Lecture Notes in Networks and Systems

Volume 230

The series "Lecture Notes in Networks and Systems" publishes the latest developments in Networks and Systems—quickly, informally and with high quality. Original research reported in proceedings and post-proceedings represents the core of LNNS.

Volumes published in LNNS embrace all aspects and subfields of, as well as new challenges in, Networks and Systems.

The series contains proceedings and edited volumes in systems and networks, spanning the areas of Cyber-Physical Systems, Autonomous Systems, Sensor Networks, Control Systems, Energy Systems, Automotive Systems, Biological Systems, Vehicular Networking and Connected Vehicles, Aerospace Systems, Automation, Manufacturing, Smart Grids, Nonlinear Systems, Power Systems, Robotics, Social Systems, Economic Systems and other. Of particular value to both the contributors and the readership are the short publication timeframe and the world-wide distribution and exposure which enable both a wide and rapid dissemination of research output.

The series covers the theory, applications, and perspectives on the state of the art and future developments relevant to systems and networks, decision making, control, complex processes and related areas, as embedded in the fields of interdisciplinary and applied sciences, engineering, computer science, physics, economics, social, and life sciences, as well as the paradigms and methodologies behind them.

Indexed by SCOPUS, INSPEC, WTI Frankfurt eG, zbMATH, SCImago.

All books published in the series are submitted for consideration in Web of Science.

More information about this series at http://www.springer.com/series/15179

Radek Silhavy
Editor

Software Engineering and Algorithms

Proceedings of 10th Computer Science On-line Conference 2021, Vol. 1

Springer

Editor
Radek Silhavy
Faculty of Applied Informatics
Tomas Bata University in Zlín
Zlín, Czech Republic

ISSN 2367-3370 ISSN 2367-3389 (electronic)
Lecture Notes in Networks and Systems
ISBN 978-3-030-77441-7 ISBN 978-3-030-77442-4 (eBook)
https://doi.org/10.1007/978-3-030-77442-4

This Springer imprint is published by the registered company Springer Nature Switzerland AG
The registered company address is: Gewerbestrasse 11, 6330 Cham, Switzerland

Preface

Software engineering research papers are presented in this volume of proceeding. This proceeding is a Vol. 1 of the Computer Science On-line Conference 2021. Papers in this part discuss intelligent algorithms, application of machine and statistical learning in software engineering research.

This book constitutes the refereed proceedings of the Software Engineering and Algorithms section of the 10th Computer Science On-line Conference 2021 (CSOC 2021), held online in April 2021.

CSOC 2021 has received (all sections) more than 390 submissions from more than 30 countries. More than 68% of accepted submissions were received from Europe, 19% from Asia, 5% from Africa, 5% from America and 3% from Australia.

CSOC 2021 intends to provide an international forum to discuss the latest high-quality research results in all areas related to computer science.

Computer Science On-line Conference is held online, and modern communication technology, which are broadly used, improves the traditional concept of scientific conferences. It brings equal opportunity to participate for all researchers around the world.

I believe that you find the following proceedings exciting and valuable for your research work.

April 2021 Radek Silhavy

Organization

Program Committee

Program Committee Chairs

Petr Silhavy	Tomas Bata University in Zlin, Faculty of Applied Informatics, Czech Republic
Radek Silhavy	Tomas Bata University in Zlin, Faculty of Applied Informatics, Czech Republic
Zdenka Prokopova	Tomas Bata University in Zlin, Faculty of Applied Informatics, Czech Republic
Roman Senkerik	Tomas Bata University in Zlin, Faculty of Applied Informatics, Czech Republic
Roman Prokop	Tomas Bata University in Zlin, Faculty of Applied Informatics, Czech Republic
Viacheslav Zelentsov	Doctor of Engineering Sciences, Chief Researcher of St. Petersburg Institute for Informatics and Automation of Russian Academy of Sciences (SPIIRAS)
Roman Tsarev	Department of Informatics, Siberian Federal University, Krasnoyarsk, Russia

Program Committee Members

Boguslaw Cyganek	Department of Computer Science, University of Science and Technology, Krakow, Poland
Krzysztof Okarma	Faculty of Electrical Engineering, West Pomeranian University of Technology, Szczecin, Poland
Monika Bakosova	Institute of Information Engineering, Automation and Mathematics, Slovak University of Technology, Bratislava, Slovak Republic

Pavel Vaclavek	Faculty of Electrical Engineering and Communication, Brno University of Technology, Brno, Czech Republic
Miroslaw Ochodek	Faculty of Computing, Poznan University of Technology, Poznan, Poland
Olga Brovkina	Global Change Research Centre Academy of Science of the Czech Republic, Brno, Czech Republic & Mendel University of Brno, Czech Republic
Elarbi Badidi	College of Information Technology, United Arab Emirates University, Al Ain, United Arab Emirates
Luis Alberto Morales Rosales	Head of the Master Program in Computer Science, Superior Technological Institute of Misantla, Mexico
Mariana Lobato Baes	Research-Professor, Superior Technological of Libres, Mexico
Abdessattar Chaâri	Laboratory of Sciences and Techniques of Automatic Control & Computer Engineering, University of Sfax, Tunisian Republic
Gopal Sakarkar	Shri. Ramdeobaba College of Engineering and Management, Republic of India
V. V. Krishna Maddinala	GD Rungta College of Engineering & Technology, Republic of India
Anand N. Khobragade (Scientist)	Maharashtra Remote Sensing Applications Centre, Republic of India
Abdallah Handoura	Computer and Communication Laboratory, Telecom Bretagne, France

Technical Program Committee Members

Ivo Bukovsky, Czech Republic
Maciej Majewski, Poland
Miroslaw Ochodek, Poland
Bronislav Chramcov, Czech Republic
Eric Afful Dazie, Ghana
Michal Bliznak, Czech Republic
Donald Davendra, Czech Republic
Radim Farana, Czech Republic
Martin Kotyrba, Czech Republic
Erik Kral, Czech Republic
David Malanik, Czech Republic
Michal Pluhacek, Czech Republic
Zdenka Prokopova, Czech Republic
Martin Sysel, Czech Republic
Roman Senkerik, Czech Republic
Petr Silhavy, Czech Republic
Radek Silhavy, Czech Republic
Jiri Vojtesek, Czech Republic
Eva Volna, Czech Republic
Janez Brest, Slovenia

Ales Zamuda, Slovenia
Roman Prokop, Czech Republic
Boguslaw Cyganek, Poland
Krzysztof Okarma, Poland
Monika Bakosova, Slovak Republic
Pavel Vaclavek, Czech Republic
Olga Brovkina, Czech Republic
Elarbi Badidi, United Arab Emirates

Organizing Committee Chair

Radek Silhavy Tomas Bata University in Zlin, Faculty of Applied Informatics, Czech Republic, email: radek@silhavy.cz

Conference Organizer (Production)

Silhavy s.r.o.
Website: https://www.openpublish.eu
Email: csoc@openpublish.eu

Conference website, Call for Papers

https://www.openpublish.eu

Contents

Multi-agent Simulation of Agile Team Dynamics: An Investigation on Team Strategies Perspective 1
Zhe Wang

SBAP: A Novel Secure Bucket Access Protocol for Leveraging User's Data Privacy in Cloud Environment 37
N. P. Hemanth Kumar and S. Prabhudeva

Energy Distribution Modeling During the Electron Beam Welding Using Dynamically Changing Thermophysical Parameters of the Product 47
Sergei Kurashkin, Vadim Tynchenko, Yuriy Seregin, Aleksandr Murygin, Vladislav Kukartsev, and Valeriya Tynchenko

Agile Architecture Frameworks: Challenges and Open Issues 59
Tumisang Akinpelu, Rene van Eck, and Tranos Zuva

Novel Approach of Video Tracking System Using Learning-Based Mechanism over Crowded Environment 67
Karanam Sunil Kumar and N. P. Kavya

A Survey on Supporting the Software Engineering Process Using Artificial Intelligence 77
Ahmed Maghawry, Rania Hodhod, Yasser Omar, and Mohamed Kholief

Automation of Electron Beam Input During the Welding of Thin-Walled Structures 88
Vadim Tynchenko, Sergei Kurashkin, Valeriya Tynchenko, Sergei Dokshanin, and Anna Leonteva

Evolution of Contrast and Illumination Adjustment Techniques for Dental Imaging to Assist Effective Implant 100
J. Shashikala and N. Thangadurai

Maturity Models, Challenges and Open Issues . . . 110
Tumisang Akinpelu, Rene van Eck, and Tranos Zuva

The Model of Parallel Convergence in the Group of Pursuers' Plane with Simultaneous Achievement of the Goal . . . 119
A. A. Dubanov

Information System for Food Origin Monitoring . . . 125
Jozef Boris, Radim Farana, and Cyril Klimeš

Novel Framework for on Demand Preprocessing Towards Facilitating Diagnosis of Diabetic Retinopathy . . . 132
M. K. Prakruthi and G. Komarasamy

Organization of Measurement Data Collection and Primary Processing Regional Environmental Monitoring Systems . . . 142
R. Y. Lapkovsky, V. A. Kushnikov, E. V. Kushnikova, A. S. Bogomolov, A. F. Rezchikov, A. D. Tsvirkun, V. A. Ivashchenko, O. I. Dranko, D. S. Fominykh, E. V. Berdnova, J. V. Lazhauninkas, and E. S. Baryschnikova

Design of an Optimal Framework to Evaluate the Throughput in Downlink Channel in LTE . . . 152
R. K. Srinivasa and A. R. Hemantha Kumar

Bioinspired Approach to Microstrip Transmission Line Design . . . 170
Danilchenko Vladislav Ivanovich, Danilchenko Yevgenia Vladimirovna, and Kureichik Viktor Mikhailovich

Analysis of Transformation Tools Applicable on NewSQL Databases . . . 180
Sarah Myriam Lydia Hahn, Ionela Chereja, and Oliviu Matei

Multi-level Trust Modelling to Resist Impact of Routing Attacks on Reliable Data/Communication Transactions in MANET-IoT Ecosystem . . . 196
G. N. Anil

Improving Determine Lexical Meanings for Mongolian Cyrillic Script and Its Used to Keyword Extraction . . . 206
Bat-Erdene Nyandag and Khishigdalai Ulaankhuu

Multidimensional Analysis of NewSQL Database Systems . . . 221
Ionela Chereja, Sarah Myriam Lydia Hahn, Oliviu Matei, and Anca Avram

A Survey on Challenges Facing Artificial Intelligence Based Pattern Recognition for Business Oriented Big Data Analytics and Solutions . . . 237
Ahmed Maghawry, Amr Elhadidi, Ahmed Alqassed, Mohamed Awad, Ayman Taha, and Mohamed Kholief

Operationalizing Analytics with NewSQL 249
Ionela Chereja, Sarah Myriam Lydia Hahn, Oliviu Matei, and Anca Avram

Grouping-Based Hybrid Channel Access Mechanism for Ultra-Dense IEEE 802.11 ax Networks 264
Jianjun Lei and Mansoor Samo

Mathematical Optimization, the Effect of the Catalyst and Solvent on the Process of Starch Sulfation with Sulfamic Acid 275
Aleksandr S. Kazachenko, Natalya Yu. Vasilyeva, Yuriy N. Malyar, and Anna S. Kazachenko

Research and Development Circuits for Error Correction 283
Sergey Gushanskiy, Alexander Gorbunov, and Viktor Potapov

Formal Development of Fault Tolerance by Replication of Distributed Database Systems 293
Rasin Katta, Ammar Alhaj Ali, Bronislav Chramcov, Said Krayem, and Roman Jasek

Fault Detection Model for Multi Robotic System Using Formal Method Event-B 307
Emad Awwama, Ammar Alhaj Ali, Roman Jasek, Bronislav Chramcov, Said Krayem, and Rasin Katta

Fault Tolerant Sensor Network Using Formal Method Event-B 317
Ammar Alhaj Ali, Bronislav Chramcov, Roman Jasek, Rasin Katta, and Said Krayem

Modeling the Temperature Field Distribution at the Stages of Input-Output of the Electron Beam 331
Sergei Kurashkin, Vadim Tynchenko, Aleksandr Murygin, Yuriy Seregin, Valeriya Tynchenko, and Aleksey Bocharov

Object Classification from a Hyper Spectral Image Using Spectrum Bands with Wavelength and Feature Set 340
Soumyashree M. Panchal and Shivaputra

Historical 3D Visualisations of Starý Světlov Castle Using Blender and Unreal Engine 351
Pavel Pokorný and David Falešník

Optimization of the Process of Obtaining Ethanol-Lignin from Abies Bark (*Abies Sibirica*) 363
Yuriy N. Malyar, Aleksandr S. Kazachenko, and Dmitry V. Zimonin

Physical Complexity of Algorithms 369
I. P. Deshko and V. Ya. Tsvetkov

The Usage of Declarative Integrity Constraints in the SQL Databases of Some Existing Software 375
Erki Eessaar

Average Consensus with Perron Matrix for Alleviating Inaccurate Sensor Readings Caused by Gaussian Noise in Wireless Sensor Networks 391
Martin Kenyeres and Jozef Kenyeres

A Conversion of 3D Graphics from Blender to Unreal Engine 406
Pavel Pokorný and Miroslav Zapletal

Technology for Dynamic Configuring of the Decision Support System for Personnel Security of the Regional Mining and Chemical Cluster 418
D. N. Khaliullina, V. V. Bystrov, and S. N. Malygina

Novel Secure Scheme for On-Field Sensors for Data Aggregation in Precision Agriculture 428
K. Vanishree and G. S. Nagaraja

Conversion Methods of Data Mining Analysis in Algorithms of Statistical and Nowcasting Forecast of Convective Precipitation 437
David Šaur and Jaromír Švejda

Interval of Spatial Correlation of Fading in a Single-Beam Decameter Radio Line 451
V. P. Pashintsev, S. A. Koval', A. F. Chipiga, and Alexander D. Skorik

Virtual Reality Model Assessment Platform: Transfer Models to a Standalone Virtual Reality Headset 469
Jakub Tomeš, Naďa Tylová, Jan Kohout, and Jan Mareš

Implementing Dynamic Mathematics Software in Calculus II for Engineering Students: Quadratic Surfaces 480
Tommy Tanu Wijaya, Jianlan Tang, Li Li, and Aditya Purnama

A Machine Learning Framework for Inter-frame Prediction for Effective Motion Estimation 492
S. K. Veena and K. Mahesh Rao

Distributed Library Model Based on Distributed Ledger Technology for Monitoring and Diagnostics System 501
E. V. Melnik, M. V. Orda-Zhigulina, and D. V. Orda-Zhigulina

Business Process Reengineering for Health Care Management and Medical Insurance: The Case of Russia 510
Yuri Antokhin, Irina Sokolova, and Anton Suprun

An Approach to Adjust Effort Estimation of Function Point Analysis 522
Huynh Thai Hoc, Vo Van Hai, and Ho Le Thi Kim Nhung

M-Learning Autonomous Model for Learning Mathematics with MatiMath Application in Students of a Confessional 538
Sol A. Mamani, Carlos R. Mamani, and Juan J. Soria

Methods and Models of Resource Allocation in Load Balancing Clusters 552
Valery Petrovich Mochalov, Gennadiy Ivanovich Linets, and Ilya Sergeevich Palkanov

Requirements Engineering in Agile Software Startups - Insights from Multiple Case Studies 564
Varun Gupta, Thomas Hanne, and Rainer Telesko

Manipulator: A Novel Collusion Attack on Trust Management Systems in Social IoT 578
Alper Ugur

Computer Science and Information Technology Students' Self-regulated Learning and Knowledge Sharing Behavior as Drivers of Individual Innovative Behavior 593
James K. Ngugi and Leila Goosen

The Geoimage Generation Method for Decision Support Systems Based on Natural Language Text Analysis 609
A. V. Vicentiy

Item-Share Propagation Link Applying for Recommendation 620
Sahraoui Kharroubi, Youcef Dahmani, and Omar Nouali

Empirical Evidence in Early Stage Software Effort Estimation Using Data Flow Diagram 632
Vo Van Hai, Ho Le Thi Kim Nhung, and Huynh Thai Hoc

Analysis of the Cybersecurity Weaknesses of DLT Ecosystem 645
Willian Dimitrov, Boyan Jekov, and Peyo Hristov

A Hybrid Approach with Machine Learning Towards Opinion Mining for Complex Textual Content 656
D. Rajeshwari and D. Puttegowda

SCRUM Framework Extended with Clean Architecture Practices for Software Maintainability 667
Eduardo C. Arango and Omar L. Loaiza

The Properties of Multistep Runge-Kutta Methods for Neutral Delay Differential Equations 682
Kewei Zhang and Haiyan Yuan

Comparison of Modified Object Detected Graph Cut and Hybrid Ant Colony Optimization - k-means for MRI Images Segmentation . . . 701
Samer El-Khatib, Yuri Skobtsov, Sergey Rodzin, and Valerii Zakharov

Robotic Sorting Line Model Using Coloured Petri Net 709
Dominik Stursa, Libor Havlicek, and Libor Kupka

Implementation of Elliptic Curve Cryptography Between Autonomous Vehicles and Control Center 718
Andrej Grnac, Frederik Valocky, and Milos Orgon

The Hybrid Methods of Object Detection and Localization in Images 730
Sergey Kuleshov, Alexandra Zaytseva, Ilya Shalnev, and Alexey Aksenov

Author Index 739

Multi-agent Simulation of Agile Team Dynamics: An Investigation on Team Strategies Perspective

Zhe Wang(✉)

Lincoln University, Lincoln, New Zealand
Zhe.wang@lincolnuni.ac.nz

Abstract. Scrum is an agile process that incrementally, iteratively and continuously delivers software based on time boxes (sprints). It comprises user stories stored in product backlogs and delivered through sprints by a Scrum teams consisting of team members, Scrum Master and Product Owner. The performance of a Scrum team is largely dependent on the team members and the technical practices that they adopt. One such practice, pair programming has been studied in a variety of contexts but not extensively in a Scrum context. Pair programming is a technique where two programmers work side by side at one computer to collaborate on the same design, algorithm, code, or test. The use of pair programming within Scrum has not been widely researched. A multi agent system is used to simulate the scrum environment where a team (with varying team members' capability) work on delivering user stories (which consists of multiple tasks with varying complexities) in multiple sprints. Using this simulated environment, various strategies are investigated. Impact is measured based on the team's work efficiency, completion time, effort time and idle time. Experiments were carried to test these strategies in varying environments and results showed that a intelligent pairing strategy performed the best in environments as it avoided any negative pairing impacts.

Keywords: Scrum · Team dynamics · Agent-based modelling · Multi-agent system · Team strategies · Solo programming · Pair programming

1 Introduction

Software is created by people using technologies and tools, which are integrated in a process known as the software development process life cycle (SDLC). They are two major approaches in software development, the waterfall model and the Agile approach. The waterfall model is a non-iterative sequential process which progresses from requirement analysis through system design, and coding, to testing and maintenance. Such an approach provides limited interactions between the customer and the software team due to its sequential nature where interactions only happen during requirement analysis.

Agile software development is an approach to creating software through an iterative and incremental process. Beck and Fowler (2001) describes Agile as: "Individuals and interactions over processes and tools; Working software over comprehensive documentation; Customer collaboration over contract negotiation; Responding to change over

R. Silhavy (Ed.): CSOC 2021, LNNS 230, pp. 1–36, 2021.
https://doi.org/10.1007/978-3-030-77442-4_1

following a plan." Scrum is one of the agile processes that realises the Agile manifesto (Beck et al. 2020). It splits users' stories into several parts and aims to achieve each part within a time period, which can last from one day to thirty days, and is known as a sprint. Within each sprint, the process is similar to the waterfall model. However, there are more opportunities for the Scrum team and the customer (also known as the product owner) to discuss the software requirements during the process because there are multiple sprints in an agile project and each sprint contains requirement analysis and this allows the requirements to be refined during each sprint. Each sprint is aimed to deliver a higher project success rate compared to the waterfall approach because the size of the software is smaller, its design goal is clearer, and the Scrum team is fully focused on the project (Schwaber and Sutherland 2017). The Scrum process is composed of user stories, tasks, sprints, sprint meetings, deliverable software and conducted by the Scrum team. User stories are designed based on the users' requirements. Each user story contains one or more tasks that need to be completed by the Scrum team during one sprint or across several. A Scrum team comprises several members who have different skills and capabilities, such as designer, developer, database administrator and tester. Scrum team members need to interact and collaborate with each other to achieve the goals of the team, incrementally or iteratively, through sprints. A Scrum team will have daily sprint meeting to discuss what has been done and what needs to be done.

Although the Scrum approach is an improvement over the waterfall model, it still suffers from several problems. One such problem is team dynamics, which largely affect the quality, risk and value of the process. Team dynamics refers to the team's composition, task allocation, interactions between team members and how they work together. Song (2015) define effective team dynamics according to the following criteria, as also indicated by Nadler et al. (1979).

Team performance (i.e., the product of teamwork meets the expectations of those who use it);

1. Member satisfaction (i.e., each team member's experience contributes to his or her personal well-being and development); and
2. Team adaptation (i.e., the team experience enhances each member's capability to work and learn together in the future).

A team comprising experienced and highly skilled members will normally perform better than a junior team that is less experienced and skilled. In a Scrum environment, the composition of the team will affect the performance of the team, because a Scrum team needs high levels of cooperation among its team members to achieve the sprint goal, and the skills, experience and capabilities of the team members affect the performance of the team. Varying methods of tasks allocation may result in different outcomes and may affect the delivery of the software.

1.1 Agent-Based Modelling

The main idea of ABM is that many real-world phenomena can be effectively modelled as an agent, an environment and as a description of agent-agent interactions and agent-environment interactions (William and Uri 2015). Modelling the real world involves developing a computational model that comprises agents and the environment in which the main actors are the agents (Mehmood et al. 2019).

Agent-based modelling can be used to model social behaviour and interactions. Boulahbel-Bachari and Saadi (2019) used agent-based modelling to model the behaviour of migration and self-selection of people. Joslin and Poole (2005), Macal and North (2006) give us an overview of how to use agents to model software project planning by modelling the developer team. In this work, agents are used to model and simulate human developers, such that an agent can have capability, skills and experience. Agents can pick up a task and work on it and deliver the product. Different agents can have different skill sets and motivation. Task modelling is important when constructing the scale of the overall project. Agents' interactions with an agent or a task can be captured using agent-based modelling, and such modelling can model the process of how each agent completes the task.

Agarwal (2007), Macal and North (2008), Phillips (2006), Yilmaz and Phillips (2007), Agarwal and Umphress (2010) indicate that agent-based modelling is suitable for undertaking team-based research to capture team behaviour and interaction when carrying out software project estimations and optimization. Team-based modelling uses multi-agents, which are a group of agents that interact with each other in a team. The shared goal of these agents is to complete all the tasks. In the real world, a software developer team usually contains more than one developer, and an agent can be used to model and simulate each developer and to group them as a team. The reasons for this are based on an agent's unique characteristics, especially its features of human behaviour and decision-making. Agent-based modelling can be used to model a group of agents and verify their impacts on team goal realization and achievement. Such features can be used to undertake team-based analysis in particular organizational-based contexts.

Abar et al. (2017) specifically reviewed all the tools that have been developed for agent-based modelling and simulation, classifying them by their functionality and scales of computing that can be applied. These tools can be distributed for various modelling purposes, such as human behaviour modelling, animal behaviour modelling and cell modelling in biology. One of the more popular tools is NetLogo (William and Uri 2015), which is used to simulate natural and social phenomena and is well suited for modelling complex systems over time. NetLogo allows developers to explore the various agents' behaviour under a variety of settings.

For the purpose of this study, a multi-agent system was used to simulate the Scrum environment to enable agents to make autonomous decisions and interact with one another using a standard agent communication language. The next sections describe the agent and the multi-agent systems in some detail.

1.2 What is an Agent and Its Characteristics?

Agent is a special software component that can make decisions autonomously based on its inner rules and logic reasoning (Wooldridge and Jennings 1995). An agent can be defined as a computer system that is situated in some environment that is capable of autonomous actions in order to meet its design objectives. Agents can be used to model human behaviour or to construct an intelligent system (Wooldridge and Jennings 1995).

Wooldridge and Jennings (1995) noted that computer agents typically have the following properties:

- Autonomy – agents operate without having direct control over their actions and internal state;
- Social ability – agents interact with other agents through some kind of 'language' (a computer language, rather than a natural language);
- Reactivity – agents are able to perceive their environment (which may be the physical world, a virtual world of electronic networks or a simulated world that includes other agents) and respond to it; and.
- Proactivity –agents are also able to take the initiative and engage in goal-directed behaviour.

Agents can communicate with each other through a common language structure, such as Foundation for Intelligent Physical Agent – Agent Communication Language (FIPA ACL), which is an agent language protocol standard, to exchange information.

1.3 What is a Multi-agent System?

As mentioned previously, agents are special software than can act autonomously on behalf of the users to achieve specific goals. Agents often need to work together to solve more complex problems. A multi-agent system (MAS) is a loosely coupled network of agents that interact to solve problems that a single agent will not be able to solve alone (Ali et al. 2019; Tyrychtr et al. 2019). Agents may need to cooperate, coordinate, collaborate and negotiate with each other to achieve their goals in a MAS environment. The ability to interact is made possible by the presence of a standard agent communication language (ACL), such as FIPA-ACL (Bellifemine 2007). Various MAS applications have been developed to solve a variety of problems in multiple domains, such as e-commerce (Tyrychtr et al. 2019), the military (Han et al. 2019), networking (Boulahbel-Bachari and Saadi 2019), logistics and supply chains (Ivanov et al. 2019) and disaster rescue (Hashimoto et al. 2019).

MAS is ideal for this research as each individual agent can be programmed to possess characteristics that embody the different roles in Scrum, such as the Scrum Master and developers. Each individual agent is able to make its own decisions about whether to take on a specific task, depending on its capability, as well as being able to make decisions about which agent to pair to gain an additional advantage. These developer agents can communicate with each other and the Scrum Master agent. They can also share information with one another either by means of communication or a shared blackboard. The Scrum MAS environment is described in the next section.

1.4 Scrum Background

This section describes Scrum and its processes in detail. There are three components in Scrum, as shown in Fig. 1, the roles, the events and the artefacts.

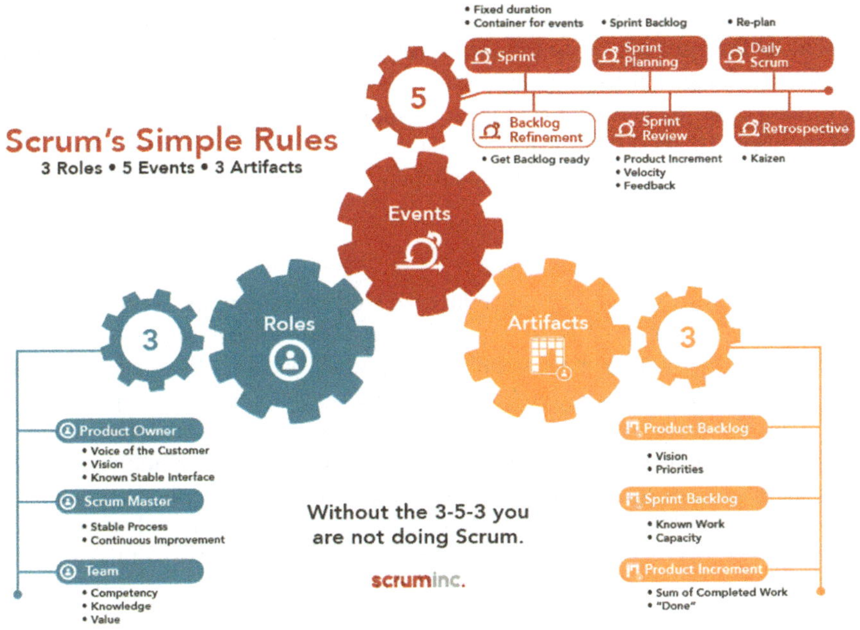

Fig. 1. Scrum model in brief (Justice 2018)

Roles

There are three roles in a Scrum team; Product Owner, Scrum Master, and Development team (including developers, designers, testers, database administrators, user experience experts). The product owner represents the stakeholders, such as the customers and the marketing manager (Stober and Hansmann 2010). The product owner must ensure that he or she is representing the interests of all stakeholders. The product owner also provides the requirements, authorises expenditure for the project, and signs off on any deliverable. The development team is responsible for developing and testing the project deliverables. Developers usually hold several key skills to solve user story problems, such as programming capability, software system design, software testing and database programming. These skills are matched with a problem-solving background. For example, an Android development would require the developer to have Java programming skills much more than C#. Finally, the Scrum Master is responsible for the Scrum process, adjusting the Scrum to best fit the project and the organisation, as well as ensuring that any issue or problem gets resolved so the Scrum team can be as effective as possible.

Scrum Events

Scrum events are referred to as Scrum ceremonies and include sprint planning, daily Scrum (stand-up), sprint review and sprint retrospective. At the beginning of the Scrum process, the user stories are defined, prioritised and estimated, then stored in the product backlog. At the beginning of each sprint, there is sprint planning where the team decides what user stories are worked on during that sprint. Each sprint is, typically, time boxed at two to four weeks. Each day, the team has a daily stand-up meeting (daily Scrum) where they discuss their progress, plan forward and identify problems that may arise. At the end of the sprint, a sprint review is conducted with the product owner to report on progress, to inspect the increments and to discuss anything (if necessary) that may require changes to the product backlog. A sprint retrospective will be conducted at the end of the sprint by the Scrum team with the Scrum Master in order to determine how the previous sprint could have been done better, the defects in the previous sprints and how to improve them.

Artefacts

The artefacts of Scrum include the product backlog (Alsalemi and Yeoh 2015), sprint backlog, and increments (Schwaber and Sutherland 2017). Requirements (user stories) are gathered and prioritised in a list called the product backlog (Stober and Hansmann 2010). The product backlog is used to hold user stories defined by the product owner and are ordered by their priority. This gives the product owner an advantage in defining the user stories (Mahnič and Hovelja 2012) iteratively and incrementally. The highest priority user story always goes to the top of the product backlog. The Scrum team will then choose the user stories with the highest priorities to work on in each sprint. A sprint can deliver one or more user stories, depending on its size, complexity of the user stories and capability of the team. Several sprints are typically necessary to complete enough user stories for the initial delivery of the software. An incremental delivery model is a key point in encouraging frequent discussions between the team and the product owner. Regular reviews help to guarantee the quality of the product and the increment delivered in each sprint. The process runs iteratively until all the user stories in the product backlog are completed by the Scrum team or the product owner decides to stop the development of the software.

1.5 Pair Programming

In pair programming, two programmers work side-by-side at one computer to collaborate on the same design, algorithm, code or test. This kind of programming has been practiced in industry with great success for years (Williams et al. 2000). Pair programming can enhance the performance of developers to deliver quality code in less time compare with solo.

1.6 Research Motivation

Teamwork is considered as the most critical factor (Silva et al. 2013) that could affect the successful rate of software projects. The Scrum guide indicates that the Scrum team plays an important role in the software development project. However, the team composition and the how the team should work together (in pair or solo) have not been fully investigated.

Pair programming is widely used in industry, because it can support two programmers working together, to share knowledge and enhance the software delivery capability. However its positive and negative impact on team performance that could affect its software delivery under scrum context has not been fully investigated. This research investigates the use of pair programming in a Scrum team and how it can affect the performance of the team through multi-agent simulations as there are limited research that focuses on the adoption of pair programming by Scrum teams. A multi-agent approach is adopted in this research to model each developer in the Scrum team and their collaboration. Moreover, there is very limited work using multi-agents system to model scrum team behaviour, especially on how to use such an approach to address Scrum team strategy design and implementation (Ramanujam and Lee 2011; Zhen et al. 2018).

2 Related Works

Task allocation was studied by Masood et al. (2017) who analysed real cases in software companies and found that the majority of developers in the software team considered their skills and matching with the task allocated (Licorish and MacDonell 2017). The goals of the whole team were often neglected by the team, and this prevented other team members from showing their options and influencing each other in a positive way. Lin (2013, 2015), Lin et al. (2014), Lin et al. (2014) proposed a task allocation algorithm for the Scrum team based on context aware methods. This method balances the workload of the task and agent status, such as the requirement for task quality, completion efficiency and an agent's psychology and pressure. Lin et al. (2014), Licorish and MacDonell (2018) conducted survey-based research about the developer's self-assessment of confidence in completing the task to find whether a correlation between the confidence and completion time of the task is correct or not for task allocation decision-making adjustments. The confidence scale was from 0 to 11 points to indicate from not confidence to very confident. Lin et al. (2014) analysed task allocation in the Scrum context by considering the developer's morale. This research had 125 undergraduate software engineering students complete a 12-week course work using the agile software development method. The key findings about the Scrum-based Adaptive Software Development (ASD) process practised by novice teams from this study were: 1) task allocation in agile teams positively correlated with the students' technical productivity; 2) collaboration is negatively correlated with the teams technical productivity, team morale, and team score; and 3) team morale was positively correlated to their technical productivity.

3 Methods

In this section, we describe a solo strategy that can be used in a situation where a single task must be worked on by a single agent. We also describe two pairing strategies where a pair of agents must work as pairs to complete a task (must pair) and where a task may be worked on by a pair of agents or a single agent (voluntary).

3.1 Strategy Design

To evaluate the impact of pairing in a Scrum software development project, we developed one solo strategy and two pairing strategies (must pair and voluntary pairing) that can be used by the scrum team to complete all the user stories in the sprint backlog. The must pair dictates that the agents must work in pair to complete a task whereas with voluntary pairing a task may be worked by a single agent or a pair of agents. The purpose of introducing these two types of pairing is to investigate the impact of compulsory pairing as opposed to voluntary pairing. Voluntary pairing is designed to maximise the benefits of pairing, so we only pair agents when it is beneficial to pair as this allows them to work solo when no acceptable pair is found. We further sub-divided the must pairs and voluntary pairs into three sub-categories based on the capability of the pair.

These strategies are designed to be evaluated in a single sprint where a sprint contains multiple user stories which are further subdivided into tasks. Each task has a size which ranges from 1 to 10, where 1 is the easiest task and 10 is the most difficult task. Tasks between 1 and 4 are categorised as easy tasks, 5–7 are intermediate tasks and 8–10 are complex tasks. These tasks will be worked on by the developer agents (the Scrum team) with varying capabilities. The developer's capability ranges from 1 to 10 where 1–4 is considered novice, 5–7 intermediate and 8–10 expert. Ideally, a developer agent should work only work on task with complexity equal to its capability. However, in situations where this is not possible, other options can be explored such as pairing with another agent or work solo on the task with penalty. It is also assumed that the team composition and the tasks are dynamics (these configurations vary in each sprint) and that certain strategy may work well for certain environment. These strategies are described in detail in the next sections.

3.2 Solo Strategy

The solo strategy allocates a task to a single agent based on the capability of the available agents. The task is assigned to the agent with the closest capability with the complexity of the task. A preference value (agent capability – task complexity) is used to determine which agents can undertake a task within the simulation (Wang et al. 2019). An agent is only permitted to work on tasks where the preference value is greater than or equal to −3. The time required to complete a given task is dependent on the agent's capability and the task's complexity. If the agent's capability is greater than or equal to the task complexity, then the time required to complete the task is equal to the task size (Eq. 1).

$$\boldsymbol{TimeRequired} = \boldsymbol{TaskSize} \tag{1}$$

otherwise, if the agent capability is lower than the task's complexity, the time required to complete a task is calculated as:

$$\boldsymbol{TimeRequired} = |\boldsymbol{AgentCapability} - \boldsymbol{TaskComplexity}| * 0.33 * \boldsymbol{TaskSize} + \boldsymbol{TaskSize} \tag{2}$$

Here, the time required to complete the task is longer since an agent with a lower capability will need more time to work on the given task. Since a gap of −3 is allowed, 0.33% is added to the time required to complete the task for each gap value (−1, −2 and −3) based on Griffith et al. (2014) where, for a given task that can be worked in a 1.0 unit of time by a middle level worker, it will take a 0.5 time unit for the senior to complete and 2.0 units of time for the junior to complete the task.

As discussed previously, a developer agent's capability can be categorised as novice, intermediate or expert agents such that same level of pairings (expert-expert, intermediate-intermediate, and novice-novice) and cross level pairings (expert-intermediate, expert-novice, and intermediate-novice) are possible. However, based on the literature (Arisholm et al. 2007; Dybå et al. 2007; Lui and Chan 2006; Nilsson 2003) the effectiveness of the pairing is dependent on the level of expertise of the individuals in the pair and the task to be worked on. The following section provides an analysis of the different pairings.

3.3 Same Level Pairing

Same level pairing refers to pairing of developers with the same capability to work together on a specific task. This is the original idea for pair programming, which aimed to enhance the quality of code cooperation between developers who enjoyed the work of coding to promote product delivery.

Expert-Expert Pair
This type of pair is considered as negative pairing as both experts may have disagreement on software design, coding and analysis. Dybå et al. (2007) supports two experts working on complex tasks as a good choice but suggests that experts should not work in pair when working on an easy task. Similarly, Arisholm et al. (2007) also supports two experts pairing to work on complex tasks but not on easy tasks. In addition, Lui and Chan (2006) shows that once a novice pair moved up to expert rank, it is no longer useful to pair them.

Intermediate-Intermediate Pair
This type of pairing is only effective when the pair is working on a complex task. Dybå et al. (2007) supports two intermediate experts working on complex tasks as a good choice, but a bad one when working on an easy task. Arisholm et al. (2007) also supported two intermediate pairs working on complex tasks but not on easy tasks. A single intermediate developer is not able to work on a complex task. However, pairing two intermediate developers can enhance the performance of each developer. This pairing is not recommended when working on intermediate tasks as an intermediate task can be worked on by a single intermediate expert (Arisholm et al. 2007; Dybå et al. 2007; Hannay et al. 2009; Lui and Chan 2006).

Novice-Novice Pair
This type of pairing is not recommended when working on complex tasks as it is not possible for two novices to solve complex problems due to gaps in knowledge. A novice-novice pairing is effective when working on easy and intermediate tasks. Based on Dybå et al. (2007), Arisholm et al. (2007) it is possible for two novices to work on all tasks.

3.4 Cross Level Pairing

The purpose of cross level pairing is to facilitate learning and knowledge transfer between the pairs (Zieris and Prechelt 2014). Good pairs can transfer knowledge, as supported by Plonka et al. (2015). Many researchers indicate that knowledge transfer between developers exists in an organization (Argote and Fahrenkopf 2016; Lin and Zhang 2019; Wang et al. 2019). By doing this, the lower level developer can learn from the high-level developer. Knowledge transfer occurs when pairings, such as when a student works together with a supervisor.

Expert-Intermediate Pair

This type of pair is effective as the intermediate can learn from the expert. This pairing is highly recommended when working on a complex task as this allows the intermediate to learn from the expert (Plonka et al. 2015). It is not recommended for this kind of pairing to work on easy and intermediate tasks as no learning can take place and the combined capabilities of the intermediate and the expert are not maximised.

Expert-Novice Pair

The expert-novice pair has the largest gap between the two developers, and this is considered an effective pairing as the novice can learn from the expert. A novice developer will not be able to work on an intermediate or complex task alone. However, by pairing with an expert, the novice has the opportunity to work on tasks at all levels.

Intermediate-Novice Pair

The intermediate-novice pair provides the opportunity for the novice to work on easy and intermediate tasks. The novice will also be able to learn from the intermediate. This pairing is not recommended when working on a complex task as the intermediate might not have the knowledge to tackle the task.

Table 1 summarises the different pairings for the varying task complexities used in this simulation. Positive pairing is the preferred pairing as it results in better code quality and allows junior developers to learn from senior developers.

Table 1. Pair programming modelling assumptions

Pair	Easy task	Intermediate task	Complex task
Novice-novice	Positive	Positive	Negative (not used)
Intermediate-intermediate	Negative	Negative	Positive
Expert-expert	Negative	Negative	Negative
Novice-expert	Positive	Positive	Positive
Intermediate-expert	Negative	Negative	Positive
Intermediate-novice	Positive	Positive	Negative (not used)

There are several situations where pairing is not recommended (negative pairing):

- Novice-novice working on a complex task
- Expert-expert (all cases)
- Intermediate-intermediate working on intermediate and easy tasks

- Expert-intermediate working on intermediate and easy tasks
- Intermediate-novice working on a complex task

In our simulation, the time required to complete a task worked on by a negative pair will be inflated by 15% to accommodate the overheads (misunderstandings and disagreements) from this pairing (Arisholm et al. 2007).

When pairing, the pair capability is based on the capability of the stronger agent (referred to as the lead agent). The time required to complete a task for a pair is half the time taken by a solo developer as there are two developers working on the same task. For negative pairing, a penalty is imposed on the pair that increases the time required for the pair by 15%. Negative pairing may also affect the quality of the completed task. In our modelling we only considered the time penalties required, as the negative pair can result in low team performance that reduces the team's efficiency. The time required to complete a given task is as follows:

$$TimeRequired = \begin{cases} TS, \ LAC \geq TC \\ (|LAC - TC| * 0.33 * TS) + TS, \ LAC < TC \end{cases} \tag{3}$$

where LAC is the lead agent capability, TS is the task size and TC is the task complexity. As before, 0.33 is the additional time imposed by a developer working on a task that has a complexity higher than the developer's capability (0.33 for a gap of 1, 0.66 for a gap of 2 and 1.0 for a gap of 3). In this scenario, the time required to work is longer since the agent with the lower capability needs more time to work on the given task.

The time required is updated based on the nature of the pairing where negative pairing has a penalty imposed as follows:

$$TimeRequired = \begin{cases} TimeRequired, \ if\ no\, penalty \\ 1.15 * TimeRequired, \ with\, penalty \end{cases} \tag{4}$$

The time required for each individual agent in the pair is:

$$TimeRequired_A = \frac{TimeRequired}{2} \tag{5}$$

as each agent can split the work equally. In Lui and Chan (2006), they indicated that pairs will use half the time to complete the same task compared with a solo.

3.5 Strategy Types

In our simulation, we modelled our pairing based on the following:

- Type 1 – the lead agent must have a capability equal to or higher than the task. In this setup, the lead agent can choose who to pair with depending on the task complexity.
- Type 2 – the lead agent must be in the same category or higher than the task. This means that a novice agent can work on any easy task, an intermediate agent can work on medium and easy tasks, and an expert agent can work on complex, medium and easy tasks.

- Type 3 – similar to type 2, but the lead agent with a gap (agent capability – task complexity) of -3 or greater can work on the task. By allowing a gap of three, it is possible for two intermediate agents to work on a complex task (Dybå et al. 2007). Also, in this setup, a pair of novice agents can work on a medium task.

These strategies were set up to observe how they affect the team's efficiency, completion time and idle time. In Type 1, the pair can only work on a task equal to or lower than the pair's capability. The Type 2 strategy allows the pair to work on a task in the same category as the capability of the pair, while Type 3, allows tasks in a higher category to be worked on by the pair (e.g. an intermediate task can be worked by a novice pair and a complex task can be worked by an intermediate pair).

3.6 Must Pair Strategy

In the must pair strategy, each task must be undertaken by a pair of agents as no agent can work on the task alone. In this strategy, each task needs to be verified by a pair of agents who have not worked on the task previously. To accommodate the must pair, a single agent must wait for another agent to pair with before working on another task. This also means that some pairings may be ineffective. The must pair strategy has three sub-types as shown in Tables 2 and 3.

Must pair Type 1 (MP1)
For easy tasks in this strategy, a novice who has a capability equal to or higher than the task complexity is the lead agent. In this situation the novice agent should pair with another novice agent. If no novice agent is found, then it may pair with any agent. However, if no novice agent is available to work on the easy task, an intermediate agent can be nominated as the lead agent and may pair with any other agent. Finally, when there are no novice or intermediate agents available, an expert, takes the role of a lead agent and can pair with any other agent.

Table 2. Must pair Type 1

Task	First choice	Second choice	Third choice	Fourth choice
Easy	Novice	Novice	Intermediate	Expert
	Novice	Any agent	Any agent	Any agent
Intermediate	Intermediate	Intermediate	Expert	Expert
	Novice	Any agent	Novice	Any agent
Complex	Expert	Expert	Expert	
	Novice	Intermediate	Expert	

For an intermediate task, an intermediate agent who fulfils the conditions will be the lead agent and must first pair with a novice (to provide an opportunity for the novice to learn), followed by any agent. If an intermediate agent is not available, the expert agent will be the lead agent who will then pair with a novice first (if available), followed by any agent.

For a complex task, an expert agent with the right capability is the lead agent and may first pair with a novice, followed by an intermediate and an expert.

Must Pair Type 2 (MP2)

The MP2 strategy is similar to MP1, so the pairing decisions are as in Table 2. However, the main difference in this strategy is that we allowed some flexibility on what task the agents can work on. A pair of agent can work on a task as long as the task complexity is in the same category as the lead agent's capability (i.e. an easy task can be worked on by a novice agent, a medium task can be worked on by any intermediate agent, and a complex task can be worked on by any expert agent).

Must Pair Type 3 (MP3)

MP3 is another variation of MP1 where, for this strategy, a novice-novice pair is allowed to work on an intermediate task and an intermediate-intermediate pair is allowed to work on a complex task.

Table 3. Must pair Type 3

Task	First choice	Second choice	Third choice	Forth choice	Fifth choice
Easy	Novice	Novice	Intermediate	Expert	
	Novice	Any agent	Any agent	Any agent	
Intermediate	Intermediate	Intermediate	Expert	Expert	Novice P-value > −4
	Novice	Any agent	Novice	Any agent	Novice
Complex	Expert	Expert	Expert	Intermediate P-value > −4	
	Novice	Intermediate	Expert	Intermediate	

This strategy allowed the intermediate and novice to work on higher level tasks and this is very useful when the team's maximum capability is lower than the task required.

3.7 Intelligent Pair Strategy

The intelligent pair strategy (voluntary pair) ensures that agents only pair if the pairing is positive and, in situations where pairing is not possible, the task will be assigned to a solo agent with sufficient capability. Each task needs to be verified by an agent who has not worked on the task. The intelligent pair strategy has three sub-types that shows in Tables 4 and 5.

Intelligent Pair Type 1 (IP1)

In this strategy (see Table 4), effective pairing has a higher priority. For an easy task, if there is a novice lead agent available, it will try to pair with another novice, followed by an intermediate and then an expert. If there are no agents available to pair with at that time tick, the novice agent will work on the task independently. If there is no novice agent, then a solo intermediate agent is preferred over an expert agent.

Table 4. Intelligent pair Type 1

Task	First choice	Second choice	Third choice	Fourth choice
Easy	Novice	Novice solo	Intermediate solo	Expert solo
	Novice			
Intermediate	Intermediate	Intermediate solo	Expert	Expert solo
	Novice		Novice	
Complex	Expert	Expert	Expert solo	
	Novice	Intermediate		

For an intermediate task, the intermediate agent is the lead agent and will pair first with a novice agent. If this pairing is not possible, a solo intermediate is the next choice, followed by an expert-novice pair and then an expert solo.

For a complex task, the expert agent is always the lead agent with the novice agent as the pair, followed by the intermediate. If novice and intermediate agents are not available, an expert agent will work on the task individually.

Intelligent Pair Type 2 (IP2)

The IP2 strategy is similar to IP1, so the pairing decision is the same as in Table 4. As before, the main difference is that in this strategy, we allow some flexibility on what task the agents can work on. An agent can work on a task as long as the task complexity is in the same category as the agent's capability (i.e. easy tasks can be worked on by a novice agent, a medium task can be worked on by any intermediate agent, and a complex task can be worked on by any expert agent).

Intelligent Pair Type 3 (IP3)

IP3 is another variation of IP1 where in this strategy, a novice-novice pair and a novice solo are allowed to work on an intermediate task and an intermediate-intermediate pair and an intermediate solo pair are allowed to work on a complex task. This strategy is similar to MP3 in that it allows intermediates and novices to work on higher level tasks.

Table 5. Intelligent pair Type 3

<table>
<tr><th>Task</th><th>First choice</th><th>Second choice</th><th>Third choice</th><th>Fourth choice</th><th>Fifth choice</th><th>Sixth choice</th></tr>
<tr><td rowspan="2">Easy</td><td>Novice</td><td rowspan="2">Novice solo</td><td rowspan="2">Intermediate solo</td><td rowspan="2">Expert solo</td><td rowspan="2"></td><td rowspan="2"></td></tr>
<tr><td>Novice</td></tr>
<tr><td rowspan="2">Intermediate</td><td>Intermediate</td><td rowspan="2">Intermediate solo</td><td>Expert</td><td rowspan="2">Expert solo</td><td>Novice
P value > −4</td><td rowspan="2">Novice solo
P value > −4</td></tr>
<tr><td>Novice</td><td>Novice</td><td>Novice</td></tr>
<tr><td rowspan="2">Complex</td><td>Expert</td><td>Expert</td><td rowspan="2">Expert solo</td><td>Intermediate
P value > −4</td><td rowspan="2">Intermediate Solo
P value > −4</td><td rowspan="2"></td></tr>
<tr><td>Novice</td><td>Intermediate</td><td>Intermediate</td></tr>
</table>

3.8 Evaluation Metrics

To evaluate the performance of the various strategies, five different metrics are defined:

Completion Time

This is the elapsed time used by the team to complete all the tasks (for the pre-defined user stories) from the start of the sprint until the end of the sprint when all tasks in the sprint backlog were completed, verified and marked as completed.

Person Hours

$$PH = CT \times NA \tag{6}$$

The person hours can be used to calculate the cost for the team to complete the product backlog. It has an economic purpose. This is calculated by multiplying the completion time with the number of agents in the team as shown in Eq. 6 where PH is person hours, CT is completion time and NA is the number of agents in the team.

Idle Time

The idle time indicates the amount of time that an agent was available but not actively working on a task. The total idle time is collected for a Scrum team in each run.

Effort Time

This indicates the amount of time an agent was working and busy. The total effort time is collected for a Scrum team in each run.

Working Efficiency

The working efficiency is the teamwork efficiency. This is obtained by dividing the completion time with the average workload of the agents (the lower the better). Here, working efficiency (WE) is defined as:

$$\mathbf{WE} = \frac{\mathbf{CT}}{\left(\frac{\mathbf{WL}}{\mathbf{NA}}\right)} \tag{7}$$

where CT is the completion time, WL is the total workload (summation of the all the tasks in the sprint backlog) and NA is the number of agents in the team.

3.9 Scrum Environment, Agents and User Stories

Figure 2 shows an overview of the Scrum simulation using a multi-agent system. In this simulation, there are two types of agents, the Scrum Master agent and the developer agents. Following the Scrum guidelines, there are between four and eight developer agents and one Scrum Master agent (Schwaber and Sutherland 2017). This team of agents work through sprints to complete all the user stories in the product backlog. In each sprint, the agents' objectives are to work collaboratively to complete the user stories (which are broken into smaller tasks) in the sprint backlog. At the beginning of each sprint, user stories from the product backlog are moved into the sprint backlog. This will be repeated until all user stories in the product backlog are completed, at which time, the software development project will be deemed to be completed. Agents share and exchange information through the use of the Scrum board where agents can view information about the status of another agent, the task and the user stories.

Scrum Master Agent

The Scrum Master agent is designed to communicate with the developer agents during task allocation, to allocate tasks and to manage the Scrum board. The Scrum Master agent communicates with all developer agents to check their status, get their preferences for particular tasks, allocate tasks to each agent, form a pair based on the task requirements and the capability of the available agents, and allocate tasks to the pairs. The number of agents and the number of user stories can be defined in the simulation. At the start of the simulation, agents (with different capabilities) and tasks (with different levels of complexities and sizes) are generated.

At the beginning of the sprint, the Scrum Master will coordinate the task allocations for the developer agents by sending a list of tasks that are available. Each agent sends its preferences to the Scrum Master and the Scrum Master decides the task allocations based on the agents' preferences[1]. At each time tick thereafter, each developer agents sends its status to the Scrum Master and in turn the Scrum Master updates the status of each agent and task. When an agent completes a task and becomes available, the Scrum

[1] This is a slight variation from the Scrum setting where developers (including the Scrum Master) negotiate on which tasks they want to work on. In this implementation, we have made Scrum Master a coordinator agent to reduce communication threads between all the agents. However, all tasks are allocated based on the agents' preferences.

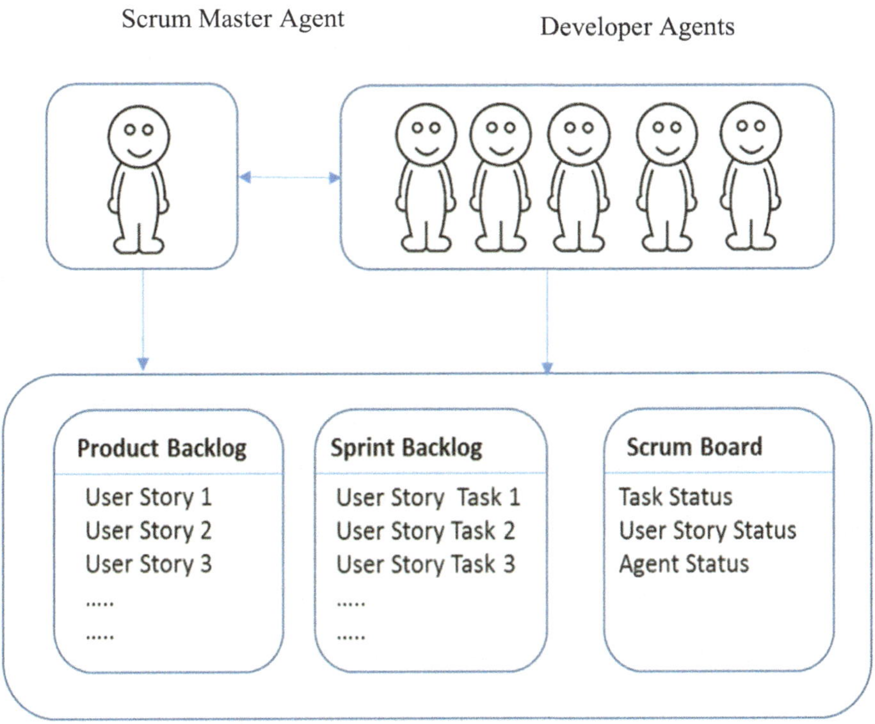

Fig. 2. Overview of Scrum simulation

Master allocates a new task to the agent, which the agent may accept or not depending on the task complexity and the agent's capability. This is repeated until all the tasks in the sprint backlog are cleared and marked as completed.

A task can be allocated to a single agent (referred to as a solo programming) or it can be allocated to a pair of agents (referred to as pair programming). For solo programming, the Scrum Master allocates the tasks to agents based on task complexity and agent preference, whereas, to allocate the task to a pair it chooses the two most appropriate agents to work on the task based on a pairing strategy.

The Scrum Master also initiates the verification of the task once it is completed by adding a verification task in the sprint backlog. When a user story is completed (all the tasks for that user stories are completed and verified), the Scrum Master initiates a user story verification task, which also gets added to the sprint backlog. The Scrum Master also manages the Scrum board where it updates all the information relating to the sprint, such as the agent's status, the agent's idle time, the agent's working time, task status, task start time, task end time, task verification time, user story status, the current time tick and the sprint completion time.

Developer Agents

The main role of the developer agents are to work on completing the tasks assigned to them. They communicate with the Scrum Master to provide information about their status, the status of the task they are working on and their task preference. Each developer agent has a name, capability and status. The capability represents technical capability ranging from 1 to 10 where 1–4 is considered novice, 5–7 intermediate and 8–10 expert[2]. This capability value is then used to determine how much time is required by the individual agent to complete a given task. Each agent has a status to indicate their current activity (working, verifying, idle). Working indicates that the agent is currently working on a task, verification indicates that the agent is verifying a task that has been completed by another agent and idle indicates that the agent is free and available to take on work.

User Stories and Tasks

Each user story is composed of multiple tasks. In this simulation, each user story is assumed to have five tasks. This is based on data collected from a number of software companies in Christchurch. Each task has a size, which indicates how much effort is needed to complete the task (the more complex the task, the larger its size). The task sizes range from 1 to 10, where 1 is the easiest task and 10 is the most difficult task. Tasks with size 1–4 are categorised as easy tasks, 5–7 are intermediate tasks and 8–10 are complex tasks. In this model, it is assumed that each task can be worked on by a single developer agent or a pair of developer agents. A task has a state with four possible values: to do (task has not been worked yet); in progress (task currently being worked); to be verified (task completed but has not been verified); and done (task is completed and verified). A task that has been completed must not be verified by the same agent(s) who worked on it.

The Scrum Board

The Scrum board shows the progress of the sprint at each time tick by showing the status of the agents, the tasks and the user stories. The Scrum board provides a platform for all agents to share and exchange information about the current sprint. The Scrum board helps the team keep all the progress transparent and is maintained by the Scrum Master.

Task Allocation Process

At the beginning of the sprint, the Scrum Master selects tasks from the sprint backlog and sends these tasks to all the developer agents. It is assumed that the tasks in the sprint backlog are already prioritised such that the most important task is at the top of the list. Each developer agent inspects the tasks and gives its preferences for these tasks. A preference value is then used to determine which agents are able to undertake a task. This preference value is the gap between the task complexity and the agent's capability (agent capability – task complexity). Once the Scrum Master has received the developer agents' preferences, task allocation is then decided. A developer agent may work on the task independently (solo) or a pair of agents (pair) may work on the task. If the work scheme is solo, the developer agent who has indicated a preference for the task will be

[2] In the real world, a developer may have specific capabilities in programming, design, and database. In this work, a single value is used for simplicity.

selected. If more than one agent is keen to work on the task, then the task is allocated to the agent with the lowest positive preference value. If agents are working in pairs, then the agent with the lowest positive preference value can work with any other agent (as discussed in strategy design.)

If the work scheme is a pair, then a lead developer will be identified to work on the task (based on preference value) and an appropriate pair will be selected (as described in detail in Strategy Design). In both cases, the task is identified before deciding which agent(s) take(s) the task. In this work, there are three main types of working schemes: agent working on a task independently (solo); two agents must work on a single task (must pair); and agents that can work either solo or in pairs (intelligent/voluntary pair).

During the sprint, when a task is completed and agents become idle, the Scrum Master advertises the next task to the idle agents and the task allocation process is repeated. The Scrum Master updates the Scrum board at each time tick. The task verification must not be verified by the same developer agent and the user story must be verified by a product owner.

Simulating a Single Sprint

Before the start of a sprint, the agents and the user stories are generated. The user stories are then broken into smaller tasks, prioritised and stored in the sprint backlog. Assuming that a single agent always works on a single task, the single sprint works as follows: When a sprint starts, the Scrum Master agent selects tasks from the sprint backlog and allocates the tasks to the developer agents based on their preferences. Each agent updates its status at every time tick. When an agent has completed a task, a task is marked as 'to be verified' and the agent's status is set to idle. At this stage, the agent can request another task from the Scrum Master. The Scrum board is updated at every time tick to show the status of the agents and tasks. This process is repeated until the end of the sprint. During the sprint, when all tasks belonging to a user story are completed, the user story needs to be verified by the product owner who is external to the developer agent team. If the product owner is satisfied with the user story, it will be marked as completed. Otherwise, if the product owner is not satisfied with the user story, the tasks that belong to that user story will be put back into the sprint backlog. Each will then be given a remedial task, named as a review task (number) where the size of the remedial task will be smaller than its original task size. At the end of the sprint, any remaining tasks in the sprint backlog will be moved to the product backlog to be worked on in the next sprint. If agents are working in pair, an available agent has to wait for another agent to become available before they can work on the task. In this simulation, it is possible for tasks to be worked by individual agents and paired agents.

To simulate the software development project, multiple sprints will be run until all the user stories in the product backlog are verified and marked as completed.

Implementation

The simulation system is developed based on the Java Agent Development framework (JADE). This framework defines agents based on their behaviour, where each agent can be designed to have various behaviours and attributes (Bellifemine 2007). JADE is a Java-based platform that can be used to develop multi-agent system. It is widely used in telecommunication industry and is not limited to modelling and simulation only.

The communication between the developer agents and the Scrum Master agent is facilitated through the message passing using the standard FIPA-ACL. The Scrum Master agent can communicate with all working agents to identify whether their working status is busy or idle, know the preference value for the task, allocate tasks to each agent, form pairs based on task requirements and allocate a task to the pair. This simulation is flexible as it allows the designer to specify the number of agents, agent's capabilities, user stories, tasks and its complexity, duration of the sprint and whether task should be worked individually or in pair. Using this simulation, various information can be collected at each time tick such as the time taken to complete a task, who worked on the task, when the task was started, when the task was completed. Information about agent working time and idle time can also be obtained.

4 Experiments Results and Discussion

This is the detailed description of expanding fixed testing and random testing to test all the strategies (7 strategies in all). I separated the context setup for fixed context and random context, they are different setup. In the fixed context, all the team agents and tasks are normally distributed, there is no extreme tasks and agents (there is no level 1 and level 10, this is consistent with the existing fixed experiments, I need to keep the same agent team to compare with existing data), but in the random testing context, there are extreme tasks and agents (there are level 1 and level 10). This makes sure both fixed and random testing can test all strategies without being affected by the strategy scale limitation and also provide insightful discussion for us compare those strategies.

The fixed task distribution strictly obeys the proportion for easy, intermediate and complex task numbers to compose the task set, and there are six types. This setup makes the fixed testing and random testing are different in its own perspective so as to provide experiments to verifying which strategy is the best. As a fixed team and tasks are more normally distributed, but random teams and tasks are not that Normal, it is purely random (Fig. 3; Tables 6, 7, 8 and 9).

Table 6. Normal distributed team for fixed testing cases

Agent capability	5 Developer team	6 Developer team	7 Developer team	8 Developer team
1				
2		One	One	One
3	One	One	One	One
4				
5	One	One	One	Two
6	One	One	One	One
7			One	One
8	One	One	One	One
9	One	One	One	One
10				

Table 7. Task distribution in each user story for fixed testing (10 user stories for a task set)

Task complexity	Case one	Case two	Case three	Case four	Case five	Case six
1						
2	One	One	One			
3		One	One	One		
4				Two	One	One
5	One		One		One	One
6	One	One		One	One	
7			One		One	
8	One	One		One	One	Two
9	One	One	One			One
10						

Table 8. Fixed task scenario

Scenario	Description
One	10 Easy task, 20 intermediate tasks, 20 difficult tasks. Each user story with 30 size as the workload. Total of 50 tasks
Two	20 Easy tasks, 10 intermediate task, 20 difficult tasks. Each user story with 28 size as the workload. Total of 50 tasks
Three	20 Easy tasks, 20 intermediate tasks, 10 difficult tasks. Each user story with 26 size as the workload. Total of 50 tasks
Four	30 Easy task, 10 intermediate tasks, 10 difficult tasks. Each user story with 25 size as the workload. Total of 50 tasks
Five	10 Easy task, 30 intermediate tasks, 10 difficult tasks. Each user story with 30 size as the workload. Total of 50 tasks
Six	10 Easy task, 10 intermediate tasks, 30 difficult tasks. Each user story with 34 size as the workload. Total of 50 tasks

4.1 Fixed Testing

Scenario One
During each task allocation process, the team will be given the number of tasks to choose, which the number is same as the team number. Such as if the team number is 5 so that 5 tasks are given at each time, if the team number is 8, then 8 tasks are given at each time. We can then observe the time need to complete those tasks are different by using various strategies. We repeatedly run the testing on each strategy for 10 times and calculate the average completion time, average effort time and the average idle time.

Table 9. Wider distributed team for random testing to test all strategies in pure random task context.

Agent capability	7 Developer team	8 Developer team
1	One	One
2		
3	One	One
4		
5	One	Two
6	One	One
7	One	One
8	One	One
9		
10	One	One

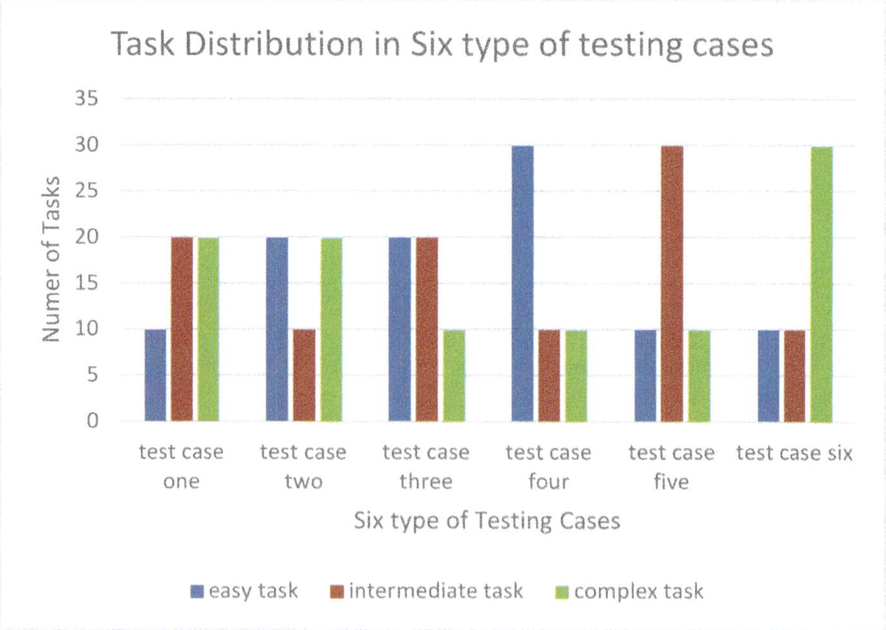

Fig. 3. Task distribution in six type of testing cases

The efficiency boxplot is widely used in this paper, the lower the value the better the strategy performance, which is useful to compare different strategies performance, the shorter the box the stable the performance.

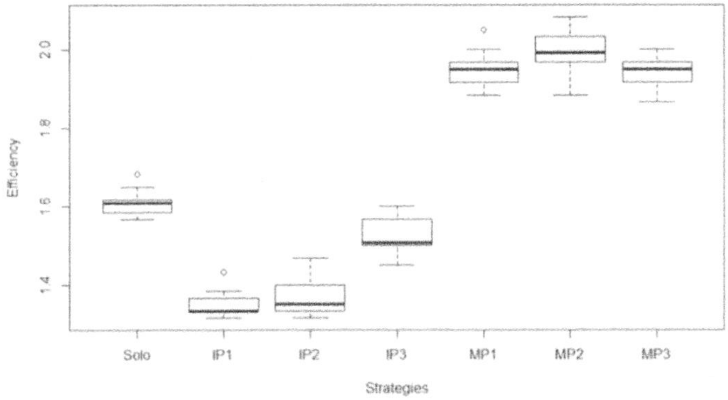

1 SOLO 2 IP-ONE 3 IP-TWO 4 IP-TRHEE 5 MP-ONE 6 IMP-TWO 7 MP-THREE

Fig. 4. 5 Agent team testing result.

We should see in Fig. 4 that intelligent pair one, two and three are very good and stable, especially IP1 and IP2 are very good. The must pair strategies are performance very worst under 5 agents as the team, even no better than solo strategy. The intelligent pair three does not performance better than intelligent pair one and two due to it allow very low agent work through pair or alone on higher level tasks.

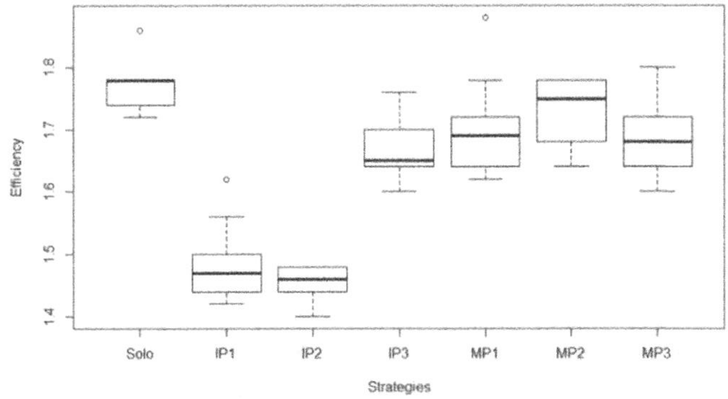

Fig. 5. 6 Agent team testing result.

We should see in Fig. 5 that intelligent pair one, two are very good and stable, especially IP1 and IP2 are very good. Under 6 number of agents as team, the must pair strategies catch up the performance of intelligent pair type three and even better than solo strategy. For the reason the task set is complex for the team to work, the must pair should performance better than solo strategy.

7 Agents Team Testing Result

Must pair performance worst under odd number of agents, even worse than solo strategy. The intelligent pair type three get the least idle time, while the must pair one get the highest idle time.

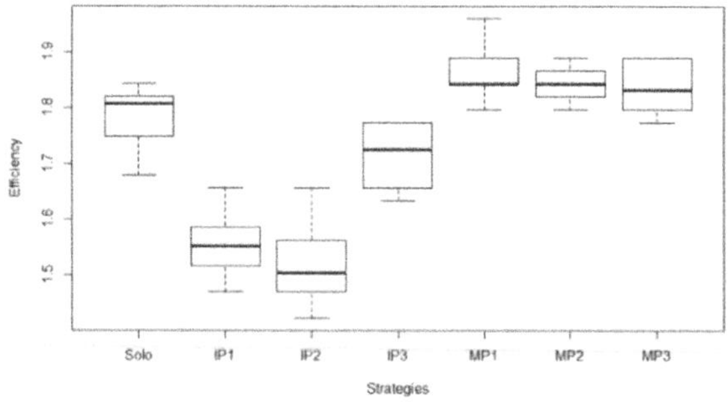

Fig. 6. 7 Agents testing result.

We should see Fig. 6 that intelligent pair one, two and three are very good and stable, specially IP1 and IP2 are very good. Under 7 number of agents as team, the must pair does not performance better than intelligent pair type three and solo strategy. Intelligent pair type two performance best at this context.

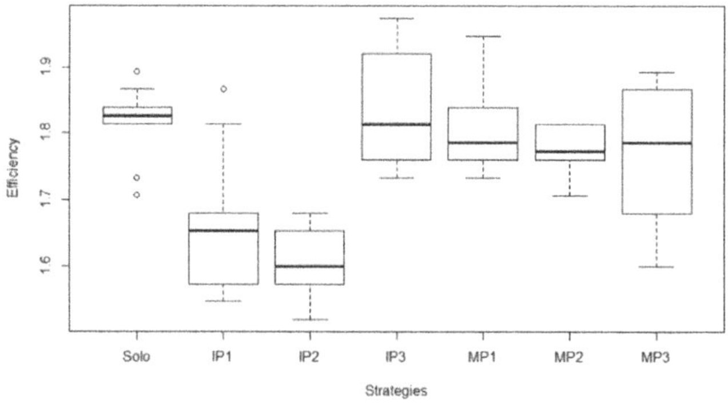

Fig. 7. 8 Agents team testing result.

We should see in Fig. 7 that intelligent pair one, two are very good and stable, specially IP1 and IP2 are very good. Intelligent pair type two performance the best under this context. Under these 8 agents as team, must pair strategies' performance catch up the performance of intelligent pair type three and solo strategy. The must pair strategy type three performance the best among all must pair strategies.

Test Case Two

The task set in this testing becomes little easier, because it has fewer intermediate tasks but more easy tasks, the complex tasks remain have level 9 tasks, which are very complex for the team to complete it. All tasks should be worked based on its priority; each five tasks compose a user story. The team need to work on the task that has higher priority than others. During each task allocation process, the team will be given the number of tasks to choose, which the number is same as the team number. Such as if the team number is 5 so that 5 tasks are given at each time, if the team number is 8, then 8 tasks are given at each time. We can then observe the time need to complete those tasks are different by using various strategies. We repeatedly run the testing on each strategy for 10 times in order to get the average complete time, to evaluate which strategy is the most effective.

Agents Team

The experiments show intelligent pair type two performance the best, followed by intelligent pair type one, the must pair type three performance better than other strategies, even better than the solo strategy.

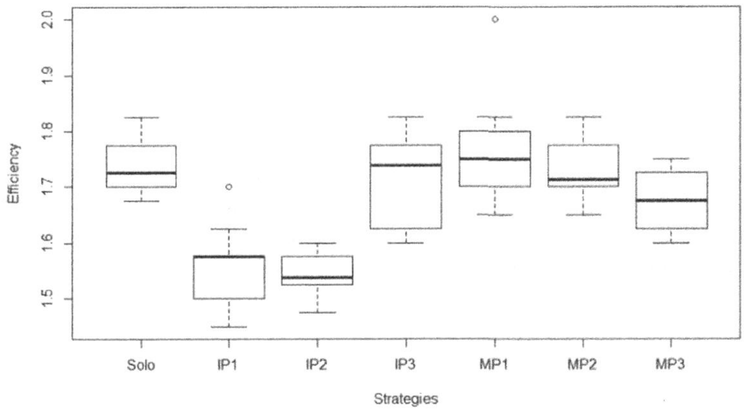

Fig. 8. 7 Agents testing result.

We should see in Fig. 8 that intelligent pair one, two are very good and stable, specially IP1 and IP2 are very good. Under the 7 agents as team, the intelligent pair type two and three performance the best. For the reason the task set complexity is high, so that must pair even catch up the performance of solo and intelligent pair type three.

Agents Team

The intelligent pair type two performance the best among all strategies, followed by must pair type three. Must pair strategy performance good when there are 8 number of agents as the team. Intelligent pair type one and must pair type two performance second best.

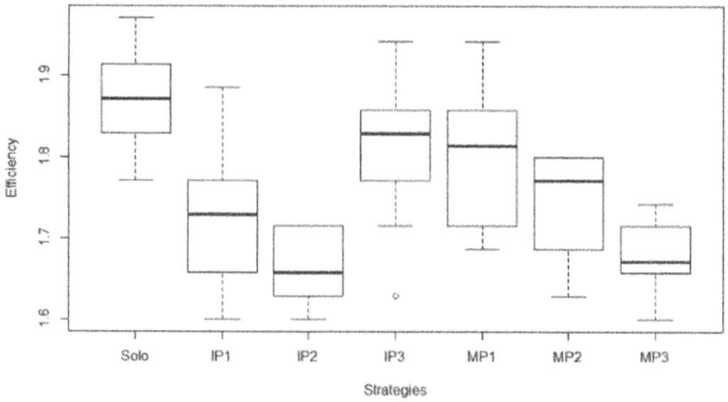

Fig. 9. 8 Agents testing result.

We should see in Fig. 9 that intelligent pair one, two are very good and stable, specially IP1 and IP2 are very good. Must pair three performance good as well. Under 8 agents as team, the must pair type three even catch up the performance of intelligent pair type two, solo performance worst under this context.

Test Case Three

The task set in this testing becomes more easier, because it has fewer complex tasks but more easy tasks and intermediate tasks, the complex tasks remain have level 9 tasks, which are very complex for the team to complete it. All tasks should be worked based on its priority; each five tasks compose a user story. The team need to work on the task that has higher priority than others. During each task allocation process, the team will be given the number of tasks to choose, which the number is same as the team number. Such as if the team number is 5 so that 5 tasks are given at each time, if the team number is 8, then 8 tasks are given at each time. We can then observe the time need to complete those tasks are different by using various strategies. We repeatedly run the testing on each strategy for 10 times in order to get the average complete time, to evaluate which strategy is the most effective.

Agents Team

The intelligent pair type two performance best among all strategies, follow by intelligent pair type three, must pair type one performance worst among all strategies, because the number of agents is odd as 7.

We should see in Fig. 10 that intelligent pair one, two and three are very good and stable, specially IP2 and IP3 are very good. In this case, the task set complexity is further reduced, which provide more opportunities for intelligent pair type three to performance better, solo can performance better than pair under such task context.

Agents Team

In this experiments, intelligent pair type two performance the best, followed by intelligent pair type three, must pair type one performance worst among all strategies.

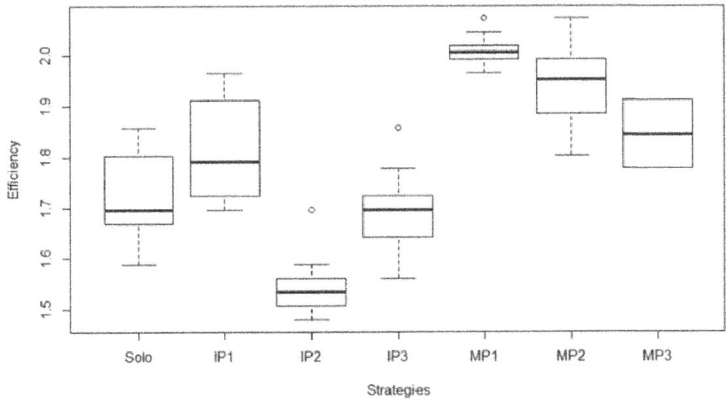

Fig. 10. 7 Agents testing result.

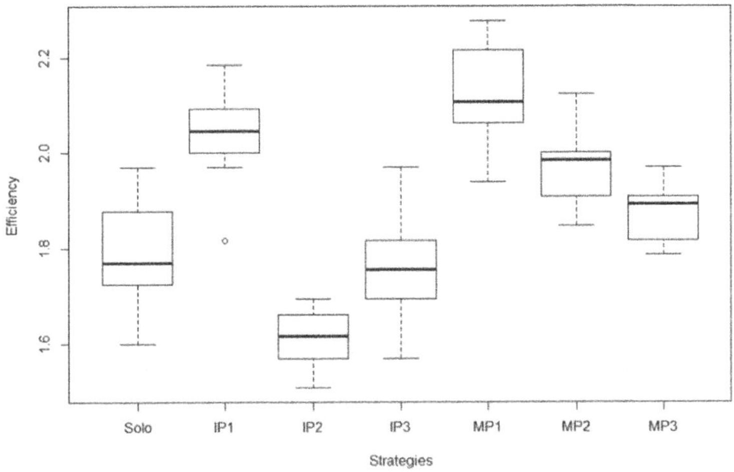

Fig. 11. 8 Agents testing result.

We should see in Fig. 11 that intelligent pair one, two and three are very good and stable, specially IP2 and IP3 are very good. Under such task set context, solo and intelligent pair type two and three all performance good.

Test Case Four

The task set in this testing becomes very easy for the team, because it has fewer complex tasks but more easy tasks and intermediate tasks, the complex tasks are only level 8 tasks, which are not complex for the team to complete it. All tasks should be worked based on its priority; each five tasks compose a user story. The team need to work on the task that has higher priority than others. During each task allocation process, the team will be given the number of tasks to choose, which the number is same as the team number. Such as if the team number is 5 so that 5 tasks are given at each time, if the team number is 8, then 8 tasks are given at each time. We can then observe the time need to complete those tasks are different by using various strategies. We repeatedly run the testing on each strategy for 10 times in order to get the average complete time, to evaluate which strategy is the most effective.

In this experiments, intelligent pair one performance the best, and intelligent pair two performance the second best among all strategies. The solo strategy performance the third best. The must pair type two and three performance the worst.

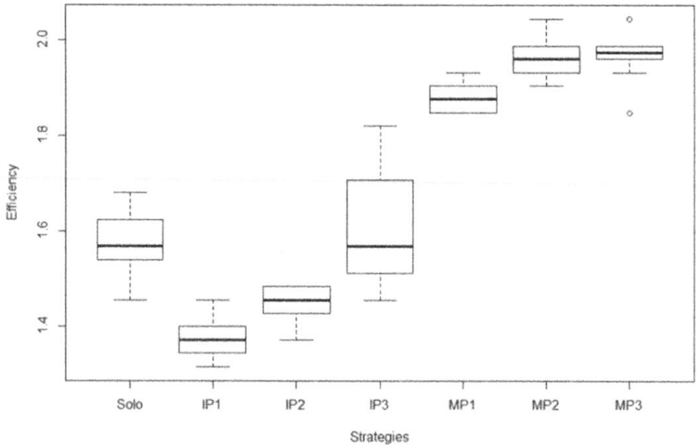

Fig. 12. 7 Agents testing result.

We should see in Fig. 12 that intelligent pair one, two and three are very good and stable, specially IP1, IP2 are very good. Under such easy task set context, the solo strategy, intelligent pair type one, two and three all performance better than any must pair strategy.

Agents Team

The intelligent pair type one performance the best among all strategies, follow by intelligent pair type two. The intelligent pair type three performance the third best, however, must pair type two performance the worst.

We should see in Fig. 13 that intelligent pair one, two and three are very good and stable, specially IP1, IP2 and IP3 are very good. Under such easy task set, the must pair strategies are not performance any better than solo strategy. The intelligent pair type one, two and three all performance good.

Test Case Five

Under such context in Fig. 14, the intelligent type one, two, three and solo performance good. The intelligent pair one performance the best. Must pair performance not good.

Under such context in Fig. 15, the intelligent type one, two performance good. The intelligent pair one performance the best. Must pair performance not good.

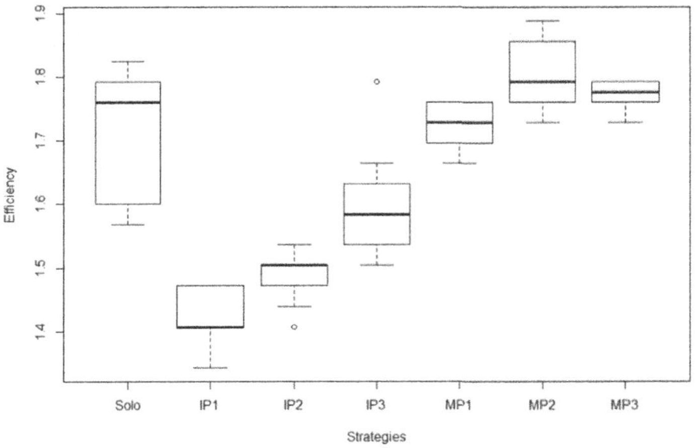

Fig. 13. 8 Agents testing result.

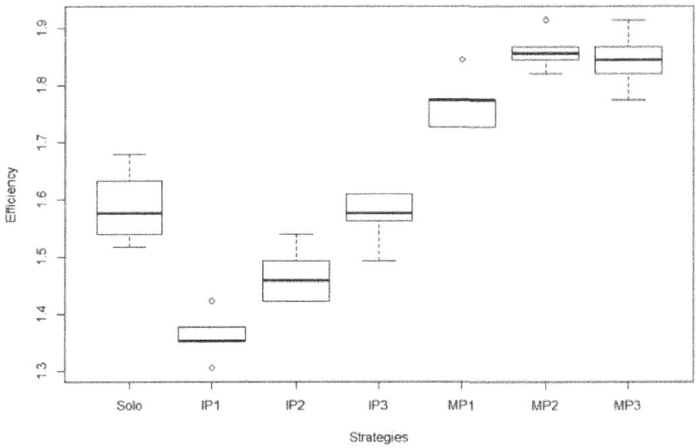

Fig. 14. 7 Agents testing result.

Test Case Six

Intelligent pair type one and two performance the best as shown in Fig. 16.

In Fig. 17, Intelligent pair type one and must pair type three performance the best, because the task set is the most complex, so that must pair works under such context. As we know, the higher the task complexity, the less chance the pair get penalty based on the rules of must pair, so that must pair can performance good when task set is very complex and there are even number of agents.

Discussion

The intelligent pair one, two and three can be used in any circumstance and performance better than solo and must pair. Its advantage comes from the design of intelligent pair

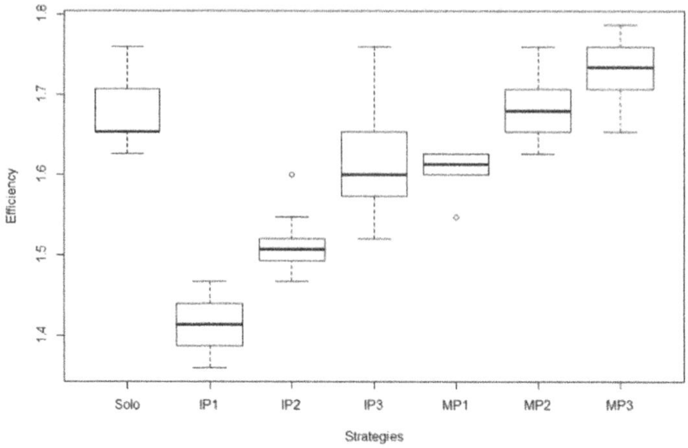

Fig. 15. 8 Agents testing result.

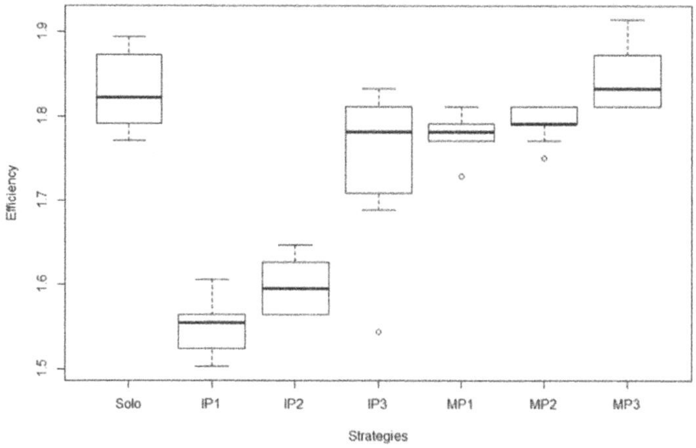

Fig. 16. 7 Agents testing result.

which using the advantage of both solo and pair and ignore the negative side of solo and pair. Intelligent pair one and intelligent two performance much better than intelligent pair three, because it follows the rules of remove the negative side of solo. However, intelligent pair three has the largest application range that other strategies does not have, intelligent pair three is more elastic to the real world when a team is composed with novice and experts, it provides the largest chance for novice to get a job to do, which is elastic for small start-up enterprise. More detailed discussion is provided in the summary of the chapter by crossing analysis.

When task complexity changed, what is the best strategy for different number of agents?

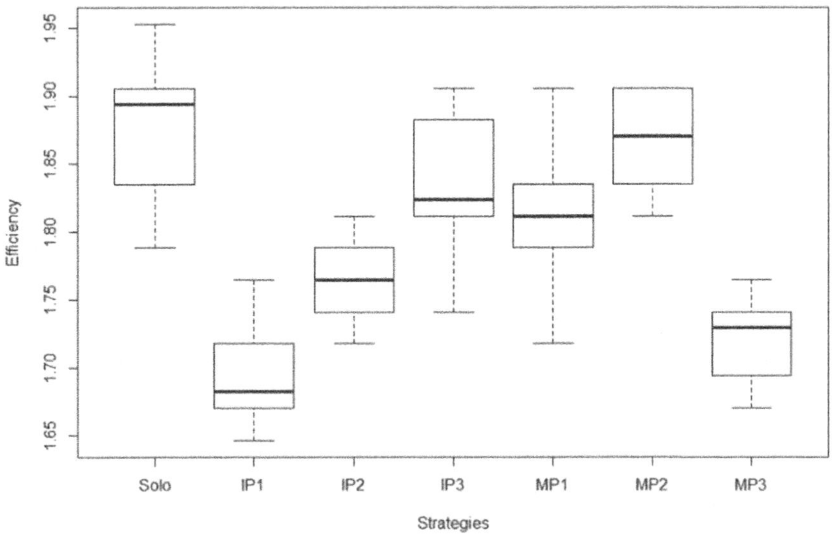

Fig. 17. 8 Agents testing result.

Table 10. 5 Agents testing.

Task complexity	Best performance	Efficiency value	Second performance	Efficiency value	Third performance	Efficiency value
Test case one	IP1	1.35	IP2	1.37	IP3	1.52

In Table 10, In the test case one, under the 5 agents as team context, the best performance is intelligent pair type one strategy, the second is intelligent pair type two strategy and the third is intelligent pair type three strategy. Intelligent pair performance overwhelmingly good than any other strategy, which the top three are all occupied by the intelligent pair family.

Table 11. 6 Agents testing.

Task complexity	Best performance	Efficiency value	Second performance	Efficiency value	Third performance	Efficiency value
Test case one	IP2	1.45	IP1	1.48	IP3	1.67

In Table 11, In the test case one, under the 6 agents as team context, the best performance is intelligent pair type two strategy, the second is intelligent pair type one strategy and the third is intelligent pair type three strategy. Intelligent pair performance overwhelmingly good than any other strategy, which the top three are all occupied by the intelligent pair family.

Table 12. 7 Agents testing.

Task complexity	Best performance	Efficiency value	Second performance	Efficiency value	Third performance	Efficiency value
Test case one	IP2	1.52	IP1	1.55	IP3	1.71
Test case two	IP2	1.54	IP1	1.56	MP3	1.68
Test case three	IP2	1.55	IP3	1.69	SOLO	1.72
Test case four	IP1	1.38	IP2	1.45	SOLO	1.58
Test case five	IP1	1.36	IP2	1.46	IP3	1.57
Test case six	IP1	1.55	IP2	1.60	IP3	1.75

In Table 12, In the 7 agents' context, through test case one to test case two, all the best performance strategies are come from intelligent pair family, intelligent pair type two occupy three cases, intelligent type pair one occupies one case. The second best are all come from intelligent pair family as well, intelligent pair type one occupies two cases, intelligent pair type three occupy one case, intelligent pair type two occupy one case. The third best are coming from intelligent pair type three, must pair type three and solo strategy. The intelligent pair family overwhelmingly performance the best in all cases.

Solo strategy can only performance better when task complexity is reduced, as from test case one to test case four, the task complexity is reducing. Must pair family be not suitable when team numbers are odd number.

In Table 13, In the 8 agents' context, through test case one to test case four, all the best performance strategies are come from intelligent pair family, intelligent pair type two occupy three cases, intelligent type pair one occupies one case. The second best are coming from both intelligent pair family and must pair family, there are intelligent pair type one, must pair type three, intelligent pair type three and intelligent pair type two. The third best are must pair type three, intelligent pair type one, solo and intelligent pair type three. The intelligent pair family overwhelmingly performance the best in all cases.

Solo strategy can only performance better when task complexity is reduced, as from test case one to test case four, the task complexity is reducing. Must pair family be suitable when team numbers are even number.

Table 13. 8 Agents testing.

Task complexity	Best performance	Efficiency value	Second performance	Efficiency value	Third performance	Efficiency Value
Test case one	IP2	1.60	IP1	1.66	MP3	1.77
Test case two	IP2	1.67	MP3	1.68	IP1	1.73
Test case three	IP2	1.61	IP3	1.75	SOLO	1.78
Test case four	IP1	1.42	IP2	1.49	IP3	1.60
Test case five	IP1	1.42	IP2	1.51	IP3	1.61
Test case six	IP1	1.69	MP3	1.72	IP2	1.77

5 Summary

The intelligent pair is the best strategy, especially intelligent par type one and two.

Based on those experiments, we can observe that the intelligent pair always performance the best in any circumstance and context, which significantly better than solo strategy and must pair strategies. The reason is because the intelligent pair use the benefit of solo and pair at the same time but avoided the negative aspect of solo and pair programming. Such intelligent feature makes the intelligent pair the best among all strategies.

The working efficiency is really a balance between give more change for agents to work or limited some agents to work on tasks. To make all novice busy or more chance to work seems to be not a good ideal, even it can reduce idle time, but will decrease the overall team performance.

References

Abar, S., Theodoropoulos, G.K., Lemarinier, P., O'Hare, G.M.P.: Agent based modelling and simulation tools: a review of the state-of-art software. Comput. Sci. Rev. **24**, 13–33 (2017). https://doi.org/10.1016/j.cosrev.2017.03.001

Agarwal, R.: A flexible model for multi-agent based simulation of software development process. Auburn University (2007)

Agarwal, R., Umphress, D.: A flexible model for simulation of software development process. Paper presented at the meeting of the Proceedings of the 48th Annual Southeast Regional Conference, Oxford (2010). https://doi.org/10.1145/1900008.1900064

Ali, S., et al.: Multi-agent system using scrum methodology for software process management. In: Bajwa, I.S., Kamareddine, F., Costa, A. (eds.) INTAP 2018. CCIS, vol. 932, pp. 787–792. Springer, Singapore (2019). https://doi.org/10.1007/978-981-13-6052-7_68

Alsalemi, A.M., Yeoh, E.: A survey on product backlog change management and requirement traceability in agile (Scrum). In: 2015 9th Malaysian Software Engineering Conference (MySEC), pp. 16–17, December 2015. https://doi.org/10.1109/MySEC.2015.7475219

Argote, L., Fahrenkopf, E.: Knowledge transfer in organizations: the roles of members, tasks, tools, and networks. Organ. Behav. Hum. Decis. Process. **136**, 146–159 (2016). https://doi.org/10.1016/j.obhdp.2016.08.003

Arisholm, E., Gallis, H., Dyba, T., Sjoberg, D.I.K.: Evaluating pair programming with respect to system complexity and programmer expertise. IEEE Trans. Softw. Eng. **33**(2), 65–86 (2007). https://doi.org/10.1109/TSE.2007.17

Beck, K., Fowler, M.: Planning Extreme Programming (2001)

Beck, K., Grenning, J., Martin, R.: Manifesto for Agile Software Development (2020). https://agilemanifesto.org/

Bellifemine, F.: Develpoing Multi-Agent Systems with JADE (2007)

Boulahbel-Bachari, S., El Saadi, N.: Migration and self-selection: an agent-based model. In: Silhavy, R., Silhavy, P., Prokopova, Z. (eds.) CoMeSySo 2019 2019. AISC, vol. 1047, pp. 288–303. Springer, Cham (2019). https://doi.org/10.1007/978-3-030-31362-3_28

da Silva, F.Q.B., et al.: Team building criteria in software projects: a mix-method replicated study. Inf. Softw. Technol. **55**(7), 1316–1340 (2013). https://doi.org/10.1016/j.infsof.2012.11.006

Dybå, T., Arisholm, E., Sjøberg, D.I.K., Hannay, J.E., Shull, F.: Are two heads better than one? On the effectiveness of pair programming. IEEE Softw. **24**(6), 12–15 (2007). https://doi.org/10.1109/MS.2007.158

Griffith, I., Taffahi, H., Izurieta, C., Claudio, D.: A simulation study of practical methods for technical debt management in agile software development. In: Proceedings of the Winter Simulation Conference, pp. 7–10, December 2014. https://doi.org/10.1109/WSC.2014.7019961

Han, J., Liu, J., Lei, H.: Agent-based simulation of emergency response of urban oil and gas pipeline leakage. In: Proceedings of the 11th International Conference on Computer Modeling and Simulation, North Rockhampton (2019).https://doi.org/10.1145/3307363.3307396

Hannay, J., Dyb, T., Arisholm, E., Sjobery, D.: The effectiveness of pair programming: a meta-analysis. Inf. Softw. Technol. **51**(7), 1110–1122 (2009). https://doi.org/10.1016/j.infsof.2009.02.001

Hashimoto, A., Takata, K., Ito, N., Matoba, R., Tani, K., Maeda, Y.: Study of the influence of an obstacle on the evacuation behavior using multi-agent simulation where the intimate space around each agent is considered. In: Proceedings of the 11th International Conference on Computer Modeling and Simulation, North Rockhampton (2019). https://doi.org/10.1145/3307363.3307369

Ivanov, D., Kapustyan, S., Kalyaev, A., Korovin, I.: Decision support systems for the oil fields with cloud multiagent service. In: Silhavy, R., Silhavy, P., Prokopova, Z. (eds.) CoMeSySo 2019 2019. AISC, vol. 1047, pp. 16–23. Springer, Cham (2019). https://doi.org/10.1007/978-3-030-31362-3_3

Joslin, D., Poole, W.: Agent-based simulation for software project planning. In: Proceedings of the Winter Simulation Conference, p. 4, December 2005. https://doi.org/10.1109/WSC.2005.1574359

Justice, J.: The 3–5–3 of Scrum (2018)

Licorish, S.A., MacDonell, S.G.: Exploring software developers' work practices: task differences, participation, engagement, and speed of task resolution. Inf. Manage. **54**(3), 364–382 (2017). https://doi.org/10.1016/j.im.2016.09.005

Licorish, S.A., MacDonell, S.G.: Exploring the links between software development task type, team attitudes and task completion performance: insights from the Jazz repository. Inf. Softw. Technol. **97**, 10–25 (2018). https://doi.org/10.1016/j.infsof.2017.12.005

Lin, J.: Context-aware task allocation for distributed agile team. In: Proceedings of the 28th IEEE/ACM International Conference on Automated Software Engineering, Silicon Valley (2013). https://doi.org/10.1109/ase.2013.6693151

Lin, J.: Human Factors in Agile Software Development. arXiv:1502.04170 (2015)

Lin, J., Yu, H., Shen, Z.: An empirical analysis of task allocation in scrum-based agile programming. arXiv:1411.6201 (2014)

Lin, J., Yu, H., Shen, Z., Miao, C.: Studying task allocation decisions of novice agile teams with data from agile project management tools. In: Proceedings of the 29th ACM/IEEE International Conference on Automated Software Engineering, Vasteras (2014). https://doi.org/10.1145/2642937.2642959

Lin, M., Zhang, Q.: Time scales of knowledge transfer with learning and forgetting. Phys. A Stat. Mech. Appl. **525**, 704–713 (2019). https://doi.org/10.1016/j.physa.2019.03.084

Lui, K.M., Chan, K.C.C.: Pair programming productivity: novice–novice vs. expert–expert. Int. J. Hum. Comput. Stud. **64**(9), 915–925 (2006). https://doi.org/10.1016/j.ijhcs.2006.04.010

Macal, C.M., North, M.J.: Tutorial on agent-based modeling and simulation part 2: how to model with agents. In: Proceedings of the 2006 Winter Simulation Conference, pp. 3–6, December 2006. https://doi.org/10.1109/WSC.2006.323040

Macal, C.M., North, M.J.: Agent-based modeling and simulation: ABMS examples. In: Proceedings of the 40th Conference on Winter Simulation, Miami (2008)

Mahnič, V., Hovelja, T.: On using planning poker for estimating user stories. J. Syst. Softw. **85**(9), 2086–2095 (2012). https://doi.org/10.1016/j.jss.2012.04.005

Masood, Z., Hoda, R., Blincoe, K.: Motivation for self-assignment: factors agile software developers consider. In: 2017 IEEE/ACM 10th International Workshop on Cooperative and Human Aspects of Software Engineering (CHASE), p. 23, May 2017. https://doi.org/10.1109/CHASE.2017.18

Mehmood, S., Ahmed, S., Kristensen, A.S.: Application of integrated model of evacuation psychology in an agent-based simulation. In: Proceedings of the 11th International Conference on Computer Modeling and Simulation, North Rockhampton (2019). https://doi.org/10.1145/3307363.3307389

Nadler, D., Hackman, J.R., Lawler, E.E.: Managing Organizational Behavior (1979)

Nilsson, K.: Increasing Quality with Pair Programming - An Investigation of Using Pair Programming as a Quality Tool (2003)

Phillips, J.: Team-RUP: An Agent-Based Simulation Study of Team Behavior in Software Development Organizations (2006)

Plonka, L., Sharp, H., van der Linden, J., Dittrich, Y.: Knowledge transfer in pair programming: an in-depth analysis. Int. J. Hum. Comput. Stud. **73**, 66–78 (2015). https://doi.org/10.1016/j.ijhcs.2014.09.001

Ramanujam, R., Lee, I.: Collaborative and competitive strategies for agile scrum development. In: 7th International Conference on Networked Computing and Advanced Information Management, pp. 21–23, June 2011

Schwaber, K., Sutherland, J.: The Scrum Guide, pp. 21–23, June 2011

Song, H., et al.: Development and validation of the primary care team dynamics survey. Health Serv. Res. **50**(3), 897–921 (2015). https://doi.org/10.1111/1475-6773.12257

Stober, T., Hansmann, U.: Considerations on teaming and leadership. In: Stober, T., Hansmann, U. (eds.) Agile Software Development, pp. 75–92. Springer, Berlin (2010). https://doi.org/10.1007/978-3-540-70832-2_5

Tyrychtr, J., Pelikán, M., Kvasnička, R., Ander, V., Benda, T., Vrana, I.: Multi-agent system in smart econometric environment. In: Silhavy, R., Silhavy, P., Prokopova, Z. (eds.) CoMeSySo 2019 2019. AISC, vol. 1046, pp. 434–442. Springer, Cham (2019). https://doi.org/10.1007/978-3-030-30329-7_38

Wang, X., Wang, J., Zhang, R.: The optimal feasible knowledge transfer path in a knowledge creation driven team. Data Knowl. Eng. **119**, 105–122 (2019). https://doi.org/10.1016/j.datak.2019.01.002

William, R., Uri, W.: An Introduction to Agent-Based Modeling: Modeling Natural, Social, and Engineered Complex Systems with NetLogo (2015)

Williams, L., Kessler, R.R., Cunningham, W., Jeffries, R.: Strengthening the case for pair programming. IEEE Softw. **17**(4), 19–25 (2000). https://doi.org/10.1109/52.854064

Wooldridge, M., Jennings, N.R.: Agent theories, architectures, and languages: a survey. In: Wooldridge, M.J., Jennings, N.R. (eds.) ATAL 1994. LNCS, vol. 890, pp. 1–39. Springer, Heidelberg (1995). https://doi.org/10.1007/3-540-58855-8_1

Yilmaz, L., Phillips, J.: The impact of turbulence on the effectiveness and efficiency of software development teams in small organizations. Softw. Process: Improv. Pract. **12**(3), 247–265 (2007). https://doi.org/10.1002/spip.318

Zhen, Y., Wanpeng, Z., Hongfu, L.: Artificial intelligence techniques on real-time strategy games. In: Proceedings of the 2018 2nd International Conference on Computer Science and Artificial Intelligence, Shenzhen (2018). https://doi.org/10.1145/3297156.3297188

Zieris, F., Prechelt, L.: On knowledge transfer skill in pair programming. In: Proceedings of the 8th ACM/IEEE International Symposium on Empirical Software Engineering and Measurement, Torino (2014). https://doi.org/10.1145/2652524.2652529

SBAP: A Novel Secure Bucket Access Protocol for Leveraging User's Data Privacy in Cloud Environment

N. P. Hemanth Kumar[1](✉) and S. Prabhudeva[2]

[1] Department of Computer Science and Engineering, Alvas Institute of Engineering and Technology, Mijar, Moodbidri, India

[2] Department of Information Science and Engineering, JNNCE, Shivamogga, India

Abstract. With large number of inter-connected of the smart appliances over distributed cloud environment, the user's private information is exposed to higher degree of vulnerability. Existing applications that executes over such devices are mandatorily require an access of user's private information in order to authorize the user to experience the privilege of cloud services rendering declination of data privacy. Therefore, proposed system introduces Secure Bucket Access Protocol (SBAP) that not only offers a secure storage of user's private information but also perform lightweight encryption of such data. The contribution of proposed SBAP is the migration of access control policy from device to cloud to offer higher scope of data privacy. The study outcome exhibits faster response time ideal to secure a communication channel over lethal threats in contrast to frequently used encryption scheme.

Keywords: Device security · Data privacy · Smart devices · Internet-of-Things · Cloud security

1 Introduction

The massive adoption of Internet of Things (IoT) collaboration of systems with the cloud is obviously a domain to be fact in the future [1]. These applications raise an associated concern towards the security aspects of the data and more importantly the specific concern of the security called privacy preservation [2]. In the eco system of the IoT and cloud computing apart from other security threats such as defense against specific attack, cryptographic approach towards data security, ensuring the privacy of the system paradigm features to be preserved which might be very sensitive to the losses concerns [3–5]. There is need of having all the associated devices to be intelligent or capable enough to identify the user's identity, along with provisioning of the data storage as well as sharing and synchronization in most effect way. The very basic agenda to build all the eco-system of IoT is to make the things more autonomous and archive an objective of fast, efficient, ubiquitous, robust, scalable and moreover environmental friendly technology provisioning. The Fig. 1 typically illustrates the future domain where such eco-system and need of data sensitivity preservation is needed.

R. Silhavy (Ed.): CSOC 2021, LNNS 230, pp. 37–46, 2021.
https://doi.org/10.1007/978-3-030-77442-4_2

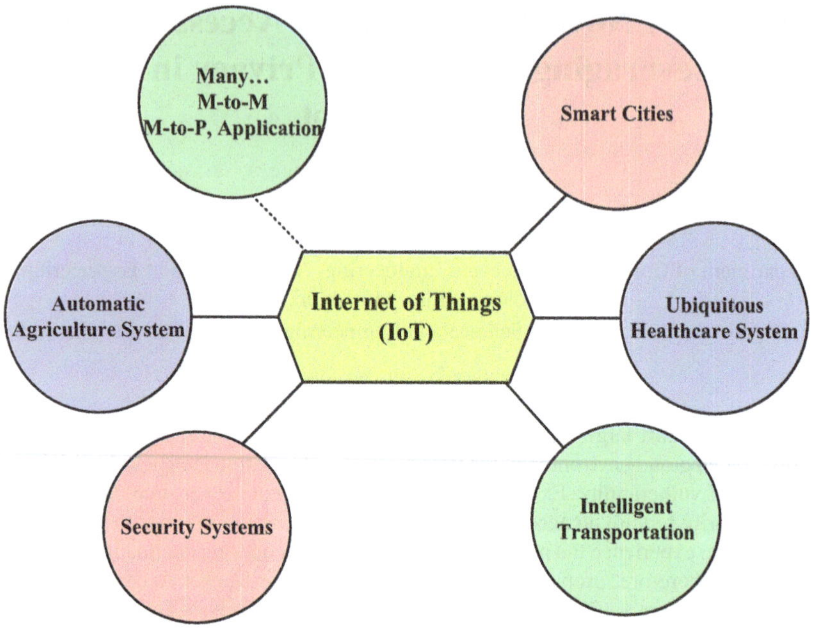

Fig. 1. IoT eco-system based application domains

In the Fig. 1, it is clearly that in order to achieve the user control the eco-system of IoT need to be seamlessly synchronized with such a device with the user that it can sense, store, process and communicate and the era of smart phone came into existence. The very basic philosophy behind such smart device is to avoid the use of the multifold keys for identity for different services. The initial steps to be registered with such system requires to provide the personal information, financial details, password to get authenticated etc., by every users and these information's are very vital which need to be preserved by any kind of leakage [6–10]. The existing mechanism considers the usage of smart phones/ devices which generates as well as stores the data.

The most frequent data from most of the applications requires is the user location. Though there are provisions by the user to permit that whether they want to be their location to be known to the application or not but in most of the cases user permits their location to be known to the application. These data's are continuously monitored unless it is interrupted by the user and the logs are deleted from the device. The sensitivity of the data leakage becomes an responsibility of the device manufacturer or the service provider of the applications. The work done by Enck and Gilbert [11] reveals most of the smart devices do not provision a transparency of the way the user's private data is used. In case of user location data leakage it could be exploited to compromise even the physical security of the user could be on stake apart from other associated risk concern with users' location. There exist much application which requires images, videos, photos etc., and the smart device of today's worlds is equipped with the camera as well as many social network applications facilitate. The authorization process for sharing these contents is very essential from the security view point. The images, photos and video contain much more vital information by which users personal and private details can be easily interpreted and personal and professional losses can occur. In case of smart phones, tablets etc., there is important role of the address book where today's for the ease of communication the user stores contact number, email ID, professional details into the advance address book. Leakage of address book poses huge and potential loss to one's wrongful exploitation of privacy data. A similar utility in the smart phones called scheduler or appointment application, where the user keeps the contact details along with the plans and schedules of the meeting with the addressee. In case of corporate user there will be huge loss by means of compromising these details by the competitor. The very vital information namely Mobile station equipment identity (IMEI) and unique device identifier (UDID) assigned uniquely to one device and most often is used to track and identify the user. Leakage of this information can extrapolate the user behaviour. When these data's are collaborated then it could be more potential threats to the application build in the eco-system of IoT. Therefore, a method needs to develop where in-spite to residing these data into local storage of the smart devices, it should be security stored and organized into the cloud infrastructure.

This paper presents design of novel and very simple solution which emphasize on the data security to strength the data privacy in cloud environment. The organization of the paper is as following: The discussion of existing approaches has been carried out in Sect. 2 followed by briefing of problem description in Sect. 3. The proposed methodology of the system model is discussed in Sect. 4 that is followed by Algorithm description in Sect. 5. Result discussion has been carried out in Sect. 6 while summary of proposed contribution in carried out in Sect. 7 in form of conclusion.

2 Related Work

This section upgrades the findings of the existing literatures from our prior review work [12] towards exploring privacy preservation techniques on cloud. Discussion of various forms of adversaries on smart devices over cloud is carried out by Liu and Sun [13], Zhou et al. [14], and Szymanski et al. [15] where it is found that the security problems are still at large. Liu et al. [16] presents an *authentication* scheme using *multiple factors* for secure data storage over cloud. Authentication scheme for smart grids have been investigation by Sha et al. [17] and Zhang et al. [18] using similar *multi-factor scheme.* Existing system has witnessed adoption of *identity-based encryption* as one of the potential solution to address data security over cloud as witnessed in Baek et al. [19]. There are certain approaches that explored deeper information for developing a security protocol. Study in similar direction is being carried out by Guan et al. [20] using *attribute-based encryption.* Adoption of *key agreement protocol* towards privacy protection was carried out by Hu et al. [21] considering fog computing. Key establishment scheme has been reported to offer a secure generation of secret key as noticed in the work of Li et al. [22]. Similar direction of investigation has been also initiated by Lu et al. [23] using *homomorphic encryption.* Literature has witnessed usage of *adaptive methodologies* for exploring data injections over cloud. The work of Jiang et al. [24] addresses latent intrusion using data injection using such technique. Integration of adaptivity and hierarchical approach was also proven to be effective against data privacy issues as reported in work of Lee et al. [25]. *Mutual-authentication*-based approach using *public-encryption scheme* was implemented by Li et al. [26] for securing smart appliances. Study towards optimizing communication link along with security has been investigated by Shin et al. [27] using *mutual-authentication scheme*. Analytical-based approaches e.g. *semantics* and *context* was used for exploring more critical intrusion issues in smart devices as seen in work of Celdran et al. [28]. Another unique form of technique has been presented by Chang and Cheng [29] where *smart cards* were used for resisting adversaries over cloud. A typical approach of traffic provisioning for reducing data leakage is studied in the work of Liu et al. [30]. A very unique form of modeling has been carried out by Sui et al. [31], by introducing a solution towards data anonymity problem using *correlation factor* between the trajectories of positions of data origination. Study towards resisting security risk during mobile offloading mechanism has been carried out by Xiao et al. [32] using *game theory* concept.

3 Problem Description

The problem description inferred after reviewing the existing approaches of data privacy over cloud environment are:

- Existing approaches of security solutions are found to offer robust cryptographic deployment to over better data privacy; however, the protection is offered for all the data without any discrimination of its sensitivity level. Such approaches offer cloud security at the cost of storage cost.
- One of the biggest challenges in existing approaches are lack of emphasis of the user's private information over the smart appliances. Present technique of application execution depends on such device-related information without which the accessibility cannot be offered for any cloud-based services.
- Existing identification and service relaying operation is initiated from the smart devices where the user's private information is highly vulnerable for man-in-middle attack as well as against rogue codes. The worst part is apart from user's private information; the intrusion can equally affect any other information that is stored within the smart devices.
- Existing authentication-based principle offers multiple layers of encryption but within the same actors. However, in reality there are presence of multiple components in clouds which are having equal probability of getting compromised. Hence, there is a need of authentication mechanism that offers mandatory encryption of data for any actors involved in data transmission or storage in cloud environment.

Therefore, the problem statement of the designed system model is "*To offer a faster and cost effective data privacy exclusively for user's private information hosted within user device over distributed cloud environment.*" The next section 4.0 the proposed solution to address the above mentioned issues is described.

4 System Model: Proposed Methodology

The prime objective of the designed system model is to preserve the private data in the IoT eco-system based applications, whereas the existing data storage mechanism in smart device having treats and means available to leak it. In order to design an efficient and robust mechanism it is essential to understand the vulnerability by which the data sharing especially image sharing application exploits the means of providing permission to share the data from one user to other user. This process can be termed as "data overhearing or data crowding", such events even occur while paying utilities or commodities by credit/debit cards, where while registration process to such application. Another terminology called threat for security can be considered as the harm to the user by means of data theft and this harm can have its own weightage. As per the weightage of the probable threats the degree of permission and cryptographic approaches need to be provisioned but these may provide a tradeoff between the level of complexity of the cryptography algorithms and the computational overhead, which is a difficult task to handle as the smart devices used by the user do not have as much higher capabilities in terms of storage, processing and computational efficiencies along with consumptions of power require for each operations. The relationship between Threat for Security Weightage (TSW), Total Time for Execution (TTE), Power Requirement (P), and the Computational Complexity (CC) may be use some field parameter £ and μ where, the Eq. (1) describes.

$$\text{£} = \text{TSW}/\ \text{TTE} \tag{1}$$

$$\&,\ \mu = \text{P} / \text{CC} \tag{2}$$

$$\text{If TTE} = \text{CC, then TSW} = (\text{£}/\mu)\ \text{x} \tag{3}$$

4.1 Authorization Model

Most of the existing devices having only dual mode mechanism for the apps to accessing or getting permission and those are either all the data or none of the data. This mechanism provides the data overhearing or data crowding possibility if the permission is set to all. Therefore, the authorization model proposes different level of permission depending upon the threat for security weightage. The minimization process is achieved on the basis of Number of Users (NU) and Data of Each User (ND), Threat for Security Weightage (TSW), and a real time constant α.

$$\text{Permitted Authorization (PA)} = [(\text{NU}/\ \text{ND})\ \text{X}\ \text{TSW}\ ] + \alpha \tag{4}$$

4.2 System Architecture

The possibilities are very rare that the user's privacy data (UPD) can be restricted to the applications as well as, at the same time expecting from the end user to be cautious enough due to uniform user behavior to control the access and authorization tracking of these private data. These compulsorily unavoidable vulnerability protects require a new pattern to deal with these data. This paper aims to develop a smart device – Cloud infrastructure to ensure optimal security of user's private data. The conceptual architecture is shown in the figure below:

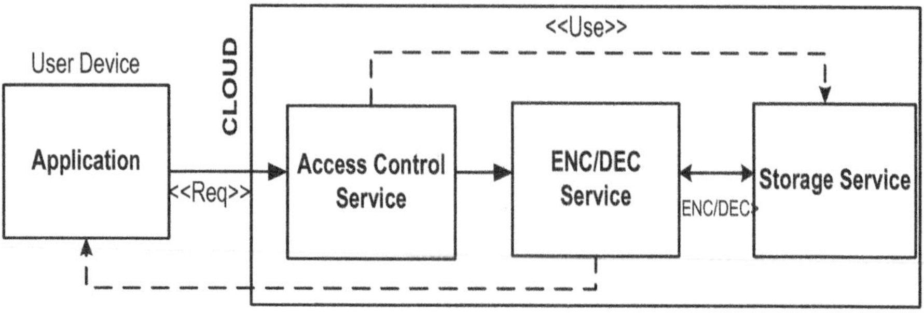

Fig. 2. Proposed Architecture

In the above proposed architecture of Cloud, following provisioning is done: i) The user data will be stored into the cloud to a place called secure bucket, ii) Create user, iii) Register user with cloud, iv) Create a secure bucket for the user, and v) Store all the private data of the user in the secure bucket in encrypted format so that the user device will have reduced task to an extent of local computation etc. This approach makes the user device scalable and it could be used for many more applications without having botheration of grooming big data as well assure security to the data to a great extent. It is an open research problem to investigate the effectiveness and reliability of existing security measure for the data security on cloud; even though as compared to device level vulnerability, cloud level data security is more preferable. The light weightness into the operation at device level is brought as all the cryptographic functions executes on the cloud infrastructure. The application device will only request to the cloud. The proposed system uses parametric encryption mechanism that uses some specific user-defined attributed to perform the process of the encryption. The system model of the proposed system uses this algorithm perform the process of the encryption of data followed by multiple checks on the legitimacy of any form of request. The significant contribution of the designed system model is that it is independent of any types of the of device-level intrusion as the complete operation of accessibility and authorization of secured data is carried out over cloud that runs some robust network security algorithm that is not possible to be implemented within the device. The next section briefs about algorithm description.

5 Algorithm Implementation

In the process of individual client application registration or installation, the application requires the private data information which is kept at secure bucket of the cloud. So, on the moment of data requirement, the user service initiates data access request to the cloud. The Secure Bucket Access Protocol (SBAP) is designed to provision a strong authentication as well authorization of every user service. In the model a secure bucket stores the data as well provides the security by optimal cryptographic approach as per the level of sensitivity of the data and its importance. The validation of the proposed prototype is illustrated with a custom application, the operational aspects of the security is divided into two categories 1) Primary Data (P_D) which user generates and stores into the secure bucket and the data out of (P_D) which the user wants it to share to the application (PDs).

Algorithm: User service (U_s) initiate primary data (P_D) to the secure bucket (B_s) .

Input: service ID, device ID, data

1. Identify the service ID access request type by the device ID
2. IF ART () == same device then
3. Service request authorized the permission
4. ELSE
5. Send the service request to secure bucket access protocol including <<service ID , device ID , data>>.
6. SBAP decides where this application has permission by service ID , device ID , data and access not get (Per).
7. IF Per == true then
8. Encrypt data
9. Store into secure bucket with service ID
10. ELSE
11. Return
12. End IF
13. End IF

The access operation for the user works in following manner: The user will be require to forward its information related to the identity of service and application to the protocol which further checks their legitimacy. In case of valid authorization, the protocol forwards the access request to the cloud bucket which further search the ciphered information and then forwards it to the respective device of the user where the decryption is carried out. A second layer of authorization is constructed which is evaluated for its legitimacy by the decryptor module to assess the authorization of decryption operation of the ciphered date followed by ultimately successful decryption. One of the interesting parts of this algorithm is that it doesn't leak any form of private information at any stage to any third party other than the final authentication of user. The second contributory factor in this algorithm is that it has to go for a series of progressive legitimacy check for accessing any data as well as storing in cloud bucket.

6 Results Discussion

As the proposed system uses a non-conventional mechanism to introduce a unique parametric-based encryption over a distributed system like cloud, hence, it is essential to evaluate the response time of the security process. Using MATLAB, the evaluation of the proposed system is carried out by forwarding 2000 bytes of information for both storage and accessing the encrypted information over cloud. We also evaluate the existing encryption algorithms e.g. Data Encryption Standard (DES), Secured Hash Algorithm (SHA), and Advanced Data Encryption Standard (ADE), which we term as *Existing system* in this analysis.

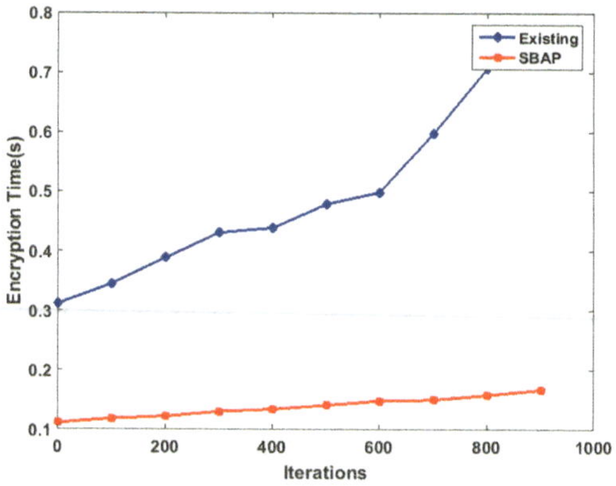

Fig. 3. Encryption Time (DT) vs number of Iteration

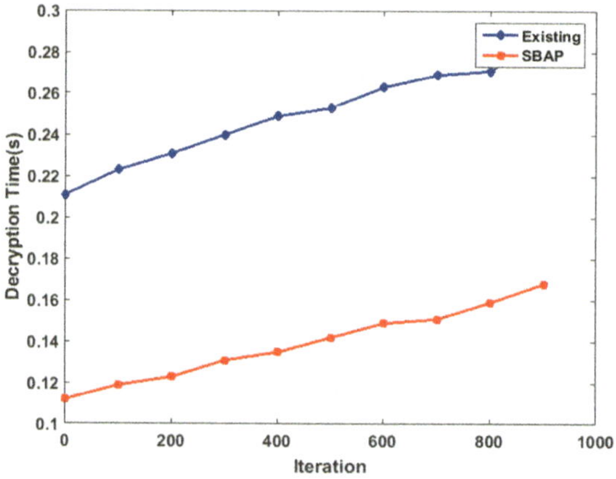

Fig. 4. Decryption Time (DT) vs number of Iteration

The Fig. 3 and Fig. 4 exhibits both the encryption time (ET) and decryption time (DT) in the respective graphs with respect to the iteration of the simulation cycle, the pattern shows that the proposed SBAP offers more than 50% of faster response rate compared to the existing system as its encryption time and decryption time is nearly similar whereas it significantly differs for existing system where 'ET' is comparatively higher than the DT.

Faster encryption time is required to be ensured as the contents are stored over distributed cloud interfaces which shouldn't introduce time lags while performing device authentication. Hence, SBAP offers cost effective data security in IoT.

7 Conclusion

This paper addresses a problem which is related to insecure exposition of private information stored within the device. The novelty in proposed solution is that authentication of such private information is migrated from device to cloud which has multiple advantages viz. i) the device becomes completely free from executing any authentication that offers more device security, ii) it makes the device free from any form of node capture attacks, iii) it also offers more flexibility to add more complex and sophisticated cryptographic approach as the algorithm runs on cloud. The response time exhibited by the proposed system is lower as compared to the existing model of consideration.

References

1. Tripathy, B.K., Anuradha, J.: Internet of Things (IoT): Technologies, Applications, Challenges and Solutions. CRC Press (2017)
2. Lopez, D., Lopez, D., Saleem Durai, M.A.: HCI Challenges and Privacy Preservation in Big Data Security. IGI Global (2017)
3. Mosenia, A., Jha, N.K.: A comprehensive study of security of Internet-of-Things. IEEE Trans. Emerging Topics Comput. **5**(4), 586–602 (2017)
4. Yang, Y., Wu, L., Yin, G., Li, L., Zhao, H.: A survey on security and privacy issues in Internet-of-Things. IEEE Internet Things J. **4**(5), 1250–1258 (2017)
5. Zhang, K., Liang, X., Lu, R., Shen, X.: Sybil attacks and their defenses in the Internet of Things. IEEE Internet Things J. **1**(5), 372–383 (2014)
6. Munir, K.: Cloud Computing Technologies for Green Enterprises. IGI Global (2017)
7. Fei, H.: Security and Privacy in Internet of Things (IoTs): Models, Algorithms, and Implementations, CRC Press (2016)
8. Kolhar, M., Alameen, A., Dhupia, B., Rubab, S., Gulam, M.: Cloud Computing Data Auditing Algorithm. Notion Press (2017)
9. Rosado, D.G.: Security Engineering for Cloud Computing: Approaches and Tools: Approaches and Tools. IGI Global (2012)
10. Russell, B., Van Duren, D.: Practical Internet of Things Security, Packt Publishing Ltd. (2016)
11. Enck, W., et al.: TaintDroid: an information-flow tracking system for realtime privacy monitoring on smartphones. ACM Trans. Comput. Syst. (TOCS) **32**(2), 5 (2014)
12. Hemanth Kumar, N.P., Prabhudeva, S.: Security and privacy preservation on Cloud-based Big Data Analysis (CBDA): a review. Commun. Appl. Electron. **6**(1), 35–42 (2016)
13. Liu, J., Sun, W.: Smart attacks against intelligent wearables in people-centric Internet of Things. IEEE Commun. Mag. **54**(12), 44–49 (2016)
14. Zhou, J., Cao, Z., Dong, X., Vasilakos, A.V.: Security and privacy for cloud-based IoT: challenges. IEEE Commun. Mag. **55**(1), 26–33 (2017)
15. Szymanski, T.H.: Security and privacy for a green Internet of Things. IT Professional **19**(5), 34–41 (2017)
16. Liu, Y., Zhong, Q., Chang, L., Xia, Z., He, D., Cheng, C.: A secure data backup scheme using multi-factor authentication. IET Inf. Secur. **11**(5), 250–255 (2017)
17. Sha, K., Alatrash, N., Wang, Z.: A secure and efficient framework to read isolated smart grid devices. IEEE Trans. Smart Grid **8**(6), 2519–2531 (2017)
18. Zhang, L., Zhang, Y., Tang, S., Luo, H.: Privacy protection for e-health systems by means of dynamic authentication and three-factor key agreement. IEEE Trans. Industr. Electron. **65**(3), 2795–2805 (2018)

19. Baek, J., Vu, Q.H., Liu, J.K., Huang, X., Xiang, Y.: A secure cloud computing based framework for big data information management of smart grid. IEEE Trans. Cloud Comput. **3**(2), 233–244 (2015)
20. Guan, Z., Li, J., Wu, L., Zhang, Y., Wu, J., Du, X.: Achieving efficient and secure data acquisition for cloud-supported Internet of Things in smart grid. IEEE Internet Things J. **4**(6), 1934–1944 (2017)
21. Hu, P., Ning, H., Qiu, T., Song, H., Wang, Y., Yao, X.: Security and privacy preservation scheme of face identification and resolution framework using fog computing in Internet of Things. IEEE Internet Things J. **4**(5), 1143–1155 (2017)
22. Li, Z., Pei, Q., Markwood, I., Liu, Y., Zhu, H.: Secret key establishment via RSS trajectory matching between wearable devices. IEEE Trans. Inf. Forensics Secur. **13**(3), 802–817 (2018)
23. Lu, R., Heung, K., Lashkari, A.H., Ghorbani, A.A.: A lightweight privacy-preserving data aggregation scheme for fog computing-enhanced IoT. IEEE Access **5**, 3302–3312 (2017)
24. Jiang, J., Qian, Y.: Defense mechanisms against data injection attacks in smart grid networks. IEEE Commun. Mag. **55**(10), 76–82 (2017)
25. Lee, Y.T., Hsiao, W.H., Lin, Y.S., Chou, S.C.T.: Privacy-preserving data analytics in cloud-based smart home with community hierarchy. IEEE Trans. Consum. Electron. **63**(2), 200–207 (2017)
26. Li, N., Liu, D., Nepal, S.: Lightweight mutual authentication for IoT and its applications. IEEE Trans. Sustain. Comput. **2**(4), 359–370 (2017)
27. Shin, D., Sharma, V., Kim, J., Kwon, S., You, I.: Secure and efficient protocol for route optimization in PMIPv6-based smart home IoT networks. IEEE Access **5**, 11100–11117 (2017)
28. Celdrán, A.H., Clemente, F.J.G., Pérez, M.G., Pérez, G.M.: SeCoMan: a semantic-aware policy framework for developing privacy-preserving and context-aware smart applications. IEEE Syst. J. **10**(3), 1111–1124 (2016)
29. Chang, C., Cheng, T.F.: A secure diverse ticket-sale system in a distributed cloud environment. Comput. J. **57**(10), 1441–1459 (2014)
30. Liu, E., Cheng, P.: Achieving privacy protection using distributed load scheduling: a randomized approach. IEEE Trans. Smart Grid **8**(5), 2460–2473 (2017)
31. Sui, P., Li, X., Bai, Y.: A study of enhancing privacy for intelligent transportation systems: k-correlation privacy model against moving preference attacks for location trajectory data. IEEE Access **5**, 24555–24567 (2017)
32. Xiao, L., Xie, C., Chen, T., Dai, H., Poor, H.V.: A mobile offloading game against smart attacks. IEEE Access **4**, 2281–2291 (2016)

Energy Distribution Modeling During the Electron Beam Welding Using Dynamically Changing Thermophysical Parameters of the Product

Sergei Kurashkin[1], Vadim Tynchenko[1,2,3](✉), Yuriy Seregin[1], Aleksandr Murygin[1], Vladislav Kukartsev[1,2], and Valeriya Tynchenko[1,2]

[1] Reshetnev Siberian State University of Science and Technology, Krasnoyarsk, Russia
[2] Siberian Federal University, Krasnoyarsk, Russia
[3] Marine Hydrophysical Institute, Russian Academy of Sciences, Sevastopol, Russia

Abstract. The article is devoted to the problem of modeling the energy distribution in the welded joint zone of a product made of VT14 titanium alloy, obtained using the method of electron beam welding. To ensure high accuracy of modeling, it is proposed not to use static values of thermophysical parameters of a product, but their dynamically changed values depending on the current temperature at the point of calculation. Regression models are used to restore the dependence of thermophysical parameters. As a result of experimental studies, it was revealed that a linear regression equation is effective to restore the dependence of the specific heat capacity; for the coefficient of thermal expansion and the heat capacity parameter - the cubic regression equation; for density - quadratic regression. The constructed regression equations allow obtaining the smallest approximation error of the existing tabular values. To solve the problem of modeling the energy distribution, a calculation algorithm was developed for the MATLAB system, considering the dynamic change in thermophysical parameters during the simulation. The conducted experimental study showed that the application of the proposed approach allows one to obtain a more natural model of energy distribution. #CSOC1120.

Keywords: Electron-beam welding · Mathematical modelling · Technological parameters · Electron beam · Optimization of process parameters · Input-output of an electron beam · Control automation

1 Introduction

Regression is the dependence of the mathematical expectation (for example, the mean) of a random variable on one or more other random variables (free variables). Regression analysis is the search for a function f that describes this dependence. Linear regression assumes that the function f depends on the desired parameter (for example x) linearly [1, 2].

The essence of regression analysis is to find the most important factors that affect the dependent variable [1, 2].

Examples of application of regression analysis [1–3]:

R. Silhavy (Ed.): CSOC 2021, LNNS 230, pp. 47–58, 2021.
https://doi.org/10.1007/978-3-030-77442-4_3

- Modeling the university enrollment rate to better understand the factors that keep children in the same institution.
- Modeling migration flows depending on factors such as average wages, availability of medical and school facilities, geographic location, etc.
- Simulation of road accidents as a function of speed, road conditions, weather, etc.; Modeling fire loss as a function of variables such as number of fire stations, call handling time, or property price.

Among the advantages of regression models are the following [4–6]:

- Simplicity of computational algorithms.
- Visibility and interpretability of results (for a linear model).

However, a number of disadvantages of such models can be identified [4–6]:

- Low forecast accuracy (mainly data interpolation).
- The subjective nature of the choice of the type of specific dependence (formal adjustment of the model to the empirical material).
- Lack of an explanatory function (impossibility of explaining a causal relationship).

Electron beam welding - welding, in which heating and melting of products is carried out by a stream of high-speed electrons moving under the action of an electric field in a vacuum. Electron beam welding is used in the aerospace industry to join various structures and assemblies made of alloys of high purity and hardness; therefore, special requirements are imposed on the selection and compliance with the technological regime [7–9].

Electron beam welding can also join dissimilar metals. This process can also be applied to so-called black and white metals. Similar statements apply to various non-ferrous metals. However, the situation can be more complicated when completely different metals (for example, chemical composition, thermal conductivity, solidification properties, coefficient of thermal expansion, etc.) must be connected to each other [10–12].

An important place is given to the choice of the optimal value of the welding speed, which depends on the accelerating voltage, and the stages of beam input and output are not given due attention. Also, often, when modeling, changes in thermophysical parameters are not considered, and tabular data are presented in an incomplete volume [13–15].

This approach is not entirely correct, therefore, in this work, it is proposed to supplement the missing values of thermophysical parameters using the methods of regression analysis and to develop an algorithm for calculating the temperature at the stages of introducing an electron beam into the water.

2 Reconstruction of Regression Dependencies for Thermophysical Parameters of the Product

During the input and output of the beam, the temperature of interest to us varies from 20 °C to melting, therefore, to improve accuracy, it is necessary to consider changes in the thermophysical parameters of the product material, such as: Heat capacity of the material, Coefficient of thermal expansion, Specific heat of the material, Density of the material.

At present, such parameters are presented in handbooks for certain material temperatures, shown in Table 1.

Table 1. Physical properties for grade VT14

Heat capacity of material, W/(m * °C)	Thermal expansion coefficient, 10^6, 1/°C	Specific heat of material, J/(kg * °C)	Density of material, kg/m^3	Temperature, °C
8.37	–	–	4520	20
9.21	8	–	–	100
10.47	8.2	0.544	–	200
11.72	8.5	0.586	–	300
12.92	8.8	0.628	–	400
13.82	8.9	0.67	–	500
–	8.7	0.712	–	600
–	–	–	–	700
–	–	–	–	800
–	–	–	4320	900

However, for more accurate modeling, it is not enough to use tables with such a discretization grid for thermophysical parameters. To restore the dependence of such values, the use of regression analysis methods is required in order to be used in the process of mathematical modeling. Since each thermophysical parameter can have its own form of dependence on the temperature of the material, the most popular approximation methods were used to restore the regression: linear, quadratic, cubic, exponential, logarithmic. Below is Table 2, showing the values of the correlation coefficient for all types of regression for each physical parameter.

Table 2. Correlation coefficient

Regression type					
	Linear	Quadratic	Cubic	Exponential	Logarithmic
Thermal expansion coefficient	0.8896	0.9695	0.9999	0.8837	0.9369
Heat capacity of material	0.9989	0.9992	0.9999	0.9938	0.9160
Specific heat of material	1	1	1	0.9992	0.9871
Density of material	−0.9999	1	–	0.9999	0.9486

Since the correlation coefficient for some parameters turned out to be equal to or close to one, even with an accuracy of 13 decimal places, the authors decided to use such a parameter as the average approximation error to select the optimal regression. Below is Table 3, which shows the average approximation error for all types of regression for each physical parameter.

Table 3. Average error of approximation

Average error of approximation					
	Linear	Quadratic	Cubic	Exponential	Logarithmic
Specific heat of material	0	$3.2 * 10^{-12}$	$1.43 * 10^{-8}$	$3.62 * 10^{-8}$	1.4335
Thermal expansion coefficient	1.5381	0.9183	0.0493	1.5676	1.1503
Heat capacity of material	0.6393	0.66020	0.0819	1.6460	6.9289
Density of material	0.0091	0	69.6297	0.0021	0.5175

Thus, as a result of the studies carried out to restore the values of thermophysical parameters for different temperatures, the cubic approximation is better, while the average error tends to zero, that is, the dependence is almost ideal.

To restore the values, a corresponding formula was drawn up for each of the parameters. So, based on Tables 2 and 3, for the specific heat capacity, a linear regression was chosen and the following formula was written:

$$C = 0.00042 * T + 0.46, \tag{1}$$

where C – specific heat used for energy distribution calculations; T – current product temperature at the point of calculation.

For the coefficient of thermal expansion, cubic regression was chosen and the following formula was written:

$$\alpha = -2.407 * 10^{-8} * T^3 + 2.02778 * 10^{-5} * T^2 - 0.002446 * T + 8.0667, \quad (2)$$

where α – coefficient of thermal expansion used for energy distribution calculations.

For the heat capacity parameter, based on Table 3, a cubic regression was chosen and the corresponding formula was drawn up:

$$\lambda = -2.58964 * 10^{-8} * T^3 + 1.789 * 10^{-5} * T^2 + 0.00880269 * T + 8.184994, \quad (3)$$

where λ – heat capacity used for energy distribution calculations.

Based on Tables 2 and 3, a quadratic regression was chosen for such a parameter as density and to restore the values, the corresponding formula was drawn up:

$$\rho = 4.73485 * 10^{-6} * T^2 - 0.231629 * T + 4524.6307, \quad (4)$$

where ρ – density used for energy distribution calculations.

Therefore, on the basis of the obtained formulas (1)–(4) and Tables 2 and 3, to restore the dependence of the thermophysical parameters, the authors selected several regressions for each of the parameters (cubic, quadratic and linear). The obtained values of physical properties for material VT14 are presented in Table 4.

Table 4. Recovered thermophysical parameters of material VT14

Heat capacity of material, W/(m * °C)	Thermal expansion coefficient, 10^6, 1/°C	Specific heat of material, J/(kg * °C)	Density of material, kg/m^3	Temperature, °C
8.368	8.026	0.468	4520	20
9.218	8.001	0.502	4501.515	100
10.454	8.196	0.544	4478.494	200
11.737	8.508	0.586	4455.568	300
12.911	8.792	0.628	4432.737	400
13.822	8.904	0.670	4410	500
14.314	8.700	0.712	4387.358	600
14.231	8.035	0.754	4364.811	700
13.418	6.764	0.796	4342.358	800
11.720	4.743	0.838	4320	900

Also, based on Table 1 and the obtained formulas (1)–(4), graphs were built for each thermophysical parameter, reconstructed using regression analysis. The results shows in Figs. 1, 2, 3 and 4.

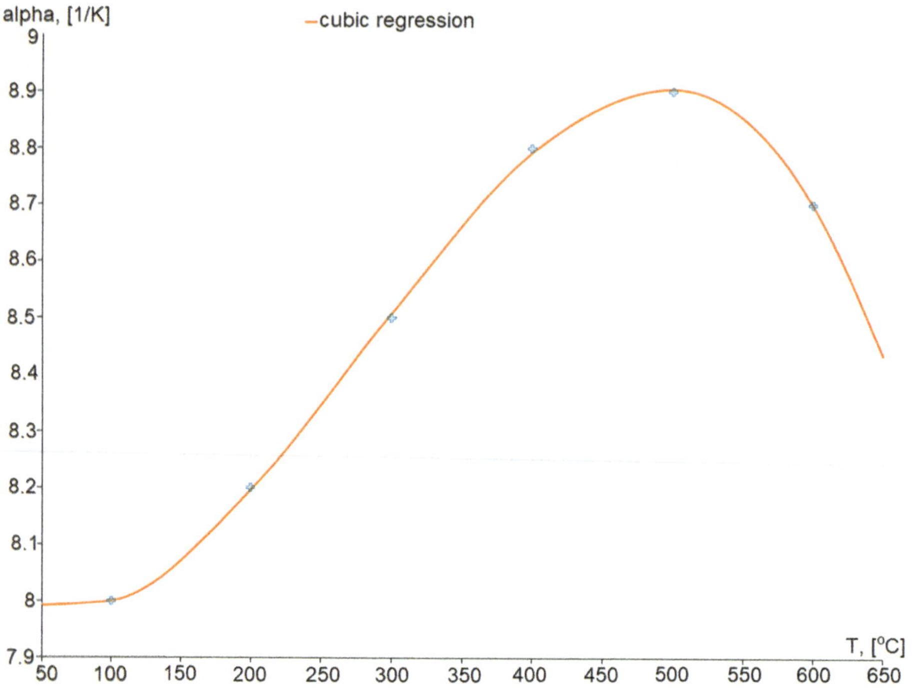

Fig. 1. Reconstructed values of coefficient of thermal expansion (α) versus temperature using cubic regression.

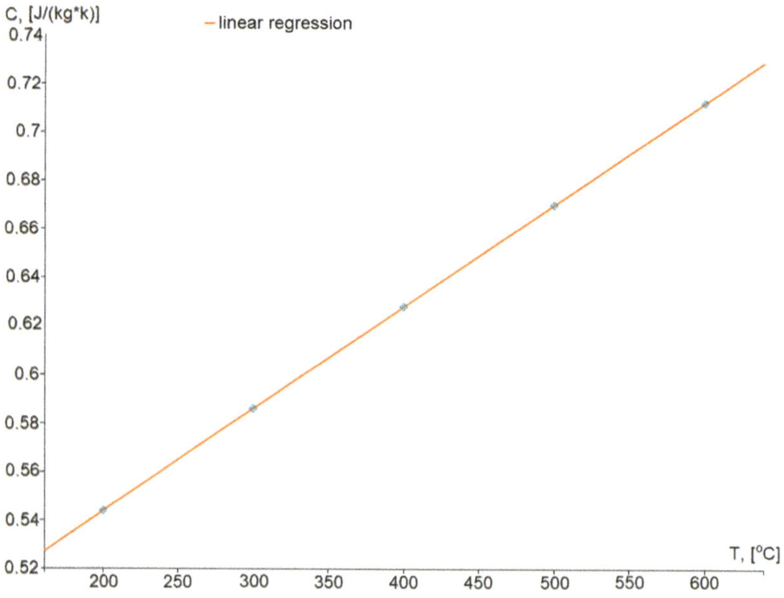

Fig. 2. Reconstructed values of material specific heat (C) versus temperature using linear regression.

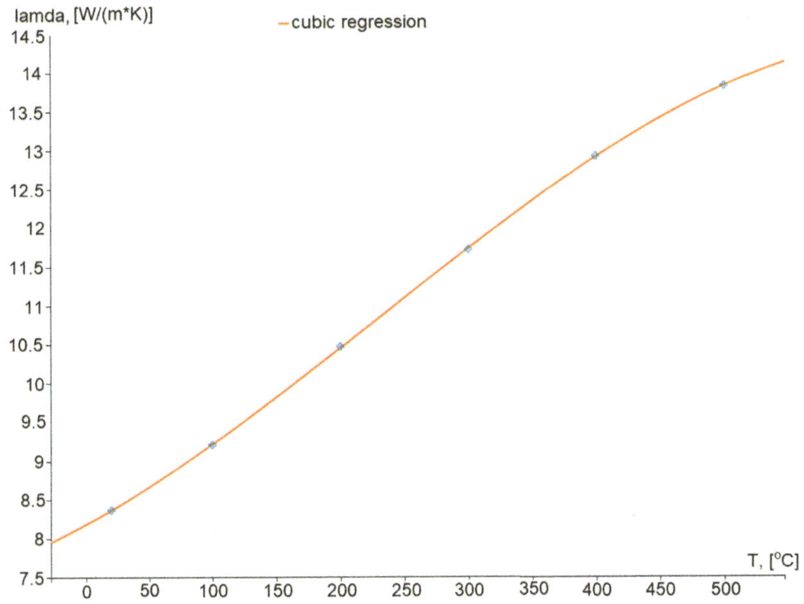

Fig. 3. Reconstructed material heat capacity (λ) versus temperature using cubic regression.

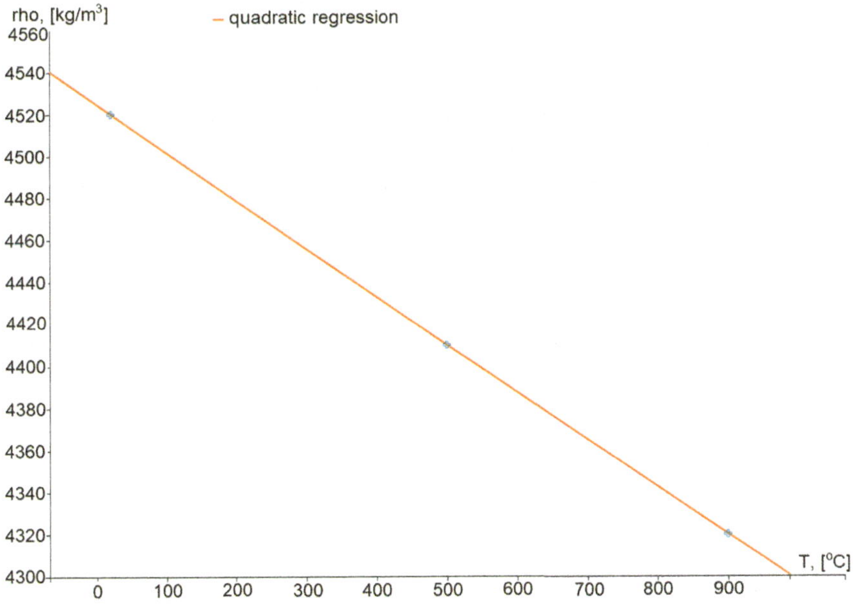

Fig. 4. Reconstructed material density (ρ) values versus temperature using quadratic regression.

3 Development of an Algorithm for Calculating the Thermal Field

In this work, the temperature is calculated at different coefficients and a non-steady-state process. Since the tabular approach is not convenient, the authors proposed an algorithm for modeling the heating process at the stage of beam input, based on the data obtained, and developed a block diagram of the algorithm for calculating the temperature distribution when the thermophysical parameters change from temperature, considering the results in previous works.

The block diagram of the algorithm for calculating the thermal field with changing thermophysical parameters shown in Fig. 5. Modeling is performed according to the algorithm in accordance with the formulas for calculating the temperature distribution [13–15].

At the first stage, the thermophysical parameters are set as: Heat capacity of the material (λ), Coefficient of thermal expansion (α), Specific heat of the material (C), Density of the material (ρ).

In the next step, the welding parameters are entered: Power (Q), Welding speed (υ), Time (t), product thickness (h), Focusing current (I_F), Beam diameter (D_B). Power - effective power, that is, the product of the efficiency of the gun by the beam current and the accelerating voltage. The initial conditions are set Q – heating source energy equivalent to the minimum value; T_{tp} – transition process temperature; Q_{sg} – size grid on power; T_{sg} – size grid on temperature; D_{BX}, D_{BY} – diameter of the beam on x and y coordinates; D_{BXsg}, D_{BYsg} – size grid on diameter of the beam on x and coordinates.

In the third stage, sub-processes are executed. First, according to formulas (1)–(4) and the given initial values of the parameters, the thermophysical parameters are calculated. Since a discrete model is used, that is, a finite number of steps is known, we can calculate thermophysical parameters, for the next step, knowing the previous temperature values, this is possible based on the developed mathematical models in the previous work. In the second subprocess, the normal distribution law of the electron beam is calculated. In the third and fourth subprocesses, the thermal field of a fast-moving linear source and the thermal field of a point source on the body surface are calculated.

At the last stage, in the fifth subprocess, the temperature is calculated, when calculating the temperature distribution, to obtain more accurate results, the current calculated temperature is checked, thus, when the temperature changes, during the calculation, the values of thermophysical parameters are calculated and substituted, with application of regression analysis methods for the current temperature.

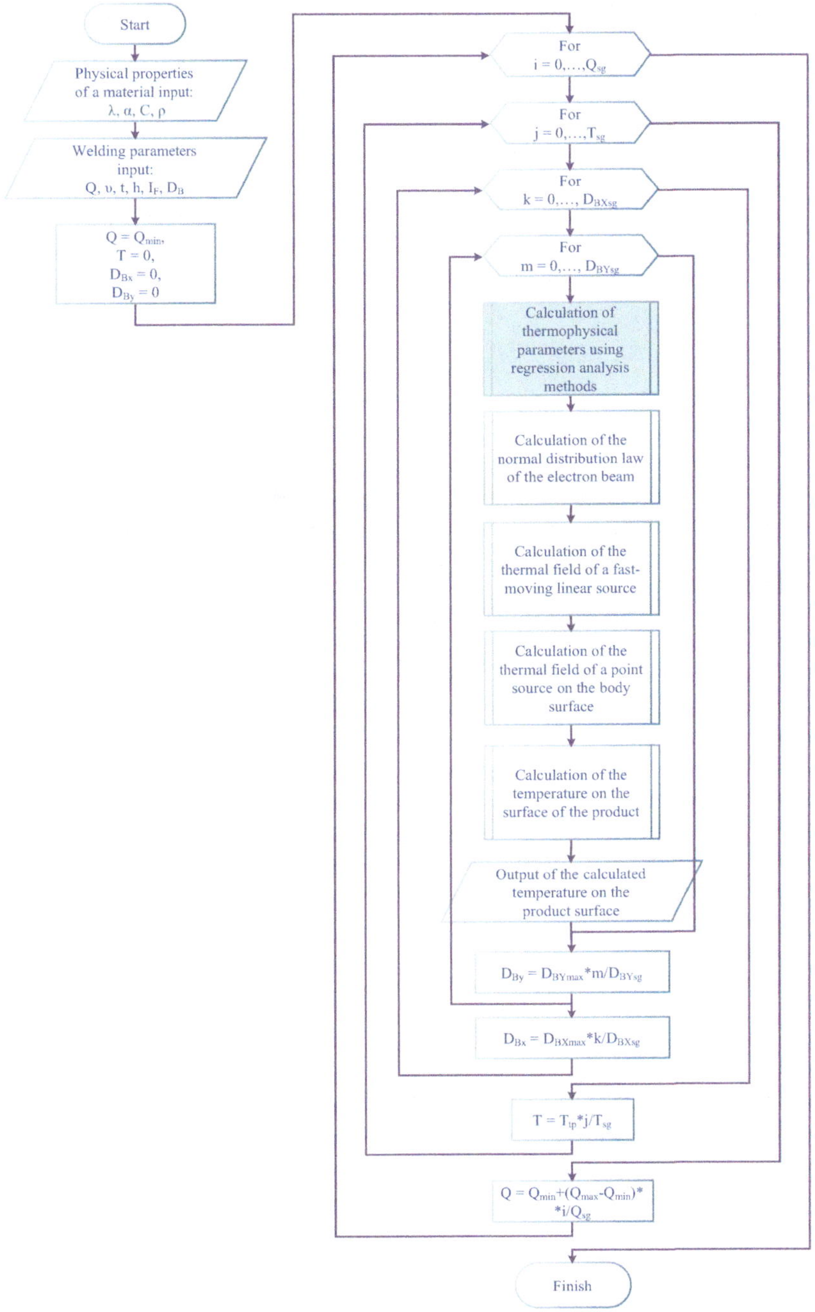

Fig. 5. Block diagram of the algorithm for calculating the thermal field when changing thermophysical parameters.

4 Experimental Study

Simulation the beam input process using the found thermophysical parameters of the product material for different temperatures.

To compare the proposed approach to modeling the process, as well as the proposed approaches in previous works [16–18], where the values of thermophysical parameters were used at the melting temperature of the product material, heating curves were plotted at several points of the weld. The resulting graphs are presented below in Figs. 6 and 7.

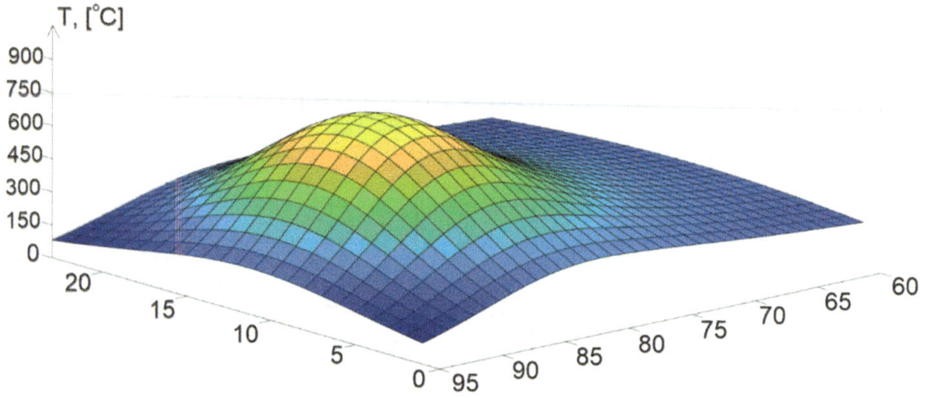

Fig. 6. Distribution of the thermal field when the beam is introduced without changing the thermophysical coefficients.

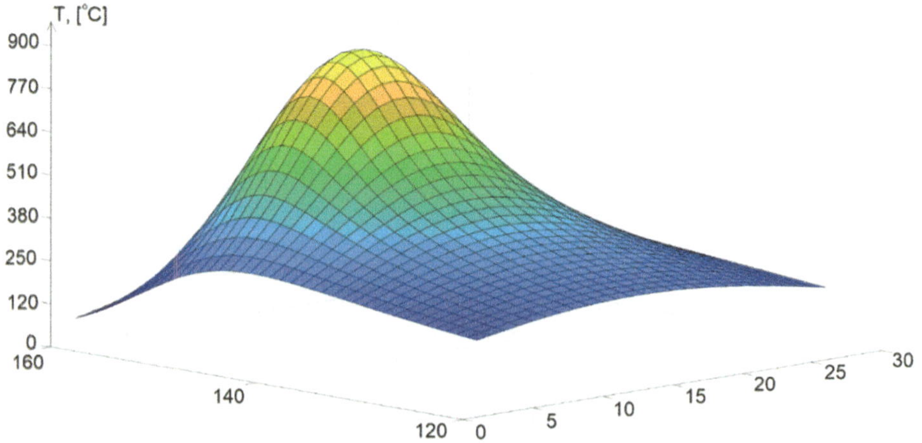

Fig. 7. Temperature distribution of titanium alloy VT14 when changing thermophysical coefficients depending on temperature.

As can be seen from the graphs, the application of the proposed approach allows one to consider the thermophysical characteristics at each stage of heating the product and, therefore, to obtain the energy distribution function in the heating zone in a more natural form, when the heating of colder zones does not give a high heat distribution to adjacent areas. Energy is concentrated in the soldering zone, in contrast to models that do not consider the dynamics of change in thermophysical characteristics (Fig. 7).

5 Results Discussions

In this work, the dependence of the thermophysical parameters of the VT14 titanium alloy was reconstructed in order to further calculate the distribution of temperature fields at the stage of beam input during electron-beam welding.

It is important to note that in this work, the calculation of the temperature when the beam was introduced was carried out not on a full-size product, but at the junction of the product.

The use of regression analysis made it possible to supplement the missing values of thermophysical parameters, and thus the authors, based on previous works, developed an algorithm for calculating the temperature when the beam was introduced, considering the temperature change during the heating process, thus, the existing set of approaches was supplemented. This made it possible to obtain more accurate values of the temperature fields and in the future will allow to reduce defects in the root of the weld seam and obtain a more stable seam quality in the process of electron beam welding.

Thus, the developed algorithm, based on a previously made mathematical model with the use of regression analysis, assumes a promising application at the stage of the introduction of an electron beam, and subsequently at the output of the beam.

Comparison of the results obtained above in a graphical form gives an idea of the feasibility of using the algorithm and shows more accurate results of the temperature distribution in the heating zone at the stage of electron beam input.

6 Conclusion

The developed algorithm for calculating the distribution of the thermal field on the basis of mathematical models of optimization when introducing an electron beam considers the nature of the distribution during heating of the product, the established technological parameters of the electron-beam welding process, the geometry of the product, as well as the thermophysical parameters restored using regression analysis. Such an algorithm for calculating the temperature distribution on the part of the product when the beam is introduced on the basis of the developed mathematical models, considering the physical properties of the material, will make it possible to create a mathematical justification for the choice of technological parameters for different modes of input and output of the beam.

In the future, the results obtained can be applied when creating models for optimizing the input and output modes of EBWs that allow you to form a better connection at the so-called zero point.

Acknowledgments. The reported study was funded by RFBR, Government of the Krasnoyarsk Territory and the Regional Science Foundation, project number 19-48-240007 "Mathematical and algorithmic support of the electron-beam welding of thin-walled aerospace structures".

References

1. Afiqah Muhamad Jamil, S., Asrul Affendi Abdullah, M., Kek, S.L., Nor, M.E., Mohamed, M., Ismail, N.: Detecting overdispersion in count data: a zero-inflated Poisson regression analysis. J. Phys. Conf. Ser. **890**, 012170 (2017)
2. Mustain, R., Vita, R., Nyoman, B.I.: Estimation of matrix variance-covariance on nonparametric regression spline truncated for longitudinal data. J. Phys. Conf. Ser. **1562**, 012014 (2019)
3. Prahutama, A.: Model Regresi Nonparametrik Polynomial Lokal Birespon Pada Data Longitudinal Tesis. Institut Teknologi Sepuluh Nopember, Surabaya (2013)
4. Suparti, Santoso, R., Prahutama, A., Devi, A.R., Sudargo: Modeling longitudinal data based on Fourier regression. J. Phys. Conf. Ser. **1217**, 012105 (2019)
5. Drapper, N.R., Smith, H.: Applied Regression Analysis 2. Marcel Dekker, New York (1992)
6. HSoE: National Research University, Regression Analysis. https://srosovet.ru/content/editor/Metod/regression/9lekciyaKolcov.pdf
7. Sitnikov, I.V., Belenkiy, V.Ya., Olshanskaya, T.V.: Study of the effect of focusing and oscillation of electron beam on the structure and properties of welded seams. IOP Conf. Ser. Mater. Sci. Eng. **611**, 012009 (2019)
8. Kasitsyn, A.N., Agafonov, R.Yu., Leonov, A.V., Grigoriev, G.V.: Control of electron beam wielding parameters based on the gap scanning system data during the welding process. IOP Conf. Ser. Mater. Sci. Eng. **759**, 012013 (2019)
9. Kharitonov, I.A., Dragunov, V.K., Goncharov, A.L., Martynov, V.N.: Electron beam welding features of cermet membranes. J. Phys. Conf. Ser. **1109**, 012022 (2018)
10. Tsonevska, T.S., Koleva, E.G., Koleva, L.S., Mladenov, G.M.: Modelling the shape of electron beam welding joints by neural networks. J. Phys. Conf. Ser. **1089**, 012008 (2018)
11. Belenkiy, V.Ya.: Scanning of an electron beam along an x-shaped trajectory as a means of reducing defects in the root of a weld in EBW. Autom. Weld. **9**, 35–37 (1986)
12. Akopianc, K.S., Emchenko-Ribko, A.V., Neporohin, V.Yu., Shilov, G.A.: Prevention of the formation of root defects in EBW with non-penetration of up to 60 mm depth. Autom. Weld. **4**, 30–34 (1989)
13. Seregin, Y.N., Murygin, A.V., Laptenok, V.D., Tynchenko, V.S.: Modeling of electron beam distribution in electron beam welding. IOP Conf. Ser. Mater. Sci. Eng. **450**(3), 032036 (2018)
14. Tynchenko, V.S., Murygin, A.V., Seregin, Y.N., Milov, A.V.: Modelling of the welded seam parameters at electron-beam welding. IOP Conf. Ser. Mater. Sci. Eng. **450**(2), 022019 (2018)
15. Laptenok, V.D., Seregin, Y.N., Murygin, A.V., Tynchenko, V.S.: Method for determination of technological mode parameters of electron-beam welding based on the application of optimality criterion with the view on the weld pool uniform heating. IOP Conf. Ser. Mater. Sci. Eng. **467**(1), 012013 (2019)
16. Seregin, Yu.N., Kurashkin, S.O.: Modeling of the EBW mode for predicting the parameters of the weld. In: Proceedings of the MPEI: Collection of Materials of 2nd International Conference, Moscow, pp. 26–36 (2017)
17. Kurashkin, S.O., Laptenok, V.D., Murygin, A.V., Seregin, Yu.N.: Analytical characteristics of the electron beam distribution density over the heated spot for optimizing the electron-beam welding process. IOP Conf. Ser. Mater. Sci. Eng. **681**, 012021 (2019)
18. Seregin, Yu.N., Kurashkin, S.O.: Modeling the thermal process using the temperature functional by electron beam welding. IOP Conf. Ser. Mater. Sci. Eng. **734**, 012003 (2020)

Agile Architecture Frameworks: Challenges and Open Issues

Tumisang Akinpelu[1,2](✉), Rene van Eck[1,2], and Tranos Zuva[1,2]

[1] Vaal Univesity of Technology, Vanderlbijl Park 1900 ZA, South Africa
[2] Springer Heidelberg, Tiergartenstr. 17, Heidelberg, Germany

Abstract. Agile architecture frameworks are flexible, adaptable to change and allows for a faster delivery of projects; these make them very important in delivery of projects. Agile Architecture frameworks are therefore of great advantage when implemented in Information and Communication Technology (ICT) environments as technology is ever changing and evolving. Although agile architecture frameworks have been explored, there is still a need for more research to be done on their challenges. This paper aims to explore the challenges and open issues for the agile architecture frameworks, various frameworks such as Agile Architecture Framework (AAF) and distributed agile or adaptive development (DAD). Agile architecture frameworks have a lot of challenges varying from communication, cost, and agile architecture processes being neglected. Even though agile architecture frameworks have a lot of challenges, there are various ways I which these can be overcome .

Keywords: Agile architecture framework · Agile architecture challenges · Archile architecture open issues · Agile architecture advanages

1 Introduction

Architecture is amongst some of the most important processes in the development of information technology systems. An agile architecture "is both an architecture that supports a team's agility by being easily modifiable and tolerant of change, and the outcome of an agile process that has a more emergent design with a shorter planning period" [1]. The advantages of agile architecture frameworks are that they allow faster development and improved delivery of projects at a lower cost [2], agile architecture is also flexible [3] and don't restrict teams with too many processes and procedures [1].

For an agile architecture to be successful, an architect is required to understand agile development, communicate with the team, must be influential to the team and use critical skills, the architect is also required to have experience with different approaches and apply functions for architecture that are independent from the methodology being used in the project [4].

Researchers [3–5] have researched different angles on agile architecture frameworks however agile architecture challenges were not explored into details, which left a gap in the study of agile architecture frameworks. This study aims to explore the challenges

R. Silhavy (Ed.): CSOC 2021, LNNS 230, pp. 59–66, 2021.
https://doi.org/10.1007/978-3-030-77442-4_4

that come with implementing agile architecture frameworks. The paper is arranged as follows: introduction, background, agile architecture frameworks, challenges and open issues and conclusion.

2 Background

Enterprise Architecture is required in all enterprises and it is vital to be applied by formal stakeholders' satisfaction in all projects. An appropriate framework for the enterprise requirements can be planned by making use of agile methods and its practices, which do not only increase the percentage of success but also returns the asset to stakeholders as soon as possible [2].

Before we can fully understand the outcome, Agile development starts to build, it also alters the designs and plans as practical knowledge is increased while building is in progress, those that are working closely to the project are trusted to have better judgment, and persistent cooperation with the eventual consumers is encouraged consumers. "Architecture establishes a technology stack, creates design patterns, enhances quality attributes, and communicates to all interested parties. The combination of these two spaces is agile architecture—an approach that uses agile techniques to drive toward good architecture" [4].

It is considered wasteful to have too much documentation in agile development processes. However, completely eliminating documentation leads to lack of architectural knowledge and information. Thus, when developing complex software systems architecture documentation is required, strategy placed between the described extremes [6].

To constantly communicate and share knowledge among members of the team are dependencies when using agile methods in order to produce working software systems of high quality in short times [5]. "The agile body of knowledge mainly discusses some technology and non-technology solutions and strategies to mitigate the communication challenges from a project management perspective" [5].

In the research by [7], there are six forces that make up the perspective of the teams and system that have been found, The team's agile architecture designs can be affected by this set of conditions. The six forces touched on in the research of [7] are the instability of requirements, technical risks, early value, culture of the teams, agility and experience of customers.

The following are the six forces they explored in their research explained in more detail:

1. Instability of requirements: is about the impact of unsound requirements on the plans that are done in advance. The ability to respond to changes in what is required is among critical values in agile environments [7].
2. Technical risks: is the impact potentially negative exposure outcome would have on the team's effort that is put up-front [7].
3. Early Value: is about the client needing to get the system and products' worth prior to implementing the functionality, perhaps a demo of the system [7].

4. Culture of the team: The effect the culture of the team has on billet to adopt agile and effort put to plan upfront is described. Culture that is focused on people and co-operative is a vital factor in communication ability of the team [7].
5. Agility of the customer: the ways of the organization of the customer are described and the enormous effect of team doing architecture design up-front has. Agile architecture culture of the customer must possess qualities like that of the culture of the team weather the team is an internal or external team, in order for the team to be completely agile. A team that has high agility does not blend in with a team that is process-oriented and prefers planning and communication which is formal [7].
6. Experience: the impact of knowledge and ability to make decision has on the time spent by an agile team to do upfront design of an architect that is highly experienced is described. Architects with experience possess knowledge, they are likely to have knowledge of options that are suitable for implementing a solution better with an understanding of what is likely to work and what is likely to fail [7].

According to [14], some common problems with current approaches are: enterprise architecture effort is not there, focus is skewed, the enterprise architecture existence is not known to project teams, the enterprise architecture is not followed by project teams, project teams don't involve enterprise architects or work with them, architecture is outdated, architecture models are narrowly focused, "charge back" schemes that are dysfunctional, and A "do all this extra work because it's good for the company" attitude. Focusing on processes and tools above personnel and communication is a common thread behind these problems, this is conflicting the first value of the Agile Alliance. İf any of these problems are experienced in an organization, it is best to take into consideration an agile approach to the enterprise architecture [14].

3 Agile Architecture Frameworks

In this section, a review of related literature relevant to the study by other researchers is briefly discussed to provide required background and clear understanding necessary for the research presented. The review of the literature mainly puts emphasis Agile Architecture Frameworks . Architecture framework a set of rules used to define different forms an architecture can be implemented; models and drawings are usually included in explanation" [2].

Study conducted by [4], in order to provide clarity on the interaction between agile and architecture. The study reveals that architects can combine agility and architectural practices to practically balance priorities of business and architectural while delivering the two with agile processes. The paper gives a clear understanding of how agility and architecture come together to function effectively however does not fully explore the challenges that come with combining the two functions.

With agile development, architects are given the opportunity to function intimately with business and technical teams while they continue to usher systems to good architecture. In so doing, challenges are presented, some inherent in the constraints of archiving architecture that is good irrespective of the methodology, some are caused by being required to produce long-lasting results making use of minimal input. "A skilled architect can adapt to agile development while staying focused on the core architectural work.

This will ensure that both individual systems and their aggregate enterprise behavior meet the needs of the business today, and are technically sustainable for years to come—an architectural value proposition that's independent of delivery methodology" [8].

A study done [9] found the simplified hybrid of scrum, extreme programming and sequential project management most successful in conducting architectural work on fourteen agile projects studied over eight years and explain the architectural functions are the ones that are generally performed on projects by architects although the listed details are not comprehensive. They also show architecturally how functions interface with interface points and an architect's main concern at that interface. They took the three categories with their four items to produce a framework that is useful for comprehension and guidance to the agile architecture that is able to extend by adding extra groupings or items based on other prioritization or preference.

A study conducted [7] on grounded theory of agile architecture and described how agile software teams and described the amount of up-front agile architecture required. They conducted the study based on grounded theory, the study involved forty-four contributors, tabled six forces that have an impact on the context of the team and five strategies that can be used by teams to assist them in determination of the amount of effort required be to put up-front into the design.

In their research they came to a realization that if too little time is spent on up-front design of the architecture, a riskier and advanced probability of failure will be faced by the team. If there is a delay in delivery of value to the customer, responding to a change can be found to be more difficult. They revealed that too much reduction in up-front design is also not good as it could result in unintended architecture that doesn't back the enablement to the development functionality of the team and unable to satisfy requirements. In order for agility to be maximized, the team is required to use a suitable trade-off between an entire up-front architecture designs and to totally develop a design. An accidental architecture has potential to lead to steady failure of the project [7].

A study [3] proposes a new architecture framework termed the Agile Architecture Framework (AAF), this framework meets the needs of the digital enterprise. The proposed framework is performed in a way that can change the supervision of the stake holders into compassion and collaboration in order to make them feel that they are capable of affecting the project and to make them more excited to finish that project; so they entirely give support to project teams.

A study [3] proposed the following in the research:

- Modularizing monolithic systems should be the focus of the Enterprise Architects, because it is the number one pre-condition for agility
- In order to remain relevant in an organization that practice agile ways, existing architecture practices and roles need to evolve
- In order to meet of the digital enterprise needs, the body of knowledge of architects should be accomplished
- When shifting from programs that are large toward multiple autonomous teams, classical architecture governance models' relevance is lost.

The proposed AAF follows the following process:

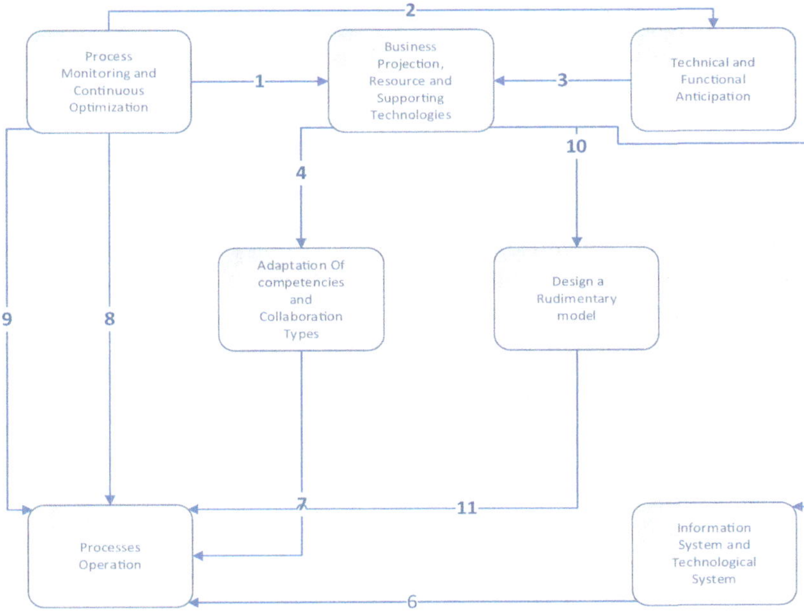

Fig. 1. Agile Enterprise Architecture Framework (AEAF) [3]

A group of researchers [5] conducted a study that explored agile Enterprise Architecture (EA) effects on Distributed Agile Development (DAD) communication and proposed development of an approach that is agile EA driven from the architecture body of knowledge for handling communication challenges of the DAD that have not been comprehensively investigated before.

EA provides a rounded overview and conditions of the entire environment in which numerous projects are managed and developed, for that reason exploration of the effect pf agile Enterprise Architecture (EA) is attempted on the DAD communication.

According to [5], active communication and effective sharing of knowledge among members of the team is a dependency for high quality of software systems which are working to be produced in short periods of time. They established that many challenges are faced with DAD projects which are mainly related to communication and that communication costs are increased in DAD than local setting due to inefficiencies of communication.

There are various frameworks that support Enterprise Architecture, the following are more of the frameworks [5]:

1. Zachman
2. Department of Defense Architecture Framework (DoDAF)
3. The Open Group Architecture Framework TOGAF

Zachman Framework "is a framework borrows from business design principles in architecture and manufacturing and provides a way of viewing an enterprise and its information systems from different perspectives, and showing how the components of the enterprise are related" [10].

"The Department of Defense Architecture Framework (DoDAF), Version 2.0 is the overarching, comprehensive framework and conceptual model enabling the development of architectures to facilitate the ability of Department of Defense (DoD) managers at all levels to make key decisions more effectively through organized information sharing across the Department, Joint Capability Areas (JCAs), Mission, Component, and Program boundaries" [11].

TOGAF "is a proven enterprise architecture methodology and framework used by the world's leading organizations to improve business efficiency"; it also assists specialists to avoid being tied down to methods that are proprietary, assists in ensuring that resources are utilized more efficiently and effectively, and greater return on investment is therefore realized [12].

The following needs are required to be met by each Agile EA approach [5]:

1. Must focus on the people,
2. Be as simple,
3. Short releases and iterations must be supported in delivering value,
4. Problems must be dealt with and addressed immediately in agile EA due to daily communications between customers and developers
5. Must work with developers that are hands-on in order to gain improved understanding of what is needed in agile EA
6. Promotes having members that are both self-organized and self-directed.

The research papers discussed in this section give a detailed explanation and provide understanding of agile architecture frameworks and how they can be implemented; however in all the research papers reviewed, one common factor is that they all provide some of the disadvantages are given briefly however do not give a detailed explanation of the challenges of implementing agile architecture in any organization.

4 Challenges and Open Issues

Many large organizations experience difficulties when responding to change in nowadays multifaceted business environments. A portion of the difficulties are caused by the lack of understanding of the complex structure and constituents in the diverse areas of the organization, where bequest information about the business is inaccessible or kept known to specific employees or business units without that being made obvious [10].

The organizational pre-requisites are often neglected and that leads to the efficiency of agile processes being put at risk [3] and Agile Architecture approaches may not be direct and have too many challenges [13]; one of the main challenges being the challenges related to communication [13]. It takes approximately 2.5 times more using DAD projects than it does the same project with local agile context [13].

As an architect, hopping into these interaction points can be a difficult experience. The developers tend to view architects with cynicism and business priority that justifies the bypass of good architecture always seems to be there. Reducing the challenges requires numerous delicate skills that only extensive experience can enhance [4].

Increased risk and chances of failure are often the results of spending too little time and effort designing architecture upfront; and if too much time is spent by the team the customer value is delayed, and it can be extremely difficult to respond to change [7].

It is not straightforward to tailor capabilities of EA adoption. The known difficulties for agile EA are as follows; satisfying the stakeholders, requirements of the customer, commitment of the stakeholders and the requirement to be flexible to changes. They argued that not even one of the conventional EA (e.g., Zachman, DoDAF, and TOGAF) is capable of dealing with the difficulties experienced in agile software development enterprise [5].

There are two organizational structures in agile EA [5]:

1. "Formal structure: this form is documented by organization such as chart, and"
2. Informal structure: developers use this form to complete tasks and is not documented, and this requires other developers with crucial skills and knowledge to be recruited into the team. "A common problem or thread of using traditional EA is the focus on tools and processes over stakeholders' interactions"

5 Conclusion

The agile architecture framework allows projects to finish better, quicker, with less cost and completion and delivery of projects is achieved with less complications [2]. It eliminates a lot of red tapes that come with processes and other methods of implementing architecture and therefore more and more organizations should explore it. Agile architecture can come in different frameworks, organizations can choose the framework most suitable for them. Even though there are various issues and challenges that often a come with implementing or choosing agile architecture, the challenges can be overcome and every framework comes with solutions to its own challenges. Communication within the architects is best achieved if the EA practice and formal EA is centralized. Even though agility in architecture reduces planning time, reducing upfront architecture could lead to a complete project failure and it is therefore necessary to have pre-design planning.

References

1. Waterman, M.G.: Reconciling agility and architecture: a theory of agile architecture, p. 348 (2014)
2. Shirazi, H.M., Rouhani, B.D., Shirazi, M.M.: A framework for agile enterprise architecture. Int. J. Intell. Inf. Technol. Appl.**2**(4), 182–186 (2009) . http://ezproxy.library.capella.edu/login?url=http://search.ebscohost.com/login.aspx?direct=true&db=iih&AN=44900091&site=ehost-live&scope=site%5Cnhttp://content.ebscohost.com/ContentServer.asp?T=P&P=AN&K=44900091&S=R&D=iih&EbscoContent=dGJyMNHr7ESeprM4yOvs
3. Dragičević, Z., Bošnjak, S.: Agile architecture in the digital era: trends and practices. Strateg. Manag. **24**(2), 12–33 (2019). https://doi.org/10.5937/straman1902011d
4. Madison, J.: Agile architecture ınteractions (was: agile architecture ınsertion points), IEEE Softw. 1–1 (2010). https://doi.org/10.1109/ms.2010.27

5. Alzoubi, Y.I., Gill, A.Q., Al-Ani, A.: Distributed agile development communication an agile architecture driven framework. J. Softw. **10**(6), 681–694 (2015). https://doi.org/10.17706/jsw.10.6.681-694
6. Maric, M., Matkovic, P., Tumbas, P., Pavlicevic, V.: Documenting agile architecture: practices and recommendations. In: Wrycza, S. (ed.) SIGSAND/PLAIS 2016. LNBIP, vol. 264, pp. 56–71. Springer, Cham (2016). https://doi.org/10.1007/978-3-319-46642-2_4
7. Waterman, M., Noble, J., Allan, G.: How much up-front? A grounded theory of agile architecture. In: Proceedings – International Conference Software Engineering, vol. 1, pp. 347–357 (2015). https://doi.org/10.1109/ICSE.2015
8. Jaeger, B., Halse, L.L., Wiig, H., Bö, A.H.: IoT technological maturity model and assessment of Norwegian manufacturing companies. In: 23rd EurOMA Conference, pp. 1–10 (2016)
9. Ramakrishnan, S.: On integrating architecture design into engineering agile software systems agile methods in practice – combination of techniques within an agile method. Issues Informing Sci. Inf. Technol. **7**, 9–25 (2010)
10. Origins, T., Zachman, J., Open, T., Architecture, G.: The Zachman Enterprise Framework. Cambridge Technical Communication, pp. 1–7 (2007)
11. Department of Defense Deputy Chief Information Officer: "The DoDAF Architecture Framework Version 2. 02," DoDAF J., p. 289 (2011). http://dodcio.defense.gov/Portals/0/Documents/DODAF/DoDAF_v2-02_web.pdf
12. Weisman, R.: An overview of TOGAF Version 9.1. The open group (2011)
13. Alzoubi, Y.I., Gill, A.Q., Al-Ani, A.: Distributed agile development communication: an agile architecture driven framework. J. Softw. **10**(6), 681–694 (2015). https://doi.org/10.17706/jsw.10.6.681-694
14. Agile Enterprise Architecture. http://agiledata.org/essays/enterpriseArchitecture.html. Accessed 13 July 2020

Novel Approach of Video Tracking System Using Learning-Based Mechanism over Crowded Environment

Karanam Sunil Kumar(✉) and N. P. Kavya

Department of Computer Science and Engineering, RNSIT, Bengaluru, India

Abstract. Video tracking is an essential operation in the area of the surveillance system. The model must possess an efficient object detection and recognition capability to facilitate a better video tracking form. Existing video tracking approaches show that they emphasize accuracy without considering the larger proportion of dynamics in mobile objects' movement patterns. Therefore, this work presents a novel learning-based approach that facilitates video tracking by differentiating mobile objects specific dynamics from other mobile objects. The approach extracts significant cells from the video feed. It subjects them to a learning-based approach in contextual information of all moving objects and dictionaries to extract dominant features. A study outcome shows higher accuracy and faster tracking in contrast to the frequently used approach.

Keywords: Video tracking · Object detection · Object recognition · Learning · Training · Accuracy

1 Introduction

With the multimedia surveillance system's rise, video tracking is an essential application that contributes to this visual surveillance system [1]. One of the essential parts to carry out an effective video tracking system is to carry out a precise object detection and recognition of it [2, 3]. There are various possibilities of this object detection method, i.e., i) static object with a static background, ii) moving object with a static background, and iii) moving object with moving background. All these are the practical cases required to be considered while developing any object detection and recognition method that directly influences the video tracking process. Hence segmentation plays a critical role in the segregation of an object's foreground with the background [4, 5]. At present, various approaches contribute to the object detection and recognition process [6–8]; however, the scenario of all the cases considered for evaluation is more or less the same. One of the different forms of a case could be to differentiate a specific form of a moving object from other moving objects for a given scene. This is highly challenging in computer vision as it is quite difficult to formulate such target rules of differentiation. Another more significant challenge is to ensure that video tracking operation is very smooth and spontaneous without any lags. Out of various scenarios, the occlusion of multiple types

R. Silhavy (Ed.): CSOC 2021, LNNS 230, pp. 67–76, 2021.
https://doi.org/10.1007/978-3-030-77442-4_5

(complete and partial) is another inevitable problem. Therefore, there is necessary to develop a video tracking system that can identify such difference in various challenging aspects of a given scene.

The proposed system introduces a novel and simplified learning-based mechanism to facilitate faster video tracking operations addressing the issues mentioned above. The proposed paper's organization is as follows: Section 2 briefs about the existing video tracking approach followed by identified issues in Sect. 3. The proposed solution is briefed in Sect. 4; algorithm design is briefed in Sect. 5, while result analysis is discussed in Sect. 6. Finally, Sect. 7 discusses the summary of this paper.

2 Related Work

Various research works are being carried out to improve the process of a video tracking system. The work carried out by Tian et al. [11] used the Bayesian network to track an object from web-based videos that better recognize dynamic objects. Adoption of complex video sets (Satellite) and facilitating tracking systems is witnessed in the work of Guo et al. [12], where Kalman filtering has been used. A similar direction of work on similar videos was also seen in Xuan et al. [13], where the motion estimation approach is integrated with the Kalman filter. Usage of filtering operation with kernel correlation is also claimed to improve the object tracking system over Du et al. [14]. More work on satellite videos and object tracking is carried out by Shao et al. [15]. The velocity correlation-based approach is used to design the filter and uses an inertial mechanism. The existing system also reports the usage of convolution regression to carry out object tracking and usage of the gradient descent method, as seen in the work of Hu et al. [16]. It also uses a deep neural network. Tracking of moving objects and recognizing the specific event associated with sports is discussed by Kong et al. [17], where compressive tracking is utilized. A study towards detecting multiple moving objects tracking operation is carried out using a sliding window approach over Jiang et al. [18]. Existing studies have also seen the usage of sparse representation and Markov random field to carry out a target tracking system, as reported in the work of Wan et al. [19]. A unique study towards ascertaining the target's reliability is carried out by Yu et al. [20], emphasizing dimension reduction. Neural network adoption is also reported to improve video tracking performance, as claimed in McLaughlin et al. [21]. The work of Yun et al. [22] has used a semi-supervised learning approach by deep reinforcement, which targets performance improvement in real-time. Apart from this, there are various other neural network-based approaches for the video-tracking system, e.g., recurrent neural network [23], deep learning over the random field [24], in-depth tracking [25], deep convolution [26], hierarchical particle filter [27], deep neural network [28], deep correlation filter [29], deep network [30]. The extensive work is done on video tracking technique by Karanam Sunil Kumar and N P Kavya [31].

3 Problem Description

After assessing the outcomes and approaches from the existing video tracking system, the following open-end problems are explored: (i) the existing system has emphasized mainly on a single moving object and not much on the context attributes associated with its dynamic environment, (ii) less number of studies are carried out towards extracting specific moving object in the presence of multiple other moving objects, (iii) majority of the approaches use are highly complex in its operation and demands higher resources to carry out video tracking, and hence they are not computationally efficient despite the better performance, (iv) maximum utilization of learning-based approach has been found frequently to be used; however, their computational performance has never been judged concerning tracking speed and spontaneity.

4 Proposed Methodology

The proposed study aims to introduce a novel approach to carrying out a unique form of video tracking system to identify an explicit dynamic mobile object from the moving scene. The flow of the proposed system is shown in Fig. 1.

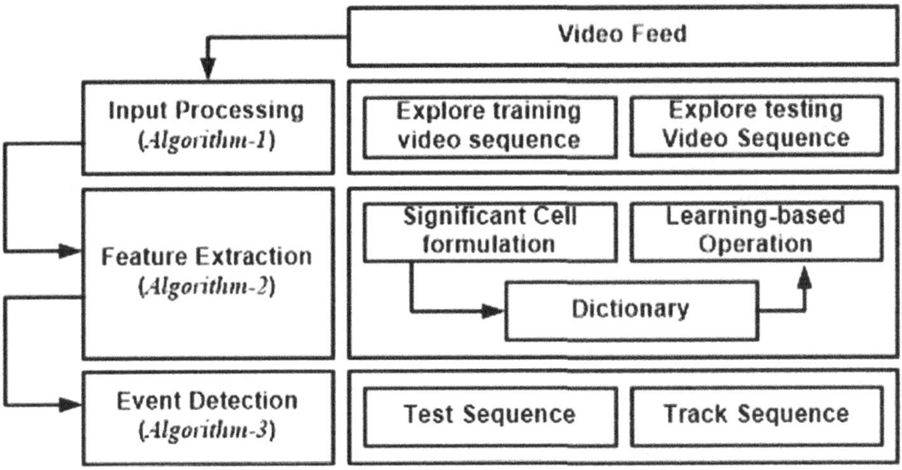

Fig. 1. Operational flow of proposed methodology

The complete implementation is classified into three stages of operation, i.e., input processing, feature extraction, and event detection using three discrete algorithms. The core idea is to apply a learning approach to evolving to offer a higher degree of accuracy. Simultaneously, the proposed system also emphasized achieving a faster process of identifying objects in the video tracking system. This is done to ensure its broader scope of applicability towards practical application. The elaboration of each block of operation concerning its algorithm is discussed in the next section.

5 Algorithm Implementation

The proposed algorithm aims to develop a novel video tracking approach that can sense and capture all the video feed granularity aspect. The complete implementation is carried out using three sequential algorithms responsible for performing feature engineering, followed by identifying an event for the given video feed. The inclusion of the algorithmic steps is discussed below:

5.1 Algorithm for Extraction of Significant Cells

This algorithm takes the complete video feed, read it, and formulates an extraction of significant cells. The term '*significant cell*' will mean a dominant block created for the given feed of the video frame, subjected to further analysis of event detection by extracting essential features from it. The operational steps are as follows:

```
Algorithm for the Extraction of Significant cells
Input: frame (frames obtained from video feed)
Output: S_cell_dic (Extracted significant cell)
Start
1.   For i=1: frame
2.       S_cell → f_1(i)
3.       S_cell_dic → f_2(h, n)
4.       For j=1: size (S_cell_dic)
5.           α → f_3(g)
6.           S_cell_dic → α
7.       End
8.   End
End
```

The above algorithm's operation is as follows: Considering all the *frames* (Line-1), the algorithm extracts its size and store the respective information in a matrix with nr rows, nc columns, and several elements k. It then applies a function $f_1(x)$ considering the frames as an input to extract significant cells S_{cell} of size mxm (Line-2). The algorithm then constructs a matrix n, with two main cells n_{rd} and n_{cd} and cumulative significant cells h (Line-3). An explicit function f_2(x) is developed which performs the following operation viz. i) it applies multiplication operation to significant cells S_{cell} and obtain n_{rd}, ii) it extracts size of S_{cell} and stores it in n_{cd}, iii) and it multiplies two significant cells S_{cell1} and S_{cell2} and obtains h while iv) it the array n retains two elements, i.e., n_{rd} and n_{cd}. The proposed system constructs a dictionary S_{cell_dic} using the parameters mentioned above (Line-3). Considering all the involved dictionary of significant cells, the proposed algorithm applies a third function f_3(x) considering the input argument of a variable g (Line-5) to yield a matrix α. The input argument g is computed by the average value of two variables v_1 and v_2, where v_1 represents the product of (nn-1) and h, while v_2 represents all the elements of *nn* and *h* (Line-5). The algorithm then computes a computational vector for obtaining the updated dictionary of significant cells (Line-6).

5.2 Algorithm for Extracting Learning-Based Features

This algorithm is a continuation of the prior algorithm, and it carries out the implementation of learning-based features. Once the dictionary-based significant cells are obtained, the generated matrix is further subjected to decomposition using a statistical approach towards evaluating multiple observations simultaneously using the proposed learning-based approach. The outcome results in extracted a feature that is utilized for event detection from video feeds. The operational steps of the proposed algorithm are as follows:

```
Algorithm for Extracting Learning-based Features
Input: S_cell_dic (dictionary of significant cells)
Output: ρ (learning-based feature)
Start
1. For i=1: size (S_cell_dic)
2.         β=S_cell_dic(i)
3.         For j=1: size(β)
4.                 γ(j)← β(j)
5.          End
6.          ρ(γ)^k=f_4(γ, w_1, w_2)
7. End
End
```

The algorithm mentioned above starts by initializing the index k and obtaining the prior algorithm's outcome, i.e., S_{cell_dic}. The proposed system considers the size of all the dictionary-based significant cells S_{cell_dic} to formulate a definitive structure (Line-1). The total of dictionary elements is reposited in a newly constructed matrix β (Line-2), while this operation is also followed by constructing another index score of γ (Line-4). This index score is responsible for all the statistical information of β. The proposed system constructs another explicit function, $f_4(x)$, which performs the sparse matrix (Line-6). Applying this function yields different attributes, e.g., total matrix with all the signs γ, basis function w_1, and fine-tuning parameter w_2. The obtained set of features is utilized for identifying the events from each frame of the input video feed.

Algorithm for Tracking Event in Video Feed

This is the final algorithm responsible for identifying a significant event to facilitate video tracking of the proposed model. The operational steps of this algorithm are as follows:

```
Algorithm for Tracking Event in Video Feed
Input: S_cell_dic (dictionary of significant cells), ρ (learning-based feature)
Output: F (Successful Event tracking)
Start
1. For i=1: ρ
2.        Apply 1st Algorithm
3.        compute p→f5(σ)
4.   End
5.   Φ→f6(t, q, D)
6.   If Φ>Th
7.      F→Flag "Event Detected."
8. End
End
```

The algorithm takes the prior algorithm's outcome as an input, which, after processing, yields event tracking. Considering all the learning-based features (Line-1), the proposed system applies the first algorithm of extraction of significant cells followed by computation of probability of event p using an explicit function f5(x) whose input argument is σ (Line-3). Σ is obtained by the square root of the summation of all the probabilities by subtracting a new matrix of dictionary S_{cell_dic1} with another matrix that retains information about its size. The new matrix S_{cell_dic1} is obtained from the cumulative dictionary function and γ. The algorithm further applies another function, $f_6(x)$, which takes the input arguments of metadata t, final feature q, and dimension D (Line-5). Therefore, this function $f_6(x)$ is mainly meant for tracking any significant changes for the given video feed. However, there are possibilities of various non-significant events in this tracking process, which will affect the accuracy of an outcome. It can track any minor changes in the position of each moving object on the given frame. For example, the movement of leaves in the tree or any smaller non-significant object. Hence, assuming that the proposed system is interested in tracking a more significant or medium-sized object, the proposed approach uses a threshold to filter out the more significant object's only movement for the given video frame. Hence, the proposed system uses a simplified operation to perform tracking of video feed. The next section discusses the results obtained from the analysis of the proposed algorithm.

6 Results Analysis

The algorithm discussed in the prior section has been implemented in MATLAB, considering the dataset [32]. The dataset consists of more than 5000 image sequences of video obtained by mounting the camera to capture pedestrians' feeds. The scene consists of people walking in two directions in the crowd, where the significant events to be tracked a non-pedestrian and dynamic movement of the pedestrian. The dataset also consists of metadata annotation consisting of a binary flag for each of the video frames for signifying significant events to be tracked (e.g., a cart moving in between pedestrian, wheelchair rolling in sideways, skaters moving in between the walking way, and biker).

The idea is to assess the proposed algorithm's capability to different this set of significant events from normal events (pedestrian walking).

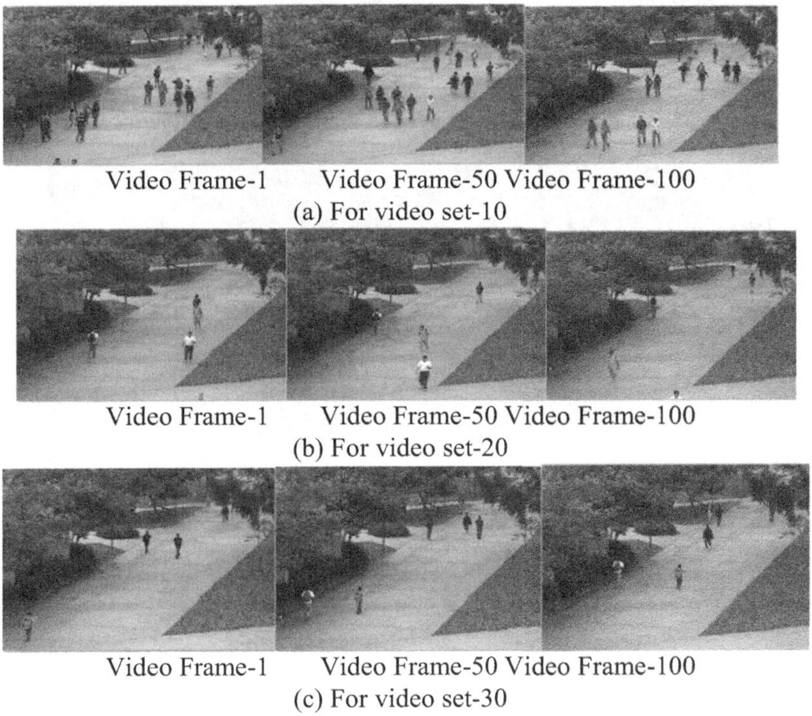

Fig. 2. Visuals of dataset used while training [32]

Figure 2 highlights the dataset's visuals of different video sets for three sample video frames subjected to learning-based feature extraction. The different set of trained datasets is mainly separately for testing. This dataset is subjected to sequential algorithmic operations discussed in the prior section. The proposed system tracked events from the video reposited and indexed discretely in different matrix compositions upon performing training. It will mean that the proposed system retains all the identified tracked events and their details in one single matrix, where the evaluation becomes quite simpler and faster. The visual outcomes of the tracked objects are shown in Fig. 3.

To carry out an effective assessment, the proposed study has compared its outcome with the related work being carried out by Fang et al. [33], where the idea was to perform a similar operation using deep learning. The assessment has been carried out over an identical test environment to obtain a numerical outcome, as shown in Table 1.

The numerical outcome in Table 1 shows that the proposed system offers better accuracy performance concerning the recall, precision, specificity, F1-score, and processing time compared to the existing in-depth learning approach. The prime justification behind this outcome is that existing deep learning has a dependency on a more considerable quantity of training dataset to retain accuracy. If the existing scheme is exposed to a lower

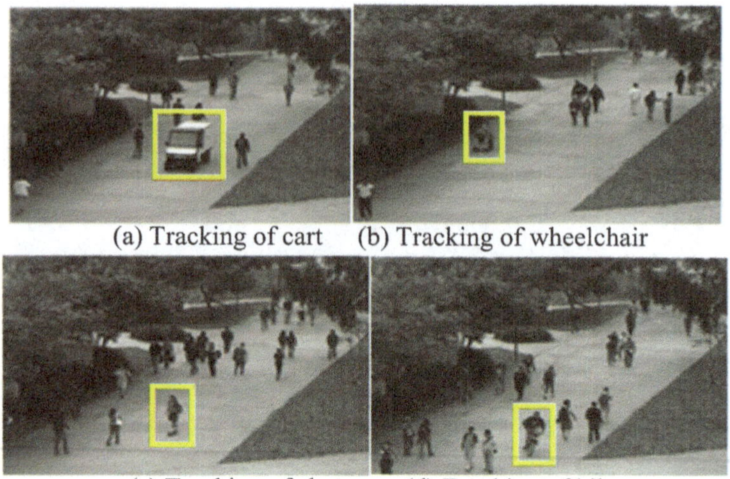

(a) Tracking of cart (b) Tracking of wheelchair

(c) Tracking of skater (d) Tracking of biker

Fig. 3. Visual outcomes of proposed video tracking

Table 1. Numerical outcomes of the proposed system

Performance parameter	Approach	Numerical value
Recall	Existing system	0.34816
	Proposed system	0.767
Precision	Existing system	0.91755
	Proposed system	0.99862
Specificity	Existing system	0.86722
	Proposed system	0.99925
F1-Score	Existing system	0.5162
	Proposed system	0.76282
Processing Time	Existing system	0.97782
	Proposed system	0.26621

scale of video sequences for training, it affects its accuracy and processing time. The proposed system does not have this dependency, and it can smartly work on a smaller set of training images, assuming that there wills not a very big significant event. For example, a truck or a jeep cannot be expected to be driving on pathways. Hence, with the progressive mode of feature extraction, the proposed system can successfully carry out video tracking operation with higher accuracy and faster processing.

7 Conclusion

The paper discusses a novel scheme of a video tracking system considering a crowd scene's standard database. The core idea is to detect a specific dynamics of a mobile object from other mobile objects as a representation of an event to be tracked. The contribution of the proposed approach are as follows: i) the proposed approach emphasizes the blocking operation by extracting a significant cell, which contributes towards higher accuracy in contrast to the existing approach, ii) the learning-based approach performs training without having an extensive dependency on video feeds, unlike another training-based approach which demands higher trained data for higher accuracy. The proposed system is capable of producing higher accuracy in the presence of low or medium training data. iii) The proposed system offers the faster tracking performance of an even in a given video feed, which showcases higher applicability in a realistic environment.

References

1. Altahir, A.A., Asirvadam, V.S., Hamid, N.H.B., Sebastian, P., Saad, N.B., Ibrahim, R.B., Dass, S.C.: Optimizing visual sensor coverage overlaps for multiview surveillance systems. IEEE Sens. J. **18**, 4544–4552 (2018)
2. Guo, Y., Bennamoun, M., Sohel, F., Lu, M., Wan, J.: 3D object recognition in cluttered scenes with local surface features: a survey. IEEE Trans. Pattern Anal. Mach. Intell. **36**, 2270–2287 (2014)
3. Bucak, S.S., Jin, R., Jain, A.K.: Multiple kernel learning for visual object recognition: a review. IEEE Trans. Pattern Anal. Mach. Intell. **36**(7), 1354–1369 (2014)
4. Chen, X., Pan, L.: A survey of graph cuts/graph search based medical image segmentation. IEEE Rev. Biomed. Eng. **11**, 112–124 (2018)
5. Xia, G., Liu, G., Yang, W., Zhang, L.: Meaningful object segmentation from SAR images via a multiscale nonlocal active contour model. IEEE Trans. Geosci. Remote Sens. **54**(3), 1860–1873 (2016)
6. Jiao, L., Zhang, F., Liu, F., Yang, S., Lim, L., Feng, Z., Qu, R.: A survey of deep learning-based object detection. IEEE Access **7**, 128837–128868 (2019)
7. Zou, X.: A review of object detection techniques. In: International Conference on Smart Grid and Electrical Automation (ICSGEA). Xiangtan, China, pp. 251–254 (2019)
8. Zhao, Z., Zheng, P., Xu, S., Wu, X.: Object detection with deep learning: a review. IEEE Trans. Neural Networks **30**(11), 3212–3232 (2019)
9. Zhou, C., Yuan, J.: Occlusion pattern discovery for object detection and occlusion reasoning. IEEE Trans. Circuits Syst. Video Technol. **30**(7), 2067–2080 (2020)
10. Kim, J.U., Kwon, J., Kim, H.G., Ro, Y.M.: BBC Net: bounding-box critic network for occlusion-robust object detection. IEEE Trans. Circuits Syst. Video Technol. **30**(4), 1037–1050 (2020)
11. Tian, S., Yin, X., Su, Y., Hao, H.: A unified framework for tracking based text detection and recognition from web videos. IEEE Trans. Pattern Anal. Mach. Intell. **40**(3), 542–554 (2018)
12. Guo, Y., Yang, D., Chen, Z.: Object tracking on satellite videos a correlation filter based tracking method with trajectory correction by Kalman filter. IEEE J. Sel. Top. Appl. Earth Obs. Remote Sens. **12**(9), 3538–3551 (2019)
13. Xuan, S., Li, S., Han, M., Wan, X., Xia, G.: Object tracking in satellite videos by improved correlation filters with motion estimations. IEEE Trans. Geosci. Remote Sens. **58**(2), 1074–1086 (2020)

14. Du, B., Sun, Y., Cai, S., Wu, C., Du, Q.: Object tracking in satellite videos by fusing the kkernel correlation filter and the three-frame-difference algorithm. IEEE Geosci. Remote Sens. Lett. **15**(2), 168–172 (2018)
15. Shao, J., Du, B., Wu, C., Zhang, L.: Tracking objects from satellite videos: a velocity feature-based correlation filter. IEEE Geosci. Remote Sens. Lett. **57**(10), 7860–7871 (2019)
16. Hu, Z., Yang, D., Zhang, K., Chen, Z.: Object tracking in satellite videos based on convolutional regression network with appearance and motion features. IEEE J. Sel. Top. Appl. Earth Obs. Remote Sens. **13**, 783–793 (2020)
17. Kong, L., Huang, D., Qin, J., Wang, Y.: A joint framework for athlete tracking and action recognition in sports videos. IEEE Trans. Circuits Syst. Video Technol. **30**(2), 532–548 (2020)
18. Jiang, Z., Huynh, D.Q.: Multiple pedestrian tracking from monocular videos in an interacting multiple model framework. IEEE Trans. Image Process. **27**(3), 1361–1375 (2018)
19. Wan, M., Gu, G., Qian, W., Ren, K., Maldague, X., Chen, Q.: Unmanned aerial vehicle video-based target tracking algorithm using sparse representation. IEEE Internet Things J. **6**(6), 9689–9706 (2019)
20. Yu, R., Cheng, I., Zhu, B., Bedmutha, S., Basu, A.: Adaptive resolution optimization and tracklet reliability assessment for efficient multi-object tracking. IEEE Trans. Circuits Syst. Video Technol. **28**(7), 1623–1633 (2018)
21. McLaughlin, N., Rincon, J.M.D., Miller, P.: Video person re-identification for wide area tracking based on recurrent neural networks. IEEE Trans. Circuits Syst. Video Technol. **29**(9), 2613–2626 (2019)
22. Yun, S., Choi, J., Yoo, Y., Yun, K., Choi, J.Y.: Action-driven visual object tracking with deep reinforcement. IEEE Trans. Neural **29**(6), 2239–2252 (2018)
23. Wang, L., Zhang, L., Yi, Z.: Trajectory predictor by using recurrent neural networks in visual tracking. IEEE Trans. Cybern. **47**(10), 3172–3183 (2017)
24. Zhou, H., Ouyang, W., Cheng, J., Wang, X., Li, H.: Deep continuous conditional random fields with asymmetric inter-object constraints for online multi-object tracking. IEEE Trans. Circuits Syst. Video Technol. **29**(4), 1011–1022 (2019)
25. Li, H., Wang, X., Shen, F., Li, Y., Porikli, F., Wang, M.: Real-time deep tracking via corrective domain adaptation. IEEE Trans. Circuits Syst. Video Technol. **29**(9) 2600–2612 (2019)
26. Li, H., Wu, S., Huang, S., Lam, K., Xing, X.: Deep motion-appearance convolutions for robust visual tracking. IEEE Access **7**, 180451–180466 (2019)
27. S. Li., S. Zhao., B. Cheng., E. Zhao and J. Chen.: Robust Visual Tracking via Hierarchical Particle Filter and Ensemble Deep Features. in IEEE Transactions on Circuits and Systems for Video Technology, vol. 30, no. 1, pp. 179–191 (2020)
28. Zhu, J., Sun, K., Jia, S., Lin, Q., Hou, X., Lin, W., Liu, B., Qiu, G.: Urban traffic density estimation based on ultrahigh-resolution UAV video and deep neural network. IEEE J. Sel. Top. Appl. Earth Observations Remote Sens. **1**(12), 4968–4981 (2018)
29. Uzkent, B., Rangnekar, A., Hoffman, M.J.: Tracking in aerial hyperspectral videos using deep kernelized correlation filters. IEEE Trans. Geosci. Remote Sens. **57**(1) 449–461 (2019)
30. Ramasinghe, S., Rajasegaran, J., Jayasundara, V., Ranasinghe, K., Rodrigo, R., Pasqual, A.A.: Combined static and motion features for deep-networks-based activity recognition in videos. IEEE Trans. Circuits Syst. Video Technol. 29(9), 2693–2707 (2019)
31. Kumar, K.S., Kavya, N.P.: Compact scrutiny of current video tracking system and its associated standard approaches. Int. J. Adv. Comput. Sci. Appl. **11**(12) (2020)
32. UCSD Anomaly Direction Dataset. http://www.svcl.ucsd.edu/projects/anomaly/dataset.htm. Retrieved on 16-10-2020
33. Fang, Z., et al.: Abnormal event detection in crowded scenes based on deep learning. Multimedia Tools and Applications **75**(22), 14617–14639 (2016). https://doi.org/10.1007/s11042-016-3316-3

A Survey on Supporting the Software Engineering Process Using Artificial Intelligence

Ahmed Maghawry[1(✉)], Rania Hodhod[2], Yasser Omar[1], and Mohamed Kholief[1]

[1] Department of Computer Science, College of Computing and Information Technology, Arab Academy for Science, Technology and Maritime Transport (AASTMT), Alexandria, Egypt

[2] Department of Computer Science, TSYS, School of Computer Science, Columbus State University, Columbus, GA, USA

hodhod_rania@columbusstate.edu

Abstract. Software engineering has been one of the highly researched sub-fields in computer science in the past years. Software engineering utilizes engineering approaches to develop software. Developing software has been facing several challenges like producing high quality software while meeting strict project time schedules with minimum resources hence maximum cost savings and of course to meet the client's requirements. Such challenges can be approached as a multi objective optimization problem. Previous literature exists that explores the potential of utilizing machine learning approaches to support the process of software engineering where it can benefit from the adaptability and learnability of such techniques. Evolutionary algorithms have been one of the highly researched machine learning approaches where it has been utilized by researchers in previous literature to solve single\multi objective optimization problems . Dealing with the challenges facing the process of software engineering as a multi-objective optimization problem promotes evolutionary algorithms as a possible solution to tackle such problems. Genetic algorithms have been among the highly researched evolutionary algorithms to tackle multi-objective optimization problems. In this paper, previous literature of techniques used to support software engineering will be reviewed as well as the possibility of using hybrid evolutionary algorithms to tackle the same problem will be discussed.

Keywords: Software engineering · Evolutionary algorithms · Multi-objective optimization

1 Introduction

As we witness the continuous surge of the digital transformation evolution in all aspects of our lives, software engineering has been one of the highly affected fields as a result of this phenomena. In the past years, we have witnessed many routine daily paper-based activities being either transformed or under intensive effort to be transformed into software that manages these activities. As a result, the process of software engineering has been exposed to various businesses where it experienced enormous pressure to adapt and to produce countless software products that serves almost all fields like military,

R. Silhavy (Ed.): CSOC 2021, LNNS 230, pp. 77–87, 2021.
https://doi.org/10.1007/978-3-030-77442-4_6

medicine, radiology, marketing, education, finance, telecommunication and more. As a result of this exposure and under this pressure, the field of software engineering has experienced intensive research to acquire techniques that will make it adapt with such business exposure and pressure, so that software engineering can evolve to meet the expectations of being at the spear head of the tools used by the digital transformation evolution. Regardless of the business, the development and maintenance of modern software have become a costly time-consuming activity. On the other hand, businesses highly require shorter time cycle with minimum resources [1].

Generally, using evolutionary algorithms to develop software raises challenges in identifying the software components while keeping the number of test cases at minimum level [2]. Approaching such challenges as a multi objective optimization problem open the doors for the use of evolutionary algorithms especially genetic algorithms in order to develop, modify and maintain software on specifications level that should lead to the production of high quality software with the least possible time consumed. Furthermore, the development and maintenance of software is usually governed by a software development life cycle (SDLC) model such as the waterfall model, iterative waterfall model, spiral model and the v-shaped model [3] as well as the agile model [4].

Regardless of which model will be used, all of them share almost the same interest of defining the software components, resource allocation and time management. Defining the software components will constitute the development and testing phase while defining the resource allocation will contribute to the cost of the software in terms of resources, and finally defining the time schedule for developing the software will contribute to the time cost of the software. For example, in the Agile SDLC, a project plan defines the project resources allocation and time schedule and can also define the software components. Defining the optimal project plan leads to the optimal component definition, resource allocation and time management, hence contributes to a high-quality software with minimum resources. Using artificial intelligence to optimize the software engineering process as a whole or one of its subcomponents has been a vast field of research in the past years [5]. In this paper we will survey literature that focused on supporting software engineering using artificial intelligence. This paper is arranged as follows: Sect. 2 is a background defining software engineering as well as SDLC models, it also defines the SDLC challenges as a multi objective optimization problem and finally defines evolutionary algorithms. Section 3 reviews software engineering challenges as well as SDLCs challenges. Section 4 discusses literature contributing to the use of artificial intelligence especially evolutionary algorithms to support software engineering in terms of effort estimation, design, and coding, testing as well as quality, reliability, and maintenance. And finally, Sect. 5 reviews the conclusion from this survey regarding using artificial intelligence to support software engineering.

2 Background

2.1 Software Engineering and the Software Crisis

Software engineering can be defined as the process of identifying and analysing user requirements and then adopt an SDLC model to design, build and test proposed software that will satisfy the gathered requirements. In the late 1960 many software projects failed as their output was an over budget unreliable software that is expensive to maintain, which started to constitute a fact that larger software is difficult and expensive to maintain [6]. In addition to this, back on this time lots of software could not serve the growing requirements of various customers. Furthermore, as the hardware capabilities increased so did the complexity of software projects. Generally, what happened back then is that the demand for software products grew faster than the ability to generate software. One way to solve such crisis was to transform unorganized coding effort into the software engineering discipline that succeeded to help companies to eliminate the causes of the software crisis by channelling their capabilities to deliver software that meets the customer requirements.

2.2 Software Development Life Cycle Models

It all starts with a business sector concerned with a business problem and willing to invest in a software that would solve this problem. A software development life cycle is the process concerned of understanding this problem to define a problem domain, then decide a plan for solution that would define the solution domain, after that a group of software developers starts to code the planned solution, then a group of software testers starts an effort to test the delivered software searching for defects or chances of improvement to feedback the software development team to start development\redevelopment according to the feedback from the software testers. Finally, a continuous effort is pursued to maintain the final product. Figure 1 shows an abstracted overview of the SDLC.

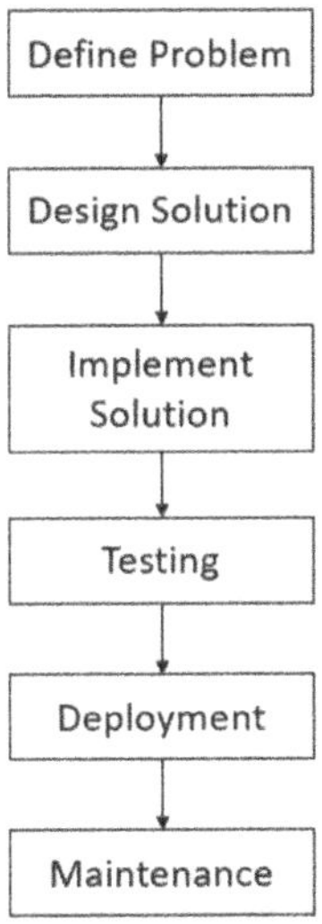

Fig. 1. SDLC Process

Once a problem is defined, a solution is designed and implemented and tested against business needs and then gets deployed. Once deployed, the maintenance effort starts to keep the software available and functional no matter the execution environment and the change in business needs. There exist different models of the SDLC al differ in the order and methodology of performing the main targets of the SDLC mentioned in Fig. 1. There exist many models that utilize different techniques with different order to fulfil the main steps of the SDLC. Examples of these models are the Waterfall model, Iterative Waterfall model, Prototyping model, Spiral model, and the V-shaped model. The Waterfall model that has many advantages like being easy to understand where the software components are defined before the design phase hence is usually used for projects with simple and strict deadlines. On the other hand, the disadvantages of this includes the assumption that the software development environment is always ideal with no surprises, while in fact the software development environment is not always like this while this model has no full tolerance to integrate risk management. It is also difficult to make changes to the software using this model. This model is also infamous of the fact that problems usually remain uncovered until testing. The Waterfall model is described in Fig. 2.

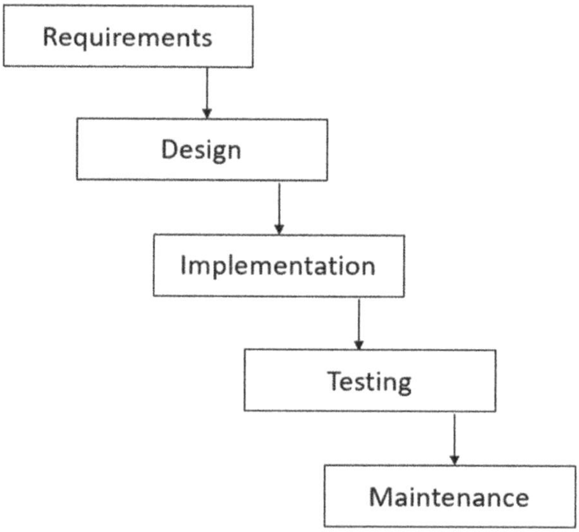

Fig. 2. Waterfall SDLC Model

The Iterative Waterfall model is a much better than the normal waterfall where the client can get continuous feedback. This model is usually used in project where requirements are not clear. However, it's not easy to manage with no clear milestones. Figure 3 shows the flow of the iterative waterfall SDLC model.

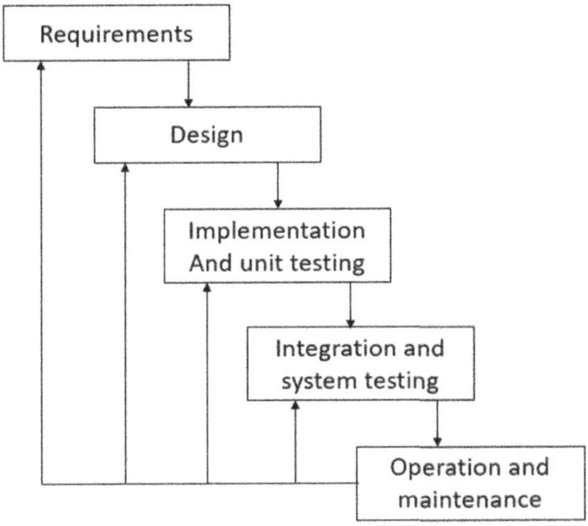

Fig. 3. Iterative Waterfall SDLC Model

The Prototyping model provides early visibility to software components and problems with higher throughput. It also provides the client with continuous feedback to the developed software and it also embraces the concept of cost reduction. On the other hand, the management of this model is complex and lacks flexibility with the possibility to system delivery failure. Figure 4 shows the Prototyping SDLC model.

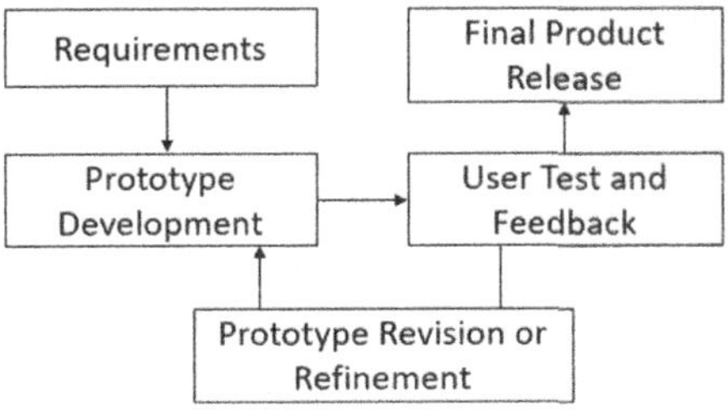

Fig. 4. Prototyping SDLC Model

The Spiral model introduces better productivity with reuse capabilities. A new prototype is obtained every iteration. In this model, errors are eliminated in early stages of development. On the other hand, the management of this model is complicated with hard estimation of cost and time. Figure 5 shows the Spiral SDLC model.

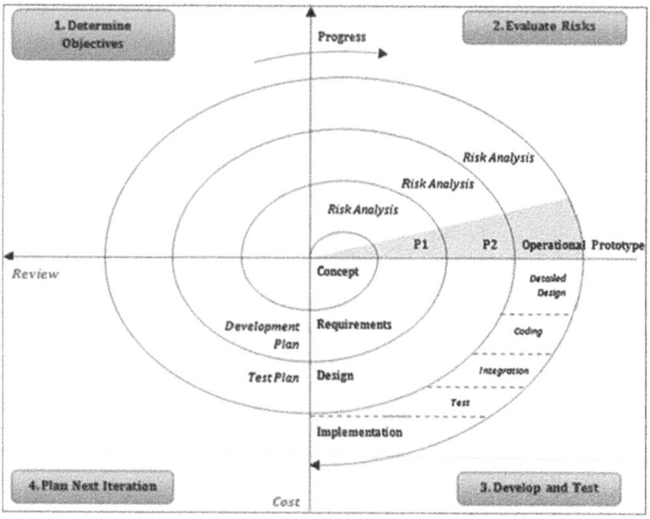

Fig. 5. Spiral SDLC model [3]

The V-Shaped model can be considered as an extended version of the waterfall model with high amount of risk analysis and better dealing with critical projects and early production. This model is also considered easy to manage and understand. However, this model is not the best for object-oriented projects and long-term projects. This model is not suitable for projects where requirements have high risk of changing.

2.3 Multi-objective Optimization

Multi-objective optimization is the approach concerned to solve problems that contains more than one optimization function. In such problems, the algorithm tackles more than one optimization function and travels through the search space to searching for a solution or a set of solutions that may achieve the optimization goal considering all the given optimization functions [7].

2.4 Evolutionary Algorithms

An Evolutionary algorithm (EAs) is a technique inspired from the biological process of evolution. Under the umbrella of artificial intelligence techniques, evolutionary algorithms target global optimization by imitating the biological process of natural evolution. EAs utilizes operators inspired from biological evolution such as procreation, crossover, mutation, and selection. EAs are based on populations where each member in an EA's population represents a possible solution to the optimization problem. The fitness of a member of population is determined by an evaluation function that measures the quality of each member in the population as a possible solution. In an evolutionary algorithm, the evolution process is fulfilled by iterating through the evolutionary operators mentioned above.

3 Challenges

3.1 Software Engineering Challenges

Developing software have become a complex resource consuming process. Complexity comes from the fact that a foggy scrambled user requirement must be restructured by analysts to be put in an organized form that can be further analysed by development teams. Then, software development teams start an effort to organize the development process of the requirements in terms of effort estimation, design and coding and maintenance while software testing teams start an effort to estimate the testing process. All the efforts mentioned above are subject to error and\or sub optimal output if it is done manually by human intervention. As a result, various researches we conducted to utilize artificial intelligence to optimize such processes [8].

3.2 SDLCs Challenges

Software engineering is the branch of science that concerned to organize the overall software production. For the actual software development effort to take place to output the final product, a software development life cycle has to commence, and for that to happen, a Software development life cycle model has to be embraced. As discussed earlier, there exist various SDLC models where each model has advantages and disadvantages and is suitable for specific type of projects and development environment.

4 Software Engineering and Evolutionary Algorithms

4.1 Effort Estimation

Software effort estimation is considered a challenging issue in software development. In order to produce an accurate effort estimation, many SDLC models have been proposed, however, none of the models proposed so far has proven efficiency in a development environment with high uncertainty.

Brajesh et al. introduced a model for the estimation of software development effort for projects sponsored by NASA by utilizing a binary genetic algorithm. Also, a modified version of the constructive cost model (COCOMO) model was utilized to consider the impact of methodology in effort estimation. NASA software project data was subject to the performance quality testing of the developed model were the developed models has provided promising estimation capabilities [9].

Vishaliet et al. introduced a genetic algorithm that has been tested and the output results were compared against those obtained using the existing COCOMO model coefficients. The final experimental results elaborate that in most test cases the results obtained using the coefficients optimized by the proposed algorithm are close to the ones obtained using the current coefficients. Comparing organic and semi-detached COCOMO model modes, it can be stated that use of the coefficients optimized by the GA and ACO in the organic mode produces better results in comparison with the results obtained using the current COCOMO model coefficients [10].

4.2 Design and Coding

Software development metrics are numeric values related to the software development process. They have traditionally been output through equations, but such technique is constrained by fully understanding the parameters interrelationships. This research aimed to find alternative methods to generate software metrics. GAs were proposed to derive metrics were it proved to have several advantages [11].

Seyed Mohammed et al. proposed a novel GA-based algorithm that proved to be a powerful optimization algorithm utilized for search space exploration, called SCI-GA (Software Component Identification using Genetic Algorithm), to identify components from analysis models. The SCI-GA utilizes software cohesion, complexity measurements and coupling to define its evaluation function. The algorithm has been evaluated using three real-world cases. The final experimental results have proven that SCI-GA succeeded to identify correct suboptimal software components and was superior to alternative heuristics like k-means and FCA-Based methods [2].

Kwong et al. has introduced an optimization model for the selection of software components for components-based software development (CBSD). The proposed model objectives are to maximize the functional performance of the CBSD as well as the cohesion while minimizing the software modules coupling. A genetic algorithm was used to enhance the search for the optimal selection of software components for CBSD development. Its efficiency was also proven by giving an example of developing a financial system for small-size and medium-size enterprises that illustrated the proposed methodology's efficiency [12].

Saxsena et al. targeted the major issue of component-based software engineering concerned with the "Component Selection". A Genetic Algorithms based technique has been utilized for component selection to minimize the gap between selected components [13].

4.3 Testing

Software testing is the process of targeting a program for execution with the intention of finding bugs, defects of missing features. Software testing is a resource consuming process in terms of effort, time in the software product's lifecycle. Constructing test cases and test data is considered a key problem in software testing, while automating such activities will improve the efficiency and effectiveness and will reduce the overall cost of software testing. The generation of test data by utilizing random, symbolic or dynamic approaches are considered not enough to generate suitable amount of test data. Some other problems, like failing to recognize the existence of infinite loops and inefficiency to generate test data for complex programs makes these techniques fail to be suitable for generating test data. Therefore, there is urgent need for generating test data using search-based techniques. In addition to that, there is also a need to generate test cases that focuses on error prone areas of code [15–18].

Utilizing Genetic Algorithms to support Software Testing is an emerging area of research that benefits from ideas across two domains. A Genetic Algorithm is utilized to generate unique test cases. It maximizes the coverage for the generated test cases. To improve the effectiveness of the generated test cases and test data, the needed measurement, quantification and the optimal modelling is done by using the accurate combination of software test metrics. The test metrics measures complexity and quality. Abhishek et.al applied the optimization study of the test case generation based on the Genetic Algorithm and generates test cases which are far more reliable [19, 20].

Aladeen et al. have investigated using genetic algorithms to generate test data for automatic path coverage against generating test data of path testing using Yong's methodology. It was observed that GAs has significantly reduced the required time for prolonged testing by restricting the generating process for the meaningful test cases for path testing. [16, 21].

Roy et al. Introduced a technique that utilizes genetic algorithms to automate test – data generation. Genetic algorithms are heuristic techniques that mimics the process of natural evolution searching for the optimal solution to the targeted problem. The technique used for test data generation was implemented using a tool called TGen where parallel processing was used to improve the performance of the searching process. To experiment with TGen, a random test data generator with the name "Random" was also implemented. Both TGen and Random were utilized to generate test data for statement and branch coverage of six programs [22].

Rajappa et al. Introduced a graph theory constituted on genetic approach for software test case generation. In this approach they created a directed graph of all possible intermediate states of the system for the expected behaviour, the initial population of the GA is generated by randomly creating a population of all the nodes of the graph. Then the elements of the genetic algorithm go through an iterative process of genetic operators. The process goes on until convergence where all nodes are covered. The proposed technique provides better performance in terms of accuracy in case of network testing or any other system testing where the predictive model-based tests are not optimized to produce the output [17].

K Singh used Genetic algorithm in scheduling of tasks to be executed on a multiprocessor system. Genetic algorithms are well suited to multiprocessor scheduling problems. As the resources are increased available to the GAs, it can find better solutions in short time. GAs performs better as compared to other traditional techniques. So GAs appears to be the most flexible algorithm for problems using multiple processors. It also indicates that the GAs can adapt automatically to changes in the problem to be solved [23].

5 Conclusion

As the need to digitalize more aspects of our lives increase, so did the pressure on the software engineering field. Since then, the software engineering field has gone through various researches introducing techniques that may support the software engineering process. The practical implementation of the concepts of software engineering is done through embracing a software development life cycle model (SDLC model). It was witnessed from previous literature that AI has been applied to support the field of software engineering on different phases of the SDLC with promising results. Artificial intelligence especially evolutionary algorithms have been utilized to optimize projects effort estimation that in return lead to significant cost reduction in terms of resources and time consumption. Evolutionary algorithms were also used to optimize the design and coding efforts as well as software testing efforts in terms of automated test cases and test data generation. In the Agile SLDC, a project plan alongside configuration management plan decides different parameters that will decide the shape of an upcoming development cycle of a project. The optimality of these plans decides the success of the software development cycle(s). Utilizing artificial intelligence techniques to attempt the optimization of these plans would lead to maximizing cost reduction and minimizing wasted effort. Choosing the suitable technique for this optimization target depends on the existence of historical data of previous plans or not. The automation of the process of generating project plan and configuration management plan will lead to great benefit to assure sustainable software development.

References

1. Vijayalakshmi, K., Ramaraj, N., Amuthakkannan, R.: Improvement of component selection process using genetic algorithm for component-based software development. Int. J. Inf. Syst. Change Manage. **3**, 63–80 (2008)
2. Hasheminejad, S.M.H., Jalili, S.: SCI-GA: SCI-GA: software component identification using genetic algorithm. J. Object Technol. **12**, 1–3 (2013)
3. Bhatnagar, V.: A comparative study of SDLC model. IJAIEM **4**, 23–29 (2015)
4. Stoica, M., Mircea, M., Ghilic-Micu, B.: Software development: agile vs. traditional. Informatica Economica 17, 64–76 (2013). https://doi.org/10.12948/issn14531305/17.4.2013.06
5. Ammar, H., Abdelmoez, W., Hamdi, M.S.: Software engineering using artificial intelligence techniques: current state and open problems (2012)
6. Report on a conference sponsored by the NATO SCIENCE COMMITTEE Garmisch, Germany, 7th to 11th October 1968
7. Li, H.-R., He, F.-Z., Yan, X.: IBEA-SVM: an indicator-based evolutionary algorithm based on pre-selection with classification guided by SVM. Appl. Math. J. Chin. Univ. **34**(1), 1–26 (2019). https://doi.org/10.1007/s11766-019-3706-1
8. Ordoñez-Ordoñez, P.F., Quizhpe, M., Cumbicus-Pineda, O.M., Herrera Salazar, V., Figueroa-Diaz, R.: Application of genetic algorithms in software engineering: a systematic literature review. In: Botto-Tobar, M., Pizarro, G., Zúñiga-Prieto, M., D'Armas, M., Zúñiga Sánchez, M. (eds.) CITT 2018. CCIS, vol. 895, pp. 659–670. Springer, Cham (2019). https://doi.org/10.1007/978-3-030-05532-5_50
9. Singh, B.K., Misra, A.K.: Software effort estimation by genetic algorithm tuned parameters of modified constructive cost model for NASA software projects. Int. J. Comput. Appl. **59**(9), 22–26 (2012)
10. Dhiman, A., Diwaker, C.: Optimization of COCOMO II effort estimation using genetic algorithm. Am. Int. J. Res. Sci. Technol. Eng. Math. **3**(2), (2013)
11. Ghiduk, A.S., Girgis, M.R.: Using genetic algorithms and dominance concepts for generating reduced test data. Informatica **34**(3), 377–385 (2010)
12. Tang, J.F., Mu, L.F., Kwong, C.K., Luo, X.G.: An optimization model for software component selection under multiple applications development. Eur. J. Oper. Res. **212**, 301–311 (2011)
13. Dixit, A., Saxena, P.C.: Software component retrieval using genetic algorithms. In: International Conference on Computer and Automation Engineering, ICCAE 2009, pp. 151–155. IEEE (2009)

14. Parnami, S., Sharma, K.S., Chande, S.V.: A survey on generation of test cases and test data using artificial intelligence techniques. Int. J. Adv. Comp. Network. Secure. **2**(1), 16–18 (2012)
15. Koboldt, D.C., Steinberg, K.M., Larson, D.E., Wilson, R.K., Mardis, E.R.: The next-generation sequencing revolution and its impact on genomics. Cell **155**(1), 27–38 (2013)
16. Mohi-Aldeen, S.M., Mohamad, R., Deris, S.: Automatic test case generation for structural testing using negative selection algorithm. (2009)
17. Rajappa, V., Biradar, A., Panda, S.:Efficient software test case generation using genetic algorithm-based graph theory. In: 2008 First International Conference on Emerging Trends in Engineering and Technology, pp. 298–303. IEEE (2008)
18. Fuqing, Y., Hong, M., Keqin, L.:Software reuse and software component technology. Acta Electr. Sin. **2** (1999)
19. Sabharwal, S., Sibal, R., Sharma, C.: Prioritization of test case scenarios derived from activity diagram using genetic algorithm. In: 2010 International Conference on Computer and Communication Technology (ICCCT), pp. 481–485. IEEE (2010)
20. Pargas, R.P., Harrold, M.J., Peck, R.R.: Test-data generation using genetic algorithms. Softw. Test. Verification Reliab. **9**(4), 263–282 (1999)
21. Kaur, A., Goyal, S.: A genetic algorithm for regression test case prioritization using code coverage. Int. J. Comput. Sci. Eng. **3**(5), 1839–1847 (2011)
22. Srivastava, P.R., Kim, T.: Application of genetic algorithm in software testing. Int. J. Softw. Eng. Appl. **3**(4), 87–96 (2009)
23. Singh, K., Rani, R.: Effective software testing using genetic algorithms. J. Glob. Res. Comput. Sci. **2**(4), 76–80 (2011)

Automation of Electron Beam Input During the Welding of Thin-Walled Structures

Vadim Tynchenko[1,2,3](✉), Sergei Kurashkin[1], Valeriya Tynchenko[1,2], Sergei Dokshanin[2], and Anna Leonteva[2]

[1] Reshetnev Siberian State University of Science and Technology, Krasnoyarsk, Russia
[2] Siberian Federal University, Krasnoyarsk, Russia
[3] Marine Hydrophysical Institute, Russian Academy of Sciences, Sevastopol, Russia

Abstract. In the article is considered automation task entering the beam at electron beam welding . During the course of research was developed flow chart to manage it entering an email address a button that shows how it works the process itself management. Developed flow chart optimizations input mode the area in which it is located subprocesses are described optimizations welding equipment. Submitted by calculation algorithm distributions temperatures during the input stage e-mail address beam, based on which one and developed by the entire system. Also developed and presented structural and functional model flow chart programs management process welding and window software version management systems. Solution of the task will allow you to not just lower it expenses for working out technological solutions parameters' electron beam welding of products with a new geometry either of the new one's materials, but also support decision making technologist at optimizations manufacturing process existing one's products. In process experimental data research, based on developed by automated systems, carried out modeling distribution energy consumption at entering an email address beam for structures from two different sources materials with different geometries. Developed by automated system the system in the future allows you to consider product dimensions, type of material, spread by temperature, technological features parameters and get more information stable weld quality seam with less quantity root defects.

Keywords: Electron beam welding · Mathematical modeling · Thermal processes · Automation management · Optimization parameters process · Input electron beam

1 Introduction

Electron beam welding is used in the aerospace industry for connections various structures and nodes manufactured by from metals alloys to which submitted by increased costs requirements to quality welded joint [1–3]. At the same time, for thin-walled structures submitted by increased costs requirements, in particular, to the depth melting point, which is even better when configured technological equipment the process can go out for the value boundaries, installed regulatory documents [4–6]. But how shows practice

R. Silhavy (Ed.): CSOC 2021, LNNS 230, pp. 88–99, 2021.
https://doi.org/10.1007/978-3-030-77442-4_7

the greatest quantity defects such as a non-producer, and other root functions defects occur in the input area and output of electronic the beam [7, 8].

On this site, be important is the definition of intensity zoom in such a characteristic as an electronic token the beam. Wrong one setting up, which can lead to "spilling" material products and services gas pores, as well as distortion paths the weld. For correct use speed selection and growth forms beam current necessary conduct mathematical description modeling distributions input power beam with a goal reducing the number of users economic factors development costs technological equipment process electron beam welding equipment. When this implementation this mathematical method security features in the automated form management systems it will allow you to significantly raise your profile repeatability process electron beam welding, as well as reduce your impact human click on quality of manufactured products by enterprises products.

2 Mathematical Description Software Automation Tasks

The main phase welding in steady-state mode is given great attention while the stages input and output beam remain in the shadows, although enough a large number of defects occurs in the same way called zero-point welding – Tom the area of the seam, where were input and output beam [7, 9, 11]. Enter and beam output complicated possible appearance such defects like: gas powered pores, non-steam, root defects, when selecting wrong one speeds and management forms with the beam current on at this stage [8, 10, 12]. In in the course of the work done mathematical analysis simulations with the use of softwarethe MATLAB product during the input stage electron beam therapy welding, were received graphs distributions temperatures for two alloys VT-14 thin-walled structures. Received results show developed mathematical models for calculate temperature on the site, complement complex available mathematical models, as well as the opportunity their application to calculate and optimization the welding process, given the excess temperature on the site input electronic beam.

For calculation temperatures in progress welding at the stage input and output beam, is offered use the formula is as follows called maximum value temperatures, submitted by in the theory of welding processes [13]. This approach allows you to calculate temperature control anywhere products even in progress I know that allow correct it temperature control at the I/o stage the beam.

Formula to calculate the maximum value temperatures when in action of a fast-moving linear heat source in an infinite plate with heat transfer [13]:

$$T_m(y, t, q, v) = T_i(x, y, z, v, t, q) + \frac{\left(\frac{2}{\pi e}\right)^{1/2} * \frac{q}{v}}{2\delta c \rho y}\left(1 - \frac{b * y^2}{2a}\right), \quad (1)$$

where T_i – initial temperature of the product;

$a = \lambda / c\rho$;

b – plate heat transfer coefficient;

v – welding speed;

q – effective beam power;

y – distance from the axis of the seam to the heating point;
cρ – heat capacity of the steel.

Based on developed by the algorithm temperature calculation was conducted modeling with the use of software security features MATLAB. As a result, for titanium alloy W-14 thick 0.12 cm is carried out modeling process electron beam welding, when set modes input and output e-mail address the beam. As a result, simulations getting a graph distributions temperature, with electron beam therapy welding of the material VT-14, pictured in Fig. 1.

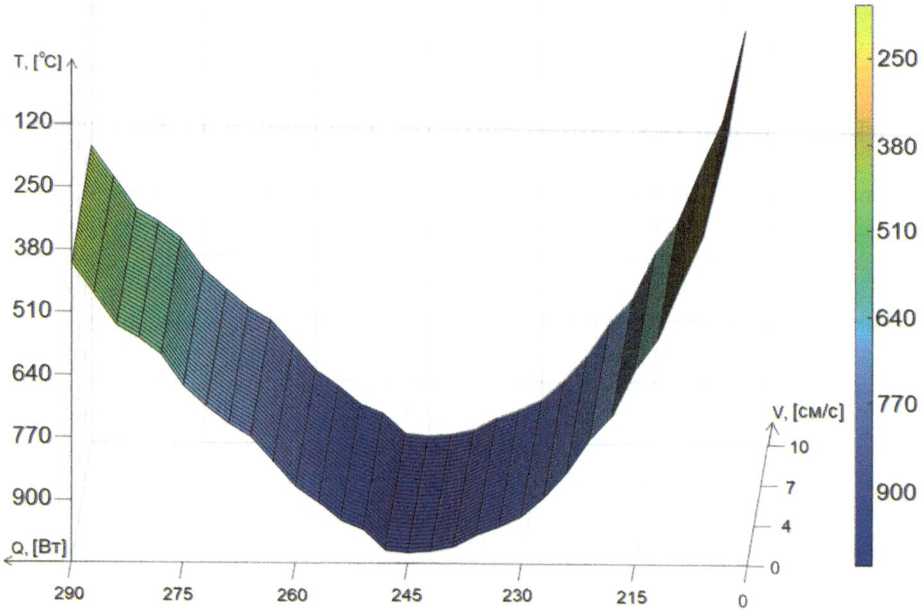

Fig. 1. Graph of temperature distribution for titanium alloy VT-14 at given modes of input-output of an electron beam

3 Algorithmic Support of the Automated System

The diagram of the electron beam injection control is shown in Fig. 2.

At the first stage, are sets the thermophysical parameters. At the next stage, it is necessary to enter the welding parameters obtained at the previous stages of modeling and tested in field experiments. At the fourth stage, the optimal value of the function I (x) is calculated - the function according to which the electron beams will be gradually introduced. Within the framework of this study, this function in general form can have one of three forms: linear, parabolic, and inverted parabolic. In accordance with the calculated I(x), a discrete input of energy is performed, and the discretization is ensured by using pauses for the time Δt.

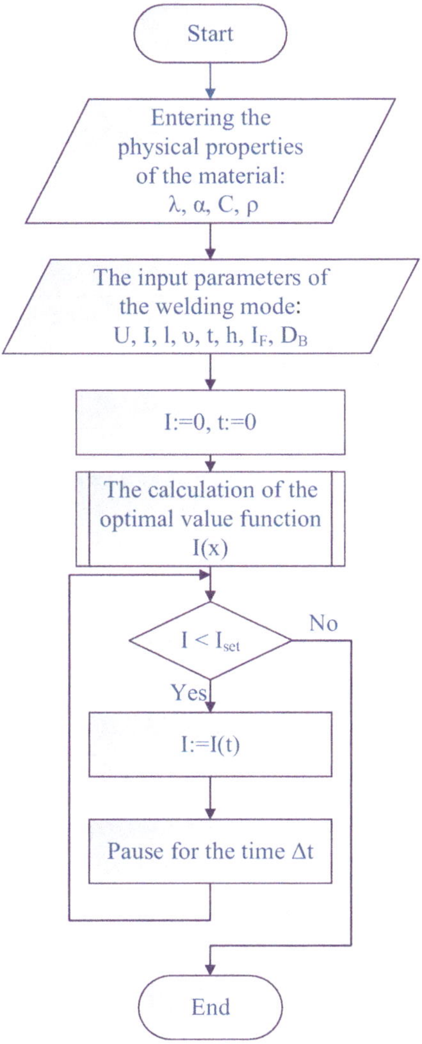

Fig. 2. Block scheme for electron beam injection control

Figure 3 shows a flowchart of the optimization the beam input mode.

On the first floor the stage is being produced mathematical description modeling calculation of heat consumption fields. This one subprocess in expanded mode the view is shown in Fig. 4. Further on the scheme is presented as follows three subprocesses, in which cases produced search for values t_{set}, an exactly t_{set}^{line} - steady state linear time form functions input, t_{set}^{quad1} – steady state time of the quadratic function form functions entries with a reduction intensity, t_{set}^{quad2} – steady state time of the quadratic function form functions input with an increase intensity.

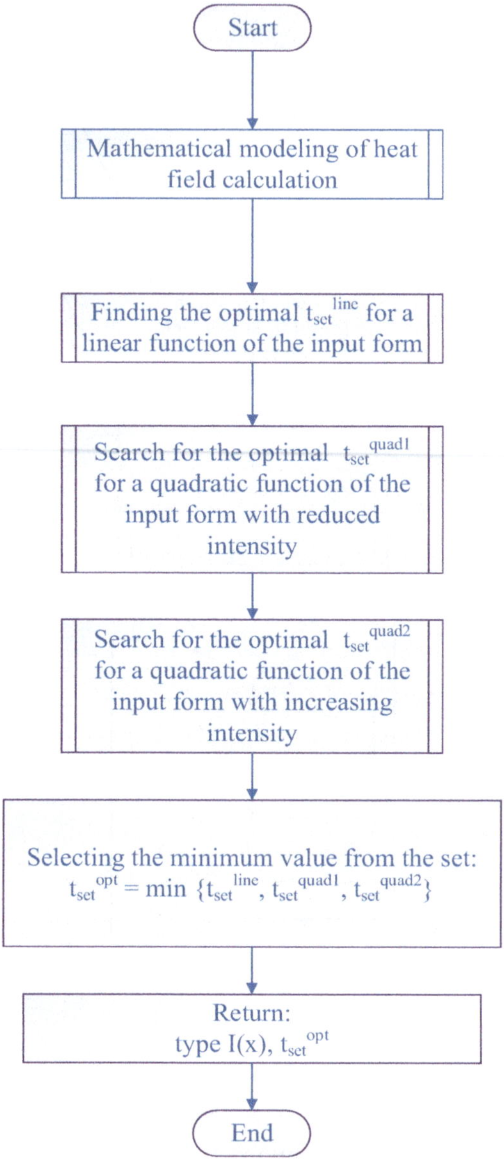

Fig. 3. Block scheme of the optimization of the beam injection mode

After that, the minimum value is selected, which is taken as t_{set}^{opt} based on the results obtained, and then the function I(x) and the optimal steady-state time t_{set}^{opt} are returned.

In order to ensure the best quality of the weld in the insertion area, three conditions must be met:

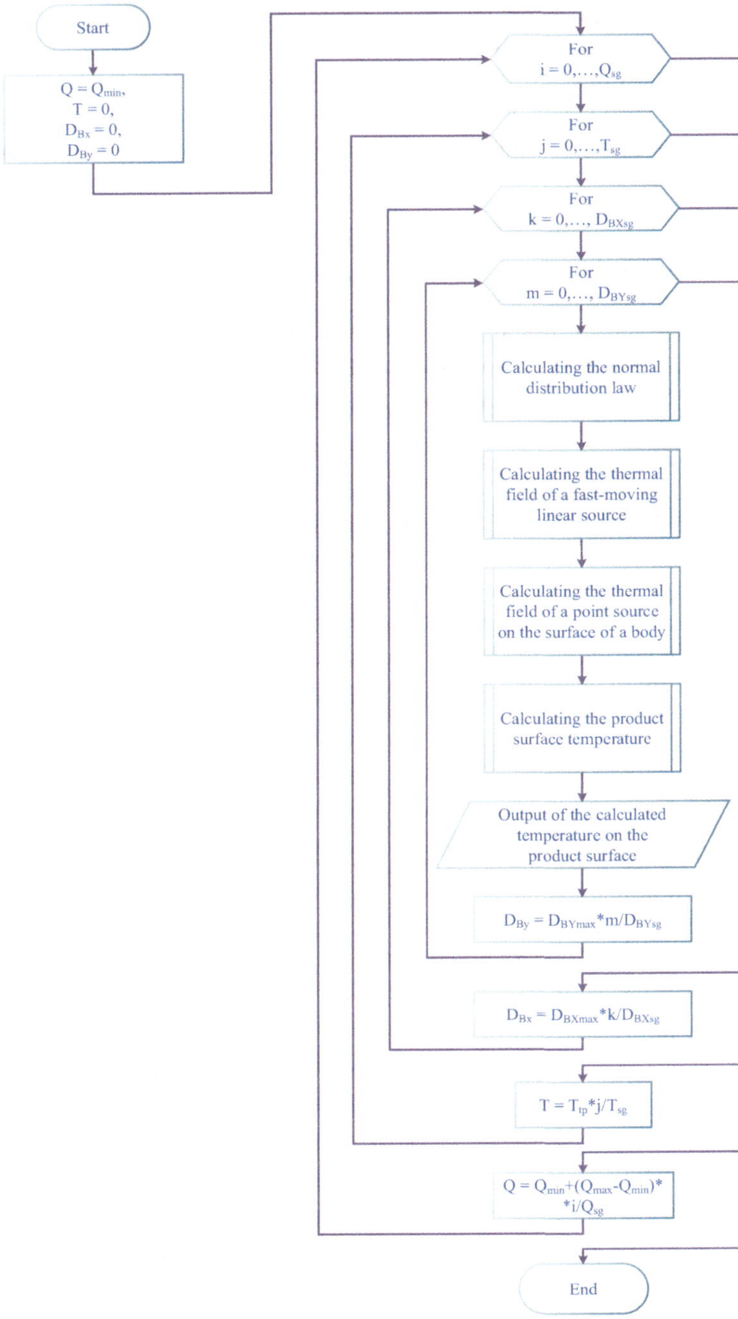

Fig. 4. Algorithm for calculating the temperature distribution at the stage of input and output of the electron beam

- input time it must be minimal with the goal of quicker exit to the management system electron beam welding on the established operating mode.
- the percentage of energy converging at the points of the product falling into the intended zone of the weld (S_{in}), must be maximum.
- the percentage of energy converging at the points of the product outside the assumed zone of the weld (S_{out}), must be minimal.

Based on this, the shape of the electron beam injection should be optimized. Figure 4 shows a block scheme for calculating the power distribution. On the diagram submitted by four subprocesses. Consider each of them learn more. In the first subprocess produced calculating the normal value distribution law e-mail address the beam, at which standard deviation rejection accepted per diameter the beam. It's connected if necessary scan input e-mail address beam for prevention root defects. Thus, produced rationing applied scans in each point.

In the second case and the third subprocess in progress calculation the thermal field fast moving linear source and the thermal field a placemark source code on the site surfaces bodies. In previous versions works by authors conducted research, and were developed mathematical features based on which produced calculation thermal fields.

In the fourth subprocess in progress calculation temperatures on the surface products, on the based on the received information formulas in which considered coefficient heat saturation. Including input dimensional analysis grid, appears possibility calculate temperature control in any topic of interest points and get them three-dimensional distribution temperatures on the surface weldable material products.

4 Software

In Fig. 5 pictured structural and functional model program outline on management process electron beam welding equipment. In this category flow chart submitted by main ad blocks, of which the program consists of for optimization purposes the EBW process. This is the module itself software interface, where they are displayed current value

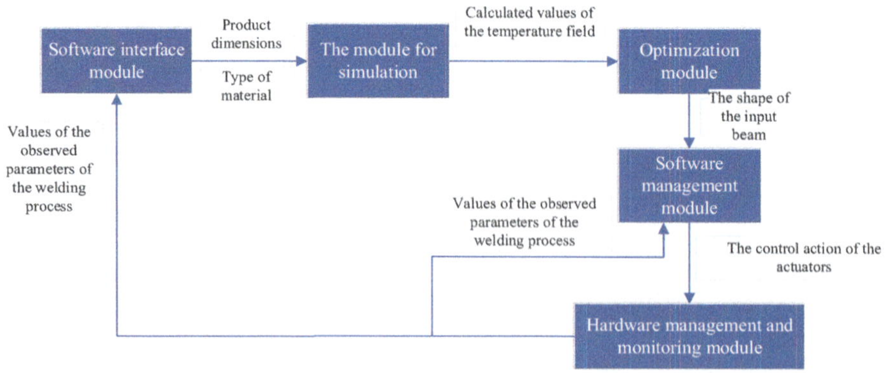

Fig. 5. Structural and functional block diagram of the welding process control program

received from sensors. This modeling module, where produced calculations temperature control fields. The same way present optimization module, who gets it data from the module simulations and already being produced bid adjustment input forms the beam. And so, on software module management and the control module monitoring hardware security features, in which cases produced software and hardware management accordingly.

Figure 6 shows the main window of the developed software system.

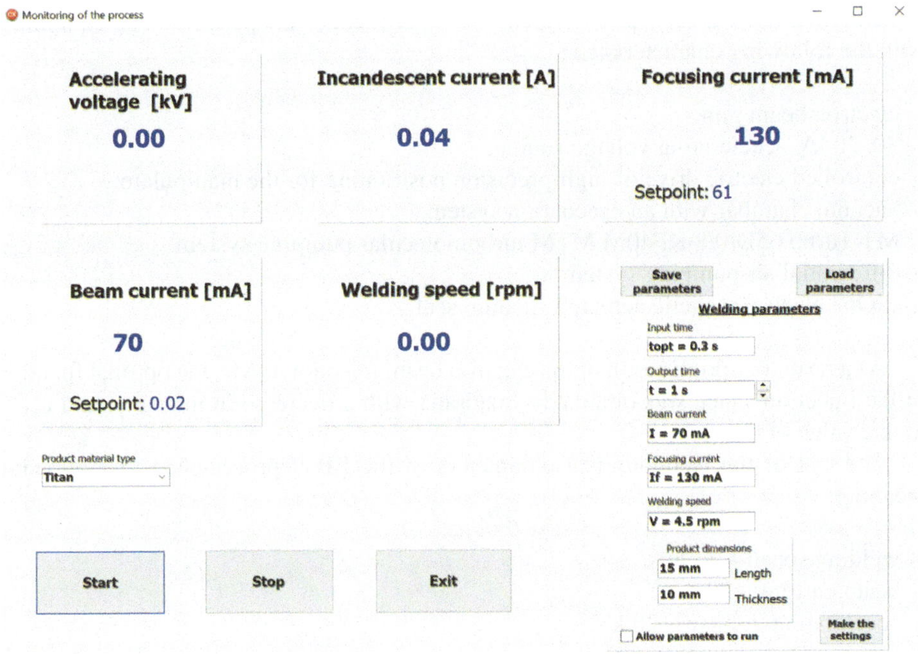

Fig. 6. Window of the software system

Current version the program allows you to set directly the technology itself parameters and save them to a separate one the file, as well as upload when if necessary. It is worth noting, what is present possibility set dimensions products, namely product length and the thickness of the product.

You can also select a type alloy material product, on the at this stage its broad used alloys in aerospace industries, namely (VT-14, AMG-6, AD-31).

Launch programs produced when clicked start buttons, a stop by clicking the "Stop" button. Exit by button «Exit». In progress welding mode the real one time, output data from measuring stations devices (acceleration device voltage, beam current, current focusing, incandescent current, welding speed).

5 Experimental Study

Using the developed approach, the electron beam input mode was optimized for a product of titanium alloy VT-14, 0.12 cm thick. The experiments were performed on equipment with the following characteristics:

- electron beam gun.
- 30, 20 kV accelerating voltage source.
- controlled electric drive of high-precision positioning for the manipulator.
- vacuum chamber with an evacuation system.
- MT-Turbo 65D/0/8 KF40M MTM turbomolecular pumping system.
- differential air pumping system.
- ISO63 electromagnetic actuated vacuum seal.

As a result of optimization of the electron beam injection mode, the optimal function of the injection shape was obtained - quadratic with a decrease in intensity, and t_{set}^{opt} took a value of 0.8 s.

The rest of the technological parameters of the EBW process had the following meaning:

- welding speed – 3 cm/s.
- beam current – 70 mA.
- accelerating voltage – 20 kV.
- focusing current – 130 mA.
- beam diameter – 0.015 cm.

As a result of metallographic inspection, the following microsections of the welded seam were obtained: in the middle of the electron beam input zone ($t_{set}^{opt}/2$) – Fig. 7; at the point of completion of the introduction of the electron beam (t_{set}^{opt}) – Fig. 8.

As you can see from the received data micrographs, in the middle of the zone entering the beam, width and depth welded joint about 1 mm, which is it speaks about a uniform distribution energy; and as seen on microchip at the end point input width and seam depth a little more 1 mm, and the shape is close to "dagger", what does he say about adequacy developed by optimization criteria, and about the future application received results.

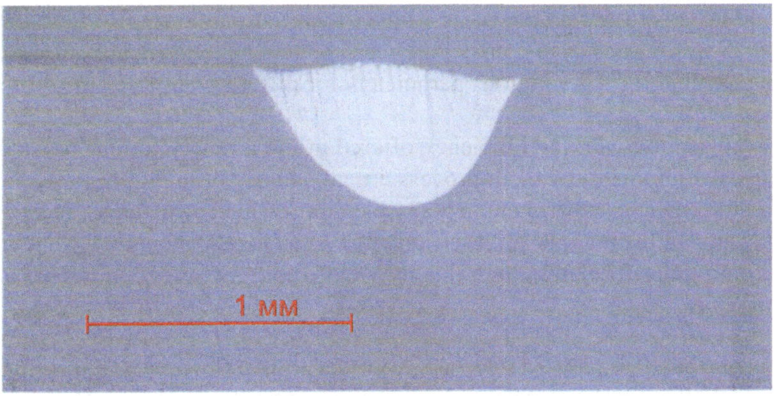

Fig. 7. Microsection of the weld in the middle of the beam entry zone

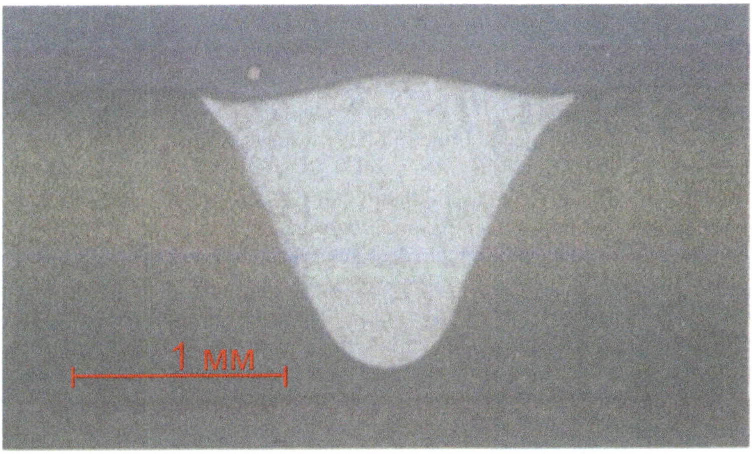

Fig. 8. Microsection of the weld at the point of completion of the introduction of the electron beam

6 Conclusion

Developed by software version automation system entering an email address beam in progress automation tools management takes into account product dimensions, namely the thickness and length, as well as the type of material. In addition, there is a possibility choose an alloy weldable material products and services technological features parameters.

Applied in the system automated management mathematical model optimization model input mode e-mail address the beam takes into account character distributions heat in the zone melting point, installed technological features parameters process electron beam welding, geometry products, as well as required parameters welded joint thin-walled structures. Usage such a model allowed create a mathematical field justification selection of technological solutions parameters for different purposes input modes the beam.

For confirmation working efficiency offered in the article solutions on automation technological equipment the input process e-mail address beam in progress there were welds conducted field experiments on samples thin-walled structures. Received weld joint in the area of the point beam insertion allows you to talk about how what is offered the system provides formation high-quality welded joint, what is confirmed submitted by micrographs.

Acknowledgments. The reported study was funded by RFBR, Government of the Krasnoyarsk Territory and the Regional Science Foundation, project number 19-48-240007 "Mathematical and algorithmic support of the electron-beam welding of thin-walled aerospace structures".

References

1. Mladenov, G., Koleva, E., Belenky, V. Ya., Trushnikov, D.N.: Modeling and optimization electron beam welding of steels. Bulletin of Perm state university national research center polytechnic university. Mech. Eng. Mater. Sci. 4, 7–21 (2014)
2. Salomatova, E.S., Trushnikov, D.N., Tsaplin, A.I.: Modeling processes evaporation rate at electron beam welding with dynamic positioning e-mail address the beam. Izvestia Tula region state-owned enterprise university. Tech. Features Nauki (6–2), 124–133 (2015)
3. Permyakov, G.L., Olshanskaya, T.V., Belenky, V. Ya., Trushnikov, D.: Modeling of electron beam welding to determine the parameters of welded joints of dissimilar materials. Bulletin of the Perm national research polytechnic university. Mech. Eng. Mater. Sci. (4), 48–58 (2013)
4. Raj, R.A., Anand, M.D.: Modeling and prediction of mechanical strength in electron beam welded dissimilar metal joints of stainless steel 304 and copper using grey relation analysis. Int. J. Eng. Technol. **7**(1), 198–201 (2018)
5. Kanigalpula, P.K.C., Jaypuria, S., Pratihar, D.K., Jha, M.N.: Experimental investigations, input-output modeling, and optimization of spiking phenomenon in electron beam welding of ETP copper plates. Measurement **129**(1), 302–318 (2018)
6. Vil'shans'ka, T.V., Salomatova, E.S.: Overview modern ones management methods electronic with the beam at electron beam welding. Bulletin Perm region national research center polytechnic university. Mech. Eng. Mater. Sci. (4), 169–187 (2016)
7. Ananthamoorthy, N., Baskaran, K.: Modelling, simulation and analysis of fuzzy logic controllers for permanent magnet synchronous motor drive. Int. Rev. Model. Simul. **6**(1), 75–82 (2013)
8. Wei, H., Yunhui, D., Lianqing, Z., Mingli, D.: Angular sensing system based on Y-type twin-core fiber and reflective varied line spacing grating fabricated by electron beam lithography. Results. Phys. **18**, 103193 (2020)
9. Peter, S.E., Kulkarni, S., Raglend, I.J., Simon, S.P.: Wavelet based spike propagation neural network (WSPNN) for wind power forecasting. Int Rev Model Simul (IREMOS) **6**(5), 1513–1522 (2013)
10. Farahat, M.A., Abd Elgawed, A.F., Mustafa, H.M.M., Ibrahim, A.: Short term load forecasting using BP neural network optimized by particle swarm optimization. Int. Rev. Model. Simul. **6**(2), 450–454 (2013)

11. Sitnikov, I.V., Belenkiy, V.Y., Olshanskaya, T.V.: Study of the effect of focusing and oscillation of electron beam on the structure and properties of welded seams. IOP Conf. Ser.: Mater. Sci. Eng. **611**(1), 012009 (2019)
12. Kasitsyn, A.N., Agafonov, R. Yu., Leonov, A.V., Grigoriev G.V.: Control of electron beam wielding parameters based on the gap scanning system data during the welding process. IOP Conf. Ser.: Mater. Sci. Eng. **759** (1), 012013 (2019)
13. Konovalov, A.V., Kurkin, A.S., Makarov, E.L., Nerovny, V.M., Yakushin, B.F.: Theory of Welding Processes. MSTU Publishing house named after N.E. Bauman, Moscow (2007)

Evolution of Contrast and Illumination Adjustment Techniques for Dental Imaging to Assist Effective Implant

J. Shashikala[1,2](✉) and N. Thangadurai[3]

[1] Department of Electronics and Communication Engineering, JAIN (Deemed-to-be University), Bengaluru, Karnataka, India
shashikala_j@bmsit.in

[2] BMS Institute of Technology and Management, Bengaluru, Karnataka, India

[3] Department of Electronics and Communication Engineering, Faculty of Engineering and Technology, Centre for Emerging Technologies, JAIN (Deemed-to-be University), Bengaluru, Karnataka, India

Abstract. Dental radiography often suffers from problems associated with low contrast and uneven illumination. The contrast enhancement over low-resolution and poor-quality dental images provide a better interpretation of hidden diagnostic information for carrying out precise pre-operative assessments. However, image-quality is a subjective measure. Therefore, the selection of contrast and illumination adjustment techniques depends on the image of interest and treatment objectives. The proposed study introduced an efficient computational model that automates the process of addressing image quality issues using different pre-processing techniques gamma correction, histogram equalization, linear and non-linear filtering, ROI extraction, and low light image enhancement techniques. The proposed system offers a flexible approach that would enable dental practitioners to select the most suitable enhanced radiographic images to carry out the effective pre-operative assessment, identifying complications and treatment planning towards appropriate implantation. The efficiency of enhancement techniques is validated based on qualitative and quantitative analysis through, PSNR as a performance metrics.

Keywords: Dental images · Contrast enhancement · Gamma correction · Histogram · Filtering · RoI extraction

1 Introduction

Image enhancement has a significant impact on the analysis and diagnosis of the patient. The prime goal of image enhancement is to improve the appealing aspect of a human organ structure without compromising the loss of important information in the input image [1, 2]. Image enhancement is generally performed to adjust the contrast and brightness factor so that the region of interest (RoI) can be evidently analyzed and discrete from the image background. Different imaging modalities belong to the medical field,

R. Silhavy (Ed.): CSOC 2021, LNNS 230, pp. 100–109, 2021.
https://doi.org/10.1007/978-3-030-77442-4_8

such as computed tomography (CT), CBCT, X-ray, magnetic resonance imaging (MRI), etc. Among them, digital X-rays are widely used in medical imaging due to its availability and its cost-effectiveness [3, 4]. This study focuses on enhancing dental X-ray images widely used in dental practices to carry out a pre-operative assessment and analyze the maxillofacial structure and abnormal conditions [5]. The manual interpretation of dental images requires subjective knowledge and physician experience. Analyzing dental images with poor quality is always difficult for the physician and dentists due to the intricate structure of the human mouth and maxillofacial region [2]. Also, the physician in dental disciplines often has to analyze many dental images to perform a comprehensive assessment to examine the conditions and surroundings of the implant location and treatment stability, which is tedious and time-consuming [6]. The adoption of automated tools can practically ease the manual process of imaging analysis. This automation involves various progressive operations such as RoI segmentation, feature selection, and classification of the RoI. However, such automation processes require the input image with high quality to provide accurate results [7]. This paper aims to introduce a cost-effective computational model that supports different image enhancement techniques for dental imaging.

In this work, contrast enhancement and illumination adjustment techniques are evaluated to support clinicians and physicians to carry out the appropriate enhancement of dental images based on the performance analysis of contrast and illumination techniques. The rest of the structure of this paper is formulated as follows: Sect. 2 discusses related work. Section 3 highlights the research problem. The evolution of different enhancement techniques is elaborated in Sect. 4, and performance analysis is carried out in Sect. 5. The proposed work is summarized in Sect. 6.

2 Related Work

A massive amount of literature is available for image Enhancement in the field of medical imaging. The proposed study focuses on the enhancement of dental images. This section discusses the existing literature available for enhancing such images. The authors in the study of Choi et al. [8] carried an analysis towards exploring the impact of different enhancements techniques on the periapical dental image. A CLAHE-Rayleigh method is adopted in the research work of Suprijanto et al. [9] to enhance the panoramic dental image. The study carried out by Kamezawa et al. [10] introduced a multiple noise filtering approach for CBCT imaging. Supriyanti et al. [11] used the point processing technique contract stretching of a dental image. Yin et al. [12] introduced noise filtering based on thresholding mechanisms and wavelet transform for improving the visual aspect of the CBCT image. An adaptive gamma correction is used in Amiri and Moudi [13] and Kandan and Kumar [14] to eliminate the noisy signal caused by positioning error in the digital panoramic dental image. In the study, Hao et al. [15], a mean filtering approach is introduced to enhance the visual quality of the CBCT image. The results are validated using PSNR and MSE. Naik et al. [16] adopted histogram equalization contrast improvement of the dental radiograph for proper bone structure and quality analysis. The work carried out by Qassim et al. [17] introduces the brightness preserving technique using entropy and histogram analysis. This study highlighted that histogram entropy and flatness are the significant metric on which quality of enhancement depends. Georgieva et al. [18] considered CLAHE, morphological technique, and homomorphic filter to enhancement dental x-ray image for caries identification.

3 Problem Description

Much research has been done in the field of medical image enhancement. However, the research works on dental image enhancement is limited. In most work, histogram equalization and gamma correction have been widely adopted. Most image enhancement methods are in the transform domain. Some artifacts, mostly over-enhancement, may appear in the output image. Most of the existing research works follow a similar approach to execute pre-preprocessing operations. However, it would be more practical if the adjustment of contrast and illumination is selected based on enhancement techniques' performances. However, there is a lack of literature in this regard. There is a requirement to design the evolution model where there is scope to evaluate the enhancement of images with different techniques.

4 Proposed System

This paper aims to introduce an evolution model for contrast adjustment and brightness preservation in dental imaging to effectively assist implant therapy. The contribution of this work is also to facilitate physicians to select appropriate contrast enhancement technique by taking into account the performance of enhanced output image metrics. The study considers various enhancement techniques for contrast adjustment non-linear enhancement operation using gamma correction, histogram, and CLAHE. Brightness preservation and illuminance adjustment using filtering and blurred image enhancement technique. The performance of the contrast enhancement techniques is evaluated concerning PSNR. The proposed system facilitates a flexible approach that meets the enhancement requirements needed to improve dental imaging quality. The contrast and illumination adjustment approaches introduced in the proposed study considers dental image **dI(x,y)** characterized by the two fundamental components illumination **Iu(x,y)** and reflectance **rf(x,y)**. Therefore, the input dental image can be described as:

$$dI_{(x,y)} = Iu(x, y) \cdot rf(x, y) \tag{1}$$

The faulty image acquisition process and motion artifact introduce a non-linear effect on luminance, and the images captured in low-lighting conditions turned out an image with poor contrasted. Therefore, considering this fact, the adopted contrast enhancements techniques aim to adjust the contrast ratio and minimize the impact of non-linear illuminance associated with dI.

4.1 Evolution of Contrast Adjustment Techniques

Gamma correction (**Γ** C) mechanism refers to non-linear operation, which uses a function of raise to power in which input pixels values elevated to a constant power to code the brightness-level of the input dental image ('d I') to construct a contrast-enhanced image. The function of this non-linear operation is given by:

$$dI_E(x, y) = C * dI(x, y)^{\Gamma} \quad (2)$$

Where, $dI_E(x, y)$ is the contrast-enhanced dental image. C refers to a constant, which can be as $1 > C > 1$ *or* $= 1$. Γ is the gamma value as encoding and decoding factor, ranging between $0 \leq \Gamma \leq 1$. If the value of $\Gamma > 1$, then the contrast adjustment is weighted towards higher resulting values, i.e., brighter and vice versa for the value of $\Gamma < 1$. Thus, the suitable gamma values for contrast adjustment is decided based on the resulting enhanced image performance metric. In the same evolution process, the next operation for dental image (dI) contrast adjustment is grayscale image histogram equalization (**Heq**), which maintains intensity distribution of the gray values in dynamic range to improve dI contrast. The histogram-equalization uses transfer mapping function $\mathbf{f_{Tm}}()$ to ensures a uniform intensity distribution followed by the resulting dI. The mapping function $\mathbf{f_{Tm}}()$ in Heq is described as follows:

$$\mathrm{dI_E(x, y)} \leftarrow \mathrm{f_{Tm}(dI)} = \mathrm{dI_0} + \left(\mathrm{dI_{g-1}} - \mathrm{dI_0}\right) \cdot \sum\nolimits_{\mathrm{i=0}}^{\mathrm{k}} \rho(\mathrm{dI_k}) \quad (3)$$

Where, $\mathrm{dI_E(x, y)}$ is the enhanced image obtained by **Heq**, $\mathrm{f_{Tm}(dI)}$ refers transfer mapping function over input grayscale dI, g refers to levels of grayscale of dI such that: gray levels $\mathrm{g} = \left\{\mathrm{dI_0}, \mathrm{dI_1}, \ldots \mathrm{dI_{g-1}}\right\}$, $\mathrm{k} = \{0, 1, \ldots \mathrm{g} - 1\}$, and ρ is the probability distribution function that defines the cumulative distribution of pixel appearance. The correction factor $\mathbf{\Gamma C}$ is used to control the overall luminance factor by taking into account the actual appearance of the input image and enhances the contrast of the input image degraded by the gamma distortion. However, Heq cannot enhance the input image, which suffers from gamma distortion. The primary function of the Heq technique is to maintain even distribution of histogram using $\mathrm{f_{Tm}}()$ of the input image. It mainly focuses on the histogram of the image rather than the appearance of the input image. To deal with Heq limitations**,** the proposed study considers the evaluation of contrastlimited adaptive-Heq (**CLAHE**). It is an advanced version of the adaptive Heq, which overcomes over-amplification of noise introduced by the Heq technique. In CLAHE $\mathbf{f_{Tm}}()$ is achieved using **a** contrast amplification limiting mechanism with neighboring pixels to reduce the issue of noise. In this process, the input dI has to be separated uniformly in non-overlapping regions/blocks, and for each uniform region, the histogram has to be computed. A cutoff-value based on user requirement is defined for the contrast adjustment, and a clip limit ι is obtained for the redistribution of histogram underι**.** The numerical formulation for ι is described as follows:

$$\iota = \frac{\boldsymbol{P}}{\boldsymbol{R}}\left(\mathbf{1} + \nu/\mathbf{10^2} \cdot (\boldsymbol{\theta_{max}} - \mathbf{1})\right) \quad (4)$$

Where, P is the dI block pixel with dynamic range R, and ν refers to clip-factor ranging between [0:100] and θ_{max} indicates histogram slope height. The value of ν is equal to 0, then ι will be equal to (P/R), indicating the uniform distribution of block pixel into possible grayscale; otherwise, no distribution of pixel will occur. The improved, enhanced image is obtained using bilinear interpolation and the cumulative distribution of neighboring pixels from the clipped histogram.

4.2 Evolution of Brightness and Illuminance Adjustments Techniques

In order to adjust the illuminance and brightness of dI at frequency levels, two kinds of filtering techniques are considered in the proposed study. The first filtering technique (**filter-type1**) is considered for addressing the visual quality issues caused due to dark regions in the dI using linear operation. The filtering design is performed by initializes the frequency co-efficient **cE** such that:

$$\mathrm{cE} = \{\mathrm{hF},\ \mathrm{lF},\ \mathrm{c1},\ \mathrm{c2}\} \tag{5}$$

Where hF is high-frequency cE, and lF is the low-frequency cE. Apart from this, the other two co-efficient c1 and c2 are initialized for frequency tuning based on brightness adjustment requirements. Further, all **cE** is decomposed with the pixel value (x, y) of dI to construct filter as follows:

$$[\vec{F}] \leftarrow (\mathrm{hF},\ \mathrm{1F})\mathrm{x}\ (1 - \mathrm{e}^{-\mathrm{c1}\left(\sqrt{((\mathrm{x}-\mathrm{q1}/2)^2+(\mathrm{x}-\mathrm{q2}/2)^2/\mathrm{c2})}\right)}) \tag{6}$$

Where, F is the filter, x ← [1:m] and y ← [1: n], and q1 ← [h/2], q2 ←between pixels and function [w/2] where 'h' and 'w' are height and width of the dI respectively. A logarithmic function is applied to perform input image frequency domain transformation followed by fourier transformation. The filtered dI is obtained from the following equations as follows:

$$F\mathrm{dI} = \mathrm{ZdI}(\mathrm{x},\mathrm{y}) \odot \vec{F} \tag{7}$$

$$\mathrm{dI}_E = \exp\left(f^{-1}(F\mathrm{dI})\right) \tag{8}$$

Where ZdI(x, y) is the image in the frequency domain, FdI is the filtered image in the frequency domain and dI_E is the outcome in terms of enhanced dI obtained from inverse log function (exp) and inverse Fourier transform. Another filtering technique, **filter-type 2** is considered to overcome noise, and over brightness, unlike filter-type 1, preserves detailed information in the output image. Filter-type 2 uses non-linear function over maximum intensity value of the input dI, which computes maximum color value (Imax) in the dI, the size of the image row 'r' and column 'c' is computed as follows:

$$[r, c] \leftarrow f2\left(\mathrm{dI}_{\mathrm{x,y}}\right) \tag{9}$$

$$Is = \sum_{r,c=1}^{N-1} 2^{round(log(f3(r,c)))-1} \tag{10}$$

Where Is ← initial spacing between pixels and function f3 () computes the smallest element from r and c. Initialize reference pixel (Rp) to normalize boundary conditions towards computing new reference pixels (nRp) for the enhanced image as follows:

$$\mathrm{Rp} = \mathrm{Imax}\left(\sum_{\mathrm{rc}}^{\mathrm{N}} \mathrm{qrc}\right) \tag{11}$$

Where, q = 1, to perform the computation of transitional product (Tp) with the Rp

$$Tp = \left(\sum_{1}^{N-r}\sum_{1}^{N-c} Rp(r,\ c)\right) + \left(\sum_{r+1}^{N}\sum_{c+1}^{N} \mathrm{dI(r,c)}\right) - \left(\sum_{1}^{N-r}\sum_{1}^{N-c} dI(\mathrm{r,c})\right) \tag{12}$$

Update: Is ← Is/2 and compute nRp

$$\mathrm{nRp} = {}^{(\mathrm{Tp}+\mathrm{Rp})}/_{2} \tag{13}$$

nRp ←the transmission medium Output enhanced image $\mathrm{dI_E}$.
In blurred image enhancement, the input image is first inverted as given below:

$$\mathrm{dI_{inv}} \leftarrow 255 - \mathrm{dI(x, y)} \tag{14}$$

The enhanced image is given by the following function numerical expression

$$\mathrm{dI_E} = \frac{\mathrm{I - Ag}}{\max(tr'(\mathrm{x}), \mathrm{tr_0})} + \mathrm{Ag} \tag{15}$$

Where I is the intensity in the input image, tr is the transmission medium, and Ag global atmospheric light.

Algorithm Description

Algorithm-1 Contrast adjustment in Dental Image

Input: dI (dental image)
Output: $\mathrm{dI_E}$ (Enhanced dental image)
Start

1. Init ΓC
2. $\mathrm{dI_E} \leftarrow \mathrm{C} * dI(x,y)^{\Gamma}$
3. $\mathrm{dI_{HE}} \leftarrow \mathrm{f_{Tm}}(dI)$
4. $\mathrm{dI_{ClahE}} \leftarrow \mathrm{f_l}(dI)$
5. $\mathrm{dI_{FType1}} \leftarrow exp(f^{-1}(FdI))$
   a. Init $hF, lF, c1, c2$
   b. Construct $[\vec{\mathrm{F}}]$
   c. $ZdI(x,y) = \mathrm{f_{ft}}\sum(log(dI+1))$
   d. $\mathrm{FdI = ZdI(x,y)}\odot\vec{\mathrm{F}}$
6. $\mathrm{dI_{FType2}} \leftarrow f2(\mathrm{dI_{x,y}})$
7. $\mathrm{dI_{blurE}} = f3(\mathrm{dI_{inv}})$

End

Initially, the algorithm initializes a variable **ΓC** to take the input value of gamma values from the user (Line1). In the next operation, gamma correction is performed to cod luminance factor with for a given value of **ΓC** (Line2) to compute enhanced image. The value of **ΓC** is considered most suitable once the satisfactory output metric is obtained for the contrast adjusted image. If the value of **ΓC** is considered to be less than one, then it would mean that the contrast adjustment is made towards darker, and if it is considered to be greater than one, then it will mean that adjustment of contrast is required towards brighter. However, ΓC is equal to one. It is a linear operation, which means no enhancement is required over the input image. In the next line, the algorithm also implements a histogram equalization Heq technique using a transform mapping function f_{Tm} to compute enhanced image dI_{HE} (Line3). However, the enhancement of gamma degraded image cannot be efficiently achieved by Heq. Therefore, the study considers other various forms of enhancement technique. The study implements the CLAHE to compute a better form of the output image by addressing the issue of noise amplification in uniform distribution of grayscale pixel and preserves detail information in the output image dI_{ClahE} (Line4). The algorithm implements a filtering approach to address issues of illuminance and brightness adjustment (Line5). The type-1 filter performs the linear operation at frequency levels, for which the system initializes the parameters hf (high frequency), lF (low frequency), c1 (slop constant), and c2 (cut-off frequency). In the next step, a decomposition operation is performed to construct a linear filter [$\overrightarrow{F}$] considering all parameters, which is further processed with the input image in the frequency domain using Fourier and inverse Fourier transform function. An inverse logarithmic function exp() is applied to transform the enhanced image into its original domain from the frequency domain. The next operation is associated with illuminance adjustment in the input image. The system implements filter-type 2, which executes the non-linear operation on the frequency components of the input image (Line-6). A non-linear function $f2$ () is applied with an input argument of the logarithmic single-color channel input image. In this process, initially, the size with the maximum color value of the image is computed. The algorithm also computes initial spacing between pixels using the round function and logarithmic function with an input argument of row size and column size. A reference pixel (Rp) is initialized to normalize boundary conditions towards computing new reference pixels for an enhanced image. Further, the initial spacing operator is updated based on the computation of transitional product with the reference pixel by following horizontal and vertical steps. Finally, the shift operator updates, and a new reference pixel is computed as an enhanced image. The next part of the algorithm focuses on implementing the different types of brightness adjustment technique using function $f3$ (). This enhancement mechanism addresses image quality issues degraded from the blurriness; haziness due to an image captured in low-lighting conditions. In this process, the image is first inverted by computing the complement of the input image. An explicit enhancement function of haze removal is applied over an inverted image to compute the image with enhanced brightness and preserving more information inside the output image (Line7).

5 Result Analysis

The implementation of the proposed model is carried on a numerical computing tool. The performance evaluation is carried out for two different dental x-ray images. The performance for each technique is analyzed in terms of PSNR. The qualitative assessment is tabulated in Tables 1 and 2 for both contrast adjustment and illumination adjustment respectively.

Table 1. Analysis of contrast adjustment

Input Image	Gamma Correction	Histogram Equalization	CLAHE
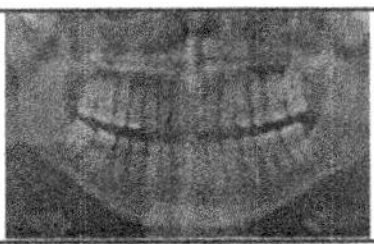	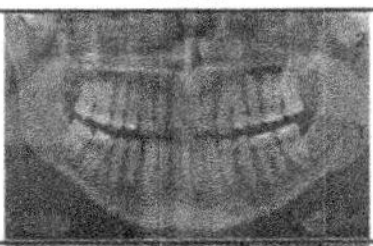	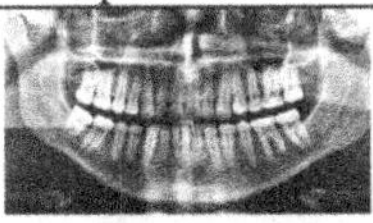	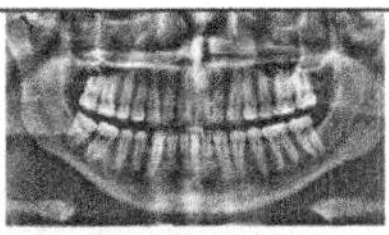
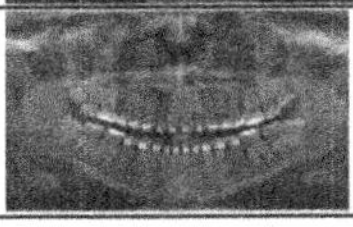	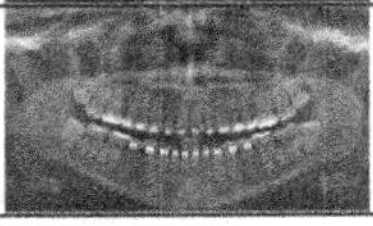	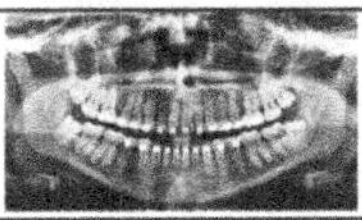	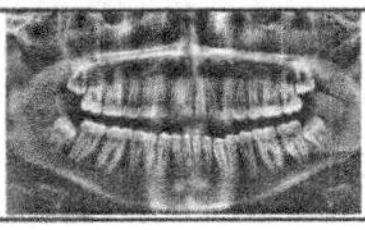

Table 2. Analysis of illuminance adjustment

Image Type	Filter-Type1	Filter-Type2	Blur Image Enhancement
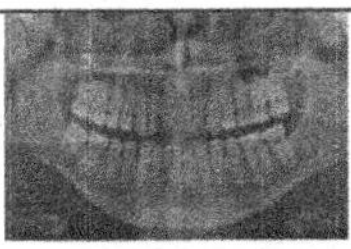		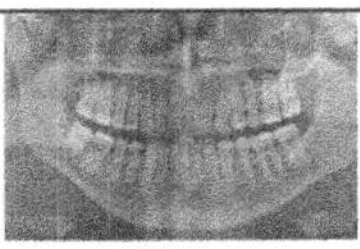	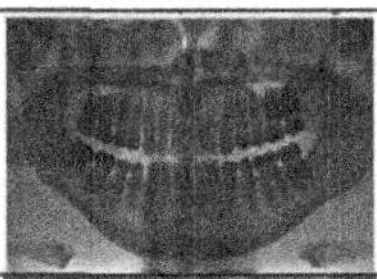
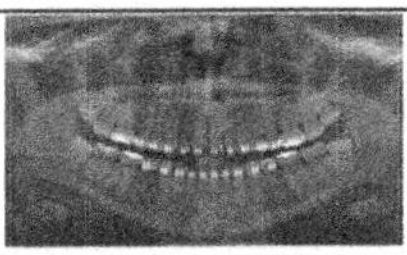		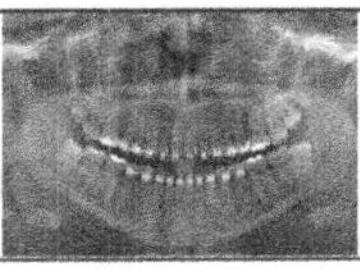	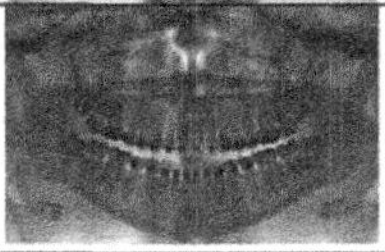

Table 1 shows the qualitative analysis of contrast adjustment technique, the visual outcome shows better output image is obtained by the gamma correction, histogram-equalization and CLAHE. The suitable value of gamma correction factor is achieved as 1.02 for the enhancement of input image based on PSNR value 27.2369 for image-1 and PSNR value 27.3825 for image-2. In Table 2 the qualitative analysis is carried for illuminance adjustment techniques. It can be seen that better performance is achieved by the filter-type2 enhancement technique compared to other two techniques. filter-type1, and the blur enhancement technique, has not shown effective performance and but poor performance is exhibited by filter-type1 due that fact it has over enhancement issues. Figure 1 shows quantitative assessment for the comparison of all enhancement technique in terms of PSNR. It is to be noted that the performance of all these techniques depends on the fact that for different images, the performance in varies to a certain extent depending on the input image visual characteristics. The major contribution of this paper is that, it introduces a modelling of unified evaluation framework with existing enhancement technique and two filtering techniques that handles enhancement issues at frequency level. The proposed study provides a flexible mechanism for dental image enhancement. This model can be also applicable to other imaging modalities for carrying out effective contrast enhancement and illuminance adjustment.

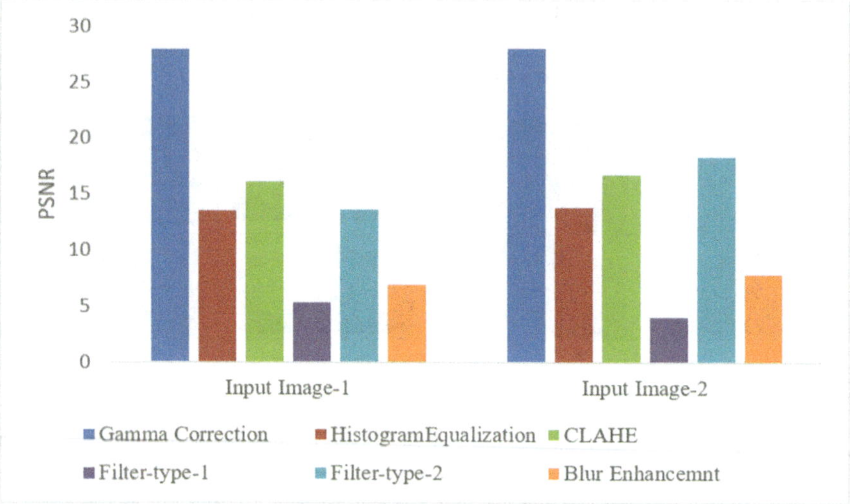

Fig. 1. Quantitative analysis

6 Conclusion

Physicians may misdiagnose medical images due to the ambiguity in the dental images. Therefore, enhancing contrast and adjusting the illuminance of input images before a diagnosis is an important step in clinical practices. This paper has presented a unified approach that integrates different image enhancement techniques for improving interpretation of dental imaging to benefit implant success. CLAHE achieves better performance for the image enhancement with a well-distributed histogram. On the other hand, the performance achieved by linear filter type-2 also exhibited better form of enhanced image and preserves more detailed information in the resulting image. The proposed work provides an evolution model of image pre-processing techniques that can offer a physician to select an enhancement technique flexibly based on their requirements.

References

1. Fu, S., Zhang, M., Mu, C., Shen, X.: Advancements of medical image enhancement in healthcare applications. J. Healthc. Eng. **29**(2018), 7035264 (2018)
2. Rahmi-Fajrin, H., Puspita, S., Riyadi, S., Sofiani, E.: Dental radiography image enhancement for treatment evaluation through digital image processing. J. Clin. Exp. Dent. **10**(7), e629–e634 (2018)
3. Shah, N., Bansal, N., Logani, A.: Recent advances in imaging technologies in dentistry. World J. Radiol. **6**(10), 794–807 (2014)
4. Bercovich, E., Javitt, M.C.: Medical imaging: from Roentgen to the digital revolution, and beyond. Rambam Maimonides Med. J. **9**(4), e0034 (2018)
5. ozylo-Kalinowska, I.: Introduction to dental radiography and radiology. In: Imaging Techniques in Dental Radiology. Springer, Cham (2020). https://doi.org/10.1007/978-3-030-41372-9_1

6. Altaf, F., Islam, S.M.S., Akhtar, N., Janjua, N.K.: Going deep in medical image analysis: concepts, methods, challenges, and future directions. IEEE Access **7**, 99540–99572 (2019)
7. Ker, J., Wang, L., Rao, J., Lim, T.: Deep learning applications in medical image analysis. IEEE Access **6**, 9375–9389 (2018)
8. Choi, J.-W., Han, W.-J., Kim, E.-K.: Image enhancement of digital periapical radiographs according to diagnostic tasks. Imaging Sci. Dent. **44**(1), 31–35 (2014)
9. Juliastuti, E., Epsilawati, L.: Image contrast enhancement for film-based dental panoramic radiography. In: 2012 International Conference on System Engineering and Technology (ICSET), IEEE, pp. 1–5 (2012)
10. Kamezawa, H., Shirieda, K., Arimura, H., Kameda, N., Ohki, M.: An approach of exposure dose reduction of cone-beam computed tomography in an image guided patient positioning system by using various noise suppression filters. In: 2014 Joint 7th International Conference on Soft Computing and Intelligent Systems (SCIS) IEEE (2014)
11. Zhang, Y., Tehrani, J.N., Wang, J.: A biomechanical modeling guided CBCT estimation technique. IEEE Trans. Med. Imaging **36**(2), 641–652 (2016)
12. Yin, Y., Yu, G., Wang, H., Liu, Z., Li, D.: CBCT image denoising based on multi-scale wavelet transform. In: 2010 3rd International Conference on Biomedical Engineering and Informatics, IEEE, vol. 1, pp. 150–153 (2010)
13. Amiri, S.A., Moudi, E.: Image quality enhancement in digital panoramic radiograph. J. AI Data Min. **2**(1), 1–6 (2014)
14. Somas Kandan, R., John Sanjeev Kumar, A.: An improved contrast enhancement approach for panoramic dental x-ray images. ARPN J. Eng. Appl. Sci. **10**, 1897–1901 (2015)
15. Hao, J., Zhang, L., Li, L., Kang, K.: An improved non-local means regularized iterative reconstruction method for low-dose dental CBCT. In: 2012 IEEE Nuclear Science Symposium and Medical Imaging Conference Record (NSS/MIC), IEEE, pp. 3422–3425 (2012)
16. Naik, A., Tikhe, S.V., Bhide, S.D.: Histogram equalization for class-identification of dental disease using digital radiography. In: Das, V.V., Vijayakumar, R., Debnath, N.C., Stephen, J., Meghanathan, N., Sankaranarayanan, S., Thankachan, P.M., Gaol, F.L., Thankachan, N. (eds.) BAIP 2010. CCIS, vol. 70, pp. 144–151. Springer, Heidelberg (2010). https://doi.org/10.1007/978-3-642-12214-9_25
17. Qassim, H.M., Basheer, N.M., Farhan, M.N.: Brightness preserving enhancement for dental digital X-ray images based on entropy and histogram analysis. J. Appl. Sci. Eng. **22**, 187–194 (2019)
18. Georgieva, V.M., Mihaylova, A.D., Petrov, P.P.: An application of dental X-ray image enhancement. In: 2017 13th International Conference on Advanced Technologies, Systems and Services in Telecommunications (TELSIKS), Nis, pp. 447–450 (2017)

Maturity Models, Challenges and Open Issues

Tumisang Akinpelu(✉), Rene van Eck, and Tranos Zuva

Vaal Univesity of Technology, Vanderlbijl Park 1900, South Africa

Abstract. Maturity models enable evaluation of the business process management maturity in organizations and assist to benchmark by revealing the opportunities to improve business processes. A very large number of academics have researched the maturity models but there is a need to correlate the challenges and open issues. The maturity models have challenges that need to be explored in detail in order for organizations to be able to properly implement the models properly to enjoy the benefits thereof. This paper reviews the maturity models in literature and highlights the challenges . The research found that the use of different models produce different results in the same organization and some of the existing maturity models are complicated and require specialized skills. It has been found that it is imperative to standardize the models so that they are able to measure the maturity of organizations in an understandable and reliable way. Despite the challenges that may exist in using maturity models, the organizations can still measure the maturity levels of their organization using the models available.

Keywords: Maturity models · Maturity model challenges · Maturity model open issues · Maturity model challenges and open issues

1 Introduction

Maturity models are ways in which organisations can measure different business processes maturity. They are used to represent the orderly and effective manner of conducting businesses in organisations [1]. Maturity models enable teams to evaluate the maturity of an organisation or department in relation to certain benchmarks [2]. Maturity is defined as a collection of business process competencies that give an indication of anticipated presentation of business processes [3].

Some of the main benefits of maturity models are that they assist to benchmark by revealing the opportunities to improve the business process. Maturity models assist in the determination of readiness or completeness [3]. They are very useful for assessing quality and how effective the existing business processes are. Maturity models also enable the organisations to come up with roadmap development as a way to enhance the level of maturity [4].

Numerous researchers like [5] and [6] have conducted studies on maturity models which provide detailed explanations and understanding of what maturity models are and how they are supposed to be used in order to archive desired maturity by different organisations. The overly exhausted studies of maturity models are unfortunately very limited to explanations and do not address in details the challenges that can be experienced in

R. Silhavy (Ed.): CSOC 2021, LNNS 230, pp. 110–118, 2021.
https://doi.org/10.1007/978-3-030-77442-4_9

the use of these models. There is a need to explore challenges maturity models in order to ensure that maturity models are used and applied correctly and their effectiveness is fully realised in different organisations.

This study aims to explore the challenges and open issues that come with making use of maturity models in different aspects of organisational processes. The paper is arranged as follows: introduction, background, related work on maturity models, challenges and open issues and conclusion.

2 Background

The maturity and capability of business processes is recognized as the main method of determining if the organization is able to adjust and react accordingly to developing threads and opportunities [6]. In the development of a roadmap, it is important to select the most acceptable maturity level as desired by the organisation even if it is not the highest. The most important thing is for the maturity level to be suitable for the organisation and for that to be achieved it does not necessarily have to be the highest maturity level [7].

The maturity models appropriately cover the elementary principles for design and descriptive purpose for use; however the design principles for prescriptive use are barely archived. Consequently maturity models deliver inadequate guidance for identifying required maturity levels and for employing measures required for improvement. Currently, there is a very high number of existing maturity models that practitioners and academics could lose track. More and more maturity models are being developed. A description of models with stages, levels of granularity that exist and maturity extends are described in organisations' maturity documents [5].

Determination the level of process maturity is important in order to ensure the business becomes more stable, improves and becomes sustainable. A maturity model is an instrument to assist organisations evaluate strong points and weak points of the business processes [8]. There are models that are capable of measuring nearly all aspects of an organisation's maturity but does not focus on the aspect relating to managing data and information [1].

Numerous organisations have invested a very large amount of money in the renewal of the business processes and improving of the Information systems in order to have advantage over other organisations in similar industries or for cost reduction. By making use of maturity models, minimal time is required to discover improvement areas [9].

It is generally assumed that organisations that have more maturity levels are more likely to archive success in the project being more effective and efficient, these organisations are also expected to have an upper hand in the market as compared to the ones with less maturity levels [10].

3 Related work on Maturity Models

In this section, a review of related literature relevant to the study by other researchers is briefly discussed to provide required background and clear understanding necessary for research presented. The review of the literature mainly puts emphasis Maturity Models.

A study conducted by researchers [3] in which maturity of Business Process Management (BPM)/Business Process Orientation (BPO) was measured using different Maturity models. Maturity models used in are McCormack (MCC), Rummler-Brache Group (RUM) and Hammer (HAM-Org). The study compared the results of maturity from different levels of function from participating organisations: Top, Middle, Low management and operational staff. The study revealed that in an organisation, maturity results can be different based on the who is assessing, this is due to the fact that some individuals tend to over-esitamte while others under-estimate. Differnet management levels and operational staff often have different views of maturity on BPM/BPO; the respondents filled out maturity questionnaire. It is revealed that different maturity score can be obtained for the same organisation if different maturity models are used in measuring maturity. The study gives a good understanding of different maturity models, however does not indicate which of the explored maturity models gives the best results or is the most accurate. In their study they recommend that external stakeholders such as service provider and customers be included in the evalution of the maturity level as they may have a great impact in providing a holistic view and it is recommended that questionnaire be set up per level, function and responsibility.

A group of researchers [5] conducted a study that gives orderly and detailed review of various Business Process Management (BPM) Maturity Models. The study identified ten maturity models which are discussed, and an attempt to discover additional models is made. The models discussed differ in their scope; some focus on the manners of BPM practices while others focus on the condition of processes or the two. A study conducted by Manteen [11] in which Project Management Maturity was measured across two department in an organisation. The study used Project Management Maturity Model (PMMM) to evaluate the maturity and found that the maturity of the project management processes have a direct impact on project performance.

Another group of researchers [8] conducted a study that gives a detailed review, comparison and evaluation of similarities in the maturity models that are currently in existence. The researched focused on reviewing the most common maturity models which include Capability Maturity Model (CMM) and Capability Maturity Model Integration (CMMI). The results revealed in the study are that most of the maturity models are not realistic and lack experiential basis that there are not enough maturity models for Business Process excellence and more maturity models need to be developed as such.

Researchers [9] conducted a study that analyses different types of maturity models for business intelligence system assessment. The study reveals that majority of the maturity models do not cover the entire domain, they however forces on certain portions of the domain and reveals that when using maturity models very minimal time is required in order to discover areas that require special attention and areas with potential for improvement. In the study it is revealed that for organisations to get more precise maturity level suitable for the organisation and results, numerous maturity models should be used to expand on areas to be covered and get more information on existing state. Potential issues to be conquered in order to archive higher maturity level and archive increased business value. The highest maturity level is not necessarily the best maturity level for the organisations.

A study conducted by [7] introduces an independent SOA Maturity Model (iSOAMM) with five maturity models. The maturity model is not dependent on the use of technology and products. iSOAMM is able to list all the challenges, advantages and the risks that are related with individual maturity levels, enabling the organisations to choose the most appropriate maturity level, not regarding how or low it is. The iSOAMM grants organisations the ability to identify the satisfactory areas that will enable them to adopt the Service-Oriented architecture (SOA) and assist in the development of the roadmap for SOA introduction. iSOAMM is independent, for that every organisation have the freedom to select technologies standards and products.

A study conducted by [12] collected and analysed the practices of maturity models from data and information management perspective. This was done by analysing and reviewing of maturity models and by collecting maturity models from similar research articles. The study reveals that there are different type of maturity models, models that can measure every aspect of the organisation without focusing on aspects relating to the management of data and information. It is revealed that the currently existing maturity models and assessment methods focus on very complicated and particular tasks that capable assessors perform in organisations. The study developed ways to perform maturity assessment procedures automatically.

Some of the most common maturity models that have been developed are as follow:

Capability Maturity Model (CMM) can be defined as a maturity model that provides systematic building set of apparatuses, this includes maturity questionnaire which are very important in improving the software processes. This model is foundation for improvement of software processes and has maturity levels [13].

Capability Maturity Model Integration (CMMI) is a more recent and improved model, basically just CMM with enhancements and framework to resolve challenges related to project management and process in the developing products and services [8]. The CMMI maturity model is commonly known for software development environment [14]. Both CMM and CMMI are maturity models that dominate when it comes to maturity models and have maturity levels [15].

Rummler-Brache Group (RUM) maturity model is a Business Process Management maturity model developed by Rumler-Brache Group (RUM) that is capable of measuring both maturity plus a process-oriented culture. Most maturity models cannot measure maturity on numerous functionalities within an organisation, this makes RUM maturity one of the best maturity models available [3].

Gartner Enterprise Information Management Maturity Model – Gartner defines Enterprise Information Management (EIM) as the organisation committing to "structure, secure and improve the accuracy and integrity of enterprise information; solve semantic inconsistencies across boundaries and; support the objectives of enterprise architecture and the business strategy". This maturity model has six levels of maturity. Gartner has however not provided a technique and strategies of assessing maturity using this maturity model [1].

Service Oriented Architecture Maturity Model is a model that is used in assessment of the present state of Service-Oriented Architecture adoption and maturity in organisations. The model is used to measure current state of things and is also used in development of change that will lead to the desired state [16].

Knowledge management maturity model (KMMM) is a maturity model that is used to measure maturity of Knowledge Management business processes. It indicates how well the Knowledge Management function is well thought-out and how mature it is [17].

ISACA's COBIT5 this is a maturity model that is more focused on measuring Information Technology governance, it measures aspects of data and information in the information technology space without focusing too much on data and information management [1].

ARMA Information Governance Maturity Model (2010) is a maturity model by ARMA3 that has principles providing high level procedures that should be practiced in record keeping but not detailed on the principles of implementing and no policies, technology, roles and procedural details. The maturity model has 5 levels like most of the maturity models available, and organisations can chose the maturity level they desire to be on. The steps are informal and have simplified statements of what is to be done but does not define how it should be done [1].

As stated, there is an ever increasing number of maturity models, there are many more maturity models that have not been discussed in this section. The studies discussed in this section give a better understanding of maturity models however do not discuss or give awareness of the challenges that are often faced when working on maturity models in Business Process Management (Table 1).

Table 1. Table below is summarized models that have been looked at in this study

Maturity model	Source	Advantage	Disadvantage
CMM	Software Engineering Institute	The affords software organizations with direction on ways to increase control of processes for development and maintenance of software	It makes it difficult for management and staff to reach an agreement on the improvement activities to tackle first
CMMI	Software Engineering Institute	The model contributes to the performance of the service provider and the assists in satisfying the customer	It is more suitable for larger organizations and the success of the evaluation depends on the evaluator
RUM	Rummler-Brache Group	Capable of measuring both maturity plus a process-oriented culture	Model is based on 10 statements which may not be enough.
iSOAM	Rathfelder and Groenda, 2008	It makes it easy to choose the most suitable maturity level	Maturity model has five models however the 5th level has not yet been fully developed

(*continued*)

Table 1. *(continued)*

Maturity model	Source	Advantage	Disadvantage
SOAMM	The Open Group	Provides an overview over SOA and to assist during the entire process of SOA adoption	The adoption of SOAMM is complicated
KMMM	KMMM	Provides a comprehensive guidance required for making decisions and planning knowledge management activities in the organization.	Provides a limited view of the vision, treating the organization as a product regardless of that fact that it is a social construct made of organisms that are alive
COBIT5	ISACA	Assists with Information Technology governance	Does not focusing on data and information management and there is a lack of important documentation regarding this model
EIMMM	Gartner	Improves the accuracy and integrity of enterprise information	Gartner has not provided technique and strategies of assessing maturity using this maturity model
ARMA	ARMA	Simplified steps defining how it should be implemented, making it easy to implement	The steps are informal statements of what is to be done but does not define how it should be done

3.1 Challenges and Open issues

There are challenges in implementing Maturity Models in an organization. The maturity models appropriately cover the elementary principles for designing and descriptive purpose for use; however the design principles for prescriptive use are usually not archived. Consequently Business Process Management maturity models deliver inadequate guidance for identifying required maturity levels and for employing measures required for improvement. As maturity models are continuously being improved and new models are being developed, there is a high number of existing maturity models that practitioners and academics could lose track [5]. Some maturity models have a very complicated way of measuring and assessing maturity [8]. Not many people understand or have knowledge and experience of the implementation of maturity model and the practice thereof is minor [18].

To get a more accurate level of maturity in an organization, different maturity models must be used and the results from using various maturity models are also not directly comparable. This could lead to measuring maturity being more complicated. When using multiple models, one must bear in mind that results from different models are not directly comparable. The reason lies in the fact that metrics, areas, levels and criteria are not standardized [9].

There is currently a high number of existing maturity models, this brings concerns that quantity is considered more than quality and makes it more difficult to choose the suitable maturity model. [5]. It is extremely difficult and impost not possible to track progress along the SOA Maturity model difficulty of the organizations business increases. A lot of maturity models have the similarities like, people, process and technology [16].

When managers are on the phase of deploying maturity models, they are usually faced with a situation where employees are resistant to the implementation [18]. There are models that can measure maturity but do not focus on the data and information management [1]. Some maturity models have informal steps and statements of what needs to be done are too simplified but these maturity models do not guide on how the steps should be followed [1].

The currently existing maturity models and maturity assessment methods are very complicated and require highly skilled and specialized assessors in organizations. The maturity models require a lot of manual intervention in the data and evidence collection process in attempt to determine the suitable maturity level. The complicated maturity models makes maturity assessment costly and processes complicated and become a burden for organizational staff [1].

Maturity can be perceived differently within the same team based on who is evaluating. Low level, Middle level, top level managers and operational staff can have different views of the maturity level. This makes the measurement of maturity level very inaccurate and unreliable. Their experiences are different and therefore they are reporting maturity from different perspectives. Some maturity models give more accurate results if there is a balance between the overestimations and underestimations [3].

4 Conclusion

There are many different types of maturity models to measure different business processes within organizations. Since the development of maturity models, more and more models are being developed; however CMM and CMMI are still the most used and accurate maturity models. There are various challenges that come with using maturity models, however the challenges can be easily overcome. Some of the challenges are that in order to archive the most accurate maturity levels, different maturity models need to be implemented. For accurate maturity results, maturity needs to be measured per department or business process as it cannot be measured for the entire organization at a time. There are so many other open issues that require researchers to standardize on, including providing clear guidance of each model and details steps of implementation. Despite the challenges that may exist in using maturity models, the organizations can still measure the maturity levels of their organization's maturity models using the models available. Using maturity models is the best way to generally improve business processes and to ensure maturity is met.

References

1. Proença, Di., Borbinha, J.: Maturity model architect: a tool for maturity assessment support. In: Proceeding of the 2018 20th IEEE International Conference on Business Informatics, CBI 2018, vol. 2, pp. 42–51 (2018). https://doi.org/10.1109/CBI.2018.10045
2. Seidel-Sterzik, H., McLaren, S., Garnevska, E.: A capability maturity model for Life Cycle Management at the industry sector level. Sustain. **10**(7), 1–19 (2018). https://doi.org/10.3390/su10072496
3. Yadava, R.N., Jain, S.: A new bioactive flavone glycoside from the seeds of Melilotus indica All. J. Asian Nat. Prod. Res. **7**(4), 595–599 (2005). https://doi.org/10.1080/10286020310001608949
4. Marcovecchio, I., Thinyane, M., Estevez, E., Fillottrani, P., Marcovecchio, I., Estevez, E.: Capability maturity models as a means to standardize sustainable development goals indicators data production. J. ICT Stand. **6**(3), 216–244 (2018). https://doi.org/10.13052/jicts2245-800x.633
5. Röglinger, M., Pöppelbuß, J., Becker, J.: Maturity models in business process management. Bus. Process Manag. J. **18**(2), 328–346 (2012). https://doi.org/10.1108/14637151211225225
6. Janeš, A., Novak, R.: Process maturity orientation measurement. In: Management International Conference, Wilson 2015, pp. 153–165 (2017). http://www.hippocampus.si/ISBN/978-961-7023-71-8/16.pdf.
7. Rathfelder, C., Groenda, H.: ISOAMM: an independent SOA maturity model. In: Meier, R., Terzis, S. (eds.) DAIS 2008. LNCS, vol. 5053, pp. 1–15. Springer, Heidelberg (2008). https://doi.org/10.1007/978-3-540-68642-2_1
8. Albliwi, S.A., Antony, J., Arshed, N.: Critical literature review on maturity models for business process excellence. In: IEEE International Conference on Industrial Engineering and Engineering Management, vol. 2015, pp. 79–83, Janua (2014). https://doi.org/10.1109/IEEM.2014.7058604
9. Kawamura, K., Schultz, A.C.: RO-MAN 2005 welcome message. In: Proceedings of the - IEEE International Workshop on Robot Human Interactive Communication, vol. 2005 (2005). https://doi.org/10.1109/ROMAN.2005.1513741
10. F. Backlund, D. Chronéer, E. Sundqvist: Project management maturity models – a critical review. Procedia – Soc. Behav. Sci. **119**, 837–846 (2014). https://doi.org/10.1016/j.sbspro.2014.03.094
11. Mateen, M.: Measuring project management maturity -a framework for better and efficient projects delivery, vol. 2, no. 4, p. 23 (2015). http://publications.lib.chalmers.se/records/fulltext/224129/224129.pdf
12. Méndez, E., Crestani, F., Ribeiro, C., David, Gabriel, Lopes, João Correia. (eds.): TPDL 2018. LNCS, vol. 11057. Springer, Cham (2018). https://doi.org/10.1007/978-3-030-00066-0
13. Berztiss, AT: Capability maturity for software development. Handbook of Software Engineering and Knowledge Engineering: Vol 3: Recent Advances, pp. 85–118 (2005). https://doi.org/10.1142/9789812775245_0004
14. Meier, F.: Service oriented architecture maturity models : a guide to SOA adoption? Spring (2006). http://his.diva-portal.org/smash/get/diva2:2439/FULLTEXT01&sa=X&scisig=AAGBfm0KJoWMWfrPvwBxKq_VQleCaxkngA&oi=scholarr&ei=V_WwT4nFMMrB0QWJvIS6CQ&ved=0CBsQgAMoAjAA.
15. Becker, J., Niehaves, B., Poeppelbuss, J., Simons, A.: Association for ınformation systems AIS electronic library (AISeL) maturity models in IS research. Matur. Model. IS Res. 42 (2010). http://aisel.aisnet.org/ecis2010/42%0Ahttp://aisel.aisnet.org/ecis2010%0Ahttp://aisel.aisnet.org/ecis2010/42

16. Inaganti, S., Aravamudan, S.: EA maturity models. In: The SIM Guide to Enterprise Architecture, pp. 89–95 (2017). https://doi.org/10.1201/9781439811146-12
17. Mthembu, M., du Plessis, T.: Maturity mapping for continuous improvement: a case study of a revenue services institution. S. Afr. J. Econ. Manag. Sci. **21**(1), 1–10 (2018). https://doi.org/10.4102/sajems.v21i1.2044
18. Blondiau, A., Mettler, T., Winter, R.: Design and implementation challenges of maturity models for healthcare organizations: an experience report. In: Proceedings of the 16th International Symposium on Health Information Management Research, January, pp. 1–9 (2013)

The Model of Parallel Convergence in the Group of Pursuers' Plane with Simultaneous Achievement of the Goal

A. A. Dubanov(✉)

Buryat State University, Ulan-Ude, Russian Federation

Abstract. This article discusses the kinematic model of a group's pursuit of a single goal using the parallel convergence method. The model is based on the fact that the pursuers try to adhere to pre-designed trajectories. The pursuer's trajectories have curvature restrictions. The initial velocity directions of the pursuers are arbitrary, which makes changes to the known method of parallel approach. In our model, the pursuers achieve the goal simultaneously. This is due to changing the lengths of the predicted trajectories in such a way as to synchronize the time to reach the goal. The change in length is due to an increase in the radius of curvature at the initial section of the trajectory.

Keywords: Pursuit · Goal · Achievement · Synchronization · Convergence

1 Introduction

A feature of the parallel approach method on the plane [1–3] is that the speed of the pursuer is directed to a point on the Apollonius circle. In Fig. 1, this is point K_i, and point P_i is the position of the pursuer, and point T_i is the position of the target at I time.

The iterative scheme of the parallel convergence method is shown in Fig. 1. The next iteration step is as follows:

$$P_{i+1} = P_i + V_P \cdot \frac{P_iK_i}{|P_iK_i|} \cdot \Delta T, T_{i+1} = T_i + V_T \cdot \frac{T_iK_i}{|T_iK_i|} \cdot \Delta T.$$

The radius R_i and the center of the Apollonius circle Q_i are calculated as follows:

$$R_i = \frac{V_T^2}{V_P^2 - V_T^2} \cdot |T_i - P_i|, Q_i = T_i + \frac{V_T^2}{V_P^2 - V_T^2} \cdot (T_i - P_i).$$

The coordinates of the point K_i are the result of solving a system of equations with respect to a parameter t:

$$\begin{cases} (K_i - Q_i)^2 = R_i^2 \\ K_i = T_i + V_T \cdot \frac{T_{i+1} - T_i}{|T_{i+1} - T_i|} \cdot t \end{cases}.$$

R. Silhavy (Ed.): CSOC 2021, LNNS 230, pp. 119–124, 2021.
https://doi.org/10.1007/978-3-030-77442-4_10

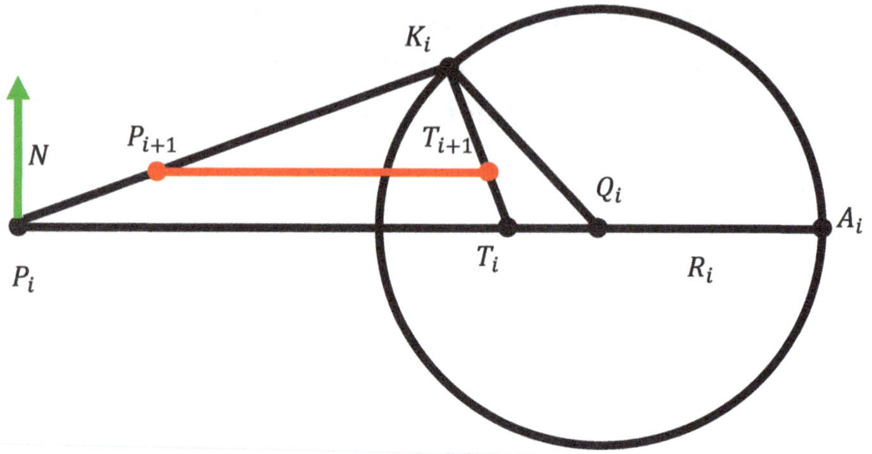

Fig. 1. Parallel approach of the pursuer and the target.

2 The Problem Statement

The description of the parallel approach method shows that the initial velocity of the pursuer cannot have an arbitrary direction in the initial statement of the problem.

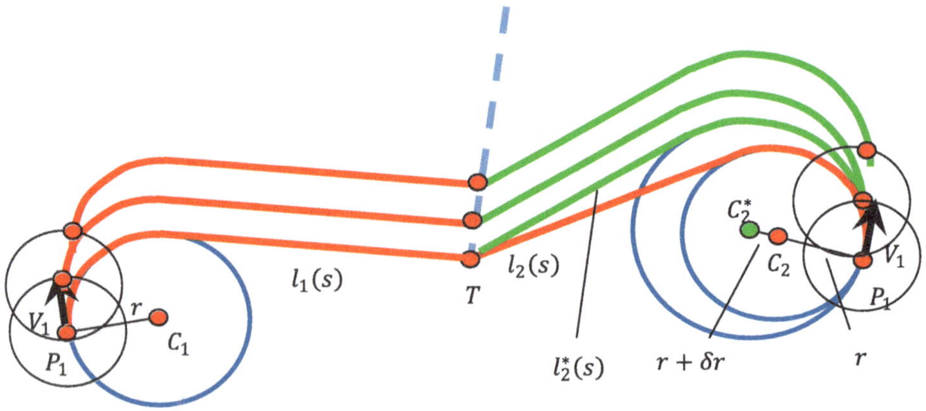

Fig. 2. Kinematic model of the parallel group approach.

In this article, we want to implement a method that is close to the parallel approximation method. In this method, the speeds V_1, V_2 of the pursuers P_1, P_2 are directed arbitrarily (Fig. 2).

Our task is to form a one-parameter set of congruent lines (Fig. 2) for each pursuer, which will be the predicted trajectories at a certain point in time. The position of the predicted trajectories at a time is determined by the position of the target T.

Our goal is to modify the projected trajectories so that the movement of the pursuers at the moment of reaching the goal is completed simultaneously. Figure 2 shows an iterative scheme for calculating the next one for each of the pursuers. This is the intersection point of the circle of the discrete step radius centered at the pursuer's current location with the line corresponding to the next position of the target.

3 Methodology for Solving the Problem

To solve this problem, we need to model a composite curve for each of the pursuers.

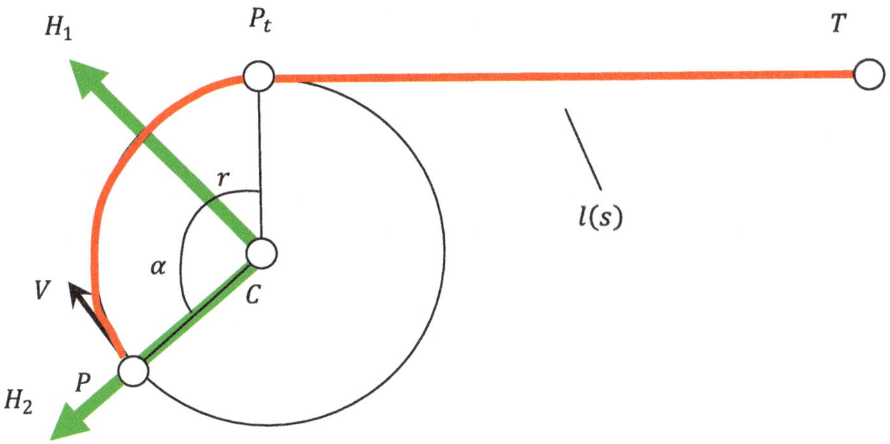

Fig. 3. Modeling many parallel lines.

Due to the restriction on the trajectory curvature of all participants in the pursuit task, the pursuer P in the predicted trajectory (Fig. 3) will pass along the arc $(\overset{\frown}{PP_t})$, then will go to the straight section $[P_tT]$ to the goal T. The radius of curvature r of the circle (C,r) in our model is assumed to be set and may increase. The center C of the circle (C,r) satisfies the system of equations:

$$|C - P| = r$$
$$V \cdot (C - P) = 0$$

In the local coordinate system (H_1, H_2) centered at point C, the arc equation $(\overset{\frown}{PP_t})$ will be:

$$L_{circle}(\alpha)^* = r \cdot \begin{bmatrix} \cos(\frac{\pi}{2} - \alpha) \\ sin(\frac{\pi}{2} - \alpha) \end{bmatrix}.$$

Where α takes values from 0 to $arccos\left(\frac{(P-C)*(P_t-C)}{|P-C|*|P_t-C|}\right)$. The basis vectors (H_1, H_2) are equal to:

$$H_1 = \frac{V}{|V|}, H_2 = \frac{P - C}{|P - C|}.$$

The conversion of the $l_{circle}(\alpha)$ line to the world coordinate system will be as follows:

$$L_{circle}(\alpha) = \begin{bmatrix} L_{circle}(\alpha)^* \cdot E_1^* \\ L_{circle}(\alpha)^* \cdot E_2^* \end{bmatrix} + C$$

$$E_1^* = \begin{bmatrix} E_1 \cdot H_1 \\ E_1 \cdot H_2 \end{bmatrix}, E_2^* = \begin{bmatrix} E_2 \cdot H_1 \\ E_2 \cdot H_2 \end{bmatrix}$$

$$E_1, E_2 - basicvectorsoftheworldcoordinatesystem$$

The equation for a straight section $[P_t T]$ is:

$$L_{line}(\varepsilon) = (1 - \varepsilon) \cdot P_t + \varepsilon \cdot T.$$

The resulting line segments $L_{circle}(\alpha)$ and $L_{line}(\varepsilon)$ form a single composite line, parametrized by the arc length.

In a test program based on the article, we obtained the combined arrays of coordinates $\{X_i, Y_i\}$, i ∈ 0..N of our composite curve. We introduce a formal parameter τ, which continuously runs through the values from 0 to N. After the cubic spline interpolation procedure, we will have continuous coordinate functions X(τ) and Y(τ) of the formal parameter τ.

From the equation for the total differential of the arc length $ds^2 = dX^2 + dY^2$ we get the differential equation of the first order for further transmission to the built-in solvers of the Cauchy task:

$$D(\tau, s) = \frac{ds}{d\tau} = \frac{1}{\sqrt{\frac{dX}{d\tau}^2 + \frac{dY}{d\tau}^2}}, \tau(0) = 0.$$

Thus, we obtained the dependences of X(s) and Y(s) on the arc length parameter. If the parameter length will satisfy the ratio $s = V \cdot t$, where t is real time, then we will receive dependence X(t) and Y(t) are coordinate functions of the baseline l(t).

The line corresponding to the target position T(T) can be obtained by adding to the basic equation of a line l(t) the vector T(T)-T(0) (Fig. 2).

If we need to increase the length of the baseline, we do the following. The length of the baseline depends on the following parameters: target coordinates T, pursuer coordinates P, pursuer velocity vector V, and minimum radius of curvature r (Fig. 2, 3).

Let a function of the path length be $L = L(T, P, V, r)$. If the pursuers P_1 and P_2 reach the goal T in the time intervals t_1 and t_2, then we choose the largest of these intervals and denote it as t_f. It would be incorrect to calculate time intervals in our model like this

$$t_1 = \frac{L(T, P_1, V_1, r_1)}{|V_1|} \text{and} t_1 = \frac{L(T, P_2, V_2, r_2)}{|V_2|}.$$

Since the movement of the pursuers is not moving along a single line at a constant speed, but is moving along lines from a one-parameter set. The next position of the pursuer is the intersection point of a circle of a radius equal to a discrete step with the line corresponding to the next position of the target (Fig. 2).

Since the presented model of group pursuit is quasi-discrete, the problem of synchronizing the time intervals t_1 and t_2 is reduced to programmatically increasing the radius of curvature r until the smallest of the time intervals is equal to the largest.

3.1 Conclusion

Based on the materials of the article, a test program is made in which two pursuers with an arbitrary speed direction begin to pursue a target moving in a straight line at a constant speed.

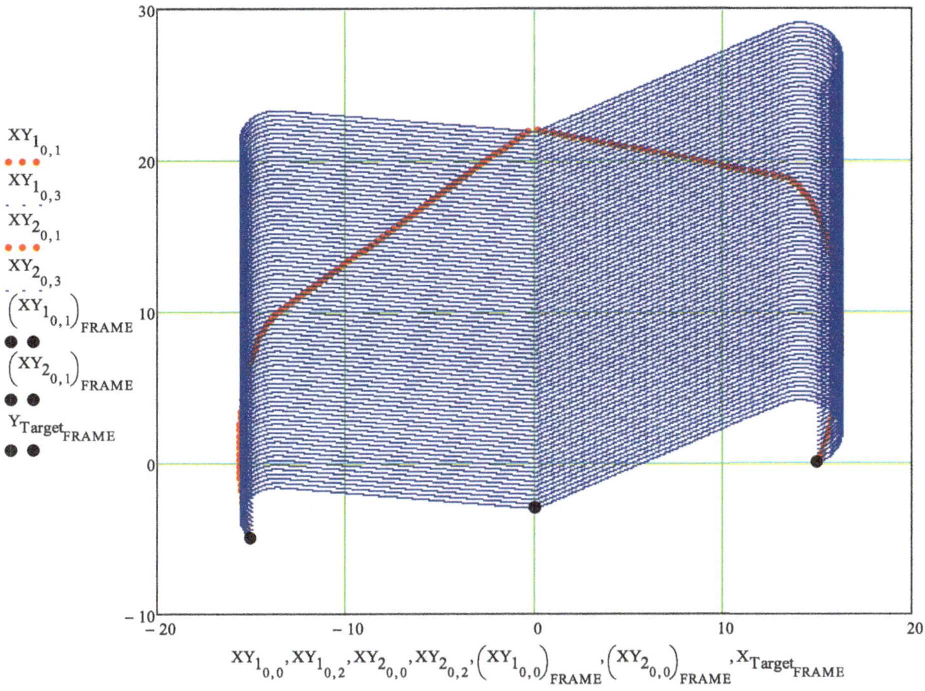

Fig. 4. Results of the program simulating the simultaneous achievement of the goal.

Figure 4 shows the first frame of the program. Figure 4 is supplemented with a link to the animated image [8].

Note that the text of the program can be found on the author's website [9]. In our work, we relied on the results obtained in [4–7]. The works [10–13] were also taken into account.

In the test program, we selected the rectilinear nature of the target movement because any curved trajectory can be interpreted as a set of rectilinear sections. And the choice of the next steps of the pursuers with synchronization of reaching the goal can be modeled as if the goal was moving in a straight line. On the next straight section, everything will be repeated.

We have considered the pursuit of only one goal, but a group of several pursuers can also apply the developed methods and algorithms in the case of simultaneous achievement of a group of several goals.

This test program considers a predefined target trajectory. In the future, as a perspective, we will consider the adaptive behavior of the goal with different strategies. This article is a continuation of the research outlined in works [14, 15].

Acknowledgement. The article was funded by innovative grant of the Buryat State University in 2021 "Control of a four-link manipulator based on signals received from the neurointerface".

References

1. Isaacs, R.: Differential games. Mir, Moscow (1967)
2. Pontryagin, L.: Linear differential game of evasion. Tr. MIAN SSSR, 30–63 (1971).
3. Krasovsky, N., Subbotin, A.: Positional differential games. Nauka, Moscow (1974)
4. Zhelnin, Y.: Linearized pursuit and evasion problem on the plane. TsAGI Sci. J. 88–98 (1977).
5. Burdakov S., Sizov, P.: Algorithms for motion control by a mobile robot in the pursuit problem. Scientific and technical Bulletin of the Saint Petersburg state Polytechnic University. Comp. Sci. Telecommun. Manag. 49–58 (2014).
6. Simakova, E.: On a differential game of pursuit. Automatics Telemech, 5–14 (1967).
7. Algorithm for following predicted paths in the pursuit problem. http://dubanov.exponenta.ru. Accessed 2019/07.22.
8. Results of the program simulating the simultaneous achievement of the goal, https://www.youtube.com/watch?v=7VNHNwCbWrg.
9. http://dubanov.exponenta.ru, section "Group pursuit with different strategies for a single goal".
10. Vagin D., Petrov N.: the Task of chasing tightly coordinated escapees. Izvestiya RAS, 75–79 (2001).
11. Bannikov A.: Some non-stationary problems of group pursuit. Proc. Inst. Math. Comp. Sci. UdSU, 3–46 (2013).
12. Bannikov A.: Non-Stationary task of group pursuit. Proceedings of the Lobachevsky Mathematical Centre, pp. 26–28. Publishing house of the Kazan mathematical society, Kazan (2006).
13. Izmestev I., Ukhobotov V.: The problem of chasing small-maneuverable objects with a terminal set in the form of a ring. Proceedings of the international conference "Geometric methods in control theory and mathematical physics: differential equations, integrability, qualitative theory", Ryazan (2018).
14. Dubanov A.A.: Modeling the behavior of objects in the pursuit problem. Adv. Intellig. Syst. Comput. 2019. T. 984. pp. 259–274.
15. Dubanov, A.A.: Trajectory Modeling in a Pursuit Problem with Curvature Restrictions. In: Silhavy, R. (ed.) CSOC 2020. AISC, vol. 1224, pp. 226–232. Springer, Cham (2020). https://doi.org/10.1007/978-3-030-51965-0_19

Information System for Food Origin Monitoring

Jozef Boris, Radim Farana(✉), and Cyril Klimeš

Mendel University in Brno, Zemědělská 1, 613 00 Brno, Czech Republic
{radim.farana,cyril.klimes}@mendelu.cz

Abstract. Food origin monitoring is a current requirement both of customers and European Union directives, which relates to food safety. The developed information system not only records data on particular kinds of foods, including their origin and easy QR-code identification, but it also enables a simple web presentation application on ingredients and origin of the food itself. The article also presents a mobile application primarily designed for small and medium-sized producers, which easily enables the producers to put together the product information by scanning QR-codes of the used ingredients. This se-cures the consistency of information about food, including all-embracing in-formation on the origin of the ingredients. Therefore, customers acquire complete information not only about ingredients but primarily about the food origin as well. Hereby, the system prevents the misuse of brands, such as FairTrade, organic food, local food, etc.

Keywords: Information System · QR-code · Food · Origin · Safety · Production

1 Introduction to the Problem

Food origin monitoring is a current requirement both of customers and European Union directives, which relates to food safety [1] as well as to widespread aware-ness of food origin [2]. Widespread awareness of the ingredients and food origin is hampered by the fact that food production uses too many ingredients - their batches, i.e. composition and origin, can differ. Thus, it is highly essential to register them in an information system and then unambiguously identify each batch separately. This also holds for each food product batch. This requirement is not new, it has been required for many years. However, it poses high requirements on food identification systems, primarily on the used range of identifiers. Current commonly used EAN codes do not meet these requirements anymore. Therefore, the developed system uses QR-codes for food identification as many modern applications regularly use such codes. A significant advantage is the support of QR-codes in smartphones and direct opening of the given web page by scanning the QR-code according to the ISO/IEC 18004:2015 standard [3]. The standard defines 40 code sizes, which enable to store from 26 to 3706 information Bytes. The code construction contains protection against incorrect reading. Nevertheless, we expect that food identification will use a MOD 97-secured digit code according to ISO 7064 [4]. Such a code enables to secure even a long sequence of digits and it is used, among others, to secure the International Bank Account Number (IBAN). Using a web page address

R. Silhavy (Ed.): CSOC 2021, LNNS 230, pp. 125–131, 2021.
https://doi.org/10.1007/978-3-030-77442-4_11

to show the output from the database as the application identifier then enables its direct opening in a browser and dis-playing all information about the food.

Similar applications to monitor food origin are not commonly available. They are not usually so ambitious to provide such a complex approach. For instance, application BitChecker reads a bar code in a retail chain shop and checks for the code correctness and for the country of origin [5]. Typical representatives of available applications are shopping applications of various retail chains, such as Scan&Shop [6]. Another group of applications focuses on food safety, such as Kawi [7], which reads the code and then assesses the food from the point of view of nutritional quality (70%), bad additives (20%), and organic dimension (10%). Specific applications are offered by some food marking providers, such as Fér potravina [8]. However, the producer must register its products and provide all information. Usually, the applications are of a national or even local nature, regarding the support extent of the provided brand.

Please note that the first paragraph of a section or subsection is not indented. The first paragraphs that follows a table, figure, equation etc. does not have an indent, either.

2 Food Database

The database for storing records of food products, sellers, and manufacturers is not complicated. Its basic structure is depicted in Fig. 1. It was realized in the environment of the NoSQL database – Firebase Realtime Database – for wider possibilities offered by NoSQL databases. Basic structured data can be easily added with non-structured text descriptions with the option of their further processing.

```
"Producers" : {
  "id_producer" : {
    "title" : "..."
  }, ...
}

"Users" : {
  "id_user" : {
    "producerId" : "..."
  }, ...
}

"Products" : {
  "id_product" : {
    "batch" : "...",
    "dateExpiration" : "...",
    "dateOfMade" : "...",
    "decription" : "...",
    "producerId" : "...",
    "products" : [ "...", ... ],
    "title" : "..."
  }, ...
}
```

Fig. 1. Database main scheme.

Figure 2 shows the principle of storing data and registration of its mutual connections as one type of food contains other registered foods as ingredients. This ensures information continuity and the possibility of acquiring detailed and real composition and primarily the origin of each registered food item. At the same time, Fig. 2 shows the possibility of simple use of the database to support food producers.

Fig. 2. Hierarchical data structure

3 Application to Support Food Producers

It is possible to build a number of various applications above the database for food origin monitoring. For the demonstration purposes, we have developed a mobile application in the Android environment for small and medium-sized food producers [9]. It enables to:

- register entrepreneurs,
- form a warehouse and maintain stocks using food QR-codes,
- create a recipe of a product,
- keep a list of recipes,
- use a recipe and warehouse stocks to create a food product, identify its batch, and assign it a QR-code.

The application can also be used without a registration. In such a case, the application enables to scan the food QR-code and display product information, including the possibility to go through the semi-product tree, as depicted in Fig. 2. The QR-code reading then uses the Xzing library, which is recommended by experts [10]. A demonstration of the system functioning is shown in Fig. 3. It displays basic information about the scanned food as well as a list of the used ingredients, which can be further searched through for information of the used basic ingredients. Thus, the application meets the basic idea of the whole system of food origin monitoring.

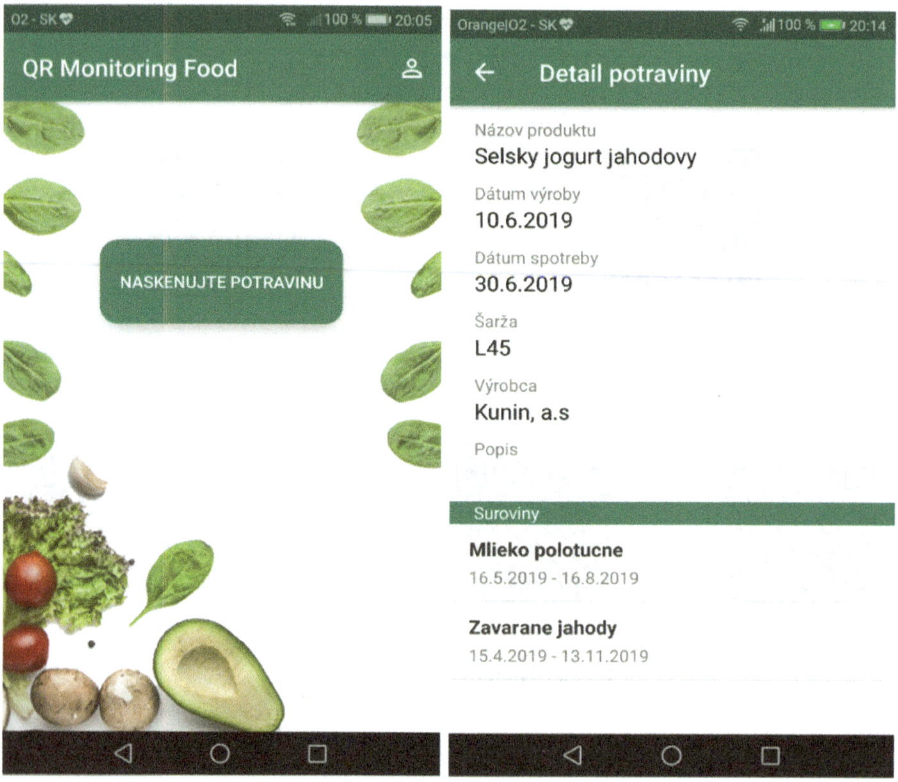

Fig. 3. Application activity without being logged in.

Having registered and logged in the application, the user acquires several other possibilities. Figure 4 depicts the basic work with the food stock. It enables to insert various types of foods into the stock, either by reading a QR-code (left) or manually in case the food or ingredient has not been inserted and does not have an assigned QR-code yet (right). The list provides the possibility to work with it, to share stored items, and to remove stored items, of course.

The main activity is the creation of new types of foods, i.e. production of food products. Figure 5 shows the creation of a set of ingredients from the stock and work with them.

The resulting product is then filed into the database and a QR-code is assigned to it. This QR-code can be sent to another application for its further use, primarily printing out and placing on the product.

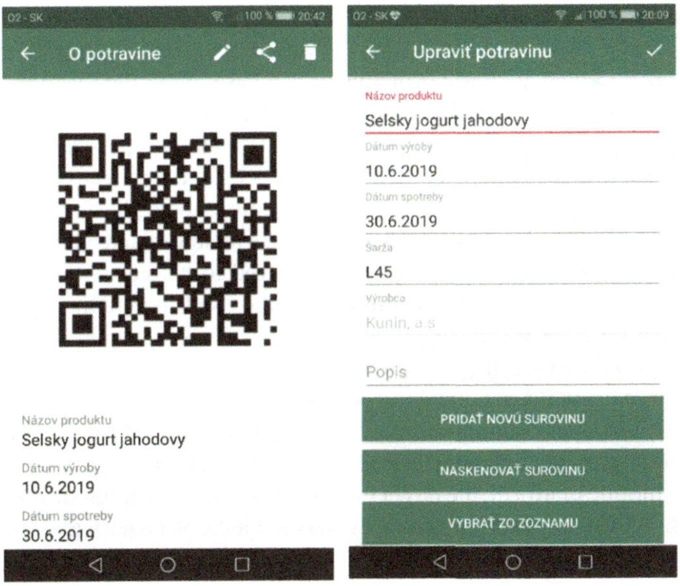

Fig. 4. Inserting items of foods into the stock.

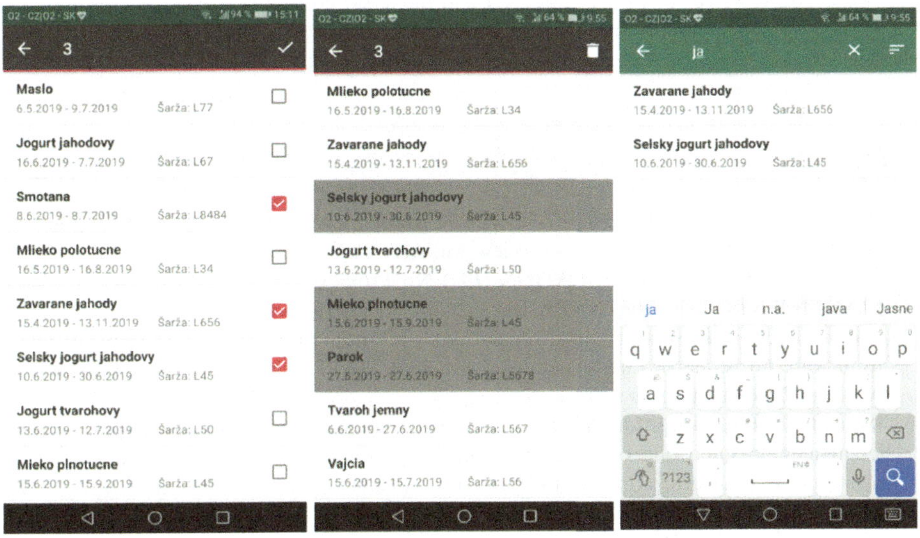

Fig. 5. Selection of basic ingredients for a food product and its naming.

The developed application has been tested in the laboratory of information systems at the authors' workplace, where it proved the correctness of the selected approach.

4 Discussion and Conclusions

The article has presented the basic principle of database operation for food origin monitoring. The database enables to monitor individual types of foods as well as their individual ingredients and composition. This results in the creation of a food composition tree, leading to the basic ingredients and their origin. Each food item is assigned a QR-code and a customer can easily access detailed information about the food item by simple scanning. When creating the database, the possibility of automated Nutrition Facts determination [11] was also analysed. However, it proved to be technically unfeasible, not even when knowing all the Nutrition facts of all the used ingredients. During the production of most types of foods, many components undergo chemical changes that are not described yet, therefore they cannot be algorithmized and included in the system.

It is possible to build a number of various applications above the database. An example of such an application is a mobile application designed for small and medium-sized producers enabling efficient use of the food database to create food products. Thanks to the mobile platform, the target group of small and medium-sized entrepreneurs can easily use it. The application primarily verified the correctness and usability of the database concept for food origin monitoring.

Acknowledgment. The development of the presented application used the equipment purchased within the project: Research Infrastructure for Young (abbreviated as Research Infrastructure), reg. No. CZ.02.1.01/0.0/0.0/16_017/0002334 and the acquired results will be used by students within the project: Support of Multidisciplinary Research Study Programmes for Young Scientists (abbreviated as Young Scientists), reg. No. CZ.02.2.69/0.0/0.0/16_018/0002333.

References

1. European Commission: Food Safety: Overview [on-line]. An Official website of the European Union. https://ec.europa.eu/food/overview_en. Accessed 20 Oct. 2020
2. European Union Law: REGULATION (EU) No 1169/2011 OF THE EUROPEAN PARLIAMENT AND OF THE COUNCIL of 25 October 2011 on the provision of food information to consumers, amending Regulations [on-line]. European Union Law. Available from web: https://eur-lex.europa.eu/legal-content/EN/ALL/?uri=CELEX%3A32011R1169. Accessed 20 Oct. 2020
3. ISO/IEC 18004:2015: Information technology — Automatic identification and data capture techniques — QR Code bar code symbology specification [on-line]. International Organization for Standardization. https://www.iso.org/standard/62021.html. Accessed 20 Oct. 2020
4. ISO/IEC 7064:2003: Information technology — Security techniques — Check character systems [on-line]. International Organization for Standardization. https://www.iso.org/standard/31531.html Accessed 20 Oct. 2020
5. Irony Software: BitChecker. Version: 1.4.7 [online]. https://play.google.com/store/apps/details?id=com.zanyatocorp.checkbarcode (2016). Accessed 20 Oct. 2020
6. Tesco Stores: TESCO Scan&Shop mobile. Version: 2.4.17 [online]. https://itesco.cz/scan-and-shop/ (2020). Accessed 20 Oct. 2020
7. Pyxweb Inc.: Kawi. Verzia: 1.1.9 [online]. https://www.kawi.io/ (2018). Accessed 20 Oct. 2020

8. Apps Dev Team Fér potravina Version: 3.5 [online]. https://www.ferpotravina.cz/ (2020). Accessed 20 Oct. 2020
9. Boris, J.: Information system for monitoring origin food (in Slovak), 41 pp. Mendel University in Brno (2019)
10. Herbst, B.: Android Barcode Scanning Library Landscape [online]. https://medium.com/bherbst/android-barcode-scanning-library-landscape-109292b81b65. Accessed 20 Oct. 2020
11. Hrádela, S.: Information system for food monitoring (in Slovak), 52 pp. Mendel University in Brno (2018)

Novel Framework for on Demand Preprocessing Towards Facilitating Diagnosis of Diabetic Retinopathy

M. K. Prakruthi[1(✉)] and G. Komarasamy[2]

[1] Jain University (Deemed to be University), Bengaluru, India
[2] Department of Computer Science and Engineering, GITAM (Deemed to be University), GITAM School of Technology, Bengaluru, India

Abstract. Identifying diabetic retinopathy in its early stage using fundus imaging is the active area of biomedical research. However, fundus images are usually related to uneven lighting and some visual distortion factors. Choosing a suitable preprocessing mechanism is crucial for improving fundus image quality to highlight hidden information and weigh the essential features such as retinal blood-vessels and exudates to analyze diabetic retinopathy. However, selecting an appropriate preprocessing mechanism is challenging to perform a comprehensive enhancement operation. It cannot be achieved using single techniques because each denoising and enhancement techniques are associated with a few advantages and limitations. Therefore, the resulting images may associate with other different artifacts, even after preprocessed. To address such issues, modeling a universal framework is designed to integrate different preprocessing techniques to evaluate fundus image enhancement. Apart from this, the proposed study also explores the applicability of optimization in image enhancement by using a swarm-based stochastic optimization algorithm. The performances of preprocessing techniques are evaluated against PSNR, and brightness is quantified in the enhanced image. This paper's main contribution is that it offers a flexible design of preprocessing framework that can meet on-demand image enhancement based on different preprocessing requirements to standardize the quality and interpretation of fundus imaging for diabetic retinopathy analysis.

Keywords: Fundus image · Diabetic retinopathy · Pre-processing · Enhancement · Denoising · Filtering · Optimization

1 Introduction

Medical imaging plays a vital role in the early identification of diseases. To date, several types of medical imaging are available, among which fundus imaging (FI) is gaining popularity among researchers. The FI provides a wealth of pathological information, indicating significant features such as exudates, blood vessels, and hemorrhages to identify the level of Diabetic Retinopathy (DR) and other diseases, like hypertension, stroke, and circulatory disease [1]. However, retinal fundus images are typically accompanied

R. Silhavy (Ed.): CSOC 2021, LNNS 230, pp. 132–141, 2021.
https://doi.org/10.1007/978-3-030-77442-4_12

by uneven illumination, noises, and low-contrast factor due to complex imaging environments, instigating difficulty for an ophthalmologist to diagnose lesions in such quality degraded FI. This may lead to affect the diagnostic process and their outcome in the context of DR identification. Therefore, effective image enhancement techniques are required to improve the FI's contrast and brightness to reveal significant features. There are enormous medical image enhancement methods introduced in the existing literature [2, 3]. The power law-based gamma correction method has been extensively adopted in literature to adjust medical images' contrast. However, power law-based enhancement is only much suitable for gamma distorted images [4]. In the same direction, the adoption of Histogram equalization (HE) is also widely considered to improve image contrast. However, a reduction in the gray-level by him is generally associated with loss of details. One of the effective enhancement techniques CLAHE is introduced to overcome the limitation of HE. However, CLAHE can be subjected to noise issues such as artificial boundaries and unwilling distortions in the resulting enhanced image [5]. Due to special visual features and particularity, FI needs a comprehensive design of enhancement mechanism. Therefore, this paper aims to introduce a universal framework modeling that integrates different techniques needed to address FI enhancement issues as per requirements. The remaining part of this paper is organized as follows: Sect. 2 presents related work. Section 3 presents the implementation of the proposed model. Result and outcome analysis are discussed in Sect. 4, and finally, a summary of the proposed work is illustrated in Sect. 5.

2 Related Work

Over the past decades, many contributions have been made in the domain of medical imaging enhancement. This section presents related work in the context of fundus imaging enhancement. The authors in the study of Yadav et al. [6] reported that image enhancement based on histogram equalization (HE) techniques is associated with the amplification of noise and suffers from computational time. Shamsudeen et al. [7] introduced exposure oriented HE technique to improve enhancement rate, adjustment of brightness, and significant noise level reduction. The work done by Datta et al. [8] used a fuzzy mechanism with a green channel of the input image, and contrast enhancement was done considering intensity level with a reduction of the mean brightness. However, higher brightness luminosity in enhanced images is not reduced. An adoption of recursive filtering-based image enhancement is considered by Cevahir Çığla and Alatan [9]. The authors have designed a cost aggregation recursion mechanism, where mapping filtered-values in the input image range is used to improve image quality. A joint approach of CLAHE and Wiener filter is considered by Santos et al. [10] for blood vessel recognition. An adaptive denoising approach using median filtering is suggested in the study of Pavel et al. [11] and Deng et al. [12] to eliminate impulse and salt and pepper noise from the images. Li et al. [13] presented an integrated approach to FI image enhancement. Mathematical morphology and multi-tree complex wavelet-transform technique. Dai et al. [14] used a normalized convolution technique for improving the contrast of retinal vasculature, which may appropriate for highlighting blood vessels. Zhou et al. [15] suggested illumination adjustment using power-law V-channel in HSV and CLAHE on the L channel of Lba color space. The research work carried out by Palanisamy et al. [16] introduced the joint mechanism of CLAHE and power-law and singular value equalization to highlight significant features in the fundus image.

3 Proposed System

The medical images are mostly preprocessed by considering specific visual quality issues such as uneven illumination and particular noise. However, other different factors and unwilling distortions also exist that degrade the intelligibility of the information presented in an image. On the other hand, the objective of performing the preprocessing operation is also to highlight the critical hidden information, tiny organs, and tissues of the area under consideration, which cannot be effectively accomplished by considering single preprocessing techniques. In this regard, the proposed system focused on designing a universal model to address various preprocessing issues, loss of details even after the image is preprocessed under limited enhancement operation scope. Therefore, the primary purpose is to introduce a computing modeling that integrates multiple enhancement mechanisms in a single-framework to benefit ophthalmologists with the most appropriate solution to perform on-demand preprocessing operations. The outline of the proposed model is exhibited in Fig. 1.

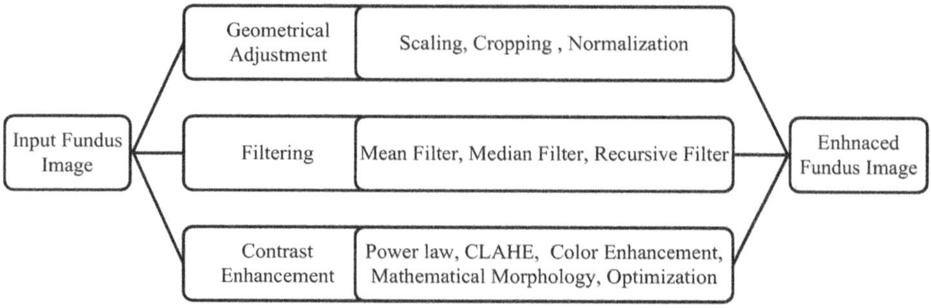

Fig. 1. Outline of the proposed pre-processing techniques evaluation model

The proposed system is designed based on different preprocessing operations requirements to standardize visual quality and highlight significant hidden features infundus imaging.The core components of the proposed framework are mentioned in Fig. 1 as i) Geometrical Adjustment, ii) Filtering for denoising, iii) Contrast enhancements techniques to adjust contrast, brightness, and address uneven illumination in the FI. The study evaluates the few most known preprocessing techniques. On the other hand, the proposed work also intended to explore the optimization in the image enhancement process using a swarm-based stochastic approach.

3.1 Geometrical Adjustment

The core function of geometric adjustment is to overcome the deformation, RoI extraction, normalization and eventually yield a revised version of the image with linearity and symmetrical. The system considers three techniques to meet the real-time preprocessing requirements of geometrical image adjustments. The first is the scaling, which is an important process of the image preprocessing application, where using the interpolation method, images are scaled up using the pixel replication process. The study considers nearest-neighbor interpolation (NNI) to enlarge input image by integer factor nR and nC using rounding to closer integer numerically expressed as:

$$I(m,n) = fx(\lfloor \mathrm{m/nR} \rceil . \lfloor \mathrm{n/nC} \rceil) \tag{1}$$

Where, m, n is the original row and column size respectively, and nR, nC is the target image size, and $\lfloor \cdot \rceil$ is the rounding to closer integer, and the function used in interpolation is given by

$$J_o(m,n) = \begin{cases} 1/(nR, nC), \ -nR/2 \le m < nR, \ -nC/2 \le n < nC/2 \\ 0 \quad else \end{cases} \tag{2}$$

The next operation in this preprocessing part is the cropping, which allows extracting the interesting part of the image from the input image. This process saves computational requirements and processes only those parts where the user is interested. The study introduces a user interactive image cropping tool, which allows users to crop the image with one click using a rectangular mask. In the next operation is the normalization is considered for the denoising input image as well as bring it in a range of normal distribution of pixel intensities given by:

$$f(y) = ((x - min) \times 255)/(xmax - xmin) \tag{3}$$

y is the normalized imaged, x is the input image, xmin, and xmax is the minimum and maximum scale of x.

3.2 Filtering Mechanism

The medical images are generally affected by the external environmental factors during acquisition, transmission resulting in quality degradation, in contrast, deteriorating the geometric aspect of objects in the image. In this regard, filtering is one of the significant preprocessing steps to eliminate different noises. In the proposed framework, three different filtering techniques are evaluated for image denoising. The mean filtering approach is considered to eliminated local variations in the image pixels. The mean filter performs a linear operation, where resulting pixels are a linear grouping of the mean value computed from the input image neighboring pixel by lowering the pixel's intensity value one after the other. The first process in this filtering operation is the computation of the mean of input image FI and further removal of image offset by subtracting computed mean values from the input image FI. The image denoising using a mean filter is given:

$$Io[m,n] = fx(I, \mathrm{F}, 3) \tag{4}$$

Io is the filtered image obtained using linear filtering function fx() with the input I image and averaging filter F of size specifying the number of rows and columns in vector. The mean filter is useful in image smoothing but lacks the edge details. Therefore, the median filtering approach comes into the picture that does a better job than mean based noise filtering. The median filter uses a non-linear operation where the pixel of output images is replaced by the median of all pixels in a neighborhood pixel of the input image instead of the taking average like in the mean filtering case. The non-linear operation of filtering is given as:

$$Io[m,n] = fx\left(x\left[i,j\right], i, j\right) \in N \tag{5}$$

Fx () is the function referring to median computation, and N is the user based neighborhood value. However, the median filtering operation is not effective in addressing all types of noises. The applicability of the recursive filter RF is also evaluated towards edge-preserving in the FI. The RF is computationally efficient since the computation of filter weights is carried out recursively. The denoised image is computed using the first-order filter using recursion as follows:

$$FIo[m,n] = fr(FI) \tag{6}$$

FIo[m, n] is the filtered image obtained from the recursive function $fr()$ over input fundus image FI. The recursion operation of $fr(FI)$ is given as follows:

$$fr(FI) \leftarrow \left(1 - c^{g[n]}\right)FI[n] + c^{g[n]} \times FIo[n-1] \tag{7}$$

where c is the constant equal to $exp\left(-\sqrt{2/Sd_{spatial}}\right)$ and Sd is the user-defined parameter.

3.3 Contrast Enhancement

Different enhancement techniques are evaluated to adjust contrast and non-linear illumination in the input image. The power law-based contrast correction is extensively adopted in the existing literature. It has the ability of contrast adjustment by controlling the gamma correction factor and constant C. For image enhancement, the power law-based contrast correction mechanism is given as:

$$FIs = C \times FI^{\gamma} \tag{8}$$

Where, FIs is the high contrast intensity output image obtained from power law factor γ, where if $\gamma < 1$ *mapping towards high brightness*, $\gamma > 1$ *towards darker*. The next technique evaluated is CLAHE ("Contrast Limited Adaptive Histogram Equalization"). It is an extended version of AHE and widely accepted in the domain of medical imaging preprocessing. This allocates the exact brightness and limits the over smoothing and noise effect in the resultant enhanced image. The optical limiting factors in fundus images capturing the environment may cause a haze in the resulting FI. The fundus images suffer from the low-lightening condition and lack the essential details inside the images' object. The haze removal algorithm based on the dark channel prior concept is evaluated in this regard. The closer association between the hazy input image and dehazed output image is given as follows:

$$FI(x, y) = FIE(x, y) \times t(x, y) + A(1 - t(x, y)) \tag{9}$$

Where FI(x,y) refers to the input image, FIE(x, y) is an enhanced image. A is the global atmospheric light and transmittance t(x,y) to a particular pixel position in the image. A contrast enhancement based on mathematical morphology is evaluated against power law, color enhancement, filtering technique, and CLAHE. The algorithm first computes the green color (Gc) component, which is further subjected to closing operation using bottom hat filtering function $fx()$ with an input argument of structuring element type disk and the image's Gc. This process results in the removal of blobs in Gc, which is again processed with the filter top-hat transform to extract essential details from the processed Gc. Further, both filtered images' differences are binarized and mapped with the input image as an enhanced output image, highlighting significant features in the FI. The study also explores the applicability of the optimization mechanism in the FI enhancement. In order to do this, a swarm-based stochastic algorithm is presented. The significant steps of optimization-based enhancement are described as follows:

Algorithm: Swarm based stochastic optimization approach for image enhancement

Input: FI (Fundus Image)
Output: FIE (Enhanced Fundus image)
Start

1. Init,Max-Iteration, Imax, Imin, v1, v2, Psize
2. Construct: empty vector ← [$\overrightarrow{Fv}$, $\overrightarrow{Pb}$, $\overrightarrow{gb}$]
3. Convert rgb IF → gIF(mxn) ← *f*x1(IF, m x n x D)
4. Particle Initialization
5. For i = 1 to Max-iteration
    a. Compute: I ← $Imax - (Imax - Imin) \times \left(\frac{i}{Max-iteration}\right)$
    b. For j = 1 to Psize
        i. Compute FIE: ← *f*x2(gFI, size)
        ii. Compute: Fv → fx3(FIE, n, m)
        iii. Fvmax ← *f*4(Fv)
        iv. Pb = P(x) → check:fv< P(x)
        v. Otherwise: Pb = Fv

    End

 End

6. Update gb ← *f*4(Pb)
7. For each P
    a. Compute: P(v)
    b. Compute: P(x)
8. Continue until the required criteria are not met.
9. Stop

End

The algorithm takes input value as FI (fundus image) and generates EFI after execution with optimization algorithm (enhanced fundus image).In this direction, the algorithm initializes the variables as max-Iteration Imax (maximum Inertial weight), Imin (maximum Inertial weight), v1, and v2 as acceleration coefficients, Particle size (Line1). The empty vector construction is carried out to store Fitness value Fv, local best Pb, and global best gb position of the particles (Line2). The input image is first converted to grayscale gFI, and the size is determined as the number of row 'n' and number of column 'c' (line3). Imax (maximum Inertial weight), Imin (maximum Inertial weight), v1, and v2 as acceleration coefficients, Particle size (Line1). The construction of an empty vector is carried out to store Fitness value Fv, local best Pb, and global best gb position of the particles (Line2). The input image is first converted to grayscale, and the size of grayscale image size is determined as the number of row 'n' and number of column 'c' (line3). For each iteration, the algorithm computes Inertial weight I (Line4 & 5). For each particle size, computation is carried to generate the enhanced FI using function fx2 () with an input argument of gFI and initial particle position. Here the function refers to a transformation operation to carry out enhancement over the input image. In the next step of the algorithm, the system computes Fv using discrete function fx3() with an enhanced image, m, and n. The other maximum value in fitness and local best position Pb is computed. The gb is computed as the maximum value of the Pb (Line6), and updating of particle new position P(x) and velocity P(v) is continued till the required criteria are matched (Line7 to 9).

4 Result Analysis

The implementation of the proposed system for fundus image preprocessing is performed on the numerical computing tool MatLab. This section discusses the outcomes achieved by different preprocessing techniques for fundus image denoising and enhancement. Each technique's performances are evaluated over three fundus images from the datasets concerning qualitative and quantitative analysis considering PSNR and quantified brightness metrics.

In Table 1, the visual outcome is presented for each technique concerning different fundus images. In Fig. 1 and Fig. 2, quantified analysis is exhibited in terms of PSNR and brightness.

From both Fig. 2 and Fig. 3, it can be analyzed that the performances achieved by each technique vary to some extent; therefore, image enhancement considering single or few techniques may not be much appropriate for the significant enhancement. The proposed swarm-based optimization technique exhibits better performance compared to other techniques. However, better performance can be furthermore exhibited by fine-tuning processes with PSO parameter initialization. Each technique's performance depends on the input image nature and visual characteristics; therefore, this work has presented a set of algorithms that can cost-effectively meet different enhancement requirements.

Table 1. Visual outcome of geometric adjustment techniques

Input images	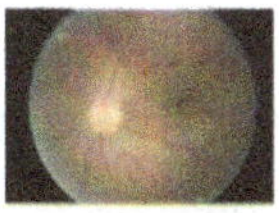	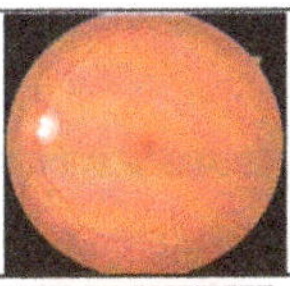	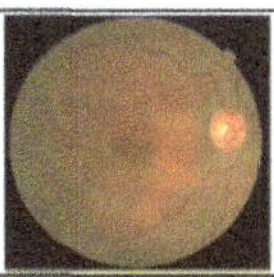
NNI	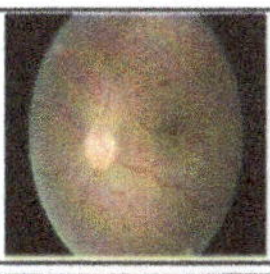	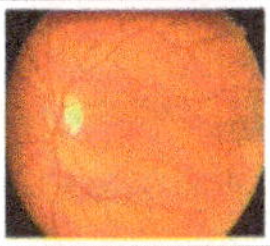	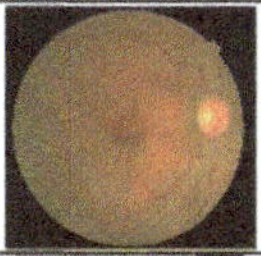
Cropping	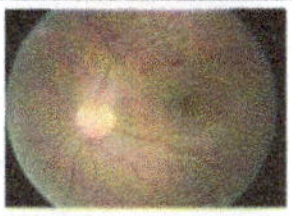	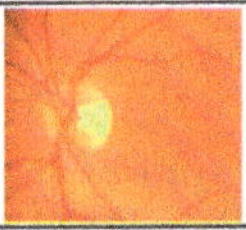	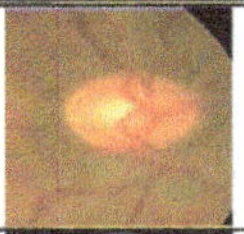
Normalization		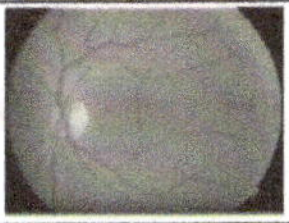	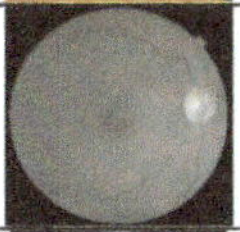
Mean Filtering	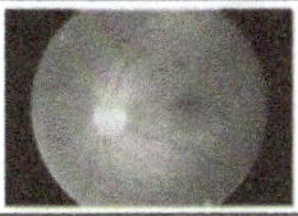	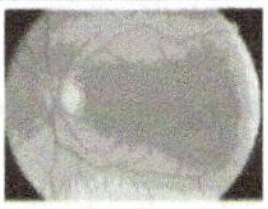	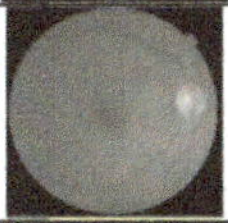
Median Filtering	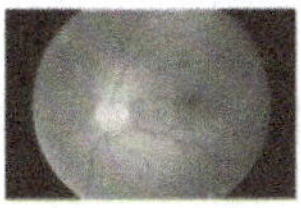	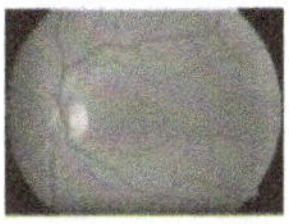	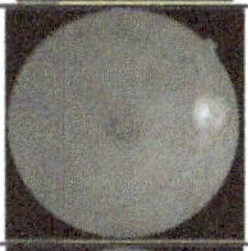
Recursive Filtering	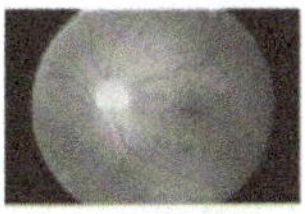	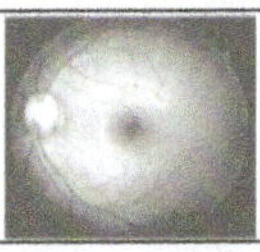	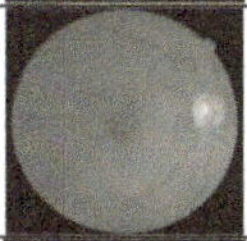
Power law	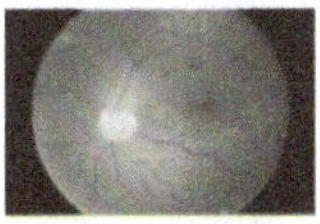	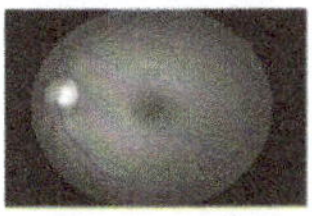	
CLAHE	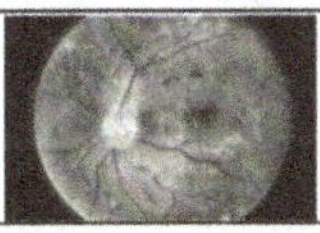	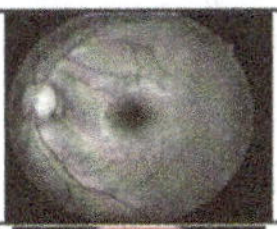	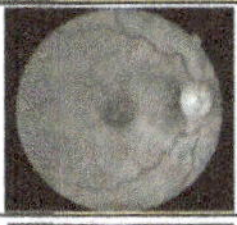
Low-light Enhancement	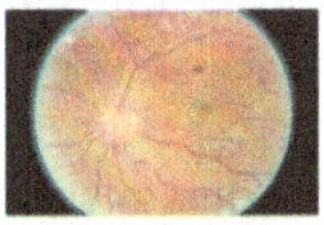	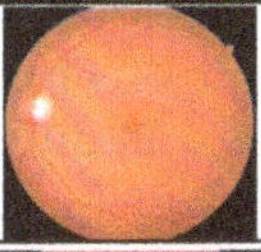	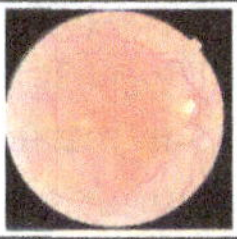
Mathematical Morphology	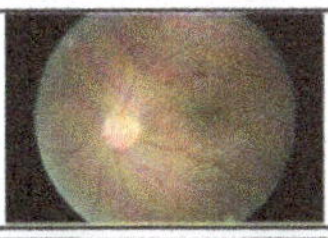	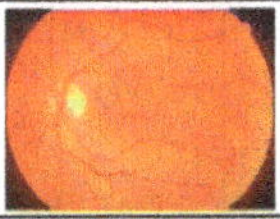	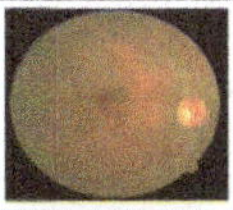
Optimization-based Enhancement	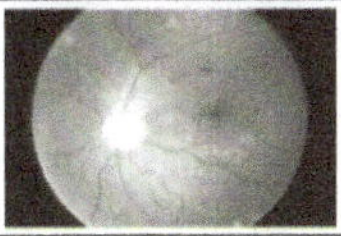	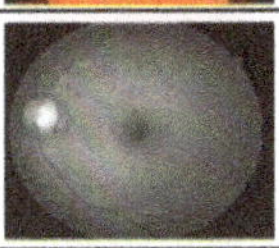	

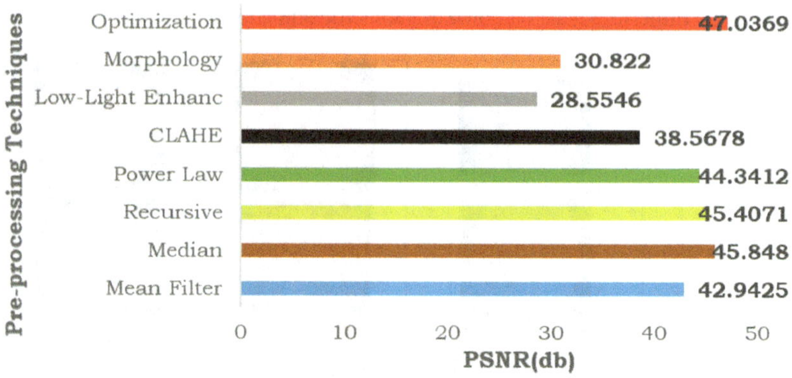

Fig. 2. Analysis of Pre-processing techniques in terms of Quantified PSNR

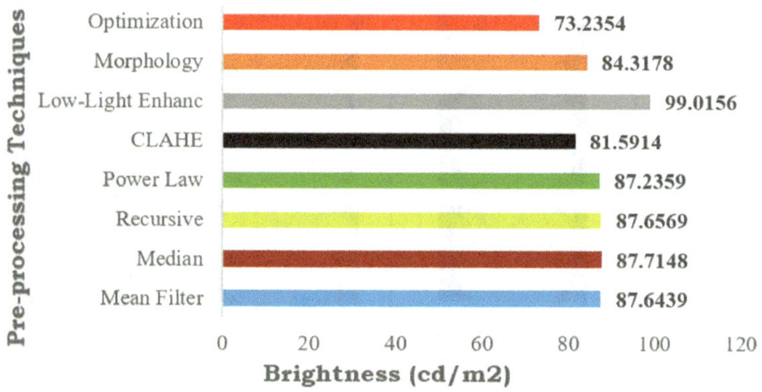

Fig. 3. Analysis of Pre-processing techniques in terms of Quantified Brightness

5 Conclusion

In this paper, multi-enhancement modeling is carried out towards which offers a flexible solution to overcomes the limitations associated with the single image enhancement technique. This work's main motive is to introduce a universal framework that can support different requirements of preprocessing in medical imaging analysis, and the user can efficiently perform appropriate enhancement over an image based on the evaluation of a particular medical image with different preprocessing options. The performance of each technique is validated based on qualitative and quantitative analysis. The proposed study can be extended to perform segmentation and diabatic retinopathy disease classification in future research work.

References

1. Tan, C.S., Chew, M.C., Lim, L.W., Sadda, S.R.: Advances in retinal imaging for diabetic retinopathy and diabetic macular edema. Indian J. Ophthalmol. **2**(5), 99–110 (2016)

2. Goel, N., Yadav, A., Singh, B.M.: Medical image processing: a review. In: Second International Innovative Applications of Computational Intelligence on Power, Energy, and Controls with their Impact on Humanity (CIPECH), Ghaziabad, pp. 57–62 (2016). https://doi.org/10.1109/CIPECH.2016.7918737.
3. Perumal, S., Velmurugan, T.: Preprocessing by contrast enhancement techniques for medical images. Int. J. Pure Appl. Math. **118**(18), 3681–3688 (2018)
4. Aziz, M.N., Purboyo, T.W., Prasasti, A.L.: A survey on the implementation of image enhancement. Int. J. Appl. Eng. Res. **12**(21), 11451–11459 (2017)
5. Mustafa, W.A., Kader, M.M.: A review of histogram equalization techniques in image enhancement application. J. Phys.: Conf. Ser. **1019**(1), 012026 (2018)
6. Yadav, S.K., Kumar, S., Kumar, B., Gupta, R.: Comparative analysis of fundus image enhancement in detection of diabetic retinopathy. In: 2016 IEEE Region 10 Humanitarian Technology Conference (R10-HTC), Agra, pp. 1–5 (2016). https://doi.org/10.1109/R10-HTC.2016.7906814
7. Shamsudeen, F.M., Raju, G.: Enhancement of fundus imagery. In: 2016 International Conference on Next Generation Intelligent Systems (ICNGIS), Kottayam, pp. 1–5 (2016). https://doi.org/10.1109/ICNGIS.2016.7854022
8. Datta, N.S., Saha, P., Dutta, H.S., Sarkar, D., Biswas, S., Sarkar, P.: A new contrast enhancement method of retinal images in Diabetic Screening System. In: 2015 IEEE 2nd International Conference on Recent Trends in Information Systems (ReTIS), Kolkata, pp. 255–260 (2015). https://doi.org/10.1109/ReTIS.2015.7232887
9. Çığla, C., Alatan, A.: An efficient recursive edge-aware filter. Signal Process.: Image Commun. **29**(9), 998–1014 (2014). https://doi.org/10.1016/j.image.2014.07.005
10. Santos, J.C.M., Carrijo, G.A., Cardoso, D.F.D.S.: Fundus image quality enhancement for blood vessel detection via a neural network using CLAHE Wiener filter. Res. Biomed. Eng. **36**, 107–119 (2020)
11. Lyakhov, P.A., Orazaev, A.R., Chervyakov, N.I., Kaplun, D.I.: A new method for adaptive median filtering of images. In: IEEE Conference of Russian Young Researchers in Electrical and Electronic Engineering (EIConRus), pp. 1197–1201 (2019)
12. Deng, X., Ma, Y., Dong, M.: A new adaptive filtering method for removing salt and pepper noise based on multilayered PCNN. Pattern Recognit. Lett. **79**, 8–17 (2016). https://doi.org/10.1016/j.patrec.2016.04.019
13. Li, D., Zhang, L., Sun, C., Yin, T., Liu, C., Yang, J.: Robust retinal image enhancement via dual-tree complex wavelet transform and morphology-based method. IEEE Access **7**, 47303–47316 (2019). https://doi.org/10.1109/ACCESS.2019.2909788
14. Dai, P., Sheng, H., Zhang, J., Li, L., Wu, J., Fan, M.: Retinal fundus image enhancement using the normalized convolution and noise removing. Int. J. Biomed. Imag. (2016)
15. Zhou, M., Jin, K., Wang, S., Ye, J., Qian, D.: Color retinal image enhancement based on luminosity and contrast adjustment. IEEE Trans. Biomed. Eng. **65**(3), 521–527 (2018). https://doi.org/10.1109/TBME.2017.2700627
16. Palanisamy, G., Ponnusamy, P., Gopi, V.: An improved luminosity and contrast enhancement framework for feature preservation in color fundus images. Signal Image Video Process. (2018)

Organization of Measurement Data Collection and Primary Processing Regional Environmental Monitoring Systems

R. Y. Lapkovsky[1], V. A. Kushnikov[1,2,4], E. V. Kushnikova[4(✉)], A. S. Bogomolov[1,2], A. F. Rezchikov[3], A. D. Tsvirkun[3], V. A. Ivashchenko[1], O. I. Dranko[3], D. S. Fominykh[1], E. V. Berdnova[5], J. V. Lazhauninkas[5], and E. S. Baryschnikova[1]

[1] Institute of Precision Mechanics and Control of RAS, 24 Rabochaya Street, Saratov 410028, Russia

[2] Saratov State University, 83 Astrakhanskaya Street, Saratov 410012, Russia

[3] V. A. Trapeznikov Institute of Control Sciences of RAS, 65 Profsoyuznaya Street, Moscow 117997, Russia

[4] Yuri Gagarin State Technical University of Saratov, 77 Politechnicheskaya Street, Saratov 410054, Russia

[5] Saratov State Vavilov Agrarian University, 1 Teatral'naya pl. Street, Saratov 410012, Russia

Abstract. The article analyses the main features of the construction of regional environmental monitoring systems. New approach to construction of distributed database of these systems is considered, structure and composition of data base of separate nodes of network of observation stations and algorithms of synchronization of data updating in them are proposed. Models and methods of primary processing of measurement information are proposed. Analysis had been carried out on how to address the problem of finding sources of pollution on the basis of operational data from monitoring systems.

Keywords: Control · Mathematical model · Unstable state · Ecology · Monitoring · Information and measurement system · Information processing

1 Introduction

The Russian state system of environmental monitoring (ESEM) was created on the basis of the Resolution of Council of ministers of the Government of the Russian Federation of 24.11.93 No. 1229 "About creation of a uniform state system of ecological monitoring" works. The main function of the system is to observe and forecast the environment so that the quality of the environment could be properly managed. The ESEM Regulation works at four main levels: a federal, a regional, a territorial and a local. At this time, the work on the formation of ESEM is at the stage of pilot projects of regions.

Operational monitoring of the state of the environment was needed to improve the effectiveness of environmental observations. This is currently being addressed through special environmental monitoring systems. There are a network of specialized stations

R. Silhavy (Ed.): CSOC 2021, LNNS 230, pp. 142–151, 2021.
https://doi.org/10.1007/978-3-030-77442-4_13

distributed over the area under research and providing monitoring of environmental quality indicators.

There are researches [1–18] on the subject of complicated systems, environmental problems. They were extremely acute for further research. This is largely due to insufficient development of regional monitoring and identification systems for pollution sources.

The used approaches did not pay sufficient attention to the primary development of measurement information and its interpretation. The calculation and linking of measurement results to the time grid is not sufficiently accurate. The calculation does not take into account the peculiarities of the sources of measured data, does not consider the analysis and analysis of data for the elimination of emissions, etc. All this directly affects the accuracy of the information received.

In addition, existing approaches did not take fully into account particularly the use of such systems in terms of the organization of a distributed database. The data are dispersed across a large number of remote nodes, with some of the links being unstable. There is no affect on the data processing process that occurs in different parts of the system. In these circumstances, existing database replication methods and the use of locks are often not effective.

Insufficient attention is also paid to the identification of sources of pollution based on environmental monitoring data. Modern monitoring systems tended to work with isolated cases of exceeding the maximum permissible concentration only. They had a negative impact on the speed and quality of management decisions taken.

2 Task Setting

The aim of the study is to modernize the standard structure of the subsystem for collecting and primary measuring information processing in automated systems for regional environmental monitoring.

3 Problem-Solving Method

In order to achieve the main aim, it is proposed to achieve the following objectives:

- To analyze the typical structure of the measurement information collection and primary processing subsystem in regional environmental monitoring systems;
- To define the measurement procedure;
- To describe the procedure for collecting and processing primary measurement information;
- To design the method for measurement results storage;
- To develop a methodology for data exchange between data processing nodes;
- To propose an approach to checking correctness of incoming single data, interpretation of obtained values, their storage and averaging;
- To create data update synchronization algorithm.

4 Typical Structure of the Subsystem for Measurement, Information Collection and Primary Processing in Regional Environmental Monitoring Systems

The environmental monitoring system has a distributed structure that includes various data processing units. They are stationary and mobile environmental stations (ESS, PES), environmental servers (POS). Environmental stations have in-stalled analytical equipment. It measures environmental quality indicators. For example, concentrations of pollutants in the atmospheric air, meteorological parameters, etc. Measurements by various means of communication are sent to environmental servers where data are stored and analyzed.

The monitoring station is a complex of interconnected equipment. It performs qualitative and quantitative chemical analysis of ambient atmospheric air, power supply systems and maintenance of operating conditions, systems for monitoring the state of the plant and checking the validity of equipment readings, as well as other modules necessary for monitoring.

These components are controlled by the software and hardware complex (PAK), which occupies the central link of the station system. This complex performs many functions, the main ones being:

- Control of measuring system equipment operation;
- Obtaining measurement results from the equipment of the measuring complex;
- Monitoring of the station state and control of its operating conditions maintenance system, power supply control;
- Processing, interpretation and accumulation of received measurement results;
- Communication with the server by means of a communication system.

For this system, an approach is proposed to improve the efficiency of the mechanism for synchronizing changes. It uses geographical and administrative characteristics. This is necessary because environmental data is a trade secret and the data-base is designed to use unstable and low-speed links. The proposed approach enables more complete and accurate collection and initial processing of measurement information in regional environmental monitoring systems.

One of the most important aspects of the operation of the station is the process of obtaining measurement results. Their processing and interpretation, as measurement data form the basis of analytical work of specialized programs and experts. On the basis controlling decisions are made.

In systems of known environmental stations, data are collected by simply storing the values obtained from the equipment in the data bases with the date and time of their receipt. However, this approach has a number of disadvantages. Each measurement result describes the state of atmospheric air at a certain point in time, which almost never coincides with the time it was received from the equipment. Therefore, it is incorrect to use the result moment as the time to which the measurement relates. Moreover, the principle of operation of an overwhelming number of measuring devices did not allow for the measurement of pollutant concentrations at point moments. An air sample is taken in a certain time interval $\left(t^1; t^2\right)$ (usually the time of pumping the air sample through a meter or sampling to a buffer storage), then (or simultaneously) an analysis of the sample is performed $\left(t_0^1; t_0^2\right)$, after which the result is transmitted from the equipment to the collection system (tr).

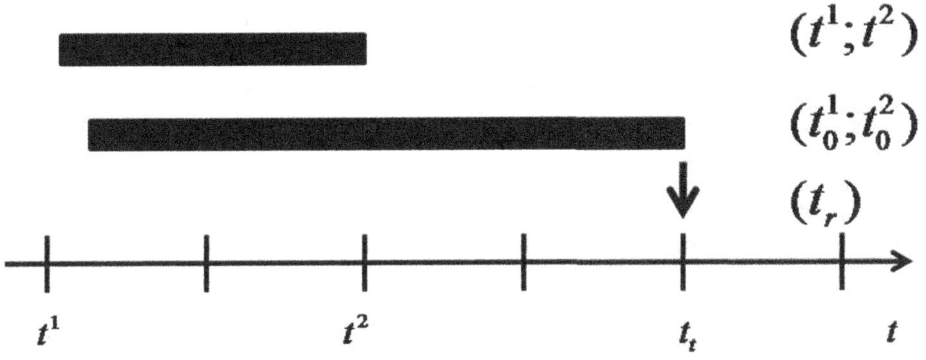

Fig. 1 Measurement performance order

In this connection, in addition to obtaining the measurement result itself, data on the period of time to which the result relates are proposed, and it is necessary to use the period $\left(t^1; t^2\right)$ (Fig. 1).

However, many instruments do not provide information on the sampling time to which the result relates. In this case, it is proposed to calculate the time required. Calculation is carried out based on the principle of equipment measurement. You can highlight some of the most common solutions.

In this research the measured values obtained obtain a clear correspondence at the point in time. At the time the atmospheric air was characterized by said measurement result.

For further analysis, measurement data obtained from analytical equipment must be stored. For this purpose, DBMS are traditionally used (SQL Server, Oracle, etc.). In modern systems, the values obtained from instruments are generally either directly stored, or the values are grouped into groups (for example, values in 5 min), and the average value of the group measurements is stored in the data base.

The choice of the preferred approach was determined by the purpose of the stored data and the nature of their further processing. In order to develop a system for monitoring the environment, means of searching for and identifying sources of pollution. The obtained data are intended to be used for two main tasks - rapid conversion to sudden excess of maximum permissible concentrations and statistical examination of accumulated data. It is usually carried out at large intervals of time (significantly larger than the interval of one measurement). For these purposes, the use of single data from instruments in their original form is not reasonable. Since the excess estimate can only be performed for data averaged in at least 20 min, and for statistical studies such a huge amount of data will significantly increase the processing time.

It is proposed to use an approach in which a sample is taken as the main unit of measurement results. A set of single measurements taken over a certain period of time at a certain geographical point. Thus, the sample comprises a set of average data measurements over the relevant period. Further data processing is carried out only with these averaged data. In this case, the unit measurements are also stored in the sample, but serve only to assess the average data and are not used mediocre in calculations. This approach greatly simplifies the processing, analysis and maintenance of data.

The fact is that the unit values do not refer to moments. But to time intervals (the air sampling interval), which may not completely cover the averaging interval; Their lengths may vary; They may fall within the averaging interval partially (Fig. 2).

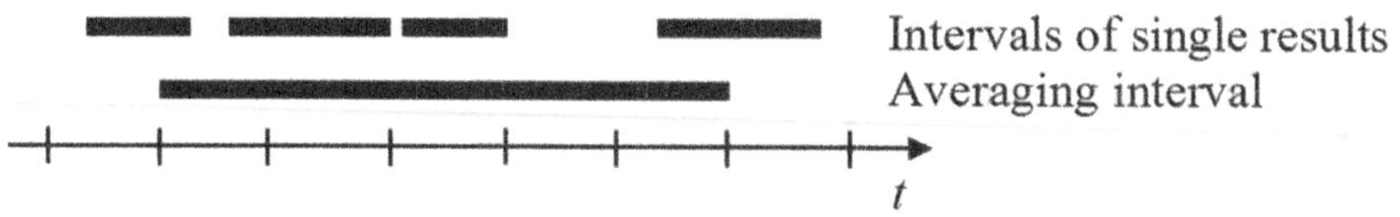

Fig. 2 Ratio of unit measurement intervals with an averaging interval

In many systems, the same approach is used to establish average values of different indicators. Calculation of arithmetic mean, which is not quite correct, because for indicators of different nature the method of calculation may be different. A classic example is the calculation of the mean wind direction. There is the calculation of the mean arithmetic in many cases yields a completely incorrect result. The main purpose of averaging wind indicators is to obtain an average value of air mass variation, which can be achieved only when processing wind direction with its speed.

When calculating average values of key figures, especially for large time intervals, the question of completeness of these data becomes relevant. As noted earlier, there are situations where the intervals for single measurements do not completely cover the averaging interval. For certain points in time, the values of some key figures remain unknown.

The methodological and technical side of the process of measuring many parameters of the surrounding environment is complex. Carried out on complex equipment, in which errors sometimes occur. In terms of measurement results, this means incorrect values. They can subsequently have an extremely negative impact on research results.

Therefore, the collection system should monitor incoming data for correctness and adequacy. Many equipment manufacturers equip their devices with means to monitor their correct operation and allow them to stop transmitting incorrect data to the collection system in case of errors in the operation of the devices. In some cases, the equipment does not provide information on its status or is unable to detect errors. In such cases, it is necessary to analyze the flow of data from the equipment for adequacy and compliance with certain conditions. Measurement results that do not meet these conditions should be screened and notified to the responsible persons when faults occur.

5 Data Exchange Between Data Processing Nodes

In known environmental monitoring systems, after performing analyses and measurements at environmental stations. The results are transmitted to an environmental server, where they are analyzed by specialists and appropriate decisions are made. When operating such systems within a settlement or industrial enterprise. This scheme is successfully operated and has no significant disadvantages. However, the operation of such systems at the regional and federal levels had significant limitations and shortcomings.

The main specificity of such large-scale environmental monitoring systems is the separation of the areas of responsibility for the operation of the system by geographical and administrative characteristics. Environmental stations were served by different groups of people in various localities. Data are also available at various levels, within human settlements and within the region, with a multiplicity of human settlements. At the same time, the same data is processed simultaneously in different parts of the system, which requires synchronization of changes made from both high and low level nodes according to their authorities and areas of responsibility. At the same time, the nature of information requires its prompt transmission between the system nodes, as well as the possibility of organizing backup transmission channels. It was important to bear in mind that data received with high speed might later be incorrect, and that information should also be promptly received by all interested persons.

Large-scale environmental monitoring systems tend to allocate multiple levels of data processing and storage. The regional system often has a three-level structure (Fig. 3).

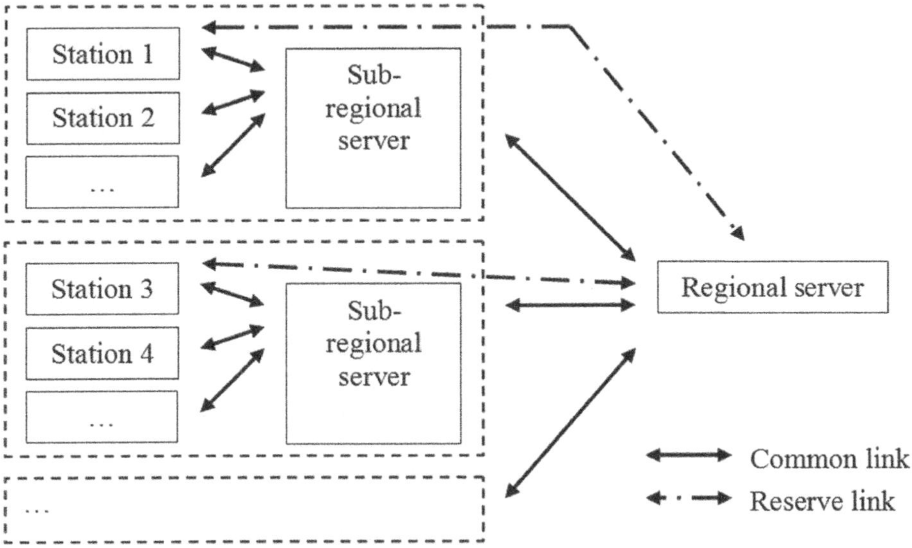

Fig. 3 Structural diagram of the regional environmental monitoring system

The region had a number of district offices, each with environmental stations and an environmental server attached to them. The personnel shall ensure the capacity of their environmental monitoring system section, perform the processing of measurement results and make certain management decisions at their level. The regional environmental server exchanged data with district servers, with staff working with already confirmed data. They take controlling decisions at the regional level, and setting global parameters for the entire monitoring system to be transmitted to district servers and monitoring stations.

These features of large-scale systems impose additional requirements on the environmental monitoring system, in which data must be stored in a distributed data-base, in which each node contains only data of lower-level nodes. They are stations contain only their own data, servers also contain data on connected stations. Changes can be made at all nodes and levels of the system according to their areas of responsibility, and any changes must be promptly agreed with other nodes. In order to implement redundant communication channels, the synchronization mechanism must allow cyclic communication structures, taking into account large volumes of data, only changed information must be transmitted.

During data exchange, nodes exchange records from the specified tables. Information on locations and sampling points, equipment, measurement results, samples and events is mainly transmitted from the lower to upper levels. Information about objects, key figures and regulations is transmitted from the upper to lower levels. In a number of cases, information is also transmitted in the reverse direction. For example, when the results of measurement and event data processing are to be transmitted from the upper levels to the lower levels or to make adjustments to the parameters of the measuring equipment.

To uniquely identify data objects within a distributed database, enter the Globally Unique Identifier (GUID) field for each record. A statistically unique identifier generated at the time the data object was created. Thus, an entity created at one node can be freely transmitted to other nodes without colliding with record identifiers at different posts, as in the case of a simple numeric identifier.

As a result of the exchange of information, the data object is distributed to the distributed database. This object may then change at one or more locations, resulting in several different versions of the same object. It is necessary to quickly synchronize data between nodes. Selection the last term of the object and transfer it to all necessary nodes. Since if you have different versions of an object, you must select the newest version of the object, the field with the date and time of its modification, as well as the field with the ID from which these changes were made, is used to determine the version of the data object.

The data update synchronization algorithm for one object is shown in Fig. 4.

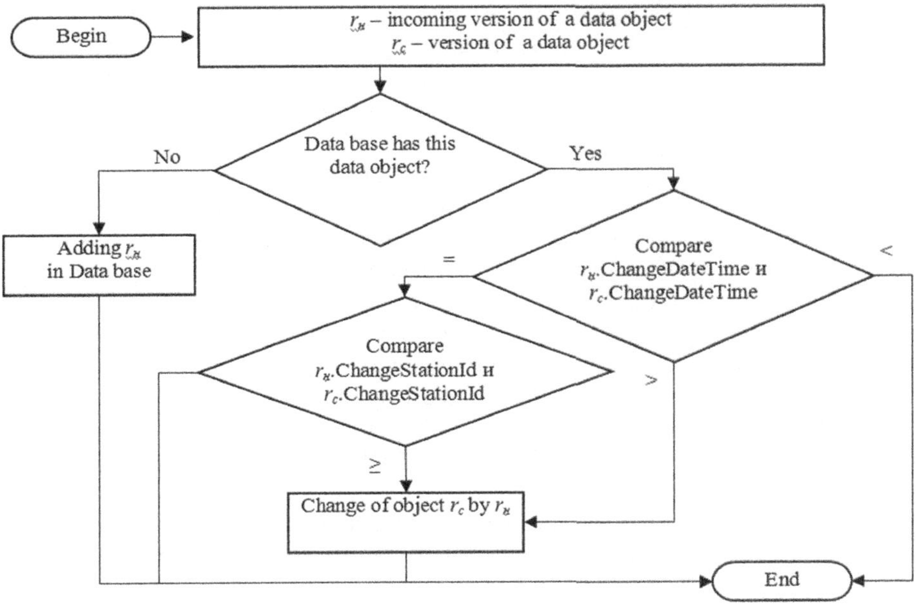

Fig. 4 Block diagram of data synchronization for one object

It should be noted that this approach requires completely synchro-reduced time on all nodes, which in practice cannot be achieved - usually there is always a time difference. This may lead to an incorrect definition of a newer object while changing it on different nodes, but in practice it is very rare in case of conflicting actions on the part of the user. Also, even with sync-down time, there may be a situation of simultaneously changing the same object - in this case, a variant with a large node number is selected. This is necessary in order to solve the conflict equally on all nodes, otherwise infinitely updating data objects may arise.

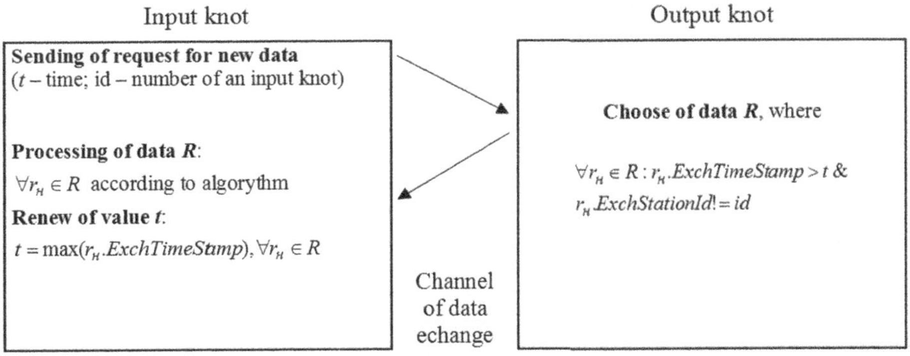

Fig. 5 Block schema of data exchange between knots

In some cases, with a significant time difference, there may be a situation where when an object changes. They appear to have a change time greater than the time. In this case, it is necessary to record in the change time field not the current time, but a value increased by some amount, thereby stressing that the record was up-dated and has a newer version.

It is proposed to arrange data exchange by means of periodic communication sessions. In it the receiving node requests new data from the transmitting node. The frequency of these sessions is selected based on the capabilities of the data link - more frequent sessions provide more efficiency, but require more bandwidth.

In practice, very many data objects are stored in the database of one node. It is not possible to execute an algorithm for updating all these objects at each communication session. This will require the transmission of almost all the DB each time, which will overload the communication channels and take a very long time.

For solution of this problem it is necessary to renew the data of objects, which changed since the previous time (Fig. 5). Special slots «ExchTimeStamp» и «Ex-chStationId» are used for this solution.

In the next session, only data records that have appeared or have been modified since the previous session are requested from the transmitting node. This eliminates records that were previously obtained from the receiving node. It allows to avoid looping of the data exchange when the same records are infinitely transmitted between the same nodes.

The continuous measurement data of the system contain, among other things, information on the actual values of air quality indicators, as well as on the physical parameters of the air medium over a certain time interval at given sampling points. These data themselves play an important role in monitoring the state of the environment. These data are not sufficient to form measures to protect the environment. In the case of significant contamination detected by the system, it was the primary task to identify the source of the pollution, which would allow for measures to be taken to develop control effects for its elimination or to bring the level of emissions to normal.

6 Conclusion

The article Key aspects of the construction of regional environmental monitoring systems (environmental monitoring) were considered. Basic standards and algorithms for primary processing of measurement data at stations of environmental monitoring systems have been formulated, allowing for more accurate, complete and adequate processing of data. Approaches are proposed to check correctness of incoming single data, interpretation of obtained values, their storage and averaging. Proposed is method of quantitative assessment of completeness of measurement results.

Based on operational data from monitoring systems, pollution sources are searched, environmental change is assessed and predicted, and management solutions are developed.

References

1. Dhillon, B.S., Chanan, S.: Engineering reliability New Techniques and Applications. John Wiley & Sons, NY (1981)

2. Guck, D., Spel, J., Stoelinga, M.: DFTCalc: reliability centered maintenance via fault tree analysis (tool paper). In: Proc. 17th Int. Conf. Formal Engineering Methods (ICFEM). LNCS 9407, pp. 304–311 (2015)
3. Ruijters, E., Reijsbergen, D., de Boer, P.-T., Stoelinga, M.: Rare event simulation for dynamic fault trees. Reliabil. Eng. Syst. Saf. **186**, 220–231 (2019)
4. Fominykh, D.S., et al.: Problem of quality assurance during metal constructions welding via robotic technological complexes. Int. Conf. Inform. Technol. Bus. Ind. **1015**(3), 032169 (2018)
5. Spiridonov, A.Y., et al.: Prediction of main factors' values of air transportation system safety based on system dynamics. Int. Conf. Inform. Technol. Bus. Ind. **1015**(3), 032140 (2018)
6. Radosteva, M., Soloviev, V., Ivanyuk, V., Tsvirkun, A.: Use of neural network models in the market risk management. Adv. Syst. Sci. Appl. **18**(2), 53–58 (2018)
7. Rezchikov, A.F., Kushnikov, V.A., Ivashchenko, V.A., Bogomolov, A.S., Filimonyuk, L.: Models and algorithms of automata theory for the control of an aircraft group. Autom. Remote Control **79**(10), 1863–1870 (2017)
8. Zheng, Y., Jia, B., Li, X.-G., Zhu, N.: Evacuation dynamics with fire spreading based on cellular automaton. Phys. A: Statis. Mech. Appl. **390**(18–19), 3147–3156 (2011). https://doi.org/10.1016/j.physa.2011.04.011
9. Alexandridis, A., Russo, L., Vakalis, D., Bafas, G., Siettos, C.I.: Wildland fire spread modelling using cellular automata: evolution in large-scale spatially heterogeneous environments under fire suppression tactics. Int. J. Wildland Fire **20**(5), 633–647 (2011)
10. Baranov, V.V., Tsvirkun, A.D.: Development control: structural analysis, problems, stability. Autom. Remote Control **79**(10), 1780–1796 (2018)
11. Filimonyuk, L.: The problem of critical events' combinations in air transportation systems. Adv. Intell. Syst. Comput. **573**, 384–392 (2017)
12. Dugan, J.B., Bavuso, S.J., Boyd, M.A.: Fault trees and sequence dependencies. In: Proc. Annu. Reliability and Maintainability Symp. IEEE, pp. 286–93 (1990)
13. Ruijters, E., Guck, D., Drolenga, P., Stoelinga, M.: Fault maintenance trees: reliability centered maintenance via statistical model checking. In: Proc. IEEE 62nd Annu. Reliability and Maintainability Symposium (RAMS), 2016978-1-5090-0248-1
14. Siergiejczyk, M., Rosinski, A.: Reliability and exploitation analysis of power supply of alarm systems used in railway objects. In: 26th Conference on European Safety and Reliability (ESREL). Glasgow, Scotland (2016)
15. Fominykh, D.S., et al.: Problem of quality assurance during metal constructions welding via robotic technological complexes. J. Phys.: Conf. Ser. **1015**, 032169 (2018). https://doi.org/10.1088/1742-6596/1015/3/032169
16. Rezchikov, A., Kushnikov, V., Ivaschenko, V., Bogomolov, A., Filimonyuk, L., Kachur, K.: Control of the air transportation system with flight safety as the criterion. Adv. Intell. Syst. Comput. **466**, 423–432 (2016)
17. Laskowski, D., Lubkowski, P., Pawlak, E., Stanczyk, P.: Anthropo-technical systems reliability. Safety and Reliability: Methodology and Applications, pp. 399–407 (2015)
18. Pilo, E.: Power Supply, Energy Management and Catenary Problems, pp. 1–191. WIT Press, UK (2010)

Design of an Optimal Framework to Evaluate the Throughput in Downlink Channel in LTE

R. K. Srinivasa[1,2](✉) and A. R. Hemantha Kumar[2]

[1] Visvesvaraya Technological University, Belagavi, India
[2] Department of Electronics and Communication Engineering, BIT, Bangalore, India

Abstract. Long-term Evolution (LTE) based network envisioned for maximizing the connectivity among user base, where it can effectively fulfill the demand of each dynamic user equipment (UE) with fairness to network resource accessibility. Addressing the limitation associated with the traditional scheduling approaches in the context of LTE where most of the approaches do not guarantee high throughput with lesser delay during packet transmission and also have design limitations owing to higher operational challenges, the study proposes an optimized analytical scheduler for LTE downlink communication which aims to optimize the throughput performance of the network cost-effective resource scheduling mechanism. For this purpose, the system adopts a set of traditional approaches such as spectrum flexibility, diversity, spatial multiplexing and many others which enhances the design of the proposed Hybrid Automatic Repeat Request (HARQ) assisted scheduler for better resource management in LTE environment. The promising outcome further illustrated that the system design is fruitful and suitable in uncertain traffic condition where the fairness of resource allocation can be maintained for each dynamic UE.

Keywords: Long Term Evolution (LTE) · HARQ · Downlink scheduler

1 Introduction

In LTE, the main payload or the data traverses through the shared channel and this channel allocation takes place in a very dynamic fashion depending upon the real-time demand. The channel allocation to the specific user does not adopt either deterministic or probabilistic approach, whereas it adopts an opportunistic approach [1]. In the 4th generation of network system like Long Term Evolution (LTE), which is devised by the combined effort of both 3GPP and IEEE 802.16 a wireless broadband standard bodies uses Orthogonal Frequency-Division Multiple Access (OFDMA) as a radio access technology in their downlink as shown in the Fig. 1. Basically, this communication channel which carries the data packet assigned to user considering opportunistic and dynamic basis. The information associated with System Information Blocks (SIB) is also carried by this shared downlink channel and used for transmitting application data too. It should be noted that this channel performs broadcasting of paging message and it also carries signaling message within itself.

R. Silhavy (Ed.): CSOC 2021, LNNS 230, pp. 152–169, 2021.
https://doi.org/10.1007/978-3-030-77442-4_14

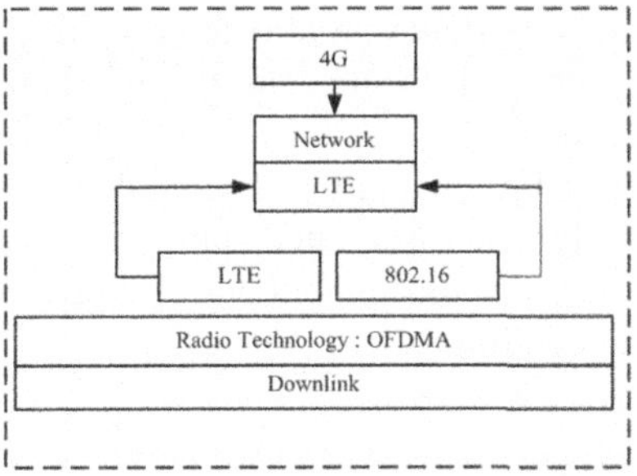

Fig. 1. Radio access technology in downlink of 4G-LTE

Taking the strength factors of the traditional approaches of Reference Measurement Channel Number (RMCN), Spectrum flexibility, diversity, Spatial Multiplexing, eNodeB set-up and many more the study design an optimal resource scheduling framework to maximize the throughput performance in LTE-downlink transmission. The framework also adopts the concept of HARQ for erroneous data packet correction and coding which also maximize the data transfer rate with lesser delay. The system is analytically constructed referring various significant components of traditional 4G-LTE where the prime emphasize is laid towards optimizing the throughput performance with higher fairness index. The supporting technologies for the proposed approach consider OFDMA as radio access technology and also it reduces the possibility of multi-path fading and also incorporates spatial multiplexing for higher data transfer rate at LTE physical and MAC layers. The system model also constructed on the basis of 16/64-QAM and QPSK modulation scheme which ensure better propagation of OFDM signal though the channel. And the system also ensures the consistency in allocation of frequency of Transport Blocks (TB) which is associated with each frame. The test evaluation to validate the performance of the proposed approach is assessed in terms of different reference channels and transmission schemes such as TX Diversity and Spatial Multiplexing (SpatialMux) and Port-0. The outcome shows that the formulated approach is suitable for uncertain traffic scenario and also provides superior throughput performance with different transmission scheme in different channel conditions.

Taking the strength factors of the traditional approaches of Reference Measurement Channel Number (RMCN), Spectrum flexibility, diversity, Spatial Multiplexing, eNodeB set-up and many more the study design an optimal resource scheduling framework to maximize the throughput performance in LTE-downlink transmission. The framework also adopts the concept of HARQ for erroneous data packet correction and coding which also maximize the data transfer rate with lesser delay. The system is analytically constructed referring various significant components of traditional 4G-LTE where the prime emphasize is laid towards optimizing the throughput performance with higher fairness index. The supporting technologies for the proposed approach consider OFDMA as radio access technology and also it reduces the possibility of multi-path fading and also incorporates spatial multiplexing for higher data transfer rate at LTE physical and MAC layers. The system model also constructed on the basis of 16/64-QAM and QPSK modulation scheme which ensure better propagation of OFDM signal though the channel. And the system also ensures the consistency in allocation of frequency of transport blocks (TB) which is associated with each frame. The test evaluation to validate the performance of the proposed approach is assessed in terms of different reference channels and transmission schemes such as TX Diversity and Spatial Multiplexing (SpatialMux) and Port-0. The outcome shows that the formulated approach is suitable for uncertain traffic scenario and also provides superior throughput performance with different transmission scheme in different channel conditions.

2 Review of Literature

The emerging trend in the field of mobile broadband communications has come up with various solutions to make the data transfer rate faster with a multiple efficient accessing mechanism in the context of cellular networks. The evolution of third-generation (3G) cellular networks initially extended its operational features targeting the maximum number of simultaneous users within a CDMA-based system. The prime emphasis was laid upon cost-effective channel utilization and maximized rate of transmission while maintaining the desired quality of services (QoS) aspects such as efficient throughput and delay performance. The communication paradigm in 3G has substantially increased the data transfer rate (~10 megabits per second (Mbps) by increasing the network capacity in contrast with previous 2G/2.5G networks. The study of (W. L. Tan, F. Lam and W. C. Lau, 2008) performed an empirical study to evaluate the performance of conventional 3G networking features in terms of different aspects such as delay, latency, data throughput and video and voice call handling capabilities. The core findings of the study outline the limitations that arises during diverse mode of resource allocation and also it shows that traditional theoretical models lack efficiency to predict the cell-capacity with ease of resource scheduling. And this do not ensure better retrieval of signal-to-noise ratio [1]. There have been extensive research effort in similar course of direction devoted by Pentikousis et al. [2], Reichl et al. [3], Kohlwes et al. [4], Cano-Garcia et al. [5], Maderet al. [6], Joyce et al. [7] to improvise the limitations that arises during the channel allocation, scheduling and approximation. The contributions are mostly towards enhancing the voice and data traffic performance under several controlled environments. To address lower scope of multi-antenna transmission and the issue associated with spectrum efficiency in 3G supported broadband communications, Long Term Evolution (LTE) had been derived and realized by combine effort on 3rd Generation partnership project (3GPP) execution model and IEEE 802.16a. This broadband bodies utilizes the standard features of OFDMA as radio-access technologies in the context of downlink transmission. LTE offers a highly flexible and robust interface which can support peak rate of 300 Mb/s with higher spectrum efficiency and minimum delay constraints (Dahlman et al. [8]), 3GPP TS36.300 [9]). The scientific communities promoted LTE as 4G-LTE and in many applications commonly also adopted as 3.95G in many cases. The study of Astely et al. [10] collectively explores the potential aspects of LTE networks and exhibited a better understanding of its features as exhibited in the following Fig. 2.

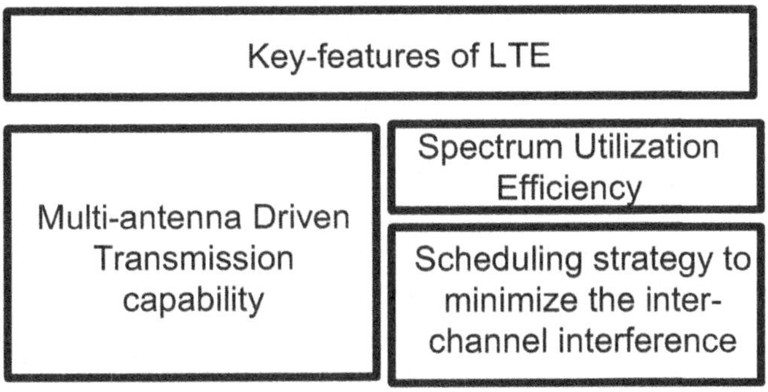

Fig. 2. Key-attributes of LTE

The study also explores the strength factors of LTE and demonstrated extensively about its spectrum flexibility attributes in the context of transmission bandwidth utilization. It clearly shows that through LTE duplex scheme, efficient spectrum regulations in a single radio-active channel can be achieved through Frequency Division-Duplex (FDD) and Time Division Duplex (TDD) both. It also talked about the essentiality of LTE-advanced and how it provides a very well foundation for further evolution [10]. The search result obtained from IEEE Xplore digital library with the key-word: Long term evolution networks shows that this area is extensively studied since past one decade. The record of the statistics in digital library database is retrieved can justify the fact that the research in LTE is drastically evolving with many terms. The following Fig. 3 shows the trend of publications of last one decade in IEEE Xplore.

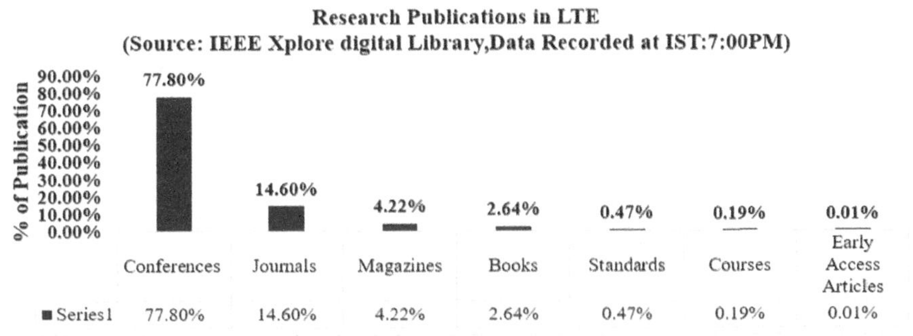

Fig. 3. Research publication relevant results from IEEE Xplore (dated: 25th Nov, 2019)

A clear analysis of Fig. 3 shows that out of total number of publication count = 11864, higher percentage of conferences and research proposals in relevant keyword is obtained. The quantified outcome also illustrates that 14.60% highly cited journals are in existence till date which talk about LTE networks. The following are the relevant studies which focuses on LTE research evolution and also talks about how the promising aspect of LTE can offer seamless data connectivity by bridging the potential gap in between physical world and virtual world. Kotagi et al. [11] have studied LTE-A in the context of futuristic Internet of Things (IoT) eco-system and derived its scope of applicability to offer seamless connectivity among a large-scale of IoT devices while balancing the energy efficiency. The study thereby introduced a novel schema to handle the energy consumption at base stations in the context of advanced mode of LTE. A resource scheduling algorithm thereby introduced for downlink transmission which can optimize the energy performance in IoT communication operations. The outcome is found promising and superior as compared to other state-of-the art baselines.

3 System Model

The proposed system initiates with developing a waveform for the downlink transmission. The conceptualization of eNodeB is the unit of hardware which is latched with the mobile network which is and associated top-level configuration. This is provisioned to communicate with the User Equipment (UEs) similar to the transceiver in the case of the networks based on GSM. The essential components are as follows:

- *Reference Measurement Channel Number (RMCN):* For RMC-s, which at the nominal target coding rate do not cover all the possible UE categories for the given modulation, reduce the target coding rate gradually (within the same modulation), until the maximal possible number of UE categories is covered.
- *DL Reference Channel Options:* The output configuration structure is initialized in accordance with the reference channels defined in TS 36.101, Annex A.3. Initialization choices available for the downlink reference channel (Table 1).

Table 1. DL reference channel configurations

RMC No	Transmission scheme	RB	Modulation	Cell ref P	R
R.0	Port 0	1	16QAM	1	½
R.1	Port 0	1	16QAM	1	½
R.2	Port 0	50	QPSK	1	1/3
R.3	Port 0	50	16QAM	1	½
R.4	Port 0	6	QPSK	1	1/3
R.5	Port 0	15	64QAM	1	¾
R.6	Port 0	25	64QAM	1	¾
R.7	Port 0	50	64QAM	1	¾
R.8	Port 0	75	64QAM	1	¾
R.9	Port 0	100	64QAM	1	¾
R.10	TxDiversity|SpatialMux	50	QPSK	2	1/3
R.11	TxDiversity|SpatialMux|CDD	50	16QAM	2	½
R.12	TxDiversity	6	QPSK	4	1/3

(continued)

Table 1. (*continued*)

RMC No	Transmission scheme	RB	Modulation	Cell ref P	R
R.13	SpatialMux	50	QPSK	4	1/3
R.14	SpatialMuxlCDD	50	16QAM	4	½
R.25	Port 5	50	QPSK	1	1/3
R.26	Port 5	50	16QAM	1	½
R.27	Port 5	50	64QAM	1	¾
R.28	Port 5	1	16QAM	1	½

- *Physical Cell ID*: The cell at the physical layer is identified by the Physical-Cell-ID(PCID) works in the same way as the primary scrambling code (PSC) in the typical UMTS cell [AR], the computation of the PCID takes place using the sync-signals of both primary and secondary(PSS, SSS) using Eq. (1)

$$\text{PCID} = 3 \times \alpha + \beta \tag{1}$$

Where α = cell group number (168 types of SSS) and β = cell number (3 types of PSS). While decoding the data transmitted at the physical layer (PL), the PCID is uses so the assignment of PCID should not be for the multiple cells as the cells and UE interfere and poses challenges in the process of cell detection [12].

- *Frame and Subframe*: The LTE data is carried out in the frames and subframes similar to the previous communication systems (Fig. 4, Fig. 5, Fig. 6). The radio network gets enables to grip the data appropriately by synchronization using frames (*f*) and subframes (*sf*) [13].

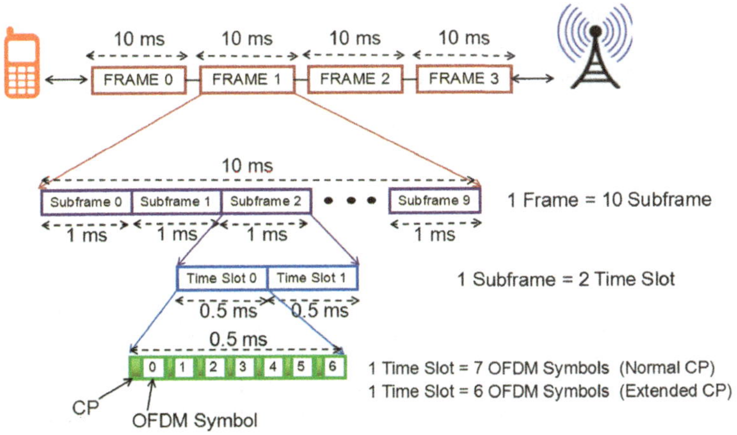

Fig. 4. Type 1 used for the LTE FDD mode systems

The frame structure differs based on whether its operating in FDD or TDD Modes. There are two types of LTE frame structure:

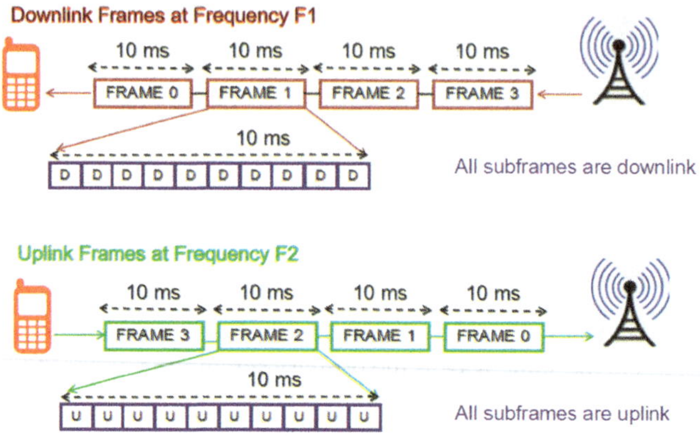

Fig. 5. Type 2 used for the LTE TDD systems

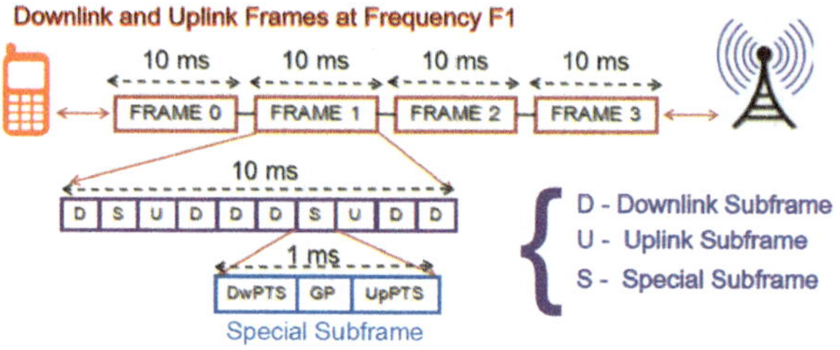

Fig. 6. Combined downlink and uplink frame transmission

One frame is divided into 10 subframes (1 ms each), and that subframe can be either downlink, uplink or special subframe. The sequence of these subframes has been defined by 3GPP with the name TDD Frame Configurations. There are fixed patterns of these configurations and network operator has to choose out of these defined patters. There are total 7 TDD configurations and typical deployment configuration is 2 (Table 2).

Table 2. Configurational detailing about TDD

config-uration	3GPP release	Downlink to uplink switch point periodic-ity (*ms*)	Subframe number										Number of Subframes / frame		
			0	1	2	3	4	5	6	7	8	9	D[DL]	U[UL]	S[SSF]
0	8	5	D	S	U	U	U	D	S	U	U	U	2	6	2
1	8	5	D	S	U	U	D	D	S	U	U	D	4	4	2
2	8	5	D	S	U	D	D	D	S	U	D	D	6	2	2
3	8	10	D	S	U	U	U	D	D	D	D	D	6	3	1
4	8	10	D	S	U	U	D	D	D	D	D	D	7	2	1
5	8	10	D	S	U	D	D	D	D	D	D	D	8	1	1
6	8	5	D	S	U	U	U	D	S	U	U	D	3	3	2

- *Release wise Evolutions of LTE and Transmission Schemes:* The advance -LTE evolution comes with the Release-10(R-10) to achieve the peak data rate as 3Gbps at downlink and 1.5 Gbps at the uplink and the maximization of the efficiency if the spectrum from 16 bps/Hz in Release-8 to 30 bps/Hz in Release-10 to coverup more number of the active users at a time along with the cell edge performance improvement along with downlink of 2 × 2 MIMO at the minimum of the 2.40 bps/Hz/cell by introducing a aggregation mechanism of aggregation to meet the multiple antenna to strengthen the relay nodes.
- *Spatial Multiplexing:* The improvement in the bit rate requires the transmission of the distinguished data stream over different antennas separated by the different reference signals in both time and frequency with common channels as shown in Fig. 7.

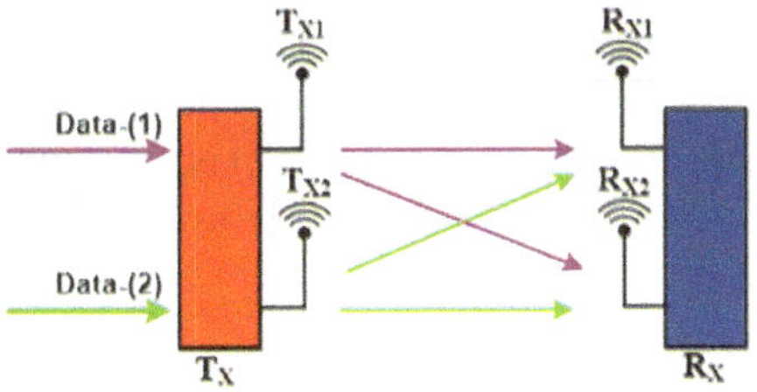

Fig. 7. Spatial multiplexing of two data streams transmission of two TX and RX using the same frequency and time, separated only by the use of different reference signal

But, the introduction of the higher Tx-Rx pair in LTE at both downlink and uplink provides capacity of handling at the higher Signal-To Noise (SNR) ratio to achieve a very channel of high quality, whereas in the case of lower SNR a transmission diversity is used as shown in the Fig. 8.

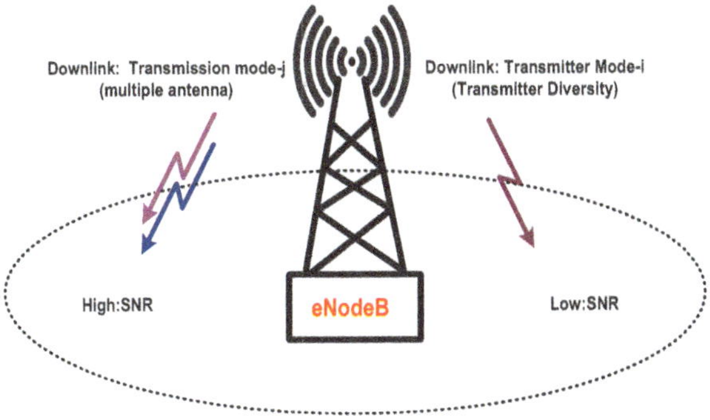

Fig. 8. Multiple antenna for higher SNR and TX diversity: lower SNR

The transmission modes are used for varied environments of the radio in order to adjust the schemes of the multiple antennas and it is informed to the User Equipment's

(UEs) through the RRC signaling. The classification of transmission mode is on the basis of antenna port used, layers, reference signal type, pre-coding and reference signals for the cell and the demodulations. As soon as the power is on and channel allocation, the User Equipment's (UEs) find the details of the antenna port of the eNodeB by a CRC mask of 16 bits of broadcast channel.

- *Downlink Power Allocation*: The downlink power allocation (ρ) is dependent of cell and the user device, there is a definite impact on throughput. The power level ranges from −60 to +50 dBm/15 kHz and the eNodeB transmits the reference signals consistently to avoid the power diversity at the user equipment's.
- *Channel State Information (CSI)*: The eNodeB controls the four coded indicators namely: coded rank, precoding matrix, precoding type and channel quality in order to report the resources used by the user equipment's.
- *Automatic Repeat Request (ARQ) and Hybrid-ARQ (HARQ)*: The reliability is achieved in the wireless communication by means of repetition. The ARQ transmits the packets in case of the failure of the previous transmission attempts and with the Forwarded Error Coded (FEC) bits this protection is achieved. To tackle the inefficient transmission, FEC and ARQ provide Hybrid-ARQ (HARQ) (Fig. 9).

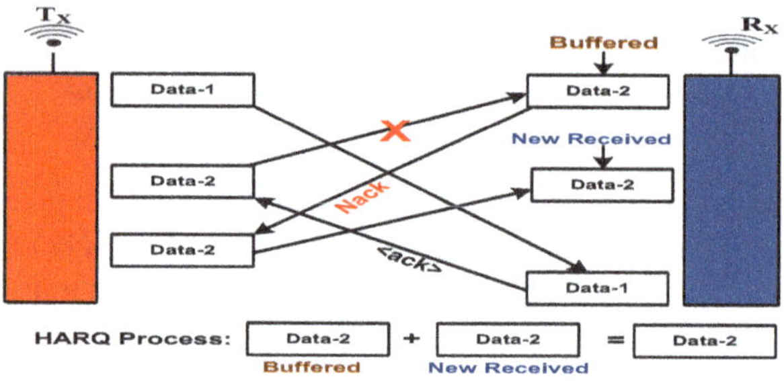

Fig. 9. Stop and wait process (SAW-process) in LTE: HARQ

The processing of the HARQ depends upon the ACK of packets with respects to the packets sent on the basis of the stop and wait sequence and increase the round-trip time which includes: a) processing time for sending and receiving and b) delay in propagation. This stop-and-wait (saw) process in LTE is a HARQ process. HARQ-RVSeq maintains one or two codewords. The number of columns in RVSeq equals the number of transmissions of the transport blocks associated with a HARQ process. The RV sequence specified in each column is applied to the transmission of the transport blocks.

4 Implementation

The system model is setup for the eNodeB configuration, initializes the different values of the releases along with the number of physical cell ID, total numbers of the subframes (fs) as shown in the Table 3.

Table 3. eNodeB configuration: Initialization parameters.

Sl. no	Setting parameters	Values
1	RMC number	R.8, R.9, R.10, R.11, R.12, R.13
2	No. of physical cell ID	10, 20, 30, 40…
3	Total numbers of subframes (fs)	10, 20, 30, 40…

The parameters of the downlink shared channels are tabulated with the initialization value in the Table 4.

Table 4. Downlink shared channel properties

Sl. no	Setting parameters	Values
1	Transmission scheme	Portx, TxDiversity, Spatial Mux, CDD
2	CRC mask	1 (16 bit mask)
3	Power allocation	−3 to +3
4	Channel state information	0.5
5	Redundancy version for HARQ	0,1

The model simulation takes place at the various values of the number of frames (f) and the varies range of the signal to noise ratio (SNR) from worst to best condition. The downlink reference measurement channel for designed eNodeB, structure is tabulated in the Table 5.

Table 5. Downlink shared channel properties

Sl. no	Components/parameters	Values
1	Cell ref point	1
2	Cyclic prefix	Normal
3	CFI	3
4	Ng	Sixth
5	Duplex mode	FDD

Table 6 and Table 7 highlight channel configuration and channel estimation considered for implementation in proposed system (Fig. 10).

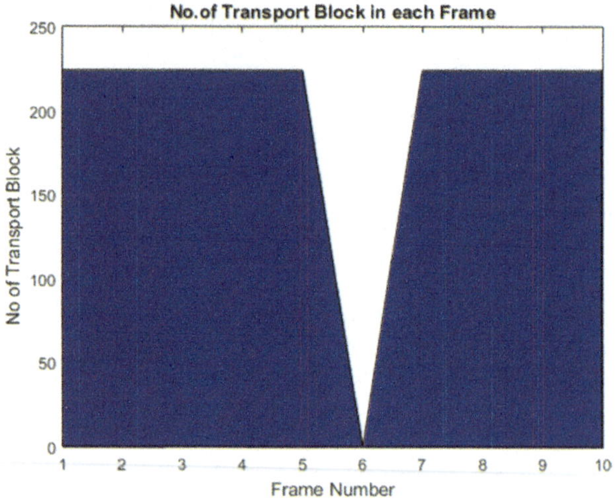

Fig. 10. Number of transport block (Tb) each frame (f)

It is being observed that the frequency of number of transport block is consistent in each frame (fs), whereas in the frame '6' it is zero and the computed code words is found = '1'. The resource grid dimension includes: a) number of subcarrier (P), b) number of OFDM symbols (Q) per subframe, and c) number of transmit antennas (R). The instantaneous resource grid dimension = {P = 180, Q = 14, R = 10}.

Table 6. Configuration of the channel

Sl. no	Components/parameters	Values
1	Seed	2
2	Number of receive antenna	2
3	Propagation/delay profile	EPA
4	Doppler frequency (Hz)	5
5	MIMO correlation	Medium
6	Oscillators in fading model	16
7	Rayleigh fading model type	GMEDS
8	Random initial phases	Random
9	Normalize delay profile	On
10	Normalize for transmit antennas	On

The analysis is also carried out with respect to transport block and throughput, which are briefed as follow.

Table 7. Channel estimation

Sl. no	Components/parameters	Values
1	Pilot symbol type averaging	Custom
2	Frequency windows size in RE	15
3	Time window size in REs (resource elements)	141
4	Cubic interpolation	Cubic
5	Interpolation of subframe	3
6	Interpolation window method	Centered

i) *Number of Transport block:* The number of transport blocks in the carrying subframes(fs) in a given data frame is computed by the equation

$$\text{Nfs} \leftarrow \Sigma(\text{Tbs} \neq 0) \quad (2)$$

The typical value of the number of transport block in R.12 is 9 from the 10 subframes in the frame number 5, which does not carry any kind of the data.
Total number of CRC block and the throughput: The CRC block and the throughput values contain the number of elements as a vector with the total number of elements as 'N' element of SNR × (Total number of transport block (Tb) × Total number of frames (f)).

5 Result and Discussion

The study basically performed an extensive numerical analysis considering the analytical framework which is mechanized to accomplish the intended solution for the addressed problem. The further section illustrates the outcome of the system evaluation which is carried out considering different test instances.

5.1 Numerical Outcome

The numerical outcome obtained by simulating the proposed approach in a numerical computing environment where the evaluation process is carried out considering different test instances of throughput measurement for different reference channels where the Table 8 shows the test evaluation outcome for R.8 where transmission scheme is used as port-0. It shows that with variable SNR which ranges between −5 to 5 with step size 1 for 20 frames the optimal throughput in port-0 obtained as 0% only (Fig. 11).

Table 8. Test evaluation for R.8 using transmission scheme port 0

Reference channel	Transmission scheme					Number of frames					
R.8	Port 0					20					
SNR	−5	−4	−3	−2	−1	0	1	2	3	4	5
Throughput-(kbps)	0	0	0	0	0	0	0	0	0	0	0
Throughput -(%)	0	0	0	0	0	0	0	0	0	0	0

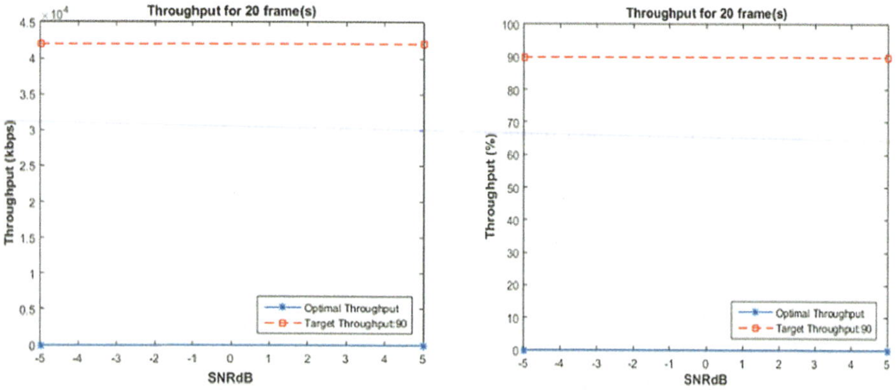

Fig. 11. Throughput analysis of R.8 using transmission scheme port 0

A similar test scenario is again considered in Table 9 for R.9 and port-0 as transmission scheme for number of frame size 20. It also exhibits the similar trend of outcome as the previous test instance.

Table 9. Test evaluation for R.9 using transmission scheme port 0

Reference channel	Transmission scheme							Number of frame size			
R.9	Port 0							20			
SNR	−5	−4	−3	−2	−1	0	1	2	3	4	5
Throughput-(kbps)	0	0	0	0	0	0	0	0	0	0	0
Throughput-(%)	0	0	0	0	0	0	0	0	0	0	0

In Table 10 reference channel R.10 is utilized with transmission scheme TxDiversity where for SNR = 5 (dB) maximum number of optimal throughput achieved is 4400 and which is approximately 100%. The trend of the curve in Fig. 12 also exhibits the same (Fig. 13).

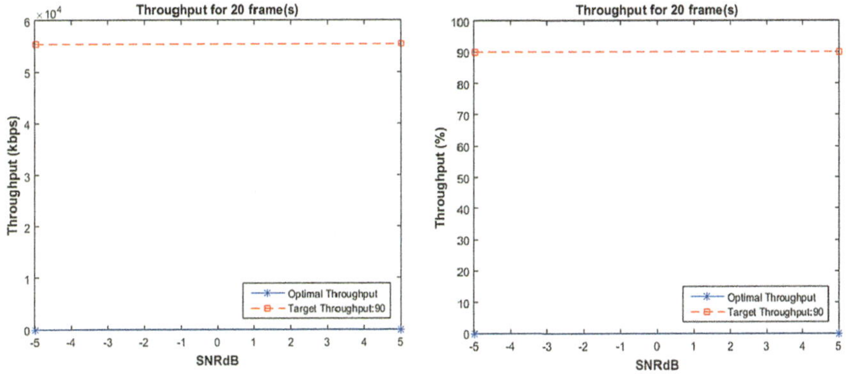

Fig. 12. Throughput analysis of R.9 using transmission scheme Port 0

Table 10. Test evaluation for R.10 using transmission scheme TxDiversity

Reference Channel		Transmission Scheme								Number of Frames	
R.10		TxDiversity								20	
SNR	5	-4	-3	-2	-1	0	1	2	3	4	5
Throughput-(kbps)		480	900	1300	1750	2200	2600	3050	3500	3995	4400
Throughput-(%)		10	20	30	40	50	60	70	80	90	100

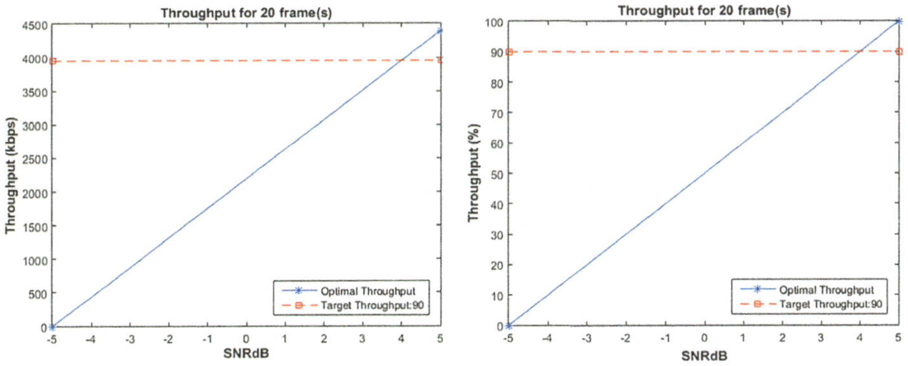

Fig. 13. Throughput analysis of R.10 using transmission scheme TxDiversity

The Table 11 and Fig. 14 shows the test outcome of the proposed scheme when applied with the transmission scheme SpatialMux. It also shows the superior trend of throughput performance with SNR ranges between −5 to 5 when reference channel R.10 is considered.

Table 11. Test evaluation for R.10 using transmission scheme SpatialMux

Reference channel		Transmission scheme							Number of frame		
R.10		SpatialMux							20		
SNR	−5	−4	−3	−2	−1	0	1	2	3	4	5
Throughput-(kbps)	0	480	900	1300	1750	2200	2600	3050	3500	3995	4400
Throughput-(%)	0	10	20	30	40	50	60	70	80	90	100

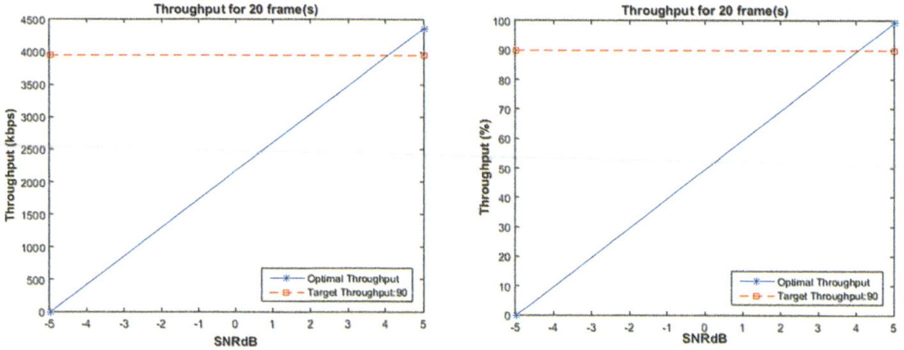

Fig. 14. Throughput analysis of R.11 using transmission scheme SpatialMux

Another test instance is evaluated considering the reference channel R.11 where the proposed approach is combinedly applied with TxDiversity mode of transmission (Fig. 15). The outcome shows that it achieves 0% throughput with every SNR value (Table 12).

Table 12. Test evaluation for R.11 using transmission scheme TxDiversity

Reference channel		Transmission scheme							Number of frames		
R.11		TxDiversity							20		
SNR	−5	−4	−3	−2	−1	0	1	2	3	4	5
Throughput-(kbps)	0	0	0	0	0	0	0	0	0	0	0
Throughput-(%)	0	0	0	0	0	0	0	0	0	0	0

Table 13 shows that with the reference channel R.12 and transmission scheme TxDiversity when the SNR is set to 5 for 20 number of frames of TB then the maximum throughput is achieved by the proposed system which is 375 kbps. The visual outcome of the quantitative data is further represented with Fig. 16.

The analysis shows that the proposed approach attains optimal throughput outcome with both SpatialMux and TxDiversity as compared to Port-0.

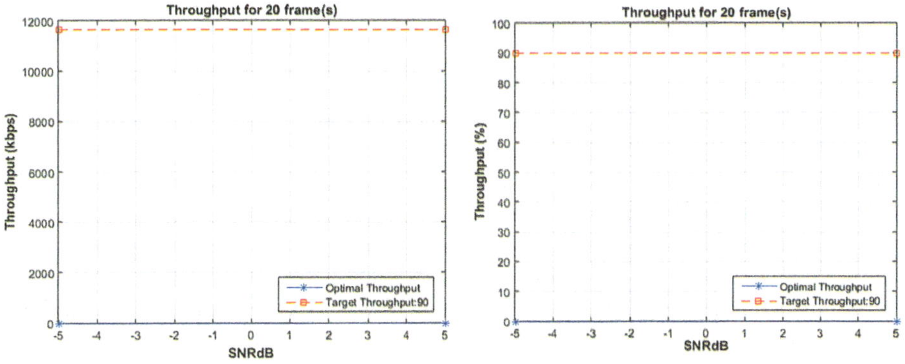

Fig. 15. Throughput analysis of R.11 using transmission scheme TxDivesrsity

Table 13. Test evaluation for R.12 using transmission scheme TxDiversity

Reference Channel		Transmission Scheme							Number of Frame		
R.12		TxDiversity							20		
SNR	5	-4	-3	-2	-1	0	1	2	3	4	5
Throughput-(kbps)		35	75	120	150	180	225	260	300	340	375
Throughput-(%)		10	20	30	40	50	60	70	80	90	100

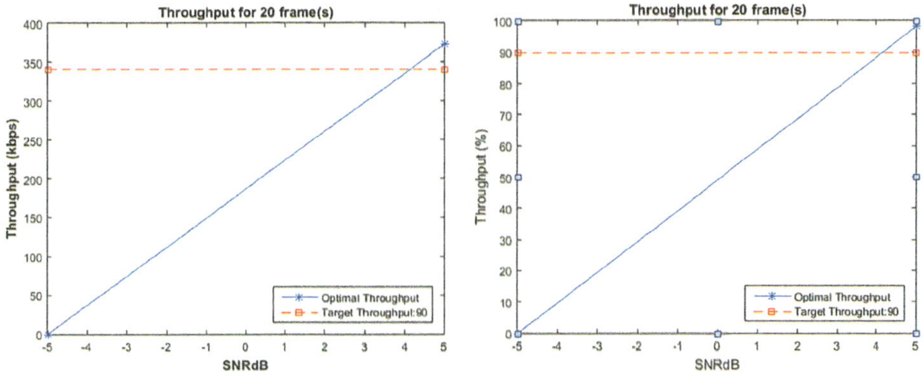

Fig. 16. Throughput analysis of R.13 using transmission scheme SpatialMux

6 Conclusion

The study introduces an optimal framework which intends to maximize the throughput during LTE downlink transmission. The framework adopts analytical methodology to achieve higher data transfer rate in different channel condition with variable dynamic traffic condition. The scheduling process basically designed considering the strength factors and component from the traditional state-of-the art communication technologies from LTE such as OFDMA, Spatial multiplexing and eNodeB and many others. The experimental analysis considers throughput estimation for variable reference channel conditions such as R.8, R.9, R.10, R.11, R.12, R.13 with respect to the increasing SNR in dB which ranges from [−5 to 5] with step size 1. The outcome is found quite acceptable and different transmission scheme such as SpatialMux, TxDiversity and port-0. The optimal throughput obtained for different channel condition suggests that the system can be adopted into the futuristic LTE based systems for efficient downlink scheduling between eNodeB and UE.

References

1. Tan, W.L., Lam, F., Lau, W.C.: An empirical study on the capacity and performance of 3G networks. IEEE Trans. Mob. Comput. **7**(6), 737–750 (2008). https://doi.org/10.1109/TMC.2007.70788
2. Pentikousis, K., Palola, M., Jurvansuu, M., Perala, P.: Active goodput measurements from a public 3G/UMTS network. IEEE Comm. Lett. **9**, 802–804 (2005)
3. Reichl, P., Umlauft, M., Fabini, J., Lauster, R., Pospischil, G.: Project WISQY: a measurement-based end-to-end application level performance comparison of 2.5G and 3G networks. In: Proc. Fourth Ann. Wireless Telecomm. Symp. (FTS) (2005)
4. Kohlwes, M., Riihijarvi, J., Mahonen, P.: Measurements of TCP performance over UMTS networks in near-ideal conditions. In: Proc. 61st IEEE Vehicular Technology Conf. (VTC Spring) (2005)
5. Cano-Garcia, J.M., Gonzalez-Parada, E., Casilari, E.: Experimental analysis and characterization of packet delay in UMTS networks. In: Koucheryavy, Y., Harju, J., Iversen, V.B. (eds.) NEW2AN 2006. LNCS, vol. 4003, pp. 396–407. Springer, Heidelberg (2006). https://doi.org/10.1007/11759355_36
6. Mader, A., Wagner, B., Hossfield, T., Staehle, D., Barth, H.: Measurements in a laboratory UMTS network with time varying loads and different admission control strategies. In: Proc. Fourth Int'l Workshop Internet Performance, Simulation, Monitoring and Measurement (IPS-MoMe) (2006)
7. Joyce, R.M., Graves, B., Griparis, T., Osborne, I.J., Lee, T.M.: Case study: the capacity of a WCDMA Network—Orange, UK. In: Proc. Fifth IEE Int'l Conf. 3G Mobile Comm. Technologies (3G) (2004)
8. Dahlman, E., et al.: 3G evolution: HSPA and LTE for mobile broadband, 2nd ed. Academic Press (2008)
9. 3GPP TS36.300: Evolved Universal Terrestrial Radio Access (E-UTRA) and Evolved Universal Terrestrial Radio Access Network (E-UTRAN): Overall Description.
10. Astely, D., Dahlman, E., Furuskär, A., Jading, Y., Lindström, M., Parkvall, S.: LTE: the evolution of mobile broadband. IEEE Commun. Mag. **47**(4), 44–51 (2009). https://doi.org/10.1109/MCOM.2009.4907406

11. Kotagi, V.J., Thakur, R., Mishra, S., Murthy, C.S.R.: Breathe to save energy: assigning downlink transmit power and resource blocks to LTE enabled IoT networks. IEEE Commun. Lett. **20**(8), 1607–1610 (2016). https://doi.org/10.1109/LCOMM.2016.2570224
12. Oppolzer, J., Bestak, R.: Physical cell identifier assignment in LTE-advanced networks. In: 2012 5th Joint IFIP Wireless and Mobile Networking Conference (WMNC), pp. 70–74. Bratislava (2012). https://doi.org/10.1109/WMNC.2012.6416150 [AR-2.] http://downloads.hindawi.com/journals/tswj/2014/685102.pdf
13. Wannstrom, J.: LTE-Advanced. An Article Extracted from https://www.3gpp.org/technologies/keywords-acronyms/97-lte-advanced. Accessed: 14 Dec 2020

Bioinspired Approach to Microstrip Transmission Line Design

Danilchenko Vladislav Ivanovich(✉), Danilchenko Yevgenia Vladimirovna, and Kureichik Viktor Mikhailovich

Federal State-Owned Autonomy Educational Establishment of Higher Vocational Education, "Southern Federal University", Taganrog, Russia
vdanilchenko@sfedu.ru

Abstract. A bioinspirated approach to the design of a microstrip line based on a genetic algorithm (GA) is described, which allows creating an algorithmic environment in the field of genetic search for solving NP complete problems. The purpose of this work is to find ways to synthesize a microstrip line based on the bioinspirated theory. The scientific novelty lies in the development of a modified genetic algorithm for bioinspirated computer-aided design of a microstrip line. The statement of the problem in this work is as follows: to optimize the design of a microstrip line by using a modified GA. A fundamental difference from the known approaches in the use of new modified modified structures in bioinspirated computer-aided design, in addition, a new method for calculating a microstrip line based on a modified GA is presented. Thus, the problem of software of algorithms and software for automated design of a microstrip line is currently of particular relevance. Its solution will improve the quality characteristics of the designed devices, reduce the design time and costs, and reduce the requirements for the developer's qualifications.

Keywords: Genetic algorithms · Graphs and hypergraphs · Evolutionary computations · Automation of microstrip line synthesis · CAD · Schematic diagram · Topology

1 Automated Synthesis of a Microstrip Transmission Line Using (GA)

1.1 Genetic Algorithm

To solve many optimization problems, you can use some classes of algorithms called "greedy". Such an algorithm makes a locally optimal choice at each step, in the hope that the final solution will also be optimal. This is not always the case, but for many problems such heuristic algorithms give an optimal result. It is said that the principle of "greedy" choice is applicable to an optimization problem if a sequence of locally

R. Silhavy (Ed.): CSOC 2021, LNNS 230, pp. 170–179, 2021.
https://doi.org/10.1007/978-3-030-77442-4_15

optimal "greedy" choices gives an optimal solution. The "greedy" operator is a language construct that allows you to create new solutions based on partial selection at each step of transformation of the locally optimal value of the objective function.

The crossing-over operator based on the "greedy" principle can be implemented on two or more chromosomes, and, in the limit, on the entire population. The proposed algorithm for the "greedy" crossing over operator on the example of automation of the synthesis of a microwave module, taking into account one criterion, for example, the distance between the strips:

- For all chromosomes of the population, the target function is calculated (the cost of the distance between the stripes). A given number of parental chromosomes is selected and the point of the "greedy" crossing-over operator is determined at random on one of the chromosomes;
- In the selected chromosome for the i-th gene located to the left of the "greedy" crossing-over operator, the value of the partial objective function is determined. For example, this is the cost of the path from the selected gene to the gene that is closest to the right. Similar actions are performed to determine the cost of the path from the crossing over operator;
- In the chromosome "descendant" select the gene, which value of the objective function is higher (lower) when maximizing (minimizing) the values of the objective function;
- The process continues in the same way, until the construction of the "descendant" chromosome. If in the process of implementation a cycle or deadlock occurs, then the unconsidered genes with the best value of the objective function are selected.

The greedy crossover operator strategy can be performed in a variety of ways. It should be noted that the search for universal and specialized operators focused on solving a given class of problems continues [1].

There are various options for choosing a pair of chromosomes with the worst value of the objective function, which should decrease during the evolution of a generation.

The probability of crossing the best chromosomes should increase in the last stages of the operator to fix the desired traits in the chromosomes.

1.2 Microstrip Line Development

Since microwave semiconductor radio transmitters are manufactured, as a rule, using integrated technology, we will focus on calculating the parameters of only an asymmetric microstrip line (see Fig. 1, a) as the main type of feeder line for microwave integrated circuits.

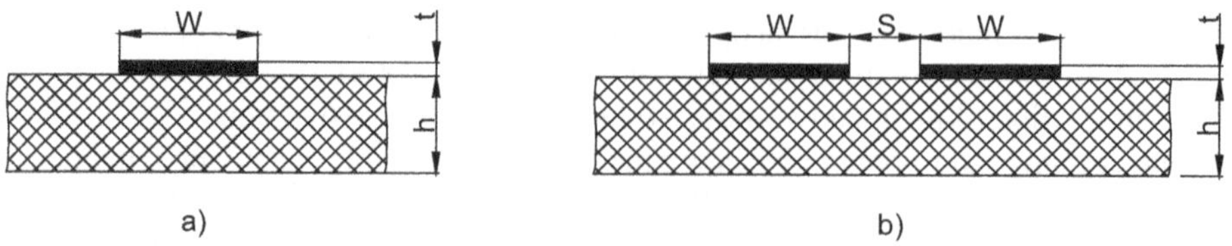

Fig. 1. Asymmetrical microstrip line.

To calculate the wave impedance ρ of such a line, you can use a subroutine - the VOL function (program 1), the effective value of its permittivity $\varepsilon_{\text{э}\phi}$ and tgθ, where $\theta = 2\pi l \cdot \sqrt{\varepsilon_{\text{э}\phi}/\lambda}$ is the phase angle of the line - the RAN subroutine function (program 2). In these subroutines-functions, the following designations are accepted: VOL - line impedance, in ohms; E is the dielectric constant of the substrate; H is the thickness of the substrate, in millimeters; WO is the width of the microstrip line, in millimeters; YO is the length of the line segment, in millimeters; FO - frequency, in gigahertz; RAN is the $tg\theta$ value.

```
Program 1:
     FUNCTION VOL (E, H, WO)
     VOL=376.8/((SQRT€)*(WO/H)*(1+1.735*
     *((E*0.0724)*((WO/H)*0.836))))
     RETURN
     END
Program 2:
     FUNCTION RAN (F0, E, H, Y0, W0)
     P=3.14159265
     ER=(E+1)/2+(E-1)/2/SQRT(1+10*H/W0)
     RAN=TAN(P*F0*Y0*SQRT(ER)/150)
     RETURN
     END
```

Consider three elementary types of links made up of microstrip lines (Fig. 2):

- In the form of two segments, one of which is short-circuited;
- In the form of two segments, one of which is open;
- As a line segment with a container [4].

The action of the previous circuit and the inhomogeneity at the junction of the line segments will be taken into account using the complex resistance ZP = RP + jXP. The input impedance of the entire link is denoted by ZB = RB + jXB. The active RB and reactive XB components can be calculated using the FAD subroutine (program 3), the designations in which coincide with those adopted in VOL and RAN, and their meaning is clear from considering Fig. 2. In addition, it should be taken into account that KB is an integer that determines the version of the circuit, and at KB = 0 W2 is the capacity.

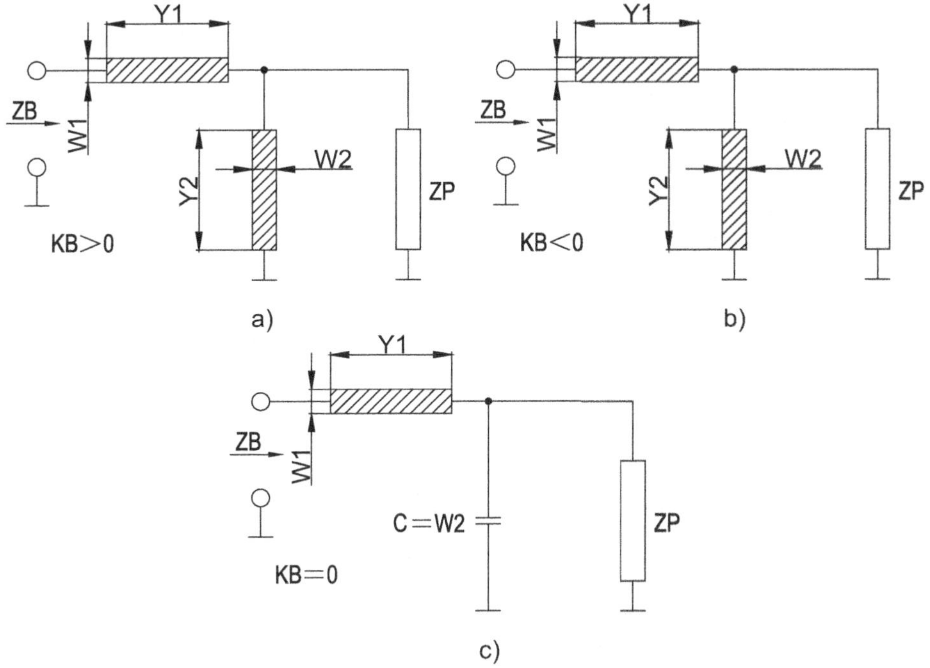

Fig. 2. Calculation of the phase angle of the microstrip line.

For a complex chain composed of the same type of links shown in Fig. 2, the FAD subroutine can be included in the chain of recurrence relations, with the help of which the parameters of the entire connection are calculated. This example will be discussed later [5].

```
Program 3:
      SUBROUTINE FAD (F0, RP, XP, RB, XB, Y1, Y2, W1, W2,
E, H, KB)
      COMPLEX ZP, Z2, ZH, ZB, TK1
      ZP=CMPLX (RP, XP)
      T1=RAN (F0, E, H, Y1, W1)
      V1=VOL (E, H, W1)
      V2=VOL (E, H, W2)
         IF(KB) 1, 2, 3
        T2=-V2/RAN (F0, E, H, Y2, W2)
        GO TO Z2
        T2=-159.24/F0/W2
        GO TO Z2
        T2=V2*RAN (F0, E, H, Y2, W2)
        Z2=CMPLX (0.0, T2)
        ZH=ZP*Z2/(ZP+Z2)
        TK1= CMPLX (0.0, T1)
        ZB=V1*(ZH+V1*TK1)/(V1+ZH*TK1)
        RB=REAL (ZB)
        XB=AIMAG (ZB)
      RETURN
  END
```

When using asymmetric microstrip lines in the upper part of the microwave range, one should take into account first of all two types of in homogeneities: associated with the open end of the microstrip line (Fig. 3, a) and determined by a sharp, stepwise change in the line width (Fig. 3, b) [7].

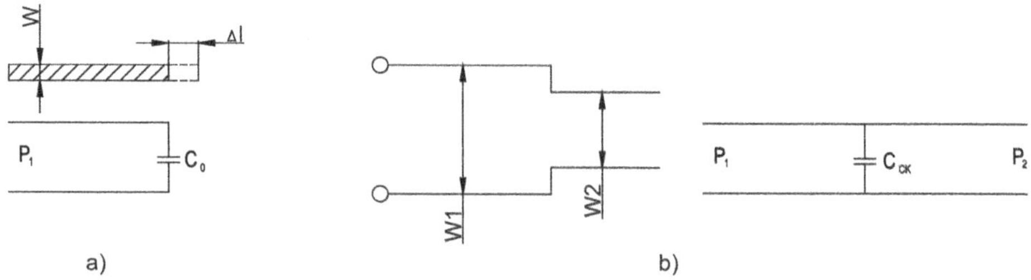

Fig. 3. Asymmetrical microstrip lines.

Both in homogeneities are accounted for by means of equivalent containers. In the first case, the action of the capacitance of the open end of the line C_0 is carried out by shortening the line by Δl [7]. In this case, in programs, you can use the following operator to calculate the value of Δl at ε_r, close to 9.6 and $0, 1 \leq W/h \leq 10$:

$$\mathrm{DY} = \emptyset 43 * \left(1 - \emptyset 64 * \mathrm{EXP}\left(-\emptyset 92 * \frac{W}{H}\right)\right) * H * \mathrm{SQRT}\left(\frac{E}{9,6}\right). \tag{1}$$

The capacity of the "jump" C_{ck} (Fig. 3, b) can be calculated using the gram function subroutine), and the effect of conductivity $Y_{ck} 1/\omega C_{ck}$ on the T-parameters of the connection of two lines - using the TR subroutine (program 4) [8].

```
Program 4:
    FUNCTION CD (W1, W2, E, H)
    B=W1/W2
    IF(B-1) 1, 1, 2
    A=B
    W=W2
    GO TO 3
    A=1/B
    W=W1
    CD=0.04*((EXP(-4.62A))-0.01)*W*E/9.6/SQRT(H)
    RETURN
    END
```

The effective value of the dielectric constant of an asymmetric microstrip line is calculated using a subroutine function (program 4) [9].

```
Program 5:
    FUNCTION DP(E, H, WO)
    DP=(E=1)/2+(E-1)/2/SQRT(1+10*H/WO)
    RETURN
    END
```

The coupled microstrip lines (Fig. 3, b) at $\varepsilon_r = 9.6$ and the characteristic impedance of the lines, all communication areas $\rho_0 = 50$ Ohm according to [5] can be calculated using the operators:

$$S = H * 2.241 * (\exp(-6.76 * CT)); \tag{2}$$

$$W = H * (-2.222 * (T * 2) + \theta CT \tag{3}$$

where, S, W, H - geometric dimensions of the lines, in millimeters (Fig. 5.5, b), CT - coupling coefficient. At 0.1 $\leq CT \geq 0.45$, the accuracy of the formulas is not worse than 3… 5%.

The dispersion properties of a microstrip line at $\varepsilon_r = 9,6$, $h = 1$ mm and $f \leq 12$ GHz can be taken into account using the empirical dependence $\varepsilon_f(f) = \varepsilon_f + 0,042(2 - \exp(-\frac{W}{h}))f$, $\varepsilon_f = ER$ is determined by the program 3 or 4.

The program includes subroutines and subroutines of the function: TL calculation of T - parameters of the line segment; TR calculation of T - parameters of the "jump"; HY calculation of conductivity Y_k; TMT multiplication of two matrices; PTS recalculation of T - parameters into S - parameters; VKF calculation of the modulus and phase of the reflection coefficient; VOL calculation of wave resistance; FAZ calculation of the phase angle of the line segment;

The following identifiers are used in the calculation: FH — initial frequency value, GHz; DF — step of the calculated frequency grid, GHz; N is the number of frequencies; K $\leq$ 19 - number of four-port networks; Y (M) - length of the M-th line segment, in mm; W (M) - when M is odd, the width of the M-th line segment, in mm; with even M and KB = − 1 and 1 - the width of the loop, in mm; at KB = 0 capacitance, in pF; RKM - transmission coefficient of the entire connection in terms of power; VBM and FVB - module and phase of the input reflection coefficient; VHM and FVH - module and phase of the load factor; UKM and FUK - module and phase (in degrees) of the voltage transfer coefficient of the circuit [10].

The active RH and reactive XH components of the load ZH = RH + jZH, depending on the frequency, are given as: $RH = AO + A1F + A2F2 + A3F3$, $XH = BO + B1F + B2F2 + B3F3$, where F is the frequency, GHz.

The program first calculates T - the parameters of the entire connection, and then determines the S - parameters, amplitude and phase-frequency characteristics. An example of the calculation will be given in Table 1.

This paper presents the results of experiments to determine the quality of the solution, the algorithm for generating the starting population, and genetic operators. Based on the research results, the optimal control parameters for a complex of algorithms for the synthesis of passive and active microwave circuits have been determined.

A hybrid genetic algorithm program was developed for experimental research. The program is written in Borland® C++ for operating systems of the Windows 98, NT, 2000, XP family. Experiments conducted on an IBM® compatible computer with an Intel® Xeon processor E5–2690, 20 MB cache, 2.90 GHz Turbo Boost 3.80 GHz (for deploying large populations in GA), 8.00 GT/s Intel ® QPI Number of cores - 8, Number of threads - 16, server RAM 32 GB. The program is registered in the register of computer programs by a certificate [13].

Table 1. An example of calculating a microstrip line.

	Y	W	V		FH = 0.2	DF = 0.2	E = 10	H = 2	N = 11
1	12.5	2.7	41.1		A0 = 15	A1 = −15	A2 = 0	A3 = 0	K = 11
2	12.2	2.0	45.2		B0 = 0	B1 = 15	B2 = 0	B3 = 0	
3	12.5	0.5	83.9						
4	14.3	2.0	48.2						
5	14.4	0.9	68.5						
6	19.4	2.0	48.2						
7	9.2	3.0	38.8						
8	19.1	2.0	48.2						
9	7.9	2.9	39.5						
10	7.4	2.0	48.2						
11	11.6	2.9	39.5						
	F0	S11		S12		S21		S22	
1	0.20	−0.31	−0.03	0.38	−0.86	0.38	−0.86	−0.23	−0.21
2	0.22	−0.33	0.01	0.28	−0.89	0.28	−0.89	−0.26	−0.20
3	0.24	−0.34	0.06	0.18	−0.91	0.18	−0.91	−0.29	−0.19
4	0.26	−0.34	0.12	0.08	−0.92	0.08	−0.92	−0.31	−0.18
5	0.28	−0.33	0.18	−0.1	−0.92	−0.01	−0.92	−0.33	−0.17
6	0.30	−0.30	0.24	−0.11	−0.91	−0.11	−0.91	−0.35	−0.15
7	0.32	−0.26	0.29	−0.21	−0.89	−0.21	−0.89	−0.37	−0.14
8	0.34	−0.21	0.34	−0.30	−0.85	−0.30	−0.85	−0.38	−0.12
9	0.36	−0.15	0.38	−0.40	−0.81	−0.40	−0.81	−0.39	−0.11
10	0.38	−0.08	0.41	−0.48	−0.76	−0.48	−0.76	−0.40	−0.09
11	0.40	−0.01	0.42	−0.56	−0.70	−0.56	−0.70	−0.41	−0.08
	F	PKM	VBM	FVB	VHM	FVH	UKM	FUK	
1	0.20	0.78	0.46	75.69	0.61	172.71	0.37	−69.14	
2	0.22	0.79	0.44	62.01	0.62	172.01	0.35	−70.28	
3	0.24	0.81	0.43	47.37	0.63	171.31	0.33	−71.15	
4	0.26	0.83	0.41	31.69	0.63	170.62	0.33	−71.84	
5	0.28	0.84	0.39	14.95	0.64	169.93	0.33	−72.43	
6	0.30	0.85	0.38	−2.74	0.65	169.24	0.33	−73.00	
7	0.32	0.85	0.37	−21.25	0.66	168.56	0.34	−73.64	
8	0.34	0.86	0.37	−40.29	0.67	167.88	0.35	−74.47	
9	0.36	0.86	0.37	−59.56	0.68	167.20	0.37	−75.64	
10	0.38	0.85	0.38	−78.80	0.68	166.53	0.41	−77.39	
11	0.40	0.84	0.39	−97.83	0.69	165.86	0.45	−80.09	

2 Conclusion

The article presents a bioinspirated approach to the design of microwave circuits based on GA, which makes it possible to simultaneously obtain a practically realizable circuit design, taking into account the peculiarities of the manufacturing technology and the initial version of the topology. The software calculations for the implementation of the method are described. The presented example of calculating a microwave amplifier confirms its effectiveness.

Acknowledgments. The reported study was funded by RFBR, project number 20-37-90151. The research is supported by Foundation for Assistance to Small Innovative Enterprises (FASIE) number 437GUTSES8-D3/62058 dated October 05, 2020.

References

1. Danilchenko, V.I., Kureichik, V.M.: Genetic algorithm for planning the placement of VLSI//Izvestiya SFU **2**, 75–79 (2019)
2. Lebedev, B.K., Lebedev, V.B.: Planning based on swarm intelligence and genetic evolution//Izvestia SFedU. Technical science. **4**(93), 25–33 (2009)
3. Zaporozhets, D.Y., Kravchenko, Y.A., Lezhebokov, A.A.: Methods of data mining in complex systems//Bulletin of the Kabardino-Balkarian Scientific Center of the Russian Academy of Sciences. 3, 52–54 (2013)
4. Kokolov, A.A., Dobush, I.M., Sheerman, F.I., Babak, L.I., et al. Complex functional blocks of broadband radio frequency amplifiers for single-chip L- and S-band receivers based on SiGe technology. In: 3rd Intern. Scientific. Conf. "Electronic Components and Electronic Modules" (International Forum "Microelectronics-2017"), Alushta, October 2017, pp. 395–401. Technosphere (2017)
5. Babak, L.I., Kokolov, A.A., Kalentyev, A.A.: A New Genetic-Algorithm-Based Technique for Low Noise Amplifier Synthesis, pp. 520–523. European Microwave Week 2012, Amsterdam The Netherlands (2012)
6. Kalentyev, A.A., Garays, D.V., Babak, L.I.: Genetic-Algorithm-Based Synthesis of Low-Noise Amplifi ers with Automatic Selection of Active Elements and DC Biases, pp. 520–523. European Microwave Week 2014, Rome Italy (2014)
7. Zhabin, D.A., Garays, D.V., Kalentyev, A.A., Dobush, I.M., Babak, L.I.: Automated synthesis of low noise amplifiers using S-parameter sets of passive elements. In: Asia-Pacific Microwave Conference (APMC 2017), Kuala Lumpur, Malaysia (2017) (accepted for publication).
8. Kurokawa, K.: Power waves and the scattering matrix. IEEE Trans. Microw. Theory Tech. **13**(2), 194–202 (1965)
9. Bockelman, D.E., Eisenstadt, W.R.: Combined differential and common-mode scattering parameters: theory and simulation. IEEE Trans. Microw. Theory Tech. **43**(7), 1530–1539 (1995)
10. Kokolov, A.A., Dobush, I.M., Sheerman, F.I., Babak, L.I., et al.: Complex functional blocks of broadband radio frequency amplifiers for single-chip L- and S-band receivers based on SiGe technology. In: 3rd Intern. Scientific. Conf. "Electronic Components and Electronic Modules" (International Forum "Microelectronics-2017"), Alushta, October 2017, pp. 395–401. Technosphere (2017)

11. Wenyuan, L., Qian, Z.: A 0.7–1.9GHz broadband pseudo-differential power amplifier using 0.13-um SiGe HBT technology. In: 2012 Int. Conf. on Microwave and Millimeter Wave Technology (ICMMT), pp. 1–4, July (2012)
12. Kokolov, A.A., Salnikov, A.S., Sheyerman, F.I., Babak, L.I.: Broadband double-balanced SiGe BiCMOS mixer with integrated asymmetric MBaluns. In: Int. Conf. "Dynamics of Systems, Mechanisms and Machines" (Dynamics-2017), Omsk, Russia, 2017 (accepted for publication).
13. Goryainov, A.E.: Construction of parametric models of passive components of microwave monolithic integrated circuits using the extraction-P program/A.E. Goryainov, I.M. Dobush, L.I. Babak//Nast. Sat. pp. 94–99.

Analysis of Transformation Tools Applicable on NewSQL Databases

Sarah Myriam Lydia Hahn(✉), Ionela Chereja, and Oliviu Matei

Computers and Information Technology, Technical University Cluj Napoca, Cluj-Napoca, Romania
sarah-hahn@gmx.net, oliviu.matei@holisun.com

Abstract. With the introduction of NoSQL databases the new opportunities in the field of cloud technology like scalability and performance were utilized. In fact not NoSQL databases could not satisfy all requirements towards some applications like consistency and ACID transactions. For this reason NewSQL databases were developed - a mixture of relational databases and NoSQL databases with the aim to get the advantages of both. Besides, presenting the most known NewSQL databases and the requirements of transformation tools, the most popular transformation tools in the field of NewSQL databases were introduced. An analysis shows how the comply with the requirements.

Keywords: NewSQL · Transformation · ETL · ELT · Cloud technology · ACID model · CAP theorem · Scalability · Consistency

1 Introduction

Not only caused by trend of big data but also caused by new possibilities in the range of cloud computing like resource pooling and rapid elasticity [39] variable new database management systems were developed. So called NoSQL databases ("not only SQL") try to eliminate the disadvantages of RDBMS by leveraging cloud technology. They go along with fast I/O, supporting big data, scalability and low cost [29]. Furthermore they are schema free to handle a variety of data. In the beginning they had not a structured query language. That is one of the reasons NewSQL database came up. These databases want to pair the advantages from both systems together - relational database management systems (RDBMS) and NoSQL [14]. Just as the mode of operation of the databases changed also the way of data load processing changed. Typical for RDBMS were Extract-Transform-Load (ETL) processes with the following steps [38]:

- Extract: First the necessary data is extracted from different data sources into a working area. The data itself is not adapted in any way.
- Transform: In the working area the data is transformed to make it usable in the target database.
- Load: In the last step the transformed data is loaded into the target database.

R. Silhavy (Ed.): CSOC 2021, LNNS 230, pp. 180–195, 2021.
https://doi.org/10.1007/978-3-030-77442-4_16

With NoSQL and NewSQL databases ETL processes changed to Extract-Load-Transform (ELT) processes. Data is first loaded as a whole into the DWH without any transformations. Whenever a transformation is needed the loading stage has not to be changed. Only new transformation has to be set up [12]. Transformations can be precomputed or calculated in real-time. Caused by new database management systems (DBMS) and processes there are new requirements toward transformation tools. This paper analyses the most popular transformation tools in the field of NewSQL databases. Until now there are several analysis regarding NewSQL databases [19,20,31]. But there is no analysis concerning transformation tools in the field of NewSQL databases. By analysing the most important properties, the advantages and disadvantages of each tool are shown and a comparison in table form can be found. All in all it should support the decision-making to find the best database system and transformation tool for each process. It also represents a market overview on the most popular tools in the context of NewSQL databases, which is existing yet. The structure is the following: the following section "related work and explanations" provides concepts and further definition as the basis for the following. In the third chapter "methodology" the approach of the analysis itself is described. Following the "analysis" of the several transformation tools can be found in chapter four. Last but not least the "conclusion" of the analysis is shown.

2 Related Work and Explanations

The section explains related work and explanations for the following analysis. First the basic of NewSQL databases - cloud computing - is introduced. Afterwards NewSQL databases and the concepts of ACID and CAP are illustrated.

2.1 Cloud Computing

There are many different definitions for the term "cloud computing" [16]. In the following the term is defined as parallel and distributed system based on various, virtualized computers provided by a cloud vendor [9]. The cloud vendors ensure the operation by providing the hardware, a simple access and professional skills. It is based on a pay per use model [16]. Concerning to the National Institute of Standards and Technology cloud computing has the following characteristics [39]:

- On-demand self-service - a customer can access the services without a human interaction with the service vendor
- Broad network access - thick and thin client applications have admission to the services through a standard mechanism
- Resource pooling - the capability of the vendor can be served to multiple consumers
- Rapid elasticity - the provision of the resources is automated, rapid and elastic
- Measured service - resource usage is monitored, controlled, and reported automatically for transparency

These characteristics has several advantages for the user such as a lesser up-front investment, lower operating costs, higher scalability and elasticity, easy access trough the internet, reduced business risks and maintenance expenses [53].

2.2 NewSQL

NewSQL databases were first mentioned in 2011 by Matthew Asslet [1]. These kind of databases bring the advantages from RDBMS and NoSQL databases together. With the new kind of databases online transaction processing (OLTP) workloads should be migrated to NewSQL as an alternative to NoSQL because of the disadvantages. In contrast to NoSQL databases it is scalable, has a high performance and availability. On the other side like RDBMS it guarantees ACID transactions and supports (partial) SQL queries. SQL is also the main language for applications using NewSQL. A non-locking concurrency control mechanism supports several read operations and writes without any locks [2,31]. Depending on the NewSQL database it uses known features from NoSQL databases such as column-oriented data model, distributed nodes or in-memory processing with the objective to process big data in the field of OLTP applications [40]. Roughly, NewSQL databases can be divided into using data sharding layer and distributed storage. A data sharding layer is used when there are multiple independent instances of RDBMS [4,11]. For each database it has to checked in detail how it how it works between the different instances especially concerning ACID transactions and availability. In contrast distributed storage has one single logical RDBMS. There is only one API endpoint which is called. The data itself is distributed on several nodes [33].

2.3 ACID Model

ACID defines the four characteristics atomicity, consistency, isolation, durability. If a database has these properties requests are also called transactions. Atomicity means that each transaction whether it is completely successful or not. If there is a power failure or any other error in between the execution of a transaction the whole transaction will fail and the data will be corrected as it was before the transaction. The second property consistency says that the database always must be in a valid state. Data sets are always consistent concerning constraints, cascades and triggers. Isolation results in that each transaction is viewed separately. Even if the execution was done in parallel it has the same result as if the execution was run separately. To guarantee the isolation tables can be locked if there are concurrent transactions. Durability ensures that once a transaction was successful it will also be like that after a failure, like a power failure [21]. This can be implemented with a transaction log on a hard disk.

2.4 CAP Theorem

The CAP theorem is a theorem for distributed systems and their characteristics. It was published in 2002 and is explained next [17]. It is focused on the three characteristics consistency, availability, and partition tolerance. Consistency means that all nodes in a distributed system look at the same data at the same time. When data is inserted, changed or deleted all nodes must have the the same data up to the last write operation. A read operation has the same result independently from the node which is queried. The definition of consistency is not the same as the consistency in the ACID model. The definition in ACID describes the consistency within a data set. Availability describes the up time of the systems which is 100%. Consequently the system is always accessible. Every request at any time gets a response within a acceptable response time. The response must not be an error. Availability does not define that the response must contain the correct or latest value. The third characteristic is partition tolerance. It is characterized by a high fault tolerance independent from loss of connection between nodes, dropped messages or other events. The theorem indicates that a database cannot satisfy all of these three characteristics. It is always at maximum a combination of two.

3 Methodology

3.1 Databases and Transformation Tools

There are many surveys which compare the different existing NewSQL databases and their properties [19,20,31]. But they do not focus how data can be transformed. In this survey selected transformation tools are evaluated. Due to the fact that several databases come along with several transformation tools the most popular databases are the basis. In the following passages it is described how the most popular databases were evaluated. There is no ranking for NewSQL databases thus the most common NewSQL databases were identified by a literature research - ClustrixDB, CockroachDB, Drizzle, GridGain IMDB, MemSQL (now SingleStore), NuoDB, Spanner, SQLFire, TransLattice Elastic Database, VoltDB [20,22,31,33]. By reason that some of the databases cannot be found in DB-Engine Ranking the databases of Google Scholar and Web of Science were used to identify their popularity. Web of Science is one of the "most trusted publisher-independent global citation database" [10]. Google Scholar provides scholarly literature containing articles, theses, books, abstracts and court opinions [36]. The most search results had the databases CockroachDB, MemSQL/SingleStore, VoltDB, NuoDB and Spanner - even the numbers were not that high. The objective of this article is to have an overview about the most important transformation tools in the context of NewSQL databases. Caused by the fact that there is no ranking yet three different approaches were used to identify them. A research were made for scientific articles about big data or transformation tools. Furthermore Gartners magic quadrant for data integration was used. For each type of technology there is a separate quadrant. The quadrant contains the most popular players divided into four quadrants - leaders, visionaries, niche players and challengers. Leaders have a leading position in the market and also a good position for the future. Visionaries have a vision for the upcoming future but they do not have a good position yet. Niche players are not that popular in the market because of a small target group or missing innovation and focus. Challengers are doing well in the moment but they are not prepared for the future [15]. As a third approach the recommendations of the database developer website will be considered. Generally third party components or internal developments are not mentioned. Coding interfaces like SQL were not mentioned.

Databases. In the following passages the chosen databases are shortly introduced as well as their transformation tools are part of the following analysis.

CockroachDB offers distributed SQL and is strongly oriented at Google's database Spanner. In its basic version it is free of charge. Hosting in AWS or Google cloud is possible [34]. Regarding to the research methodology there was no transformation tool found which has a standard connector for CockroachDB. Therefore the database is not evaluated in the following parts.

NuoDB is a distributed SQL database which can be used in several enviroments such as Google Cloud, OpenShift, AWS, Azure, docker, kubernetes, Linux or VMWare [41]. Pentaho offers a connector to NuoDB [3].

SingleStore is a distributed SQL system which can be hosted at different cloud vendors like AWS, Google Cloud, Microsoft Azure and RedHat OpenShift or in a private cloud. It supports different types of data like relational, JSON, geospatial, key-value and time series [45]. The researched transformation tools are HVR, Informatica and Talend [50].

Spanner is classified as a relational database but in the context of NewSQL databases and provided by Google [24]. Spanner is supported by Informatica and Safe Software [24,50].

VoltDB is a in-memory database written in Java and C++. It aims to bring data storage in form of a database and streaming together. Furthermore it can be used for OLTP and OLAP applications [49]. Within the scope of this research several Analysis and Self-Service Reporting tools were found but not a transformation tool [49]. It is not considered furthermore.

Transformation Tools. In the next sections the chosen transformation tools are shortly presented.

HVR is a data integration software. It runs in a separate distributed environment and uses log-based change data capture (CDC) technology. There are over 1500 deployments and clients in over 30 countries [23].

Informatica provides several data integration products for cloud as well as for on premise solutions. These products include solutions for advanced data transformation, B2B data exchange, ultra messaging and enterprise data catalog [26]. The focus of the following evalution will be set on Informatica Power Center.

Pentaho is a data integration tool by Hitachi Vantara. There are two versions of it - Pentaho Community Project and Pentaho Enterprise Edition. With the free version a codeless pipeline development is possible. All other features like load balancing, scheduling, streaming, spark or the data science tool kit require the enterprise version [37].

Safe Software offers the tool Feature Manipulation Engine (FME). A data integration tool for all industrial sectors like local government, energy, transportation, utilities, education, natural resources, AEC and commercial [25].

Talend offers not only a tool for ETL processes but also for data integrity, governance and application and API integration. In the scope of data integration it offers a pipeline desginer, data inventory, data preparation and stitch [47].

Table 1 gives a shortly overview which transformation tool can be used with which database. If a transformation tool is not mentioned in the focus of a specific database it can also support the database. This can be seen in the overview.

Table 1. Combination possibilities of NewSQL databases and transformation tools

	HVR	Informatica	Pentaho	Safe software	Talend
NuoDB	No	No	Yes	No	No
SingleStore	Yes	Yes	No	Yes	Yes
Spanner	No	Yes	No	Yes	No

3.2 Evaluation Criteria

The presented transformation tools will be reviewed in relation to various criteria. In the subchapters these criteria will be explained.

Consistency. NewSQL databases are geared to RDBMS and offer ACID transactions. For a higher availability and fault tolerance the standard properties can be changed. This results in possible stale data [24,41,45].

Scalability. Scalability of a system is relevant if there are much more or less data in a data flow. Whether a system is scalable or not depends on the underlying database and not on the transformation tool. In the evaluation it is shown what kind of scalability the databases support. Referring to the evaluation the following kind of scaling are of note:

Partitioning. There are two ways of partitioning - horizontal and vertical partitioning. Horizontal partitioning means that data sets of one entity are stored in different partitions on different nodes. Vertical partitioning describes the storage of one or more columns in one partition. This way of partitioning is very common for column-oriented data stores. Horizontal partitioning splits up into the two most common techniques: range partitioning and consistent hashing. Range partitioning creates the partitions on a partition key. Each partition on each server has a defined range of the partition key. By querying data using the partition key the query is very fast because only a few partitions are needed. On the other side there can be data hot spots and load-balancing issues when loading data of only one partition. Further a routing server is needed. Consistent hashing builds the partition based on the hash of each data set. The hash values are approximative uniformly distributed. The advantage is that the partitions have nearly the same size and there is no need for restructuring of the partition borders. Even if there is a new partition the movement of the data is easy because only the partitions next to the new one has to be split up. Queries instead can be slow when an access on each partition is needed [19].

Replication. A replication of data enlarges not only read and write operation performance but also system reliability, fault tolerance, and durability. In the following it is differed between master–slave and multi-master replication. In the master-slave model always one node represents the master and the other nodes are so called slaves. All writes are committed at the master node. Afterwards they are replicated to the slaves. The master node is consistent. All the nodes can be used for read requests to have a performance increase. By reading from a slave node the data can be stale. In the opposite using multi-master replication means that all nodes are masters and can receive write operations. The replication is happening between all nodes [19]. There is not one leading node which is consistent.

Licence and Costs. Many of the new databases are open source [35]. But there are also proprietary software which goes along with licence costs such as Google Spanner. In the analysis it is not only mentioned what kind of licence and concomitant costs the transformation tool has but also the underlying database. This information supports to calculate the whole resources needed for the development of an use case.

Transformation Types. A data flow contains one ore more data sources with data, one ore more transformation steps and at least a data target for the data output [5]. Concerning to the ELT approach data source and data target is the same database instance. The focus is on the transformation tool thus the data source and target will not be considered. There are many common transformation types. To implement known transformation types in the environment of NewSQL databases the transformation tool has to provide at least these common functionalities:

- Additional columns/Expression [6,7,52]
- Aggregation [6,13,46,52]
- Conversion [6,13,46]
- Data Cleansing [7,46]
- Filter [6,13,46,52]
- Join/Lookup [6,13,46,52]
- Merge/Union [6,13,46]
- Sequence/Surrogate Key [13,52]
- Update [52]

It is measured how many of these transformation types are supported by each tool. Also if there are special transformation types they will be highlighted.

Standard Components. The transformation types listed before can often be implemented in a user-defined code like a query language or with user interface. With such an interface data flows can be build. In a transform editor the result can be seen in real-time [5]. There are several benefits of using low-developing platforms such as no technological knowledge is needed and the development is faster. Both decreases the price of a new development. Also complexity is reduced and the maintenance is more easy [42,43]. It is measured how many of the provided transformation types mentioned before can be implemented via the interface of the transformation tool as a standard component. The advantages of standard components in the case of transformation tools are that the development of new transformations and the data preparation itself is much easier and can be made without a deep technological and development knowledge. The time needed for new development decreases as well as the implementation costs. Furthermore the maintenance of transformation processes is not only easier but also can be made by users without a technical or development background. This is especially important in case of limited (developer) resources.

Usability, Popularity and Stability. ISO 9241 was published in 1999 containing seventeen parts of requirements for office work with visual display terminals. The requirements range from menu dialog, colours to usability [28]. In part eleven the term of usability is defined. It is the extent to which a software can be used by specified users to fulfill given tasks with effectiveness, efficiency and satisfaction in a specified context of use [27]. Depending on other use cases he conclusion can be made that there is a correlation between usability and popularity [44]. To measure the popularity of each tool on the basis of DB-Engine Ranking the following metrics are used [18]:

- Number of search results in Google and Bing
- Frequency of searches by means of Google Analytics
- Amount of questions in Stack Overflow
- Number of job advertisements on Indeed and Simply Hired
- Sum of LinkedIn accounts with reference in the profile
- Total amount of Tweets

- Number of search results in Web of Science and Google scholar

The popularity is an indication by choosing a transformation tool how common the usage of the tool and how big the community is. Depending on the project this can be very relevant. It is not only important how popular a tool is but also how long it is on the market. To appreciate the popularity better the year of the launch is mentioned. This is also an indicator how stable a software is. The stability of a software which is available over years can be ranked higher than a new one. The stability of a new software is vague.

Performance. Performance concerning time, hardware resources, capacity and modes is also an important factor for data integration processes. These factors can be measured as [48]:

- Time efficiency: grade of low response and processing times
- Resource utilization: resources utilized by the process differentiate between the type of resource and the quantity
- Capacity: demand that can be executed without missing the other requirements
- Modes: modes of data processing (batch, real-time etc.)

Within the scope of this article the performance of each tool is not measured. If available there will be a reference to other studies or benchmarks.

Competitive Advantages. If there are any competitive advantages they will be also shown in the evaluation. These can also be features which are not in the context of transformation and the underlying database itself. Depending on the use case these features can be helpful for the implementation. It has to be decided if it is useful in a project and if the additional value is high enough to choose over another transformation tool.

4 Analysis

This sections gives an overview about the databases in the context of the presented, important criteria for transformation tools followed by the criteria which are specific for transformation tools. A summary of the criteria on database level can be found in Table 2. Also the consistency is specified there. It is not analysed in detail because all of the NewSQL databases offer ACID transactions. The properties of the transformation tools are shortly summarized in Table 3.

Scalability. Singlestore creates replicas if it is requested at table creation or the cluster is in high availability or redundancy-2 mode. If a replication is necessary master-slave replication is used. By default synchronous mode is chosen. This means that all replicas of a table are locked until a write transaction is finished. This is important to guarantee consistency. Also an asynchronous mode can be chosen. In this mode the replicas are not locked [45]. Google Spanner does not have a partition model. But it divides data depending on the primary key in different separated ranges. The ranges are set automatically for balancing. Google Spanner's replication also has a master (Google calls it a leader) node and several slave nodes. The implementation is based on the Paxos protocol. It is another way of defining the agreed value [24]. A different way of partitioning is used by NuoDB. The storage is maintained by storage managers. Each storage manager has different access storage groups. The connection between the storage manager and a storage group is a many to many relationship. A storage group is a physical storage. Such a storage group contains a number of partitions. A partition is always is stored in one storage group. Replication is made by NuoDB automatically in an asynchronous multi-master replication [41].

Table 2. Database properties

	Consistency	Scalability	Licence and costs
NuoDB	ACID	Storage manager, Multi-master	Open source (Proprietary)
SingleStore	ACID	Master-slave replication	Proprietary
Spanner	ACID	Automated range partitioning, Paxos protocol replication	Proprietary

Licence and Costs. Google is a cloud vendor and offers based on cloud technology its database Spanner. There are not fixed costs. Instead they depend on the usage of the database. The measurement for the cost units are reading and writing operations, storage, back-up, replication, in-memory cache, streams and data transfer. This price model has a correlation with the usage [24]. In this analysis the use cases are not further specified. Hence a fixed price cannot be assumed. A similar price model has Singlestore. The price depends on the cloud, region and resources chosen [45]. NuoDB's community version is free but limited and only usable for low storage. The enterprise version goees along with a licence to pay [41]. HVR is a proprietary software but the pricing is not revealed. In advance an three hour trial can be done [23]. Also Informatica does not have a public price listing. It is a proprietary software [26]. Pentaho also has two versions - the open source Pentaho Enterprise Edition and the proprietary Pentaho Community Project [37]. Talend has a basic open source software called Talend Open Source. For the extended version there is a licence per user e.g. for using cloud integration [47]. A more special pricing concept has Safe Software. There are free licenses for students, graduates, research, non-profit organizations, teaching and trials. The license fee has not been raised since 2002. The license depends on the environment like on premise or cloud and on the kind of organization [25].

Transformation Types. In HVR there are so called actions for data transformation. In the restrict action data can be filtered, joined and duplicates can be removed. Within the action of ColumnProperties that data type can be changed and a surrogate key can be used. The integrate action can also merge data in the target. A derived column can be created via code in the ColumnProperties action. An aggregation can not be made out of the box [23]. With Power Center and Data Quality - part of Informatica's product range - all the relevant transformation types are supported [26]. In Pentaho the transformation types are called Data Integration Steps. They provide all necessary transformations [37]. Also the Safe Software's operators support all defined transformation types [25]. In Talend there are many components which can be used in the transformation process. Also code can be used e.g. for querying the database. The derivation of columns or expression should be made in code. Finally all transformations can be implemented [47].

Standard Components. Actions in HVR are based on labels and text boxes. There are predefined parameters in the restrict action for filtering, lookup and removing duplicates. They are called "CaptureCondition", "HorizLookupTable" and "SelectDistinct". With the parameters "Datatype" and "SurrogateKey" in the action "ColumnProperties" data conversion can be made and a surrogate key deposited. The property "Resilient" in action "Integrate" allows to make a merge in the target database. It can also be used for updates [23]. In Informatica Power Center there are several transformation types which can be added to the graphical data flow like "aggregator", "filter", "sorter", "expression", "joiner", "lookup", "union", "sequence generator" and "update". With the "sorter" transformation also duplicates can be removed [26]. In Pentaho there are different Data Integration Steps which can be used for the implementation of the different transformation types such as "group by" for aggregations, "filter rows", "unique rows", "select values", "concat fields", "merge join", "append streams", "add sequence" and "combination lookup/update" [37]. In Safe Software's FME the transformation types are called transformers. The most important ones are "Aggregator", "AttributeFilter", "DuplicateFilter", "BaseConverter", "AttributeCreator", "StringConcatenator", "DatabaseJoiner", "FeatureMerger", "Counter" and "DatabaseUpdater" [25]. Talend has several components in its Talend Open Studio and expanded modules for transformations. The most necessary are "tAggregateRow", "tFilterRow", "tUniqRow", "tConvertType", "tMap" and "tUnite". There are database specific components for update and merge in the target [47]. In many parts there is not only one standard component available for a certain transformation type. However only the most popular one is shown in this article.

Usability, Popularity and Stability. Analysis of the transformation tools Pentaho and Talend shows that they are easy to use [8,32]. In total there is no comparison to all of the focused transformation tools. To classify the popularity of a tool the measures were related to each other. The greatest popularity has

Informatica. Almost 40 % less is the popularity of Talend followed by Pentaho. The tools with the lowest popularity are HVR and Safe Software. After first explorations HVR was deployed in 2012 [23]. Informatica, as a software company, is on the market since 1993 [26]. Pentaho was developed in 2004. In 2015 the software was bought by Hitachi Vantara [37]. Safe software began their company history in 1993 helping forestry companies [25]. Talend was founded in 2005. Since then they enlarged their software solution [47].

Performance. HVR does not support complex ETL processes very well [51]. Talend improves their performance by supporting parallel processing [30,47]. Due to the fact that there is no study comparing the performance of all tools, the tools cannot be ranked by performance. Also not for every transformation tool information was found.

Competitive Advantages. HVR supports many different topologies such as uni-directional, bi-directional, multi-directional, broadcast, integration and cascading [23]. Informatica provides several products which supports the data integration process. These support by ultra messaging, build a data hub or the data exchange between companies [26]. With the Lumada data services Hitachi Vantara offers a wide product range for Analytics, Governance and Ops Agility containing multi cloud solutions such as data catolog, data optimizer for Hadoop and a data lake [37]. Talend has a strong focus on data integration processes. This results in over 900 free components and connectors which can be used [47].

Table 3. Transformation tool properties

	Licence and costs	Transformation types	Standard components	Popularity	Founding year	Competitive advantages
HVR	Proprietary	8/9	7/9	Not popular	2012	Supporting many topologies
Informatica	Proprietary	9/9	8/9	Popular	1993	Continuative data integration products
Pentaho	Open source/ Proprietary	9/9	9/9	Less popular	2004	Lumada data services
Safe Software	Proprietary	9/9	9/9	Not popular	1993	
Talend	Open source/ Proprietary	9/9	8/9	Popular	2005	Over 900 components & Connectors

5 Conclusion and Future Research

Because NoSQL databases cannot satisfy all requirements an application has towards a database NewSQL databases are getting more popular. This analysis gives an overview about the most popular transformation tools on the market for NewSQL databases including their advantages and disadvantages. In fact there are not that many tools that support NewSQL databases. There are also popular NewSQL databases which are not supported by any tool. This shows how restricted the selection is. Instead code or a user specific connection has to be used. The tools which have connectors for NewSQL databases are not new on the market. Instead they are existing ETL tools which can be now also user for transformations in NewSQL datbases. In this study fundamental concepts regarding the change of NewSQL databases and their properties were presented. Using a self-developed selection method the most popular NewSQL databases and transformation tools were chosen. The most important properties of a transformation tool were introduced and discussed for each database and tool. In overview tables the advantages and disadvantages can be seen quickly and clearly. The analysis is not only a market overview but a decision guidance for the implementation of specific use cases. It is shown that a choice of a transformation tool always includes the choice of a database system. The research is an overview about different transformation tools in the context of NewSQL databases. There are more NewSQL databases on the market which are not mentioned here because of their less popularity. There is not only a change of the popularity over time also depending on the use case other databases could fit better. Continuing specific use cases can be implemented with the several tools and be measured and analyzed. Within the scope of the study not every property could be analyzed in detail. There are further studies necessary. The performance of standard components have to be analyzed in relation to the performance of custom code. Furthermore the usability of each tool has to be measured with a representative group of people. Besides that it is partly possible to connect a database with a transformation tool without a standard connector.

Acknowledgments. This research was made possible by funding from the ICT-AGRI-FOOD 2020 Joint Call. This work was supported by a grant of the Romanian National Authority for Scientific Research and Innovation, CCCDI - UEFISCDI, project number COFUND-ICT-AGRI-FOOD-GOHYDRO 200/2020, within PNCDI III.

References

1. Anslett, M.: Nosql, newsql and beyond: the drivers and use cases for database alternatives. In: The Drivers and Use Cases for Database Alternatives, NoSQL, NewSQL and Beyond (2011)
2. Aslett, M.: How will the database incumbents respond to NoSQL and NewSQL (2011)
3. Atriwal, L., Nagar, P., Tayal, S., Gupta, V.: Business intelligence tools for big data. J. Basic Appl. Eng. Res. **3**(6), 505–509 (2016)

4. Authors, V.: The vitess documentation. https://vitess.io/docs/
5. Belyy, A., et al.: Dataset previews for ETL transforms, 28 May 2013. US Patent 8,452,723
6. Bergamaschi, S., Guerra, F., Orsini, M., Sartori, C., Vincini, M.: A semantic approach to ETL technologies. Data Knowl. Eng. **70**(8), 717–731 (2011)
7. Bhide, M.A., Bonagiri, K.K., Mittapalli, S.K.: Column based data transfer in extract transform and load (ETL) systems, 20 August 2013. US Patent 8,515,898
8. Biplob, M.B., Sheraji, G.A., Khan, S.I.: Comparison of different extraction transformation and loading tools for data warehousing. In: 2018 International Conference on Innovations in Science, Engineering and Technology (ICISET), pp. 262–267. IEEE (2018)
9. Buyya, R., Yeo, C.S., Venugopal, S., Broberg, J., Brandic, I.: Cloud computing and emerging it platforms: vision, hype, and reality for delivering computing as the 5th utility. Future Gen. Comput. Syst. **25**(6), 599–616 (2009)
10. Clarivate: Web of science. https://clarivate.com/webofsciencegroup/solutions/web-of-science/
11. Citus Data: Citus documentation. https://docs.citusdata.com/
12. Davenport, R.J.: ETL vs ELT: a subjective view. In: Insource Commercial Aspects of BI Whitepaper (2008)
13. El Akkaoui, Z., Zimanyi, E., Mazón, J.N., Trujillo, J.: A model-driven framework for ETL process development. In: DOLAP 2011: Proceedings of the ACM 14th International Workshop on Data Warehousing and OLAP, pp. 45–52 (2011)
14. Fatima, H., Wasnik, K.: Comparison of SQL, NoSQL and NewSQL databases for Internet of Things. In: 2016 IEEE Bombay Section Symposium (IBSS), pp. 1–6. IEEE (2016)
15. Gartner, I.: Magic quadrant research methodology (2020). https://www.gartner.com/en/research/methodologies/magic-quadrants-research
16. Geelan, J., et al.: Twenty-one experts define cloud computing. Cloud Comput. J. **4**, 1–5 (2009)
17. Gilbert, S., Lynch, N.: Brewer's conjecture and the feasibility of consistent, available, partition-tolerant web services. ACM SIGACT News **33**(2), 51–59 (2002)
18. solidIT consulting & software development GmbH: Db-engines (2020). https://db-engines.com/
19. Grolinger, K., Higashino, W.A., Tiwari, A., Capretz, M.A.: Data management in cloud environments: NoSQL and NewSQL data stores. J. Cloud Comput. Adv. Syst. Appl. **2**(1), 22 (2013)
20. Gurevich, Y.: Comparative survey of NoSQL/NewSQL DB systems. The Open University of Israel, Department of Mathematics and Computer Science (2015)
21. Haerder, T., Reuter, A.: Principles of transaction-oriented database recovery. ACM Comput. Surv. (CSUR) **15**(4), 287–317 (1983)
22. Hajoui, O., Dehbi, R., Talea, M., Batouta, Z.I.: An advanced comparative study of the most promising NoSQL and NewSQL databases with a multi-criteria analysis method. J. Theoret. Appl. Inf. Technol. **81**(3), 579 (2015)
23. HVR Integration Software: enterprise data integration software — HVR (2020). https://www.hvr-software.com/
24. Google Inc., Cloud spanner — google cloud (2020). https://cloud.google.com/spanner
25. Safe Software Inc.: Safe software — fme — data integration platform (2020). https://www.safe.com/
26. Informatica: Enterprise cloud data management — informatica deutschland (2020). https://www.informatica.com/de/

27. ISO: Ergonomische Anforderungen für Bürotätigkeiten mit Bildschirmgeräten - Teil 11: Anforderungen an die Gebrauchstauglichkeit - Leitsätze. Beuth Verlag, Berlin (1999)
28. ISO: Ergonomische Anforderungen für Bürotätigkeiten mit Bildschirmgeräten - Teil 1: Allgemeine Einführung (ISO 9241–1:1997) (enthält Änderung AMD 1:2001); Deutsche Fassung EN ISO 9241–1:1997 + A1:2001. Beuth Verlag, Berlin (2002)
29. Han, J., Haihong, E., Le, G., Du, J.: Survey on NoSQL database. In: 2011 6th International Conference on Pervasive Computing and Applications, pp. 363–366 (2011)
30. Katragadda, R., Tirumala, S.S., Nandigam, D.: ETL tools for data warehousing: an empirical study of open source talend studio versus microsoft ssis. In: Computing Conference Papers [147] (2015)
31. Kaur, K., Sachdeva, M.: Performance evaluation of NewSQL databases. In: 2017 International Conference on Inventive Systems and Control (ICISC), pp. 1–5. IEEE (2017)
32. Kherdekar, V.A., Metkewar, P.S.: A technical comprehensive survey of ETL tools. In: Advanced Engineering Research and Applications, p. 20 (2016)
33. Kumar, R., Gupta, N., Maharwal, H., Charu, S., Yadav, K.: Critical analysis of database management using NewSQL. Int. J. Comput. Sci. Mob. Comput. **3**, 434–438 (2014)
34. Cockroach Labs: Cockroach labs, the company building cockroachdb (2020). https://cockroachlabs.com/
35. Leavitt, N.: Will NoSQL databases live up to their promise? Computer **43**(2), 12–14 (2010)
36. Google LLC: about google scholar. https://scholar.google.com/intl/de/scholar/about.html
37. Hitachi Vantara LLC: Pentaho enterprise edition — hitachi vantara (2020). https://www.hitachivantara.com/
38. Mehra, K.K., Kumar, P., Choudhury, P., Lakkur, A.L., Samanta, S.: Extract, transform and load (ETL) system and method, 5 April 2017. US Patent 9,633,095
39. Mell, P., Grance, T., et al.: The NIST definition of cloud computing. NIST special publication 800–145 (2011)
40. Moniruzzaman, A.B.M.: NewSQL: towards next-generation scalable RDBMS for online transaction processing (OLTP) for big data management. Int. J. Database Theory Appl. **7** (2014)
41. NuoDB: Nuodb home — nuodb (2020). https://nuodb.com/
42. OutSystems: the state of application development: Is it ready for disruption? Technical report, Department of Computer Science, Michigan State University, OutSystems: Boston, MA, USA, September 2019
43. Richardson, C., Rymer, J.R.: New Development Platforms Emerge for Customer-Facing Applications. Forrester, Cambridge (2014)
44. Scowen, G., Regenbrecht, H.: Increased popularity through compliance with usability guidelines in e-learning web sites. Int. J. Inf. Technol. Web Eng. (IJITWE) **4**(3), 38–57 (2009)
45. SingleStore, I.: Singlestore is the database of now™ powering modern applications and analytical systems (2020). https://www.singlestore.com/
46. Song, X., Yan, X., Yang, L.: Design ETL metamodel based on UML profile. In: 2009 Second International Symposium on Knowledge Acquisition and Modeling, vol. 3, pp. 69–72. IEEE (2009)
47. Talend: Talend - a cloud data integration leader (modern etl) (2020). https://www.talend.com/

48. Theodorou, V., Abelló, A., Lehner, W.: Quality measures for ETL processes. In: International Conference on Data Warehousing and Knowledge Discovery, pp. 9–22. Springer (2014)
49. VoltDB I.: Home - voltdb (2020). https://www.voltdb.com/
50. Zaidi, E., Thoo, E., Heudecker, N., Menon, S., Thanaraj, R.: Gartner magic quadrant for data integration tools. Gartner Group (2020)
51. Zaidi, E., Thoo, E., Heudecker, N., Menon, S., Thanaraj, R.: Magic quadrant for data integration tools **2020** (2020)
52. Zamanian, K., Nesamoney, D.: Apparatus and method for performing data transformations in data warehousing, 15 January 2002. US Patent 6,339,775
53. Zhang, Q., Cheng, L., Boutaba, R.: Cloud computing: state-of-the-art and research challenges. J. Internet Serv. Appl. **1**(1), 7–18 (2010)

Multi-level Trust Modelling to Resist Impact of Routing Attacks on Reliable Data/Communication Transactions in MANET-IoT Ecosystem

G. N. Anil(✉)

Department of Computer Science and Engineering,
BMS Institute of Technology and Management, Bengaluru, India

Abstract. The latest technological advances in the field of futuristic technology haveled to communication conceivable through MANET. This provides an attractive and flexible service to the end-userin terms of faster communication at a large scale and interaction with physical objects in the realworld. However, due to malicious attacks by an adversary in the MANET ecosystem, ensuring reliable communication and data security is quite a challenging task. This paper has presented a multi-level trust model based on the behavioral analysis of sensor nodes. Different trust-values are computed for each node, and the computed trust values are accumulated based on probability distribution to compute the final trust-value of the nodes. In routing operation, the proposed scheme maintains the trust records for each node, and the route is established between trusted intermediate nodes. The proposed scheme offers a secure and reliable routing process against different routing attacks in the IoT-MANET ecosystem. The design and implementation of the proposed methodology are carried on a numerical computing tool. The performance of the proposed system is assessed based on comparative analysis. The simulation outcome and comparative demonstrate the effectiveness of the proposed system in terms of reduced network overhead, higher packet delivery rate, and higher energy efficiency.

Keywords: MANET · Internet of Things · Routing · Trust model

1 Introduction

Recent advances in communication technologies and integration of sensory devices opened up various opportunities for industry transformation and a wide range of services for mobile users [1, 2]. Internet of Things (IoT) provides a unified association between physical objects and the rapid exchange of information [3]. The convergence of MANET in the Internet of Things opens up new ways to provide smarter, safer, and more reliable services, making the system fully automated in a smart environment [4, 5]. However, this also brings many challenges to the network. One of the critical factors in such ecosystem lof IoT-MANET is the security concerns due to resource constraints and the absence of physical security of nodes [6]. Over the past several years,

R. Silhavy (Ed.): CSOC 2021, LNNS 230, pp. 196–205, 2021.
https://doi.org/10.1007/978-3-030-77442-4_17

various research efforts have been made to improve security MANET [7, 8]. Since the network is open and nodes are limited in resources, ensuring a comprehensive security level is a challenging task. Robust and complex cryptographic techniques are not suitable for such resource constraint networks [9]. Therefore, many researchers are still focusing on designing energy-efficient and efficient security solutions to extend network service time and provide a comprehensive security level. This paper presents a novel design of a security solution that supports energy efficiency and provides better security against routing attacks. In this paper, multi-level trust modeling is proposed based on the behavioral analysis of sensor nodes. Different trust-values are computed for each node, and the computed trust values are accumulated based on probability distribution to obtain the nodes' final trust value. During data transmission, secure routing decisions are made via trusted intermediate nodes between the source and the destination node. The rest of this paper isarranged in the following manner: Sect. 2 presents related work. Section 3 describes the proposed system followed by schematic architecture and methodologies adopted. Section 4 discusses the simulation outcome and performance analysis of the proposed multi-level trust scheme. Finally, the summarization of the contribution of the proposed study is presented in Sect. 5.

2 Related Work

In this section, a review of the state-of-the-art is carried out in the context of MANET-IoT. The work carried out by Khanna et al. [10] carried extensive analysis for investigating responsible parameters used in the trusted-based mechanism against routing attacks in the MANET. The authors have highlighted various researches issues and provided a significant recommendation as well. A secure mechanism for data security based on a strong authentication scheme in MANET-IoT is presented in the study of Anil [11]. In this, an efficient ciphering mechanism is introduced for the communication process, and access-permission is designed to ensure authorized nodes' participation. Desai et al. [12] suggested a predictive approach for secure routing against different attacks. The work of Nehra et al. [13] investigated the impact of distributed denial of service (DDoS) attacks in the MANET. The authors have also explored the possible solution based on the analytical discussion to defend against DDoS attacksearly. Kaur et al. [14] investigated the performance of various existing routing schemes, namely DSR, AODV, and ZRP, in the presence of wormhole attacks. A robust security scheme to address security vulnerabilities in IoT is given by Anil [15]. In this, SDN technology is used to resist attacks. Another work carried out by Anil [16] presented an energy-aware routing mechanism based on optimization of the PHY/MAC layer in the IoT ecosystem. The work carried out by Ghosh et al. [17] presented a trust-based clustering mechanism to address different security issues and achieve higher energy-efficiency in MANET. The authors in the study of Merlin and Ravi [18] presented a trust-oriented energy-aware path selection mechanism against blackhole attacks in the MANET.

3 Proposed Model

This section discusses the design and implementation of the methodology adopted for developing the proposed multi-level trust modeling. The proposed study considers the network scenario of the MANET convergence in the IoT-ecosystem. In this, networking scenario the sensed packets via sensory nodes are transmitted through secure routing by the MANET to the Gateway nodes. Therefore, the prime aim of the proposed scheme is to provide a trust-oriented security mechanism to carry out energy-aware and secure data transmission between the trusted node and the IoT gateway. The core methodology adopted for secure data transmission in the proposed model is based on the computation of each sensory node's trust-degree. The trust-degree of nodes is carried out in a multi-level approach that includes computation of primary trust value, secondary trust value, and mutual trust value. A probabilistic distribution function is used to combine primary secondaryand mutual trust parameters to achieve the final trust value of each node in the network. The computation of node trust values is carried by cluster-head. It establishes the secure route between the trusted intermediate nodes between the source and destination (IoT gateway node) during packet transmission. The schematic architecture of the proposed model is depicted in Fig. 1.

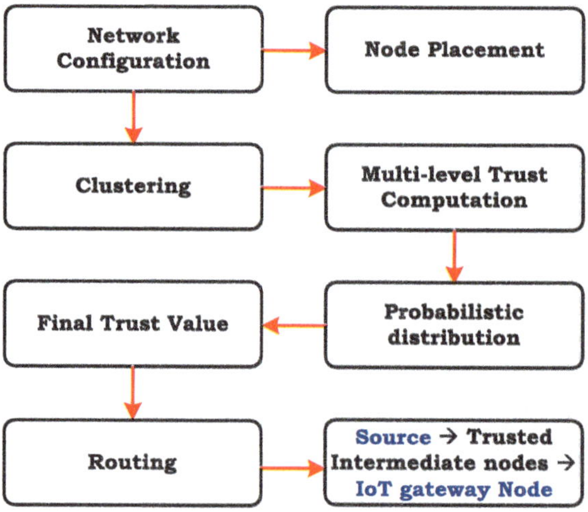

Fig. 1. The schematic architecture of the proposed scheme for data transmission in MANET-IoT

Figure 1 exhibits the core attributes and design of the proposed security model to resist the impact of routing attacks by discarding the participation of malicious nodes in the network communication and data transaction processes. The trust degree of each node is computed by a cluster head (CH) based on each node's behavioral factors under consideration. The computed behavioral parameters nodes are categorized into different categories: true nodes and suspected nodes. The trust computation of the nodes is carried out periodically and updated to other clusters over a certain period of time.

3.1 Network Configuration

This is the first module of the proposed system, which refers to a network set-up process. In this phase of the model, the placement of sensor nodes (S_N) in the simulation region S_R

is carried out. On the other hand, clustering operation is performed after the placement of S_N, which is responsible for executing the process of node trust degree computation. The proposed system considers N numbers of mobile S_N randomly placed in the fixed S_R with their respective positioning coordinates [lx, ly] given as follows:

$$S_N\left(l_{x,y}\right) = l_{x,y} \times f_{rand}\ (S_N)\ for\ \forall\ 1 \leq S_N \leq \mathrm{N} \tag{1}$$

Where, f_{rand} denotes a random generator function. In the network scenarioconsidered in the proposed study, each S_N is assigned with the unique-id provided by the gateway node. The clustering mechanism is considered based on the hop distance. Once the S_N are distributed in the simulation boundary, and powered up, it initiates initial cluster set-up followed by searching for their nearby nodes. If any nearby S_N is identified, it sends cluster formation requests. The nearby S_N upon receiving the cluster formation request, also add their neighbor to join in the clustering operation. After the cluster formation, S_N from each cluster begins intra-communication process to select the cluster head-(CH). In this process most stable S_N is selected as CH In the clustering operation, the S_N can leave or join another cluster due to their mobility factor. The cluster is also not static; it can be divided and merged at any point in time. It is to be noted that each CHis also assigned with their unique id provided gateway node.

3.2 Multi-level Trust Computation

The trust-degree of each S_N is carried out by CH for their particular cluster. The computed trust parameter of S_N further distributed to other clusters in the network by the CHs. The computation of trust degree is performed based on the periodic interval. The logic behind this is that, in different communication cycles, the topology of the network often changes, and single-time computation of trust-value of nodes is not an appropriate choice concerning security vulnerabilities. Therefore, the proposed multi-level trust model is based on the nodes' behavioral analysis in different communication cycles over a certaintime. This process makes network security for a longer period of time. The trust-degree of each $\mathrm{S_N}$ in a particular cluster is computed based on the behavioral parameters of nodes such as packet forwarded rate, misrouted rate, dropped rate, and falsification of the packets. In this process, $\mathrm{S_{N-A}}$ analyzes the behavior of the adjacent sensor nodes ($\mathrm{AS_{N_S}}$) of $\mathrm{S_{N-B}}$, based on which the class of $\mathrm{S_N}$ is determined as a true node (α) or suspected node (β). The following are the responsible behavior factors that decide the α and β category of the nodes.

$S_{N-\alpha} \rightarrow$ data packets are forwarded with honesty.
$S_{N-\beta} \rightarrow$ data packets are misrouted, dropped, or packets injected falsely.

Figure 2 demonstrates the generic represents of sensor node monitoring for behavioral analysis-based trust analysis of the S_N.

Figure 2 demonstrates the generic scenario of the node trust factor computation. Considering the resource constraints of $\mathrm{S_N}$ behavior monitoring is carried for a fixed interval of time T. After expiration of T the trusted parameters are achieved, and the trust-value of a node is computed using a probability distribution. Since the proposed

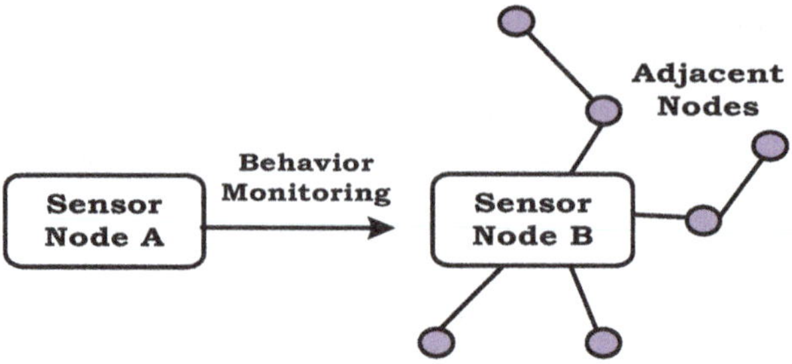

Fig. 2. A generic representation of sensor node behavior analysis for trust computation

study considers multi-level trust modeling, the behavior analysis of nodes is carried for three trust factors described as follows:

3.2.1 Primary Trust Value

In this, the trust is computed for particular S_N. The computation of trust value is given as follows:

$$p(A, B) = \frac{w(A, B)}{w(B)}, \; where \; P(A, B) \in [0, 1] \tag{2}$$

where $p(A, B)$ is the primary trust value between the nearby sensor node A and B $S_{N(A-B)}$. $w(A, B)$ denotes the level of strength between nearby or neighboring $S_{N(A-B)}$ which are at neighboring hop distance.

3.2.2 Secondary Trust Value

In this, trust is computed for S_{N-A} to analyze the influence caused by S_{N-B}. The computation of secondary trust value is given as follows:

$$S(A, B) = \begin{cases} mt^{\frac{d_{max} - d_{A,B} + 1}{d_{max}}} & if \; d_{A,B} \le d_{max} \\ 0 & if \; d_{A,B} > d_{max} \end{cases} \tag{3}$$

Where, mt denotes the distance of the intermediate path between $S_{N(A-B)}$ given by

$$mt = min\left(d\left(S_{N(A)}A,\; S_{N(A1)}\right),\; d\left(S_{N(A1,A2)}\right),\; \ldots d\left(S_{N(An,\;B)}\right)\right) \tag{4}$$

and, d_{max} trust level between $S_{N(A-B)}$ with the maximum distance.

3.2.3 Mutual Trust Value

In this, trust value is estimated to analyze trust association between S_{N-A} and S_{N-B} or vice versa. The mutual trust relationship of $S_{N(A-B)}$ given by following numerical expression.

$$\mathrm{m(A, B)} = \begin{cases} \mathrm{min\;(T(A, B))\; if\; min\;(T(A, B))} \ge \tau \\ 0 \qquad\qquad\qquad \mathrm{else} \end{cases} \tag{5}$$

where τ denotes trust is the trust acceptance factor for controlling the leastacceptable trust degree. Each S_N computes the trust-level of its nearby nodes that provides behavior evidence α (trust node) and β (suspicious behavior of nodes). Based on these multi-layer trust values, final trust values are be computed using bets probability distribution considering variables α and β computed for each trust value. Once the multi-level trust value is accumulated and the final trust value of S_N is computed, CH node broadcasts to the other cluster so that reliable routing is carried out only via a trusted node.

3.3 Trust-Oriented Routing

In order to perform secure data transmission, routing operation is carried out only between the S_N whose trust value is computed. In this process, the proposed scheme maintains the records of trusted intermediate S_N, along with their trust-values and routing is carried between the trusted nodes only. In addition, each S_N during the active phase of the communication and transmission process, periodically monitors the transmission link and assists the data transmission process if any link failure occurs due to mobility of S_N. Figure 3 exhibits the routing scenario for the secure data transmission between the source S_N and IoT gateway.

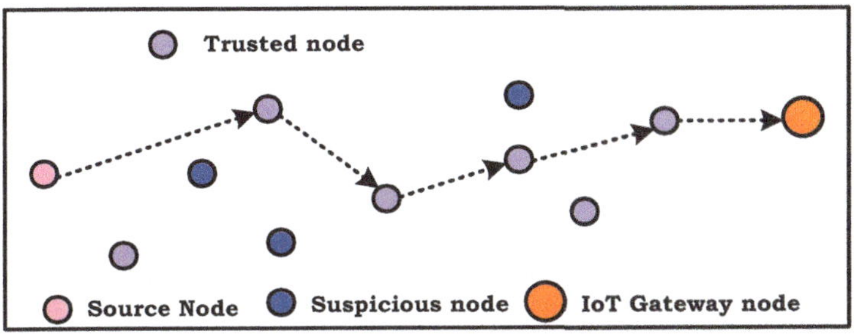

Fig. 3. Routing via trusted node

Figure 3 exhibits the secure routing process via trusted nodes between the source node and the IoT gateway nodes. This process offers secure and reliable transmission to end-user via the gateway node even if there are malicious nodes in the network. The proposed scheme's significance is that it offers an effective mechanism for secure routing and reliable data-packet transmission in the IoT-ecosystem. Another interesting factor of the proposed multi-level trust model is that it also offers a mechanism of malicious node identification, i.e., if the node's trust factor is low compared to a certain cut-off value. The node is regarded as suspicious and not allowed to take participate in the routing process. However, the suspicious node will not immediately be declared as a malicious node. Trust computation will be computed periodically. If the node's unexpected behavior is computed multiple times, it will be declared as a malicious node. The next section discusses the performance and outcome of the proposed scheme.

4 Result Analysis

The design and development of the proposed multi-level trust modeling are carried on a numerical computing tool. The proposed trust-level-based scheme's performance is analyzed based on the comparative analysis considering energy consumption, network overhead, and packet delivery ratio. The performance and outcome of the proposed scheme is compared with the existing approaches, namely, CBRP [19], CBTRP [20]. The description of parameters considered in the simulation is described in Table 1.

Table 1. Simulation parameters

SI. No	Parameters	Values
1	Simulation region	100×100 m^2
2	Packet size	512 *bytes*
3	Number of nodes	500
4	Number of malicious nodes	5 (%)
5	Mobility of nodes	5 m/s
6	Simulation time	2000

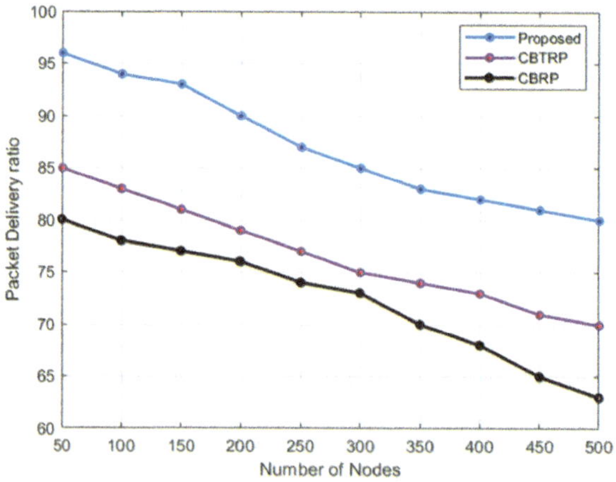

Fig. 4. Analysis of packet delivery ratio

Figure 4 shows the performance analysis of the proposed multi-level trust model in terms of packet-delivery-ratio. It can be seen that the graph trend exhibits a better rate of packet-delivery ratio achieved by the proposed system compared to existing approaches.

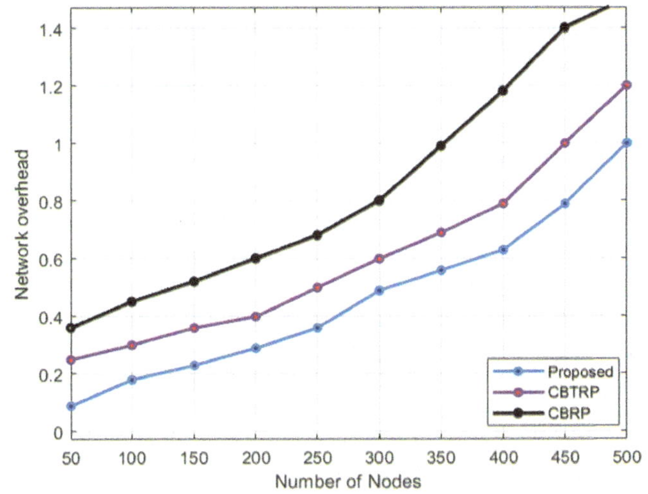

Fig. 5. Analysis of network overhead

Figure 5 shows the performance analysis of the multi-level trust model in terms of network overhead. The proposed system offers low overhead factors regarding communication cost and network congestion during the data transmission process.

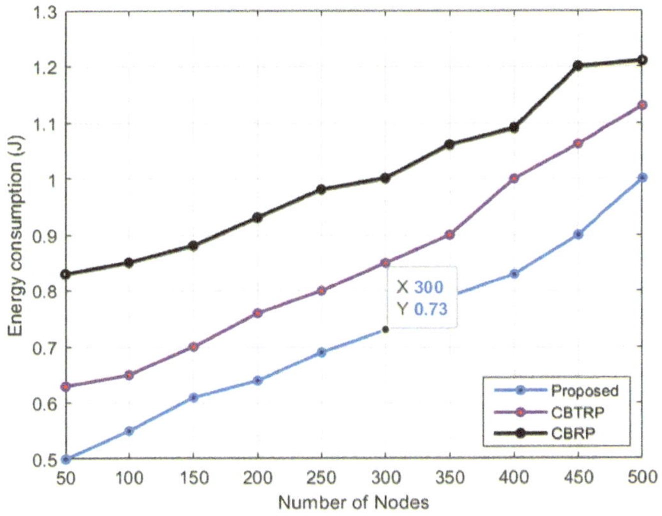

Fig. 6. Analysis of energy consumption

Figure 6 shows the performance analysis of the proposed multi-level trust model in terms of energy consumption. The proposed system offers a better energy-conservation over increasing participation of nodes in the communication process in the network from the graph trend. The prime reason behind achieving better performance by the proposed system is the simple and effective implementation mechanism for trust computation and routing operation. Another reason is that less chance of participation of malicious nodes can cause sensor nodes in the network to deplete their energy. The energy will use only in the process of communication and data transmission. The uniqueness of the proposed system is that it provides secure data transmission against routing attacks as the routing is carried between trusted intermediated nodes.

5 Conclusion

This paper has introduced a novel design of a security mechanism for secure data transmission against routing attacks. An approach of multi-level trust modeling is implemented, which computes the trust-level of nodes based on the behavioral analysis of nodes based on which accumulated trust value is computed for each node in the network. During data transmission, the routes are establishing via trusted intermediate nodes between source and destination nodes. The effectiveness of the proposed multi-level trust model is assessed based on simulation outcome and comparative analysis. The study outcome proves the stability and scope of the proposed system in terms of secure and energy-efficient data transmission.

References

1. Jan, M.A., Khan, F., Alam, M. (eds.): Recent Trends and Advances in Wireless and IoT-enabled Networks. EICC, Springer, Cham (2019). https://doi.org/10.1007/978-3-319-99966-1
2. Qin, W., Chen, S., Peng, M.: Recent advances in Industrial Internet: insights and challenges. Digital Commun. Netw. **6**(1), 1–13 (2020)
3. Kumar, S., Tiwari, P., Zymbler, M.: Internet of Things is a revolutionary approach for future technology enhancement: a review. J. Big Data **6**(1), 1–21 (2019). https://doi.org/10.1186/s40537-019-0268-2
4. Bellavista, P., Cardone, G., Corradi, A., Foschini, L.: Convergence of MANET and WSN in IoT urban scenarios. IEEE Sens. J. **13**(10), 3558–3567 (2013)
5. Mukherjee, S., Biswas, G.P.: Networking for IoT and applications using existing communication technology. Egyptian Inform. J. **19**(2), 107–127 (2018)
6. Raj, N., Bharti, P., Thakur, S.: Vulnerabilities, challenges and threats in securing mobile ad-hoc network. In: 2015 Fifth International Conference on Communication Systems and Network Technologies, Gwalior, pp. 771–775 (2015) https://doi.org/10.1109/CSNT.2015.101
7. Boulaiche, M.: Survey of secure routing protocols for wireless ad hoc networks. Wireless Pers. Commun. **114**(1), 483–517 (2020). https://doi.org/10.1007/s11277-020-07376-1
8. Amiri, E., Keshavarz, H., Heidari, H., Mohamadi, E., Moradzadeh, H.: Intrusion detection systems in MANET: a review. Procedia Soc. Behav. Sci. **129**, 453–459 (2014)
9. Hameed, S., Khan, F.I., Hameed, B.: Understanding security requirements and challenges in Internet of Things (IoT): a review. J. Comput. Netw. Commun. **2019**, 1–14 (2019)
10. Khanna, N., Sachdeva, M.: Study of trust-based mechanism and its component model in MANET: current research state, issues, and future recommendation. Int. J Commun. Syst. **32**(12), e4012 (2019)
11. Anil, G.N.: FAN: framework for authentication of nodes in mobile adhoc environment of Internet-of-Things. In: Silhavy, R. (ed.) CSOC2018 2018. AISC, vol. 763, pp. 56–65. Springer, Cham (2019). https://doi.org/10.1007/978-3-319-91186-1_7

12. Desai, A., Jhaveri, R.: Secure routing in mobile ad hoc networks: a predictive approach. Int. J. Inf. Technol. **11**, 345–356 (2019)
13. Nehra, D., Dhindsa, K.S., Bhushan, B.: Clustering based model to provide defense against DDoS attacks in MANETs. J. Commun. Eng. Syst. **9**(1), 7–13 (2019)
14. Kaur, P., Kaur, D., Mahajan, R.: Simulation based comparative study of routing protocols under wormhole attack in MANET. Wirel. Pers. Commun. **96**, 47–63 (2017)
15. Anil, G.N.: RF-IoT: a robust framework to optimize Internet of Things (IoT) security performance. In: Computer Science On-line Conference, pp. 213–222. Springer, Cham, April 2019
16. Anil, G.N.: Designing an energy efficient routing for subsystems sensors in Internet of Things eco-system using distributed approach. In: Computer Science On-line Conference, pp. 111–121. Springer, Cham, July 2020
17. Ghosh, U., Datta, R.: A secure addressing scheme for large-scale managed MANETs. IEEE Trans. Netw. Serv. Manag. **12**, 483–495 (2015)
18. Merlin, R.T., Ravi, R.: Novel trust based energy aware routing mechanism for mitigation of black hole attacks in MANET. Wireless Pers. Commun. **104**(4), 1599–1636 (2019). https://doi.org/10.1007/s11277-019-06120-8
19. Chiang, C.-C., Wu, H -K., Liu, W., Gerla, M.: Routing in clustered multihop, mobile wireless networks with fading channel. In: IEEE Singapore International Conference on Networks, SICON'97, pp. 197–211, April 1997
20. Safa, H., Artail, H., Tabet, D.: A cluster-based trust-aware routing protocol for mobile ad hoc networks. Wireless Netw. **16**, 969–984 (2010)

Improving Determine Lexical Meanings for Mongolian Cyrillic Script and Its Used to Keyword Extraction

Bat-Erdene Nyandag[1(✉)] and Khishigdalai Ulaankhuu[2]

[1] Department of Mathematics, Physics, and Information Technology, School of Engineer and Technology, Mongolian University of Life Sciences, 17024 Ulaanbaatar, Mongolia
baterdene@muls.edu.mn

[2] School of Business, Mongolian University of Humanities, 14200 Ulaanbaatar, Mongolia
khishigdalai@humanities.mn

Abstract. Improving determine lexical meanings for Mongolian Cyrillic scripts with agglutinative languages are essential tool to information retrieval. This method had integrated the following methods such as to tag part of speech through creating lexicon and to determine lexical meanings through parsing word structure according to orthography of Mongolian Cyrillic script. We evaluate the effectiveness of our method and its contribution the keyword extraction from text materials of Mongolian Cyrillic scripts.

Keywords: Lexical meaning · Parsing word structure · Orthography of Mongolian · Mongolian cyrillic script · Mongolian word structure

1 Introduction

This Mongolian is the official language of Mongolia. The Mongolian language is used two different alphabets, Mongolian Cyrillic and traditional Mongolian. The Mongolian Cyrillic script is used in Mongolia. But the traditional Mongolian script is used in Inner Mongolia Autonomous, People Republic of China. Both scripts are different characterization and rules. This study has focused on Mongolian language, which is used in Mongolian Cyrillic script.

Nowadays, the electron texts materials are written in various languages at network are growing very rapidly [1–3]. One part of those large amounts of text materials is written in Mongolian text materials. Therefore, it is increased requirements for data extraction and information retrieval from Mongolian text materials through the network.

The lexical meanings are used for searching materials from the system. It is increased chance to extract optimized information from the texts materials database.

One or more suffixes are concatenated to word stem in word structure of Mongolian Cyrillic script. The lexical meanings can change to concatenate suffixes. Therefore, word stems extraction and lexical meanings determinations are essential tool for information extraction. The lexical meanings are determined extract word stem through to parse word structure from inflected word structure.

R. Silhavy (Ed.): CSOC 2021, LNNS 230, pp. 206–220, 2021.
https://doi.org/10.1007/978-3-030-77442-4_18

For example: "**Цахилгаанжуулагчдынхантайгаа**" is one of the longest words at Mongolian Cyrillic script.

A word stem is '**цахь**' (to fireflaught); seven lexical meanings are '**цахь**' (to fireflaught), '**цахил**' (to flash), '**цахилгаан**' (electric), '**цахилгаанж**' (to electrify), '**цахилгаанжуул**' (to electrize), '**цахилгаанжуулагч**' (electrician), '**цахилгаанжуулагчид**' (electrician's); four word-formation affix are (**-ж, -уул, (а), -гч, -д**), five inflected suffix are (**-ын, -хан, тай, (г), -аа**) are composed this the longest word. This word between sentences can be translated the following way.

- In Mongolian Cyrillic script: "**Би цахилгаанжуулагчдынхантайгаа явсан.**"
- In English: "I went with professional team of electricians."
- In Chinese: "我和电工人员一起走了。".

A research work and experiments related to determining lexical meanings, extracting keyword, and searching by content are not conducted yet. Therefore, it is a basic research of keyword extraction and optimized content extraction.

We had evaluated and detected their errors for previous studies related to word stem extraction. We had integrated methods based on lexical meanings and orthography of Mongolian Cyrillic script and had improved results. The advantages of our method were classified typical and untypical inflected words during the parsing words structure and machine had learned methods based on orthography of Mongolian Cyrillic script.

We had used to Mongolian laws and e-learning materials for the experimental data and evaluated our methods based on results.

2 Related Work

First, To determine word stem is the first significant for indexing and exploring the information [4, 5]. The research works related to stemming methods for agglutinative languages are studied the following scientists. Thai et al. (2000) for Malay language [6], Vega et al. (2001) for Indonesian language [6], Laurikkala et al. (2004) for Finnish [7], Larkey et al. (2002) for Arabic language [8], Carlberger et al. (2001) for Swedish language [9], Popovic et al. (1992) for Slovene language [10], Ekmekçioglu et al. (1996) for Turkish language [11].

Ehara et al. (2004) and Jaimai et al., (2005) are proposed method that morphological rules are processed by manually and are produced Japanese- Mongolian bilingual dictionary [12]. Purev et al., had suggested that POS is tagged using PC-KIMMO (Antworth, 1990) at Mongolian [13]. PC-KIMMO was based on finite-state two-level. Purev et al. (2005) method accuracy was 60.5%. The reasons for any failures are found for manually rules. Sanduijav et al. (2005) analyzed noun and verb by morphological analysis and proposed a lemmatization method for Mongolian. This method accuracy was 56.5%. [14]. It was limited lexicon.

Khaltar et al. (2009) proposed a lemmatization method based on finite-state and state-of-the-art for Mongolian. Distinguish of previous methods are used solely verbs. This method accuracy was 84.2% **байсан**[15]. The reasons for any failures, and found that for 12 conventional nouns and 11 loanwords, the suffixes were incorrectly segmented.

Chimeddorj et al. (2012) integrated methods of Khaltar et al., and part of speech (POS), used statistical machine translation for Mongolian. In this method, entering words are tagged by POS and parsing words by Khaltar et al. (2009) method. This method accuracy was 93.5% [16]. But failures were indicated on the POS. They used to Khaltar et al. (2009) method directly, so same failures was again indicated.

Previous research works had their own advantages. However, the following errors were typical. Here:

The first: In the methods of Khaltar, Fujii et al., and Chimeddorj, Fujii et al., noun, verb, numbers were identified incorrectly and word structure were pursed also in correctly. For example: 'уран' is a verb with suffixes and integrated by 'ур' (to tear) 'ан' (suffix for verb); 'уран' (uranium) is a noun and 'уран' (art craftsman) is an adjective. The examples of those mistakes are shown in (1), (2) and (3) respectively.

(1) **Дүү цамцаа**'ур + ан' (Verb + Suffix) хаяв.
(My brother is tearing his T-shirt.)
(2) **Зарим улс**'уран' - (Noun) **олборлодог**.
(The uranium is mined in some countries).
(3) **Миний дүү**'уран' (adjective) **дархан болно**.
(My brother will be art craftsman).

The correction of mistakes: To do correct tag for POS and to form appropriate lexicon.

The second: This method did not calculated chance to add double suffixes. The examples of those mistakes are shown in (1), (2) and (3) respectively.

(1) Input word form: noun with suffixes is a 'ахынхаа' (brother) + '-ын' (genitive suffix) + '(х)' + '-аа' (possessive suffix).
(2) Incorrect parsing form: 'ахын' (brother + genitive suffix).
(3) Correct form: 'ах' (brother).

The correction of mistakes: To calculate versions of added double suffixes according to the orthography of Mongolian Cyrillic script and to develop algorithm.

The third: To calculate incorrectly that the letter 'ь' is converted to 'и', and the vowel is eliminated. The examples of those mistakes are shown in (1), (2) and (3) respectively.

(1) Input word form: noun with concatenated suffixes '**сургуулийн**' or '**сургууль**' (school) + '-ийн' (genitive suffix).
(2) Incorrect parsing form: '**сургуули**' (school + и did not change the converted vowel).
(3) Correct form: '**сургууль**' (school).

The correction of mistakes: To calculate rules of the letter "ь" is converted to "и" and last letter of word stem according to the orthography of Mongolian Cyrillic script and to develop algorithm.

The fourth: Did not calculate fully that rules for insertion of eliminated vowel. The examples of those mistakes are shown in (1), (2) and (3) respectively.

(1) Input word form: noun with inflected suffixes is '**үнийг**' or '**үнэ**' (price) + '-ийг' (accusative suffix).
(2) Incorrect parsing form: '**үн**' (price + last vowel had missed).
(3) Correct form: '**үнэ**' (price).

The correction of mistakes: To calculate rules for refill of eliminated vowel.

The fifth: The same forms of words and suffixes mistakes are found. For example: words and suffixes' of similar forms are became error conditions. For example: '-aa' (possessive suffix) and '-aa' (ablative case) and '-д' (plural suffix) and '-д' (dative case).

The correction of mistakes: To calculate characterization of concatenated suffixes and develop morphotactic database.

3 Methodology of Parsing Word Structure, Determining Lexical Meanings

This research work aims are determining lexical meanings through extracting word stem and separate inflectional suffixes from word structure of Mongolian Cyrillic script.

This research work results would be solve the five failures for previous studies. For solving the first and fifth failures, we tagged POS correctly and lexicon created by fully and half automatically. For solving the second failure, we calculated versions of concatenated suffixes according to the orthography of Mongolian Cyrillic script and created suffix database. For solving the third and fourth failures, we calculated typical and untypical changes for end of words, developed rules of the letter "ь" is converted to "и" and last letter of word stem according to the orthography of Mongolian Cyrillic script.

This method is consist of three parts. POS should be marked on the words of input text materials. After we had parsed word structure of words in the input sentence and determined lexical meanings by extracting word stem. At last, we had created new text materials with same meanings from the lexicon. We showed the examples step by step.

Input text. "**Багш сайн найздаа уран зургийн ном бэлэглэсэн.**" (Teacher good friend gradino book give.)

Step 1. Багш сайн найздаа зургийн ном бэлэглэсэн.
ns11 ns3 ns11 ns3 ns11 vs1
Step 2. Багш сайн найз+даа зург+ийн ном бэлэглэ+сэн.
'найз' (friend) + '-д' (locative dative case) + '-aa' (reflexive suffix), 'зураг' (gradino) + '-ийн' (genitive case), 'өгөх' (give) + '-сэн' (past suffix)
Step 4. Багш сайн найз зураг ном бэлэглэ.
ns11 ns3 ns11 ns3 ns11 vs1

In the first step, input words were compared with lexicon, determined POS and tagged certain words. In the second step, word structure was parsed and lexical meanings were determined by extracting word stem. In the third step, new text materials were created by word first form with lexical meanings. Here: ns was represented nouns, ns1 was represented anonym, ns11 was represented proper name, ns3 was represented adjectives, and vs1 was represented verbs.

3.1 Lexical Meaning of Mongolian Cyrillic Script /Lexical Meanings at Written Language/

Words meaning are classified for direct and indirect meanings based on characterization of named. Direct meanings should be pointed exact things. Indirect meaning should be pointed things compared with similar things [17]. Both types are saved for lexicon and expressed lexical meanings.

New unit of lexicon would be created during the process of word formation. If word became new word of database, it would be independent and represent to express only one meaning. The second word unit with new lexical meaning would be created by suffixes attaching to unit of lexical meanings. For example: "мал + чин = малчин" consist of word stem "мал' + 'чин" suffixes for profession. Two unit of lexical meanings and two meaning are created here as 'мал' and 'малчин'.

малчинд = малчин+дwould be created by 'малчин' word stem attached with '-д' (dative case). (Here: '=' symbol indicate new word with stem attached to suffixes. This symbol will used following parts).

3.2 Part of Speech at Mongolian Cyrillic Script

Modern Mongolian Cyrillic script words are classified into basic and auxiallary words depends on their meaning, form, function. The basic words can express independent meaning, have word building methods, inflect by suffixes, can part of a sentence independently. For example: 'хүн' (person), 'мал' (animal), 'өндөр' (high), 'нам' (low), 'сайн' (good), 'муу' (bad), 'хоёр' (two), 'би' (i), 'чи' (you), 'яв' (go), 'үз' (see) etc. Auxiallary words cannot express independent meaning, have not generating methods, do not inflect by suffixes, can not part of a sentence independently. For example: 'ба' (and), 'бөгөөд' (or), 'тул' (to), 'магад' (may), 'ялангуяа' (in particular), 'ёстой' (should), etc.

We had classified noun (ns) and verb (vs) according to characterization of attaching suffixes. We can distinguish word stem for noun and verb from lexicon. For example: "хар + ыг" (black + Accusative case) [ns3 + Fc311], "хар + сан" (see + Identinite Past) [vs1 + Fd11] etc. But noun can be distinguish from lexical meanings based on their inflecting. For example: "нар + анд" (sun + Locative-dative case), "нар + т" (Plural suffix + Locative-dative case), "баг + ын" (small + Genitive case), "баг + ийн" (mask + Genitive case) etc.

Therefore, If words had parsed word structure, it can be expressed ambiguous meanings. In that case, other meanings could be compared with morphotactic (Stem + suffix). For example: үе+ийнshould be separated the following formula as ns11 + Fc112, ns1 + Fc112 etc.

3.3 Orthography of Mongolian Cyrillic Script and Types of Inflecting Word

Inflecting word, parsing word structure, үеэр таслах, orthography rules, morphological rules and specific symbols etc. are used in Mongolian Cyrillic script. It is also including 2 rules for Mongolian Cyrillic script alphabet, 17 rules for vowel, 14 rules for consonants, 1 rules for hard and soft signs, 11 rules for orthography of word stem, affix and suffixes, 14 rules for morphology, 5 rules for others [18].

These rules are classified in groups and taught for computer.

- Word stem 'ном' (book) is added by '-оор' (Instrumental case) suffixes directly attached to word stem according to vowel harmony rule.
- Word stem **'ажил'** (work) is added by '-аас' (ablative case) when suffixes attached to word stem, vowel will reduce according reduced vowel rule.
- Word stem 'ах' (brother) is added by **'-д'** (locative-dative case) when suffixes attached to word stem, vowel will insert according insertion vowel rule.
- Word stem **'харь'** (back) is added by '-в' (past suffix), 'ь' converted to 'и' according to hard and soft sign rule.
- Word stem **'мал'** (animal) is added by '-ын' (genitive suffix) word stem will not convert according to no inflection rule.
- Word stem **'үнэ'** (price) is added by '-ийг' (accusative suffix) word stem will convert according to replacement rule.
- Word stem 'сур' (study) is added by '-х' (Participial + x suffix) vowel added according to Participial + x suffix rule.
- Word stem **'ард'** (public) and **'хүн'** (person) 'ы' and 'ий' suffixes are attached according to distinguishing rules of 'ы' and 'ий' rule.

3.4 Changes of Word Stem

In At Mongolian Cyrillic script, suffixes and suffixes are attached with word stem. "Word stem is part of the word, which is possible to take suffixes [19]. Word stem of Mongolian Cyrillic script are have following structure as Э, ЭГ, ЭГЭ, ЭГЭГ, ЭГГЭ, ГЭ, ГЭГ, ГЭГЭ, ГЭГЭГ, ГЭГГЭ [20]. We also discovered ГЭГГ version during our studies. For example: 'ганц' (one), 'харц' (eye). It is also happened in foreign words at Mongolian Cyrillic script. The following versions are also noted as ЭГ = Э + Г, ЭГЭ = Э + Г + Э,..., ГЭГГ = Г + Э + Г + Г. 13 vowels (which will be noted 'Э' hereafter) are divided 7 basic (which will be noted 'YЭ' hereafter) and 6 auxiliary (which will be noted 'TЭ' hereafter). 20 consonants (which will be noted 'Г' hereafter) are divided 7 vowel consonants (which will be noted '7' hereafter), 9 voiced consonant (which will be noted '9' hereafter), 4 loanword consonants (which will be noted '4' hereafter).

Table 1. Typical changes of word stem

	Final Letter of word	Word Stem	Inflecting Type	Translation
0	ГГЭ	дарга		director
1	ГГ		дарг+ыг	director + accusative case
0	7ь	морь		horse
1	7и		мори+ор	Horse + Instrumental case

Word stem are classified the following versions as one versions and several versions. Word stem with one versions will not change depends on suffixes [21]. Typical changes of word stem are shown in Table 1.

Word stem with several version will depends on suffixes and form several word stem [21]. Some word stem had changed depends on suffixes except morphology of Mongolian Cyrillic script. For example: word 'би' (I) have several forms as 'мин' (I), 'над' (I), 'намай' (I). It is named untypical changes of word stem and is shown in Table 2.

Table 2. Untypical changes of word stem

Tag	Word Examples			
	Mongolian Cyrillic script	**English script**	**Mongolian Cyrillic script**	**English script**
0	Би	I	чи	you
1	Мин + ий	I + genitive case	чин+ий	you + genitive case
2	Над + ад	I+ locative-dative case	чам+д	you + dative case
3	Намай + г	I + accusative case	чамай+г	you + accusative case

Word stem with 0 form can be independent word. But 1, 2, 3, forms can not be express lexical meaning.

3.5 Morphotactic (Word Stem + Suffix)

We had studied word inflecting rules for morphology of Mongolian Cyrillic script and developed word inflecting rules based on last letters of word. When we developed word inflecting rules, registered all suffixes containing word structure. Suffixes versions are shown in Table 3.

Table 3. Untypical changes of word stem

Conjugations suffix form	Suffix code	Conjugations suffix form	Suffix code
nc1|0 + genitive case	nc2	nc1|0 + ablative case	nc5
nc1|0 + Locative-dative case	nc3	nc1|0 + Instrumental case	nc6
0 + Accusative case	nc4	nc1|0 + Cooperative case	nc7

Where: nc1|0 for noun stem 0 for word stem, 1 for inflected word stem, nc2 for genitive case suffixes, nc3 for dative case suffixes, nc4 for accusative case suffixes, nc5 for ablative case suffixes, nc6 for instrumental case suffixes, nc7 for cooperative case suffixes.

Examples of created morphotactic (word stem + suffix) are shown in Table 4.

Table 4. Morphotactic

Parts of Speech	Word Stem	Morphotactic (word stem + suffix)
Nc0	санал	1 + np61, 1 + np64, 0 + nc21, 1 + nc311, 1 + nc42, 1 + nc51, 0 + nc61, 1 + nc71

3.6 Parsing Word Structure

Verbs are inflected verb suffixes at Mongolian Cyrillic script. Those inflections are shown following items: "'verb stem' + 'Voice suffix' + 'Aspect suffix' + 'Conjunctive suffix' and 'Finite suffix'".

For example: "'**явуулчихсан**' (to send) = 'яв' (go) + '-уул' (Voice suffix) + '**-чих**' (Aspect suffix) + '-сан' (Finite suffix)".

When the parsing word structure and extracting word stem of Mongolian Cyrillic script, we had followed three requirements:

- If words are written according to ortography of Mongolian Cyrillic script, parsing word structure would be based on ortography of Mongolian Cyrillic script and morphotactic (stem + suffix).
- Possible all suffixes contained morphotactic (suffix + suffix) database. Concatenated suffixes had with constant step. So it was calculated on program.
- It was provided to identify and parse for correct version 700 words, which are written with mistake on Mongolian Cyrillic script.

We had followed rules for some suffixes attached only certain suffixes. It is depends on front limb at Mongolian Cyrillic script like other agglutinative languages. Therefore we had created conjugations suffix form for every suffix. For example: verb personal ending suffixes at suffixes database are shown in Table 5.

Table 5. Suffix example database of inflectional suffix

v	Verb suffix	Final letter of word	Possibility of attaching inflectional suffix
vp4	**Personal ending suffix**		
vp41	+аач	Г(м,г,л,в,р,д,т,з,с,ц,х,ж,ч,ш)	vs,vv21,vv3,vv4,vv5,va1,va2
vp42	+ач	Э (а,э,о,ө,и,я,е,ё)	vs,vv11,vv221,va4
vp43	+гаач	УЭ болон ХЭ (й,ау,уа,аа,ээ,оо,өө,уу,үү,яа,яу,е э,еү,еө,ёо,ёу,юу,юү,иа,ио,иу)	vs,vv12,va3,va5

Where: The suffixes form at Table 5. are available attach to verb stem (vs), verb voice suffix (vv), aspects suffix (va) and other suffixes.

When the suffixes attach to the word stem, we need calculate to add last letters. It is shown in final letter of word column. Here we noted as following: 'vp4' for personal ending suffix, 'vp41' for '-аач' suffixes. In the conjugation possibilities column, 'vs' for verb stem, 'vv21' verb voice suffix as '-уул' and **'-үүл'**, 'va1' for verb aspect suffix as **'-чих'** etc. **'-аач'**, **'-ач'** and **'-гаач'** suffixes can attach after the suffixes at conjugation possibilities column.

When the parsing word structure, word suffixes are separate from behind to front. If word contain several suffixes, suffixes would be separate one by one depends on their conjugation characterization under calculation as a "'Voice suffix' + 'Aspect suffix' + 'Conjunctive suffix' and 'Finite suffix'". Example of parsing word structure is shown in Fig. 1. such as "**'явуулчихсан'** = 'яв' (go) + '-уул' (Voice suffix) + **'-чих'** (Aspect suffix) + '-сан' (Finite suffix)".

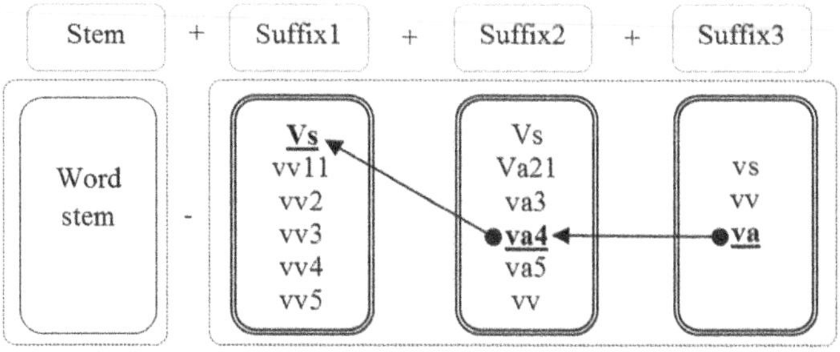

Fig. 1. Calculation of suffixes conjunction characterization

The method of parsing word structure are based on suffixes conjugation possibilities and defined by phonetics principle.

3.7 Algorithm for Parsing Inflectional Suffixes

The process for parsing inflectional suffixes of Mongolian Cyrillic script is shown in Table 6. If signs are finished the following type as a parsing rule column, we would be perform approticate process. For example: verb **'бодов'** (think + b past) in that case, remove suffix 'в' front of vowel 'Э' and extract word stem **'бод'** (think).

According to our developed parsing suffixes rules such as noun suffix '-ын' (genitive case) and verb suffix '-в' (b past) are shown in Table 7.

For example: noun 'номын' is If (end of * = Г + ын | ийн) true (book + Genitive case) in that case, suffix '-ын' (Genitive case) was removed and 'ном' (book) is extracted. But verb 'явав' in that case, If (end of * = 7 + (+Э) + в) true (go + b past) 'Э' + 'в' is removed and 'яв' (go) is extracted.

Table 6. Orthography and types of inflecting word in mongolian cyrillic script

Suffixes	Parsing the Rules	Word Examples	Suffix Code
		Mongolian Cyrillic script	English script
Noun suffixes	If ((end of * = ØЭ + (н) + ы) | (end of * = ØГ + ын)) The suffixes started with 'ы' | 'ын' will be separated.	ямааны = ямаа - ны номын = ном - ын	goat + genitive case book + genitive case
	If ((end of * = ØЭ + (г) + УЭ) | (end of * = ØГ + УЭ)) The suffixes started with 'УЭ' will be separated.	эмээгээр = эмээ - (г) - ээр ахаас = ах – аас	grandmother + instrumental case brother + ablative case
	If ((end of * = ØЭ + ГЭ) | (end of * = ØГ + ГЭ)) The suffixes started with 'ГЭ' will be separated.	тэмээтэй = тэмээ - тэй сартай = сар – тай	camel + cooperative case moon + cooperative case
Verb suffixes	If ((end of * = ØЭ + (г) + Э) | (end of * = ØГ + Э)) The suffixes started with 'Э' will be separated.	орлогоос = орло - гоос өгөөд = өг - өөд	to substitute + ablative case give + dative suffix
	If ((end of * = ØЭ + ГЭ) | (end of * = ØГ + 9Э)) The suffixes started with 'ГЭ' will be separated.	зөвлөжээ = зөвлө - жээ тохирсон = тохирох – сон	to indicate + identinite present suffix to belong + identinite past suffix
	If ((end of * = ØЭ + ГГ) | (end of * = ØГ + Э + ГГ)) The suffixes started with 'ГГ' will be separated.	бүжиглэвч = бүжиглэ – вч гэгээрэгтүн = гэгээр – (+э) - гтүн	to dance + repetitive suffix to dawn + appeal suffix

Table 7. Morphotactic

Suffixes	Parsing the rules
'-ын' genitive case	If ((end of * = Э + (н) + ын|ийн)) true Remove '-ын' suffix
	If ((end of * = Г + ын|ийн)) true Remove '-ын' suffix
	If ((end of * = ь + (н) + ы)) true Remove '-ы' suffix
'-в' past	If ((end of * = 9 + (+Э) + в)) true Remove 'Э + в' suffixes
	If ((end of * = 7 + (+Э) + в)) true Remove 'Э + в' suffixes
	If ((end of * = УЭ + в)) true Remove '-в' suffix
	If ((end of * = Г + и + в) true Remove 'и + в' suffix and add '-ь' soft sign

4 Experiments

We conducted two separate experiments to evaluate our parsing word structure and extracting word stem methods for Mongolian Cyrillic script. In Sect. 4.2, our method is evaluated on parsing word structure and extracting word stem. In Sect. 4.6, we evaluate the effectiveness of our methods in keyword extraction.

4.1 Assessment Used Datas

The parsing word structure and extracting word stem methods are assisted on the Mongolian laws and e-learning text materials. We collected 1.780.968 word, 560 law documents with 22.234 words from http://www.legalinfo.mn/ [22]. Mongolian laws are contained common used words. We also collected 178.448 words and 75 e-materials with 17.709 words from Mongolian universities distance learning text materials [23]. Distance learning text materials are contained terminology of several fields.

4.2 Result Evaluation of Parsing Word Structure and Extracting Word Stem

The sets with 1.000 words are randomly selected 20 times from every material to assess the results of extracted word stems. Using the classification types of the correctly extracted, incorrectly extracted, non-extracted, and corrected errors, the results are as follows: 91.5% of e-materials are correctly extracted, 0.0% are incorrectly extracted, 6.1% are non-extracted and 2.4% are corrected errors; In e-materials of laws, 93.7% are correctly extracted, 0.0% are incorrectly extracted, 4.4% are non-extracted and 1.9% are corrected errors.

We had corrected previous research works failures, which is had done by Khaltar et al. (2009) and Chimeddorj et al. (2012). In practice: The first, when we developed lexicon, POS was noted half automatically and manually. The second, form of suffixes doubled possibilities was fully calculated and entered morphotactic database by algorithm. The third, rule which is hard and soft signs converted to "и" was calculated correctly. The forth, we improved rules for replacing reduced vowels and developed insertion vowel rule. The fifth, we calculated attaching types of words and suffixes with same forms. We also improved discernment results. But we still find failures depends on following three factors. Example: Ancient words, Direct writing of foreign words and Words with professional meaning.

4.3 Evaluating the Contribution of Parse Word Structure to Keyword Extraction

Evaluating the contribution of parse word structure to keyword extraction.

In this experiment, we evaluated results of extracting word stem during key word extracting. We compared and extracted key words from two different text materials. Including: New text material (NTM) is composed of meaningful words of lexicon, which is parsed word structure and extracted word stem from Mongolian text materials or Original Text Material (OTM) is not parsed word structure of Mongolian text materials.

4.4 Evaluation Methods

Many measures for evaluating the performance of information retrieval systems have been proposed, the performance of the extractor system is measured by comparing the generated keywords by the system for each document with the extracted suggested keywords by an expert. The performance measure is based on the number of matches between the system generated keywords and the human generated keywords.

We had used following parameters as Recall (R), Precision (P) and F-measure (F). Recall is the fraction of keywords made by the human that are extracted by the system [24]. Precision is the fraction of keywords extracted by the system that are made by the human. The above two measures clearly trade off against each other, therefore F-measure can be used to measure the trade off of precision versus recall [25].

4.5 Evaluation Datasets

We evaluated our keyword extraction method using two different text materials in Mongolian Cyrillic script, e-learning text materials and law text materials. We collected 250 e-learning text materials (hereafter "e-Documents") from the "http://muls.edu.mn/ [23]" and 1450 law text materials (hereafter "Laws") from the "http://www.legalinfo.mn/ [22]". Each corpora consists of documents collected from Mongolia laws websites and Mongolian distance learning websites.

The collected text materials may be management, economical, tourism, education and law, where a number of keywords are assigned manually to each of the collected text materials. These text materials are used to test the effectiveness of the proposed system to extract proper and significant keywords.

4.6 Evaluation Results

The experiment was conducted on the collected text materials; where the results for the WF-IWF method are shown in Table 8 and Table 9 respectively.

Table 8. Performance of the OTM

Documents type	P	R	F
e-Documents	0.61	0.73	0.68
Laws	0.64	0.78	0.70
Average	0.63	0.77	0.69

Table 8 shows the performance of the OTM. Results showed that the e-document set achieved the highest F-measure with a value of 0.68, followed by the laws document set with an F-measure value of 0.70.

Table 9 shows the performance of the NTM. Results showed that the laws document set achieved the highest F-measure with a value of 0.87 changed at 0.17%, followed by the e-document set with an F-measure value of 0.74 changed at 0.06%.

Table 9. Performance of the NTM

Documents type	P	R	F
e-Documents	0.69	0.81	0.74
Laws	0.82	0.93	0.87
Average	0.75	0.87	0.81

Figure 2 and Fig. 3 shows the comparison between of results for the two methods in the e-documents and laws.

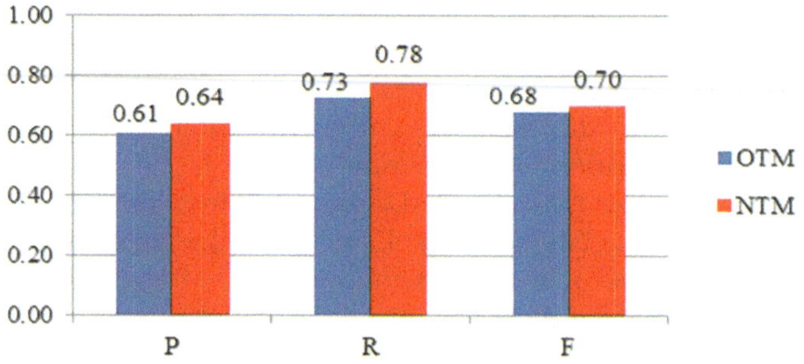

Fig. 2. Performance of WF-IWF among e-Documents

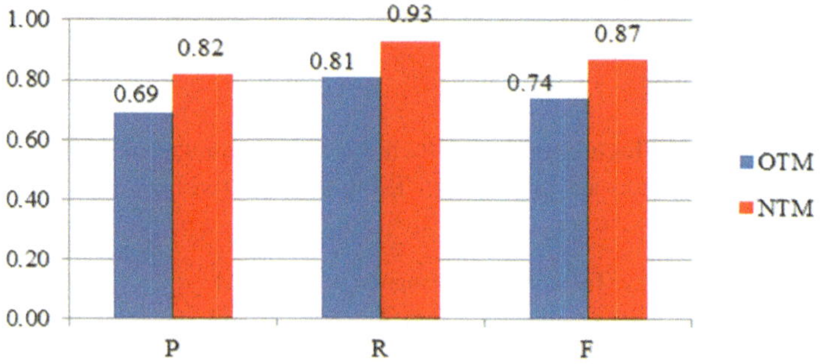

Fig. 3. Performance of WF-IWF among Laws

The above figures show that the NTM performed better than OTM when dealing with Laws, the highest value was for Recall with a value of about 0.93. Also the value of NTM was better than the value of OTM within e-Documents result 0.78 Recall.

This percentage gives us a good indication about the capability of the system to capture important concepts from a set of Mongolian Cyrillic script text materials that belongs to local database.

Figure 3 shows the accuracy of the two methods to capture important concepts within specific domain. The NTM method results better results than OTM. The NTM method results 88% accuracy while the OTM method results 69% accuracy.

This percentage gives us a good indication about the capability of the system to capture important concepts from a set of Mongolian Cyrillic script text materials that belongs to specific domain.

The experiments results give a positive indication about of the system to extract optimized keywords from text materials of Mongolian Cyrillic script, as well as the capability of the system to extract optimized words that reflect the domain of each text materials.

5 Conclusion

After In this paper, we had performed that parsing word structure, extracting word stem and determining lexical meanings. We also had checked and developed new algorithm for parsing word structure based on lexicon and orthography of Mongolian Cyrillic script. Advantages of this method:

- To improve possibilities for parsing word structure of inflected words with correct version when we parse the word structure. Also it can distinguish word stem meanings.
- To improve possibilities for parsing word structure of inflected words all versions which are followed rules on "Word structure form depends on last letter of word at Mongolian Cyrillic script."

Parsing word structure and extracting word stem algorithm assisted a set with 1000 words for each text materials of Mongolian Cyrillic script. Percentage of parsing word structure and extracting word stem was 92.6%.

The experimental results were shown that system abilities of extracting keyword from text materials of Mongolian Cyrillic script was positive.

Further we need to create conjuctions database, proper name database, terminology database. MS Word Authors: please try to use the paragraph styles contained in this document.

Acknowledgements. This work is supported by the School of Engineering, Mongolia University of Life Sciences, Ulaanbaatar, Mongolia. The authors thank U. Khishigdalai for their helps, suggestions and comments.

References

1. Yang, W.: Chinese keyword extraction based on max-duplicated strings of the documents. In: Proceedings of the 25th Annual International ACM SIGIR Conference on Research and Development in Information Retrieval, Tampere, Finland, pp. 439–440 (2002). ISBN1-58113-561-0
2. Timonen, M., Toivanen, T., Teng, Y., Chen, C., He, L.: In formativeness-based keyword extraction from short documents. In: International Conference on Knowledge Discovery and Information Retrieval (KDIR 2012), Spain, pp. 411–421 (2012)
3. Choi, Y.L., Jeon, W.S., Yoon, S.H.: Improving database system performance by applying NoSQL. J. Inf. Process. Syst. (JIPS) **10**, 355–364 (2014)

4. Hull, D.A.: Stemming algorithms – a case study for detailed evaluation. J. Am. Soc. Inform. Sci. Technol. **47**(1), 70–84 (1996)
5. Porter, M.F.: An algorithm for suffix stripping. Program **14**(3), 130–137 (1980)
6. Berlian Vega, S.N, Bressan, S.: Indexing the Indonesian web: language identification and miscellaneous issues. In: Proceedings of Tenth International World Wide Web Conference, pp. 46–47 (2001)
7. Korenius, T., Laurikkala, J., Järvelin, K., Juhola, M.: Stemming and lemmatization in the clustering of Finnish text documents. In: Proceedings of the 13th Association for Computing Machinery International Conference on Information and Knowledge Management, pp. 625–633 (2004)
8. Larkey, L.S., Ballesteros, L., Connel, M.E.: Improving stemming for Arabic information retrieval: light stemming and co-occurrence analysis. In: Proceedings of the 25th Annual International ACM SIGIR Conference on Research and Development in Information Retrieval, pp. 275–282 (2002)
9. Carlberger, J., Dalianis, H., Hassel, M., Knutsson, O.: Improving precision in information retrieval for Swedish using stemming. In: Proceedings of NODALIDA 01 – 13th Nordic Conference on Computational Linguistics (2001)
10. Popovic, M., Willett, P.: The effectiveness of stemming for natural-language access to Slovene textual data. J. Am. Soc. Inform. Sci. Technol. **43**(5), 384–390 (1992)
11. Ekmekçioglu, C.F., Lynch, M.F., Willett, P.: Stemming and n-gram matching for term conflation in Turkish texts. Inf. Res. News **7**(1), 2–6 (1996)
12. Ehara, T., Hayata, S., Kimura, N.: Mongolian morphological analysis using ChaSen. In: Proceedings of the 10th Annual Meeting of the Association for Natural Language Processing, pp. 709–712 (2004). (in Japanese)
13. Jaimai, P., Zundui, T., Chagnaa, A., Ock, C.-Y.: PCKIMMO-based description of Mongolian morphology. Proc. Int. J. Inf. Process. Syst. **1**(1), 41–48 (2005)
14. Sanduijav, E., Utsuro, T., Sato, S.: Mongolian phrase generation and morphological analysis based on phonological and morphological constraints. J. Nat. Lang. Process. **12**(5), 185–205 (2005). (in Japanese)
15. Khaltar, B.-O., Fujii, A.: A lemmatization method for Mongolian and its application to indexing for information retrieval. Inf. Process. Manag. **45**, 438–451 (2009)
16. Chimeddorj, O., Fujii, A.: Enhancing lemmatization for Mongolian and its application to statistical machine translation. In: Proceedings of the 10th Workshop on Asian Language Resources, pp. 115–124 (2009)
17. Bazarragchaa, M.: Some issues of Mongolian sentence study. Mongolian National University School of Language and Culture, Ulaanbaatar, p. 190 (2012)
18. Mongol khelnii zuv bichikh durem (buren ekheer). http://www.news.mn/r/13435. (in Mongolia)
19. Nergui, B., Korpp, O.: Mongol khelnii khel shinjleld khoer tuvshint morphologiin arga khereglekh, Ulaanbaatar (1996). (in Mongolia)
20. Nyandag, B.-E., Ru, L., Indruska, G.: Performance analysis of optimized content extraction for Cyrillic Mongolian learning text materials in the database. J. Comput. Commun. **4**, 79–89 (2016). https://doi.org/10.4236/jcc.2016.410009
21. Unurbayan, T.: Orchin tsagiin mongol khelnii ug zui, Ulaanbaatar (2004). (in Mongolia)
22. http://www.legalinfo.mn/. Accessed Dec 2016
23. http://www.muls.edu.mn/. Accessed Dec 2016
24. Nyandag, B.-E., Ru, L., Demberel, O.: Keyword extraction based on statistical information for Cyrillic Mongolian script. In: Proceedings of the 29th Chinese Control and Decision Conference (CCDC), 28–30 May 2017 (2017). https://doi.org/10.1109/CCDC.2017.7978889
25. Al-Kabi, M., Al-Belaili, H., Abul-Huda, B., Wahbeh, A.H.: Keyword extraction based on word co-occurrence statistical information for Arabic text. Basic Sci. Eng. **22**(1), 75–95 (2013)

Multidimensional Analysis of NewSQL Database Systems

Ionela Chereja(✉), Sarah Myriam Lydia Hahn, Oliviu Matei, and Anca Avram

Computers and Information Technology, Technical University Cluj Napoca, Cluj-Napoca, Romania
{ionela.chereja,anca.avram}@ieee.org, sarah-hahn@gmx.net, oliviu.matei@holisun.com

Abstract. Collecting, transforming and analyzing data is seeing an all time high interest over the last few years. Due to the large amount of emerging or enhanced data management systems, as well as the high amount of differences in features and options they support, this paper is focused on reviewing the existing systems that are part of the NewSQL category. By looking at the top nine most used and researched implementations this work investigates how these compare in terms of supported data models, transactionality features and querying abilities. Furthermore, it provides an overview of scalability implementations and functionality as well as security assets and liabilities. It also includes an analysis of costs, by looking at both licensing as well as infrastructural support. This work can be used as a guiding analysis for providing insights into what features are supported and offered by the top NewSQL systems available over multiple dimensions.

Keywords: NewSQL · Database systems · Performance · Security · HTAP · Querying capabilities

1 Introduction

The field of computer science has been at the heart of worldwide focus and development for a few decades now, having seen one of the most rapid growth a field has ever encountered. As computers and computer science has developed, so did the amount of sub-fields it has.

A new generation of databases were created as a solution to this new demand and were designed with complete new requirements in mind compared to the existing Relational Database Management Systems (RDBMS), called NoSQL [23]. NoSQL systems are non-relational databases, provide high scalability and are defined through a different set of properties such as BASE (Basically Available, Soft State, Eventual Consistency) and the CAP theorem [1]. These are implemented as opposed to ACID (Atomicity, Consistency, Isolation and Durability) properties that RDBMSs implement [16]. NoSQL systems were designed as a reaction to the relational databases as the name also suggests (referring

R. Silhavy (Ed.): CSOC 2021, LNNS 230, pp. 221–236, 2021.
https://doi.org/10.1007/978-3-030-77442-4_19

originally to "non-SQL" or "non-relational") [17]. However, doubts and concerns have been raised on how they will fit the needs of the industry due to certain limitations [32].

In 2011, a new term was coined by Matthew Aslett [5] - NewSQL - referring to a newer generation of database systems. This term intended to categorize a new set of highly scalable and highly performant products that were still SQL databases, but provided various enhancements that the existing systems were missing (such as a distributed architecture). Following the introduction of the term, more vendors and more systems have been called NewSQL. However, due to the high variety of products as well as the variety and versatility of features they provide, there is a lack of an accepted definition and standard for NewSQL databases.

This paper has the following objectives: define the database management systems available, key terms, concepts and data models. Define the taxonomy used to address the existent NewSQL systems and focus thereafter on those characteristics to create a clear selection of NewSQL products that will be analyzed more in depth in relation to their capabilities on data models supported, transactionality features and querying capabilities. Furthermore, the analysis will also include the scalability functionalities and approaches these NewSQL databases use and the security features together with their known limitations and vulnerabilities. An overview of cost and infrastructural support considerations is also provided.

2 Related Work and Terms

2.1 SQL and Relational DBMS

A relational database is a structure allowing linking of information from different data containers, or tables. Every data container must have a key or index that uniquely identifies information. Different data containers can reference that key in order to create a relation or link between the two containers [41].

There are a few key elements to the relational concept, as defined by E.F. Codd in his 1986 paper "An Evaluation Scheme for Database Management Systems that are claimed to be Relational". Codd defines twelve rules to identify relational systems. All the rules are based on one foundational rule, called Rule Zero: "For any system that is advertised as, or claimed to be, a relational database management system, that system must be able to manage databases entirely through its relational capabilities." [15]. Any system that does not implement this rule, cannot qualify as a Relational Database Management System.

2.2 NoSQL and Unstructured Data

NoSQL is a non-relational database management system that provides flexibility in schema. It is generally used with distributed data, particularly with applications that involve big data and real time data.

There are a number of existing articles providing an overview on the categories, differences between SQL and NoSQL systems and the advantages or disadvantages that each group of database management systems has [23,37,38,44,45]. In this study there will not be an in-depth comparison again, only a reference to the categories used to classify the NoSQL data stores. The classification made by Hecht and Jablonski will be used [25]: Key-value, Document, Column Family and Graph databases.

2.3 NewSQL Systems

NewSQL is a new class of database management systems leveraging scaling capabilities that are unavailable in legacy SQL systems. NewSQL seeks to supply the scalable performance of NoSQL systems while still providing the ACID guarantees. These systems attempt to attain the scalability of the NoSQL DBMSs with the relational model and transactional support of SQL DBMSs [31].

NewSQL surveys and comparisons between the different management systems have been proposed in other articles before. The articles surveyed focus mainly on comparisons between the three different data store categories: SQL, NoSQL and NewSQL [7,30]. A few articles focus on comparisons between the two latter categories [10,22]. Surveys that focus specifically on NewSQL data stores and their features are limited [3].

3 Methodology and Criteria

3.1 Taxonomy

This survey uses a systematic approach to select the solutions presently included. Due to the large number of data stores on the market that would qualify in the category of NewSQL databases, it is not feasible to try and include all of them. These systems are continuously evolving and changing, many adding support for new features or releasing new versions.

The considerations on which this filtering is done is constituted of three parts:

1. DB-Engines Ranking [19] score
2. Presence of the database in scientific papers in Google Scholar
3. Presence of the database in scientific papers in WebOfKnowledge

The method used by DB-Engines to calculate the Rank of each database includes [19]: number of mentions of the system on websites; interest in the system by frequency of searches (using Google trends); frequency of technical discussions about the system; numbers of job offers that mention the system; number of profiles that mention the system in professional networks; social networks relevance by counting the mentions of the system.

We use these three dimensions as a means to assert maturity and stability of the system, presence in the market and online, and also interest from the scientific community.

The list of databases initially surveyed are the following: VoltDB, MemSQL, TiDB, Google Spanner, Citus, Vitess, NuoDB, Clustrix DB, CockroachDB, FaunaDB, SAP HANA, Apache Trafodion, Harper DB, Altibase, GemFire XD, Actor DB, C-TreeACE, Active Spaces and TokuDB.

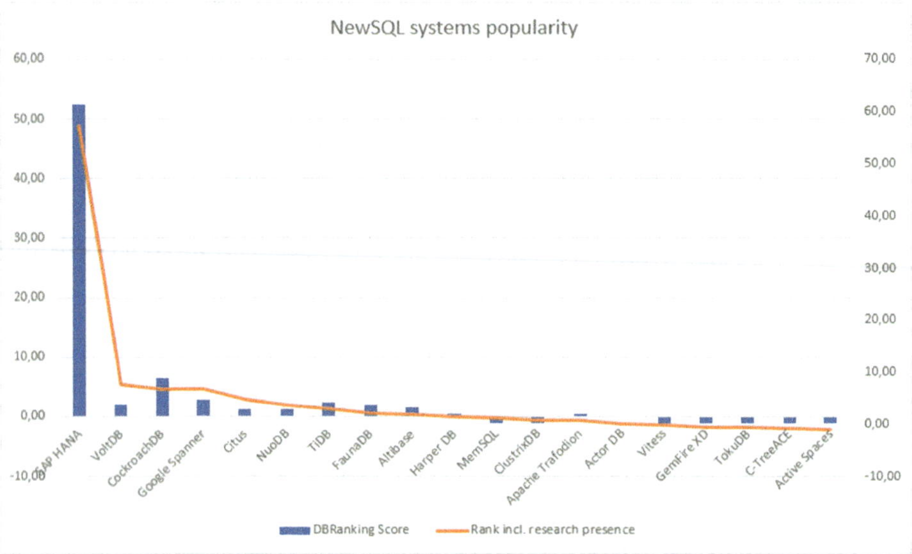

Fig. 1. NewSQL Databases Overview with three criteria selection

The systems chosen for analysis are the top 9 in the ordered list shown in Fig. 1: **SAP HANA, VoltDB, CockroachDB, Google Spanner, Citus, NuoDB, TiDB, FaunaDB and Altibase**.

3.2 Criteria

3.2.1 Data Model

Data models can be defined as entities that introduce abstraction in a database management system [14]. Database management systems support a large variety of data models. With the emergence of NoSQL and NewSQL systems, multi-model DBMSs [33] became more common. Looking at the data models supported by each system provides an overview of the systems' flexibility.

The classification in this paper includes the Relational model, Graph and Document. The analysis will also include the Time Series category. Time series data is data that consists of a series of timestamps and corresponding values [6]. The databases that use the time series data model are also called TSDBs (Time Series Databases).

Under the Data Model category this paper will also look if the surveyed databases offer predefined data types (such as float or date) under the subcategory of Typing. There is also an explicit mention if they have embedded XML support.

3.2.2 Transactionality

Relational Database Management Systems are defined by their ability to abide by the ACID properties [21]. In the criteria defined as Transactionality, we will investigate ACID compliance for the surveyed systems, including specifics that vendors mention for the system's ability to run transactions.

Consistency will also be inventoried separately. Although Consistency is part of the ACID properties, there are more definitions about what Consistency represents, including the different ACID and CAP definitions. In relation to distributed databases, consistency refers to the freshness of the data when the data is queried and it indicates if copies are the same in different sites [34]. Consistency levels are connected to the type of replication a database uses. Synchronous protocols present immediate consistency (or strong consistency) by propagating changes to all of the replicas at the same time and in the same order. Asynchronous and hybrid protocols ensure different levels of consistency by applying the updates to a replica while allowing another to still be read. This results in temporary inconsistency, also called weak consistency or eventual consistency [1].

Consistency also has the requirement of all defined user-constraints to be satisfied. Referential Integrity (Constraints) ensure that every child row has a parent row, which in turn may require that every row in the database is uniquely identified by a primary key [26]. This paper reviews the support for every NewSQL database for referential constraints, also called foreign keys.

Durability refers to the support offered by the database in making data persistent. Thus, it is the guarantee that once a transaction has been successfully ran it will survive system failures, it will not roll back or abort and also that the integrity constraints have been checked [9].

3.2.3 Querying Capabilities

To analyze the flexibility and performance of the database system with the data models available, this paper looks at the **querying capabilities** that each surveyed system supports such as SQL or SQL-like languages.

APIs (Application Programming Interfaces) in the field of databases have increased in diversity and volume with the increase of data-centric services [24]. We address, thus, what access methods and what APIs are supported by the databases under the criteria of querying capabilities, checking support for REST-based APIs in particular.

3.2.4 Scalability

An important criteria of performance for the NewSQL databases is the scalability they offer. Scalability will be assessed in terms of Partitioning, Replication, Concurrency Control and In-memory capabilities of each system.

Partitioning is mainly categorized in two ways: vertical and horizontal. While vertical partitioning is about the database's capacity to organize data into disjoint sets of columns and is generally supported, horizontal partitioning allows database objects to be partitioned into disjoint sets of rows. Two categories of horizontal partitioning are hash and range [2].

Replication is used as a mechanism for performance, scalability and fault-tolerance. This introduces a trade-off between efficiency and consistency [2], but ensures that the failure of one of the replicas does not allow for complete system failure. The requests of the failed replica will be taken over by other available replicas.

In order to provide the data copies with consistency when updates occur, the database system is responsible for replica control. This defines where the replication tool is located, where updates can occur and how they are applied, as well as how changes are propagated to the replicas [36].

With **Concurrency Control** we analyze the system's capacity to manipulate data concurrently and the specific implementation limitations that some systems have.

In-memory capabilities specifies the system's capacity to define some or all of the structures in the database to be held in memory only.

3.2.5 Security

Security is a relevant aspect of any data storing system especially as constraints regarding privacy increase. Four aspects of security are addressed for the surveyed NewSQL database systems. These aspects were used in one of the surveyed articles [20], however the paper does not focus on NewSQL data stores alone and the ones that are mentioned are scarce. The dimensions analyzed are as follow [20]:

Authentication - the ability to verify the identity of the users who are attempting to access the data stored in the database. There are multiple mechanisms supporting this, including user and password authentication, user certificates and the ability to integrate with enterprise directory systems.

Authorization - the capacity to provide access to resources through an associated set of permissions.

Encryption - the mechanisms used to encrypt the data stored so that it cannot be deciphered by unauthorized attempts to read it. Encryption usually refers to multiple levels: server to server connections, data at rest and client to server communication.

Auditing - the logged trail of performed actions on a database system. This is especially important in analyzing events that involve security.

3.2.6 Costs and Infrastructural Support

Next to the features that a system provides, one of the most important factors in deciding whether to use it or not are the costs involved and the limitations or infrastructural support they offer.

Two components are analyzed: the licensing of the surveyed system and infrastructural support. While there are a few database vendors that offer a free version, there are also multiple ones that have enterprise licensing or a mix of the two depending on how the system will be used. In terms of infrastructural support, the indicator of whether it is a Cloud-based service only or not is used. This provides insight in the required resources to be able to put the system into place.

4 Analysis

4.1 Data Model

NewSQL systems are built to support the relational data model, which includes all of the chosen database systems for this survey. However, even though all of these systems do support the relational model, some of them are multi-model databases. SAP HANA [35] and FaunaDB [18] both offer embedded support for Graph data and Document models (Table 1).

SAP HANA Graph is part of the SAP HANA core functionality, as it expands the database platform with native support for graph processing. This allows for typical graph operations to be executed directly over the data stored in the SAP HANA system. Relational storage allows all the functions to be applied to the graph data including access control and backup and recovery [35].

Similarly, FaunaDB offers a multi-model approach to the data stored allowing multiple modeling techniques (including graph, document and temporal). FaunaDB unifies all of the different interactions with the distinct data models under the same query language [18].

TiDB database runs over a TiKV server, which is responsible for storing the data. This is a distributed transactional key-value storage engine which uses regions as the basic unit to store the data [43]. Next to the relational model

Table 1. Supported data models

Database name	Relational	Graph	Document	Time series	XML support	Typing
Google Spanner	Yes	No	No	Yes	No	Yes
SAP HANA	Yes	Yes	Yes	No	No	Yes
VoltDB	Yes	No	No	No	No	Yes
TiDB	Yes	No	Yes (Key-Value)	No	Yes (experimental)	Yes
Altibase	Yes	No	No	No	No	Yes
Citus	Yes	No	No	Yes	Yes	Yes
CockroackDB	Yes	No	No	No	No	Yes
FaunaDB	Yes	Yes	Yes	Yes	No	No
NuoDB	Yes	No	No	No	No	Yes

and the key-value store (categorized here under Document data model), TiDB is introducing support for XML which is currently an experimental feature.

Temporal data, referred in the categories as Time Series, is supported by three of the surveyed engines. FaunaDB implements a Temporal Model to track changes and have auto-expiring records, for example. FaunaDB does that by adding extra dimensions in the schema that will track validity and transaction time for each record. By using a temporal snapshot storage, the engine natively provides auditing for change history, querying previous versions of data and even recovering from mistakes [18].

Citus provides support for scalable Time Series. Big tables with time-ordered data are broken down into multiple tables per time ranges, with natively distributed sharding. The partitions can also be automatically created on demand, reducing manual interaction [12]. There are some caveats to be taken into account while using this technique, such as limitations for creating an index over the partitioned table.

Google Spanner also offers a temporal option for the data through timestamps. Data is read and written with timestamps and can also be therefore read at specific timestamps from the past [3]. A mechanism called TrueTime is used by Spanner to provide a globally synchronized clock which is used to avoid communication across replicas in distributed systems [8]. This is used to generate increasing timestamps that are assigned to distributed transactions and is also used to ensure consistency.

XML support refers to the capacity of the system to natively be able to process data in XML format, for example by supporting XQuery, XPath or XSLT. This capability is embedded for Citus only and is mentioned as being experimental for TiDB.

Typing, or predefined data types, are offered by all of the systems analyzed with the exception of FaunaDB which is a schema-free system.

4.2 Transactionality

All of the NewSQL systems analyzed offer ACID properties for transactions, strong (or immediate) consistency and durability. However, one of the criteria component is not satisfied by all systems and that is referential integrity, or the ability to define foreign keys. VoltDB does not offer the possibility of defining a foreign key constraint [46]. Similarly, TiDB supports the creation and the dropping syntax for foreign keys in order to avoid migration errors, but does not perform constraint checking [43]. For this paper we will mention, thus, that it is unsupported.

Several solutions use the term "fully ACID-compliant" to define their engines. SAP HANA and VoltDB use this terminology for defining the products offered.

Spanner supports the execution of read and write-transactions at a single logical point in time through the use of timestamps. The read-write transactions provide ACID guarantees and moreover they even run with stronger guarantees. All reads within a transaction return data from the exact same timestamp and all writes in the transaction are committed at the exact same timestamp. This will only be visible when the transaction is over. Due to the usage of timestamps, Spanner offers serializability (or strict serializable isolation) and external consistency (every transaction has a commit timestamp and this timestamp is related to real time by using TrueTime [8]).

TiDB performs distributed ACID transactions through the adoption of the Percolar model used by Google in BigTable [42]. The implementation done by TiDB supports distributed transactions with both optimistic and pessimistic models, having the pessimistic transaction model as default starting from version 3.0.8.

Similarly, FaunaDB uses a storage engine that is equivalent to the primary storage engine in Google BigTable [11] to provide durability. This is implemented as a compressed log-structured merge tree [40]. The consistency model that FaunaDB uses, delivers strict serializability across multi-key transactions [18].

Altibase is a hybrid database that offers both in-memory and on-disk processing and adheres to ACID compliance. It also offers integrity checks through the usage of primary and foreign keys, unique constraints and is compatible with multiple legacy DBMSs [4].

NuoDB uses Durable Distributed Cache architecture (DDC) with Multi-Version Concurrency Control (MVCC) to ensure ACID compliance in a distributed and elastic scale-out database. Atomicity, Consistence and Isolation guarantees are managed through the Transaction Engines (TE) and Durability is implemented through Storage Managers (SM). A Transaction Engine is a NuoDB component that processes SQL commands, caches data and coordinates transactions together with other TE [39].

CockroachDB supports ACID transactions in a distributed environment across a set of arbitrary keys [13]. The engine works with transaction records in order to ensure transactional atomicity. It offers serializable isolation and synchronous writes.

4.3 Querying Capabilities

None of the surveyed articles on NewSQL systems has had a systematic look at querying languages. These are closely related to the data models supported by the engine and they are indicative of the flexibility provided and also the ease of use. As these database systems implement a scalable, distributed relational model, providing a SQL querying language or at least a SQL-like language comes as an advantage in migration, manageability and maintainability. Secondly, knowledge of SQL querying languages can be leveraged in working with a new system with a similar interface.

FaunaDB offers a flexible querying language that accommodates all the different supported data models. It runs ACID transactions and it runs with temporal retention as well. This is executed on top of distributed cloud-native storage [18]. However, FaunaDB is the only system that does not provide a SQL-like querying language.

Table 2. Querying capabilities

Database name	SQL(like)	REST APIs	APIs and other Access methods
Google Spanner	ANSI 2011	Yes	gRPC API; JDBC info
SAP HANA	Yes	No	JDBC; ODBC
VoltDB	Subset of SQL 99	Yes	Java API; JDBC; JSON API
TiDB	Yes	No	JDBC; ODBC; Proprietary protocol
Altibase	ANSI SQL-92	No	JDBC; ODBC; Proprietary protocol
Citus	Yes	No	ADO.NET; JDBC; native C library ODBC; streaming API for large objects
CockroachDB	Yes	No	JDBC
FaunaDB	No	Yes	–
NuoDB	Yes	No	ADO.NET; JDBC; ODBC

Rest API implementations are provided by Spanner, VoltDB and FaunaDB. The rest of NewSQL databases provide different APIs or access methods and an overview of those is available in Table 2.

4.4 Scalability

Concurrency control is offered by all the surveyed systems and will not be detailed into more depth, but it is still a relevant marker of a system's capacity to support multiple requests in parallel.

TiDB uses a storage engine called TiKV which uses Key-Value pairs to save the data. It is a persisted storage engine that eventually stores data on disk. However, it does not write it directly to disk but uses RocksDB as data storage in order to ensure data availability in case of a machine failure. This is achieved by replicating the data over multiple machines using the Raft algorithm. This is a consensus algorithm providing features as Leader election, log replication and membership changes like adding or removing replicas [27]. By using Raft, TiDB will have every change in a Raft log and through it data is replicated to multiple nodes in the Raft group. Moreover, the Raft protocol only requires a majority of the nodes to have the data replicated in order to consider the write successful [43]. Similar to TiDB, CockroachDB uses Raft for consensus-based replicated writes. CockroachDB is built over a distributed consistent key-value store that is split into multiple key-value ranges. These ranges are spread across several nodes [13].

FaunaDB is vertically and horizontally scalable, designed to be self-coordinating without any single point of failure [18]. This is achieved at the level of every node in the distributed cluster, each of which implements simultaneously three different roles as log replica, data replica and query coordinator. Thus every node is able to handle any request. Partitioning is done in FaunaDB across all the nodes: indexes are partitioned by term and instances by primary key [18]. They scale linearly regardless of the numbers of nodes in the cluster or their cardinality.

SAP HANA supports hash and round-robin partitioning. It also offers range partitioning, yet in this case understanding the contents of the data is needed. The database offers either Single-Level partitioning (no nested partitioning) or Multi-Level, which allows partitioning by a column that is not part of the primary key. In terms of replication, SAP HANA offers four distinct modes [35] that can be configured.

In NuoDB, partitioning is done using the concepts of table partitions, storage groups and partitioning policies. A table partition can be associated with a table as an attribute and it will contain a subset of the row versions that are stored and serviced by a storage group (a logical storage unit). A partitioning policy governs the data that is stored within a table partition [39]. NuoDB also provides multi-master replication as a default mechanism and it uses Transaction Engines and Storage Manager to achieve it.

Altibase offers three partitioning techniques: list, range and hash partitioning. In the case of list partitioning, the partitions are defined depending on the membership of the column value in one of the discrete values lists. Altibase also offers four replication models: primary-standby, active-active, primary multi standby using propagator and primary-multi standby. The details of this implementations have been described in a separate paper [29].

Citus provides Source-Replica high availability [12]. Data is written to a primary system called a Writer and is instantly replicated to a secondary node called Reader.

4.5 Security

All surveyed systems have in place some form or another of authentication, authorization, encryption and auditing mechanisms. However, there are differences in what kind of support is provided for each and to what extent.

TiDB supports password-based and certificate-based authentication, but it does not support external authentication such as LDAP (Lightweight Directory Access Protocol). It does not support password validation, password expiry, password last-changed tracking or password lifetime. It performs authorization through a role-based access control [43]. TiDB uses a non-encrypted connection by default between server and client. To avoid data leaking, it offers support for encrypted connection based on TLS (Transport Layer Security). Encryption at rest is implemented in TiDB by doing the encryption before storage. Known vulnerabilities include metadata information that is not encrypted and encryption keys and user data being present in core dumps [43].

Contrary to TiDB's approach, Citus data that is hosted in Citus Cloud is encrypted at rest including all of the backups for disaster recovery. Next to that, Citus Cloud only accepts SSL connections and all nodes also encrypt the communication with one another (also using SSL). Furthermore, to avoid man in the middle attacks, when connecting to the coordinator node a certificate for the server can be verified. Citus also accepts two factor authentication [12]. Regarding authorization Citus offers a role-based authorization and it supports granting privileges in bulk also.

FaunaDB provides security using two options for access: identity-based and anonymous. The resources that FaunaDB offers in order to support this access models are Secrets, Attribute-based access control (ABAC), Tokens, JSON Web Tokens (JWTs) and Keys. Authorization is role-based through either built-in roles or user-defined roles [18].

NuoDB provides encryption at rest using TDE. Role-based authorization is used and the latest version also comes with an addition in administrating role-based access at the Admin tier [39]. The database management system also offers support for external authentication through LDAP.

CockroachDB uses TLS 1.2 security protocol which makes use of symmetric and asymmetric encryption. CockroachDB uses asymmetric encryption to authenticate but also to establish the channel of communication, and uses symmetric encryption for protecting data in flight. It also performs encryption at rest using AES in counter mode [13]. Cockroach also supplies role-based authorization for user management and it supplies audit logging that is especially relevant in troubleshooting possible data breaching queries.

Similar to most systems, Altibase also adopts TLS protocols to ensure security. Altibase supports both server-only and mutual authentication and it offers policy based password management, object privileges and IP access control. Furthermore, it has the option to configure tracking through auditing statements [4].

Google Spanner is a Cloud-based only service and thus uses the Cloud Storage feature of always encrypting data before it is written to disk. Access is provided through the Identity and Access Management (IAM) mechanism. This allows user and group access control at the project, instance or database level. In terms of logging Spanner offers a multi-level setup which includes System Event audit logs, Data Access audit logs, Admin Activity and Policy Denied logs [28].

VoltDB provides a secure connection using hashing rather than encryption when it passes the user and password from client to server. Clients use either SHA-2 or SHA-1 hashing and passwords are also hashed in the database itself. VoltDB also provides TLS and Kerberos security. However, by default VoltDB does not perform authentication. This needs to be enabled in the configuration file. VoltDB also provides audit logging capabilities by using Log4J, an open source logging service [46].

SAP HANA provides four authentication methods: Basic (through username and password), Kerberos, SAML and Session Cookies. These methods are supported at the primary node's level and secondary nodes delegate the authentication to the primary. It also supports both server-side and client-encryption. On the server side data is encrypted at rest and it also offers the option for internal encryption. On the client side it offers the option to encrypt data that is sensitive by using an encryption key that will only be accessible to the client. In terms of authorization, SAP HANA supports a series of options that allow role based authorization, LDAP, as well as cross-database authorization in tenant databases. Furthermore, SAP HANA offers multiple auditing configuration options by using Audit Policies, Audit Trails. These can be maintained and managed through Auditing Configuration and Audit Policy Management [35].

4.6 Costs and Infrastructural Support

Finally, we find it important to create an overview of the costs involved (specifically for licensing) visible in Table 3 and also mentioning what type of infrastructure is required in order to run the NewSQL system. For assessing this we use the limitation of a system of being supported cloud-only.

Table 3. Licensing and infrastructural support

Database name	Open source license	Enterprise license	Cloud-based only
Google Spanner		Commercial	Yes
SAP HANA		Commercial	No
VoltDB	AGPL3	Commercial	No
TiDB	Apache 2.0		No
Altibase	GNU AGPLv3		No
Citus	GNU AGPLv4	Commercial	No
CockroachDB	BSL and CCL	Commercial	No
FaunaDB		Commercial	Yes
NuoDB		Commercial	No

VoltDB, TiDB, Altibase, Citus and CockroachDBs are all Open Source systems, with VoltDB, Citus and CockroachDB also offering Enterprise licensing options. In terms of infrastructural support, the two systems that are available only as cloud services are Google Spanner and FaunaDB.

5 Further Investigation and Expansion Areas

5.1 Conclusions

This research is intended to offer an overview over the generation of databases that promise to combine the strong guarantees of relational databases with the scalability and performance of NoSQL systems. While that is indeed what these new solutions offer under various implementations, they also provide in a vast majority an integration of operational and analytical workloads under a fast and highly scalable solution.

NewSQL systems come in a wide variety of implementations and they offer a strong case in supplying features that are in high demand. NewSQL systems provide high performance on a distributed architecture and come with much greater scalability than the legacy relational systems. They also keep guarantees that relational systems are still preferred for - like ACID and a relational approach to storing data. This offers NewSQL systems a considerable advantage over NoSQL stores, even more so in the case of multi-model databases that can store, process and query large amounts of data in structured or unstructured formats.

Cost considerations are also an important factor in choosing a solution, and NewSQL implementations offer both open-source and enterprise licensing options. However, in some implementations there are still limitations in terms of security functionalities they offer as well as missing a standardized query language and a standard supported API. Moreover, there is no clear overview on how a migration can be easily made from a legacy system to a NewSQL database.

5.2 Opportunities

While these systems offer multiple advantages over both legacy relational databases and NoSQL databases, they also pose a series of questions and opportunities. The analysis done over NewSQL databases exposed the following opportunities to be further researched in the field:

NewSQL systems are very diverse in what they offer and what features they implement. Even for labeling a system as NewSQL there is no accepted standard, but more a consensus based on the initial usage of the term [5]. As a result, common terminology is not yet established or agreed upon. Creating such a standard can provide an easier method in comprehending and implementing such a solution. This also is true for standards on querying languages or access methods.

As there is such a diverse offering of NewSQL databases, there is also no predefined standard or guide on how migrations are supported from legacy systems into the NewSQL systems. A survey on the migration support offered by the most widely used and recognized systems would provide a good staring point for picking the right one for a specific list of requirements.

Acknowledgments. This research was made possible by funding from the ICT-AGRI-FOOD 2020 Joint Call. This work was supported by a grant of the Romanian National Authority for Scientific Research and Innovation, CCCDI - UEFISCDI, project number COFUND-ICT-AGRI-FOOD-MUSHNOMICS 205/2020, within PNCDI III.

References

1. Abadi, D.: Consistency tradeoffs in modern distributed database system design: cap is only part of the story. Computer **45**(2), 37–42 (2012)
2. Agrawal, S., Narasayya, V., Yang, B.: Integrating vertical and horizontal partitioning into automated physical database design. In: Proceedings of the 2004 ACM SIGMOD International Conference on Management of Data, pp. 359–370 (2004)
3. Almassabi, A., Bawazeer, O., Adam, S.: Top newsql databases and features classification. Int. J. Database Manage. Syst. **10**, 11–31 (2018)
4. Altibase: Altibase (2020). https://altibase.com/
5. Aslett, M.: What we talk about when we talk about newsql (2020). https://blogs.451research.com/information_management/2011/04/06/what-we-talk-about-when-we-talk-about-newsql/
6. Bader, A., Kopp, O., Falkenthal, M.: Survey and comparison of open source time series databases. Datenbanksysteme für Business, Technologie und Web (BTW 2017)-Workshopband (2017)
7. Binani, S., Gutti, A., Upadhyay, S.: Sql vs. nosql vs. newsql-a comparative study. Database **6**(1), 1–4 (2016)
8. Brewer, E.: Spanner, truetime and the cap theorem (2017)
9. Burchall, L.D., Goodsell, A.E., Shirley, B.A.: Delaying database writes for database consistency (2014). uS Patent 8,768,890
10. Chandra, U.: A comparative study on: Nosql, newsql and polygot persistence. Int. J. Soft Comput. Eng. **7** (2017)
11. Chang, F., et al.: Bigtable: a distributed storage system for structured data. ACM Trans. Comput. Syst. **26**(2), 1–26 (2008)
12. CitusData: Citus (2020). https://www.citusdata.com/
13. Cockroachlabs: Cockroachdb (2020). https://www.cockroachlabs.com/
14. Codd, E.F.: Data models in database management. In: Proceedings of the 1980 Workshop on Data Abstraction, Databases and Conceptual Modeling, pp. 112–114 (1980)
15. Codd, E.: An evaluation scheme for database management systems that are claimed to be relational. In: 1986 IEEE Second International Conference on Data Engineering, pp. 720–729. IEEE (1986)
16. Corbellini, A., Mateos, C., Zunino, A., Godoy, D., Schiaffino, S.: Persisting big-data: the nosql landscape. Inf. Syst. **63**, 1–23 (2017)
17. nosql database.org: Nosql (2020). http://nosql-database.org/
18. Fauna: Fauna (2020). https://fauna.com
19. solidIT consulting & software development GmbH: Db-engines (2020). https://db-engines.com/
20. Grolinger, K., Higashino, W.A., Tiwari, A., Capretz, M.A.: Data management in cloud environments: nosql and newsql data stores. J. Cloud Comput.: Adv. Syst. Appl. **2**(1), 22 (2013). https://link.springer.com/article/10.1186/2192-113X-2-22
21. Haerder, T., Reuter, A.: Principles of transaction-oriented database recovery. ACM Comput. Surv. **15**(4), 287–317 (1983)
22. Hajoui, O., Dehbi, R., Talea, M., Batouta, Z.I.: An advanced comparative study of the most promising nosql and newsql databases with a multi-criteria analysis method. J. Theor. Appl. Inf. Technol. **81**(3), 579 (2015)
23. Han, J., Haihong, E., Le, G., Du, J.: Survey on nosql database. In: 2011 6th International Conference on Pervasive Computing and Applications, pp. 363–366. IEEE (2011)

24. Haselmann, T., Thies, G., Vossen, G.: Looking into a rest-based universal api for database-as-a-service systems. In: 2010 IEEE 12th Conference on Commerce and Enterprise Computing, pp. 17–24. IEEE (2010)
25. Hecht, R., Jablonski, S.: Nosql evaluation: a use case oriented survey. In: 2011 International Conference on Cloud and Service Computing, pp. 336–341. IEEE (2011)
26. Holenstein, P.J., Holenstein, B.D., Highleyman, W.H.: Referential integrity, consistency, and completeness loading of databases (2011). uS Patent 7,949,640
27. Huang, D., et al.: Tidb: a raft-based htap database. Proc. VLDB Endowment **13**(12), 3072–3084 (2020)
28. Inc., G.: Spanner (2020). https://cloud.google.com/spanner
29. Jung, K.C., Lee, K.W., Bae, H.Y.: Design and implementation of replication management in main memory dbms altibase tm. In: International Conference on Parallel and Distributed Computing: Applications and Technologies, pp. 62–67. Springer (2004). https://doi.org/10.1007/b103538
30. Khasawneh, T.N., AL-Sahlee, M.H., Safia, A.A.: Sql, newsql, and nosql databases: a comparative survey. In: 2020 11th International Conference on Information and Communication Systems (ICICS), pp. 013–021 (2020)
31. Kumar, R., Gupta, N., Maharwal, H., Charu, S., Yadav, K.: Critical analysis of database management using newsql. Int. J. Comput. Sci. Mob. Comput. **3**, 434–438 (2014)
32. Leavitt, N.: Will nosql databases live up to their promise? Computer **43**(2), 12–14 (2010)
33. Lu, J., Holubová, I.: Multi-model databases: a new journey to handle the variety of data. ACM Comput. Surv. **52**(3), 1–38 (2019)
34. Mahfoud, Z., Nouali-Taboudjemat, N.: Consistency in cloud-based database systems. Informatica **43**(3) (2019)
35. May, N., Bohm, A., Lehner, W.: Sap hana–the evolution of an in-memory dbms from pure olap processing towards mixed workloads. Datenbanksysteme für Business, Technologie und Web (BTW 2017) (2017)
36. Meyer, S., Degener, J., Michener, B., Giannandrea, J.: Database replication (2012). uS Patent 8,204,856
37. Mohamed, M.A., Altrafi, O.G., Ismail, M.O.: Relational vs. nosql databases: a survey. Int. J. Comput. Inf. Technol. **3**(03), 598–601 (2014)
38. Nayak, A., Poriya, A., Poojary, D.: Type of nosql databases and its comparison with relational databases. Int. J. Appl. Inf. Syst. **5**(4), 16–19 (2013)
39. NuoDB: Nuodb (2020). https://nuodb.com/
40. O'Neil, P., Cheng, E., Gawlick, D., O'Neil, E.: The log-structured merge-tree (lsm-tree). Acta Informatica **33**(4), 351–385 (1996)
41. Padhy, R.P., Patra, M.R., Satapathy, S.C.: Rdbms to nosql: reviewing some next-generation non-relational database's. Int. J. Adv. Eng. Sci. Technol. **11**(1), 15–30 (2011)
42. Peng, D., Dabek, F.: Large-scale incremental processing using distributed transactions and notifications (2010)
43. PingCap: Tidb (2020). https://pingcap.com/
44. Stonebraker, M.: Sql databases v. nosql databases. Commun. ACM **53**(4), 10–11 (2010)
45. Vatika, S., Meenu, D.: Sql and nosql databases. Int. J. Adv. Res. Comput. Sci. Softw. Eng. **2**(8), 20–27 (2012)
46. VoltDB, I.: Voltdb (2020). https://www.voltdb.com/

A Survey on Challenges Facing Artificial Intelligence Based Pattern Recognition for Business Oriented Big Data Analytics and Solutions

Ahmed Maghawry[2](✉), Amr Elhadidi[2], Ahmed Alqassed[2], Mohamed Awad[2], Ayman Taha[2], and Mohamed Kholief[1]

[1] College of Computing and Information Technology, Arab Academy for Science, Technology and Maritime Transport (AASTMT), Alexandria, Egypt

[2] Research and Development, Business Solutions Department, EFinance Cairo, Egypt

{Ahmed_Maghawry,Amr_ElHadidi,Ahmed_Alqassed,Mohamed_Awad, Ayman_Taha}@efinance.com.eg

Abstract. The digital transformation revolution is targeting all aspects of our lives. Soon, we will have almost all fields transformed from routine manual operation into digital operation releasing a massive amount of data as a result. Such big data can be either a burden to any institution or a gold mine depending on how any institution approaches its output data. Big Data analytics is considered the fruit of digital transformation providing insights with decision support to decision makers to direct the institution toward efficiency to achieve their business goals. Applying big data analytics on raw data is considered un efficient with insignificant outcome. Applying artificial intelligence techniques for pattern recognition before big data analytics makes it more efficient with significant outcome. Clustering based pattern recognition can be considered one of the most efficient pattern recognition support techniques for big data analytics. In this paper, we will review previous work done in this field from critical point of view while proposing possible future trends or extended research areas to achieve better business-oriented data analytics to enhance business oriented institutional decision support.

Keywords: Big data science · Big Data analytics · Artificial intelligence · Clustering

1 Introduction

Digital transformation is one of the pillars instituting the concept of sustainable development endorsed by the united nations. It can be abstractly defined as targeting almost all aspects of human lives to be transformed from manual into digital form. Such transformation introduces significant benefits including resource reduction and maximizing business goals that will provide broad spectrum development in all fields.The concept of digital transformation has surged lately to be referred to as the digital transformation evolution. This was a result of the increase in the speed of which digital transformation

R. Silhavy (Ed.): CSOC 2021, LNNS 230, pp. 237–248, 2021.
https://doi.org/10.1007/978-3-030-77442-4_20

is reaching all fields. On the other hand, the more fields that gets digitally transformed, the more data is obtained from each field. Such data can be either a burden or a gold mine to any institution as nowadays data is power.

Handling the output data of digital transformation efforts can be done on various levels starting with simple statistics, extending to visualized reports and peaking at pattern recognition and data analytics. Each level has its own requirements and benefits and is embraced according to the business goals of the institution, but none will provide significant decision support benefits such as AI based big data analytics. The existence of such systems helps different institutions either private or governmental to efficiently achieve their business goals including minimizing resources while maximizing profits. There exists a variety of big data analytics tools and systems that can be utilized to obtain decision support mechanisms, most are based on a variety of AI based techniques that are utilized for pattern recognition. Applying data analytics on raw data will be a relatively time and resource consuming process compared to applying the same on clustered smaller groups of the same data, where each members of the same group share the same pattern while members of different groups each has its unique pattern that defines all members of the same group. Once massive data is split into pattern recognition based smaller groups, questions are raised, such questions are then answered by data analysts that apply data analytics techniques that will answer different questions raised by applying for example AI based clustering on a set of massive data generated from a full-scale management system that facilitates digital payments for tax suppliers. The rest of the paper is organized as follows, Sect. 2 features a background laying foundation to terms and definitions that institute the main pillars of this paper. Section 3 focuses on challenges. Section 4 reviews related previous work, and finally conclusions are presented in Sect. 5.

2 Background

2.1 Big Data

Such term has been in use since 1990 relating to John Mashey [3, 4]. The concept of big data refers to data sets with sizes that exceed the ability of traditional software tools to deal with, hence, such concept is subjective, such that, one organization considers a multi hundred gigabyte data as big data, while for others, it requires a multi thousand gigabyte data before such organization starts to reconsider its data management tools. It is the concept of introducing methods to deal with data sets that are extremely large or complex to be approached using traditional data-processing applications. Furthermore, it also deals with analysing or even extracting information from such massive data sets. The more rows in a data set, the greater statistical power it will offer. While the higher its complexity the more possible it will lead to a higher false discovery rate (FDRs).

FDR is a method that frames the rate of type I errors in null hypothesis testing that results from conducting multiple comparisons. The less FDR the more confidence will be introduced with the results with the goal to eliminate false discoveries. Big data targets include capturing data, storing data, analysing data, search through data, data transfer, visualizing data, privacy as well as the data source itself. The concept of big data has been associated with three key terms: volume variety and velocity. Big data usually includes data with sizes that exceed the processing capacity of traditional software. "There is little doubt that the quantities of data now available are indeed large, but that's not the most relevant characteristic of this new data ecosystem" [1]. "Analysis of data sets can find new correlations to "spot business trends, prevent diseases, combat crime and so on" [2]. Different entities including scientists, business executives, medical practitioners, advertising agencies and government institutions all face the same difficulties when dealing with large datasets.

Handling big data with traditional database management systems and software packages often faces difficulties, such work may require intensive parallel software running on hundreds or even thousands of high-performance servers. Labelling a data set as big data depends on the capabilities of the users and their tools, expanding such capabilities makes big data a moving target. For some institutions, a multi hundred gigabyte data set is all what it takes to trigger the need to reconsider data management options, while for others, it requires ten times this number before data size becomes a significant consideration. Figures 1, 2 and 3 shows a visualized conceptual definition of data velocity, variety, and volume.

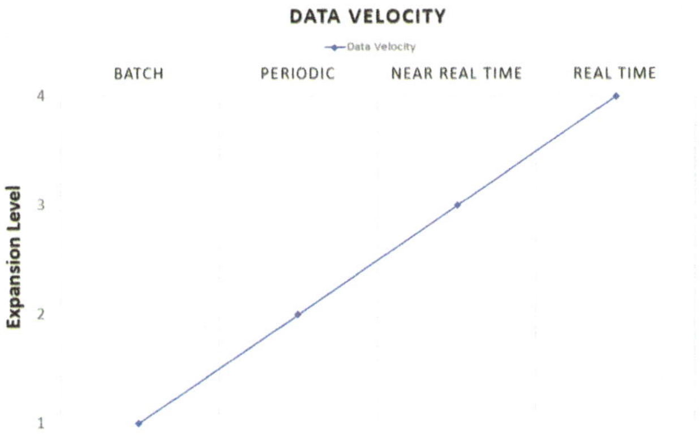

Fig. 1. Data velocity

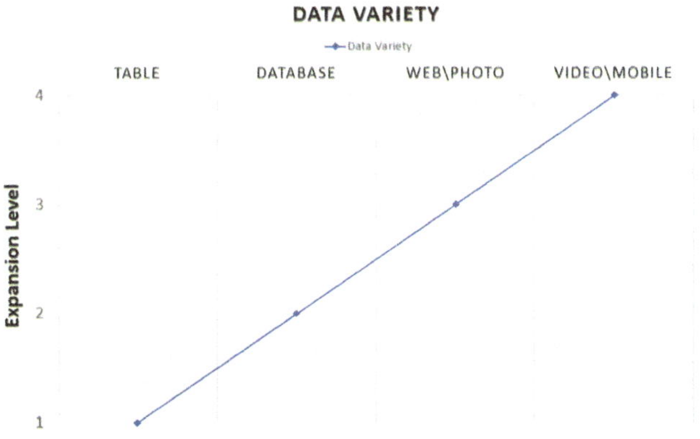

Fig. 2. Data variety

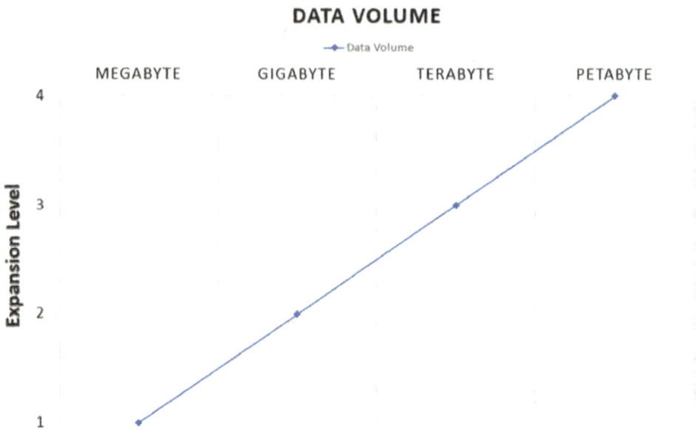

Fig. 3. Data volume

2.2 Data Science

It is an inter-disciplinary field that utilizes a combination of scientific methods, processing algorithms and systems to obtain insights and knowledge from data wither its structural or unstructured [5, 6]. Data science can be directly related to data mining and machine learning as well as big data. Data science unifies statistics and data analysis alongside their related methods in order to analyse and understand phenomena in data [7]. The concept of data science also utilizes methodologies and theories driven from a variety of fields including mathematics, statistics, computer science, domain specific knowledge as well as information science. Data science has been also portrayed as a "fourth paradigm" of science (empirical, theoretical, computational, and now data-driven), in addition to that, it was also proposed that science is changing as a result of the impact of information technology and data [8, 9]. There exist a variety of techniques and technologies that are utilized in data science including clustering, linear regression, support vector machines and generally machine learning. Implementation wise, many programming languages will support the implementation of any data science technique, however, Python gained popularity among data scientists for having simple syntax and readymade libraries including Numpy pandas and Scipy that makes data scientists life much easier. Many frameworks also exist for data scientists to use including Google's TensorFlow and Facebook's Pytorch as well as Apache Hadoop. In addition to that, a variety of visualization tools also exists including Microsoft's PowerBI and Google Charts. Not to mentions data science platforms including RapidMiner, Anaconda as well as MATLAB.

2.3 Data Analysis

Data analysis can be considered as the process of inspection, cleansing, transforming as well as modelling data with the intention of discovering useful information, conclusions that may offer comprehensive decision support. Data analysis has multiple approaches with diverse techniques under a variety of names, and its also utilized in different fields including business, science as well as social science. Nowadays, data analysis has a great impact on decision making that makes decision makers take scientific based steps allowing businesses to operate more effectively. Data Mining (ML) is considered a data analysis technique that has a predictive rather than descriptive targets by utilizing statistical modelling and knowledge. On the other hand, Business Intelligence (BI) focuses on data analysis that focuses mainly on business information [10]. Data analysis can be divided into descriptive statistics, exploratory data analysis (EDA) that focuses on the discovery of new features in a given data set, and confirmatory data analysis (CDA) that focuses on the negative or positive confirmation of an existing hypotheses. Predictive analytics targets applying statistical models to achieve predictive forecasting or classification, while text analytics utilizes statistical linguistic and structural techniques to obtain and classify useful information from text-based sources. All of the analytics types mentioned earlier can be considered a varieties of data analysis. Figure 4 shows and outline of the relation between data analysis and reality.

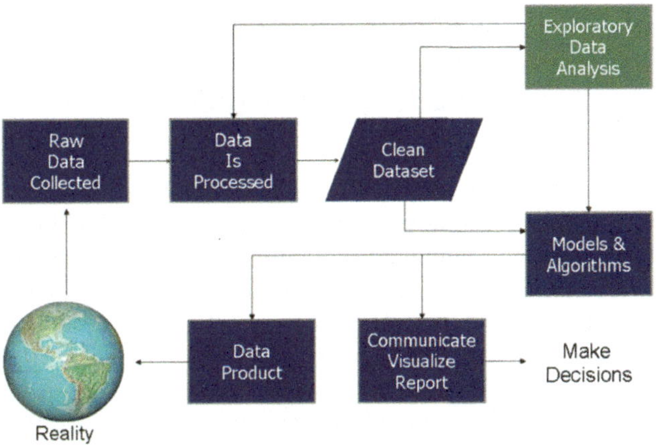

Fig. 4. Data analysis and reality [11].

Data is collected from a variety of sources, how this will happen, is communicated by analysts to database administrators such as information technology personnel within an organization. Furthermore, the data can also be collected from sensors for example in the context of internet of things (IoT). Once data is collected, it must be further processed or organized to be prepared for analysis, for example, data may be required to be put in a structured data format in rows and columns within a table. The processed data may be distorted in the form of duplicity or useless values; hence, the data may require further data cleaning on which multiple processes is executed on the processed data to further prepare it for analysis [12]. Figure 5 shows the phases of the intelligence cycle used to convert raw data into useful intelligence or knowledge that are similar to the phases of data analysis.

2.4 Artificial Intelligence

The term refers to intelligence manifested by machines. One study defines artificial intelligence (AI) as any machine that perceives its environment in such a way that enables

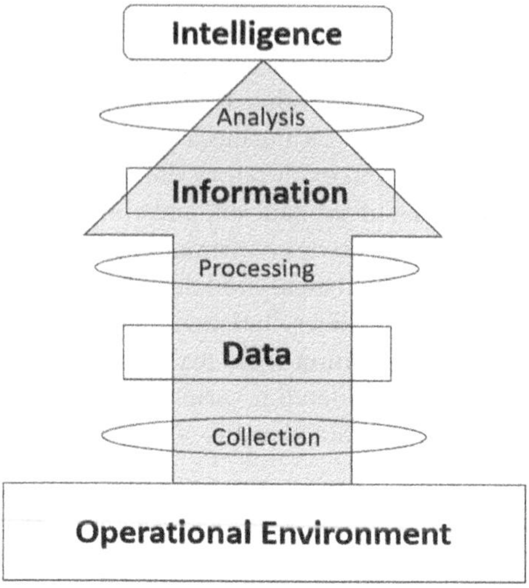

Fig. 5. Data to Intelligence

it to take actions that maximizes its chance of successfully achieving its targets [13]. On the other hand, other research defines AI by referring to machines that mimic the cognitive functions of human intelligence manifested in learning and problem solving [14]. The AI effect refers to the fact of removing tasks that no longer require intelligence to be done from the circle of AI [15]. "AI is whatever hasn't been done yet" [16]. There exist a variety of artificial intelligence techniques [19] that can be used for example to achieve pattern recognition in a given data set that can later be used for analysis to obtain decision support valuable information. Such techniques include Genetic Algorithms (GAs), Artificial Neural Network (ANNs\NNs) as well as Clustering techniques including K-Means clustering algorithm [17].

2.5 K-Means Clustering

The K-means clustering algorithm is considered one of the partitioning based, non-hierarchical clustering techniques [18], it is one of the most popular clustering algorithms that has been utilized in a variety of scientific and technological fields [20, 21]. The k-means clustering algorithm is used commonly as a result of its applicability on different data types. The algorithm starts with a set of targeted numeric objects X and an integer number k that represents the desired number of clusters. The algorithm then attempts to partition all members of X into k clusters while minimizing the sum of squared errors [22]. In the beginning, the algorithm randomly initializes the cluster centres, then the algorithm attempts to assign each member of X to its closest cluster centre according to the square of the Euclidean distance from the cluster centre [23].

Then, the value of each centre is updated by computing the mean value of each cluster, this updating process is a result of the change of membership of the cluster members [24]. The algorithm then iterates through these processes until no more changes in the cluster's membership is achieved which is known as convergence. To calculate how near a data vector is to a cluster's center, the following formula is used:

$$d\left(z_p, a_j\right) = \sqrt{\sum_{k=1}^{d}\left(z_{pk} - a_{jk}\right)^2} \tag{1}$$

3 Challenges

3.1 Data Challenges

Many organizations suffer initializing their big data projects. This is often caused by the lack of awareness of the challenges of big data, or the lack of skill and equipment needed to tackle those challenges. Challenges facing big data project include:

A) Big Data Misconception
This is usually caused by insufficient understanding of the big data concept by individuals who will have key roles in big data projects such as data base administrators. In such case, if involved data base administrators lacks the understanding of the big picture of big data, they may for example ignore keeping backups of sensitive data, they may also not use databases properly and efficiently for storage. As a result, if such important data is needed, its highly likely that it will not be retrieved easily, or in worst case scenario may not be retrieved at all.

Such problem can be tackled through big data workshops and seminars being held for all employees within the same organization despite being key persons on big data projects or not. This will secure the basic understanding of the big data concept by all levels within the same organization.

B) Data Growth
As the digital transformation peaks, so is the amount of data being stored in data centres and databases. If such data is not stored properly and efficiently, it will be extremely difficult to handle with time. Most of such data is unstructured and may come from documents, video, audio or even text files as well as other resources, which implies that they may not be found in databases. Such challenge can be tackled by opting modern techniques including data compression, tiering, and deduplication. Compression reduces the number of bits consumed by the data while deduplication removes duplicate values from a data set. While data tiering allows an organization to store data in various storage tiers including public cloud, private cloud, or even flash storage to ensure that the data resides in the most appropriate storage space.

C) Big Data Tool Selection
Most organizations may get confused while choosing the best tool for big data storage and analysis. Such confusion may lead organizations to make poor decisions and choose an inappropriate technology that will result in resource, time, and effort waste. This can be mitigated by seeking professional advice and consultation. Consultants will give recommendation of the best tools that will meet the targeted business needs with maximum efficiency.

D) Unskilled individuals
Organizations need skilled data professionals to run modern technologies and big data tools. Such group includes data scientists, analysts and engineers who are experienced on running such tools to make sense out of massive data sets.

Such challenge can be tackled by providing training for existing individuals while heiring skilled professional at the same time depending on the organization's needs. In addition to that, organizations can purchase AI based data analytics solutions that can be run by professionals who may not be data science experts but have enough knowledge to run such applications. Such step can help organizations to save a lot of recruitment resources.

E) Data Security
Organizations usually get busy understanding, storing, and analysing large data sets that they push security concerns for late stages which is not a smart thing to do. Unprotected data repositories can become breeding ground for malicious hackers. Data breaches may lead to catastrophic losses. To mitigate data security concerns, a collection of data security concepts must be applied on large data sets including encryption that makes stolen data useless, Data segregation in which data is gathered into clusters with homogenous properties that makes it easy to access to serve your needs efficiently. Identity access control will eliminate unwanted access for unauthorized individuals to protect sensitive data. Real-time security monitoring will enable security offices to mitigate threats on real-time before they even become a significant threat.

F) Data Sources Variation
In large organizations, data comes from a variety of sources like social media, Enterprise Resource Planning (ERP) applications, log files, emails, presentations and reports. Organizing such various data is a notoriously challenging task. Data integration is important for further analysis and reporting as well as business intelligence efforts. To tackle this challenge, large organizations must either purchase the right tools or even implement their own tools that will facilitate data integration efficiently.

3.2 Data Science Challenges

Many organizations across the globe are looking forward to organizing, process and uncover the value of the massive amounts of data they generate to transform them into a high value actionable business insight. Such organizations hire data scientists to do this task, and it's not an easy one, as it faces its own challenges. Such challenges include:

A) Data Preparation
Enough time should be spent to perform data preparation including cleaning and preparing data to improve its quality by making it accurate and consistent before the analysis phase. This step can be considered the worst part of this job as it will consume time and effort. It is required to go through terabytes of data with multiple formats, sources, and platforms daily while keeping a log of their activities to prevent duplication. One way to tackle this challenge is by adopting AI based data science technologies like augmented

analytics that automates the manual data cleansing and preparation tasks that will enable data scientists to be more productive.

B) Multiple Data Sources
Large organization will be generating their own data from multiple sources that may include Point of Sale (POS) machines, portals, or full-scale management systems. Such variety of data sources will make data scientists job almost impossible. One way to tackle such problem is by implementing a centralized data integration platform that will gather data from various sources into a central place that will be targeted later by data scientists.

C) Data Security
It's not an accepted practice that data scientists target large data sets for processing and analysis while compromising the data for external threats, in this case, the outcome of data science will not overcome its negative effects. Hence data security must be approached seriously in each step in the work of data scientists.

D) Understanding the Business Problem
This is probably the most important challenge facing a data scientist, which is to understand the targeted business problem. A Data scientist may be approaching data generated from any field, medicine, engineering, military or governmental. In all cases, a data scientist have to understand the business significance of the data on hand and the targeted problem to solve to reach a specific business oriented goal.

E) Dealing with Non-Technical Stakeholders
As mentioned before, all individuals must be involved in the big data concept to embrace it to be utilized and orchestrated toward the organization goals. This also includes data science. A data scientist must have enough communication skills to break down bulk scientific information to communicate with non-technical stakeholders who might find it difficult to understand. A data scientist job hence includes a communication skills part in addition to the scientific part.

D) Misconceptions About the Role
In large organizations, a data scientist is expected to clean data, retrieve data, build models, and conduct analysis. However, this is an overconsumption of a data scientists. In order to make a data science team function correctly, tasks must be distributed among individuals.

4 AI Based Big Data Analytics

There exist a variety of AI based techniques used to maximize the potential of big data analytics including data and big data mining techniques like clustering, classification and frequent patterns mining technologies [25]. In this section of the paper, we will review previous work done in this field to attempt to spot gaps that may contribute to future research directions. Performance wise, parallel computing is among the most important future trends to make data analytics work efficiently for big data. Hence, technologies of cloud computing, Hadoop, and map-reduce will play a major role to support big data analytics. This will require new methods to handle the computing power of cloud-based systems to fulfil tasks as fast as possible manifested in a scheduling problem. Such problem is considered an important axis for research.

As a result, designing solutions for big data analytics will require such solutions to basically consider parallel computing as a must. Furthermore, fulfilling the ability to display the knowledge we get from big data analytics will decide wither big data analytics is applicable to work for real world approaches, not only theoretical and lab environments. Combining information from different sources to achieve information fusion will also improve the end result of big data analytics [25]. Social networks became part of our daily lives. The data generated from social networks is also part of the big data concept. Large organizations should also consider analysing the data of social networks trying to predict for example the customer behaviour [25]. Not to mention that security and privacy is an important aspect to consider in each step toward big data analytics projects, as its no acceptable to compromise data security to achieve big data analytics.

Many tools and frameworks already exist to server big data needs, some of them are already operational while at the same time an intensive amount of research effort is done to optimize them to get the most benefit out of these frameworks [26]. As a result, this interlinked the big data field to many other fields like AI including neural networks, genetic algorithms, and clustering algorithms. As the amount of data to be handled is continuously increasing, so is the technology and research efforts continue in order to meet the expanding increase of such data. The importance of big data for large complex organizations can be concluded in data management, mining, pattern recognition and analysis. It is a big mistake to attempt data analytics without pattern recognition first. As pattern recognition is applied first, massive amounts of data is them split into smaller groups with similar features and patterns learned by the execution algorithm. Such smaller groups can then be targeted for analysis in order to visualize the recognized patterns within each group. This maximizes the benefit of data analytics and makes it efficient in the first place.

Assume a large organization with various data sources that works for various systems developed by this organization. It is possible to approach each data source independently and apply AI based big data analytics on it that will result in significant decision support benefits. However, data scientists can gather all these independent data sources, analyse them and apply pre-processing and cleansing techniques that will result in one big Datawarehouse that can be mined with the big picture with any possibility to link the various data sources. Hence, the recognized patterns will be on a greater and larger perspective that can be analysed later to introduce large scale insight for the most efficient decision support efforts.

5 Conclusion

The digital transformation revolution is surging and transforming almost all aspects of our lives into digital solutions that generates massive amounts of data. The generated data must be handled efficiently in terms of storage and security to avoid data leaks and at the same time facilitate future processing and retrieval. Big data analytics empowers businesses to reach their targeted objectives in the most efficient way by minimizing resource and maximizing benefits. Hence, large organizations have to invest in non-conventional ways to handle data including frameworks, tools as well as professional skilled human resources. Many challenges will face large organization when attempting to tackle the big data realm including data storage, security, and sources as well as frameworks and tools selection in addition to unskilled personnel ending with communication with non-technical stakeholders. All such challenges can be mitigated with decent planning in addition to consulting professionals to choose the right frameworks and tools and provide training for all human resources involved in big data project regardless of being a key person or not. Also planning and work distribution is another critical point that must be handled properly to avoid over\under fitting of big data analytics efforts.

Big data is an advancing field with a significant amount of research is yet to be done. However, we believe that more research is needed to investigate the impact of applying combinations of artificial intelligence techniques in the phase right before the actual data analytics phase. Such effort will provide comprehensive pattern recognition output that will maximize the output of big data analytics. Furthermore, we believe there is a room for investigating the application of AI techniques in the early phases of this process including data gathering and integration as well as pre-processing and cleansing. Such application will further maximize the efficiency of big data analytics while facilitating the job of later phases including pattern recognition and later analysis.

References

1. Boyd, D., Crawford, K.: Six Provocations for Big Data. Social Science Research Network: A Decade in Internet Time: Symposium on the Dynamics of the Internet and Society (2011). https://doi.org/10.2139/ssrn.1926431. S2CID 148610111
2. Data, data everywhere. The Economist. 25 February 2010. Retrieved from 9 Dec 2012
3. John, R.M.: Big Data ... and the Next Wave of InfraStress (PDF). Slides from invited talk. Usenix (1998). Retrieved from 28 Sept 2016.
4. Steve, L.: The Origins of 'Big Data': An Etymological Detective Story. The New York Times (2013). Retrieved from 28 Sept 2016
5. Dhar, V.: Data science and prediction. Commun. ACM. **56**(12), 64–73 (2013). https://doi.org/10.1145/2500499. S2CID 6107147. Archived from the original on 9 November 2014. Retrieved from 2 Sept 2015
6. Jeff, L.: The key word in "Data Science" is not Data, it is Science. Simply Statistics (2013). Archived from the original on 2 January 2014. Retrieved from 1 Jan 2014
7. Hayashi, C.: What is data science? fundamental concepts and a heuristic example. In: Hayashi, C., Yajima, K., Bock, H.-H., Ohsumi, N., Tanaka, Y., Baba, Y. (eds.). Data Science, Classification, and Related Methods. Studies in Classification, Data Analysis, and Knowledge Organization, pp. 40–51. Springer, Japan (1998). https://doi.org/10.1007/978-4-431-65950-1_3. ISBN 9784431702085.
8. Tony, H., Stewart, T., Kristin, M.T.: The Fourth Paradigm: Data-intensive Scientific Discovery. Microsoft Research (2009). ISBN 978-0-9825442-0-4. Archived from the original on 20 March 2017. Retrieved from 16 Dec 2016
9. Bell, G., Hey, T., Szalay, A.: Computer science: beyond the data deluge. Science **323**(5919), 1297–1298 (2009). https://doi.org/10.1126/science.1170411.ISSN0036-8075.PMID19265007.S2CID9743327
10. Blog: Spotless. (n.d.). Retrieved from 15 Dec 2020. https://web.archive.org/web/20171018181046/https://spotlessdata.com/blog/exploring-data-analysis

11. O'Neil, C., Schutt, R.: Doing Data Science, O'Reilly (2014). ISBN: 978-1-449-35865-5. Forbes-Gil Press-A Very Short History of Data Science-May 2013
12. Arasu, A., König, A. C., Ganjam, K., Kaushik, R., Chaudhuri, S., Narasayya, V.: Data Cleaning (2017). Retrieved from 15 Dec 2020. https://www.microsoft.com/en-us/research/project/data-cleaning/?from=http%3A%2F%2Fresearch.microsoft.com%2Fen-us%2Fprojects%2Fdatacleaning%2F
13. Poole, D., Mackworth, A., Goebel, R..: which provides the version that is used in this article. These authors use the term "computational intelligence" as a synonym for artificial intelligence (1998)
14. Russell, S.J., Norvig, P.: Artificial Intelligence: A Modern Approach, 3rd edn. Prentice Hall, Upper Saddle River, New Jersey (2009). ISBN 978-0-13-604259-4.
15. McCorduck, P.: Machines Who Think, 2nd edn. A.K. Peters, Ltd., Natick, MA (2004), ISBN 1-56881-205-1
16. Maloof, M.: Artificial Intelligence: An Introduction, p. 37 (PDF). georgetown.edu. Archived (PDF) from the original on 25 August 2018
17. Maghawry, A.M., Omar, Y., Badr, A.: Initial centroid selection optimization for k-means with genetic algorithm to enhance clustering of transcribed arabic broadcast news documents. In: Silhavy, R., Silhavy, P., Prokopova, Z. (eds.) CoMeSySo 2017. AISC, vol. 662, pp. 86–101. Springer, Cham (2018). https://doi.org/10.1007/978-3-319-67621-0_8
18. Maghawry, A., Hodhod, R., Omar, Y., Kholief, M.: An approach for optimizing multi-objective problems using hybrid genetic algorithms. Soft. Comput. **25**(1), 389–405 (2020). https://doi.org/10.1007/s00500-020-05149-3
19. Maghawry, A., Kholief, M., Omar, Y., Hodhod, R.: An approach for evolving transformation sequences using hybrid genetic algorithms. Int. J. Comput. Intell. Syst. **13**(1), 223 (2020). https://doi.org/10.2991/ijcis.d.200214.001
20. Xu, R., Wunschii, D.: Survey of clustering algorithms. IEEE Trans. Neural Netw. **16**, 645–678 (2005). https://doi.org/10.1109/tnn.2005.845141
21. An Introduction to Classification and Clustering. Cluster Analysis Wiley Series in Probability and Statistics, pp. 1–13 (2011). https://doi.org/10.1002/9780470977811.ch1.
22. Hamerly, G., Drake, J.: Accelerating Lloyd's Algorithm for k-means clustering. In: Celebi, M.E. (ed.) Partitional Clustering Algorithms, pp. 41–78. Springer, Cham (2015). https://doi.org/10.1007/978-3-319-09259-1_2
23. Shrivastava, P., Kavita, P., Singh, S., Shukla, M.: Comparative analysis in between the k-means algorithm, k-means using with Gaussian mixture model and fuzzy c means algorithm. Commun. Comput. Syst. (2016). https://doi.org/10.1201/9781315364094-186
24. Lei, X.-F.: An efficient clustering algorithm based on local optimality of K-means. J. Softw. **19**, 1683–1692 (2008). https://doi.org/10.3724/sp.j.1001.2008.01683
25. Tsai, C.-W., Lai, C.-F., Chao, H.-C., Vasilakos, A.V.: Big data analytics: a survey. J. Big Data **2**(1), 1–32 (2015). https://doi.org/10.1186/s40537-015-0030-3
26. Malgaonkar, S., Soral, S., Sumeet, S., Parekhji, T.: Study on big data analytics research domains. In: 2016 5th International Conference on Reliability, Infocom Technologies and Optimization (Trends and Future Directions) (ICRITO), pp. 200–206. Noida (2016). https://doi.org/10.1109/ICRITO.2016.7784952

Operationalizing Analytics with NewSQL

Ionela Chereja(✉), Sarah Myriam Lydia Hahn, Oliviu Matei, and Anca Avram

Computers and Information Technology, Technical University Cluj Napoca, Cluj-Napoca, Romania
{ionela.chereja,anca.avram}@ieee.org, sarah-hahn@gmx.net, oliviu.matei@holisun.com

Abstract. Operational data and analytical data are no longer two separate disciplines and discussions. Data Analysis is gaining more ground and more request from companies that begin to base their strategies - as well as decision intelligence and decision management - on factual information. In surveying the state of the data industry and the trends in data management technologies, NewSQL systems appear to be more present. They are able to answer the question of bridging operational data storage and administration with providing real-time access to analytical data. This paper aims to provide a structured look into the features and capabilities offered by NewSQL systems that can be leveraged to allow Data Analysis over a variety of data types. Furthermore, it provides an overview of Realtime Analytics offerings, Map Reduce capabilities and hybrid (transactional and analytical) features.

Keywords: NewSQL · Data analysis · Database systems · OLAP · OLTP · HTAP · Realtime analytics · Map reduce · Streaming · Timestamp

1 Introduction

Over the course of the last few years, more and more of the focus of the IT industry has turned towards data. Google trends [31] reports an all time high of the term "data science" in the year 2020, as well as the term "storage". More data is being generated currently than ever before, with an estimated number of 2.5 quintillion bytes every day [54] and the expectancy that it will grow to 463 exabytes by 2025. With this amount of available information, there has been a shift not only towards improving how we store that data, but also how we analyze it in order to generate value from the stored bytes.

Traditional database management systems have been developed for supporting transactional processing [26] by implementing the ACID (Atomicity, Consistency, Isolation and Durability) properties. Multiple solutions have been implemented and have matured within the SQL standard [9], usually called SQL databases or Relational Database Management Systems (RDBMS). The data stored in the SQL databases is therefore highly relational, having a clearly

R. Silhavy (Ed.): CSOC 2021, LNNS 230, pp. 249–263, 2021.
https://doi.org/10.1007/978-3-030-77442-4_21

defined schema and predefined sets of links between entities, also known as structured data.

The increase in the volume of data generated, the variety of data structures and lack thereof, as well as the subsequent need to store and access it, has lead to the term of Big Data. While Big Data is mainly linked to other technologies as cloud computing, Internet of Things, data centers and Hadoop [12], it is also associated with a revolution in data management systems.

With RDBMs systems being seen as unfit, NoSQL database systems were developed [28]. These systems are not relational and they provide high scalability at the expense of the ACID properties. NoSQL databases follow a set of BASE principles. BASE stands for Basic Availability, Soft state and Eventual consistency. It can be considered an alternative for ACID [15]. However, doubts appeared around how fit NoSQL systems are in providing the features required by the industry, especially for the applications and organizations that require reliability and strong transactional support [35].

Following the hype of NoSQL databases and the disillusion thereafter, a new category of database systems began to arise. These are known as **NewSQL** and they aim at combining the ACID support and highly relational data model with the scalability and performance of NoSQL stores. NewSQL has only appeared as a term since 2011 when it was used in an article trying to label a new generation of database management systems [5]. Ever since, the term has been adopted and used by different vendors, although the definition and concepts surrounding it are not standardized or widely accepted.

Simultaneously to the need of storing systems that are sufficiently capable, flexible, scalable and performant to handle Big Data, there has been an increasing interest in actually using the data for analyze. If previously the limitations in storage space, computing power and other IT resources have drawn the interest in creating a technology to eliminate those blockers, the general adoption of Cloud Computing built yet another turning point. With the arrival of Cloud Computing and its widespread use, resources were barely an issue anymore in terms of physical limitations. The term XaaS (Everything-as-a-Service) [18] points towards the availability of resources directly "as a service". From Infrastructure-as-a-Service to Software-as-a-service, storing and management of data has been simplified. This prompted more organizations to start questioning what the data they store will be used for, and what it actually contains. The combination of performant ingest, store, process and analyze flows for data became key to organizational performance. Decision intelligence and decision modelling are among the areas that require strong data management and analytics design. They bring together different disciplines and provide a framework to model and design, align and perform tuning on decision models in order to improve the understanding of outcomes and behaviours for businesses [23].

This paper has the objective to create an overview of the current state of the data industry, as well as the foreseeable directions and challenges on storing and processing information. In doing so we analyze the data management system trends over the last three Gartner hype cycles [23], correlating these with the technologies that are part of the NewSQL implementations. The approach in this work is to define the taxonomy in regards to how the systems are selected and the dimensions of the analysis. The survey will create an overview of how the analyzed systems provide storage support for different categories of data: historical, realtime, streaming and temporal. Moreover, we inquire the embedded support offered in analyzing data. This is assessed through looking at Map-Reduce capabilities and realtime analytics support offered either through native features or integration points. Lastly, we also offer a look at the systems' capacity to act as an OLTP, OLAP or HTAP system. This provides an understanding on the flexibility of the system and the advantage of having hybrid engines that would better support high volumes of data and real time querying.

2 Related Work and Definitions

2.1 Overview of the Data Industry

The amount of information generated worldwide is leading to multiple threads in the development of the data field. Along with the emergence of Big Data as a separate category of data in terms of size and management requirements, storage systems have been developing to sustain this need. However, a big focus of the data industry is the analysis of this information. Terms like Data Analysis, Data Mining, AI and Data Science are a few of the ones that got attention over the course of the last few years.

According to Gartner [23], the most notable trends of 2020 in data analytics include: AI - the prediction over AI is that more than 75% of companies will shift towards operationalizing it, as opposed to just using it as pilot. Traditional reporting and dashboarding is declining while decision intelligence is expected to be a part of more than 33% of organizations by 2023. There is an expectation that more enterprises will explore capabilities from vendors for video, audio or image data by using a varied range of structured and unstructured content. This is called 'X analytics' [23].

One of the most relevant foreseeable trend refers to data and analytics. Although previously considered two different disciplines, the increased offer of end-to-end workflows creates a less distinct split of the two capabilities and generates convergence. Analytics directions also include graph analytics, which are techniques that allow exploration of relationships in data that were not easy to analyze with traditional approaches.

Next to these, Cloud usage introduces a challenge in optimizing cost and performance for data and analytics.

The current and the predicted trends for data focus on exploring the meaning of the stored information through varied and continuously improving analytics technologies. For the emerging analytics technologies to provide value in a reliable and performant manner, the systems that store the data have to keep up with the demands of the industry [45].

2.2 Big Data

Big Data is a term increasingly utilized to refer to the amount of data enterprises store nowadays. It defines data that is too voluminous to store, process or analyze using traditional database technologies. The amount of data defined with the term Big Data has also been called "enormous datasets" [20]. Big Data often includes a big quantity of unstructured data, generally in need of more realtime analysis [37].

The term Big Data is known as a buzzword [44] and as a consequence the definitions of what it actually represents are also numerous. A work often cited on Big Data, attempts to compare different existing definitions for the term [58].

Most often Big Data is characterized by several Vs. The initial model was the 3Vs, developed by a Gartner [23] expert. The 3Vs are Volume, Velocity and Variety. Since the initial definition, the Vs model has expanded to include multiple other dimensions. These include Value, Visualisation, Veracity, Viscosity, Virality and Validity [34].

The evolution and expansion of data has lead to what is called "big data problems" [24]. The disproportionate evolution between data and the technologies to process and store it lead to the emphasis on processing capabilities. The Big Data problems underline the relationship between the growth of data and the necessity of having the proper tools to manage it and process it adequately.

2.3 Data Management and Big Data

Big Data is often used together with terms such as Cloud computing, Artificial Intelligence or Analytics and paired with notions like streaming data, social media data feeds and massive amounts of information. While in the past the widely accepted and used standard for managing data was the SQL standard - Relational Database Management Systems that handle structured information - with the increase of the data volume and the changes in its structure, the traditional database systems were no longer fitted [39].

NoSQL databases were developed alongside the development of the internet and cloud computing, with high-performance demands in mind, including low latency and high concurrency, scalability, efficient big data storage and lower operational and management costs [28]. NoSQL data stores are built on a schema-free architecture to be able to accommodate unstructured and semi-structured data. These did not have a structured query language and did not follow the same principles as SQL systems did, falling short of features that enterprises were still demanding [35].

NewSQL is a generation of database systems that put together characteristics of both SQL and NoSQL systems in order to take the advantages of each of them and minimize the disadvantages of using one or the other [8]. NewSQL systems are designed in the era of Big Data, considering that they have been acknowledged as a "new" category of databases in 2013 [5]. The focus of NewSQL databases is to leverage the scalability, high performance and versatility of NoSQL systems, while still offering the ACID properties of SQL systems.

3 NewSQL Evolution

3.1 NewSQL Data Stores

With the arrival of web applications, database management systems were faced with bottlenecks due to increasing demands in resources, but limitations in hardware at the time. To overcome this, the first attempts were to scale RDBMSs vertically. However, moving the database management system to a new server with better hardware implied processes that were complex and required downtime, which was not an option for some of the systems being migrated. Another approach was to create a horizontal scaling mechanism, called sharding. This means that there are several multiple physical nodes that store databases, but the data is presented to the application as a single logical database by using a middleware component. The management of the data across the shards would be therefore handled by this middleware component. Known examples of such systems are the Google MySQL based cluster [51] and the eBay Oracle based cluster [50]. Companies, however, moved away from this sharding mechanism due to the fact that it did not manage to operate complex joins between tables or to update multiple records in a transaction with the required performance.

To solve this demand, some vendors took a different approach to database systems. After using NoSQL systems as a way to overcome scalability limitations of RDBMs and running into durability and consistency issues, a few new systems started to arise. Some of these databases offered implementations based on the Calvin system [55]. These systems called NewSQL have one major advantage over NoSQL stores - they bring ACID complacency. The article "What's Really New with NewSQL?" published in 2016 by Matthew Aslett proposes that the DBMSs that fall under the NewSQL category are only OLTP systems. NewSQL implementations target support for applications that execute read-write transactions that are repetitive, have a short execution life and work with small amounts of data [42]. However, some of the implementations labeled NewSQL offer the possibility of acting as an OLAP database or both.

While there are a few surveys addressing NewSQL systems, the majority fall under the comparison between SQL, NoSQL and NewSQL [8,33]. A few other articles focus specifically on NoSQL and NewSQL comparisons [11,27], and a limited number of papers focus on NewSQL stores alone considering that they are a relatively recent category. One example of such a work exemplifies the systems' features [1] and proposes a categorization.

The approach we use in this work is different. The focus is specifically on the capabilities offered by NewSQL systems in the area of analysis and the presence of the systems in the industry and the scientific community. We analyze thus the data available in the Gartner [23] hype cycles and create an overview of how technologies that are supported and implemented in NewSQL database systems have evolved over the last available three years. Moreover, a selection of the systems to be analyzed is made using a systematic approach. Both are defined in the following sub-chapters.

3.2 Gartner Hype Cycle Research Methodology

Gartner hype cycles are graphical representations of the perceived value of an emerging technology trend or innovation and the market promotion it has [40]. It explains a generally applicable path of a technology on an x-y axis. It is looking at the visibility and expectations of the value of the technology on the y-axis, and how that evolves over the course of time using the x-axis. The cycle is composed of two different curves, which are merged. The first curve or equation describes the human expectations in the shape of a hype level curve. The second one refers to the technology's maturity through an S-shaped curve.

There are five stages or phases on the hype cycle [22]:

1. Technology/Innovation trigger. The cycle is triggered (usually by a public announcement or a demonstration). When awareness regarding the technology starts to spread, media coverage begins. A few companies start to use it on the premise of early adapter advantages.
2. Peak of Inflated Expectations. Expectations are inflated through further media coverage. As a result, more companies start investing in the emerging technology.
3. Through of Disillusionment. Commercial early adoption fail to meet expectations. This can be in terms of performance or revenue wise. This, in turn, brings public disappointments which are spread by the media.
4. Slope of Enlightenment. Adopters who continued to invest and work with the technology begin to experience benefits. As a result of persistence and experience with the technology there is more knowledge and understanding gained. Performance also tends to increase as a result of that acquired knowledge. The technology enters another awareness wave to the industry.
5. Plateau of Productivity. In this phase, the value of the technology is better known and evaluated. Adoption is accelerated with the raised awareness and palpable results.

'Time-to-value gap' is a term representing the time between the Peak of Inflated Expectations and the Plateau of Productivity [22]. It may differ between technologies, depending on factors such as performance constraints, penetration potential into the market and integration complexity. When looking at a specific technology, the length of the hype cycle related to it may vary between two years to twenty years. However, there are three categories acknowledged related to the time-to-value gap. One category called 'normal technologies' have a foreseen gap between five to eight years. A second category called 'fast track technologies' are anticipated to only take two to four years. The third category is called 'long fuse technologies', and it can continue for up to two decades. These might go through several hypes.

The hype cycle model has seen growing interest in the industry and lately from scholars. As a result, the model has grown in maturity, combining qualitative methods with quantitative methods focusing on specificity [17].

There are several areas over which Gartner publishes yearly hype cycles. One of the areas is Data and Analytics, which includes a subcategory of Data

Management. This subcategory helps towards a better understanding of the emergence and pace of data management technologies.

This paper looks at the presence of NewSQL in the last three available Hype Cycles for Data Management (2017, 2018 and 2019). By analyzing the Hype Cycles of 2017, 2018 and 2019 this paper investigates where in the cycle NewSQL technologies are. We use this as an indicator of the technology's maturity and adoption levels, as well as an indicator of how fast the technology moves through the cycle in a three year time span.

The 2017 Hype Cycle mentions technologies as In-Process HTAP (Hybrid Transactional/Analytical Processing) [6], Point-of-Decision HTAP, Multimodel DBMSs, Graph DBMSs and Operational and Analytical In-Memory DBMSs. The same technologies that fall under the NewSQL umbrella are still present in the 2018 Hype Cycle: HTAP (In-Process and Point-of-Decision), Multimodel DBMS and Graph and Key-Value Database Management Systems. Similarly, the hype cycle of 2019 contains Operational In-Memory DBMSs, Analytical in Memory DBMSs, Multimodel, Graph and Key-Value DBMSs. Although Graph and Key-Value Stores have primarily NoSQL implementations, there are also a few NewSQL systems that can fall under that category, such as FaunaDB [21] which supports GraphQL or Spanner [32] which evolved from a key-value store.

The comparison for the three years is made per sub-category and the changes present in the Gartner Hype Cycle, as seen in Fig. 1.

In-Process HTAP [30] systems are found only in the 2017 and 2018 Cycles. In 2017, the technology is already classified under the category that will reach the Plateau in 5 to 10 years, and it is located right at the Peak of Inflated Expectations. By 2018, it reached the beginning of the Trough of Disillusionment. In the 2019 Cycle, it is no longer mentioned.

Similar to In-Process HTAP, **Point-of-Decision HTAP** is only present in the 2017 and 2018 cycles. This sub-category is at the edge between the Peak of Inflated Expectations and the Trough of Disillusionment in 2017 and by 2018 there's only a small change in the downward slope of the latter quadrant. In both years, the technology is classified under category that would reach the Plateau in 2 to 5 years. The 2019 Cycle no longer contains the category.

Multimodel DBMSs [29,41] are present across the three analyzed cycles. In all three years, the technology is mentioned under the 2 to 5 years category. However, during the three hype graphics it only is descending from the high slope of the Peak of Inflated Expectations towards the beginning of the Trough of Disillusionment.

Examples of NewSQL implementations that are also categorized as Multimodel include Google Spanner [32], CockroachDB [14], TiDB [43] and SequoiaDB [48].

The sub-category of **Operational In-Memory DBMSs** include examples as VoltDB [57], NuoDB [38] and MemSQL [52]. This is part of the so called 'fast track technologies', as it is under the category expected to reach Plateau in less than 2 years. In 2017, it is located on the downward slope of Trough of Disillusionment. In 2018, it's already at the beginning of the upward curve

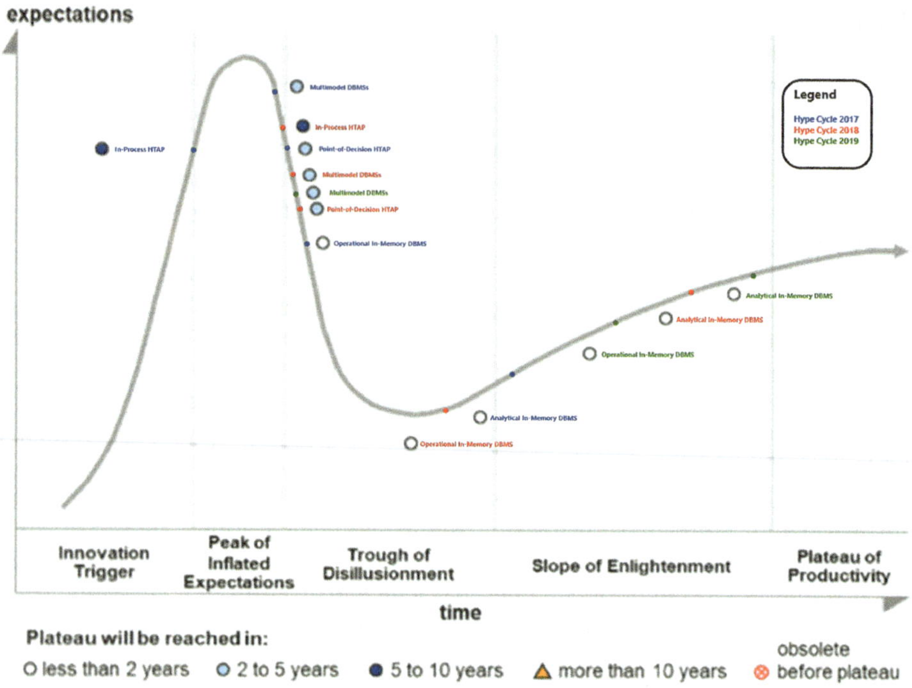

Fig. 1. Gartner Hype Cycle Comparison for Data Management 2017-2018-2019

leading to the Slope of Enlightenment, which is then reached as seen in the 2019 Hype Cycle.

Analytical In-Memory DBMS sub-category [36] is missing from the 2017 Hype Cycle, appearing only in 2018 and already on the upward curve of the Slope of Enlightenment quadrant. By 2019 it is already almost at the edge of the Plateau of Productivity. One example of NewSQL implementation is VoltDB [19].

A new emerging technology that is part of the 2019 Hype Cycle is Event Streaming Processing for which several NewSQL systems have support.

4 Analysis

4.1 Analyzed Systems

Available works that have been surveyed for this work (including, but not only [1,8,11,27,33]) include mentions of varied NewSQL implementations, such as Google Spanner, MemSQL, VoltDB, TiDB, Citus, Vitess, NuoDB, Clustrix DB, CockroachDB, FaunaDB and SAP HANA among the most recognizable and known. Other mentions of NewSQL systems include Harper DB, Apache Trafodion, GemFire XD, C-TreeACE, Altibase, Active Spaces, Actor DB and TokuDB.

As the NewSQL category is highly dynamic and volatile through emerging systems that become available or through changes in features offered by existing databases, this paper uses a two-dimensional approach to choosing which systems to analyze. The score available for the database on DB-Engines Ranking [25] and the presence of the system in the academic community. The DB-Engines score in itself is a multi-dimensional evaluation, including the number of times the system is mentioned online, the interest and frequency seen on technical websites, but also the number of job offers and mentions in the profiles available on professional networks (such as LinkedIn). This offers an overview on how known, stable and widespread the industry's awareness and usage is. Secondly, we use the number of mentions the system has in scientific papers. Both Google Scholar and WebOfKnowledge have been used.

As a result of this initial analysis we limit the systems surveyed to the top nine databases from the ones mentioned above. These are as follow: **SAP HANA** [47], **VoltDB** [57], **CockroachDB** [14], **Google Spanner** [32], **Citus** [13],**NuoDB** [38], **TiDB** [43], **FaunaDB** [21], **Altibase** [2].

4.2 Taxonomy and Analysis

The database systems were categorized first over the data types supported. The databases are categorized into Real-Time, Historical, Streaming or Temporal [1].

Real-Time data is data that gets updated with a high frequency and is intensely accessed as a result of heavy traffic. The database needs to provide the information requested usually within specified time constraints [7]. Distributed Real-Time database systems manage large volumes of data that is dispersed across multiple locations while performing real time processing over it [49].

Historical data refers to the ability of storing big volumes of data in the database. The data store holding the information is therefore the main repository of the historical data that is usually optimized for analytics by following a warehouse architecture [46].

Streaming data is data that gets ingested constantly in small batches from numerous data sources, called streams. Streaming data can have a variety of types from a wide variety of sources, for examples log records, click streams in web tracking, data feeds from sensors or network packages. The databases that support this data type need to support querying this information most often within time constraints [4].

Temporal data is defined as information that can be queried, updated and written at a specific timestamp. Data can therefore be read at a specific point in time, recreating the database's state at that moment. This is also called timestamp data or time series data [53].

An overview of the different types of data that the systems analyzed offer is presented in Table 1. All of the analyzed databases offer real-time data support, however Google Spanner together with CockroachDB and FaunaDB are not built as stores for historical data. Spanner does offer the option to query data at specific timestamps [32], yet it has limited support for high volumes of data intended for historical purposes.

Table 1. NewSQL Database data types support

Database name	Historical	Real-time	Streaming	Temporal
Google Spanner	No	Yes	No	Yes
SAP HANA	Yes	Yes	Yes	No
VoltDB	Yes	Yes	Yes	No
TiDB	Yes	Yes	No	No
Altibase	Yes	Yes	No	No
Citus	Yes	Yes	No	Yes
CockroachDB	No	Yes	No	No
FaunaDB	No	Yes	Yes	Yes
NuoDB	Yes	Yes	No	No

Streaming capabilities are offered by VoltDB under the concepts called 'streams' [57], import, export and migration. Streams operate similar to regular database tables, however they are not stored in the database but managed 'in flight' without necessarily having to persist it. SAP HANA offers an optional feature which enables collection, processing and analyzing of streaming sources in real time [47]. FaunaDB supplies a feature for real time streaming for documents. This allows changes on the stored documents to be streamed to subscribers [21].

Time Series or temporal data, is supported by three of the surveyed engines. In Google Spanner, data rows have a timestamp attached [1]. Spanner uses TrueTime to offer a globally synchronized clock, used to avoid communication issues in distributed systems [10]. It generates increasing timestamps that are assigned to transactions, creating a serializable isolation.

FaunaDB offers a Temporal Model to track changes. It adds extra columns in the schema that will track transaction time and validity for every row [21]. Citus also supports scalable Time Series. Time-ordered data in big tables is split into multiple tables according to time ranges, by using a natively distributed shading mechanism [13].

An inventory is used to provide an overview, in Table 2, of features that can facilitate analysis and warehousing. The databases are analyzed in terms of their capacity to act as OLAP, OLTP and HTAP. We are using the capacity of the systems that can act as both OLAP and OLTP databases as a performance and efficiency measurement. If a system can perform both use cases, it eliminates the need of Extract-Transform-Load (ETL) operations thus being more time efficient.

Within the criteria of Data Analysis support, **Realtime analytics** options within the database are surveyed. Realtime analytics refers to the ability of a system to provide real-time insight into data to support decisions by generating value from the data available [56].

Lastly, **Map Reduce** facilitates processing and generating large data sets, by allowing users to specify a map function that processes key-value pairs and a reduce function that merges intermediate results. By design, running Map Reduce will be executed in parallel and over a large number of machines [16].

Table 2. NewSQL Analytics features

Database name	OLTP	OLAP	HTAP	Realtime analytics	Map reduce
Google Spanner	Yes	No (Online Analytical Processing support)	No	Yes (materialized views and optimized distributed reads)	Yes
SAP HANA	Yes	Yes	Yes	Yes (no ETL or batch processing)	No
VoltDB	Yes	Yes	Yes	Yes (materialized views and optimized distributed reads)	No
TiDB	Yes	Yes	Yes	Yes (TiFlash - extended analytical engine)	Yes (TiSpark connector)
Altibase	Yes	Yes	Yes	Yes	No
Citus	Yes	Yes	Yes	Yes	No
CockroachDB	Yes	No	No	No	No
FaunaDB	Yes	No	No	No	No
NuoDB	No	Yes	Yes	Yes	No

TiDB [43] is a HTAP storage system that offers TiSpark and TiFlash as analytics support. TiFlash offer medium-scale analytical processing, while TiSpark supports large-scale analytical processing. TiSpark is built on top of TiDB/TiKV to run Apache Spark [3] as a way to execute complex OLAP requests.

Citus is designed for scaling and for big data while maintaining a highly performant setup. Citus parallelizes SQL over multiple nodes and can in consequence provide real-time responsiveness. By using this mechanism it is able to get parallelism and concurrency for high performance ingest and for "sub-second responses" [13]. It also supports complex joins on co-located tables and reference tables, advanced subqueries and searches or funnel analytics.

VoltDB supports both OLAP and OLTP workloads and offers materialized views and optimized distributed reads as embedded features to support advanced realtime analytics [57].

SAP HANA offers, as well, a combination of OLTP and OLAP support requiring no ETL/ELT (Extract-Transform-Load/Extract-Load-Transform) or batch processing to perform the ingesting, transforming and loading of the data into a separate warehouse. This also provides instant insight over the live data. SAP also offers SAP S/4 HANA as an integrated ERP system that runs over SAP HANA database. This provides embedded analytics [47].

Altibase is another HTAP system with a built-in mechanism for real-time analytics called Altibase CEP. This is an In-Memory Middleware that provides real-time processing, querying and also analyzing data streams [2].

The last of the surveyed databases that can be categorized as HTAP is NuoDB, which also offers "real-time operational intelligence" [38]. Features that support this are its ability to execute long-running queries on dedicated hosts built over an architecture that is designed with an easy scale-out mechanism.

CockroachDB [14], FaunaDB [21] and Spanner [32] are not categorized as HTAP systems and the first two do not have embedded support for real time analytics. However, while Google Spanner does not claim to be designed for OLAP workloads, it does provide Online Analytical Processing support [32]. Spanner also offers mechanisms such as distributed data reads and materialized views, both being used in supporting realtime analytics.

Map Reduce is only available in Google Spanner and TiDB. In order to run it though, TiDB needs to integrate with Apache Spark through the native TiSpark connector [43].

5 Conclusions

This paper is aimed at providing a systematic overview of analytical features and support for NewSQL databases. Analytics, especially real-time or 'on the fly' analytics, are part of an increasing request of the industry. By looking at the emerging technologies over the course of the last three years we have identified several data management trends that point towards the rising visibility and usability of NewSQL databases. These systems are capable not only of scaling easily, but they offer mechanisms that support large amount of data and tools to process them. Moreover, NewSQL systems offer ACID guarantees with low latency and are capable to support different types of data. Multiple emerging needs (for example, streaming) are met by some of the NewSQL implementations, as well as the opportunity to store operational data and perform analytical workloads without including intricate and time consuming ingest, prepare and store processes such as ETL/ELT.

5.1 Opportunities and Expansion Areas

NewSQL technologies are still being adopted and perfected, but have a few limitations as well. As a result, several opportunities have been identified. Even though NewSQL data stores offer advantages over both legacy relational systems and NoSQL engines, they also fall short in some areas or offer opportunities to be further explored and analyzed. In this work we identify the following expansion options in the area of analysis using NewSQL:

While some of the presented solutions are part of the HTAP category, understanding how these capabilities can be put to use in improving or removing the ETL processes between source and warehouse storage can provide great insight. This can change the view on how data preparation has to happen to support processes such as Data Mining. Are there ways to use the actual data storage system to streamline and fast track towards analysis?

One of the most relevant contribution that a vast majority of NewSQL systems offer is bridging the gap between operational and analytical work. A more data oriented approach has been noticeable over the past few years, which brings analytics and real time data processing into attention. An overview of how analytical work can be performed and how different existing tools can integrate with NewSQL systems is not yet available. Next to the embedded or integrating features presented in this survey, how do strong analytics tools integrate or support NewSQL data sources? And next to the Map-Reduce capabilities presented in this work, what other integration points with artificial intelligence tools or machine learning algorithms are available? Such a structured overview is yet to be created.

Acknowledgments. This research was made possible by funding from the ICT-AGRI-FOOD 2020 Joint Call. This work was supported by a grant of the Romanian National Authority for Scientific Research and Innovation, CCCDI - UEFISCDI, project number COFUND-ICT-AGRI-FOOD-MUSHNOMICS 205/2020, within PNCDI III.

References

1. Almassabi, A., Bawazeer, O., Adam, S.: Top NewSQL databases and features classification. Int. J. Database Manag. Syst. **10**, 11–31 (2018)
2. Altibase: Altibase (2020). https://altibase.com/
3. Apache: Apache spark (2020). https://spark.apache.org/
4. Arasu, A., Babcock, B., Babu, S., McAlister, J., Widom, J.: Characterizing memory requirements for queries over continuous data streams. ACM Trans. Database Syst. (TODS) **29**(1), 162–194 (2004)
5. Aslett, M.: What we talk about when we talk about NewSQL (2020). https://blogs.451research.com/information_management/2011/04/06/what-we-talk-about-when-we-talk-about-newsql/
6. Barber, R., et al.: Evolving databases for new-gen big data applications. In: CIDR (2017)
7. Bestavros, A., Lin, K.J., Son, S.H.: Real-Time Database Systems: Issues and Applications, vol. 396. Springer, Boston (2012). https://doi.org/10.1007/b116080
8. Binani, S., Gutti, A., Upadhyay, S.: SQL vs. NoSQL vs. NewSQL-a comparative study. Database **6**(1), 1–4 (2016)
9. Bowman, J.S., Emerson, S.L., Darnovsky, M.: The Practical SQL Handbook: Using Structured Query Language. Addison-Wesley Longman Publishing Co., Inc., Boston (1996)
10. Brewer, E.: Spanner, truetime and the cap theorem (2017)
11. Chandra, U.: A comparative study on: NoSQL, NewSQL and Polygot persistence. Int. J. Soft Comput. Eng. (IJSE) **7** (2017)

12. Chen, M., Mao, S., Liu, Y.: Big data: a survey. Mobile Netw. Appl. **19**(2), 171–209 (2014)
13. CitusData: Citus (2020). https://www.citusdata.com/
14. Cockroachlabs: Cockroachdb (2020). https://www.cockroachlabs.com/
15. Corbellini, A., Mateos, C., Zunino, A., Godoy, D., Schiaffino, S.: Persisting big-data: the NoSQL landscape. Inf. Syst. **63**, 1–23 (2017)
16. Dean, J., Ghemawat, S.: MapReduce: Simplified data processing on large clusters (2004)
17. Dedehayir, O., Steinert, M.: The hype cycle model: a review and future directions. Technol. Forecast. Soc. Chang. **108**, 28–41 (2016)
18. Duan, Y., Fu, G., Zhou, N., Sun, X., Narendra, N.C., Hu, B.: Everything as a service (XAAS) on the cloud: origins, current and future trends. In: 2015 IEEE 8th International Conference on Cloud Computing, pp. 621–628 (2015)
19. Duggirala, S.: NewSQL databases and scalable in-memory analytics. In: Advances in Computers, vol. 109, pp. 49–76. Elsevier, Amsterdam (2018)
20. Emani, C.K., Cullot, N., Nicolle, C.: Understandable big data: a survey. Comput. Sci. Rev. **17**, 70–81 (2015)
21. Fauna: Fauna (2020). https://fauna.com
22. Fenn, J., Raskino, M.: Mastering the Hype Cycle: How to Choose the Right Innovation at the Right Time. Harvard Business Press (2008)
23. Gartner: Gartner (2020). https://www.gartner.com
24. Geczy, P.: Big data characteristics. Macrotheme Rev. **3**(6), 94–104 (2014)
25. solidIT consulting & software development GmbH: Db-engines (2020). https://db-engines.com/
26. Gray, J., Reuter, A.: Transaction Processing: Concepts and Techniques. Elsevier, Amsterdam (1992)
27. Hajoui, O., Dehbi, R., Talea, M., Batouta, Z.I.: An advanced comparative study of the most promising NoSQL and NewSQL databases with a multi-criteria analysis method. J. Theor. Appl. Inf. Technol. **81**(3), 579 (2015)
28. Han, J., Haihong, E., Le, G., Du, J.: Survey on NoSQL database. In: 2011 6th International Conference on Pervasive Computing and Applications, pp. 363–366. IEEE (2011)
29. Holubová, I., Svoboda, M., Lu, J.: Unified management of multi-model data. In: Laender, A., Pernici, B., Lim, E.P., de Oliveira, J. (eds.) Conceptual Modeling. ER 2019. Lecture Notes in Computer Science, vol. 11788, pp. 439–447. Springer, Cham (2019). https://doi.org/10.1007/978-3-030-33223-5_36
30. Huang, D., et al.: TIDB: a raft-based HTAP database. Proc. VLDB Endow. **13**(12), 3072–3084 (2020)
31. Inc., G.: Google trends (2020). https://trends.google.com
32. Inc., G.: Spanner (2020). https://cloud.google.com/spanner
33. Khasawneh, T.N., AL-Sahlee, M.H., Safia, A.A.: SQL, NewSQL, and NoSQL databases: a comparative survey. In: 2020 11th International Conference on Information and Communication Systems (ICICS), pp. 013–021 (2020)
34. Khurshid, K., Khan, A., Siddique, H., Rashid, I., et al.: Big data-9vs, challenges and solutions. Tech. J. **23**(03), 28–34 (2018)
35. Leavitt, N.: Will NoSQL databases live up to their promise? Computer **43**(2), 12–14 (2010)
36. May, N., Bohm, A., Lehner, W.: Sap HANA-the evolution of an in-memory DBMS from pure OLAP processing towards mixed workloads. Datenbanksysteme für Business, Technologie und Web (BTW 2017) (2017)

37. Murazzo, M., Gómez, P., Rodríguez, N., Medel, D.: Database newsql performance evaluation for big data in the public cloud. In: Naiouf, M., Chichizola, F., Rucci, E. (eds.) Cloud Computing and Big Data. JCC&BD 2019. Communications in Computer and Information Science, vol. 1050. Springer, Cham (2019). https://doi.org/10.1007/978-3-030-27713-0_10
38. NuoDB: Nuodb (2020). https://nuodb.com/
39. Ohlhorst, F.J.: Big Data Analytics: Turning Big Data into Big Money, vol. 65. Wiley, Hoboken (2012)
40. O'Leary, D.E.: Gartner's hype cycle and information system research issues. Int. J. Account. Inf. Syst. **9**(4), 240–252 (2008)
41. Özsu, M.T., Valduriez, P.: NoSQL, NewSQL, and Polystores. In: Principles of Distributed Database Systems, pp. 519–557. Springer (2020). https://doi.org/10.1007/978-3-030-26253-2_11
42. Pavlo, A., Aslett, M.: What's really new with NewSQL? SIGMOD Rec. **45**(2), 45–55 (2016)
43. PingCap: Tidb (2020). https://pingcap.com/
44. Pokorný, J.: How to store and process big data: are today's databases sufficient? In: Saeed, K., Snášel, V. (eds.) Computer Information Systems and Industrial Management, pp. 5–10. Springer, Heidelberg (2014). https://link.springer.com/chapter/10.1007/978-3-662-45237-0_2
45. Pokornỳ, J.: How to store and process big data: are today's databases sufficient? In: Saeed, K., Snasel, V. (eds.) Computer Information Systems and Industrial Management. CISIM 2015. Lecture Notes in Computer Science, vol. 8838, pp. 5–10. Springer, Heidelberg (2015). https://doi.org/10.1007/978-3-662-45237-0_2
46. Rifaie, M., Kianmehr, K., Alhajj, R., Ridley, M.J.: Data modelling for effective data warehouse architecture and design. Int. J. Inf. Decis. Sci. **1**(3), 282–300 (2009)
47. SAP: Sap hana (2020). https://www.sap.com/products/hana.html
48. SequoiaDB: Tidb (2020). http://www.sequoiadb.com/en/
49. Shanker, U., Misra, M., Sarje, A.K.: Distributed real time database systems: background and literature review. Distrib. Parallel Databases **23**(2), 127–149 (2008)
50. Shoup, R., Pritchett, D.: The EBAY architecture. In: SD Forum (2006)
51. Shute, J., et al.: F1: A distributed SQL database that scales (2013)
52. SingleStore: Memsql (2020). https://www.singlestore.com/
53. Steiner, A.: A generalisation approach to temporal data models and their implementations. Ph.D. thesis, ETH Zurich (1998)
54. TechJury: How much data is created every day (2020). https://techjury.net/blog/how-much-data-is-created-every-day
55. Thomson, A., Diamond, T., Weng, S.C., Ren, K., Shao, P., Abadi, D.J.: Calvin: fast distributed transactions for partitioned database systems. In: Proceedings of the 2012 ACM SIGMOD International Conference on Management of Data, pp. 1–12 (2012)
56. Verma, S., Kawamoto, Y., Fadlullah, Z.M., Nishiyama, H., Kato, N.: A survey on network methodologies for real-time analytics of massive IoT data and open research issues. IEEE Commun. Surv. Tutor. **19**(3), 1457–1477 (2017)
57. VoltDB, I.: VoltDB (2020). https://www.voltdb.com/
58. Ward, J.S., Barker, A.: Undefined by data: a survey of big data definitions. arXiv preprint arXiv:1309.5821 (2013)

Grouping-Based Hybrid Channel Access Mechanism for Ultra-Dense IEEE 802.11 ax Networks

Jianjun Lei(✉) and Mansoor Samo

School of Computer Science and Technology, Chongqing University of Posts and Telecommunications, Chongqing 400065, China
leijj@cqupt.edu.cn

Abstract. IEEE 802.11ax offers Uplink OFDMA Random Access (UORA) to support simultaneous multiuser operation. In dense WLANs scenario, as the stations (STAs) access the RUs (Resource Units) on a random basis, the risk that either STAs collide on the same RU, or some idle RUs are not utilized. In this paper, we present a Grouping-based hybrid channel access (HCA Grouping) mechanism, in which STAs during contention process use hybrid channel access mechanism, that consist of OFDMA based backoff and MC-CSMA-CA (Multichannel-CSMA-CA) to increase RUs utilization efficiency and throughput performance. The results justify that our proposed grouping mechanism achieves improved RUs efficiency by reducing idle RUs probability and also increase in throughput compared with conventionl UORA and adaptive grouping schemes.

Keywords: IEEE 802.11ax · MAC · Uplink OFDMA Random Access (UORA) · Grouping mechanism

1 Introduction

Wireless local area networks (WLANs), due to their high speed, flexible deployment, and low cost have become the primary traffic bearing wireless networks. The Cisco reports mentioning a sharp increase in the wireless data which will exceed 80 exabytes by 2021 [1]. However, in dense WLAN scenarios, the single-user (SU) transmission suffers from low efficiency [2, 3]. To this end, for improving channel efficiency, the IEEE 802.11 ax task group is proposing the multiuser (MU) OFDMA access technology [4, 5]. The OFDMA technology divides the whole bandwidth into multiple sub-channels, which are called resource units (RUs). Every STA in OFDMA is assigned to each RU. This OFDMA mechanism enables the scenario of simultaneous MU transmissions for uplink (UL) and downlink (DL) traffic.

In IEEE 802.11ax, the Uplink OFDMA-based Random Access (UORA) enables multiple STAs to randomly picks RUs for simultaneously transmitting uplink data to AP. In the UORA process using TF-R (Trigger Frame Random Access), AP dynamically announces some eligible random access RUs (RA-RUs) in the Trigger Frame (TF) so that the associated STAs can take part in channel competition for these RUs. After receiving

R. Silhavy (Ed.): CSOC 2021, LNNS 230, pp. 264–274, 2021.
https://doi.org/10.1007/978-3-030-77442-4_22

the TF-R, each STA decreases its OBO counter by the number of eligible RA-RUs. The low throughput in a dense environment is directly attributable to the limitation of the OBO mechanism. Figure 1 shows an example of UORA operation. After receiving TF, STA1, STA6, STA7, and STA8 gain transmission opportunity with OBO counter values of -3, -1, -6, and -2, respectively. However, STA4 and STA5 select RU4, and a collision occurs. According to the process of UORA, STA2, STA3, and STA9 could not complete their OBO counter to zero. Therefore, these STAs could not select the RUs, results in the wait for the next TF-R to resume the backoff. This kind of scenario in high-density WLANs leads to degradation of overall system performance due to high collision rate and Idle RUs.

As the STAs access the RUs on a random basis, there exist risks either STAs collide on the same RU, some idle RUs does not utilize, or both are high. Usually, UORA achieves only 37% of the maximum normalized throughput like slotted ALOHA because of the high probability of collisions at high loads. This result has an intuitive practical significance, as the RU efficiency of the UL OFDMA random access scheme is considered as the average probability that exactly one STA transmits in one RU during transmitted TFs. The existing literature on uplink OFDMA based MAC algorithms provide several insights on its performance. The first concept for uplink OFDMA based schemes introduced was based on the UORA [6]. As the de-facto in dense WLANs, the performance of uplink OFDMA based systems is mostly dependent on UORA. Initially, the authors [7–10] firstly investigated the UORA system using a bi-dimensional Markov chain. For improving optimal system efficiency by extending the contention windows, the adaptive backoff mechanism was proposed [11]. However, having the limited size of the backoff window in UORA, this mechanism cannot adopt in dense WLANs.

For handling the collision probability in dense networks, an adaptive grouping algorithm is proposed that separates the contending STAs into different groups using Target Wake Time (TWT) [12]. As a result, system performance is improved. But still, it does not consider decreasing the idle RA-RU probability for further improve system performance. The separation of the access process and data transmission process two-stage based mechanism is proposed in [13]. The authors use a traffic congestion control mechanism by p-persistent probability to mitigate the collision during the channel contention phase. A spatial clustering group division-based OFDMA access protocol [14] is proposed for reducing interference extension problem in OFDMA based WLANs and divide the basic service set into different cluster groups.

For increasing channel utilization in dense WLANs, one of the effective ways is to limit the number of simultaneous accessing multi-users (MUs) through STAs grouping. Due to severe contention of a large number of STAs, the collision probability is high. Additionally, STAs cannot achieve transmission opportunity due to idle RUs, which degrades the overall system performance. Hence we can break this limit through a mechanism in which STAs access the hybrid channel access mechanism during the channel contention phase to reducing the idle RA-RU probability without a corresponding increase in the collision probability resulting in improved RUs efficiency and system throughput.

This paper comprises of five sections: Sect. 2 is about Grouping-based HCA mechanism with the Algorithm. Section 3 described the performance analysis. Moreover, Sect. 4 is composed of simulation results. Finally, Sect. 5 gives the conclusion.

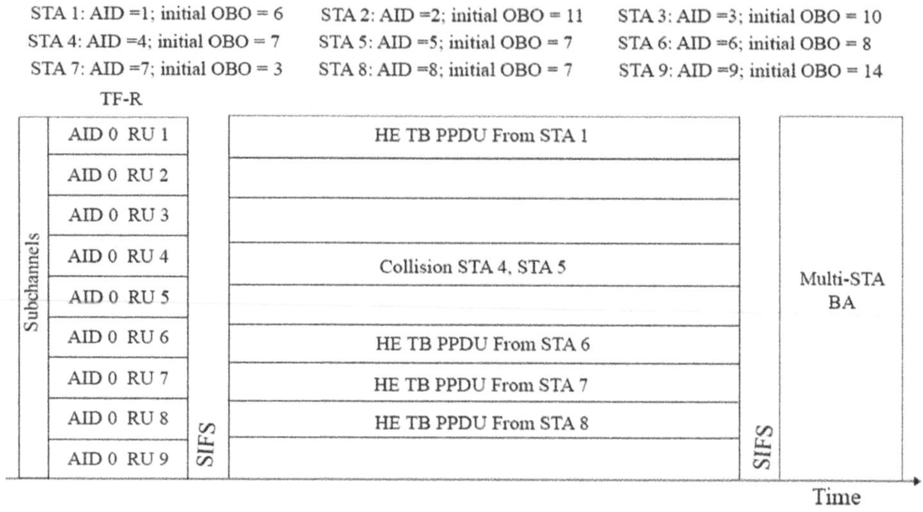

Fig. 1. Illustration of an example of UORA operation.

2 Proposed Grouping-Based Hybrid Channel Access Mechanism

2.1 Protocol Description

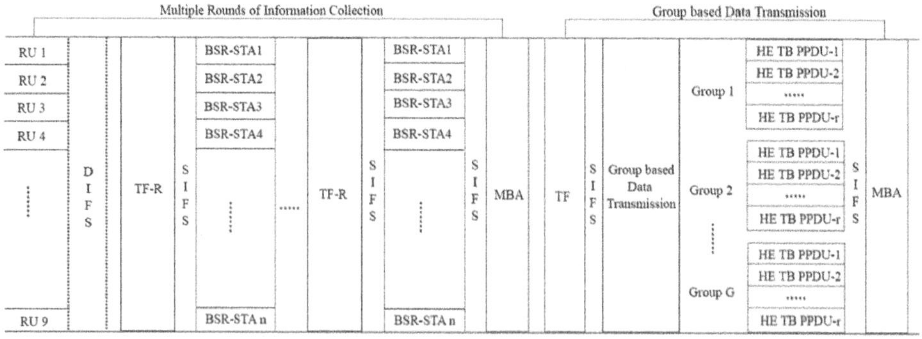

Fig. 2. Illustration of grouping mechanism.

The proposed grouping based Hybrid channel access combines the OBO based backoff and trigger-based multi-channel CSMA/CA [15] to reduce the collision and idle RA-RU probability in the first stage. In the second stage, STAs perform data transmission in the groups, as shown in Fig. 2.

The partitioning of the information collection and data transmission process efficiently improves the throughput of the whole system. Every time AP will run multiple rounds of information collection to minimize the access number of STAs for reducing collision probability. BSR delivery efficiency depends on multiple rounds of information collection, which is based on two key features: 1) AP adapt the OBO based backoff time, through which STAs completing their backoff counter equal to or less than 0 obtain transmission opportunity to send their BSR, 2) AP adapt MC-CSMA-CA during the same round of information collection in TF through which STAs did not complete their backoff counter start carrier sensing mechanism to detect idle RUs for sending their BSR. The number of information collection rounds needs to be proved by taking an experiment, as shown in Sect. 4. After a few rounds of information collection, based on maximized BSR delivery efficiency, AP sends a Multiple-Block Acknowledgement (M-BA) containing an MR_IC_offset to STA stop send BSR.

For hybrid channel access mechanism we can explain through Fig. 1, STA1, STA6, STA7, and STA8 can access the RA-RU after finishing their backoff. However, STAs did not complete their backoff use MC-CSMA-CA, to detect idle RUs using carrier sensing with fixed duration or t_{slot} slots. Length of t_{slot} is fixed for sensing the idle RUs shown in simulation parameters Table 2. After sensing that RU2, RU3 and RU9 are idle, STA2, STA3 and STA9 decrements their OBO to 0 to send their BSR at certain t_{slot}. Thus STA2, STA3 and STA9 gained transmission opportunity from the OBO-based backoff to perform p-persistent CSMA using the probabilities until they succeed or no idle RUs remain. Starting from first t_{slot}, an STA generates a random number from 0 to 1 and comparing to a p-persistent CSMA transmit probability parameter. If the generated random number is less than the p-persistent CSMA probability, the STAs start transmitting their BSR. Otherwise, it defers and repeats carrier sensing. If there are still idle RUs remaining, the STA will start again carrier sense at next t_{slot} time. In the second stage, AP will establish optimal groups for data transmission using the obtained BSR. The number of groups for data transmission depends on maximum BSR delivery efficiency (Fig. 3).

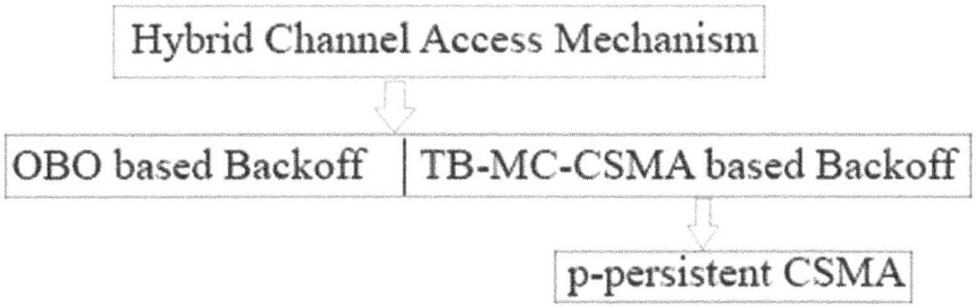

Fig. 3. Hybrid channel access mechanism for information collection stage.

2.2 Grouping Algorithm Details

In this section, we propose a grouping algorithm based on a hybrid channel access mechanism that takes fewer rounds of information collection stages and generates optimal groups of STAs that can transmit data with less collision probability and increased RUs efficiency. The proposed algorithm improves the system performance, by using maximized BSR_{opt} due to a hybrid channel access mechanism, and then divide STAs into multiple groups. The grouping algorithm is shown in Algorithm 1. AP takes a hybrid channel access mechanism using OBO and TB_MC_CSMA based backoff to start the information collection process. AP runs multiple rounds of information collection to obtain the BSR_{opt}. Maximum BSR_{opt} depends on the number of successful STAs sending their BSR to AP and N_{MR_IC}. Some STAs get transmission opportunities to send their BSR after finishing their OBO. STAs not completed their OBO get transmission opportunity through TB_MC_CSMA based backoff using p-persistent in this grouping based hybrid channel access mechanism, which certainly improves BSR delivery efficiency.

If BSR_{opt} is maximized during N_{MR_IC}, AP will continue N_{MR_IC}. If BSR_{opt} start decreasing due to a large number of N_{MR_IC}, AP sends MB_A contains an offset for STAs to stop sending the BSR. After the first stage, AP divides the overall number of successful STAs with N_{RU} into different groups for the second stage, in which STAs start data transmission to AP. The proposed grouping based HCA mechanism maximize the utilization of RUs, that certainly improves overall system performance (Table 1).

Table 1. Summary of notations

Notation	Meaning
N_{MR_IC}:	Number of Multiple Rounds of Information Collection
BSR [s]:	Successful Buffer Status Report
N_{RU}:	Number of RU
$OCW_{min}(W_0)$:	Minimum contention window size
m:	Maximum number of backoff stage
p:	Collision Probability
τ:	Transmit Probability
G:	Total Group Number
N_j:	Successful BSR during N_{MR_IC}
BSR_{opt}:	Buffer Status report delivery efficiency
RU_{idle}:	Idle RU Number
MR_IC_offset:	Multiple Rounds Information Collection offset

Algorithm 1: HCA Grouping Algorithm

Input: N_{MR_IC}, n
Output: $G, N_j, \forall j \in [1, R], G_{opt}$
1 **while** true **do**
2 $HCA \Leftarrow [OBO\ based\ backoff,\ TB\_MC\_CSMA\ based\ backoff]$
3 **if** $N_{MR\_IC} > 0$
4 **if** $STAs \Leftarrow OBO \leq 0$
5 **if** $STAs \Leftarrow RU_{idle}$
6 $STAs \Leftarrow TB\_MC\_CSMA\ based\ backoff$
7 **for** $i = 1{:}n$
8 $STAs \Leftarrow BSR[s], s = 1,2, \ldots, n$
9 $STAs \Leftarrow N_j(BSR[s])$
10 $N_j, \forall j \in [1, R]$
11 **end for**
12 **if** $N_j BSR_{opt}$ is maximized
13 $AP\ continue\ N_{MR\_IC}$
14 **else**
15 break;
16 $MR\_IC\_offset = 1$
17 **end if**
18 **end if**
19 **end if**
20 **end if**
21 $G \Leftarrow \left\lfloor \frac{STAs \Leftarrow (BSR[s])}{N_{RU}} \right\rfloor + 1$
22 $G \Leftarrow G_{opt} \in (1, G)$
21 **Repeat from Line 2**
23 **end while**

3 Performance Analysis

This section describes the performance analysis of the proposed scheme and derives expressions based on two key performance metrics: i) BSR delivery and RU Efficiency ii) Throughput.

For avoiding complexity in the performance analysis section, we are not including the MC-CSMA-CA of the channel accessing mechanism to derive the RU efficiency. We are only considering UORA performance analysis for the derivation of RU efficiency. The contention process for transmitting BSR on RA-RU is based on OBO backoff and MC-CSMA-CA. By contenting STAs picks a random integer the OBO in the range of $[0, OCW_{min}]$. According to the process of UORA, STAs decrement their OBO by the number of RU after receiving the TF. We are following the two-dimensional Markov chain model [16, 17]. The probability that an STA transmits its BSR in any of the RUs can be given as follows

$$\tau = \frac{2(1-2p)}{(1-2p)\left(\frac{W_0}{N_{RA}}+1\right)+p\frac{W_0}{N_{RA}}\left(1-(2p)^m\right)} \tag{1}$$

The probability that two STAs simultaneously selects one RU results in a collision can be computed as

$$p = 1 - \left(1 - \frac{\tau}{N_{RA}}\right)^{n-1} \quad (2)$$

The probability of successful transmitting BSR in RA RUs can be calculated as,

$$P_{tr} = 1 - \left(1 - \frac{\tau}{N_{RA}}\right)^{n} \quad (3)$$

Here, the probability P_s that BSR transmission in RA RUs is successful is given by obability of successful trans

$$P_s = \frac{n\tau(1-\tau)^{n-1}}{P_{tr}} \quad (4)$$

Now RUs efficiency in terms of the collected BSRs during MR_IC can be analyzed by BSR delivery efficiency, which is mentioned in [18]. Here we denote the buffer status report delivery efficiency by BSR_{opt} which can be calculated by combining Eqs. (3) and (4).

$$BSR_{opt} = N_{RA} P_{tr} P_s \quad (5)$$

Equation (5) maximizes the BSR delivery rate according to algorithm 1 using a hybrid channel access mechanism. AP adaptively adjust N_{MR_IC} for achieving the BSR_{opt} which is proved in Sect. 4. The collision occurs during information collection if as result no BSR is successfully transmitted to AP. Therefore, both success and failure cases must be considered with corresponding probability, Thus the transmission K_{BSR} of each round can be obtained by

$$K_{BSR} = P_s^{(n,N_{RU})}(0).K_{BSR}^{fail} + \left(1 - P_s^{(n,N_{RU})}(0).K_{BSR}^{succ}\right) \quad (6)$$

Where K_{BSR}^{fail} and K_{BSR}^{succ} are the expected cases of failure and success which are the time duration of each information collection round, respectively. $P_s^{(n,N_{RU})}(s)$ is the probability that s STAs successfully access the RU among n STAs and number of RUs N_{RU} in TF-R.

In case of failure of BSR in any round of information collection, the transmission time duration can be is calculated as:

$$t_{idle} = t_{DIFS} + T_{trigger} \quad (7)$$

For throughput, we need to express the time duration of each transmission. The transmission duration of successful data is obtained by:

$$t_{duration} = t_{DIFS} + 2T_{trigger} + 4T_{SIFS} + T_{PPDU} + 2T_{MBA} \quad (8)$$

The estimated time duration of each group of STAs can be obtained by

$$t_G = t_{idle} + t_{duration} \quad (9)$$

The throughput of each STAs group can be obtained as:

$$TH_G = \frac{BSR[s]*L_{payload}}{t_G} \quad (10)$$

In Eq. (10) $BSR[s]$ express the number of successful BSR received in each round of information collection, $L_{payload}$ is the length of payload and t_G is the time duration of each group By using Eq. (10) the average throughput of the system can be expressed as

$$TH_{avg} = \frac{1}{G}\sum\nolimits_{j=1} G\ TH_G \quad (11)$$

The average throughput of STA in each group can be obtained as

$$TH_{avg\_th} = \frac{1}{G}\sum\nolimits_{j=1} G\ TH_{avg\_th}, j \quad (12)$$

4 Results and Discussions

To justify the efficiency of the proposed method, we compared it with Adaptive grouping and conventional UORA grouping scheme. It is ensured that the RU efficiency and throughput is maximized using our proposed grouping mechanism. Moreover, experiments are conducted using Matlab, where the total number of RUs is 9, minimum OBO is set to 15 and the maximum number of backoff stage is 2. The following parameters are used for simulation in Table 2.

Table 2. Simulation parameters

Parameters	Value	Parameters	Value
Total number of RUs	9	DIFS	34 μs
W_0	15	Trigger	89 bytes
m	2	BSR	32 bytes
Payload	1000 bits	SIFS	16 μs
MCS	9	M-BA	46 bytes
GI	0.8 μs	Control Packet Physical Rate	6 Mbps
No. of MC-CSMA t_{slot}	7	MC-CSMA t_{slot} duration	16 μs

4.1 BSR Delivery BSR_{opt} and RU Efficiency

We first see the effect of the HCA mechanism for maximized BSR delivery efficiency in terms of multiple rounds of the information collection stage. Because this BSR_{opt} is directly proportional to an increase in RU efficiency. As is shown in Fig. 4, due to the utilization of the proposed HCA mechanism BSR_{opt} is maximized from 37% to 45% with $N_{MR\_IC} = 3$. In each round of information collection, through the HCA mechanism, STAs not completed their OBO backoff found transmission opportunity by MC-CSMA-CA to detect the idle RUs. STAs using p-persistent CSMA probability using a fixed number of t_{slot} as explained in Sect. 2, transmit their BSR to AP. When we increase the N_{MR_IC} from 3 to 5 during the experiment, BSR_{opt} is decreasing. This result prove that AP through the proposed HCA mechanism can adaptively adjust N_{MR_IC} to maximize BSR_{opt}. The RU efficiency of the grouping based HCA algorithm is shown in Fig. 5, compared with adaptive grouping and conventional UORA mechanism. The proposed grouping algorithm is achieving higher RU efficiency by using a hybrid channel access mechanism every time by reducing the STAs channel contention by a fixed number of N_{MR_IC}.

4.2 Throughput

In Fig. 6, the throughput of grouping based HCA mechanism is higher than Adaptive grouping and conventional UORA mechanism. The proposed grouping scheme is achieving higher system throughput when number of STAs is 200. Buffer status report delivery efficiency is decreasing due to hybrid channel access mechanism, already proved through fixed N_{MR_IC} shown in Fig. 4. Maximum BSR_{opt} leads to increased throughput when AP divide the STAs for data transmission into groups according to number of RU in the TF. In Fig. 7, the average throughput of each STA is improved than Adaptive grouping and UORA mechanism. Because AP has more efficient groups for data transmission compared to other schemes. One problem in our proposed scheme is that the average system throughput is falling from peak point of Adaptive grouping and UORA because of overhead of during the BSR delivery to AP in multiple rounds of information collection. Thus the grouping based hybrid channel access mechanism reduce the idle RU probability for STAs having opportunity for more number of STAs to transmit their data to AP in groups.

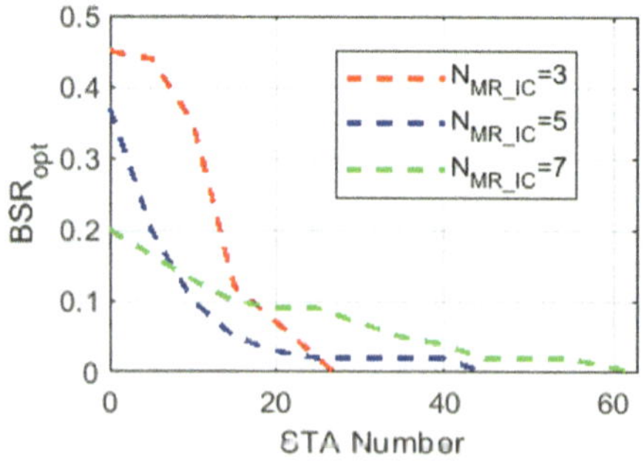

Fig. 4. Optimal BSR delivery rate of HCA

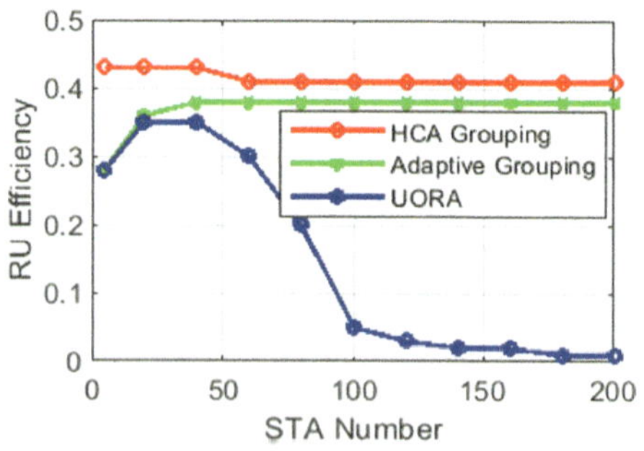

Fig. 5. RU efficiency of HCA

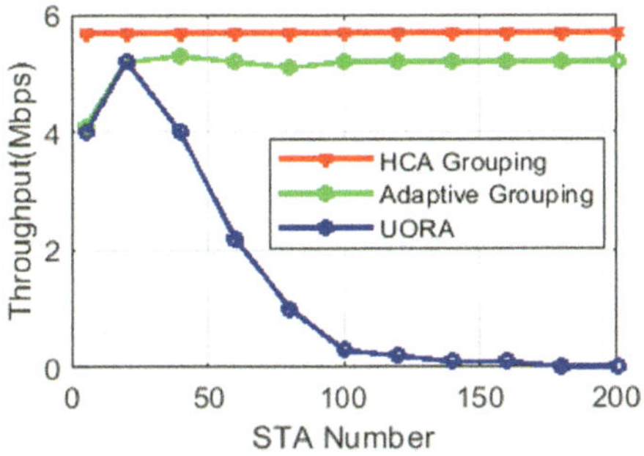

Fig. 6. System throughput versus STA number

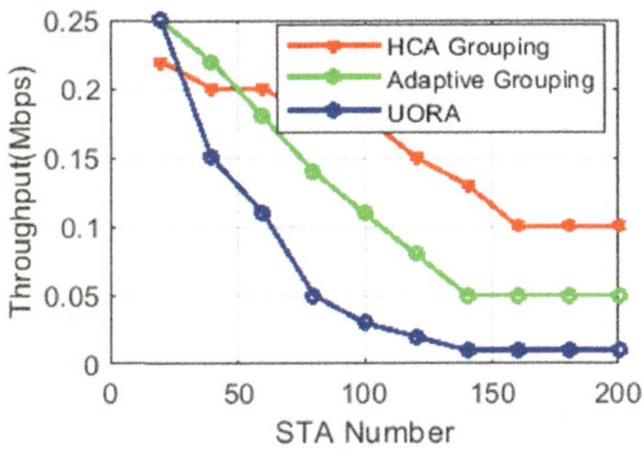

Fig. 7. Single user throughput versus STA number

5 Conclusion

In this paper, we proposed a grouping based HCA scheme that improve the RU efficiency and system throughput of UORA in ultra-dense networks. Using this mechanism, AP collect the BSR of STAs based on fixed rounds of information collection to reduce the collision probability. When AP is getting the maximized BSR delivery efficiency, STAs are divided into groups for efficient data transmission. The hybrid channel access mechanism along with OBO uses MC-CSMA-CA so STAs detect the idle RUs. STAs getting transmission opportunity by using p-persistent CSMA-CA and send their BSR to AP by successfully accessing idle RUs. The BSR efficiency is increased which also increases to RU efficiency of UORA. Based on simulation results, our proposed scheme show improved overall system performance compared to Adaptive grouping scheme and general UORA mechanism.

References

1. Global mobile data traffic forecast update Cisco Visual Networking Index: White Paper. https://www.cisco.com/c/en/us/solutions/service-provider/visual-networking-index-vni/index.html. Accessed 26 Feb 2018
2. Omar, H.A., Abboud, K., Cheng, N., Malekshan, K.R., Gamage, A.T., Zhuang, W.: A survey on high efficiency wireless local area networks: next generation WiFi. IEEE Commun. Surv. Tutorials **18**(4), 2315–2344 (2016)
3. Deng, D.J., Lien, S.Y., Lee, J., Chen, K.C.: On quality-of-service provisioning in IEEE 802.11ax WLANs. IEEE Access **4**, 6086–6104 (2016)
4. IEEE P802.11ax/D3.0, Part 11: Wireless LAN Medium Access Control (MAC) and Physical Layer (PHY) Specifications: Enhancements for High Efficiency WLAN, 802.11 Working Group of the LAN/MAN Standards Committee (2018)
5. Deng, D.J.: IEEE 802.11ax: highly efficient WLANs for intelligent information infrastructure. IEEE Commun. Mag. **55**(12), 52–59 (2017)
6. Ghosh, C.: Random Access with Trigger Frames using OFDMA. https://mentor.ieee.org/802.11/dcn/15/11-15-0875-01-00ax-random-access-with-trigger-frames-using-ofdma.pptx
7. Lanante, L., Uwai, H.O.T., Nagao, Y., Kurosaki, M., Ghosh, C.: Performance analysis of the 802.11ax UL OFDMA random access protocol in dense networks. In: 2017 IEEE International Conference on Communications, pp. 1–6 (2017)

8. Naik, G.: Performance analysis of uplink multi-user OFDMA in IEEE 802.11 ax. In: 2018 IEEE International Conference on Communications, pp. 1–6 (2018)
9. Yang, H.: Performance analysis of IEEE 802.11 ax UL OFDMA-based random access mechanism. In: 2017 IEEE Global Communications Conference, pp. 1–6 (2017)
10. Khan, P.W., Abbas, K., Shaiba, H., Muthanna, A., Abuarqoub, A., Khayyat, M.: Energy efficient computation offloading mechanism in multi-server mobile edge computing—an integer linear optimization approach. Electronics **9**(6) (2020)
11. Uwai, T., Miyamoto, T., Nagao, Y., Lanante, L., Kurosaki, M., Ochi, H.: Adaptive backoff mechanism for OFDMA random access with finite service period in IEEE 802.11ax. In: 2016 IEEE Conference on Standards for Communications and Networking, pp. 1–6 (2016)
12. Bai, J.: Adaptive Uplink OFDMA random access grouping scheme for ultra-dense networks in IEEE 802.11ax. In: 2018 IEEE/CIC International Conference on communications in China (ICCC), pp. 1–6 (2018)
13. Ahn, W.: An energy efficient multiuser uplink transmission scheme in the next generation WLAN for Internet of Things. Int. J. Distrib. Sens. Netw. **12**(7), 1512054 (2016)
14. Li, Y., Li, B., Yang, M.: A spatial clustering group division-based OFDMA access protocol for the next generation WLAN. Wireless Netw. **25**(8), 5083–5097 (2019). https://doi.org/10.1007/s11276-019-02115-2
15. Kwon, H., Seo, H., Kim, S., Lee, B.G.: Generalized CSMA/CA protocol for OFDMA systems. In: 2008 IEEE Global Telecommunications Conference, pp. 1–6 (2008)
16. Bianchi, G.: Performance analysis of the IEEE 802.11 distributed coordination function. IEEE J. Sel. Areas Commun. **18**(3), 535–547 (2006)
17. Arun, I.B., Venkatesh, T.G.: Adaptive Backoff Algorithm for IEEE 802.11 DCF under MPR Wireless Channels (2013)
18. Bhattarai, S., Naik, G., Park, J.J.: Uplink resource allocation in IEEE 802.11ax. In: 2019 IEEE International Conference on Communications (ICC), pp. 1–6 (2019)

Mathematical Optimization, the Effect of the Catalyst and Solvent on the Process of Starch Sulfation with Sulfamic Acid

Aleksandr S. Kazachenko[1,2](✉), Natalya Yu. Vasilyeva[1,2], Yuriy N. Malyar[1,2], and Anna S. Kazachenko[2]

[1] Institute of Chemistry and Chemical Technology SB RAS, Federal Research Center "Krasnoyarsk Science Center SB RAS", Akademgorodok, 50/24, Krasnoyarsk 660036, Russia
kazachenko.as@icct.krasn.ru
[2] Siberian Federal University, Svobodny Avenue, 79, Krasnoyarsk 660041, Russia

Abstract. Starch is one of the most common polysaccharides. The scope of its application is wide enough, and chemical modification expands them even more. In this work, the preparation of a biologically active starch derivative, sulfate, is considered. In addition, in this work, for the first time, the study of the effect of catalysts and solvents on the process of sulfation of starch with sulfamic acid. It was shown that the most effective catalyst is urea, and the solvent is DMSO. For this system "sulfamic acid-urea-DMSO" mathematical optimization was carried out. It is shown that the optimal conditions for obtaining starch sulfates, providing the maximum sulfur content in the resulting reaction product, are: temperature 87.1 °C and duration 2.7 h. The introduction of a sulfate group into a starch molecule was confirmed by elemental analysis and FTIR spectroscopy.

Keywords: Starch · Sulfation · Optimization · Sulfamic acid · Solvents · Catalysts

1 Introduction

Starch is a natural, renewable and biodegradable polymer produced by many plants as a stored energy source. It is the second most abundant biomass material in nature. It is found in roots, stems, seeds of crops, and major crops such as rice, corn, wheat, tapioca and potatoes [1, 2].

Starch consists of glucose residues linked by only two types of bonds: α-1,4 and α-1,6 glucosidic bonds. Despite the simple chemical composition of starch, the final starch molecules are diverse and complex [3, 4]. Biopolymers such as cellulose and starch are biodegradable materials that can be used in many areas of industry, as well as a platform molecule for their modification with various functional groups [1, 5, 6]. Due to chemical modification, starch can acquire a number of new properties, for example, hydrophilicity/hydrophobicity, etc. [7].

Among the wide range of starch derivatives, special attention can be paid to its sulfated derivatives due to their biological activity. Recently, starch sulfates have attracted

R. Silhavy (Ed.): CSOC 2021, LNNS 230, pp. 275–282, 2021.
https://doi.org/10.1007/978-3-030-77442-4_23

great attention due to their wide variety of biological activity [8–12]. In addition to their biological activity, starch sulfates are widely used in the sizing of paper, textiles, and food products, as well as in drilling fluids [13].

Traditionally, starch sulfates are prepared using highly corrosive sulfating agents such as sulfuric acid, chlorosulfonic acid, sulfuryl chloride, sulfur trioxide [14, 15], and various organic solvents such as pyridine, dimethyl sulfoxide, triethylamine, toluene were used as the reaction medium. or ethylene dichloride [16, 17]. In addition, there are known methods for sulfation of starch in the absence of a solvent [18, 19] and in aqueous media [12].

Despite the many ways to sulfate starch, the effect of the solvent on this process is poorly understood.

The aim of this work was the mathematical optimization of the process of sulfation of starch with sulfamic acid, as well as the study of the influence of solvents and catalysts on this process.

2 Experimental

We used potato starch (Khimreaktivsnab, Russia).

2.1 Features of the Starch Sulfation

2.1.1 Sulfation of Starch

Starch was sulfated by sulfamic acid in different solvents (1,4-dioxane, dimethyl sulfoxide, dimethylformamide, piperidine, pyridine, morfoline) in the presence of organic bases (urea, thiourea, pyridine, morfoline) under different conditions. For this purpose, solvents (50 ml), sulfamic acid (10.0 g or 102 mmol), and organic bases (102 mmol) were placed into a 100-ml three neck flask equipped with a thermometer, a mechanical stirrer, and a reflux condenser. The prepared mixture was heated to 90 °C under constant stirring, added with an air-dry starch (2 g), and thermostated at a specified temperature for 3.0 h. Upon completing of sulfation, the solvent was decanted, and the residue was dissolved in 25 ml of water. The excess sulfamic acid was neutralized with aqueous ammonia (25%) until the neutral reaction. To find the optimal conditions, the best solvent and the best catalyst were used with varying temperatures and process times.

2.1.2 Dialysis of Sulfated Starch

The sulfated starch ammonium salt was purified by dialysis on cellophane against distilled water in an MF-503−46 MFPI cellophane dialyzing sac (US) with a pore size of 3.5 kDa (−0.1 μm). The product was dialyzed for 10 h with changing the water in 1–2 h.

2.2 Calculation

The numerical optimization of the starch sulfation process with sulfamic acid was carried out using the Statgraphics Centurion XVI software, according to the procedure described in [20, 21].

2.3 Elemental Analysis

The elemental analysis of the sulfated starch was made on a ThermoQuest FlashEA-1112 elemental analyzer (Italy).

2.4 Fourier-Transform Infrared Spectroscopy

The FTIR spectra of the starch and sulfated starch were recorded on a Shimadzu IRTracer-100 FTIR spectrometer (Japan) in the wavelength range from 400 to 4000 cm^{-1}. The spectral data were analyzed using the OPUS software (version 5.0). Solid specimens in the form of tablets in a KBr matrix (2-mg specimen/1000 mg of KBr) were prepared for the analysis.

3 Results and Discussion

Sulfation of starch was carried out with sulfamic acid in dioxane in the presence of bases: urea, thiourea, pyridine, morpholine and in the absence of a basic catalyst. The sulfation process was carried out for 3 h at a temperature of 90 °C with an excess of sulfation complex. The sulfation results are shown in Table 1.

Table 1. Influence of the basic catalyst nature on the sulfur content (% wt.) in sulfated starch during sulfation in 1,4-dioxane at 90 °C and time of process 3.0 h.

№	Organic base	K_B*	Sulfur content, % (wt.)
1	1,4-dioxane	$1.2 \cdot 10^{-17}$	2.2
2	Urea	$1.51 \cdot 10^{-14}$	11.2
3	Thiourea	$6.35 \cdot 10^{-15}$	9.1
4	Pyridine	$1.51 \cdot 10^{-9}$	6.9
5	Morpholine	$2.14 \cdot 10^{-6}$	4.2

* - basicity constant

According to the data given in Table 1, the highest sulfur content in sulfated starch is achieved when using urea as a catalyst.

In order to find the optimal medium for carrying out the process of sulfation of starch with sulfamic acid in the presence of urea, aprotic solvents with different dielectric constant were investigated: 1,4-dioxane, pyridine, piperidine, morpholine, DMF, DMSO, and diglyme. The results are presented in Table 2.

Table 2. Influence of the solvent on the sulfur content in starch sulfates during sulfation with sulfamic acid in the presence of urea at 90 °C for 3.0 h.

№	Solvent	Dielectric constant	Sulfur content, % (wt.)
1	1,4-dioxane	*2.3*	*11.2*
2	Pyridine	*12.5*	*9.1*
3	Morpholine	*7.3*	*7.6*
4	Piperidine	*5.9*	*8.1*
5	DMF	*38*	*8.2*
6	DMSO	*46.7*	*12.1*
7	Diglyme	*7.4*	*10.3*

In contrast to the heterogeneous reaction of starch sulfation in dioxane, pyridine, etc. (Table 2), starch sulfation in dimethyl sulfoxide is a homogeneous reaction. Carrying out the process of sulfation in the medium in DMSO under homogeneous conditions makes the hydroxyl groups of starch more accessible for attack by the sulfating complex, which ultimately leads to the formation of a sulfated product with a high sulfur content. In addition, it is known about the formation of a fairly reactive but unstable complex of dimethyl sulfoxide with sulfur trioxide, which can also be the cause of the formation of starch sulfates with a high content of sulfate groups [22, 23].

In order to find the optimal conditions for the sulfation of starch with sulfamic acid in DMSO in the presence of urea, a mathematical and experimental optimization of this process was carried out.

Two factors are included in the study as independent variables: X_1 - process temperature, C; X_2 is the duration of the sulfation process, hours. The result of the sulfation process was characterized by the output parameter: Y_1 - sulfur content in sulfated starch, wt. %. Table 3 shows the experiment design matrix and its implementation.

The dependence of the output parameter (Y_1) on variable process factors was approximated by second-order regression equations. The results of analysis of variance are given in Table 4.

Analysis of variance showed that within the accepted experimental conditions, both factors make a significant contribution to the total variance of the output parameter.

The dependence of the degree of substitution in starch sulfates (Y_1) on variable process factors is approximated by the regression equation:

$$Y_1 = -155.578 + 3.53X_1 + 7.8667X_2 - 0.01933X_1^2 - 0.03X_1X_2 - 0.9833X_2^2 \quad (1)$$

The predictive model based on Eq. (1) is shown in Fig. 1. According to this model, the closer to the center line of a point, the more accurately the mathematical model describes the experimental data.

Table 3. The experiment design matrix and its implementation

№	Temperature, °C	Time, h	Sulfur content, wt.%
1	80	1	7.7
2	85	1	8.9
3	90	1	9.8
4	80	2	10
5	85	2	11.6
6	90	2	11.9
7	80	3	10.6
8	85	3	12
9	90	3	12.1

Table 4. The result of analysis of variance

Source	Mean square	F-ratio	P-value
X_1	5.04167	132.16	0.0014
X_2	11.4817	300.98	0.0004
X_{12}	0.467222	12.25	0.0395
X_1X_2	0.09	2.36	0.2221
$X_2{}^2$	1.93389	50.69	0.0057

The quality of approximation is characterized by the coefficient of determination $R^2{}_{adj}$. In our case, the value of $R^2{}_{adj} = 98.2\%$, which indicates a good quality of approximation.

The graphic display of Eq. (1) in the form of a response surface is shown in Fig. 2.

The optimal conditions for obtaining starch sulfates (within the accepted experimental conditions), providing the maximum sulfur content in the reaction product are: temperature 87.1 °C and duration 2.7 h.

The original starch and sulphated starch (obtained under optimal conditions) were analyzed by FTIR spectroscopy (Fig. 3).

The introduction of a sulfate group into a starch molecule was confirmed by IR spectroscopy. Thus, in the IR spectrum of sulfated starch, absorption bands appear at 1243, 861, and 817 cm^{-1}, corresponding to the vibrations of the sulfate group.

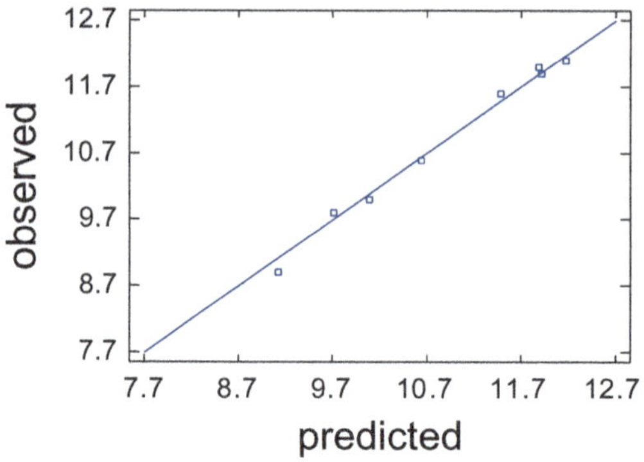

Fig. 1. The results of observations against the values of the output parameter Y1 predicted by the mathematical model (1).

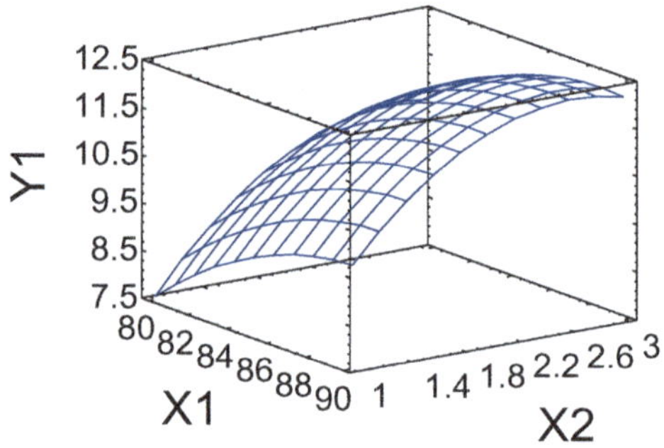

Fig. 2. Response surface the output parameter Y_1 from the variable factors X_1 and X_2.

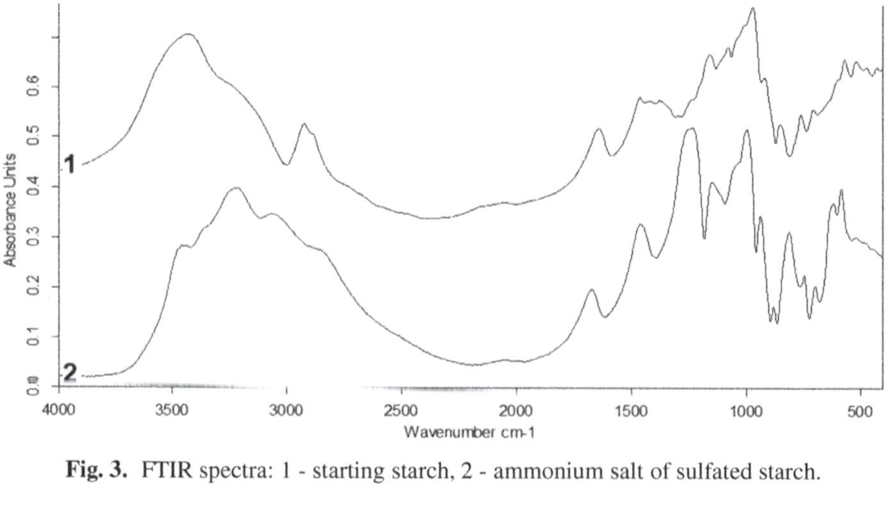

Fig. 3. FTIR spectra: 1 - starting starch, 2 - ammonium salt of sulfated starch.

4 Conclusions

The effect of catalysts and solvents on the process of starch sulfation with sulfamic acid has been studied for the first time. It was shown that urea was the most effective catalyst for this process.

When DMSO is used as a solvent for the process of sulfating starch with sulfamic acid, the highest sulfur content in the product is achieved. This is probably due to the fact that in this solvent the reaction proceeds under homogeneous conditions.

A mathematical model of starch sulfation with sulfamic acid in DMSO in the presence of urea has been constructed. It is shown that the optimal conditions for obtaining starch sulfates, providing the maximum content of sura in the resulting reaction product are: temperature 87.1 °C and duration 2.7 h.

The introduction of a sulfate group into a starch molecule was confirmed by elemental analysis and FTIR spectroscopy.

Acknowledgments. The research was funded by RFBR, Krasnoyarsk Territory and Krasnoyarsk Regional Fund of Science, project number 20-43-243001. The devices of the Krasnoyarsk Regional Center of Research Equipment of Federal Research Center «Krasnoyarsk Science Center SB RAS» were used in the work. This study was carried out within the budget project # 0287-2021-0017 for the Institute of Chemistry and Chemical Technology, Siberian Branch of the Russian Academy of Sciences.

References

1. Shaterian, H.R., Rigi, F.: Starch sulfate as an efficient and biodegradable polymer catalyst for one-pot, four-component reaction of 2H-indazolo[2,1-b]phthalazine-triones. Starch/Stärke **63**, 340–346 (2011)
2. Whistler, R.L., BeMiller, J.N., Paschall, E.F. (eds.): Starch: Chemistry and Technology, p. 220. Academic Press, Orlando, Florida (2012)
3. Yoo, S.H., Jane, J.L.: Structural and physical characteristics of waxy and other wheat starches. Carbohydr. Polym. **49**, 297–305 (2002)
4. Blennow, A., Engelsen, S.B., Nielsen, T.H., Baunsgaard, L., Mikkelsen, R.: Starch phosphorylation: a new front line in starch research. Trends Plant Sci. **7**, 448–450 (2002)
5. Molvinger, K., Quignard, F., Brunel, D., Boissiere, M., Devoissells, J.M.: Porous chitosan-silica hybrid microspheres as a potential catalyst. J. M. Chem. Mater. **16**, 3367–3372 (2004)
6. Shaabani, A., Rahmati, A., Badri, Z.: Sulfonated cellulose and starch: new biodegradable and renewable solid acid catalysts for efficient synthesis of quinolines. Catal. Commun. **9**, 13–16 (2008)
7. Tharanathan, R.N.: Starch—value addition by modification. Crit. Rev. Food Sci. Nutr. **45**(5), 371–384 (2005)
8. Cui, D., Liu, M., Wu, L., Bi, Y.: Synthesis of potato starch sulfate and optimization of the reaction conditions. Int. J. Biol. Macromol. **44**(3), 294–299 (2009)
9. Drozd, N.N., Logvinova, Y.S., Torlopov, M.A., Udoratina, E.V.: Effect of sulfation and molecular weight on anticoagulant activity of dextran. Bull. Exp. Biol. Med. **162**(4), 462–465 (2017). https://doi.org/10.1007/s10517-017-3640-2
10. Drozd, N.N., Torlopov, M.A., Kuzhim, A.A., Makarov, V.A.: Dependence of the anticoagulant activity of starch and inulin on their degree of sulfonation. Eksp. Klin. Farmakol. **75**(6), 31–35 (2012)

11. Staroszczyk, H., Fiedorowicz, M., Zhong, W., Janas, P., Tomasik, P.: Microwave-assisted solid-state sulphation of starch. e-Polymers **7**(1), (2007). https://doi.org/10.1515/epoly.2007.7.1.1635
12. Cui, D., Liu, M., Liang, R., Bi, Y.: Synthesis and optimization of the reaction conditions of starch sulfates in aqueous solution. Starch - Stärke **59**(2), 91–98 (2007)
13. Geresh, S., Lupescu, N., Arad, S.: Fractionation and partial characterization of the sulphated polysaccharide of porphyridium. Phytochemistry **13**(12), 4181–4186 (1992)
14. Schierbaum, F., Kordel, K.: Reaction of starch with the chlorosulfonic acid-formamide reagent, in carbohydrate sulfates. Am. Chem. Soc. **V11**, 173–192 (1978)
15. Guiseley, K.B.: Some novel methods and results in the sulfation of polysaccharides, in carbohydrate sulfates. Am. Chem. Soc. **V9**, 148–162 (1978)
16. Cirelli, A.F., Covian, J.A.: Effect of sulfation on the biological activity of b-(1–3)glucans from the tree fungus Cyttaria harioti Fischer. Carbohydr. Res. **190**(2), 329–337 (1989)
17. Geresh, S., Mamontov, A., Weinstein, J.: Sulfation of extracellular polysaccharides of red microalgae: preparation, characterization and properties. J. Biochem. Biophys. Methods **50**, 179–187 (2002)
18. Akman, F., Kazachenko, A.S., Vasilyeva, N.Y., Malyar, Y.N.: Synthesis and characterization of starch sulfates obtained by the sulfamic acid-urea complex. J. Mol. Struct. **1208**, 127899 (2020)
19. Kazachenko, A.S., Vasilyeva, N.Y., Malyar, Y.N., Kazachenko, A.S.: Optimization of starch sulfation process with a deep eutectic solvent-mixture of sulfamic acid:urea. J. Phys.: Conf. Ser. **1679**, 052082 (2020)
20. Kazachenko, A.S., Vasilyeva, N., Sudakova, I.G., Levdansky, V.A., Lutoshkin, M.A., Kuznetsov, B.N.: Numerical optimization of the process of abies ethanol lignin sulfation with sulfamic acid in 1,4-dioxane medium in the presence of urea. J. Sib. Fed. Univ. Chem. **13**(2), 232–246 (2020)
21. Pen, R.Z.: Planning an experiment at Statgraphics Centurion, p. 293. SibSTU, Krasnoyarsk (2014)
22. Dzhilbert, E.E.: Sulfonation of organic compounds, p. 416. Chemistry, Moscow, Russia (1969)
23. Whistler, R.L., King, A.H., Ruffini, G., Lucas, F.A.: Sulfation of cellulose with sulfur-trioxide-dimethyl sulfoxide. Arch. Biochem. Biophys. **121**(2), 358–363 (1967)

Research and Development Circuits for Error Correction

Sergey Gushanskiy, Alexander Gorbunov, and Viktor Potapov(✉)

Department of Computer Engineering, Southern Federal University, Taganrog, Russia
{smgushanskiy,avgorbunov,vpotapov}@sfedu.ru

Abstract. One of the main problems to be overcome in the near future for the practical creation of quantum computer to make quantum computing resistant to interaction with the external environment and to eliminate inaccuracy of operations. In this paper we study quantum codes correcting errors, proposed as a way to protect against the occurrence errors due to the influence of the external environment. Results of work are considered software implemented model of the simplest quantum code. A number of questions related to quantum computing are also raised. In their works, Shor (factorization algorithm), Grover (search in a disordered database) and a number of other scientists showed that a quantum computer (QC) is capable of processing a small class of functions with higher performance than any of the existing "classic" computers. This work is a study of the influence of the environment on the quantum system of qubits. The article is supposed to describe the foundations of quantum information theory, as well as the place in it of the concept of quantum entanglement. A technique for correcting two main types of quantum errors has been developed, based on the implementation and execution of certain quantum circuits. The main difficulties of ensuring the protection of a quantum channel from various types of errors are analyzed and considered.

Keywords: Qubit · Entanglement · Quantum scheme · Wave function · Quantum gate

1 Introduction

Until recently, the possibility of creating an active QC [1] (operating about 1000 qubits) was ruled out due to the strong exposure quantum particles to outside influence. However, as a solution to this problem were Quantum Error Correcting Codes [2] (QECC) have been proposed. The main difference between QECC and classic codes is to correct the error "without delving into the meaning" of a quantum state [3] is explained by the fact that measurement leads to the destruction of the quantum superposition [4] of states. It is believed that the interaction of a qubit with the environment can lead to an error of one of three possible types:

1. Bit (or, as they are also called, amplitude) errors – X, i.e. leading to a qubit [5] flip: $|0> \rightarrow |1>, |1> \rightarrow |0>$;

R. Silhavy (Ed.): CSOC 2021, LNNS 230, pp. 283–292, 2021.
https://doi.org/10.1007/978-3-030-77442-4_24

2. Phase errors – Z: $|0>\ \rightarrow |0>, |1>\ \rightarrow$ - $|1>$;
3. Bit-phase (amplitude-phase) errors (i.e., simultaneous action of errors of the first and second types) – Y: $|0>\ \rightarrow |1>, |1>\ \rightarrow$ - $|0>$.

These errors are called independent, because an error can occur in an arbitrary qubit, and it will not affect the state of other qubits. Moreover, bit errors can be turned into phase errors and vice versa by changing basis:

$$|0\rangle \rightarrow \frac{1}{\sqrt{2}}(|0\rangle + |1\rangle), \quad |1\rangle \rightarrow \frac{1}{\sqrt{2}}(|0\rangle - |1\rangle)$$

Among the existing quantum codes, there are simple ones that fix only one type of error in only one qubit. We have begun research on quantum codes. By modeling the code on a "classic" computer, we trace the work of the code step by step, then you can also get some code reliability characteristics. We started our work with the simplest code, proposed by Peter Shor. So, at the first stage, auxiliary qubits (as you can see, initially they must be in zero state) are set to the same state ($|c>$), as the first qubit, i.e. to clone the state of a qubit, you can use gate Controlled NOT (CNOT). At the second stage an error may occur. The third stage is usually called the preparation of the error syndrome. At the fourth stage, the error is corrected. We note at once two important drawbacks of this code: auxiliary qubits [6] do not return to their initial state, and next time they must be set to zero again; the scheme is "not designed" for errors in auxiliary qubits. Next is the question of the reliability of the code. In view of these shortcomings in in this case, the code is very unreliable. But the second drawback can be eliminated, if you use more auxiliary qubits. notice, that in this case, the CCNOT gate [7] must be replaced with a network that corrects the error for the most common state among the auxiliary qubits. To evaluate such codes, you need to know, for example:

- How the stability of the code depends on the probability of an error in one a qubit with a given number of auxiliary qubits;
- How the stability of the code depends on the number of auxiliary qubits with a fixed probability of occurrence of an error in one qubit;
- And most importantly, how to achieve code stability of n% for a given error probability.

According to the graph of the dependence of the number of qubits on the probability of error in one qubit to achieve ~100% code stability, one can judge the inefficiency in the essence of the classical approach when building a QECC (Fig. 1). Apart from the potentially large number of auxiliary qubits, this "code does not fix all possible types of errors. The code can be recommended for an initial acquaintance with the concepts of correcting errors in QC, not more. The code chosen by us for further research, according to its creators, corrects all possible errors, is resistant to the appearance of errors in the process of correcting the discovered. However, before embarking on further research, it is necessary to solve one problem that has arisen before us. Let's move a little away from the QECC topic, and let's turn to the theory of quantum computing. The state of a qubit can be described by an equation of the form $|\phi> = \alpha|0> + \beta|1>$, where $|\alpha|^2 + |\beta|^2 = 1$. Here, α and β, in general, are complex amplitudes [8].

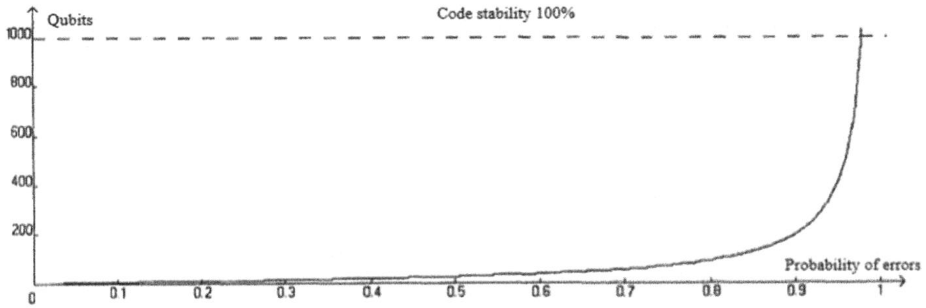

Fig. 1. The dependence of the number of qubits on the probability of error

The state of a system of n-qubits is usually described by the state vector dimension 2^n. The probability of finding the entire system in state m (m = 0 … N (in binary), where $N = 2^n - 1$ is determined by the square of the corresponding amplitude. In addition to the state vector, a density matrix is used to describe the state of a quantum system. Let us explain with the example of a qubit. If its state vector |q >, then the density matrix can be obtained as follows: $|q\rangle\,|q\rangle = \begin{pmatrix} \alpha \\ \beta \end{pmatrix} \cdot (\alpha^* \beta^*)$, where α*, β* are complex conjugate α, β numbers. The idea of fault-tolerant codes is borrowed from classical information theory, but their implementation in the case of encoding quantum states differs markedly from the classical situation. Let S be an arbitrary state in a Hilbert space M. By the code is an isometric mapping V: M → N that takes states S to coding states VS in the Hilbert [9] space N. A quantum code V can detect an error if the dimension of the Hilbert space N is greater than the dimension of the space M. in words, as in the classical case, any fault-tolerant quantum code is based on the principle of redundant coding. Let the transformation of a quantum state during encoding and subsequent occurrence of an error be described by the following scheme: S → VS → Φ(VS), where Φ ∈ E is an error operator of a certain class E. Then a code V is called an error-correcting code if there exists an operator Ψ such that Ψ[Φ(VS)] → c(Φ)S, for any Φ ∈ E. Here c(Φ) is some constant depending on the operator errors. It is called ancilla [10] of error syndrome, since its meaning depends only on the type of error and its localization. The error correction process is a two-step process:

1) by means of quantum circuits, individual for each correction code, the state of the coding ancilla is analyzed, during which the change in the state of ancilla syndrome;
2) the state of the ancilla of the syndrome is measured, and, depending on it, certain transformations are carried out with the coding ancilla.

Since the error of a single qubit is determined by a unitary transformation, then to restore the state, it is enough just to reapply to the erroneous qubit the transformation that caused the error.

By analogy with classical coding, the notation of the form [[n, k, d]] is used to denote quantum correction codes. Here n is the total number of code qubits, k is the number of information qubits, d is called the minimum code distance of the code, is defined as the minimum distance Hamming between all possible pairs of codewords (double brackets distinguish quantum code from classical). The whole variety of quantum correction codes is usually divided into three class:

1) additive stabilizing (simplex) codes. The most extensive class of codes is based on the classical correction codes, and therefore have received the greatest development. Stabilizing codes can be classified according to the minimum code distance:
a) a class of codes with a code distance of two. General notation [[n, n – 2, 2]];
b) a class of codes with a code distance of three. General designation $[[2^j, 2^j - j - 2, 3]]$;
c) a class of codes with a code distance of four. General designation $[[2^j, 2^j - 2j - 2, 4]]$.
Here j can take arbitrary integer values. This classification of stabilizing codes is not the only one and is generally accepted. More often you can come across a classification by the type of the code stabilizer matrix [11]:
- Cyclic codes;
- CSS – codes;
- Amplitude-protective codes, they are also "code over GF";
2) Non-additive codes that are not stabilizing;
3) Concatenated or Clifford codes. The probability of error for them is much lower than for the previous two, for which you have to pay a multi-level structure of the code, and as a consequence, an increase in the number of qubits. Concatenated codes are a generalization of additive stabilizing codes. In classical coding theory, they have an analogue called convolutional codes. The definition of X – error for Steen's seven-qubit code is given in Fig. 2.

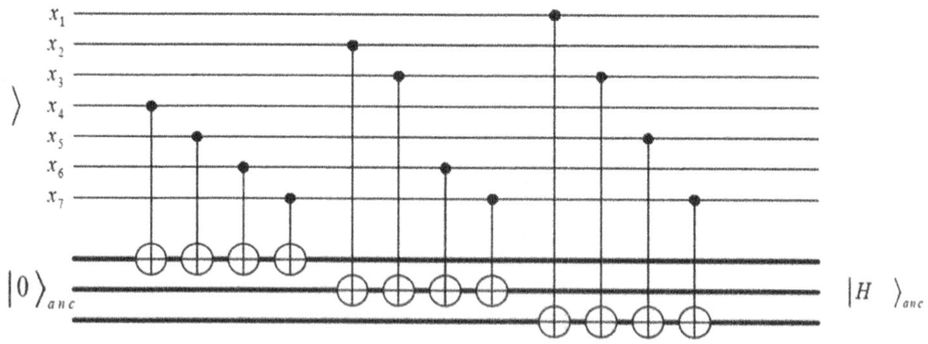

Fig. 2. X – error for Steen's seven-qubit code

We can now give a quantum circuit to detect X – an error in any of the seven code qubits (Fig. 2). The circuit uses generally accepted designations for controlled gates (the dot denotes the control qubit, and the circle with a cross is controlled). This diagram allows you to understand how the error syndrome is detected for Steen's code. Each qubit of ancilla syndrome is the result of summing modulo 2 of certain qubits code. Moreover, if we compare the matrix of checking the Hamming code and the circuit (Fig. 2), then it can be seen that it is the verification matrix that determines the numbers of the code qubits summed (modulo 2) in each qubit of the ancilla. However, it is clear that in addition to X – errors, it is necessary to fight with Z – errors (if it will be possible to eliminate both types of these errors, then Y – errors will also be automatically eliminated, since they can be represented as the result of the successive action of the first two). As you know, the concepts of bit and phase error are relevant only for the introduced basis. Changing the basis in a certain way converts the bit error into phase error and vice versa. The transformation that rotates the basis is called the Hadamard transformation

$$R = \frac{1}{\sqrt{2}}\begin{pmatrix} 1 & 1 \\ 1 & -1 \end{pmatrix}$$

Hadamard transform translates encoding states to $|0\rangle_{code}$ and $|1\rangle_{code}$ in new coding states

$$\left|\tilde{0}\right\rangle_{code} \equiv \frac{1}{\sqrt{2}}(|0\rangle_{code} + |1\rangle_{code}), \quad \left|\tilde{1}\right\rangle_{code} \equiv \frac{1}{\sqrt{2}}(|0\rangle_{code} - |1\rangle_{code}).$$

Considering that the Hadamard transformation is unitary, i.e. its double application will return the system to its original state, then a possible implementation of the circuit that allows detecting any errors in Steen's seven-qubit code will be have the form. The definition of any error for Steen's code is shown in Fig. 3.

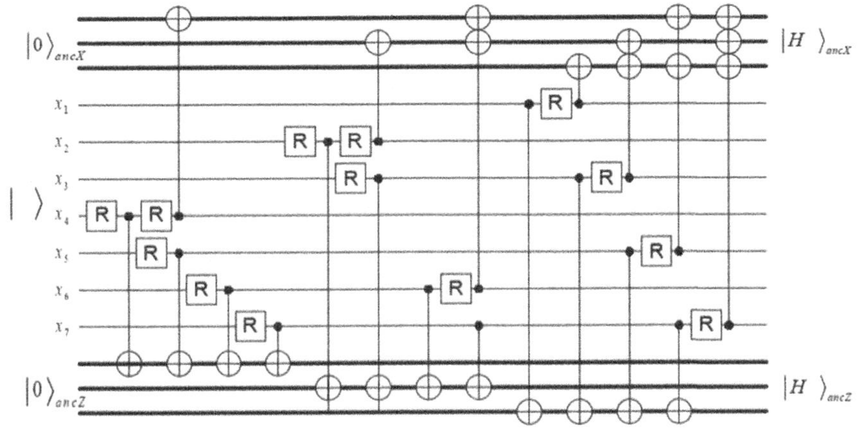

Fig. 3. The any error for Steen's code

In the proposed version of the quantum scheme (Fig. 3), two ancillas are used (ancilla X – errors and ancilla Z – errors). Since, according to our assumption, an error can occur only in one qubit, there are 4 possible outcomes of error detection:

1. Both ancilla are zero: no error occurred;
2. The ancilla X is zero, but the ancilla Z is not: there was Z – an error in the qubit, whose the number is contained in the ancilla Z;

3. The ancilla Z is zero, but the ancillus X is not: an X occurred – an error in the qubit, whose number is contained in ancilla X;
4. Both ancilla are nonzero and contain the same value: happened Y – an error in the qubit, whose number is contained in the ancilla X and Z.

If the type of error and its localization are established, then the correction process consists only in applying the appropriate transformation to the erroneous qubit: X, Y, or Z. Let's look at the error detection process using an example. Let there be an initial state of a single qubit $|\psi\rangle = \frac{1}{\sqrt{3}}|0\rangle + \frac{2}{\sqrt{3}}|1\rangle$, then the encoded state, according to the Steen code coding rules, will have the form.

$$|\psi\rangle_{code} = \frac{1}{\sqrt{3}}|0\rangle_{code} + \frac{2}{\sqrt{3}}|1\rangle_{code} = \frac{1}{2\sqrt{6}}(|0000000\rangle + |0001111\rangle$$

$$+ |0111100\rangle + |1010101\rangle + |1011010\rangle + \Big| |1100110\rangle + |1101001\rangle) + \frac{1}{2\sqrt{3}}$$

$$(|1111111\rangle + |1110000\rangle + |1001100\rangle + |1000011\rangle +$$

$$+ |0101010\rangle + |0100101\rangle + |0011001\rangle + |0010110\rangle).$$

Suppose the error is just an X – transformation, applied to the third qubit of the code, i.e. the error operator is $\Phi = I \otimes I \otimes X \otimes I \otimes I \otimes I \otimes I$, then, as a result of an error, the encoded state will take the form

$$\Phi|\psi\rangle_{code} = \frac{1}{\sqrt{3}}|0\rangle_{code} + \frac{2}{\sqrt{3}}|1\rangle_{code}$$

$$= \frac{1}{2\sqrt{6}}(|0010000\rangle + |0011111\rangle + +|0101100\rangle + |1010101\rangle)$$

$$+ |1000101\rangle + |1001010\rangle + +|1110110\rangle + |1111001\rangle$$

$$+ \frac{1}{2\sqrt{3}}(|1101111\rangle + |1100000\rangle + |1011100\rangle + +|1010011\rangle)$$

$$+ |0111010\rangle + |0110101\rangle + |0001001\rangle + |0000110\rangle$$

denote the state of the ancilla syndrome as $|c_1c_2c_3\rangle$ initially $c_1 = c_2 = c_3 = 0$. Having carried out the transformations, according to the scheme (Fig. 3), we obtain for our case

$$c_1 = x_4 \oplus x_5 \oplus x_6 \oplus x_7 = 0$$

$$c_2 = x_2 \oplus x_3 \oplus x_6 \oplus x_7 = 1$$

$$c_3 = x_1 \oplus x_3 \oplus x_5 \oplus x_7 = 1$$

It can be seen that the ancilla state of syndrome 011 unambiguously indicates X – error in the third qubit. To eliminate it, it is enough just to reapply the error operator Φ to the encoded state.

2 Modeling a Circuit for Error Correction

Some redundant information needs to be added to protect messages from noise. This procedure is called coding information. Thus, even if some of the information in the encoded message will be corrupted, redundancy will allow restoring all the original information by decoding the message. To find the results, we used computer algebra system Wolfram Mathematica.

2.1 Initial State

Formalism of quantum mechanics using density matrices allows you to describe the interaction of a system of qubits with the environment.

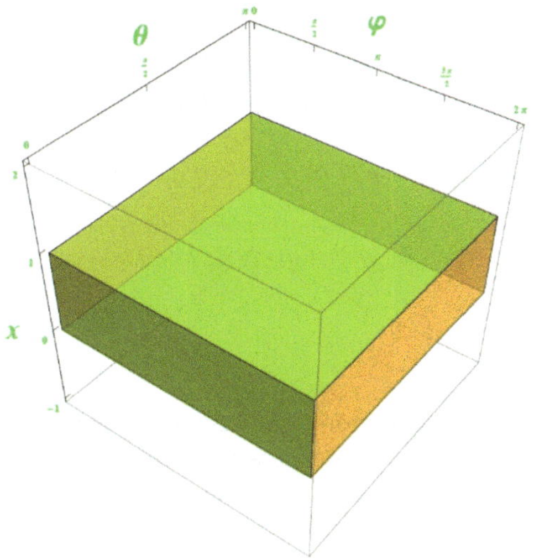

Fig. 4. Possible values of the parameter χ (*p*) of the superoperator for different density matrices. A combination of bit and phase errors.

The initial density matrix of three qubits is written in the form shown in Fig. 4.
$P = |\psi\rangle\langle\psi|\otimes|0\rangle\langle 0|\otimes|0\rangle\langle 0|$

2.2 Encoding

The CNOT element can be written in the same way through the Kronecker symbols [12] δ_{10} means that the value is 1, if the 1st qubit is in state 1, in otherwise, the value is 0.

$$CNOT = \begin{pmatrix} 1 & 0 \\ 0 & 1 \end{pmatrix} \otimes \left[\begin{pmatrix} 0 & \delta_{11} \\ \delta_{11} & 0 \end{pmatrix} + \begin{pmatrix} \delta_{10} & 0 \\ 0 & \delta_{10} \end{pmatrix} \right] = I \otimes \begin{pmatrix} \delta_{10} & \delta_{11} \\ \delta_{11} & \delta_{10} \end{pmatrix}$$

.

This result can be easily generalized to the case of N qubits, among which the first is the master, and N − 1 is the controlled.

$$CNOT(N) = I \otimes \prod_{N-1}^{\otimes} \begin{pmatrix} \delta_{10} & \delta_{11} \\ \delta_{11} & \delta_{10} \end{pmatrix}$$

The encoding operation is represented: $p_{CNOT}^{(3)} = CNOT(3) \cdot p_{initial}^{(3)} \cdot CNOT(3)$. The resulting density matrix is an entangled state matrix. We will assume that coding occurs without errors.

2.3 Overlay Noise

It is possible to compose a system of equations describing the effect of the environment on the system three qubits. For this we use the action of the environment in the form of a superoperator.

$$\$_{1p^{(3)}} = (1-p) \cdot p^{(3)} + p \cdot X \otimes I \otimes I \cdot p^{(3)} \cdot X \otimes I \otimes I$$

$$\$_{12p^{(3)}} = (1-p) \cdot \$_1 p^{(3)} + p \cdot X \otimes I \otimes I \cdot \$_1 p^{(3)} \cdot X \otimes I \otimes I$$

$$\$_{123p^{(3)}} = (1-p) \cdot \$_{12} p^{(3)} + p \cdot X \otimes I \otimes I \cdot \$_{12} p^{(3)} \cdot X \otimes I \otimes I$$

2.4 Bug Fixes

We will use syndromes to correct errors. The first syndrome determines whether the 1st and 2nd are equal, the second syndrome answers the question of equality of the 2nd and 3rd qubits. Knowing the results of the syndromes, we the most we will find out which of the three qubits is wrong. The syndromes are measurements, which leads to the loss of some off-diagonal elements of the density matrix. The reduction will eliminate all probabilities corresponding to transitions from states with equal qubits to states with unequal, as well as the probabilities of transition from states with unequal into states with equal qubits. The first/second syndrome nullifies the elements marked with X:

...	000	001	010	011	100	101	110	111
000	...	...	X	X	X	X	...	...
001	...	...	X	X	X	X	...	...
010	X	X	...	...	...	...	X	X
011	X	X	...	...	...	...	X	X
100	X	X	...	...	...	...	X	X
101	X	X	...	...	...	...	X	X
110	...	...	X	X	X	X	...	...
111	...	...	X	X	X	X	...	...

...	000	001	010	011	100	101	110	111
000	...	X	X	...	...	X	X	...
001	X	...	...	X	X	...	...	X
010	X	...	...	X	X	...	...	X
011	...	X	X	...	...	X	X	...
100	...	X	X	...	...	X	X	...
101	X	...	...	X	X	...	...	X
110	X	...	...	X	X	...	...	X
111	...	X	X	...	...	X	X	...

From this it is easy to show that these measurements lead to the loss of all matrix elements outside the main and secondary diagonal. We divide the density matrix into 4 components $\rho 1$, $\rho 2$, $\rho 3$ and $\rho 4$ according to the following principle:

1. All qubits are equal;
2. Third qubit with error;
3. Second qubit with error;
4. First qubit with error.

Having carried out such a division, it is enough to simply correct the classical error:

$$p_{corrected}^{(3)} = p_1^{(3)} + X \otimes I \otimes I \cdot p_2^{(3)} \cdot X \otimes I \otimes I + X \otimes I \otimes I \cdot$$

$$\cdot p_3^{(3)} \cdot X \otimes I \otimes I + X \otimes I \otimes I \cdot p_4^{(3)} \cdot X \otimes I \otimes I$$

...	000	001	010	011	100	101	110	111
000	1	0	0	0	0	0	0	1
001	0	2	0	0	0	0	2	0
010	0	0	3	0	0	3	0	0
011	0	0	0	4	4	0	0	0
100	0	0	0	4	4	0	0	0
101	0	0	3	0	0	3	0	0
110	0	2	0	0	0	0	2	0
111	1	0	0	0	0	0	0	1

The result of the error-correcting algorithm is the matrix density of the corrected qubit $p_{corrected}^{(1)}$. To get this matrix it is necessary to leave the basis *CNOT* and discard the second and third qubits, reducing the matrix $p_{corrected}^{(3)}$.

$$p_{corrected}^{(3)} = CNOT(3) \cdot p_{corrected}^{(3)} \cdot CNOT(3)$$

So, the above correction scheme is presented in the form of Fig. 5.

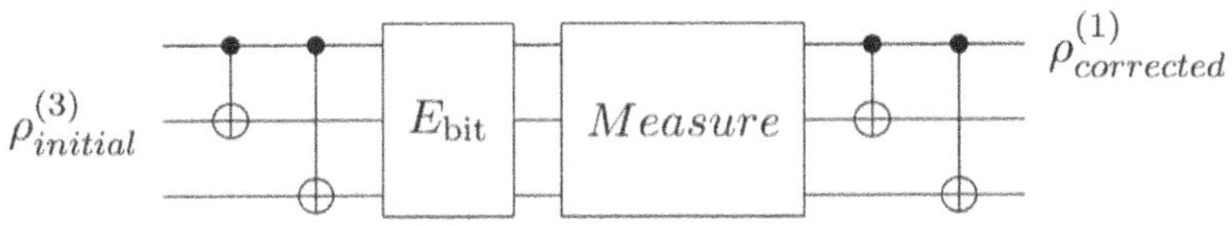

Fig. 5. The correction scheme

3 Conclusion

Correcting errors is one of the major challenges facing quantum computing devices. And without solving this problem, further successful developments in this promising area will become ineffective. In this paper, the codes for correcting various types of errors are numerically simulated. The main obstacles and difficulties in the way of protecting the channel from noise are analyzed, and some methods of overcoming them are proposed.

Implementation of schemes for correcting two main types of quantum errors has been made. The dependences of the data distortion on the noise level and the decoherence measure on the noise level in one qubit are demonstrated, as well as the dependence of the error on the measure and purity of entanglement.

Acknowledgments. The research was funded by RFBR according to the project № 20-07-00368.

References

1. Guzik, V., Gushanskiy, S., Polenov, M., Potapov, V.: Complexity estimation of quantum algorithms using entanglement properties. In: 16th International Multidisciplinary Scientific GeoConference, Bulgaria, pp. 20–26 (2016)
2. Guzik, V., Gushanskiy, S., Polenov, M., Potapov, V.: Models of a quantum computer, their characteristics and analysis. In: 9th International Conference on Application of Information and Communication Technologies (AICT). Institute of Electrical and Electronics Engineers, pp. 583–587 (2015)
3. Collier, D.: The comparative method. In: Finifter, A.W. (ed.) Political Sciences: The State of the Discipline II, pp. 105–119: American Science Association. Washington, DC (1993)
4. Williams, C.P.: Explorations in quantum computing. Texts in Computer Science. Chapter 2. Quantum Gates, pp. 51–122. Springer, Heidelberg (2011). https://doi.org/10.1007/978-1-84628-887-6
5. Olukotun, K.: Chip Multiprocessor Architecture – Techniques to Improve Throughput and Latency. Morgan and Claypool Publishers, San Rafael (2007)
6. Potapov, V., Guzik, V., Gushanskiy, S., Polenov, M.: Complexity estimation of quantum algorithms using entanglement properties. In: Informatics, Geoinformatics and Remote Sensing, Proceedings of 16-th International Multidisciplinary Scientific Geoconference, SGEM 2016, Bulgaria, vol. 1, pp. 133–140. STEF92 Technology Ltd. (2016)
7. Potapov, V., Gushanskiy, S., Guzik, V., Polenov, M.: The computational structure of the quantum computer simulator and its performance evaluation. In: Silhavy, R. (ed.) CSOC2018 2018. AISC, vol. 763, pp. 198–207. Springer, Cham (2019). https://doi.org/10.1007/978-3-319-91186-1_21
8. Milner, R.G.: A short history of Spin. In: Contribution to the XV International Workshop on Polarized Sources, Targets, and Polarimetry. Charlottesville, Virginia, USA, 9–13 September 2013 (2013). arXiv:1311.5016
9. Hallgren, H.S.: An improved quantum Fourier transform algorithm and applications. In: Proceedings of the 41st Annual Symposium on Foundations of Computer Science, Redondo Beach, CA, p. 515. IEEE (2000)
10. Potapov, V., Gushanskiy, S., Polenov, M.: The methodology of implementation and simulation of quantum algorithms and processes. In: Proceedings of 11th International Conference on Application of Information and Communication Technologies (AICT 2017), pp. 437–441. IEEE (2017)
11. Boneh, D., Zhandry, M.: Quantum-secure message authentication codes. In: Johansson, T., Nguyen, P.Q. (eds.) EUROCRYPT 2013. LNCS, vol. 7881, pp. 592–608. Springer, Heidelberg (2013). https://doi.org/10.1007/978-3-642-38348-9_35
12. Potapov, V., Gushansky, S., Guzik, V., Polenov, M.: Architecture and software implementation of a quantum computer model. In: Silhavy, R., Senkerik, R., Oplatkova, Z.K., Silhavy, P., Prokopova, Z. (eds.) Software Engineering Perspectives and Application in Intelligent Systems. AISC, vol. 465, pp. 59–68. Springer, Cham (2016). https://doi.org/10.1007/978-3-319-33622-0_6

Formal Development of Fault Tolerance by Replication of Distributed Database Systems

Rasin Katta(✉), Ammar Alhaj Ali, Bronislav Chramcov, Said Krayem, and Roman Jasek

Faculty of Applied Informatics, Tomas Bata University in Zlin, Zlin, Czech Republic
{Katta,alhaj_ali,chramcov,krayem,jasek}@utb.cz

Abstract. Data replication in a distributed database system is a crucial topic for enhancing data availability and improving the system faults tolerance, in the distributed database we think that clients use a single database yet, in fact, a set of databases are stored on multiple servers is used.

Replication often plays an important role in database management strategy; organizations use database replication for load balancing, offline processing, redundancy, and other typical administrative tasks.

In this paper we describe our experiences with using Event-B formal method and show how to conduct an assessment of system reliability at the refinement stages rather than at the implementation level; we demonstrate by case study how to obtain a mathematically proven solution using Event-B formal method. Here we propose using the Rodin modeling tool for Event-B that integrates modeling and proving together, and as a case study; we present data replication technologies in SQL Server databases.

Keywords: Distributed systems · Replication · Database · Fault tolerance · Event B · Rodin

1 Introduction

Replication is a set of technologies for copying and distributing data and database objects from one database (Source DB) to another database (Target DB) see Fig. 1, and then synchronizing between both databases to maintain consistency. Using replication, allows to distribute data to different geographical locations and to remote or mobile users over local and wide area networks, dial-up connections, wireless connections, and the Internet [1].

Replication uses distributed database technology to share data between multiple sites, but there is a difference between a replicated database and a distributed database. In a distributed database, data is available at many locations, but a particular table resides at only one location [2].

Our paper addresses layered refinement, where in each refinement, the modeler can introduce new variables, invariants, and events to enhance the system and provide general guidelines for modeling of replication of distributed database systems in formal methods,

R. Silhavy (Ed.): CSOC 2021, LNNS 230, pp. 293–306, 2021.
https://doi.org/10.1007/978-3-030-77442-4_25

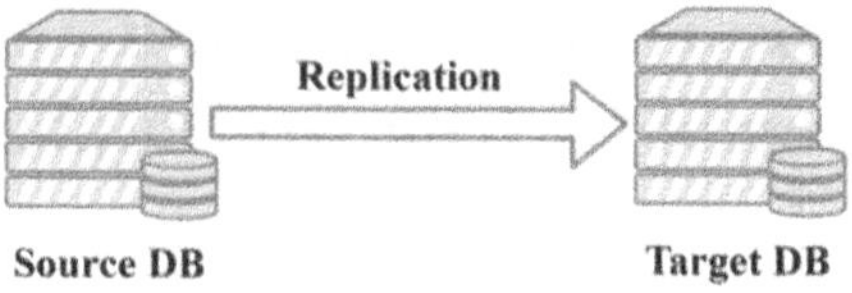

Fig. 1. Replication database

and show the suitability of formal method for developing the distributed storage systems and to motivate for developing a new methodology that improves both accuracy and reliability.

This paper is structured as follows: Sect. 2 presents related works, Sect. 3 provides reasons for using replication in database systems, Sect. 4 describes SQL server replication, Sect. 5 presents types of replication within SQL Server, Sect. 6 presents modeling and Refinement in Event B, Sect. 7 presents Formal modeling of replication within SQL Server in Event-B, Sect. 8 shows discussion and statistics, and lastly Sect. 9 concludes the paper.

2 Related Works

There is existing literature covering the concept of modeling Replication of distributed Database systems by formal methods.

Raghuraj and Divakar [3]; they have used Event-B as a formal technique for construction of their model. Event-B provides complete framework by rigorous description of problem at abstract level and discharge of proof obligations arising due to consistency checking. In this paper, they outline formal construction of model of multiversion concurrency control scheme for database transactions using Event-B.

Divakar and Butler [4]; they present a formal development of a distributed system using Event B that ensures atomic commitment of distributed transactions consisting of communicating transaction components at participating sites.

Pooja and Raghuraj [5]; In their model; a coordinator site finds a site with a largest replica number and then broadcast its updated replica to all other sites in a distributed environment.

None of these studies provided general guidelines for modeling of replication of distributed database systems in formal methods. Moreover, they did not address layered refinement, where in each refinement a modeler can introduce new variables, invariants, and events.

The approach of modeling replication of distributed database systems by gradual refinement steps is an important aspect and contribution of this research; as well as we present proofed model by Rodin platform. Where proof obligations are automatically generated after models have been created. The most powerful aspect of the Rodin platform is that a semiautomatic theorem proves is embedded in the tool, which saves lots of time when comparing to the technique of creating the proofs manually.

3 Why Replicate?

There are many rationales behind replication; however, they typically fall into four groups:

- Data Availability: Consider a case where you have a database at central office and multiple branch offices. The links between the branch offices and the center office are not stable, the network latency involved would make the application using database painfully slow, with the use of replication you could make copies of the database at branch offices and application speed would improve [6].
- Fault Tolerance: Fault tolerance means configuring your Database system so that it can withstand faults; a fault can be defined as an event that will prevent users from accessing their data [6].
- Data Distribution: Replication can be used to distribute data over multiple regional locations. Then, applications can access various regional servers instead of accessing one central server [2].
- Performance: Replication provides fast, local access to shared data because it balances activity over multiple sites. Some users can access one server while other users access other servers, thereby reducing the load at all servers [2].

4 SQL Server Replication

Microsoft SQL Server provides the following types of replication for use in distributed applications:

- Snapshot Replication.
- Transactional Replication.
- Merge Replication.

Each of these types of replication has the same common components:

1. The Publisher: is your source database that has a publication and a set of articles (the actual selections of data that will be replicated) defined on it to denote what data will be published to the Subscribers of that data.
2. The Distributor: keeps track of the subscriptions and is the hub for the activity between the Publisher and the Subscriber.
3. The Subscriber is a server that receives and maintains the published data, uses can be either push or pull, meaning data can be sent from the Publisher to the Subscriber, or the Subscriber can grab the data itself [7], see Fig. 2.

Each type of replication typically begins with an initial synchronization of the published objects between the Publisher and Subscribers. This initial synchronization can be performed by replication with a snapshot, which is a copy of all the objects and data specified by a publication [8].

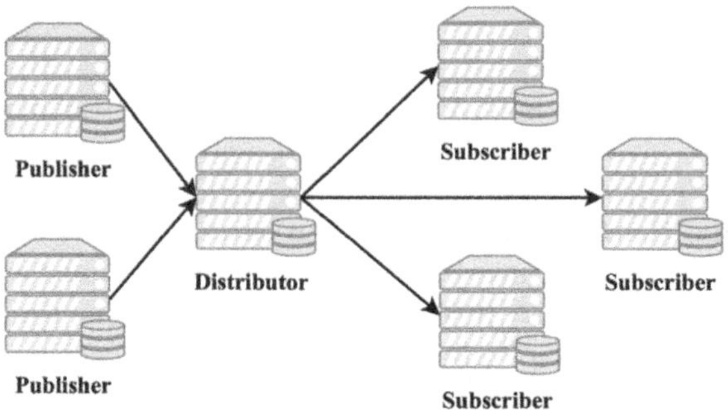

Fig. 2. Common components of SQL Server Replication.

5 Types of Replication Within SQL Server

5.1 Snapshot Replication

Snapshot replication, in its simplest form, is the easiest type of replication to set up and manage. Complete tables or table segments (for partitioned tables) are written to the subscribers during replication [9]. Snapshot replication distributes data exactly as it appears at a specific moment in time and does not monitor for updates to the data [8].

Replication is implemented through replication agents. Each agent is essentially its own, small, independent program that takes care of the tasks of monitoring transactions and distributing data as required for that particular type of agent [9].

Snapshot Agent
The Snapshot Agent supports snapshot and initial synchronization of data tables for other types of replication. In addition, it is responsible to take the "picture" of the published data and stores the files on the distributor [9].

Distribution Agent
The Distribution Agent is used for moving data for initial synchronization and snapshot

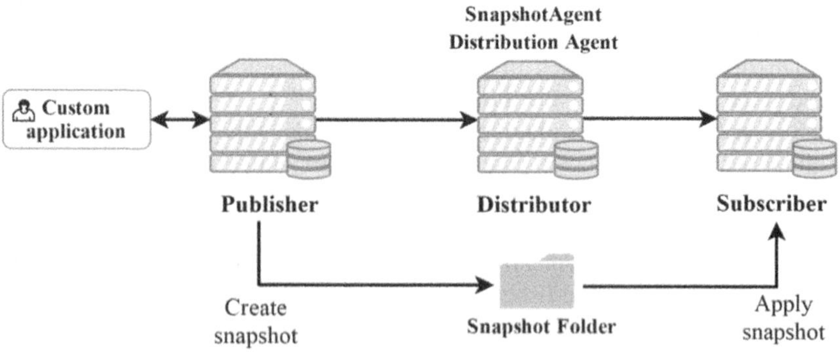

Fig. 3. Snapshot replication architecture.

replication from the publisher to the subscriber(s). For push subscriptions, the Distribution Agent typically runs on the distributor. For pull subscriptions, the Distribution Agent typically runs on the subscriber [7] [9, 10], see Fig. 3.

The Process of Snapshot Replication

1. The Snapshot Agent places a shared lock on all articles in the publication to be replicated, ensuring data consistency.
2. A copy of each article's table schema is written to the distribution working folder on the distributor.
3. A snapshot copy of table data is written to the snapshot folder.
4. The Snapshot Agent releases the shared locks from all articles in the publication.
5. The Distribution Agent creates the destination tables and database objects, such as indexes, on the subscriber and copies the data, overwriting the existing tables [9].

5.2 Transactional Replication

Transactional replication is arguably the most popular form of replication, especially when it comes to availability. It is also the most granular and gives you the lowest latency and often the best throughput [7]. The difference between transactional replication and snapshot replication is that incremental changes, rather than full tables, are replicated to the subscribers. Any changes logged to published articles, such as INSERT, UPDATE, and DELETE statements, are tracked and replicated to subscribers. In transactional replication, only changed table data is distributed, maintaining the transaction sequence. In other words, all transactions are applied to the subscriber in the same order that they were applied to the publisher [9].

Transactional Replication is implemented through replication agents, Log Reader Agent has the responsibility for copying transactions marked for replication from the publisher to the distributor as shown in Fig. 4. The Distribution Agent is also used in transactional replication and is responsible for moving transactions from the distributor to the subscriber(s) [9].

The Process of Transactional Replication

1. Modifications are posted to the publisher database and recorded in the associated transaction log.
2. The Log Reader Agent reads the transaction log and identifies changes marked for replication.
3. Changes taken from the transaction log are written to the distribution database on the distributor.
4. The Distribution Agent applies the changes to the appropriate database tables in Subscriber [9].

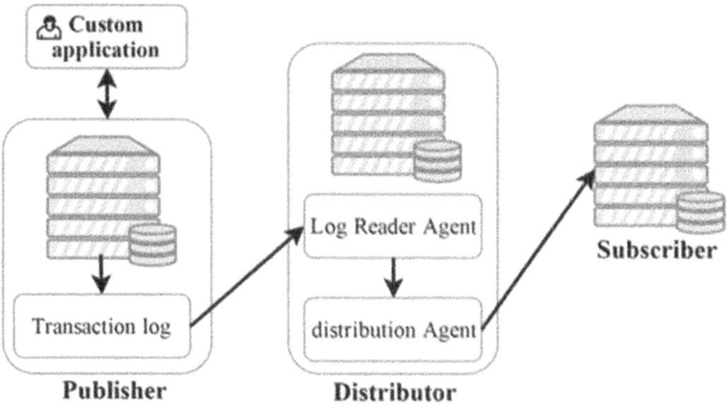

Fig. 4. Transactional replication architecture.

5.3 Merge Replication

Merge replication is similar to transactional replication, typically starts with a snapshot of the publication database objects and data. Subsequent data changes and schema modifications made at the Publisher and Subscribers are tracked with triggers. The Subscriber synchronizes with the Publisher when connected to the network and exchanges all rows that have changed between the Publisher and Subscriber since the last time synchronization occurred [12].

Merge Replication is also implemented via replication agents (the Merge Agent). As shown in Fig. 5, the agent copies changes from all subscribers and applies them to the publisher. It then copies all changes at the publisher (including those made by the Merge Agent itself during the resolution process) to the subscribers. The Merge Agent typically runs on the distributor for push subscriptions and on the subscriber for pull subscriptions [9].

The Process of Merge Replication

1. Triggers installed by SQL Server track changes to published data.
2. Changes from the publisher are applied to subscribers.
3. Changes from subscribers are applied to the publisher, and any conflicts are resolved.

6 Modeling and Refinement in Event B

Event-B is a mathematical approach for developing formal models of systems. An Event-B model is constructed from a collection of modeling elements. These elements include variables, invariants, events, guards and actions.

The modeling elements have attributes that can be based on Set Theory and Predicate Logic. Set Theory is used to represent data types and to use the data. Logic is used to apply conditions to the data.

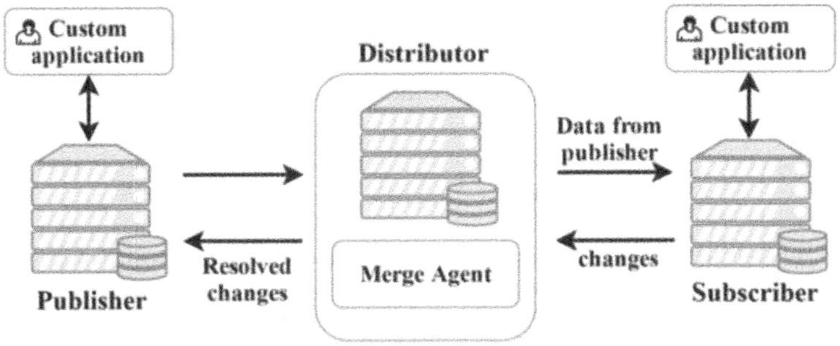

Fig. 5. Merge replication architecture.

The development of an Event-B model goes through two stages; abstraction and refinement, the abstract machine specifies the initial requirements of the system. Refinement is carried out in several steps, with each step adding more detail to the system [13].

The Rodin Platform is an Eclipse-based IDE for Event-B that provides effective support for refinement and mathematical proof. The platform is open source, contributes to the Eclipse framework and is further extendable by using plugins [14].

In Event-B there is two kinds of components machine and context, [15].

4. Context: describes the static elements of a model.
   A context has the following components:
- Sets: User-defined types can be declared in the SETS section.
- Constants: this is used to declare constants.
- Axioms: These axioms define rules that will always be the case for given elements of the context [16, 17].
5. Machine: describes the dynamic behavior of a model.

A machine has the following components:

- Variables: The variable's values are determined by an initialization event and can be changed by events.
- Invariants: These are predicates that should be true for every reachable state.
- Events: An event can assign new values to variables and the guards of an event specify the Conditions under which it can be executed [16].

7 Formal Modeling of Replication Within SQL Server in Event-B

In this paper, we model the database as a total function from objects to data,

$$\mathtt{DATABASE} \in \mathtt{OBJECT} \rightarrow \mathtt{DATA}$$

And we will use a partial database (`PartialDB`) that will be involved in an individual transaction of updates between servers,

$$\texttt{PartialDB} \in \texttt{OBJECT} \twoheadrightarrow \texttt{DATA}.$$

An update function maps a partial database (pdb1) to another partial database (pdb2) between tow servers:

$$\texttt{UPDATE} \in \texttt{PartialDB} \twoheadrightarrow \texttt{PartialDB}.$$

And the update function (update) updates the database as follows [3].

$$\texttt{database:} = \texttt{database} \lhd\!\!\!+ \texttt{update (pdb)}.$$

7.1 Snapshot Replication

Abstract Model

The abstract model of Snapshot replication is given in Fig. 6.

Below is a brief description of the machine.

- TRANSACTION, OBJECT and DATA are defined as sets.
- The databases are represented by variables Publisher_DB and Subscriber_DB.
- The variable transSA represents set of transactions from Publisher to SnapShot-Folder.
- The variable transDA represents set of transactions from SnapShotFolder to Subscriber.
- The variables transUpdateSA and transUpdateDA are total functions which maps each transaction to an object update function

$$\texttt{UPDATE} \in (\texttt{OBJECT} \twoheadrightarrow \texttt{DATA}) \twoheadrightarrow (\texttt{OBJECT} \twoheadrightarrow \texttt{DATA}).$$

Model Refinement

In snapshot replication the snapshot agent places a shared lock on all articles (objects) in the publication to be replicated, for this in model refinement we will use a new variable:

$$\texttt{ObjectStatus} \in \mathbb{P}1\ (\texttt{OBJECT}) \rightarrow \texttt{BOOL}$$

This variable will prevent making update on objects until transaction is completed, see Fig. 7.

```
machine SnapshotM0 sees SnapshotC0
variables Publisher_DB SnapShotFolder Subscriber_DB transSA
transUpdateSA transDA transUpdateDA
invariants
  @inv1 Publisher_DB ∈OBJECT→DATA
  @inv2 SnapShotFolder ⊆ (OBJECT⇸DATA)⇸(OBJECT⇸DATA)
  @inv3 Subscriber_DB∈OBJECT→DATA
  @inv4 transSA∈ℙ(TRANSACTION)
  @inv5 transUpdateSA ∈ transSA→((OBJECT⇸DATA)⇸(OBJECT⇸DATA))
  @inv6 transDA∈ℙ(TRANSACTION)
  @inv7 transUpdateDA ∈ transDA→((OBJECT⇸DATA)⇸(OBJECT⇸DATA))
events
  event INITIALISATION
    then
      @act1 Publisher_DB:∈OBJECT→DATA @act2 SnapShotFolder≔∅
      @act3 Subscriber_DB:∈OBJECT→DATA @act4 transSA≔∅
      @act5 transUpdateSA ≔ ∅ @act6 transDA≔∅
      @act7 transUpdateDA ≔ ∅
  end
  event Read_PublisherDB_By_SnapshotAgent
    any t u
    where @grd1 t∉transSA @grd2 u∈(OBJECT⇸DATA)⇸(OBJECT⇸DATA)
    then @act2 transUpdateSA(t)≔u @act1 transSA≔transSA∪{t}
  end
  event Write_to_SnapShotFolder_By_SnapshotAgent
    any t  where
      @grd1 t∈transSA
    then
      @act1 SnapShotFolder≔SnapShotFolder ∪ {transUpdateSA(t)}
      @act2 transUpdateSA≔{t}⩤ transUpdateSA
      @act3 transSA≔transSA\{t}
  end
  event Read_SnapShotFolder_By_DistributionAgent
    any t u where
      @grd1 t∉transDA @grd2 u∈SnapShotFolder
    then
      @act1 transDA≔transDA ∪ {t} @act2 transUpdateDA(t)≔u
  end
  event Write_to_SubscriberDB_By_DistributionAgent
    any t pdb
    where
      @grd1 t∈transDA @grd2 pdb∈dom(transUpdateDA(t))
    then
      @act1 Subscriber_DB≔Subscriber_DB  
transUpdateDA(t)(pdb)
      @act2 transUpdateDA≔{t}⩤ transUpdateDA @act3 trans-
DA≔transDA\{t} End end
```

Fig. 6. A specification of SnapshotM0.

```
event Read_PublisherDB_By_SnapshotAgent refines
Read_PublisherDB_By_SnapshotAgent
    any t u o
    where
      @grd1 t∉transSA
      @grd2 o∈ ℙ1(dom(Publisher_DB))
      @grd3 u∈(o⇸DATA)⇸(o⇸DATA)
      @grd4 ObjectStatus(o)=TRUE
    then
      @act2 transUpdateSA(t):=u
      @act1 transSA:=transSA∪{t}
      @act3 ObjectStatus(o):=FALSE
  End
event Write_to_SnapShotFolder_By_SnapshotAgent refines
Write_to_SnapShotFolder_By_SnapshotAgent
    any t o
    where
      @grd1 t∈transSA
      @grd2 o∈ ℙ1(dom(Publisher_DB))
      @grd3 ObjectStatus(o)=FALSE
    then
      @act1 SnapShotFolder:=SnapShotFolder ∪ {transUpdateSA(t)}
      @act2 transUpdateSA:={t}⩤ transUpdateSA
      @act3 transSA:=transSA\{t}
      @act4 ObjectStatus(o):=TRUE
  end
```

Fig. 7. A specification of SnapshotM1.

7.2 Transactional replication.

Abstract Model

The abstract model of Transactional replication is given in Fig. 8.

Below is a brief description of the machine.

- Transactions are written in the transaction log of the publication database.
- Log Reader Agent reads updates in the publication database by event `Read_TransactionLog_By_LogReaderAgent`
- Log Reader Agent writes updates in the distribution database by event `Write_to_DistributorDB_By_LogReaderAgent`.
- Distribution Agent reads from distribution database by event `Read_DistributorDB_By_DistributionAgent`.
- Distribution Agent transfers data to the subscription database by event `Write_to_SubscriberDB_By_DistributionAgent`.

Model Refinement

In model refinement of Transactional replication, we define new variable

```
  event Read_TransactionLog_By_LogReaderAgent
    any t u o where
      @grd1 t∉transLog @grd2 u∈(OBJECT⇸DATA)⇸(OBJECT⇸DATA)
      @grd3 o ∈ ℙ1(OBJECT)
    then
      @act1 transLog:=transLog∪{t} @act3 transObjectSA(t):=o
      @act2 transUpdateSA(t):=u
  End
  event Write_to_DistributorDB_By_LogReaderAgent
    any t pdb  where
      @grd1 t∈transLog
      @grd2 pdb=transObjectSA(t)◁ Distributor_DB
      @grd3 pdb∈dom(transUpdateSA(t))
    then
      @act1 Distributor_DB:=Distributor_DB • transUpdate-
SA(t)(pdb)
      @act2 transUpdateSA:={t}⩤ transUpdateSA
      @act3 transObjectSA:={t}⩤ transObjectSA
      @act4 transLog:=transLog\{t}
  End
  event Read_DistributorDB_By_DistributionAgent
 any t u o
    where
      @grd1 t∉transDA @grd2 u∈(OBJECT⇸DATA)⇸(OBJECT⇸DATA)
      @grd3 o ∈ ℙ1(OBJECT)
    then
      @act1 transDA:=transDA ∪ {t} @act3 transObjectDA(t):=o
      @act2 transUpdateDA(t):=u
  end
  event Write_to_SubscriberDB_By_DistributionAgent
    any t pdb
    where
      @grd1 t∈transDA @grd2 pdb=transObjectDA(t)◁ Subscriber_DB
      @grd3 pdb∈dom(transUpdateDA(t))
    then
      @act1 Subscriber_DB:=Subscriber_DB  
transUpdateDA(t)(pdb)
      @act2 transUpdateDA:={t}⩤ transUpdateDA
      @act3 transObjectDA:={t}⩤ transObjectDA @act4 trans-
DA:=transDA\{t}
  End end
```

Fig. 8. A specification of TransactionalM0.

$$\texttt{transstatus} \in \texttt{TRANSACTION} \rightarrow \texttt{TRANSSTATUS}$$

This variable denotes a status of transactions The TRANSSTATUS is an enumerated set containing the values INITIAL, PENDING and COMPLETED. These values are used to represent the global status of the transactions see Fig. 9.

```
event Read_TransactionLog_By_LogReaderAgent refines
Read_TransactionLog_By_LogReaderAgent
    any t u o  where
      @grd1 t∉transLog
      @grd2 u∈(OBJECT⇸DATA)⇸(OBJECT⇸DATA)
      @grd3 o ∈ ℙ1(OBJECT)
      @grd4 transstatus(t)=INITIAL
    then
      @act1 transLog:=transLog∪{t}
      @act3 transObjectSA(t):=o
      @act2 transUpdateSA(t):=u
      @act4 transstatus(t):=PENDING
  End
  event Write_to_DistributorDB_By_LogReaderAgent refines
Write_to_DistributorDB_By_LogReaderAgent
    any t pdb  where
      @grd1 t∈transLog
      @grd2 pdb=transObjectSA(t)◁ Distributor_DB
      @grd3 pdb∈dom(transUpdateSA(t))
      @grd4 transstatus(t)=PENDING
    then
      @act1 Distributor_DB:=Distributor_DB • transUpdate-
SA(t)(pdb)
      @act2 transUpdateSA:={t}⩤ transUpdateSA
      @act3 transObjectSA:={t}⩤ transObjectSA
      @act4 transstatus(t):=COMPLETED
      @act5 transLog:=transLog\{t}
end
```

Fig. 9. A specification of TransactionalC1 & TransactionalM1.

8 Discussion and Statistics

Formal methods establish correctness of a program (or a system) by mathematical methods which has independently been proved sound. Ideally, a formal verification should cover as much as possible of the system's construction: beginning with a specification that is very clear so the user can have no doubt of its meaning; and ending with an implementation that is very concrete so the manufacturer can have no doubt of how to build it.

Proof obligations are automatically generated after models have been created. The most powerful aspect of the Rodin platform is that a semiautomatic theorem proves is embedded in the tool, which saves lot of time compared to creating the proofs manually. The Rodin platform is open source, so it supports a large number of plug-in tools [17–19].

In Table 1, we can see proof statistics for Snapshot replication model and Transactional replication model using the Rodin3.2 platform, the statistics provide us the proof obligations generated and discharged by the Rodin (auto). The finial development of the Snapshot replication results in 33 POs (Proof obligations), and Transactional replication results in 51 POs that are proved automatically by the Rodin.

Table 1. Proof Obligations of Snapshot and Transactional replication models.

Snapshot replication model			
Step	Auto. (%)	Man. (%)	Total
Abstract model	16(100%)	0(0%)	16(100%)
First refinement	17(100%)	0(0%)	17(100%)
Total (%)	33(100%)	0(0%)	33(100%)
Transactional replication model			
Step	Auto. (%)	Man. (%)	Total
Abstract model	24(100%)	0(0%)	24(100%)
First refinement	27(100%)	0(0%)	27(100%)
Total (%)	51(100%)	0(0%)	51(100%)

9 Conclusion

In this paper we proposed a methodology for modeling and developing the distributed storage systems by using Event-B formal method from requirements analysis to code implementation using strict safety assessments.

The formal method is one of the mechanisms which help us to understand the complex specification of the system and how to analysis and build it, and aims to improve systems quality, provides a high degree of confidence and to produce zero-defect software by controlling the whole system-development process, from specifications to implementation.

Event-B is a formal method that is used for specifying and reasoning of complex systems and Rodin tool is a platform where verification of the program is done, it offers a reactive environment for constructing and analyzing models as most modern integrated development environments do, and provides integration between modeling and proving whereas this is an important feature for the developers to focus on the modeling task without switch between different tools to check proving in the same time.

Our paper shows the fitness of formal method for developing the distributed storage systems and it encourages the development using new methodology that improves the accuracy and reliability in the distributed storage systems.

Our final models are translated into the Event-B notation to verify required properties; finally we can conclude that Event-B allows us to define modeling methodology by writing the correct mathematical notions.

References

1. SQL Server Replication (2018). https://msdn.microsoft.com/en-us/library/ms151198.aspx. Accessed 20 Nov 2020
2. Replication Overview (2010). https://docs.oracle.com/cd/A87860_01/doc/server.817/a76959/repover.htm. Accessed 20 Nov 2020

3. Suryavanshi, R., Yadav, D.: Modeling of multiversion concurrency control system using Event-B. In: 2012 Federated Conference on Computer Science and Information Systems (FedCSIS), pp. 1397–1401. IEEE (2012)
4. Yadav, D., Butler, M.: Rigorous design of fault-tolerant transactions for replicated database systems using Event B. In: Butler, M., Jones, C.B., Romanovsky, A., Troubitsyna, E. (eds.) Rigorous Development of Complex Fault-Tolerant Systems. LNCS, vol. 4157, pp. 343–363. Springer, Heidelberg (2006). https://doi.org/10.1007/11916246_18
5. Sharma, P., Suryavanshi, R.S.: Formal verification of distributed transaction execution in replicated database system using Event-B. Int. J. Comput. Appl. **143**(6) (2016)
6. Cotter, H.: A Guide to SQL Server 2000 Transactional and Snapshot Replication, Not While the Surf's Up Press (2004). ISBN-10: 0974973602, ISBN-13: 978–0974973609
7. Hirt, A.: Pro SQL Server 2008 Failover Clustering. Apress. Springer, New York (2009). ISBN-13 (pbk): 978–1–4302–1966–8 ISBN-13 (electronic): 978–1–4302–1967–5
8. Types of Replication (2017). https://msdn.microsoft.com/en-us/library/ms152531.aspx. Accessed 20 Nov 2020
9. Vieira, R.: Professional Microsoft SQL Server 2008 Programming, Wiley Publishing, Inc. Published simultaneously in Canada (2009). ISBN: 978–0–470–25702–9
10. Snapshot Replication (2017). https://msdn.microsoft.com/en-us/library/ms151832.aspx. Accessed 20 Nov 2020
11. Transactional Replication (2017). https://msdn.microsoft.com/en-us/library/ms151176.aspx. Accessed 20 Nov 2020
12. Merge Replication (2017). https://msdn.microsoft.com/en-us/library/ms152746.aspx. Accessed 20 Nov 2020
13. Said, K.: Modern Theory of Information–Structured and Knowledge Based System Design (2016)
14. Event-B and the Rodin Platform (2018). http://www.event-b.org. Accessed 20 Nov 2020
15. Damchoom, K., Butler, M., Abrial, J. R..: Modelling and proof of a tree-structured file system in Event-B and Rodin. In International Conference on Formal Engineering Methods (pp. 25–44). Springer, Heidelberg (2008)
16. Jastram, M., Butler, M.: Rodin User's Handbook: Covers Rodin v.2.8, CreateSpace Independent Publishing Platform. (2014). ISBN 10: 1495438147 ISBN 13: 9781495438141, USA
17. Ali, A.A., Krayem, S., Chramcov, B., Kadi, M.F.: Self-stabilizing fault tolerance distributed cyber physical systems. Ann. DAAAM Proc. **29** (2018)
18. Ali, A.A., Jasek, R., Krayem, S., Zacek, P.: Proving the effectiveness of negotiation protocols KQML in multi-agent systems using Event-B. In: Silhavy, R., Senkerik, R., Kominkova Oplatkova, Z., Prokopova, Z., Silhavy, P. (eds.) CSOC 2017. AISC, vol. 574, pp. 397–406. Springer, Cham (2017). https://doi.org/10.1007/978-3-319-57264-2_40
19. Ali, A.A., Jasek, R., Krayem, S., Chramcov, B., Zacek, P.: Improved adaptive fault tolerance model for increasing reliability in cloud computing using Event-B. In: Computer Science On-line Conference, pp. 246–258. Springer, Cham. (2018)

Fault Detection Model for Multi Robotic System Using Formal Method Event-B

Emad Awwama, Ammar Alhaj Ali, Roman Jasek, Bronislav Chramcov, Said Krayem, and Rasin Katta(✉)

Faculty of Applied Informatics, Tomas Bata University in Zlin, Zlin, Czech Republic
{awwama,alhaj_ali,jasek,chramcov,krayem,Katta}@utb.cz

Abstract. Non-compliant behaviour prediction is an essential feature of open systems; this non-compliance may be due to accident, necessity, or malice. Therefore, to achieve a sustainable operation in a system we need to handle non-compliance using error detection and correction based on type and severity.

In this paper, we will present Event-B as a formal method that can be used in the development of the multi-robotic system and we propose using the Rodin modelling tool for Event-B that integrates modelling and proving. Where formal methods are traditionally used for reasoning about systems correctness. We used fault detection technique to improve the reliability of multi agent system by applying step-by-step refinement in the Event-B.

Keywords: Fault detection · Multi-robotic system · Formal method · Modelling · Event-B · Rodin

1 Introduction

Multi Agent Systems have witnessed rapid extension in the domain of physical agents since the 1980s, this expansion is performed by robots involvement in many tasks spreading from exploration [1] to warehouse-management [2]. Individual robots of the MRS (multi-robotic system), are obviously sensible to failure similar to their virtual counterparts. On the other hand, virtual agents, faults in robots could be also electromechanical which can be demonstrated in the robot's sensor and actuation devices. Furthermore, simple, and small-sized robots usually used in large scale collectives, lack the hardware capabilities to detect some of the common faults (e.g., rotary encoders to detect actuator faults). Many research has concentrated on engineering fault detection in robot collectives (e.g., see endogenous [3], and exogenous [4] fault detectors). Most of these models are developed assuming that the normal operating behaviour is identified and can be described beforehand. Consequently, the models are trained to recognize given normal behaviour (e.g., see [5]), any behaviour that is not known by the model is labelled abnormal. Nevertheless, while this approach does deliver some remarkable results of robust

R. Silhavy (Ed.): CSOC 2021, LNNS 230, pp. 307–316, 2021.
https://doi.org/10.1007/978-3-030-77442-4_26

fault detection and tolerance, the fault detection in the collective may not easily convert to different and varying characterization of normality. This is mostly relevant in scenarios in which agents change their behaviour through learning, or as a reaction to agitations in their environment. Furthermore, the prior information on the characterization of normal operating behaviour may not always be in the hand of the system designer. Briefly, the fault detection capabilities of the MRS tend to be tailored to the particular behaviour demonstrated by the target system.

In this study, we use a distributed multi agent-based model to illustrate the system behaviour. We used a technique where each agent should be able to detect other agents failure and assist its recovery. To apply this a counter is used for the number of agents agreeing that an agent task is down. All the agents which have duplicate of this task can modify the value of this counter.

The formal method Event-B is proposed as a natural choice for its formal modelling and verification. Event-B adopts an event-based modelling style that facilitates a correct-by-construction development of distributed system.

The paper is organized as follows: in Sect. 2 we briefly introduce modelling in Event-B and structure of model in Event-B. In Sect. 3 we introduce fault detection techniques. Section 4 presents modelling of fault detection in Event-B, Sect. 5 gives a brief discussion, and finally, in Sect. 6, we present our conclusions.

2 Modelling in Event-B

Event-B is a state-based formal approach that promotes the correct-by-construction development paradigm and formal verification by theorem proving. In Event-B, a system model is specified using the notion of an abstract state machine [6].

An abstract State Machine encapsulates the model state, represented as a collection of variables, and defines operations on the state, i.e., it describes the dynamic behaviour of a modelled system. The variables are strongly typed by the constraining predicates that together with other important properties of the systems are defined in the model invariants. Usually, a machine has an accompanying component, called context, which includes user-defined sets, constants and their properties given as a list of model axioms. The dynamic behaviour of the system is defined by a set of atomic events.

Generally, an event has the following forms, see Table 1.

Table 1. Event forms

Three Possible Forms of an Event
$e = begin\ R_e\ end$
$e = when\ G_e\ then\ R_e\ end$
$e = any\ a\ when\ G_e\ then\ R_e\ end$

Where *e* is the event's name, *a* is the list of local variables, the guard *Ge* is a predicate over the local variables of the event and the state variables of the system. The body of an event is defined by a multiple (possibly nondeterministic) assignment over the system variables. In Event-B, an assignment represents a corresponding next-state relation *Re*. The guard defines the conditions, under which the event is enabled, i.e., its body can be executed. If several events are enabled at the same time, any of them can be chosen for execution non-deterministically.

Event-B employs a top-down refinement-based approach to system development; Development starts from an abstract specification that non-deterministically models the most essential functional requirements. In a sequence of refinement steps, we gradually reduce non-determinism and introduce detailed design decisions. In particular, we can add new events, split events as well as replace abstract variables by their concrete counterparts, i.e., perform data refinement. When data refinement is performed, we define gluing invariants as a part of the invariants of the refined machine. These machines define the relationship between the abstract and the concrete variables. The proof of data refinement is often supported by supplying witnesses (A witness is to an event's parameter what a gluing invariant is to a variable: it is a mapping between the abstract parameter and the new parameter and allows the abstract parameter to disappear [7]). Witnesses are specified in the event clause *with*. The consistency of Event-B models, i.e., verification of well-formedness and invariant preservation as well as correctness of refinement steps, is demonstrated by discharging the relevant proof obligations. The Rodin platform [8] provides an automated support for modelling and verifying. In particular, it automatically generates the required proof obligations and attempts to discharge them.

3 Fault Detection Techniques

Any fault recovery technique is based on redundancy. So, if an agent is in failure, there should be other agents in the MAS that are ready to replace the in-fault agent. Therefore, each agent should be able to detect other agents failure and assist to recover it. Let *aj* be an agent such that *ai* has some of its tasks replicated in *aj*, i.e., *aj* ∈ *duplicate(ai)*. Variable ai can be down or some (and not all) of its tasks could be in failure.

In order for the agents in *duplicate(ai)* to take over the failed tasks, they have to recognize whether ai is down or in failure. To enable this, we propose to use a handshake protocol to ensure that coupled agents are able to identify whether their counterparts are down or not, Table 2 shows the functions used in our technique.

If *ai* is down, then each of the agents in *duplicate(ai)* will not receive a handshake from *ai* and will assume that they have to take over. Still, all these agents must agree that *ai* is down in order to take over.

Table 2. Functions used in our techniques

#	Function	Description
1	*handShake*	*handShake(s, a_i, a_j) =true if a_j receives a handshake from a_i in state s. If handShake(s, a_i, a_j) = false, then a_j knows that a_i is down.*
2	*agentDown*	*agentDown(s, a_i, g_j) = true if there is communication between a_i otherwise false.where g is one of the services the system has to provide*
3	*downDetected*	*downDetected(s, a_j) = true* if counter = *duplicate(a_j).*
4	*coupling*	*coupling(t) = (a_i, a_j)*,when task t is copied between two agents.
5	*informFailure*	*informFailure(s, a_i, t) := a_j ;agent a_i inform agent a_j to take over the task t.*
6	*abortedTasks*	*abortedTasks(s, a_j):=t; the task t is aborted*
7	*agentFailure*	*agentFailure(s, a_i, g_j) = true, agent a_i in failure.*

We propose to use a counter for the number of agents which agreed that ai is down. All the agents which are in *duplicate(ai)* are able to modify the value of the proposed counter. Each agent which detects that *ai* is down, will increment the value of this counter. If the value of the counter is equal to *duplicate(ai),* then the agents agree that *ai* is down. Otherwise, the agents that have incremented the value of the counter assume that *ai* is not able to communicate with them, and *ai* is not down. In the same way, if ai is in failure and cannot perform a task t, then it should notify a counterpart, the other agent *aj* acting as a backup agent for task *t*, i.e., *aj* in *coupling(t) = (ai, aj)*, as each task is replicated in one agent as stated earlier.

The following algorithm summarizes the fault-detection step, see Fig. 1, when the failure is detected, the system should recover from it, whenever possible.

4 Formal Modelling of Fault Detection in Event-B

In an Event-B development, instead of having a large single model, we can build the system using a series of successive layers, abstraction, and refinement [9, 10].

The abstract model should provide a simple view of the system, emphasizing on main purpose and key features of the system while ignoring how its details is achieved. Then gradually we add the functionality details of the system to the abstract model step by step. This process is called refinement [7, 11].

4.1 Abstract Model

We start by defining the context, it is the part where we declare sets and constants see Fig. 2.

```
// Let gi is one of the services provided by the system.
// we verify ai has some of its tasks replicated in aj .
if agentDown(s, ai, gi) = true then
       for each aj ∈duplicate(ai) do
           if handShake(s, ai, aj) = false then
           counter(ai) := counter(ai) + 1 ;
           // we increment the counter associated to ai.
           end if
        end for
end if
// we verify whether the agents agree over ai's failure.
if counter =  |duplicate(ai)| then
          downDetected(s, ai) = true;
end if
// if the agent is not down but in failure.
if (agentDown(s, ai, gi) = false) and (agentFailure(s, ai, gi)=
true) then
//we inform the agents in which t is replicated that t is in
failure.
for each t ∈ abortedTasks(s, ai) do
        for each aj ∈ duplicate(ai) do
             if coupling(t) = (ai, aj) then
             informFailure(s, ai, t) := aj ;
             end if
        end for
end for
end if
```

Fig. 1. The fault-detection algorithm

```
context CtxFDT0
sets S // the set of states of the system
     A // the set of agents operating in a MAS
     G // the set of goal situations
     T // the set of all tasks
constants s_initail a_initail g_initail t_initail
axioms @axm1 finite(A) @axm2 s_initail∈S
       @axm3 a_initail∈A
       @axm4 g_initail∈G @axm5 t_initail∈T
end
```

Fig. 2. A specification of abstract context

Then we move to defining the abstract machine; this is where we declare variables and invariants see code in Fig. 3, in addition two events are defined: *AgentDown* and *AgentIsNotDownButFailure*; in the first event we change the value of variable *downDetected* to True when the agent come to be down. And in the second event agent ai inform agent *aj* to take over the task *t*, when ai is not down but failure, see code in Fig. 4.

```
machine MFDT0 sees CtxFDT0
variables
s // s is one of observable states of the system.
g // g is one of the services the system has to provide.
t // t is one  of the task  of the system.
index
a_index // a  is one of  agents operating in a MAS
informFailure agentDown downDetected agentFailure
invariants
  @inv1 s ∈ S
  @inv2 g ∈ G
  @inv3 t∈T
  @inv4 index⊆1‥card(A)
  @inv5 a_index∈A→index
  @inv6 informFailure∈{s}×{a_index}×{t}→{a_index}
  @inv7 agentDown∈{s}×{a_index}×{g}→ BOOL
  @inv8 downDetected∈{s}×{a_index}→ BOOL
  @inv9 agentFailure∈{s}×{a_index}×{g}→ BOOL
```

Fig. 3. Variables and invariants of abstract machine

```
event AgentDown
    any ai where
      @grd1 ai∈a_index @grd2 agentDown(s↦{ai}↦g)=TRUE
    then
      @act1 downDetected(s↦{ai})≔TRUE
  End
event AgentIsNotDownButFailure any ai aj
    where
      @grd1 ai∈a_index @grd4 agentFailure(s↦{ai}↦g)=TRUE
      @grd2 aj∈a_index @grd3 agentDown(s↦{ai}↦g)=FALSE
    then
      @act1 informFailure(s↦{ai}↦t) ≔ {aj} end
```

Fig. 4. A specification of Events in abstract machine

4.2 Refinement Model

Last step is the refinement step, here we added new variables and invariants, see code in Fig. 5. After that we added new events see code in Fig. 6.

```
machine MFDT1
refines MFDT0
sees CtxFDT1

variables
...
    handShake task_a_index
    duplicate_Counter_a_index
    Counter_a_index abortedTasks
   coupling duplicate
invariants
...
  @inv10 handShake∈{s}×{a_index}×{a_index}→BOOL
  @inv11 task_a_index∈a_index→T
  @inv12 duplicate_Counter_a_index∈a_index→ℕ
  @inv13 Counter_a_index∈a_index→ℕ
  @inv14 abortedTasks∈{s}×{a_index}→ T
  @inv15 coupling∈{t}→{{a_index}×{a_index}}
  @inv16 duplicate∈{a_index}→{a_index}
  @inv17 ∀ai·(ai∈a_index  ∧ ∀ aj·(aj∈a_index  ∧ aj≠ai ∧
∃tt·(tt ∈ T ∧  coupling(tt)={{ai}}×{{aj}}   )⇔(
task_a_index(ai)=tt∧task_a_index(aj)=tt  )))

  @inv18 ∀ai·(ai∈a_index  ∧ ∀ aj·(aj∈a_index  ∧ aj≠ai  ∧  aj
∈duplicate({ai}))⇔( ∃tt·(tt ∈ T ∧
task_a_index(ai)=tt∧task_a_index(aj)=tt  )))
```

Fig. 5. Variables and invariants of refinement machine

4.3 Proof Statistics

In Table 3, we present the proof statistics for our model using the Rodin3.2 platform, the statistics provide us the proof obligations generated and discharged by the Rodin, the final development of our model results in 56 POs (Proof obligations), in which around (90%) have been proved automatically by the Rodin platform and the rest have been proved manually.

```
event IncrementCounter
  any ai aj
  where
    @grd1 ai∈a_index @grd2 aj∈a_index @grd3 ai≠aj
    @grd4 agentDown(s↦{ai}↦g)=TRUE
    @grd5 aj∈ duplicate({ai})
    @grd6 handShake(s↦{ai}↦{aj})=FALSE
  then
    @act1 Counter_a_index(ai)≔Counter_a_index(ai)+1
end
event GetNumberOftasksReplicated
  any ai tt
  where
    @grd1 ai∈a_index @grd2 tt∈{task_a_index(ai)}
    @grd3 ∀aj·(aj∈a_index ∧ ai≠aj ∧  aj ∈ duplicate({ai}))
  then
    @act1
duplicate_Counter_a_index(ai)≔card(task_a_index~[{tt}])
End
event AgentsAgreeOverAgentFailure refines AgentDown
  any ai
  where
    @grd1 ai∈a_index
    @grd2 Counter_a_index(ai)=duplicate_Counter_a_index(ai)
  then
    @act1 downDetected(s↦{ai})≔TRUE
end
event AgentIsNotDownButFailure refines AgentIsNotDown-
ButFailure
  any ai aj tt
  where
    @grd1 ai∈a_index
    @grd2 aj∈a_index
    @grd3 tt∈{task_a_index(ai)}
    @grd4 agentDown(s↦{ai}↦g)=FALSE
    @grd5 agentFailure(s↦{ai}↦g)=TRUE
    @grd6 tt∈{abortedTasks(s↦{ai})}
    @grd7 aj∈duplicate({ai})
    @grd8 coupling(tt)={{ai}}×{{aj}}
  then
    @act1 informFailure(s↦{ai}↦tt) ≔ {aj}
end
```

Fig. 6. A specification of Events in refinement machine

Table 3. Proof Obligations of our model

FDT Fault Detection Technique	Total 56(100%)	Auto 51 (90%)	Manual 5 (10%)
CtxFDT0	1	1	0
CtxFDT1	5	4	1
MFDT0	18	14	4
MFDT1	32	32	0

5 Discussion

Our study revealed a robust maintenance of tolerance to normal swarm behaviours, understood as being exhibited by a large proportion of the agents, irrespective of the type of swarm behaviours. Consequently, our abnormality detection system does not require looking for persistence in normal behaviour, only abundance. Importantly, the resilience to error is essential, since the normal behaviour is not just represented by a single feature-vector but distributed over a set of them. The error detection model introduced in this study highlights an important difference between fault detection and abnormality detection. The behaviours that are not exhibited by most of the agents in the swarm may be considered as abnormal, although they may not necessarily be faults.

6 Conclusion

Our fault detection model benefit from the concept of task duplicates in multi agent distributed system to create a solid background for fault tolerance distributed system in general. The behaviour of this system is verified using Rodin platform.

Improvement of the reliability of Multi-agent systems is an important issue for computing systems. Recently, formal methods have become an important approach to guarantee the reliability, the correctness and providing a high degree of confidence in Multi-agent Systems.

The used technique improved fault detection using multi agent systems. Furthermore, we used Event-B Modelling as it facilitates a correct-by-construction development of distributed system. Our paper shows the suitability of formal method for developing the multi-agent systems and to encourage for developing a new methodology that improves the accuracy and reliability in multi-agent systems.

Our final models are translated into the Event-B notation to verify required properties, this encourage us to consider that event-B allows to define a kind of modelling methodology by writing the correct mathematical notions; for this we can apply event-B in the modelling of many different complex projects, yet we should be careful during the selection of invariants and variables in order to facilitate the effort of proof.

Together with the Rodin tool that offers reactive environment for constructing and analysing models as do most modern integrated development environments and provides integration between modelling and proving as this is an important feature for the developers that allows them to focus on the modelling task without switching between different tools to check proving at the same time.

References

1. Hauert, S., Zufferey, J.C., Floreano, D.: Evolved swarming without positioning information: an application in aerial communication relay. Auton. Robot. **26**(1), 21–32 (2009)
2. Wurman, P.R. D'Andrea, R., Mountz, M.: Coordinating hundreds of cooperative, autonomous vehicles in warehouses. In: Proceedings of the 19th National Conference on Innovative applications of artificial intelligence - Volume 2, IAAI 2007, pp. 1752–1759. AAAI Press (2007)
3. Christensen, A.L., O'Grady, R., Birattari, M., Dorigo, M.: Fault detection in autonomous robots based on fault injection and learning. Auton. Rob. **24**(1), 49–67 (2008)
4. Christensen, A.L., O'Grady, R., Dorigo, M.: From fireflies to fault tolerant swarms of robots. IEEE Trans. Evol. Comput. **13**(4), 1–12 (2009)
5. Terra, M.H., Tinos, R.: Fault detection and isolation in robotic manipulators via neural networks: a comparison among three architectures for residual analysis. J. Rob. Syst. **18**(7), 357–374 (2001)
6. Abrial, J.R.: Modeling in Event-B: System and Software Engineering. Cambridge University Press (2010)
7. Jastram, M., Butler, M.: Rodin User's Handbook: Covers Rodin v.2.8, CreateSpace Independent Publishing Platform (2014). ISBN 10: 1495438147 ISBN 13: 9781495438141, USA
8. Event-B and the Rodin Platform (2018). http://www.event-b.org. Accessed 20 Nov 2020
9. Ali, A.A., Krayem, S., Chramcov, B., Kadi, M.F.: Self-stabilizing fault tolerance distributed cyber physical systems. Ann. DAAAM Proc. **29** (2018)
10. Ali, A.A., Jasek, R., Krayem, S., Zacek, P.: Proving the effectiveness of negotiation protocols KQML in multi-agent systems using Event-B. In: Silhavy, R., Senkerik, R., Kominkova Oplatkova, Z., Prokopova, Z., Silhavy, P. (eds.) CSOC 2017. AISC, vol. 574, pp. 397–406. Springer, Cham (2017). https://doi.org/10.1007/978-3-319-57264-2_40
11. Ali, A.A., Jasek, R., Krayem, S., Chramcov, B., Zacek, P.: Improved adaptive fault tolerance model for increasing reliability in cloud computing using Event-B. In: Silhavy, R. (ed.) CSOC2018 2018. AISC, vol. 765, pp. 246–258. Springer, Cham (2019). https://doi.org/10.1007/978-3-319-91192-2_25

Fault Tolerant Sensor Network Using Formal Method Event-B

Ammar Alhaj Ali, Bronislav Chramcov, Roman Jasek, Rasin Katta(✉), and Said Krayem

Faculty of Applied Informatics, Tomas Bata University in Zlin, Zlin, Czech Republic
{alhaj_ali,chramcov,jasek,katta,krayem}@utb.cz

Abstract. Cyber physical systems (CPS) are being increasingly deployed in different critical infrastructures such as transportation, healthcare, power, water, and other networks, these deployments witnessed a growing complexity of components by increased use of advanced technologies like sensors, actuators, communication networks and multicore processors: this posed reliability as a major challenge. We face many issues to ensure the reliability of the CPS, especially in the presence of system failures, any sensor fault that occurs during the readings from the physical systems manipulates the knowledge extracted from the physical systems and is most likely to cause the loss of the CPS reliability unless appropriate measures are taken.

In this paper we propose formal development of a software-based mechanism for fault tolerance in cyber physical system by the Event-B Method, in addition we present an approach to translate the Event-B model into C# code; which can be executed as a computer program or as simulation.

Keywords: Fault Tolerance · CPS · Cyber physical system · Modeling · Simulation · Event-B · Rodin · C#

1 Introduction

Cyber-physical systems (CPSs) represent a new field in automatic control that has been recently used increasingly in life critical systems, where the probability of tragic failure has to be kept below very low levels.

Cyber-physical systems represent an interconnection between processing systems and physical systems through communication, sensor and actuator technologies [1], see Fig. 1.

Where a sensor fault refers to a sensor's reading that is inconsistent with the expected behavior of the physical phenomenon, fault tolerance is used to face failures and achieve best performance of systems, thus; it becomes critical to develop models that are tolerant towards a system failure. This will make it possible for the CPS to continue its operation even in the presence of faults.

Modeling and simulation enables designers to test whether design specifications are met by using virtual rather than physical experiments. The use of virtual prototypes significantly shortens the design cycle and reduces the cost of design. It further provides the

R. Silhavy (Ed.): CSOC 2021, LNNS 230, pp. 317–330, 2021.
https://doi.org/10.1007/978-3-030-77442-4_27

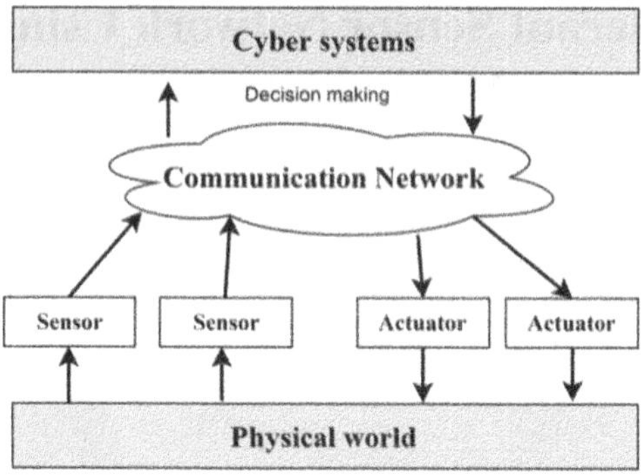

Fig. 1. Cyber physical system view

designer with immediate feedback regarding design decisions which, in turn, promises a more comprehensive exploration of design alternatives and a better performing of the final design [2].

In this paper; we present a mechanism for fault tolerance in cyber physical systems for the case where faults occur in the sensor interface of CPSs between the cyber and the physical parts, we propose Event-B as formal method for system-level modeling and analysis, and we recommend using the Rodin platform as modeling tool for Event-B that integrates modeling and proving, finally we present a method to the translation of our Event-B model into C# code that can be executed as a computer program.

The paper is organized as follows: in Sect. 2; the related works are presented, Sect. 3 gives definitions for basic terms and concepts, In Sect. 4 we briefly introduce modeling in Event-B and structure of model in Event-B, In Sect. 5 we introduce a fault tolerant model in cyber-physical systems. Section 6 presents modeling the FMS in Event-B, Sect. 7 gives a brief of transforming of Event B models into verified C# implementations, and finally, in Sect. 8, we present our conclusions.

2 Related Works

There is different literature covering the concept of Fault Tolerance in Cyber Physical Systems and formal development.

Sudeep Ghimire [3]; they presented the model for defining sensing model for representing sensors and actuators and creating virtual objects that will enable data regeneration when the physical devices fails. The data regeneration algorithm is based on the virtual instance attributed with contextual details of the deployed physical devices.

Volkan Gunes [4], in his dissertation, he focuses on sensor fault mitigation and achieving high reliability in CPS operations. He examines the falling ball example (FBE) using binary event detectors, a controller, and a camera for timely motion detection and estimation of a falling ball. Another challenge he tackle is satisfying thermal comfort and energy efficiency under faulty sensor conditions in a multi-room building incorporating temperature sensors, controllers, and heating, ventilation, and air conditioning (HVAC) systems. For both cases, he adopts a model-based design (MBD) methodology to analyze the effect of sensor faults on the system outcome. In this regard, he develops well-defined fault and system evaluation models and incorporates them into the traditional CPS model that comprises the cyber, interface (e.g., sensors and actuators) and physical models.

Dubravka Ilić [5], in their paper they propose formal development of a software-based mechanism for tolerating transient faults in the B Method. The mechanism relies on a specific architecture of the error detection actions called the evaluating tests. These tests are executed (with different frequencies) on the predefined subsets of the analyzed data.

3 Definition of Basic Terms and Concepts

System: is a construct or collection of different elements that together produces results that are not obtainable by the elements separately or alone. The elements can include people, hardware, software, facilities, policies, documents -all things required to produce a system - level qualities, properties, functions, behavior, and performance. Vitally, the value of the system as a whole is the relationship among the parts [6].

Modeling: is the process of representing a model which includes its construction and working. This model is a physical, mathematical, or otherwise logical representation of a system and it is similar to the real system, it helps the analyst to predict the effect of changes to the system. In other words, modeling is creating a model which represents a system including its properties.

Simulation of a system: is the operation of a model in terms of time or space, which helps to analyze the performance of an existing or a proposed system. In other words, simulation is the process of using a model to study the performance of a system [7].

4 Modeling in Event-B

Event-B is a formal method for system-level modeling and analysis. Key features of Event-B are the use of set theory as a modeling notation, the use of refinement to represent systems at different abstraction levels and the use of mathematical proof to verify consistency between refinement levels. Event-B is an extension of the B method, which was developed by Jean Raymond Abrial several years ago [8].

4.1 Why Event-B

- Event-B is a mature formal method which has been widely used in several industry projects in many different domains, such as automotive, transportation, space, business information, medical device and so on.
- An FP7 project supported by the European Commission, called the Deploy Project, is using Event-B to improve system dependability, to handle system complexity and to reduce the capital spent on the testing and debugging stages of software development [9].
- A recent technical report [10] written by Mery and Singh described a formal development of a Cardiac Pacemaker using Event-B, which shows the feasibility and validity of using Event-B during the development cycle of medical devices.

- Another reason for using Event-B is that it is a tool-supported formal specification language. The Rodin platform, an Eclipse-based IDE for Event-B, provides a user-friendly interface to create, refine and mathematically prove properties of models. Proof obligations are automatically generated after models have been created. The most powerful aspect of the Rodin platform is that a semiautomatic theorem proves is embedded in the tool, which saves lots of time compared to creating the proofs manually. The Rodin platform is open source, so it supports a large number of plug-in tools [9].

4.2 Event-B Structure

A model in Event-B consists of contexts and machines. Contexts contain the static part (types and constants) of a model while machines contain the dynamic part (variables and events), there are various relationships between contexts and machines [11–14]. See next Fig. 2.

5 Fault Tolerant Model in Cyber-Physical Systems

We face enormously increasing challenges to ensure the reliability of CPS, especially in the presence of system failures, as any sensor fault that occurs during the readings from the physical systems damages the knowledge extracted from the physical systems and presents the risk of reducing or even losing the reliability of CPS unless appropriate measures are taken.

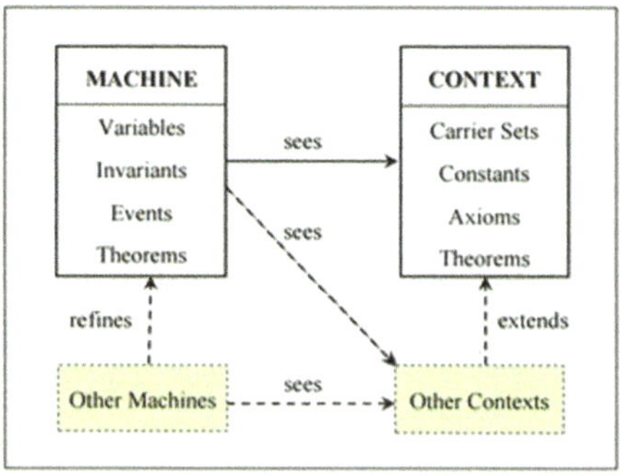

Fig. 2. Relationships between contexts and machines

Transient faults are the temporal defects within the system, they frequently occur in the hardware functioning. In our case study, we focus on designing a controller that is able to tolerate transient faults, in the complex fault-tolerant control systems; a controller largely consists of the mechanisms for implementing fault tolerance.

This is often perceived as a separate subsystem dedicated to fault tolerance, this subsystem is traditionally called Failure Management System (FMS) [5].

Obviously, correctness of the FMS itself is essential for ensuring dependability of the overall system. Formal methods are traditionally used for reasoning about software correctness; we will demonstrate how to develop the FMS by stepwise refinement in the Event-B, then translate final proven model to C# code to get the simulation.

The Failure Management System may be part of CPS as shown in Fig. 3.

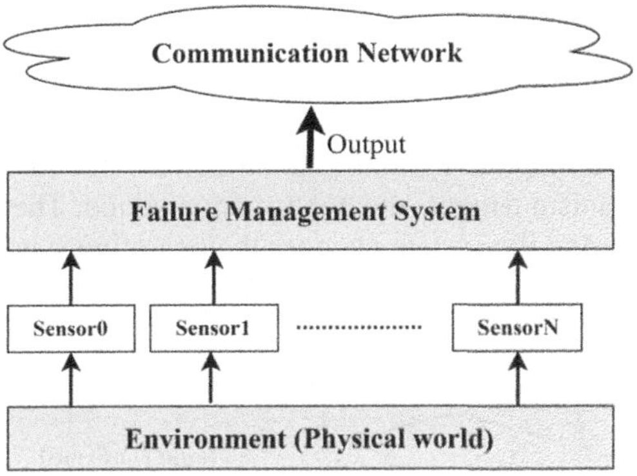

Fig. 3. Failure Management system in CPS

The sensor readings are considered as the inputs to the FMS and the outputs from the FMS are forwarded to the communication networks and the task of the FMS is to detect incorrect inputs and prevent their propagation into the communication networks, Hence the main purpose of the FMS is to supply the communication networks of the system with the fault-free inputs from the environment.

We assume that initially the system is error-free and the FMS tests the inputs by applying a certain detection procedure; As a result, the inputs are categorized as fault-free or faulty, then the FMS analyses the categorized faulty inputs to distinguish between recoverable and non-recoverable faulty inputs.

This is achieved by assigning a status to each analyzed input; the status can be [5]:

- The fault-free inputs are marked as **ok**.
- The recoverable inputs are marked as **suspected**.
- The non-recoverable inputs are marked as **confirmed failed**.

After completing analysis, the FMS takes the corresponding actions; these actions can be classified as healthy, temporary or confirmation see Fig. 4.

1. Healthy action:
   If the FMS is in the Normal state, a received input is fault-free (ok), then the input is forwarded unchanged to the communication networks and the FMS continues its operation by accepting another input from the environment.
2. Temporary action:
   If the FMS is in the Normal state and detects the first faulty input, it changes the operating state from Normal to Recover, see Figure 4, While in the Recover state,

the FMS counts the number of faulty inputs in successive operating cycles and at the same time the status of the faulty input is marked as suspected, One of the requirements imposed on the FMS is to give a fault-free output even when the input is faulty, Hence, while operating in the Recover state, the FMS calculates the output using the last good values of this inputs obtained before entering the state recover. Once a temporary action is triggered, it will keep the system in the state Recover until the counting mechanism determines whether the input (i.e., the corresponding sensor) has recovered. In this case, the system changes its state from recover to normal.

3. Confirmation action:
   If the system has been operating in the state Recover and the input fails to recover, the counting mechanism triggers the confirmation action, Then the input is marked as confirmed failed and the system changes the operating state to Failed [5].

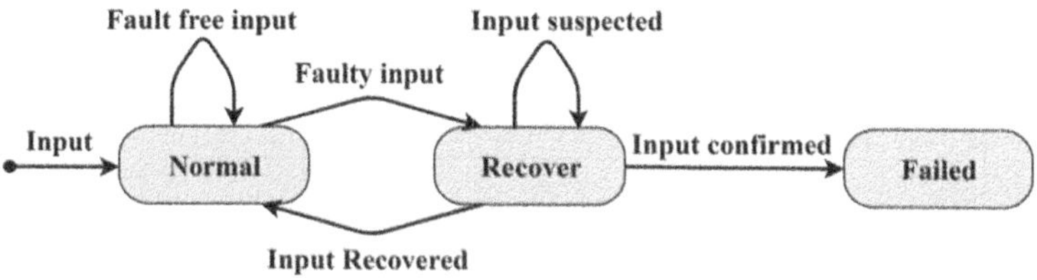

Fig. 4. Specification of the FMS behavior

6 Modeling the FMS in Event-B

The abstract specification defines the behavior of the FMS during one operating cycle is modeled using the variable `FMS_State`. The type `STATES` of `FMS_State` is defined in the context Global, as follows:

```
STATES = {env, det, anl, anlloop, act, output, freeze};
```

Where the values of FMS_`State` define the phases of the FMS execution in the following way:

- env: obtaining inputs from the environment.
- det: performing tests on the inputs and detecting erroneous inputs,
- anlloop and anl: deciding upon the input status,
- act: setting the appropriate actions,
- output: sending output to the communication network either by simple forwarding one of the obtained inputs or by calculating the output based on the last good values of the inputs,
- freeze: freezing the system

The variable `FMS_State` models the progress of the system behavior in the operating cycle. At the end of the operating cycle the system either reaches the terminating (freezing) state or produces a fault-free output. In the latter case, the operating cycle starts again.

6.1 Variables

Table 1 shows all variables in abstract, First refinement and second refinement model.

Table 1. Variables in FMS model.

No	Variables	description	invariants
abstract			
1	Indx	Index of The input values by the sensors	*0 ‥ max_indx* where *max_indx* ∈ ℕ
2	InputN	The input values produced by the sensors (N multiple homogeneous sensors.)	Indx→ ℕ
3	Input_StatusN	it is an array that for each of N inputs contains a value of the type: *I_STATUS* = {*ok, suspected, confirmed_failed*}; (previous cycle)	Indx→ **I_STATUS**
4	Input_In_ErrorN.	It is **TRUE**, if an error is detected on the sensor reading of a particular input, and **FALSE** otherwise.	Indx→**BOOL**
5	Last_Good_InputN	Last good input or last input – fault-free	Last_Good_InputN ∈ Indx →ℕ
6	Output	Out of FMS	Output ∈ ℕ
7	FMS_State	The variable *FMS_State* models the evolution of the system behavior in the operating cycle.	FMS_State ∈ **STATES** Where *STATES* = {*env, det, anl, anlloop, act, out, freeze*};
First refinement			
8	Input_StatusN1	it is an array that for each of N inputs contains a value of the type: *I_STATUS* = {*ok, suspected, confirmed_failed*};(current cycle)	Indx→ **I_STATUS**
9	Processed	It is **True**,After the operation AnalysisLoop is completed else **False**.	Processed ∈ Indx →BOOL
Second refinement			
10	cc	For N inputs, we introduce counters cc_i (i∈0.. *max_indx*), which contain accumulated values determining how trustworthy a particular input *i* is.	cc ∈ Indx ⇸ ℕ
11	num	For N inputs, the counter which counts the number of the consequent recovering cycles for each suspected input	num ∈ Indx ⇸ ℕ

6.2 Events

In abstract model we use the following events:

```
event Environment
  where
    @grd1 FMS_State = env
  then
    @act1 InputN :∈ Indx →ℕ
    @act2 FMS_State ≔ det
end
```

Fig. 5. Specification of environment Event.

1. Environment: The input values produced by the sensors are modeled by the variable InputN. The variable represents the readings of N multiple homogeneous sensors, see Fig. 5.

2. Detection: After obtaining the sensor readings from the environment, the FMS starts the operation Detection, we assign the value TRUE to the variable Input_In_ErrorN, if an error is detected on the sensor reading of a particular input, and FALSE otherwise, see Fig. 6.

```
event Detection
  where
    @grd1 FMS_State = det
  then
    @act1 Input_In_ErrorN :∈ Indx →BOOL
    @act2 FMS_State ≔ anlloop
End
```

Fig. 6. Specification of detection Event.

3. AnalysisLoop: After the detection phase, the FMS performs the operation AnalysisLoop, the operation Analysis sets the results of the analysis for all the inputs at once, see Fig. 7.

```
event AnalysisLoop
  where
    @grd1 FMS_State = anlloop
  then
    @act1 FMS_State :∈ { anlloop , anl }
End
```

Fig. 7. Specification of analysis loop Event

4. Analysis: The FMS decides upon the status of an input (Based on the results obtained at the previous state); fault-free (i.e., ok), suspected or confirmed failed. The variable Input_StatusN is an array that for each of N inputs contains a value of the type: I_STATUS = {ok, suspected, confirmed_failed}, see Fig. 8.

```
event Analysis
    where
      @grd1 FMS_State = anl
    then
      @act2 FMS_State ≔ act
      @act3 Input_StatusN :∈ { ff | ff ∈ Indx →I_STATUS ∧
            (∀ee·ee∈Indx ∧ Input_In_ErrorN(ee)=FALSE ⇒
ff(ee)∈{ok, suspected,confirmed_failed} ) ∧
            (∀ee·ee∈Indx ∧ Input_In_ErrorN(ee)=TRUE ⇒
ff(ee)∈{suspected, confirmed_failed})}
  End
```

Fig. 8. Specification of analysis Event.

5. Action: here we have four action events, the operation Action will decide if system will go to freeze status or out status, see Fig. 9.
6. Return: in this Event FMS will send output to the communication network either by simple forwarding one of the obtained inputs or by calculating the output based on the last good values of the inputs, then move to Environment event to read new values, see Fig. 10.

6.3 Refinement

First refinement: we specify in detail the operation AnalysisLoop. The operation gradually performs the input analysis, considering inputs one by one until all the inputs are processed. The information about the input status of the processed inputs is correspondingly accumulated in the variable Input_StatusN1. After the operation AnalysisLoop is completed, the value of Input_StatusN1 is assigned to Input_StatusN in the operation Analysis.

Second refinement: For each of N inputs, we introduce counters cci ($i \in 0..$ max_indx), which contain accumulated values determining how trustworthy a particular input i is. If $cci = 0$ then the input i is ok. If $0 < cci < zz$, where zz is some predefined value, the input i is suspected. Otherwise, the input i is considered failed.

At every cycle the counters cci are reevaluated depending on the detection results (suspected status). Each faulty input i increments the counter cci by a certain predefined value xx. Similarly, each fault-free input i decrements the corresponding counter cci by another predefined value yy. If at some point the value of cci reaches 0, the input i is declared ok. Similarly, if the value of cci exceeds zz, the input i is declared confirmed_failed and should be removed from the set of inputs used by the FMS.

The predefined values zz, xx and yy are set after observing the real performance of the FMS. The input may behave in such a way that the counter cc is practically oscillating between some values but never reaches the limit zz or zero. To overcome this problem, we introduce the second counter Num which counts the number of the consequent recovering cycles for each suspected input (i.e., when $0 < cci < zz$). When a certain limit for Num is exceeded, the recovery terminates and, if cc is different from zero, the input is confirmed_failed [5].

```
event ActionAllOut
    where
      @grd1 FMS_State = act
      @grd2 confirmed_failed ∉ ran (Input_StatusN)
      @grd3 Indx ≠ ∅
    Then @act1 FMS_State ≔ output
  end
  event ActionSomeOut
    where
      @grd1 FMS_State = act
      @grd2 confirmed_failed ∈ ran ( Input_StatusN )
      @grd3 Input_StatusN~[ { ok , suspected } ] ≠ ∅
    then
      @act1 Indx ≔ Input_StatusN~[ { ok , suspected } ]
      @act2 InputN ≔ Input_StatusN ~ [ { ok , suspected } ]◁
InputN
      @act3 Input_StatusN ≔ Input_StatusN ⩥{ ok , suspected }
      @act4 Input_In_ErrorN ≔ Input_StatusN ~ [ { ok ,
suspected } ] ◁ Input_In_ErrorN
      @act5 Last_Good_InputN ≔ Input_StatusN ~ [ { ok ,
suspected } ] ◁ Last_Good_InputN
      @act6 FMS_State ≔ output
  End
  event ActionAllFailed
    where
      @grd1 FMS_State = act
      @grd2 confirmed_failed ∈ ran ( Input_StatusN )
      @grd3 Input_StatusN~[ { ok , suspected } ] = ∅
    then
      @act1 FMS_State ≔ freeze
  end
  event ActionNoOut
    where
      @grd1 FMS_State = act
      @grd2 confirmed_failed ∉ ran (Input_StatusN)
      @grd3 Indx = ∅
    then
      @act1 FMS_State ≔ freeze end
```

Fig. 9. Specification of action Events.

```
event Return
    any In
    where
      @grd1 FMS_State = output
      @grd2 In = ( Last_Good_InputN ⩤ ( Input_StatusN ∼ [ {
ok } ] ◁ InputN ) )
      @grd3 ran(In)≠∅
    then
      @act1 Last_Good_InputN ≔ In
      @act2 Output :∈ ran ( In )
      @act3 Input_In_ErrorN ≔ Indx × { FALSE }
      @act4 FMS_State ≔ env
  End
```

Fig. 10. Specification of return Event.

6.4 Proof Statistics

In Table 2, we can see proof statistics for our model using the Rodin3.2 platform, the statistics give us the proof obligations generated and discharged by the Rodin. The finial development of our model results in 227 POs (Proof obligations), around (98%) of them have been proved automatically by the Rodin platform and the rest have been proved manually in the Rodin interactive proving environment [13–15].

Table 2. Proof Obligations of our model.

Reliability (Our model)	Total	Auto	Manual
	227(100%)	223 (98%)	4 (2%)
Global	0	0	0
Global1	0	0	0
FMS	24	24	0
FMS1	133	133	0
FMS2	70	66	4

7 Transforming EVENT B Models into Verified C# Implementations

The RODIN platform allows translation of Event-B to the C# programming language by use plugin tools; EB2ALL is a set of translator tools that automatically generates efficient target programming language code (C, C++, Java and C#) from Event-B formal specification related to the analysis of the complex problems. The EB2ALL contains four plugin namely EB2C, EB2C++, EB2J and EBC#. The goal of EB2ALL is to be able to generate a verified source code that satisfies behavioral properties of the develop formal system (abstractly). The EB2ALL tool is developed as a set of plugins for RODIN development tool under the Eclipse framework [16].

7.1 C# Translation Philosophy

After final refinement step; we can generate C# code by plugin tools (i.e., EB2ALL) of Rodin or we can convert manually Event B syntax to c# code [17, 18].

A top-level C# main function must be provided to call the generated functions "INITIALISATION" and "Iterate". The calling of INITIALISATION function of FMS must be called before Iterate, see Fig. 11.

```
if (INITIALISATION() == true)
{
Iterate();
}
//****************************************//
public bool Iterate()
{
if (Environment() == true) return true;
if (Detection() == true) return true;
if (AnalysisLoop() == true) return true;
if (Analysis() == true) return true;
if (Action() == true) return true;
if (Return() == true) return true;
return false;
}
```

Fig. 11. Structure of calling Events in C#

7.2 FMS Simulation in C#

Simulation is a technique in which a real-life system or process is emulated by a designed model. The model encapsulates all of the system's features and behaviors, the simulation is the execution of this system over time. Due to the fact that simulation programming can be complicated, there have been many attempts to create languages that embody all the requirements of the simulation paradigm in order to ease the development. One such language is SIMULA.

Nowadays, the focus is more on creating packages, frameworks or libraries that incorporate what programmers need when creating a simulation. These libraries are meant to be called from ordinary languages like C#, C++, Java or Python [19].

We used MS Visual studio to create our simulation with C# language, see Fig. 12.

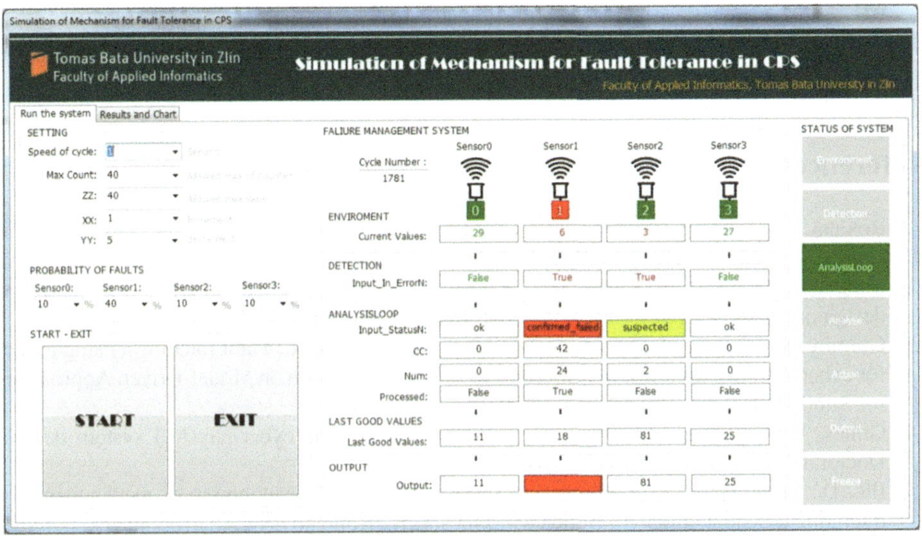

Fig. 12. FMS simulation (Run the system)

8 Conclusions

In this paper we offered a formal pattern for specifying and refining a mechanism for fault tolerance in cyber physical systems. Our formal development of the failure management system adopts the CPS approach itself.

Formal methods are traditionally used for reasoning about software correctness, we demonstrated how to develop the FMS by stepwise refinement in the Event-B, then we translated the final proven model to C# code to get the simulation. We introduced the main principles, rules and implementation solutions for the translation tools in order to generate the target programming language code (C#) from the Event-B specifications.

The translator generates for all events of concrete modules separate code methods in the target programming language this approach may be applicable to massive parallel processing. After the final refinement step, we can generate C# code by plugin tools (i.e., EB2ALL) of Rodin or we can convert manually Event B syntax to C# code.

The Rodin tool uses an internal database to handle model information which allows model generation to be based on the underlying meaning of the model and reduces the syntax dependency.

Rodin tool is a platform where verification of the program is done and it offers reactive environment for constructing and analyzing models as do most modern integrated development environments, It provides integration between modeling and proving, whereas this is an important feature for the developers that allows them to focus on the modeling task without the need to switch between different tools to check proving correctness of model at same time. Our final models are translated into the Event-B notation to verify required properties; we can say that Event-B allows us to define a kind of modeling methodology by writing the correct mathematical notions.

We verified our complete development with Rodin platform; around 98% of proof obligations have been proved automatically by the tool, the rest have been proved using the Rodin interactive proving environment.

References

1. Roxana, R.B., Eva-Henrietta, D:. Fault-tolerant control of a cyber-physical system. MS&E **261**(1), 012003 (2017)
2. Sinha, R., Paredis, C.J., Liang, V.C., Khosla, P.K.: Modeling and simulation methods for design of engineering systems. J. Comput. Inf. Sci. Eng. **1**(1), 84–91 (2001)
3. Ghimire, S., Sarraipa, J., Agostinho, C., Jardim-Goncalves, R.: Fault tolerant sensing model for cyber-physical systems. In: Proceedings of the Symposium on Model-Driven Approaches for Simulation Engineering, pp. 1–9 (2017)
4. Gunes, V.: Ensuring reliability and fault-tolerance for the cyber-physical system design, Doctoral dissertation, UC Irvine (2015)
5. Ilić, D., Troubitsyna, E., Laibinis, L., Snook, C.: Formal development of mechanisms for tolerating transient faults. In: Butler, M., Jones, C.B., Romanovsky, A., Troubitsyna, E. (eds.) Rigorous Development of Complex Fault-Tolerant Systems. LNCS, vol. 4157, pp. 189–209. Springer, Heidelberg (2006). https://doi.org/10.1007/11916246_10
6. Sokolowski, J.A., Banks, C.M.: Modeling and Simulation Fundamentals. Wiley, Hoboken (2010). ISBN 978-0-470-48674-0
7. Modeling & Simulation Tutorial (2018). https://www.tutorialspoint.com/modelling_and_simulation/index.htm. Accessed 20 Nov 2020
8. Event-B and the Rodin Platform (2018). http://www.event-b.org. Accessed 20 Nov 2020
9. Xu, H.: Model based system consistency checking using Event-B, Doctoral dissertation (2012)
10. Méry, D., Singh, N.K.: Technical report on formal development of two-electrode cardiac pacing system (2010)
11. Jastram, M., Butler, M.: Rodin User's Handbook: Covers Rodin v.2.8. CreateSpace Independent Publishing Platform. (2014). ISBN 10: 1495438147 ISBN 13: 9781495438141, USA
12. Boulanger, J.L. (ed.): Industrial Use of Formal Methods: Formal Verification. Wiley (2013)
13. Ali, A.A., Krayem, S., Chramcov, B., Kadi, M.F.: Self-stabilizing fault tolerance distributed cyber physical systems. Ann. DAAAM Proc. **29** (2018)
14. Ali, A.A., Jasek, R., Krayem, S., Zacek, P.: Proving the effectiveness of negotiation protocols KQML in multi-agent systems using Event-B. In: Silhavy, R., Senkerik, R., Kominkova Oplatkova, Z., Prokopova, Z., Silhavy, P. (eds.) CSOC 2017. AISC, vol. 574, pp. 397–406. Springer, Cham (2017). https://doi.org/10.1007/978-3-319-57264-2_40
15. Ali, A.A., Jasek, R., Krayem, S., Chramcov, B., Zacek, P.: Improved adaptive fault tolerance model for increasing reliability in cloud computing using Event-B. In Computer Science On-line Conference, pp. 246–258. Springer, Cham (2018)
16. EB2ALL - The Event-B To C, C++, Java And C# Code Generator (2011). http://eb2all.loria.fr/. Accessed 20 Nov 2020
17. Elsayed, E.K., El-Sharawy, E.: High Quality implementation for automatic generation C# code by Event-B pattern. Int. J. Softw. Eng. Appl. **5**(1), 43 (2014)
18. Degerlund, F., Grönblom, R., Sere, K.: Code generation and scheduling of Event-B models. Turku Centre for Computer Science, Technical report, 1027 (2011)
19. Discrete Event Simulation: A Population Growth Example (2016). https://msdn.microsoft.com/en-us/magazine/mt683796.aspx. Accessed 20 Nov 2020

Modeling the Temperature Field Distribution at the Stages of Input-Output of the Electron Beam

Sergei Kurashkin[1], Vadim Tynchenko[1,2,3](✉), Aleksandr Murygin[1], Yuriy Seregin[1], Valeriya Tynchenko[1,2], and Aleksey Bocharov[1,2]

[1] Reshetnev Siberian State University of Science and Technology, Krasnoyarsk, Russia
[2] Siberian Federal University, Krasnoyarsk, Russia
[3] Marine Hydrophysical Institute, Russian Academy of Sciences, Sevastopol, Russia

Abstract. In this study, the authors proposed an approach for calculating the distribution of the temperature field with simulation at the stages of input and output of an electron beam on the basis of mathematical models of the thermal field, considering the thermophysical parameters of the product material and its geometric dimensions. The approach considered by the authors for calculating the thermal field at the stages of input and output of an electron beam, at the weld seam, based on the developed mathematical models, taking into account the geometric dimensions of the product and the thermophysical parameters of the product material, will reduce the cost of setting up the technological process when welding new material alloys and other thicknesses, and will also reduce the number of defects in the used technological modes in production. Thus, the purpose of the study is to develop a new modeling approach for developing a technological process with its further application for a more conscious choice of technological parameters when developing a new technological process for electron beam welding of other materials and alloys and obtaining a stable quality of a weld. #CSOC1120

Keywords: Electron-beam welding · Modelling · Technological parameters · Electron beam · Optimization of process parameters · Input-output electron beam · Automation

1 Introduction

The equipment of modern satellite systems is high-tech and complex and consists of a large number of simple elements that require connection into a single whole [1–3]. At the same time, high requirements are imposed on the quality of such joints, which necessitate the use of modern science-intensive approaches to the formation of permanent joints [4–6]. There are several approaches that are successfully and widely used at the enterprise of the rocket and space industry, allowing to solve the task:

- soldering methods: induction and chemical soldering.

R. Silhavy (Ed.): CSOC 2021, LNNS 230, pp. 331–339, 2021.
https://doi.org/10.1007/978-3-030-77442-4_28

- welding methods: diffusion, electron-beam, argon-arc and resistance welding.

Within the framework of this study, the method of electron beam welding is considered, which is one of the most promising due to a number of its advantages, such as: the ability to weld materials with a wide range of thicknesses (from tenths of a millimeter to tens of centimeters), stable quality of the welded joint, ample opportunities for automation of technological process [7–9].

One of the disadvantages of electron-beam welding technology is the need to select the exact values of technological parameters that ensure high quality of the welded joint [10–12]. At the same time, experiments on adjusting such parameters are expensive, which indicates the expediency of replacing full-scale experiments with numerical ones using models of energy distribution in the zone of formation of a weld [13, 14]. To date, the issue of energy distribution has been widely covered in a number of studies [14–16] devoted to the construction of mathematical models of various stages of the electron-beam welding process.

In addition, a number of studies highlight the issues of simulation, as well as the issues of numerical studies of thermal processes occurring with EBW of various designs, showing the effectiveness of this approach when adjusting technological parameters [16–18]. However, existing scientific research focuses on the steady state and omits the issues of modeling the modes of input and output of an electron-beam [18–20].

2 Mathematical Model for Calculating the Thermal Field

Accurately considering the complex configuration of the body, analyzing the thermal processes during welding, is rather difficult from an analytical point of view, therefore, for calculations, it is necessary to simplify the forms of the considered schemes [21, 22]. In this work, considering the introduced restrictions, the following schemes (Fig. 1) of the heated body were chosen:

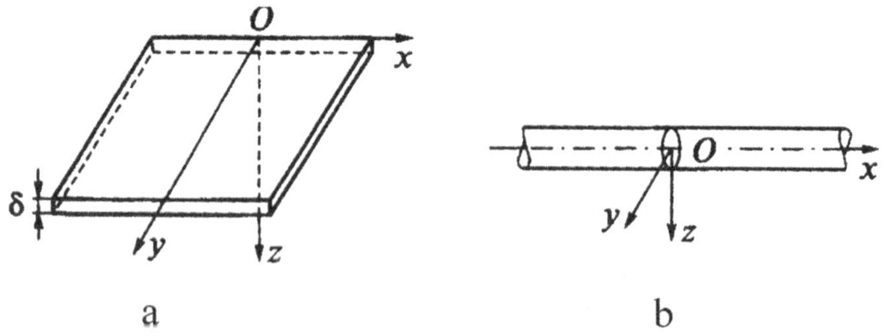

Fig. 1. Design schemes of solids: a – semi-infinite plate; b – endless rod.

- a semi-infinite plate that is a body bounded by two parallel planes $z = 0$ и $z = \delta$ (Fig. 1a).
- an endless rod, which is a body with a straight and curved axis (Fig. 1b).

To describe the temperature fields of mobile sources, the superposition principle is used, thus, a mobile heat source can be represented by a set of instantaneous sources acting sequentially in time [21, 22]. In this study is chosen a moving Cartesian coordinate system, the beginning of which coincide with the point O of the current position of the source, for a semi-infinite plate (Fig. 2a) and an endless rod (Fig. 2b).

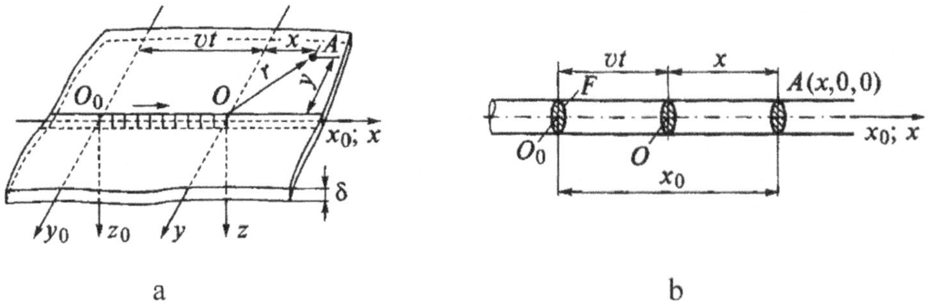

Fig. 2. Schemes for calculating temperature fields from moving sources: a – moving linear source in a semi-infinite plate; b – movable flat source in an endless rod.

The expression for the temperature field in a moving coordinate system for a semi-infinite plate has the following form [21, 22]:

$$T(x, y, z, t) = T_i + \frac{q}{4\pi\lambda\delta} exp\left(-\frac{\upsilon x}{2a}\right) \int_0^t exp\left[-\left(\frac{v^2}{4a} + b\right)\tau - \frac{r^2}{4a\tau}\right] \frac{d\tau}{\tau}, \quad (1)$$

where $\tau = t - t'$ – time variable;

$r = \sqrt{x^2 - y^2}$ – the length of the radius vector of point A in the moving coordinate system;

T_i – initial temperature of the product;

q – heating source energy;

b – plate heat transfer coefficient;

v – welding speed;

a – thermal diffusivity;

δ – thickness;

λ – coefficient of thermal conductivity.

For a movable flat source in an endless rod, in cases where the temperature field in a movable coordinate system is determined mathematically in the following form [21, 22]:

$$T(x, t) = T_i + \frac{q}{c\rho\sqrt{4\pi a}} exp\left(-\frac{\upsilon x}{2a}\right) \int_0^t exp\left[-\left(\frac{v^2}{4a} + b\right)\tau - \frac{x^2}{4a\tau}\right] \frac{d\tau}{\sqrt{\tau}}, \quad (2)$$

where $c\rho$ – heat capacity of the steel.

Thus, to simulate thermal processes and calculate the temperature field in electron beam welding, expression (1) was used on the parts of the welded products, and in the cases of a waveguide path, expression (2) was used and an additional restriction was imposed in the form of expression (1).

3 Simulation Modeling

In this work, simulation modeling was carried out, the temperature change at the stages of beam input and output was considering, which will ensure a stable quality of the weld during electron beam welding of thin-walled aerospace structures, as well as reduce the complexity of setting up the EBW technological process, and, accordingly, reduce the influence of the human factor. The task of creating such models of temperature distribution for various modes of input and output of an electron beam during welding is an urgent task. This model considers the nature of heat distribution in the melting zone.

In this work, simulation models of electron beam input-output during EBW of thin-walled aerospace structures have been developed, which make it possible to optimize the modes of input and output of an electron beam in order to improve the quality of welded joints.

Simulation modeling, using mathematical models of temperature distribution at the stages of electron beam input and output, considering the geometric and thermophysical parameters of the product, will justify the choice of specific values of technological parameters for such modes depending on the geometric dimensions of the product and the required characteristics of the welded joint. Simulation conducted in ANSYS Discovery AIM student software.

4 Experimental Study

Let us carry out simulation of the electron-beam welding process at the stages of beam input and output, considering the thermophysical parameters of the product material (density, tensile yield strength, tensile ultimate strength, coefficient of thermal expansion, specific heat, isotropic thermal conductivity, isotropic resistivity) and the geometric dimensions of the product (length, width, height).

At the beginning, simulation was carried out at the stages of beam input and output for titanium alloy VT14 with a thickness of 1.2 mm, considering the technological parameters (welding speed, beam current, focal length, focusing current, accelerating voltage). It is important to note that the modeling was carried out at the site of the product, namely at the junction of the products, since the beam input area is small and for simplicity, it was decided to minimize the area in this model. This is also due to the fact that there is no critical spread in temperature and at the stage of electron beam output, the temperature is close to melting, which is confirmed by the temperature distribution graph on the surface of the VT14 titanium alloy product, at the stages of beam input and output (Fig. 3).

The next material for which the modeling was carried out at the stages of beam input and output was an aluminum alloy AMg-6 with a thickness of 1.2 mm; during

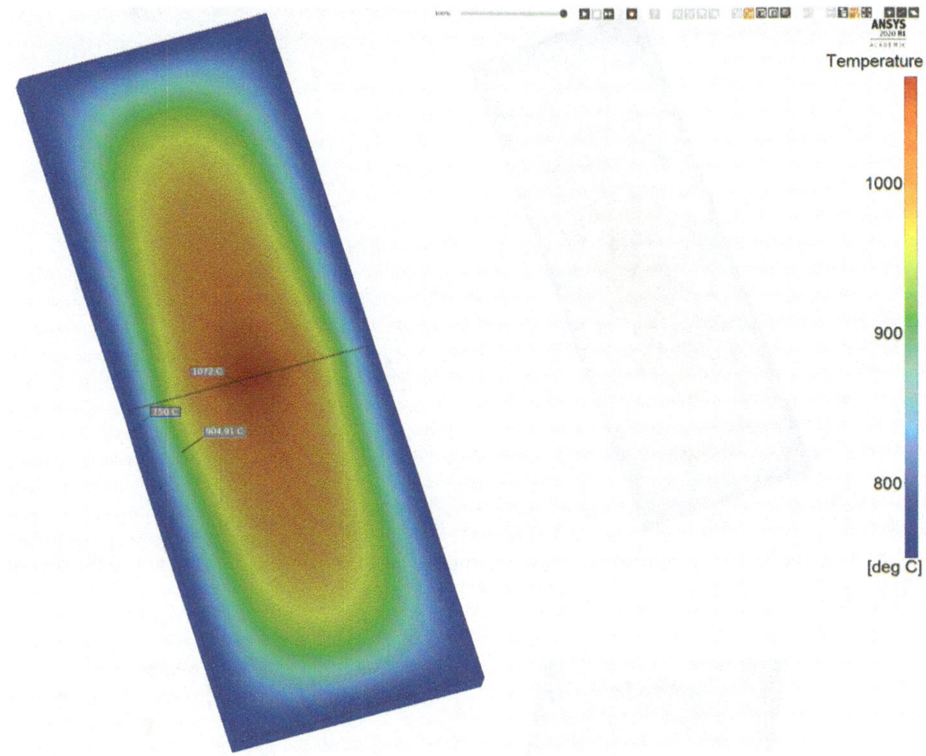

Fig. 3. Distribution of the temperature field during the input and output of the beam in the section of the titanium glory VT14 product.

the simulation, technological and geometric parameters (welding speed, beam current, focal length, focusing current, accelerating voltage, width, height, length).

As for the previous alloy, modeling was carried out on a small area of the product, namely at the junction (Fig. 4).

In conclusion, modeling was carried out for the waveguide path; the thickness of the AMg-6 aluminum alloy was reduced to 0.65 mm. Thus, as for the previous alloys, the simulation was carried out considering the technological and geometric parameters (welding speed, beam current, focal length, focusing current, accelerating voltage, width, height and length of the product).

Modeling was carried out at the site of the product, namely at their junction (Fig. 5).

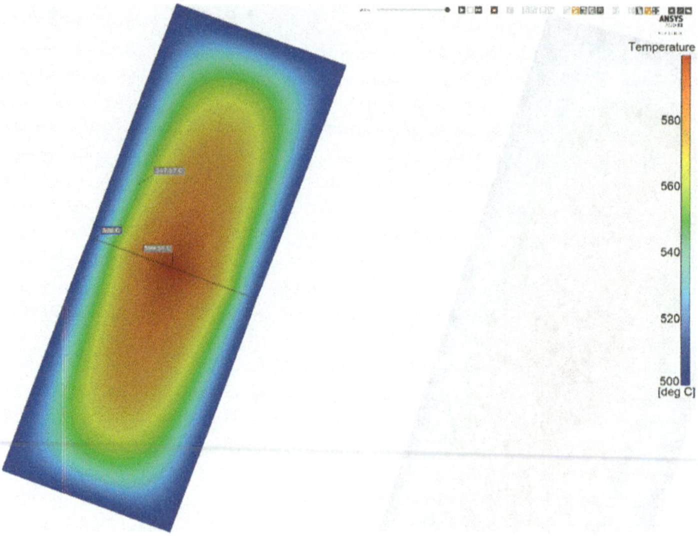

Fig. 4. Distribution of the temperature field during the input and output of the beam on the section of the product of the aluminum alloy AMg-6, 1.2 mm thick.

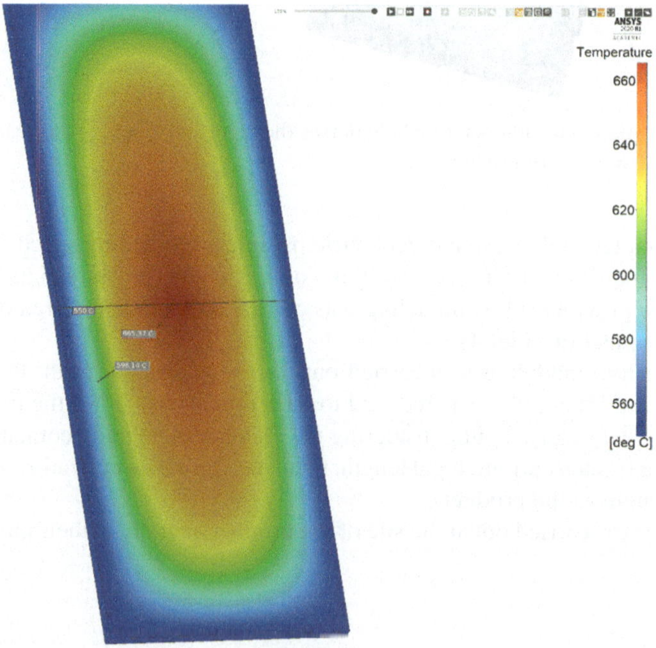

Fig. 5. Distribution of the temperature field during the input and output of the beam on the section of the product of the waveguide path, 0.65 mm thick

5 Results Discussion

In this work, simulation modeling was carried out for titanium alloy VT14, aluminum alloy AMg-6 with a thickness of 1.2 mm and a waveguide path made of aluminum alloy AMg-6 with a thickness of 0.65 mm using the ANSYS Discovery AIM software product at the stages of electron beam input and output.

The results obtained in the course of the study show that the approach proposed by the authors can be used to optimize the spread of energy on the surface of the product during the input and output of the electron beam, which will reduce the number of defects arising at these stages.

Note that in this work, in order to simplify the process of simulation at the stages of beam input and output, and since the beam input section is small, the product surface was limited by small dimensions, which did not affect the final results, but on the contrary made it possible to judge the feasibility of the proposed approach.

The application of this approach makes it possible to calculate the distribution of the temperature field at various stages of electron-beam welding, thus, it will reduce the number of defects throughout the entire process of electron-beam welding and, as a result, will allow obtaining a stable quality of the weld.

Comparison of the simulation results obtained above in a graphical form gives an idea of the feasibility of using simulation models and shows more accurate results of the distribution of the temperature field in the heating zone at the stages of beam input and output during electron beam welding.

6 Conclusion

The approach proposed by the authors in this study is used for calculating the distribution of the temperature field. It could be used when carrying out simulation at the stages of the electron beam input and output. The approach is based on the mathematical models of the thermal field, considering the thermophysical parameters of the product material and the geometric dimensions of the product. The results of the study will allow working out the welding process for new products with alloys and thicknesses differ with the known ones. Also, it could cause the reduce of the defects number.

In the future, the results obtained can be used for a more conscious choice of technological parameters in electron beam welding of materials and alloys that have not been used before and to obtain a stable quality of the weld.

Acknowledgments. The reported study was funded by RFBR, Government of the Krasnoyarsk Territory and the Regional Science Foundation, project number 19-48-240007 "Mathematical and algorithmic support of the electron-beam welding of thin-walled aerospace structures".

References

1. Mastanaiah, P., Sharma, A., Madhusudhan, R.G.: Process parameters-weld bead geometry interactions and their influence on mechanical properties: a case of dissimilar aluminium alloy electron beam welds. Defence Technol. **14**, 137–150 (2018)

2. Wei, H., Yunhui, D., Lianqing, Z., Mingli, D.: Angular sensing system based on Y-type twin-core fiber and reflective varied-line spacing grating fabricated by electron beam lithography. Results Phys. **18**, 103193 (2020)
3. Seregin, Yu.N., Laptenok, V.D., Uspenskiy, N.V., Nikitin, V.P.: Experimental research on the optimization of the technology of electron-beam welding of aluminum alloys. In: Technologies and equipment of EBW-2011 Materials of the St.-Petersburg International Scientific and Technical Conference, Russia, Saint Petersburg, pp. 71–80 (2011)
4. Kanigalpula, P.K.C., Pratihar, D.K., Jha, M.N., Derose, J., Bapat, A.V., Rudra Pal, A.: Experimental investigations, input-output modeling and optimization for electron beam welding of Cu-Cr-Zr alloy plates. Int. J. Adv. Manuf. Technol. **85**(1–4), 711–726 (2015). https://doi.org/10.1007/s00170-015-7964-7
5. Salomatova, E.S., Trushnikov, D.N., Tsaplin, A.I., Belenkiy, V.Ya.: The high performance computing for research of evaporation during electron beam welding. Bull. PNIPU Aerosp. Eng. **39**, 96–108 (2014)
6. Reich, M., et al.: Real-time beam tracing for control of the deposition location of electron cyclotron wave. Fusion Eng. Des. **100**, 73–80 (2015)
7. Trushnikov, D.N., Belen'kii, V.Y.: Investigation of the formation of the secondary current signal in plasma in electron beam welding with oscillations of the electron beam. Weld. Int. **27**(11), 877–880 (2013)
8. Poli, E., Peeters, A.G., Pereverzev, G.V.: TORBEAM, a beam tracing code for electron-cyclotron waves in tokamak plasmas. Comput. Phys. Commun. **136**(1–2), 90–104 (2001)
9. Seregin, Yu.N., Kurashkin, S.O.: Modeling of the EBW mode for predicting the parameters of the weld. In: MPEI: Collection of Materials of 2nd International Conference, Russia, Moscow, pp. 26–36 (2017)
10. Permyakov, G.L., Trushnikov, D.N., Belenky, V.Ya., Olshanskaya, T.V.: Numerical modeling of the electron beam welding process with longitudinal beam oscillation. Bull. PNIPU Aerosp. Eng. **39**, 139–145 (2015)
11. Kurashkin, S.O., Laptenok, V.D., Murygin, A.V., Seregin, Yu.N.: Analytical characteristics of the electron beam distribution density over the heated spot for optimizing the electron-beam welding process. IOP Conf. Ser. Mater. Sci. Eng. **681**, 012021 (2019)
12. Rai, R., Palmer, T.A., Elmer, J.W., Debroy, T.: Heat transfer and fluid flow during electron beam welding of 304L stainless steel alloy. Weld. J. **88**(3), 54–61 (2009)
13. Erofeev, V.A., Logvinov, R.V., Nesterenkov, V.M.: Features of the use of the equivalent heat source, taking into account the strain and stress in the process of electron beam welding. Weld. Diagn. **4**, 22–26 (2010)
14. Yazovskikh, V.M.: Model of deep penetration in electron beam welding. Weld. Eng. **20**(8), 16–20 (2005)
15. Seregin, Yu.N., Kurashkin, S.O.: Modeling the thermal process using the temperature functional by electron beam welding. IOP Conf. Ser. Mater. Sci. Eng. **734**, 012003 (2020)
16. Sudniky, W., Radajz, D., Erofeew, W.: Computerized simulation of laser beam welding, modelling and verification. J. Phys. D Appl. Phys. **29**(11), 013 (1996)
17. Cho, W.I., Na, S.J., Thomy, C., Vollertsen, F.: Numerical simulation of molten pool dynamics in high power disk laser welding. J. Mater. Process. Technol. **212**(1), 262–275 (2012)
18. Trushnikov, D., Belenkiy, V., Schavlev, V., Piskunov, A., Abdulin, A., Mladenov, G.: Plasma charge current for control and monitoring at electron beam welding with the beam oscillation. Sensors **12**(12), 17433–17445 (2012)
19. Sudnik, V.A., Erofeev, V.A., Richter, K.-H., Heins, K.-V.: Numerical Modelling of the EBW process. In: Proceedings of International Conference Computer Technology in Welding and Manufacturing & Mathematical Modeling and Information Techhnologies in Welding and Related Processes, Kiev, pp. 295–300 (2000)

20. Felici, F., et al.: The TCV team: real-time physics-model-based simulation of the current density profile in tokamak plasmas. Nucl. Fusion **51**(8), 083052 (2011)
21. Rykalin, N.N., Zuev, I.V., Uglov, A.A.: Fundamentals of electron beam processing of materials. Mashinostroyeniye, Russia, Moscow (1978)
22. Nerovny, V.M., Konovalov, A.V., Yakushin, B.F., Makarov, E.L., Kurkin, A.S.: Theory of welding processes. Bauman MSTU, Russia, Moscow (2007)

Object Classification from a Hyper Spectral Image Using Spectrum Bands with Wavelength and Feature Set

Soumyashree M. Panchal(✉) and Shivaputra

Department of Electronics and Communication Engineering, Dr. AIT, Bangalore, India

Abstract. Hyperspectral imaging is increasingly adopted for various remote sensing purposes with the increasing usage of sensors in various applications. One of the significant challenges in processing hyperspectral images is to make the data ready for carrying out an effective unmixing. However, this is yet an open-end problem as there is no apriori information to the system. Therefore, the proposed system introduces a simplified analytical model where an abundance-based methodology is used for extracting a useful feature followed by classification using a simple and non-iterative matrix-based operation. The proposed system emphasizes an unmixing process where the data is analyzed concerning data cube, wavelength, and feature set. Simulated in MATLAB, the study outcome shows that the proposed system offers higher classification accuracy and lower processing time.

Keywords: Hyperspectral image · Remote sensing · Accuracy · Classification · Wavelength · Feature extraction

1 Introduction

Hyperspectral imaging involves aggregation as well as processing sensitive data collected from the electromagnetic spectrum [1]. The core idea is to extract the spectrum information associated with all involved pixelsto explorea specific object over the given field [2, 3]. The sensors used in this process are frequently used for various applications, e.g., geoscience, biomedical imaging, surveillance, astronomy, etc. [4, 5]. Such images aim to extract fingerprint,a unique identification over the electromagnetic spectrum [1]. These are used for identification purposes [6]. There are various research trends in hyperspectral imaging (HSI) that includes i) sparse unmixing (SU), ii) dimensionality reduction (DM), iii) spectral-spatial classification (SSC), iii) localization and distribution of the subpixels [7]. This paper focuses on the spectral-spatial classification problem using abundance estimation. There exist various challenges to find the spatial location and the distribution of the subpixel because, during the process of subpixel mapping, the errors that occurred due to incorrect abundance computation are ignored while performing the SU [8]. In this process, spectral unmixing is an important phenomenon that can identify the original spectral signals from a given set of the field of materials [9]. The greatest challenge in this regard is that the system never has any apriori information about

R. Silhavy (Ed.): CSOC 2021, LNNS 230, pp. 340–350, 2021.
https://doi.org/10.1007/978-3-030-77442-4_29

this surface material, which leads to higher challenges in classification performance. Therefore, the proposed system introduces a simplified approach tothe classification of Hyperspectral Images (HIS), unlike conventional processes. The manuscript is organized as follows: Sect. 2 discusses the existing research work, followed by problem identification in Sect. 3. Section 4 discusses the proposed methodology analysis of the accomplished result is discussed in Sect. 5. Finally,the conclusion of the paper is written in Sect. 6.

2 Related Work

This section discusses diversified approaches being carried out towards processing and analyzing a hyperspectral image. The work carried out by Huang et al. [10] has addressed the challenges associated with sparse unmixing where the pure spectral abundances have been estimated using contextual spatial data. Adoptingthe firefly algorithm was also claimed to offer a better selection of the hyperspectral bands (Su et al. [10]). It has also been observed that the usage of spatial and spectral data enhances the classification of hyperspectral data using an iterative approach (Guccione et al. [11]). The work carried out by Tong et al. [12] hasused a genetic approach for addressing the issues of subpixel mapping. Uezato et al. [13] have also addressed the unmixing process's challenges using Markov random process. The work carried out by Zhou et al. [14] has also considered the unmixing problems and solved them by using a learning method over subspace structure with a matrix factorization approach. Qi et al. [15] have combined both spatial and spectral approaches for solving an unmixing problem by presenting a multi-view approach of data processing using a sparse matrix. A unique study of Chatterjee and Yuen [16] have implemented a learning process over the given dictionary for improving the classification performance. Usage of matrix factorization is also adopted in He et al. [17], where the total variation approach is used, assuming the abundance matrix to be highly sparse. Yuan et al. [18] have used k-nearest neighboring considering sparse constraint where matrix factorization process has been used for solving unmixing problems. Qu et al. [19] have used dictionary-based anomaly identification using a simplified matrix decomposition process. The adoption of matrix factorization is also seen in Wang and Qian [20], where a thresholding mechanism has been used iteratively for better unmixing consequences. Apart from this, various other approaches are being introduced in existing approaches viz. sparse regression using semi-supervised approach (Yuan et al. [21]), unmixing using apriori data (Tang et al. [22]), spatial-spectral based factorization of the matrix (Huang et al. [23]), sparse coding (Soltani-Farani and Rabiee [24]), attribute embedding (Hong and Zhu [25]), distributed unmixing using least mean p-power (Khoshsokhan et al. [26]), blind unmixing (Yao et al. [27]), spectral variability based unmixing (Borsoi et al. [28]). Hence, there are various existing schemes for improving the analysis of hyperspectral images. The next section outlines the issues associated with it, followed by a discussion of HIS's proposed classification process.

3 Problem Description

There have been various approaches towards processing HIS in the existing literature; however, it has been seen that matrix factorization is the most dominant one. Apart from this, there is also various iterative mechanism to address the unmixing performance of HSI. Unfortunately, such an approach is a computationally expensive process and doesn't emphasize much on classification errors. The classification error depends on how much error occurs for the estimation of the abundance field during spectral unmixing, which is less attended in existing studies.

4 Proposed Methodology

The proposed study presents an analytical model capable of performing the classification of an object from the given HIS using a simplified approach. The idea is to assist in performing an effective unmixing of the spectral signal to contribute to better classification performance. Figure 1 highlight the proposed model.

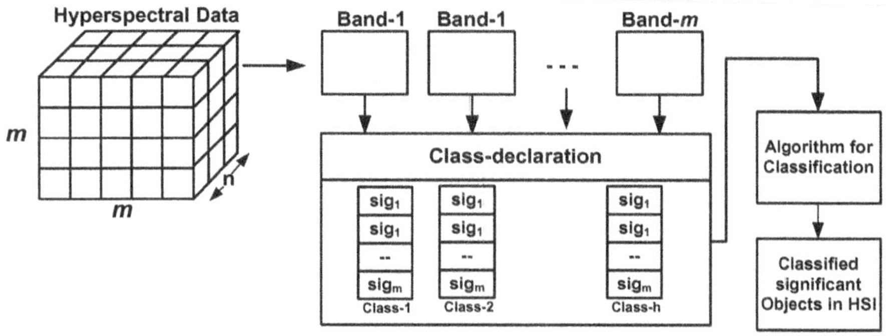

Fig. 1. Proposed model of HSI classification

The core challenge during the process of classification using the HSI feature is to handle the downsample and the high dimensionality as ∀ pixel (P) $\in\{P_p, P_m\}$, where P_p is the pure-pixel and P_m is mixed-pixel. The very objective of this paper to design a model to classify various sections in the input HSI. To handle the imbalance context of less samples and complexities poses by the P_p and P_m, an Abundance Representation Of Map (A_m) is used. A_m represents the particular classes of the objects (O_c) in the HSI by the corresponding abundance-feature set (Afs). The Afs computed from ∀ pixel (P) $\in\{P_p, P_m\} \rightarrow$ probability factor (Pf) of $(O_c)_k \in O_c$. For example, suppose an HSI has three classes in $O_c = \{$House, Water, Trees$\}$, then the value of Afs represents the percentage of probability of house, water, and tree. The model takes the respective dataset (Ds), which provides the intrinsic features as a set Fs = {sB, cS, nsB}, where sB = spectral band, cS = signature of the class and nsB is a total number of spectral bands. Generally, the minimum and the maximum ranges of wavelengths are specified with the HSI dataset. Whereas to compute the datacube (Dc) and Afs, the respective wavelength (λ) of ∀ sBk $\in$ nsB are require. Algorithm-1 describes the automated classification method.

Algorithm-1: Classification of HIS-CL based on Afs
Input: Ds
Output: HIS-CL
Start

$[F_s]=\{sB,cS,nsB, \lambda_{min}, \lambda_{max}\})] \leftarrow f(Ds)$

1. Compute, Central wavelength(λ_c) ∀ sB :
   a. $\lambda c = \frac{(\lambda max - \lambda min)}{nsB}$
2. Compute Data-cube:
   a. $Dc \leftarrow f_2(sB, \lambda_c)$
3. Colour equivalence of Dc using Contrast stretching
   a. $H_c \leftarrow f_3(Dc, P)$
4. Object Class Visualization
   a. $Plot \leftarrow f_4(Labels(O_c))$
5. Computation of Afs
   a. $Afs \leftarrow f_5(D_c, cS)$
6. Afs Visualization
   a. Plot($\forall O_c$, scaled colour equivalence of the Afs)
7. Channel -Sig-Mapping
   a. $Ch_{iD} \leftarrow (cS)I$ with max(Afs) | max(Afs)∈{Pp,Pm}.
8. Mapped HSI-Classification
9. HIS-CL<-overlaying(Afs,S_{img})

End

The computation of the wavelength (λ) in Spectral Band (Sb) of HSI is described in Sect. 4.1.

4.1 Wavelength (λ) in Spectral Band (Sb) of HSI

The HSI consists of many spectral bands (nSb) with the wavelength (λ) ranging from λ_{min} to λ_{max}. The central wavelength (λ_c) ∀ Sb is computed by considering an equal division of the range between λ_{min} and λ_{max}. Figure 2 showing the pattern of the λ_c for ∀ sB, which is computed using Eq. (1).

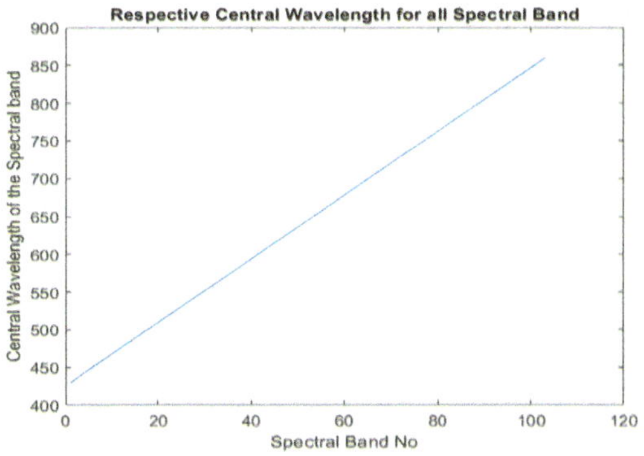

Fig. 2. Central wavelength (λ_c) of each spectral band (Sb)

$$\lambda_c = \frac{(\lambda \max - \lambda \min)}{nsB} \tag{1}$$

The explicit input for computing Afs is in the form of a hyperspectral data cube (Dc). A monochromatic image and the features set of this Dc = {S_p, S_t}, where S_p and S_t are spatial and spectral features. The representation of Dc: {x, y, z}, where x and y ∈ S_p and z ∈ S_t as wavelength. The characterization of the ground truth of the Dc is as in Table 1.

Table 1. Characterization of the hyperspectral image data cube (Dc)

No. of spectral band (nSb)	Wavelength range (Wr in nm)	Geometric resolution (Gr in m)	Spatial resolution (Sr in pixel) dimension
220	400.02–2499	1.3	145 × 145

The Dc contains 'n' classes of the objects (Oc) as ground truth (Gt).The typical *Indians pine hyperspectral image dataset* (IPHSI) is acquired through the AVIRIS sensor on North-Western Indiana's test site. It consists of sixteen different classes of the object (n = 16). These classes are a set of objects as class-number of sample pair: {Alfalfa-46, Corn-notill-1428, Corn-mintill-830, Corn-237, Grass-pasture-483, Grass-trees-730, Grass pasture-mowed-28, Hay-windrowed-478, Oats-20, Soybean-notill-972, Soybean-mintill-2455, Soyabean-clean-593, Wheat-205, woods-1265, Buildings-Grass-Trees-Drives-386, Stone-Steel-Towers-93}. Further, the transformation of Dc is performed by estimate color image equivalence of hyperspectral data, as described in Sect. 4.2.

4.2 Transformation of HSI to the Optimal RGB

A hypercube data (Dc) is formed using the original HSI as Io and the wavelength (λ) to an explicit function that converts the HSI data to the color equivalence (Hc) such that Hc ← {Dc, P}, where Dc and P is the hyperspectral data cube and the associated properties respectively. Further, the estimation of the color image of Dc takes place using the logical operation for the stretching of the contrast by using contrast limited adaptive histogram equalization when it is true (1) [29]. Figure 3 is the obtained visualization of the hyperspectral data into the visualization environment.

The ground truth labelling of the object classes and their visualization in the respective bands is described in Sect. 4.3.

Fig. 3. HSI to the optimal RGB visualization

4.3 Object Class Visualization

The ground truth of the object class (Oc) consists of sixteen different classes (n) where the descriptions of each class represented by the object or class descriptor (cS). Figure 4 illustrates these class object visualization.

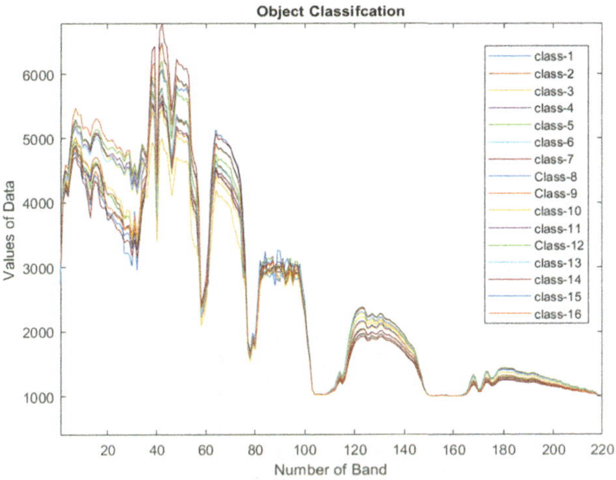

Fig. 4. Object class variation visualization

4.4 Least Square Method (LSM) for Afs

The Dc = $\{P_p, P_m\}$, the challenge is that the P_p contains only the uni-Class's intrinsic characteristics. In contrast, the P_m signifies the intrinsic characteristics of the spectral of multi-class.

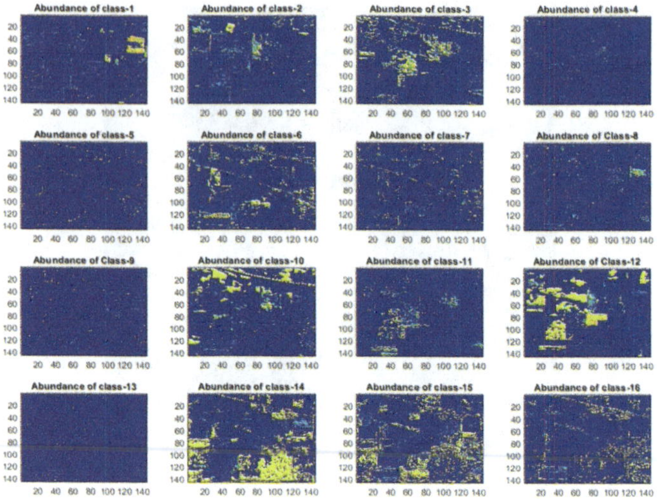

Fig. 5. Abundance of the class

The cS of P_p represents the unique class in the Dc; however, the cS of P_m represents linearly correlated objects (Oc_k) ∈ Oc. The typical behavior of the Afs ∀ pixel (P) ∈{P_p, P_m} signifies the share of the specific classes $(Oc)_k$ ∈ Oc based on the probability factor (Pf). The computation of the Afs occurs using a Fully Constrained Least-Squares Method (FC-LSM) method by taking the input set of {Dc, cS}. The computation of Afs decomposes the spectra's of {P_p, P_m} of ∀ Oc, popularly known as spectra-unmixing. Figure 5 illustrates displays ∀ Oc in the scaled color equivalence of the Afs.

4.5 Classification Based on Afs

In this stage of computation, the computation of the Channel-ID (Ch_{iD}) takes place based onmax (Afs) ∈ {P_p, P_m}. The mapping of the Ch_{iD}: → (cS)I with max (Afs). Figure 6 illustrates all the heat-map code of only high Afs (8-Classes).

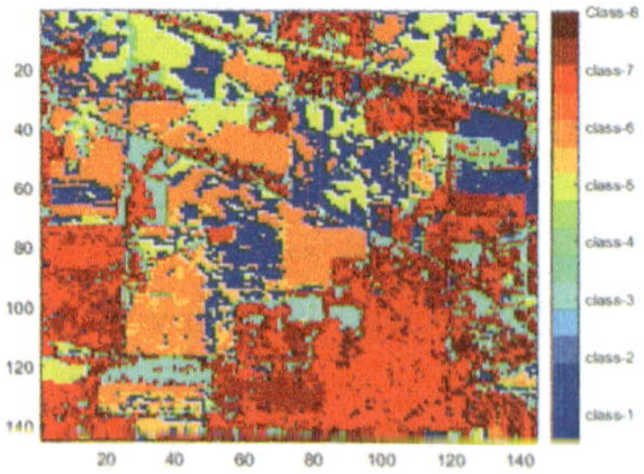

Fig. 6. Heat map of the classes

The identified regions of the classes (Oc) undergo overlaying the equivalent image with solid colorcodes (S_{img}) based on the binary masking's mapping function. Figure 7 illustrates the mapped classification Oc representation with the HIS-zones along with the class labels (HIS-CL).

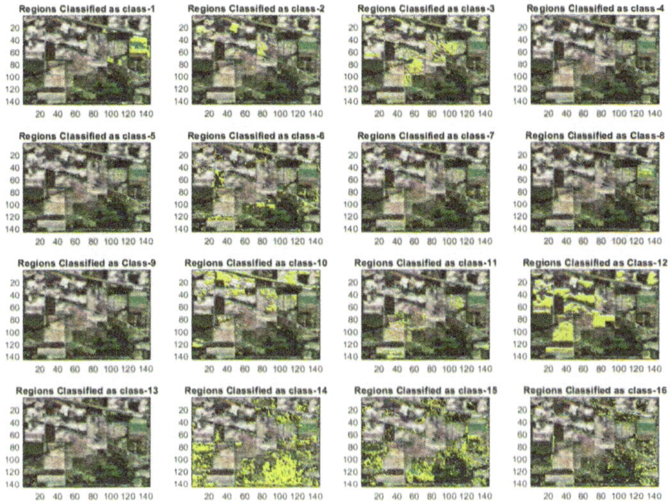

Fig. 7. Classified region

5 Result Analysis

The proposed study's implementation is carried out considering the typical *Indians pine hyperspectral image dataset* (IPHSI) dataset [30] in MATLAB. The outcome has been

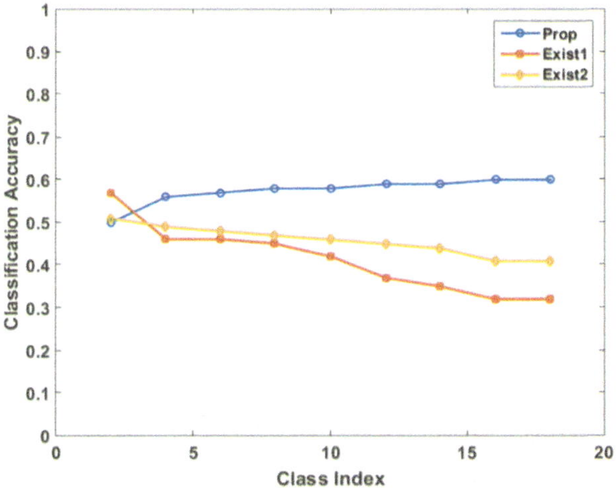

Fig. 8. Comparative analysis of classification accuracy

compared with the relevant work of Sun et al. [31], represented as Exist1, and Sun et al. [32], represented as Exist2.

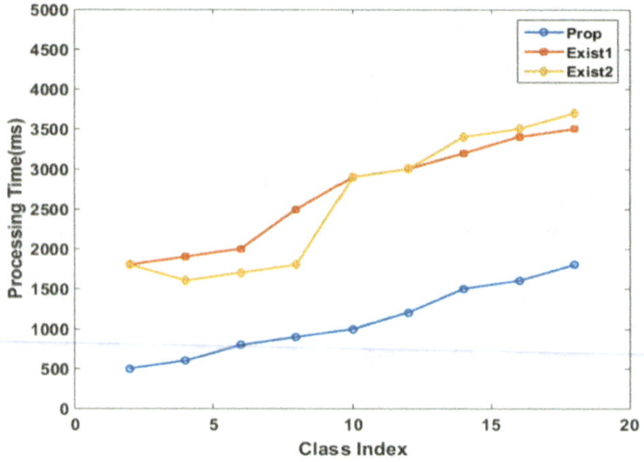

Fig. 9. Comparative analysis of processing time

A closer look into Fig. 8 and Fig. 9 showcase that the proposed system offers significantly higher classification accuracy and reduced processing time, which results in a simplified classification model in HSI. The prime reason behind this is less inclusion of sophisticated operations, unlike the conventional approach, while performing an unmixing operation.

6 Conclusion

Processing hyperspectral images is one of the challenging tasks due to the complexity involved in it—existing literature reviews showcase sophisticated operations' inclusion to obtain classification accuracy. Although classification accuracy is obtained, the models are not found to be addressing the problems of unmixing spectral images in a simplified manner. Therefore, the proposed study implements a technique capable of performing an effective classification from the Indiana Pines dataset. The study outcome shows a proposed system offers better classification accuracy and reduced processing time thanthe existing approach.

References

1. Nasrabadi, N.M.: Hyperspectral target detection: an overview of current and future challenges. IEEE Signal Process. Mag. **31**(1), 34–44 (2013)
2. ElMasry, G., Sun, D.-W.: Principles of hyperspectral imaging technology. In: Hyperspectral Imaging for Food Quality Analysis and Control, pp. 3–43. Academic Press (2010)
3. Ravikanth, L., Jayas, D.S., White, N.D.G., Fields, P.G., Sun, D.-W.: Extraction of spectral information from hyperspectral data and application of hyperspectral imaging for food and agricultural products. Food Bioprocess Technol. **10**(1), 1–33 (2017)

4. Calin, M.A., Coman, T., Parasca, S.V., Bercaru, N., Savastru, R., Manea, D.: Hyperspectral imaging-based wound analysis using mixture-tuned matched filtering classification method. J. Biomed. Opt. **20**(4), 046004 (2015)
5. Johnson, W.R., Wilson, D.W., Fink, W., Humayun, M.S., Bearman, G.H.: Snapshot hyperspectral imaging in ophthalmology. J. Biomed. Opt. **12**(1), 014036 (2007)
6. Martin, G., Plaza, A.: Spatial-spectral preprocessing prior to endmember identification and unmixing of remotely sensed hyperspectral data. IEEE J. Sel. Top. Appl. Earth Obs. Remote Sens. **5**(2), 380–395 (2012)
7. Bioucas-Dias, J.M., et al.: Hyperspectral unmixing overview: Geometrical, statistical, and sparse regression-based approaches. IEEE J. Sel. Top. Appl. Earth Obs. Remote Sens. **5**(2), 354–379 (2012)
8. Ji, R., Gao, Y., Hong, R., Liu, Q., Tao, D., Li, X.: Spectral-spatial constraint hyperspectral image classification. IEEE Trans. Geosci. Remote Sens. **52**(3), 1811–1824 (2013)
9. Keshava, N.: A survey of spectral unmixing algorithms. Lincoln Lab. J. **14**(1), 55–78 (2003)
10. Su, H., Yong, B., Du, Q.: Hyperspectral band selection using improved firefly algorithm. IEEE Geosci. Remote Sens. Lett. **13**(1), 68–72 (2016)
11. Guccione, P., Mascolo, L., Appice, A.: Iterative hyperspectral image classification using spectral-spatial relational features. IEEE Trans. Geosci. Remote Sens. **53**(7), 3615–3627 (2015)
12. Tong, X., et al.: A new genetic method for subpixel mapping using hyperspectral images. IEEE J. Sel. Top. Appl. Earth Obs. Remote Sens. **9**(9), 4480–4491 (2016)
13. Uezato, T., Murphy, R.J., Melkumyan, A., Chlingaryan, A.: Incorporating spatial information and endmember variability into unmixing analyses to improve abundance estimates. IEEE Trans. Image Process. **25**(12), 5563–5575 (2016)
14. Zhou, L., et al.: Subspace structure regularized nonnegative matrix factorization for hyperspectral unmixing. IEEE J. Sel. Top. Appl. Earth Obs. Remote Sens. **13**, 4257–4270 (2020)
15. Qi, L., Li, J., Wang, Y., Huang, Y., Gao, X.: Spectral–spatial-weighted multiview collaborative sparse unmixing for hyperspectral images. IEEE Trans. Geosci. Remote Sens. **58**(12), 8766–8779 (2020)
16. Chatterjee, A., Yuen, P.W.T.: Sample selection with SOMP for robust basis recovery in sparse coding dictionary learning. IEEE Lett. Comput. Soc. **2**(3), 28–31 (2019). https://doi.org/10.1109/LOCS.2019.2938446
17. He, W., Zhang, H., Zhang, L.: Total variation regularized reweighted sparse nonnegative matrix factorization for hyperspectral unmixing. IEEE Trans. Geosci. Remote Sens. **55**(7), 3909–3921 (2017)
18. Yuan, Y., Fu, M., Lu, X.: Substance dependence constrained sparse NMF for hyperspectral unmixing. IEEE Trans. Geosci. Remote Sens. **53**(6), 2975–2986 (2015)
19. Qu, Y., et al.: Hyperspectral anomaly detection through spectral unmixing and dictionary-based low-rank decomposition. IEEE Trans. Geosci. Remote Sens. **56**(8), 4391–4405 (2018)
20. Wang, W., Qian, Y.: Adaptive ${L}_{\bf 1/2}$ sparsity-constrained NMF with half-thresholding algorithm for hyperspectral unmixing. IEEE J. Sel. Top. Appl. Earth Obs. Remote Sens. **8**(6), 2618–2631 (2015)
21. Yuan, Y., Feng, Y., Lu, X.: Projection-based NMF for hyperspectral unmixing. IEEE J. Sel. Top. Appl. Earth Obs. Remote Sens. **8**(6), 2632–2643 (2015)
22. Tang, W., Shi, Z., Wu, Y., Zhang, C.: Sparse unmixing of hyperspectral data using spectral a priori information. IEEE Trans. Geosci. Remote Sens. **53**(2), 770–783 (2015)
23. Huang, R., Li, X., Zhao, L.: Spectral–spatial robust nonnegative matrix factorization for hyperspectral unmixing. IEEE Trans. Geosci. Remote Sens. **57**(10), 8235–8254 (2019)
24. Soltani-Farani, A., Rabiee, H.R.: When pixels team up spatially weighted sparse coding for hyperspectral image classification. IEEE Geosci. Remote Sens. Lett. **12**(1), 107–111 (2015)

25. Hong, D., Zhu, X.X.: SULoRA: subspace unmixing with low-rank attribute embedding for hyperspectral data analysis. IEEE J. Sel. Top. Signal Process. **12**(6), 1351–1363 (2018)
26. Khoshsokhan, S., Rajabi, R., Zayyani, H.: Sparsity-constrained distributed unmixing of hyperspectral data. IEEE J. Sel. Top. Appl. Earth Obs. Remote Sens. **12**(4), 1279–1288 (2019)
27. Yao, J., Meng, D., Zhao, Q., Cao, W., Xu, Z.: Nonconvex-sparsity and nonlocal-smoothness-based blind hyperspectral unmixing. IEEE Trans. Image Process. **28**(6), 2991–3006 (2019)
28. Borsoi, R.A., Imbiriba, T., Bermudez, J.C.M.: A data dependent multiscale model for hyperspectral unmixing with spectral variability. IEEE Trans. Image Process. **29**, 3638–3651 (2020)
29. Su, H., Qian, D., Peijun, D.: Hyperspectral image visualization using band selection. IEEE J. Sel. Top. Appl. Earth Obs. Remote Sens. **7**(6), 2647–2658 (2014). https://doi.org/10.1109/JSTARS.2013.2272654
30. Hyperspectral Remote Sensing Scene. http://www.ehu.eus/ccwintco/index.php/Hyperspectral_Remote_Sensing_Scenes. Accessed 13 Jan 2021
31. Sun, W., Tian, L., Xu, Y., Zhang, D., Du, Q.: Fast and robust self-representation method for hyperspectral band selection. IEEE J. Sel. Top. Appl. Earth Obs. Remote Sens. **10**(11), 5087–5098 (2017)
32. Sun, W., Zhang, L., Du, B., Li, W., Lai, Y.M.: Band selection using improved sparse subspace clustering for hyperspectral imagery classification. IEEE J. Sel. Top. Appl. Earth Obs. Remote Sens. **8**(6), 2784–2797 (2015)

Historical 3D Visualisations of Starý Světlov Castle Using Blender and Unreal Engine

Pavel Pokorný(✉) and David Falešník

Department of Computer and Communication Systems, Tomas Bata University in Zlín, Faculty of Applied Informatics, Nad Stráněmi 4511, 760 05 Zlín, Czech Republic
pokorny@utb.cz, david.falesnik@seznam.cz

Abstract. The main task of this paper is to describe a visualisation method for Starý Světlov Castle. This castle was probably founded in the first half of the 14th Century when the massive castle core was built. Until the mid of 15th Century Starý Světlov Castle was an important residence of the local aristocracy. At the end of the 15th Century, the castle lost its importance and was gradually abandoned. Research by historians has made it possible to create sketches and paintings of the castle at the time of its greatest importance – in the first half of the 15th Century. The 3D complex model of the castle was created based on mentioned sketches and other resources from this period. All buildings and accessories were separately modelled and textured using UV mapping techniques. The rendered outputs are performed by a visualization application. The Blender software creation suite was used for modelling. Quixel Mixer was used for texturing and the final application was developed in Unreal Engine.

Keywords: 3D visualisation · Modelling · Historical visualiszation

1 Introduction

The visualisation process is usually defined as the communication of information using graphical representation. This communication mechanism was formed much earlier than the formalisation of written language. A single picture can contain a wealth of information, and can be then processed much more quickly than a comparable page full of words. Pictures can also be independent of local language - just as a graph, chart or map may be understood by a group of people with no common tongue [1].

Nowadays, visualisation is very often associated with data obtained from various fields. Visualisation Research at SCI [2] has focused on applications spanning computational fluid dynamics, weather data analysis, medical imaging and analysis, poetry, biomedical data analysis, healthcare data analysis, network and graph analysis, and financial data analysis, etc. This data generally has numerical representation in nature – they have a value and meaning when the data is processed and analysed. Nowadays, these processes are performed by computers [3].

The Computer Graphics field is a typical example of the creation of successful visualisations. In line with developments and improved computer technologies performance, the limits and possibilities of computer graphics continue to increase. The core

R. Silhavy (Ed.): CSOC 2021, LNNS 230, pp. 351–362, 2021.
https://doi.org/10.1007/978-3-030-77442-4_30

3D graphics system – the rendering pipeline – moved on from software to raster graphics hardware via the development of geometry – and pixel–processing chips; and demanding computing operations gradually took over from graphics cards (GPU) [4].

One of these fields is historical visualisations. Based on historical documents, maps, drawings, photographs or plans; using 3D models, one can create reconstructions of virtual objects that no longer exist in their original form.

This article describes the 3D visualisation method used for Starý Světlov Castle. Using modern 3D software modelling tools, we have tried to achieve a very credible appearance of the castle buildings and its vicinity in the first half of the 15th Century.

2 A Short History of Starý Světlov Castle

The Starý Světlov ruins are those of a castle that stood (and still stands) in Eastern Moravia in the Czech Republic. Its exact location was 10 km southwest of Vizovice Township and about a km northwest of the Podhradí village near Luhačovice. It has been protected as a cultural monument since 1964 [5].

The foundations of this castle - with its original name Světlov, date back to the second half of the 13th century. The oldest mention of the castle - later called Starý Světlov, comes from 1275; although there is no mention of the castle's owner. Another historical document, dating from 1360, mentions that Světlov Castle was then in the possession of the Lords of Šternberk. At that time, the castle had a strategic importance and housed a large military garrison [6].

Based on the findings, we can get a pretty clear idea of the historical castle's form and its surroundings in this period. The castle core was built in the 14th Century, and surrounded by a fence and a massive moat and ramparts during the ownership of the first Šternberk family. Part of these fortifications was the Upper Fort - built of quarried stone.

From 1423, the castle was held by King Zikmund and, in the fourteen forties, (15th) Century, the mortgaged Světlov estate passed into the hands of the lords of Cimburk, who undertook marauding raids in a wide surrounding area. At this time, the Světlov Estate was probably densely populated by a large group of inhabitants - poor families, merchants, and craftsmen. The castle itself probably housed the Lord of the Castle, with some of their most loyal warriors; and the lower sub-castle was built and inhabited by other warriors.

During the Czech-Hungarian wars, the castle faced several devastating raids, during which at least a dozen villages around the castle disappeared. In the course of these wars, the castle itself was severely damaged in the years 1469–1470. However, Starý Světlov remained inhabited at least until the early 80s of the 15th Century.

After the end of the Czech-Hungarian wars, the location of Světlov Castle began to prove more and more unfavourable. This castle found itself on the very edge of the manor and relatively far from new important roads. Its then owners were the Šternberk brothers, who decided to build a new seat there and moved to Bojkovice in 1490. The Castle was then left to its fate and ever since then, Starý Světlov was always mentioned as desolated in period documents [6].

The abandoned ruin was mainly damaged by its dismantling for building material from the second half of the 18th Century. Although the castle has become a ruin, it is still a bearer of interesting information and there is evidence of remarkable architectural details - such as, for instance, the unique construction of the outer castle gate, the castle walls' surface finish or several construction peculiarities in the palace.

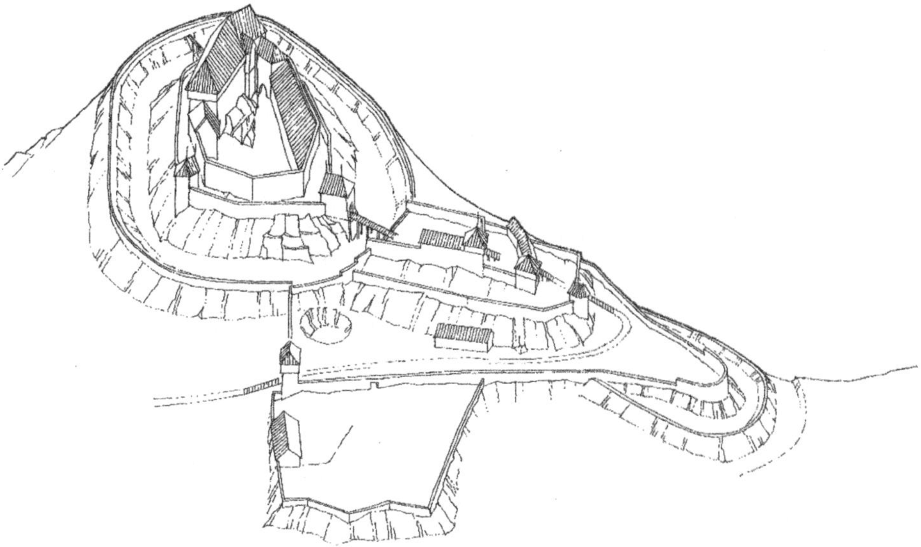

Fig. 1. Starý Světlov castle - artwork from the mid-15th century; Author: Radim Vrla [7]

3 Resources and Software Used

The first phase was to collect all suitable, available historical materials related to Starý Světlov Castle and to select the adequate visualisation creation software.

3.1 Acquiring Resources

This work's overall progress was the collation of any available historic materials and information about Starý Světlov Castle, down through the centuries. The main resources were found in the National Heritage Institute in Kroměříž [7], the Castle's webpage [6], as well as a book on this subject [5].

Its attention was mainly focused on drawings and sketches. We also visited the castle ruins and created a number of photographs to accurately capture its current appearance.

The most useful drawings obtained were those sourced from Radim Vrla, a Kroměříž National Heritage Institute employee. Based on archaeological excavations, he created nice drawings of Starý Světlov Castle's appearance in the mid-15th Century. Some of these drawings are shown in Figs. 1 and 2.

The castle 3D models and created were mainly based on these drawings. Since there is no exact form of the castle from the past and these drawings capture only the main shapes, the details of these shapes were a matter of improvisation. Since it was well-known the castle was built in the Gothic Style, we attempted to retain all this of architectural style's characteristics. In addition – so as to avoid any eventual historical inaccuracies, all works on these models were consulted with Radim Vrla - the above-mentioned historian.

Fig. 2. More detailed Starý Světlov castle artwork; Author: Radim Vrla [5]

3.2 Software Used

We compiled the requisite equipment to create high-quality visual outputs in order to select applications that we wanted to use. The whole process includes 3D model creation, texture mapping correction, and then to create a visualisation application that imports the prepared models with their textures. All 3D models with appropriate UV maps were made in Blender. Textures were created in the Quixel Mixer software tool and the final application was developed in Unreal Engine.

Blender [8] is a free and open-source 3D creation suite. It supports the entirety of the 3D pipeline—modelling, rigging, animation, simulation, rendering, compositing and motion-tracking, video-editing and a 2D animation pipeline. This software is also available under GNU GPL License and users can use it on most common platforms – Windows, Linux, Mac OS, etc.

Blender opens up a world of creativity, which has been traditionally exclusive to those able to afford high-end graphics software. Its GUI is the arrangement of windows and panels containing the controls for operating the programme. Blender supports many tools that allow one to create and animate models and characters in scenes combined with stunning visual effects. Scenes may be rendered into photo-realistic images and animations into video clips. Video may be edited and compiled into movies [9].

Quixel Mixer [10], this represents a revolutionary approach to asset texturing. One can texture one's own assets with multi-channel 3D painting, scan-data, real-time 3D curvature, displacement sculpting, ID masking, procedural noises, seamless texture projection, Smart Materials and much more. It also provides instant access to the entire Megascans library – which guarantees photorealistic base textures or their effortless transformation into fantastic stylised assets. This software is extremely easy to use, and free forever, for everyone [11].

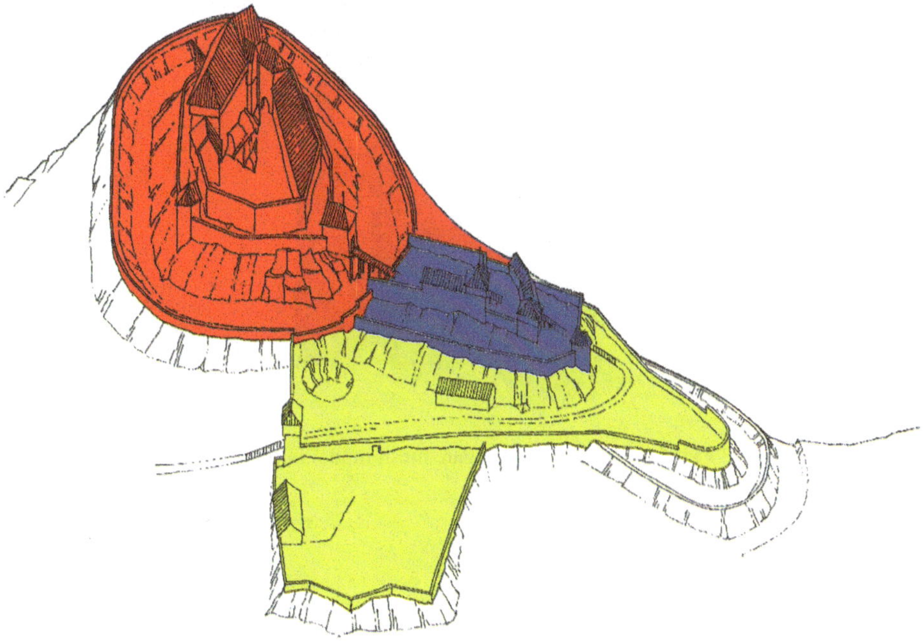

Fig. 3. Colorized divisions of the castle parts for the modelling phase

Unreal Engine [12] is the world's most open and advanced real-time 3D creation tool. Continuously evolving to serve not only its original purpose as a state-of-the-art game engine; today, it gives creators across industries the freedom and control to deliver cutting-edge content, interactive experiences, and immersive virtual worlds.

Unreal Engine is a complete suite of development tools for anyone working with real-time technology. From design visualisation and cinematic experiences to high-quality games across PC, consoles, mobiles, VRs, and ARs, Unreal Engine gives one everything they need to create very high-quality applications. This engine - with its own editor, offers complex tools for photorealistic rendering, dynamic physics and effects, life-like animation, robust data translation - and much more. A license for creators is free-to-use and 100% royalty-free; users can use it to create internal or free projects, or to develop linear content or custom projects for clients; but not for publishing off-the-shelf offerings [13].

Fig. 4. The outer-castle gate mesh model in blender

4 Modelling

The Modelling phase began after gathering a sufficient amount of information and resources. Due to the size of the whole model, the castle was divided into three parts – the Sub-Castle, the Outer-Castle and the Inner-Castle itself. These parts are colour-coded in Fig. 3. It was then possible to model and place the corresponding objects within these parts. Timbered house and farm building models were created for the Sub-Castle; while the Outer-Castle was formed by models of a palisade tower and a wooden gate. The Inner-Castle was composed of models of a girder bridge - (that connects the Outer-Castle and Inner-Castle), towers, ramparts and the Castle building itself. Each main object was further split into smaller parts, which were modelled separately and later joined. For example, a house was separated into main walls, doors, windows and roof.

Only Polygonal Representation was used for all models – i.e. all models were recomposed from vertices, edges and faces. This representation is performed using the Mesh objects in Blender.

The Castle Model Creation Process always proceeded in the same way. In dependency of basic shape, appropriate mesh primitives - (e.g. Planes, Circles or Cubes), were inserted into the 3D scene. This object was then transformed in order to reach the correct position, orientation and sizes. After that the model was switched into the Edit Mode, where is possible to modify the object geometry – user can transform any vertex, edge of face. Additionally, Blender offers many editing tools (Extrude, Knife, Subdivide, Loop Cut, etc.) which improve model geometry – they add new vertices, edges and faces and then is possible to model smaller and more detailed models' parts.

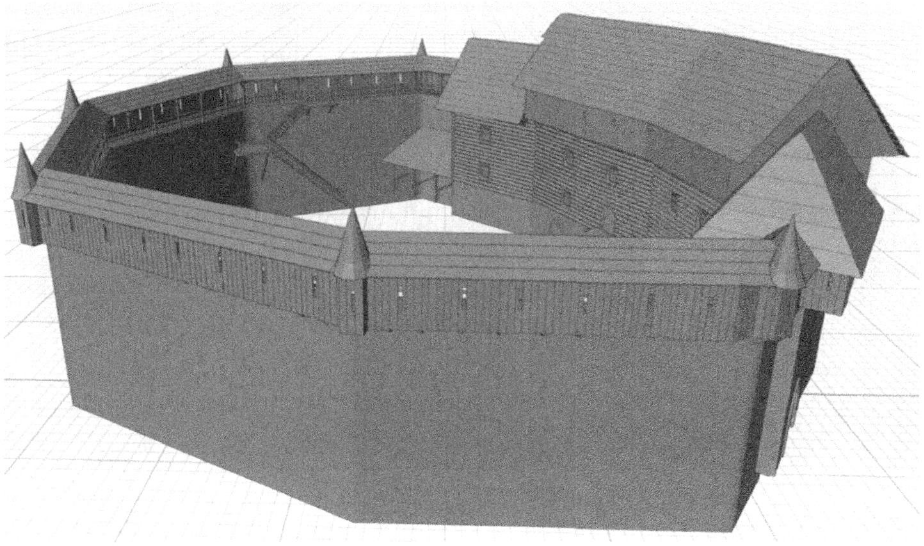

Fig. 5. The inner-castle mesh model in blender

Modifiers are another class of Blender tools. They can improve/transform any shapes of selected objects into other ones. The Boolean Modifier was often used - which can make new shapes using logical operations among selected objects (Union, Intersect, and Difference). Finally, the Array Modifier was used for the creation of repeat objects - (it can rapidly make any number of transformed copies).

Figure 4 shows a 3D model of the Outer-Castle gate with a girder bridge created in this way. This whole model contains approximately thirty thousand vertices. Another example is Fig. 5 – this is a model of the whole Inner Castle and contains approximately one hundred and forty thousand vertices.

5 Texturing

The Texturing Phase began when all of the models had been created. UV-mapping techniques were used for texture mapping in Blender. This process begins with the decomposition of each object into 2D sub-surfaces - (e.g. a UV map). Blender supports the Unwrap tool with several settings for these purposes.

After completing the UV mapping process, the models are ready to apply textures. The above-mentioned Quixel Mixer programme was used for this process. Exporting an object from Blender to Quxiel Mixer was performed by means of the FBX graphic format [14]; which Quxiel Mixer supports.

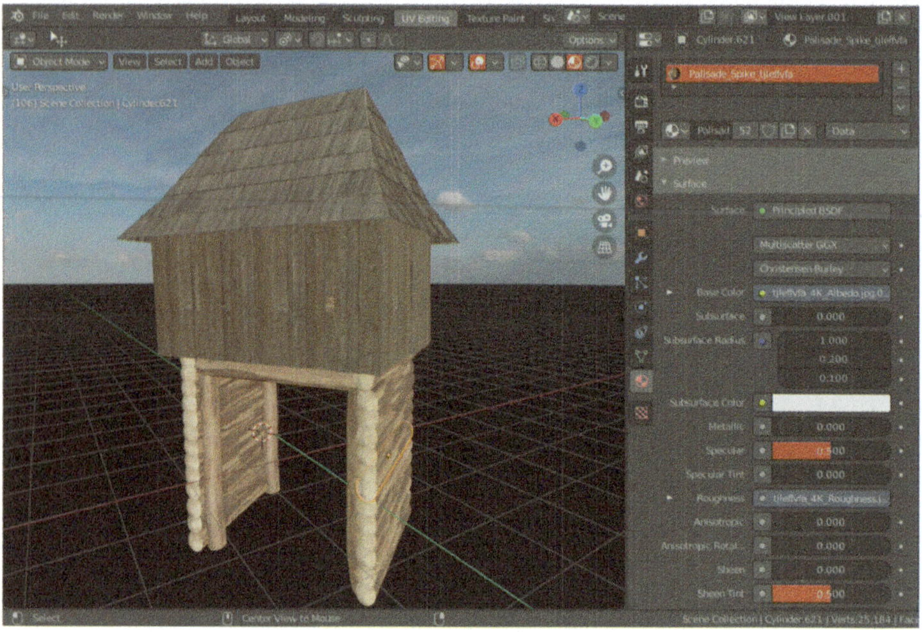

Fig. 6. Textured wooden-gate model in blender

Quixel Mixel software was used in a new project to texture each model - a corresponding model was then imported into this project. The texture process always proceeded in the same way. At first, suitable textures were selected – they were performed via previously mentioned Megascans library or Internet resources [15]. These textures were then mapped on imported object surfaces using various types of brushes.

Three texture layers were used for the models' surfaces – one affected the colours - (Diffuse Map); the second glossiness - (Roughness Map); and the third, irregularities - (Normal Map).

The mapping process was complete when all layers were integrated into corresponding contacts, and thus work in the Quixel Mixer programme was concluded. The resulting textures were then exported via the standard raster graphic format – JPG, which can easily be imported into Blender and Unreal Engine environments.

An example of a 3D model with mapped textures in Blender is shown in Fig. 6. This is the model of a wooden gate that is placed between the Outer-Castle and the Sub-Castle.

6 Application Creation

This visualisation application was created in Unreal Engine. In this application, it was first necessary to create a new 3D project. In the following step, the whole environment should be created. The Starý Světlov Castle - with textures model was ready, but the main object was missing - the terrain model. It was not created in Blender since Unreal Engine contains all of the necessary tools for its creation.

Unreal Engine provides the Landscape command in the Modes menu in order to create a landscape model. Here it is possible to set the size of the model, as well as the number of segments from which the landscape model is to be created - or the position where model will be generated. After generation, it is also possible to make manual modifications in order to attain the required shape. To do this, a user can use the three graphic tools - Sculpt, Flat and Flatten; which modify the shapes according to the set parameters. The final shape of the landscape was therefore, created manually according to the real altitudes obtained from the source [16].

The above-mentioned FBX graphic format was used to import castle objects since, in addition to the model's geometric shapes, it also supports all other requisite information - like UV coordinates for the correct application of textures [17]. After the importation of all partial models of the castle, these models were then placed in the correct position and then the terrain model was slightly modified in order to allow all models to follow each other correctly and the most realistic form be achieved.

Another task was to add materials - (textures), to the landscape model. Unreal uses so-called "Smart Materials". Using them, it is possible to link more textures to these materials; the implemented algorithm can recognise inequalities on the object and then automatically applies a suitable texture according to these inequalities. Grassy textures are then automatically applied to flat surfaces, while stony textures are applied to inclined surfaces. Afterwards, it is then possible to draw in various details - paths, and so on, directly on the surface of the object using special brushes.

In order to further increase the realism, it was necessary to add some vegetation to the scene - trees, grasslands, etc. These types of objects do not need to be modelled, they can be obtained from the Unreal Marketplace or Quixel Bridge.

Some of them are free - others for a fee. We selected free models and after integrating them into our project, it was possible to apply this to the 3D environment by means of special brushes. These brushes can have settings for many parameters like size, density, location; or variability of objects to be generated. These objects were added to the scene to selected places, and in the same way, a wooden palisade was created in the vicinity of the castle grounds.

The next step was to illuminate the whole scene in a suitable way. Unreal engine offers several types of light sources with adjustable parameters that can be combined in any way. In order to achieve the most realistic simulation of sunlight, a combination of sky light with directional light, which can also create shadows, was chosen for the castle visualisation scene.

To create rendered images and animations, there was still a need to add extra cameras to the scene. These images and animations are then rendered according to their location and direction-of-view. The Render command is located in the Viewport window under the "High Resolution Screenshot" button. Here, users can select the size of the generated image and then create the render. Examples of these rendered images are shown in Figs. 7 and 8. In the case of animation, the camera moves and this movement is set by the so-called "keying frames" technique. This technique means the setting of the camera position in the selected animation frames.

Fig. 7. Rendered visualisation of Starý Světlov castle in unreal engine

Fig. 8. Rendered visualisation of Starý Světlov sub-castle in unreal engine

The application itself was created from the 3rd perspective. This type of application is provided by means of a selected blueprint in Unreal Engine, which ensures the basic movement of the character. The camera is linked to this character and their movement is performed based on user input from keyboards and peripherals. After setting up the camera, it was still necessary to set where the character can go - for example, it cannot pass through the solid walls of objects. So-called "Collision Objects" solve this. They are basically cube-type objects that are not visible but specify places that cannot be entered.

After the application testing phase in Unreal Engine, a separate project for the Windows operating system (64-bit) was then generated and can be run on all common personal computers and laptops with an "average" current hardware configuration.

7 Conclusion

This article briefly describes the visualisation method used for depicting Starý Světlov Castle artwork from the mid-15th Century – at the time of its greatest glory. Although the castle is currently a mere ruin, historians' research has helped to create the probable form of the castle at the time. The results of this research have become the main source for creating this visualisation.

All models and their UV maps were created in the Blender software suite and the textures were prepared and mapped in Quixel Mixel. When all these assets had been completed, they were then imported into Unreal Engine, where - as mentioned above, the visualisation application was made.

The future goal is to improve and expand these models. This process will include modelling more detailed parts of the castle and adding period accessories in order to make the scene more realistic. Further improvements can be performed by application optimisation, improved interactions and customisation for the Virtual Reality suite. This will enable any users to get to know the historic environment of this castle.

Acknowledgments. This work arose from collaboration with the Czech National Heritage Institute and the Kroměříž Regional Council, which provides both methodological and professional assistance to heritage-site owners and managers in the Zlín Region.

References

1. Ward, M.: Interactive Data Visualization: Foundations, Techniques, and Applications, 1st edn. A K Peters, Ltd., India (2010). ISBN 978-1568814735
2. Visualization. http://www.sci.utah.edu/research/visualization.html. Accessed 18 Jan 2021
3. Srinivasa, K.G., Siddesh, G.M., Srinidhi, H.: Introduction to data visualization. In: Srinivasa, K.G., Siddesh, G.M., Srinidhi, H. (eds.) Network Data Analytics - A Hands-On Approach for Application Development, pp. 321–331. Springer, Cham (2018). https://doi.org/10.1007/978-3-319-77800-6_16
4. Hughes, J.F., Van Dam, A., Foley, J.D., McGuire, M., Feiner, S.K., Sklar, D.F.: Computer Graphics: Principles and Practice, 3rd edn. Addison-Wesley, Boston (2014). ISBN 978-0-321-39952-6
5. Michálek, L.: Starý Světlov, 1st edn. Putujme (2014). ISBN 978-80-87891-06-3

6. Historie hradu Starý Světlov: Podhradí. https://www.obec-podhradi.cz/historie-hradu-stary-svetlov/ds-1085. Accessed 18 Jan 2021
7. National Heritage Institute in Kroměříž. https://www.npu.cz/cs/uop-kromeriz. Accessed 18 Jan 2021
8. Blender.org – Home of the Blender project. https://www.blender.org/. Accessed 18 Jan 2021
9. Blain, J.M.: The Complete Guide to Blender Graphics, 3rd edn. CRC Press (2016). ISBN 978-1-4987-4647-2
10. Quixel Mixer. https://quixel.com/mixer. https://www.textures.com/. Accessed 18 Jan 2021
11. Quixel ships Mixer 2020.1.6. http://www.cgchannel.com/2020/11/quixel-ships-mixer-2020/. Accessed 18 Jan 2021
12. The most powerful real-time 3D creation platform - Unreal Engine. https://www.unrealengine.com/en-US/. Accessed 18 Jan 2021
13. Plowmanm, J.: 3D Game Design with Unreal Engine 4 and Blender, 1st edn. Packt Publishing (2016). ISBN 978-1-78588-146-6
14. FBX – Adaptable File Formats for 3D Animation Software. https://www.autodesk.com/products/fbx/overview. Accessed 18 Jan 2021
15. Textures – Textures for 3D, graphic design and Photoshop! https://www.textures.com/. Accessed 18 Jan 2021
16. Daftlogic – Google Maps Find Altitude. https://www.daftlogic.com/sandbox-google-maps-find-altitude.htm. Accessed 18 Jan 2021
17. Blender to Unreal: How to Import Models for Unreal. https://all3dp.com/2/blender-to-unreal-engine-import-models-unreal-ue4/. Accessed 18 Jan 2021

Optimization of the Process of Obtaining Ethanol-Lignin from Abies Bark (*Abies Sibirica*)

Yuriy N. Malyar[1,2(✉)], Aleksandr S. Kazachenko[1,2], and Dmitry V. Zimonin[1,2]

[1] Institute of Chemistry and Chemical Technology SB RAS, Federal Research Center "Krasnoyarsk Science Center SB RAS", Akademgorodok, 50/24, Krasnoyarsk 660036, Russia
malyar@icct.ru

[2] Siberian Federal University, Svobodny av., 79, Krasnoyarsk 660041, Russia

Abstract. Bark is a large-scale waste in the wood processing industry. The methods of organosolvent processing of lignocellulosic raw materials are now widely used. In this work, the numerical optimization of the process of water-ethanol extraction of abies bark with the production of ethanol-lignin is carried out. It is shown that the optimal conditions for obtaining ethanol-lignin from abies bark are: time 4.5 h, temperature 185 °C and ethanol/water ratio 70/30. The resulting ethanol-lignin can be widely used as a raw material for the production of bio-medical preparations and components of motor fuels. #COMESYSO1120

Keywords: Abies Sibirica · Abies extraction · Ethanol-lignin · Ethanol

1 Introduction

Renewable plant biomass, in particular wood, is an alternative to traditional petroleum feedstock for obtaining a variety of chemical products [1–3]. The composition of woody biomass, which consists of three main components - cellulose, lignin, and hemicelluloses [4] makes it difficult to chemically process it into demanded chemical compounds [5].

Traditional industrial technologies are focused on processing only the cellulosic component of woody biomass [4]. Although lignin is one of the main components of biomass and a potential source of valuable aromatic substances [6, 7].

Technical lignins formed in traditional processes of processing lignocellulosic raw materials are not widely used in practice. The complexity of chemical processing of technical lignins is due to their low reactivity due to the high degree of polymerization and the presence of sulfur. By treating lignocellulosic raw materials with aliphatic alcohols, or their mixtures with water at temperatures of 180–200 °C, organosolvent lignins are isolated, which do not contain sulfur and have increased reactivity [8, 9].

Organosolvent lignins can find wide application for their chemical modification [10, 11], as well as depolymerization to obtain components of motor fuels [12, 13].

Bark-wood waste is the main solid carbon-containing waste of modern woodworking, pulp and paper industries [14]. Waste amounts to 13–21 wt.% of woody biomass [15], and only in Europe and Canada, they are generated annually in the amount of 38, 17 million m^3, respectively [16].

R. Silhavy (Ed.): CSOC 2021, LNNS 230, pp. 363–368, 2021.
https://doi.org/10.1007/978-3-030-77442-4_31

The aim of this work was to find the optimal conditions for the release of ethanol-lignin from the abies bark with the maximum product yield.

2 Experimental

In this work, the abies bark (Ábies sibírica) was used.

Ethanol-lignin was obtained according to the procedure [17].

An autoclave with a capacity of 2.5 L was loaded with crushed air dry abies bark (100 g), 1000 ml of a water-ethanol mixture (ethanol/water ratio 40/60, 60/40 or 80/20%). Then the autoclave was sealed. Extraction of ethanol lignin was carried out at temperatures of 160, 185 or 210 °C and time of 1, 3.5 or 6 h. After the process, the autoclave was cooled, the resulting mixture was filtered on a paper filter, and the solid product was washed with a water-ethanol mixture.

A fourfold volume of cooled (up to 4 °C) distilled water was added to the filtrate, after which a precipitate of ethanol-lignin of the abies bark precipitated. The resulting mixture was placed in a refrigerator (temp. 10 °C) for 24 h for more complete precipitation of ethanol-lignin. The resulting precipitate was filtered off on a paper filter and washed with distilled water, after which it was dried in air at room temperature for 2 days. After that, the ethanol-lignin obtained was weighed and the yield (wt.%) was determined.

The numerical optimization was carried out using the Statgraphics Centurion XVI software, according to the procedure described in [18, 19].

3 Results and Discussion

Three factors were included in the study as independent variables (in brackets - the levels of their variation): X_1 - time (1, 3.5, 6 h); X_2 - process temperature (160, 185, 210 °C) and X_3 - alcohol content in the water-ethanol solution (40, 60, 80%). The result of the ethanol-lignin extraction process was characterized by the output parameter: Y_1 - the yield of ethanol-lignin (wt. %). Each experiment was conducted in duplicate. The Box-Behnken Design (BBD) experiment plan was used. Variables are designated in Table 1.

Table 1. Designations of independent variables (factors) and output parameters (experimental results)

Factors and parameters	Notation in equations	Range of variation
Time, h	X_1	*1–6*
Temperature, °C	X_2	*160–210*
Alcohol content in the water-ethanol solution, %	$X3$	*40–80*
Yield, % wt.	Y_1	

The experimental results are shown in Table 2.

Table 2. Yield of ethanol lignin from Abies bark

№	Time, h	Temperature, °C	Alcohol content in the water-ethanol solution, %	Yield, % wt.
1	3.5	185	60	17.2
2	1	160	60	5.1
3	6	160	60	13.1
4	1	210	60	4.0
5	6	210	60	13.5
6	1	185	40	5.7
7	6	185	40	11.9
8	3.5	185	60	17.1
9	1	185	80	5.2
10	6	185	80	16.7
11	3.5	160	40	12.4
12	3.5	210	40	9.7
13	3.5	160	80	15.1
14	3.5	210	80	13.2
15	3.5	185	60	17.2

The results of analysis of variance are given in Table 3.

Table 3. The result of analysis of variance

Sources variance	Statistical characteristics	
	F-ratio	*P-value*
X_1	327.56	0.0000
X_2	7.43	0.0415
X3	29.15	0.0029
X_12	234.79	0.0000
X_1X_2	1.19	0.3252
X_1X_3	14.85	0.0120
X_2^2	59.41	0.0006
X_2X_3	0.34	0.5860
X_3^3	25.54	0.0039
Df	14	
R^2	99.1	
R^2_{adj}	97.9	

The dependence of the yield of ethanol lignin abies bark (Y1) on variable process factors is approximated by the regression equation:

$$Y_1 = -152.207 + 5.2013X_1 + 1.5614X_2 + 0.44138X_3 - 0.877333X_1^2 + 0.006X_1X_2 + 0.027X_1X_3 - 0.004413X_2^2 + 0.0004X_2X_3 - 0.0045208X_3^2 \quad (1)$$

Analysis of variance showed within the accepted experimental conditions, the greatest contribution to the total variance of the output parameter is made by the factor - the time of the ethanol-lignin extraction process. This is indicated by the high values of the F-dispersion coefficients for the main effects. The information contained in the P columns of the table is interpreted similarly (Table 3). The influence of the source of variance on the output parameter is considered statistically significant if the significance level is less than the specified critical value (in our case, 0.05).

A graphical display of the dependence of the output parameter (Y_1) on the variable factors X_1, X_2 and X_3 in the form of a response surface is shown in Fig. 1.

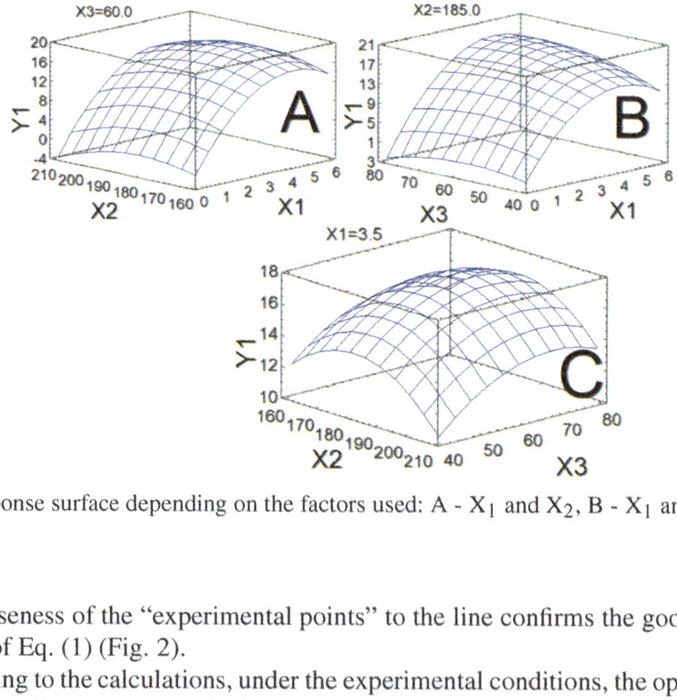

Fig. 1. Response surface depending on the factors used: A - X_1 and X_2, B - X_1 and X_3, C - X_2 and X_3.

The closeness of the "experimental points" to the line confirms the good predictive properties of Eq. (1) (Fig. 2).

According to the calculations, under the experimental conditions, the optimal conditions for obtaining ethanol-lignin from abies bark are: time 4.5 h, temperature 184.2 °C and alcohol content in water-ethanol solution 70.5%.

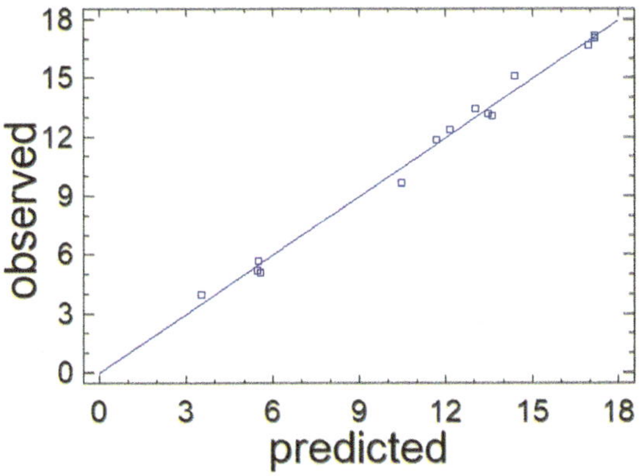

Fig. 2. The results of observations against the values of the output parameter Y_1 predicted by the mathematical model (2).

4 Conclusions

For the first time the process of thermal extraction of the bark of Abies (Abies Sibirica) with the production of ethanol-lignin has been studied. It has been shown that the optimal conditions providing the maximum yield of the target product are: time 4.5 h, temperature 184.2 °C and alcohol content in water-ethanol solution 70.5%.

The obtained ethanol-lignin of Abies bark can be used as a source for obtaining both components of motor fuels and biomedical preparations.

Acknowledgments. This study was carried out using the equipment of the Krasnoyarsk Regional Centre for Collective Use, Krasnoyarsk Scientific Center, Siberian Branch of the Russian Academy of Sciences.

Funding. The research was funded by RFBR, Krasnoyarsk Territory and Krasnoyarsk Regional Fund of Science, project number 20-43-242906. This study was carried out within the budget project # 0287-2021-0012 for the Institute of Chemistry and Chemical Technology, Siberian Branch of the Russian Academy of Sciences.

References

1. Luterbacher, J.S., Alonso, D.M., Dumesic, J.A.: Targeted chemical upgrading of lignocellulosic biomass to platform molecules. Green Chem. **16**(12), 4816–4838 (2014)
2. Gallezot, P.: Conversion of biomass to selected chemical products. Chem. Soc. Rev. **41**(4), 1538–1558 (2012)
3. Zhao, X., et al.: Biomass-based chemical looping technologies: the good, the bad and the future. Energy Environ. Sci. **10**(9), 1885–1910 (2017)
4. Rowell, R.M.: Handbook of Wood Chemistry and Wood Composites, 2nd edn. CRC Press: Taylor and Francis Group (2012)
5. Popa, V.I.: Pulp Production and Processing: From Papermaking to High-Tech Products. Smithers Rapra Technology, Shropshire (2013)

6. Amen-Chen, C., Pakdel, H., Roy, C.: Production of monomeric phenols by thermochemical conversion of biomass: a review. Biores. Technol. **79**(3), 277–299 (2001)
7. Gazi, S.: Valorization of wood biomass-lignin via selective bond scission: a minireview. Appl. Catal. B: Environ. **257**, 117936 (2019)
8. Kim, J.-Y., et al.: Effects of various reaction parameters on solvolytical depolymerization of lignin in sub- and supercritical ethanol. Chemosphere **93**(9), 1755–1764 (2013)
9. Zhang, K., Pei, Z., Wang, D.: Organic solvent pretreatment of lignocellulosic biomass for biofuels and biochemicals: a review. Biores. Technol. **199**, 21–33 (2016)
10. Kuznetsov, B.N., et al.: Optimization of the process of abies ethanol lignin sulfation by sulfamic acid–urea mixture in 1,4-dioxane medium. Wood Sci. Technol. **54**(2), 365–381 (2020). https://doi.org/10.1007/s00226-020-01157-6
11. Vasilyeva, N.Yu., et al.: Synthesis and study of copper-containing polymers based on sulfated ethanol lignin of abies wood. J. Sib. Fed. Univ. Chem. **13**(1), 142–152 (2020). https://doi.org/10.17516/1998-2836-0170
12. Sharypov, V.I., Beregovtsova, N.G., Baryshnikov, S.V., Miroshnikova, A.V., Lavrenov, A.V., Kuznetsov, B.N.: Conversion of pine wood ethanol-lignin in supercritical ethanol in the presence of catalysts based on borate-containing alumina. J. Sib. Fed. Univ. Ser. Chem. **11**(1), 81–92 (2018)
13. Sharypov, V.I., Beregovtsova, N.G., Baryshnikov, S.V., Taran, O.P., Lavrenov, A.V., Kuznetsov, B.N.: Hydrogenation of larch wood ethanol-lignin in supercritical ethanol in the presence of sulfated ZrO2 and Pt/ZrO2 catalysts. J. SFU. Chem. **11**(4), 593–603 (2018)
14. Best Available Techniques (BAT) Reference Document for the Production of Pulp, Paper and Board. Industrial Emissions Directive 2010/75/EU (Integrated Pollution Prevention and Control). Publications Office of the European Union (2015)
15. Feng, S., Cheng, S., Yuan, Z., Leitch, M., Xu, C.: Valorization of bark for chemicals and materials: a review. Renew. Sustain. Energy Rev. **26**, 560–578 (2013)
16. Miranda, I., Gominho, J., Mirra, I., Pereira, H.: Chemical characterization of barks from Picea abies and Pinus sylvestris after fractioning into different particle sizes. Ind. Crops Prod. **36**(1), 395–400 (2012)
17. Quesada-Medina, J., López-Cremades, F.J., Olivares-Carrillo, P.: Organosolv extraction of lignin from hydrolyzed almond shells and application of the δ-value theory. Bioresour. Technol. **101**(21), 8252–8260 (2010). https://doi.org/10.1016/j.biortech.2010.06.011
18. Kazachenko, A.S., Vasilyeva, N., Sudakova, I.G., Levdansky, V.A., Lutoshkin, M.A., Kuznetsov, B.N.: Numerical optimization of the process of abies ethanol lignin sulfation with sulfamic acid in 1,4-dioxane medium in the presence of urea. J. Sib. Fed. Univ. Chem. **13**(2), 232–246 (2020)
19. Pen, R.Z.: Planning an experiment at Statgraphics Centurion. Krasnoyarsk: SibSTU **293** (2014)

Physical Complexity of Algorithms

I. P. Deshko[1] and V. Ya. Tsvetkov[1,2](✉)

[1] Russian Technological University (RTU MIREA), Vernadsky Prospekt, 78, 119454 Moscow, Russia
dip@mirea.ru

[2] Research and Design Institute of Design Information, Automation and Communication on Railway Transport, 27, bldg 1 Nizhegorodskaya Str., Moscow, Russia

Abstract. The article analyzes the complexity of the first kind algorithms. The first kind' algorithms features are investigated. A new concept is introduced in the theory of algorithmic complexity "algorithm efficiency" and "physical computation time". A new concept in the algorithmic complexity theory "physical complexity of the algorithm" is introduced. The difference between the first and second kind algorithms is shown on the deterministic and non-deterministic Turing machines example. A first kind algorithms generalized model is given. The conventionality of complexity' some types are noted. It consists in the fact that algorithms that belong to the same complexity class or type of complexity physically use different computation times. A situation is noted in which a complexity simpler type spends more time on computations than a more complex one. The algorithmic complexity existing theory lack is noted - the calculation result and cognitive complexity quality exclusion from the complexity analysis. The complexity' existing theory focuses on computation and computation time. The main thing in the computations is the result. If the computations' result is not high-quality or unclear, then the computation time loses its importance. Accordingly, the complexity classes assessment should be tied not only to time but also to the resulting quality. The algorithm physical complexity takes into account the computation result quality #CSOC1120.

Keywords: Algorithms · Complexity · Algorithmic complexity · Computational complexity · Time complexity · Complexity classes · Algorithm physical complexity · Computational model · Turing machine

1 Introduction

The algorithmic complexity theory current state is associated with a computational model based on the Turing machine. To estimate, for example, the time complexity, typical procedures are used: estimate the time spent for this algorithm operation; compare the known classified algorithm and the new algorithm; if there is a similarity or proportionality, the new algorithm is assigned to a known class. The algorithms complexity classes are considered as algorithms set that solves commensurate computational problems. Essentially, the algorithms' classification is based on comparative analysis and an asymptotic complexity model.

R. Silhavy (Ed.): CSOC 2021, LNNS 230, pp. 369–374, 2021.
https://doi.org/10.1007/978-3-030-77442-4_32

Information algorithmic processing corresponds to the input information transformation into the processing result [1]. In computations, the algorithm plays an intermediary role in the information transformation. The main factor in assessing complexity in the complexity' existing theory [2, 3] is the algorithm running time, and not its actions result. The information processing experience shows that according to the information processing and the computation trajectory type, algorithms two types can be distinguished: the first and the second kind [4]. With a first kind algorithm help [5], the problem solution can be obtained using one computational trajectory, even if there are several possible trajectories. The first kind most algorithms involve solving problems that can be solved by a deterministic Turing machine (DTM) with limited time or space resources. The second kind' algorithms are solutions to problems that can be solved using a non-deterministic Turing machine (NTM). In this paper, we investigate the algorithm of the first kind as simpler. There are different types of algorithms' complexity. Time complexity is traditionally used as a basis [6, 7], which is called computational complexity, linking the computation complexity with the computation time. The computation time is denoted by $T(n)$. Time complexity is usually expressed using the capital *O*: $O(n)$, $O\ (n\ log(n))$, $O(n^a)$, $O(2^n)$ and others, where n is the size of the input data in bits *log required* for presentation. For example, a time complexity algorithm is a linear time algorithm, and time complexity algorithm > 1 is a constant 1 is an algorithm with polynomial time.

2 Algorithm Performance

To assess complexity, the term "Time complexity class" is often used, which is estimated by the time resources required to solve the problem. The complexity class is usually determined by three factors: the computational problem type, the computation model, and the limited computational resource. For example, the complexity class *P* is defined as the set of problems whose solution can be obtained by a deterministic Turing machine in polynomial time. Figure 1 shows a generalized algorithm of the first kind.

The first kind' all algorithms (Fig. 1) are characterized by one input *X* and one output Y. The first kind' all algorithms can be simplified and combined into one block with an input (*X*) and an output (*Y*) (Fig. 1). Figure 1 shows that the complexity is estimated according to the algorithm, and not according to its work result. This does not take into account the algorithm physical running time (*PhT*) and the resulting quality (*QR*) or the algorithm effectiveness. The same algorithm (*PhT*) physical running time varies depending on the data amount

$$PhT = f(N) \tag{1}$$

In expression (1) *N* is the data amount. It is fundamental that

$$at\ N1 > N2 \rightarrow PhT1 > PhT2 \tag{2}$$

Expression (2) is valid for time complexity any class or it does not depend on the complexity class.

The resulting quality or the effectiveness (*QR*) can be assessed by the accuracy (accuracy) and by the resulting optimality. Depending on the data quality (Fig. 1), the result may have high accuracy (*A1*), satisfactory accuracy (*A2*), and low accuracy (*A3*). This takes place on one algorithm and one algorithm complexity class.

Depending on the data type and quality (Fig. 1), when solving optimization problems, the result can be optimal (*Op1*), close to optimal (*Op2*), and not optimal (*Op3*). This takes place on one algorithm and one algorithm complexity class. For example, discrete optimization [8, 9] on integers gives different values than optimization on real numbers.

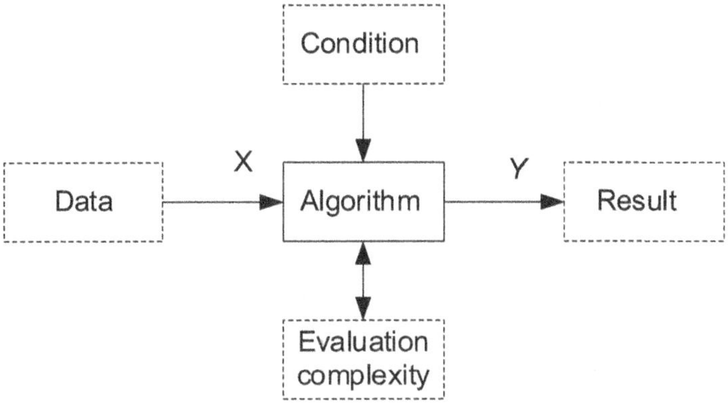

Fig. 1. The first kind Generalized algorithm.

When inverting large matrices, the problem of "machine zero" may arise [10]. If the matrix elements differ from each other by many orders of magnitude, then when the matrix is inverted, a machine zero occurs. Small values of elements that differ from each other and are commensurate with each other, but not commensurate with zero, when divided into large elements become commensurate with zero or turn into machine zero. This creates computational complexity that is not accounted for in time complexity classes. A similar computational situation takes place with large numbers. When calculating the large numbers' powers, the disproportionate numbers may become commensurate to a certain extent.

When inverting matrices, the problem of "resonance" or instability of the solution may arise. If the matrix elements or data are in certain quantitative relationships, then a resonance effect appears, in which the solution is not obtained, while it is obtained on other data. For example, when solving an inverse photogrammetric intersection (determining the photographing point by the image points coordinates and the corresponding points on the terrain), the solution is possible if the control points do not lie on a second-order surface. If this condition is not met, then the solution fails. The resonance effect can occur in many algorithms in which inverse matrices are calculated or problems of successive approximation methods are solved. The convergence problem [11] is also algorithmic but is not taken into account in complexity classes. If you set certain sizes in successive approximations' area, then the solution can be obtained with the required accuracy. If the approximation area is larger, then the solution will not be obtained at all. Therefore, in addition to the theoretical computational complexity, it is advisable to introduce the algorithm physical complexity concept (*PCA*) Fig. 2.

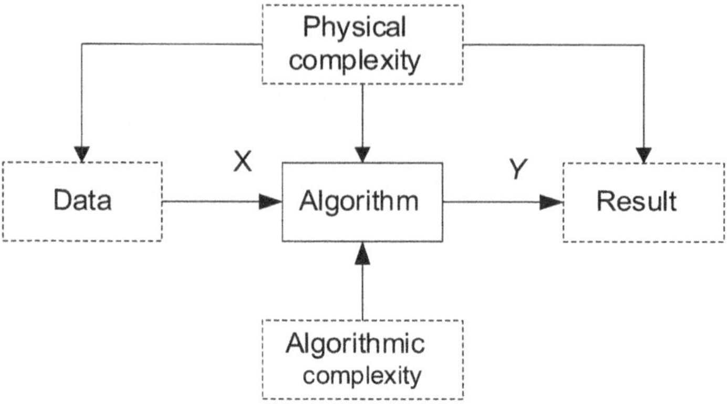

Fig. 2. The algorithms physical complexity.

The algorithm physical complexity covers more parameters, but these are physical parameters. The algorithm physical complexity can be represented as a complex system.

$$PCA = \langle N,\ TD,\ PhT,\ A,\ Op,\ RC,\ Deg,\ Con\rangle \tag{3}$$

In expression (3), N is the input data amount, TD is the input data types, PhT is the physical computation time estimate, A is the result accuracy, Op is the solution optimality, RC is the resonance conditions, Deg is the order between small and large values of numbers, Con - convergence area.

Expression (3) can have a qualitative (logical) notation and quantitative). The logical notation has two meanings for each parameter, 0 and 1. Zero means no influence of this factor on physical complexity. For example, for the complexity of constant time $O(1)$ parameter $N = 0$. This means that it does not affect the algorithm operation. And vice versa, for the linear time complexity $O(n)$ $N = 1$, since in this algorithms kind there is an influence of the data volume on the computation time. If the algorithm has absolute convergence, then $Con = 1$. If the algorithm has no resonance effects, it means $RC = 0$, otherwise, $RC = 1$. If the algorithm provides the required accuracy, then $A = 1$. If the algorithm provides the required optimality, then $Op = 1$. An ideal or simple algorithm, in this case, has the following qualitative characteristics.

$$PCA = \langle 0,\ 0,\ PhT,\ 1,\ 1,\ 0,\ 0,\ 1\rangle \tag{4}$$

Expression (4) suggests that the algorithm physical complexity understudy is such that: there is no influence of the input data volume on the physical computation time; data types ($TD = 0$) do not affect the result; the algorithm provides the required accuracy of calculations ($A = 1$); the algorithm ensures the optimal solution ($Op = 1$); the algorithm has no resonance effects or loops ($RC = 0$); the algorithm has no significant difference between the data ($Deg = 0$); the algorithm has absolute convergence ($Con = 1$).

The algorithm proposed physical complexity has access to the analysis' cognitive area and requires the cognitive logic and cognitive modelling connection. Many algorithms have instability in operation, which is revealed during operation. During operation, the experience is accumulated in identifying algorithm defects and experience in eliminating defects of individual algorithms or a group of algorithms is accumulated. During operation, the algorithm reliability increases and additional reliability "accumulated reliability" appears [12]. It does this by connecting the person's cognitive area.

3 Conclusion

The disadvantage of the indicator physical computation time is its dependence on the computing system technical characteristics. On the other hand, it gives the real computation time, while the computational complexity focuses on the computation time asymptotic bounds. The an algorithm physical complexity proposed concept is not an alternative to computational time complexity but supplements it. The physical complexity proposed concept is not fully formalized but depends on expert knowledge. The physical complexity concept and physical computation time make it possible to compare time complexity different classes algorithms. In practice, a high complexity class algorithm shorter running time is possible in comparison with a lower class algorithm due to the data used a smaller amount. For example, numerical algorithms that use tabular data can have a simple analytical form but require a lot of computation time. Analytical algorithms based on strict dependencies can use more complex formulas, but they can get a solution in one computation cycle. They make regular attempts to construct complexity classes general hierarchies. For example, an attempt is known to link temporal and spatial complexity classes (in bold) as follows: **L** ⊂ **NL** ⊂ P ⊂ NP ⊂ **PSPACE** ⊂ EXPTIME ⊂ **EXPSPACE**. Physical complexity complements this hierarchy. In the research development, it should be recommended to use the information units concept as computing units and time units generalization. The Turing machine steps and cells are such information units. The first kind algorithms are relatively simple classes algorithms. This makes it possible to study them in more detail, and this article is one of such research stages.

References

1. Kudzh, S.A., Tsvetkov, V.Y., Rogov, I.E.: Life cycle support software components. Russian Technol. J. **8**(5), 19–33 (2020)
2. Hidary, J.D.: Complexity theory. In: Hidary, J.D. (ed.) Quantum Computing: An Applied Approach, pp. 37–44. Springer, Cham (2019). https://doi.org/10.1007/978-3-030-23922-0_4
3. Doerr, C.: Complexity theory for discrete black-box optimization heuristics. In: Doerr, B., Neumann, F. (eds.) Theory of Evolutionary Computation, pp. 133–212. Springer, Cham (2020). https://doi.org/10.1007/978-3-030-29414-4_3
4. Tsvetkov, V.Ya.: Incremental solution of the second kind problem on the example of living system. Biosci. Biotechnol. Res. Asia **11**(S), 177–80 (2014)
5. Shchennikov, A.E.: Direct algorithms models. Slavic Forum **4**(18), 103–109 (2017)
6. Xia, T.: A constant time complexity spam detection algorithm for boosting throughput on rule-based filtering systems. IEEE Access **8**, 82653–82661 (2020)
7. Consolini, L., et al.: Optimal time-complexity speed planning for robot manipulators. IEEE Trans. Rob. **35**(3), 790–797 (2019)

8. Kasperski, A., Zieliński, P.: Robust discrete optimization under discrete and interval uncertainty: a survey. In: Doumpos, M., Zopounidis, C., Grigoroudis, E. (eds.) Robustness analysis in decision aiding, optimization, and analytics. ISORMS, vol. 241, pp. 113–143. Springer, Cham (2016). https://doi.org/10.1007/978-3-319-33121-8_6
9. Tilahun, S.L., Ngnotchouye, J.M.T.: Firefly algorithm for discrete optimization problems: a survey. KSCE J. Civ. Eng. **21**(2), 535–545 (2017). https://doi.org/10.1007/s12205-017-1501-1
10. Nishikawa, H.: On Estimating Machine-Zero Residual. arXiv preprint arXiv:2007.06394 (2020)
11. Zhang, X., et al.: The convergence analysis and uniqueness of blow-up solutions for a Dirichlet problem of the general k-Hessian equations. Appl. Math. Lett. **102**, 106124 (2020)
12. Titov, E.K., Tsvetkov, V.Ya.: Accumulated reliability of information hardware and software systems. In: IOP Conference Series: Materials Science and Engineering, vol. 919, p. 022055 (2020). https://doi.org/10.1088/1757-899X/919/6/022055

The Usage of Declarative Integrity Constraints in the SQL Databases of Some Existing Software

Erki Eessaar(✉)

Department of Software Science, Tallinn University of Technology, Tallinn, Estonia
erki.eessaar@taltech.ee

Abstract. Enforcing integrity constraints declaratively in databases gives numerous advantages like protecting the database against recording certainly incorrect propositions and giving information about the database and data to different programs, including the database management system itself, as well as human users. We investigated the enforcement of declarative integrity constraints in the PostgreSQL databases of four programs, three of which were open source. We automated the work by using system catalog-based queries. We found serious design problems in all the databases. For instance, only one of the databases had check constraints and even there many checks were probably missing. In all the databases many uniqueness, foreign key, and not null constraints were missing. We explain different SQL database design problems and how to detect the occurrences of these by using system catalog-based queries. The queries are available through a webpage (https://github.com/erki77/database-design-queries).

Keywords: SQL · Database design · Integrity constraint · PostgreSQL · Design smell · Automated quality assessment

1 Introduction

It is possible to define different types of integrity constraints (primary key, unique, not null, foreign key, check) in SQL databases according to the SQL standard. In addition, determining that a column has a certain data type and possibly a maximum field size is also an act of specifying a constraint (an attribute constraint) [1]. Declaring constraints means that the database management system (DBMS) checks propositions against the constraints at write time (schema-on-write) and rejects those that do not conform to the constraints. Integrity constraints cannot ensure that the recorded propositions are correct. However, the constraints ensure that at least some certainly incorrect propositions are immediately rejected by the system. Thus, later, one has to spend less resources for trying to correct mistakes. It could be that the information sources based on which to correct the data are later not available. The constraints give information and assurances about data to the DBMS and thus help it to optimize the execution of data manipulation statements. The constraints can be used by external programs that, for instance, generate user interfaces of database applications, generate test data, or reverse engineer the database structure into a conceptual data model. The constraints help developers to understand the nature

R. Silhavy (Ed.): CSOC 2021, LNNS 230, pp. 375–390, 2021.
https://doi.org/10.1007/978-3-030-77442-4_33

of data and database. In this way the constraints are a kind of executable specification. Kapfhammer et al. [2] note that mistakes made in database schema definition, including in the definition of integrity constraints, propagate to the database applications and are later costly to fix.

Enforcing constraints in the database ensures that these cannot be bypassed in any way. Enforcing the constraints in database applications does not ensure this. Without considering that the same data element can be used concurrently by different users and by overlooking DBMS-specific properties of concurrency control, the applications could permit registration of data that violates the constraints. For instance, Bailis et al. [3] demonstrate, based on applications written in Ruby on Rails, how constraint (uniqueness, referential integrity, and built-in or custom validations) checks in applications fail to stop violations in case of concurrent (parallel) deletions or insertions of propositions. No violations were observed in case of constraints that were declared in the database. The lack of the constraints is a part of database technical debt that makes it gradually more difficult to use, understand, manage, and evolve the database.

The goal of the paper is to investigate how well the integrity constraints are implemented in existing SQL databases. We analyzed the databases of four programs [4–7] by using our developed software for detecting occurrences of database design problems in PostgreSQL databases [8]. The software queries the system catalog part of database to find information about the database schema. We have cataloged, implemented, and published a large number of checks [8]. For this paper, we applied a subset of these for automating the analysis of database schemas. We explain the identified problems in designs and how we detected the possible occurrences of these.

The rest of the paper is organized as follows. Section 2 introduces some related work. Section 3 explains the analysis method. Section 4 presents the results of the analysis. Its subsections are dedicated to different types of constraints. In Sect. 5 we give some additional thoughts. Finally, we conclude and point to further work.

2 Related Works

Balogh et al. [9] note that the structure analysis of large SQL databases it is a complicated undertaking and requires supporting software. Their industrial partner has started development of a static database analysis software. It captures knowledge about database design by using rules. Balogh et al. [9] mention five rules about integrity constraints, four of which have been determined as important or highly important by developers and managers. The catalog of queries that we used for our research [8] also captures database design knowledge in the form of rules and the use of the queries is a variant of static analysis. An alternative method to find missing or incomplete integrity constraints is to generate test data that should be rejected by the DBMS [2].

Blaha [10] reverse engineered 35 SQL databases and found the lack of declared primary keys and especially foreign keys in these. Blaha and Premerlani [11] present a catalog of SQL database design idiosyncrasies based on their experience of reverse engineering real world databases. The catalog includes a number of problems with the integrity constraints. Karwin [12] has SQL database design antitpatterns about the lack of declared foreign keys, a naming convention of the primary key columns, and listing permitted column values in a check constraint. Weber et al. [13] write about technical debt in SQL databases with respect to undeclared foreign keys. They point out the relative absence of foreign key declarations in OSCAR medical record system. Al-Barak and Bahsoon [14] present a taxonomy of SQL database design debt that includes the lack of declared foreign key constraints. Factor [15] presents a catalog of 150 SQL code smells that include four smells about integrity constraints. Delplanque et al. [16] present a prototype tool for analyzing SQL data definition statements to find the occurrences of eleven design problems, three of which are about integrity constraints. Sharma et al. [17] present a catalog of 13 SQL database design smells, which includes missing foreign key constraints, unnecessary surrogate keys, and defining in the database the list of values that are permitted in a column (possibly by using a check constraint). The lack of foreign key declarations is among those where the biggest number of respondents to developers' survey (52 complete answers) always consider it as a smell. Still some consider it to be a universal or context-dependent recommended practice. Defining permitted values is among those where the smallest number of developers always consider it as a smell. Both [16, 17] evaluate open-source programs in terms of smells as a part of their research. Definition of the list of permitted values is in open-source projects the best indicator that there are also other design problems in the database design [17]. Dintyala et al. [18] introduce a toolchain for detecting and fixing antipatterns (26 in total) in SQL database applications. Five of these are about integrity constraints. They found that four of these were common in various databases, including open-source systems in GitHub. Piattini et al. [19] propose a number of metrics for evaluating the maintainability of SQL databases. Maintainability depends on the complexity of schema and some metrics consider the declared foreign key constraints in the database.

Foidl et al. [20] propose a model of technical debt in data intensive systems. They point out that the debt in one system component may increase the debt in another. For instance, the absence of foreign key constraints can decrease the data quality and thus increase the complexity and decrease the maintainability of the entire system.

An observation about the related work is that the missing constraints usually mean missing foreign key constraints and only [18] mentions missing check constraints.

3 The Method

We utilized our catalog of system catalog-based SQL SELECT statements, which are meant for identifying the possible occurrences of database design problems in PostgreSQL databases (https://github.com/erki77/database-design-queries) [8]. The queries automate searching and solving the problems and are thus especially useful for investigating databases with a large number of database objects or a large number of databases. We use PostgreSQL because it is open source, popular, and conforms well to the SQL standard, i.e., offers a lot of design options for developers. Although many of the queries are specific to the system catalog structure of PostgreSQL, most of the problems that these queries codify can appear in any SQL database. Moreover, a number of queries are solely based on the views of the standardized *information_schema* schema.

Our present investigation involved queries from the categories about missing data, uniqueness, relationships between tables, and check constraints. Due to the space restrictions, we did not consider problems related to data types and field sizes. At the time of conducting the research the categories contained 240 queries in total, including 178 problem detection queries, which results directly point to possible problem occurrences. We have collected the queries that returned results that are presented in this paper to the collection "Find problems about integrity constraints". It consists of 93 queries, 69 of which are problem detection queries. Thus, we checked the databases against a bigger number of problems and presented only these that appeared at least once.

We investigated the database of four programs that use a PostgreSQL database and are from different domains. We analyzed the database structure but we did not have documentation about the meaning of the data. We did not consider enforcement of integrity constraints by the applications. All the programs were started a long time ago but were still actively developed and used at the time of the investigation, meaning that there had been enough time to accumulate problems as well as fix these.

FusionForge is an open-source project-management and collaboration software that can aid development teams during the entire development cycle. The first commit in GitHub was done in August 13, 2001 and we used the version that was committed in May 12, 2019 [4]. LedgerSMB is an open-source accounting and enterprise resource planning (ERP) software that is aimed at small and mid-size businesses. The first commit in GitHub was done in September 1, 2006 and we used the version that was committed in November 24, 2020 [5]. OTRS Community Edition is an open-source software for managing tickets and facilitating communication. The first commit in GitHub was done in December 1, 2001 and we used the version that was committed in October 24, 2019 [6]. Stansoft is a Linux financial accounting software [7]. We used the demo database of the version 7.20.

For the analysis purposes, we created the databases on a server with the latest PostgreSQL version (13) at the time. In general, we executed the SQL scripts as-is, i.e., we ignored the errors. Thus, if some database object was not created due to an error, we ignored it. However, in case of FusionForge and Stansoft, we found that in case of some base tables the creation script specified WITH OIDS property, which caused an error. We removed the property specification from the scripts because version 13 of PostgreSQL did not support it anymore. In case of FusionForge some foreign keys were implemented by declaring foreign key constraints whereas others were implemented with the help of constraint triggers that invoked certain internal functions that are used to enforce referential integrity. Moreover, some constraint triggers of the same table had the same name (in particular, sometimes the name was "<unnamed>"). Therefore, some of the triggers were not created because the trigger name must be unique in case of the same table. In case of counting the existing foreign key constraints and searching the missing ones, we only considered declaratively specified constraints.

The driving business need is to have databases where the set of declaratively enforced integrity constraints is complete, i.e., all the relevant constraints for the database domain are defined. In addition, all the defined constraints should be correct and relevant in terms of the database domain. The work was inspired by the research of integrity constraints in existing databases that at one hand point to the design problems but at the same time are old [10, 11] or cover only one type of constraints like, for instance, foreign key constraints [13, 14]. A goal of the work was to analyze a set of databases in terms of the state of declaratively enforcing integrity constraints by considering different types of constraints. We covered classes of problems that a constraint could be missing, redundant, have wrong properties, or has been defined in an inconsistent manner, i.e., the same thing is done differently in different places. Constraints and indexes belong to the conceptual and internal level of a database, respectively [1]. Because in some cases unique indexes were used to ensure uniqueness of rows in a table instead of declaring a constraint, we were forced to cover unique indexes as well in the research.

Another goal was to further tune the queries that helped us in this process in order to improve these and to demonstrate the feasibility of using the queries for detecting occurrences of database design problems. The method of investigation was action design research [21]. We continuously improved and extended the queries as a by-product of the process. For instance, problems with FusionForge led us to the creation of queries about enforcing referential integrity with constraint triggers that invoke internal functions. After improvements of the queries the databases were always re-checked.

The results can be used to improve the particular databases but more importantly it should raise the awareness of the extent of problems.

4 Analysis Results

Table 1 contains some general numeric values that characterize the databases. In the brackets is the percentage of columns covered by the constraints.

Table 1. The total number of some types of database elements and the percentage of columns covered by the constraints.

Software measure	FusionForge	LedgerSMB	OTRS	Stansoft
Number of base tables (tables)	206	162	116	174
Number of columns of the tables	1097	978	962	1931
Number of primary key constraints	144 (16.5%)	151 (24.1%)	82 (8.5%)	0 (0%)
Number of unique constraints	17 (2.5%)	64 (7.3%)	56 (7.3%)	0 (0%)
Number of foreign key constraints	172 (15.8%)	270 (27.6%)	264 (27.4%)	0 (0%)
Number of check constraints	0 (0%)	65 (7.2%)	0 (0%)	0 (0%)

In the following sections, we firstly present the proportion (percentage) of elements (tables, columns, or constraints) with certain properties from all the elements of this type (see Table 1) and after that the total number of elements with the properties.

In case of a simplistic SQL database design each base table (table in short) has a simple primary key that is a surrogate key. In addition, few (if any) natural keys (which often involve more than one column) are enforced, there are few (if any) composite foreign keys (although these could reduce the need of join operations), the checks (if these are even used) involve mostly one column, and many columns are optional, i.e., permit NULLs. Table 2 shows some simplicity characteristics of the designs. We found information about the keys based on the constraints that were declared in the database.

The depth of referential degree of schema is defined as the longest referential path between tables in the database schema by Piattini et al. [19]. The designs of OTRS and Stansoft seem to be more simplistic than the other two based on these values. Next, we write about problems with different types of integrity constraints.

Table 2. Numeric characteristics about the simplicity of the database design.

Software measure	FusionForge	LedgerSMB	OTRS	Stansoft
% of tables with a simple (involves one column) natural primary key	7.8% (16)	21% (34)	1.7% (2)	(0)
% of composite primary keys	22.2% (32)	38.4% (58)	0% (0)	(0)
% of composite unique keys	41.2% (7)	10.9% (7)	16.1% (9)	(0)
% of composite foreign keys (FK)	0.6% (1)	1.9% (5)	0% (0)	(0)
% of FK constraints that implement an identifying relationship	29.1% (50)	38.5% (104)	0% (0)	(0)
Depth of referential degree of a schema	6	8	6	0
% of checks that involve more than one column	(0)	18.5% (12)	(0)	(0)
% of optional columns	33.8% (371)	43.9% (429)	20% (192)	96.8% (1869)
% of tables where all the non-primary key columns are optional	24.8% (51)	25.3% (41)	0.9% (1)	78.2% (136)
% of optional foreign key columns from all the foreign key columns	24.3% (42)	34.8% (94)	5.7% (15)	(0)

4.1 Key Constraints

UQ1: A table does not have any primary key and unique constraints, i.e., the table does not have any declared keys. The biggest proportion of such tables was in Stansoft – 100% (174). It was followed by OTRS with 25% (29) and FusionForge with 24.3% (50). The smallest proportion of this kind of tables was in LedgerSMB – 1.9% (3).

UQ2: A table does not have any means for preventing duplicate rows. A DBMS may offer alternatives for declaring the primary key or a unique constraint. In PostgreSQL one could define an exclusion constraint, which is a generalization of unique constraint or create a unique index. In LedgerSMB was an exclusion constraint that implemented conditional uniqueness. Stansoft used unique indexes as the only database-level mean for ensuring uniqueness but not all the tables had these. Still, in Stansoft 32.8% (57) and in FusionForge 19.9% (41) of tables were without any uniqueness enforcement.

UQ3: Using unique indexes instead of uniqueness constraints. It is a sign of missing key constraints (see UQ14) and lowers the abstraction level because the creation of a uniqueness constraint in PostgreSQL would cause the automatic creation of a supporting index. In Stansoft 67.2% (117) and in FusionForge 4.4% (9) of tables had a unique index as the only mean for ensuring uniqueness of rows. In Stansoft 143, in FusionForge 14, and in LedgerSMB 2 unique indexes were created instead of a unique constraint. For the schema explorers, it obfuscates information about unique identifiers.

UQ4: Missing identity. If each unique constraint and index of a table covers at least one optional column, which permits NULLs, then there can be duplicate rows in the table and thus there could be rows without any unique identifier. In Stansoft 54.6% (95) and in FusionForge 1.9% (4) of tables were such. The proportion of tables where all the unique columns were optional was 51.1% (89) in Stansoft and 1.9% (4) in FusionForge.

UQ5: Duplication in case of ensuring uniqueness of rows. No table had more than one uniqueness constraints declared to the same set of columns. However, in LedgerSMB 1.9% (3) of tables had duplicate unique indexes that covered the same set of columns. In one table a unique index was defined to the columns of the primary key. In one table two indexes implemented the same conditional uniqueness rule and in one table three identical function-based unique indexes were defined.

UQ6: A table has a unique constraint but not the primary key. In FusionForge 5.8% (12), in LedgerSMB 4.9% (8), and in OTRS 4.3% (5) of tables were such. Firstly, it is a question of consistency because all databases except Stansoft had both primary key and unique constraints. A convention is to have the primary key in each table. Secondly, because unique constraints do not prevent NULLs in the covered columns it may lead to the missing identity (see UQ4) if NOT NULL constraints have been forgotten.

UQ7: A table has a surrogate key as the primary key and does not have any unique constraints. Surrogate key involves one column, which has an integer type and the column values are generated by the system (by using a sequence generator in case of PostgreSQL). In such tables substantive duplicates can still be registered because natural uniqueness raising from the database domain has not been identified and enforced. For instance, slow network connection and resulting multiple commands to register the same entity could lead to the registration of duplicates. The proportion of such tables was in FusionForge 39.8% (82), in OTRS 26.7% (31), and in LedgerSMB 25.3% (41).

UQ8: One size fits all. This is a SQL-database design antipattern [12] according to which tables have the surrogate primary key where the key column has name *id*. In OTRS 68.1% (79), in LedgerSMB 24.7% (40), and in FusionForge 8.3% (17) of tables followed the naming convention. Although the convention has problems like making join queries more complicated, the value of following a convention comes from its universal usage. In case of these databases the usage was not universal. A problem of following the convention is that it may lead to omitting other uniqueness constraints from the database (see UQ7). In FusionForge 100% (17), in LedgerSMB 82.5% (33), and in OTRS 39.2% (31) of the tables with "id" did not have any unique constraint.

UQ9: Unnecessary surrogate keys. In LedgerSMB we found that 11.1% (18) of tables had the primary key that is a natural key and the table also had a column that formed the surrogate key. It means that the column had an integer type, it had an associated sequence generator, and it had been declared unique. However, with the presence of the natural key the surrogate key column is superfluous.

UQ10: Definition of a non-minimal superkey instead of a candidate key. A superkey can contain columns that are not needed for ensuring uniqueness of rows. One should declare primary key and unique constraints based on candidate keys, i.e., minimal superkeys where any proper subset of columns does not have the uniqueness property. Thus, constraints that enforce non-minimal superkeys are superfluous. Firstly, in LedgerSMB 1.2% (2) and in FusionForge 1% (2) of tables had a primary key or a unique constraint that columns were a proper subset of columns of another primary key or unique constraint. Secondly, in Stansoft 4.6% (8 – in total 12 pairs of indexes) of tables had a non-conditional unique index that proper subset of columns had another non-conditional unique index and at least one of the indexes in the pair was not automatically created to support a constraint. Thirdly, in FusionForge 1% (2) and LedgerSMB 0.6% (1) of tables had a column with associated sequence generator that was a proper subset of some declared key. The column apparently contains unique values. This query did not reveal any new non-minimal superkeys. In all the cases the sets of columns that constitute the proper subsets are the keys that should be declared in the database.

Fourthly, in LedgerSMB 6.2% (10) of tables had the primary key where a proper subset of columns could contain only one value due to the check constraint on a column. The tables were created by using table inheritance and the check constraints were used to restrict values in the subtables. In this case the primary key should cover only columns without such constraints.

UQ11: Case sensitive and case insensitive uniqueness of the same simple key. In PostgreSQL one can implement case insensitive uniqueness check by using an exclusion constraint or a function-based unique index where the function is either *Lower* or *Upper*. In LedgerSMB 0.6% (1) of tables had a column that was covered by the primary key, which ensures case insensitive uniqueness as well as a unique function-based index. In this case the unique index that supports the primary key is supefluous. In this example a better approach is to set another key of the table as the primary key.

UQ12: Candidate key columns that probably have a static default value. Key values must be unique in case of each row. In FusionForge 11.8% (19) of uniqueness constraints consisted of only columns that had a static default value (an integer or string literal). In FusionForge 3.6% (40) and in LedgerSMB 0.1% (1) of columns participated in a uniqueness constraint and had a static default value. Such default value could be a mistake but it could also indicate that these columns should not constitute the key.

UQ13: A uniqueness constraint is not redefined in a subtable. An object-relational feature of PostgreSQL allows defining new tables based on existing tables by using inheritance. In PostgreSQL (13) the subtables do not inherit unique, exclusion, primary key, and foreign key constraints. Thus, these constraints should be redefined. In LedgerSMB the primary key constraint of a table was not redefined in one of its subtables.

UQ14: Missing key constraints (based on foreign keys, sequence generators, or column names). There can be tables that implement many-to-many relationship types but do not have a unique or the primary key constraint over the set of foreign key columns. It means that duplicate relationships are not prevented. In LedgerSMB 2.5% (4), in OTRS 1.7% (2), and in FusionForge 1% (2) of tables had such property.

A sign of a missing constraint is that a column has an associated sequence generator but it does not belong to any uniqueness constraint. In Stansoft 20.1% (35) of tables had such a column. The proportion of such tables was 2.9% (6) in FusionForge and 1.2% (2) in LedgerSMB. In the databases each such table had one such column.

One could consider a convention to name primary key columns *id* or *code* and search such columns that do not belong to any primary key or unique constraint. In FusionForge 3.9% (8), in Stansoft 1.7% (3), and in LedgerSMB 1.2% (2) of tables had such a column. In the databases each such table had one such column.

Usernames in tables of user accounts must be unique. E-mail address can be used as the username. One can search textual columns that name refers to username and that are not covered by a primary key, unique, or exclusion constraint or a unique index that implements case insensitive uniqueness. In addition, either the table name should point to user accounts or the table must contain a column that name refers to passwords. In Stansoft 2.3% (4 – one such column in each), in OTRS 0.9% (1), and in FusionForge 0.5% (one table with two such columns) of tables had such columns.

4.2 Foreign Key (Referential) Constraints

FK1: Missing foreign key constraints. We searched tables that do not participate in any foreign key constraint – neither as the referencing table nor as the referenced table. In Stansoft 100% (174), in FusionForge 31.6% (65), in OTRS 15.5% (18), and in LedgerSMB 4.9% (8) tables were such. However, sometimes such "islands" may be legitimate (tables for storing parameter values or intermediate results). We excluded the tables that have subtables because these supertables could be used as abstract blueprints for defining other tables and in PostgreSQL foreign key constraints are not inherited.

Another way of finding missing foreign key constraints is to search columns that based on their names (the name has the prefix or suffix *code* or *id*) and non-coverage by uniqueness constraints could be foreign key columns but the columns are not covered by any foreign key constraint. In FusionForge 7.9% (87), in Stansoft 4.4% (84), in OTRS 3.7% (36), and in LedgerSMB 3% (29) of columns were such.

In LedgerSMB we found two non-key and non-foreign key columns that had a textual type with a small field size (i.e., the columns potentially contain codes) and there was also a table with a similar name (in this case with the name *currency*).

FK2: Triggers are used to enforce referential integrity. In FusionForge 18.4% of tables (38) had constraint triggers that invoked system-defined functions for ensuring referential integrity. There were also declared foreign key constraints in the database (see Table 1). Thus, in addition to unnecessary complexity it also means inconsistency.

FK 3: A foreign key column is associated with a sequence generator. Foreign key values must be selected amongst the values of the referenced candidate key (or the value can be missing, i.e., NULL, in case of SQL). The foreign key value should not be generated when a new row is added to the child table. However, in LedgerSMB three foreign key columns (from three different tables) had an associated sequence generator.

FK4: Duplicate foreign key constraints. In LedgerSMB two tables had two constraints that covered the same column and referenced to the same table and its key. The constraint definition statements did not specify the constraint name and were duplicated in the script. Thus, the execution of the script caused the duplication. A name that follows a convention would have caused a name conflict and prevented the duplication.

FK5: A foreign key does not reference the primary key. Delplanque et al. [16] (rule 2) call it a code smell because uniqueness of the reference is not guaranteed. It is the case only if the foreign key references to a unique constraint where at least one column is optional, i.e., permits NULLs. Still, the standard practice is to reference to the primary key columns. Although the databases mainly followed the practice in LedgerSMB 34.8% (94) and in FusionForge 5.2% (9) of foreign key constraints referenced to the columns of a unique constraint. Sometimes there was no primary key in the referenced table. However, in LedgerSMB 32.6% (88) of foreign keys referenced to the unique constraint columns while there was also the primary key in the referenced table. This is a question of design consistency. In LedgerSMB 1.5% (4) and in FusionForge 0.6% (1) foreign key constraint referenced a set of columns that was covered with a unique index but not with a unique constraint. In addition to inconsistent design, while permitted in PostgreSQL it may not be permitted in other DBMSs and thus it reduces portability.

FK6: A referenced column is optional, i.e., permits NULLs. This may happen if a foreign key does not reference the primary key columns (see FK5). In FusionForge 0.6% (1) and in LedgerSMB 0.4% (1) of foreign key constraints where such.

FK7: Parent table's key column and child table's foreign key column have a different type or field size. In OTRS 18.6% (49) and in LedgerSMB 4.4% (12) of foreign key constraints were such. If the data type/field size of a foreign key column (for instance, INTEGER or VARCHAR(6)) permits a larger set of values than the data type/field size of the refenced key column (for instance, SMALLINT or VARCHAR(4)), then it is just an inconsistency. However, if the data type/field size of a foreign key column permits a smaller set of values than the data type/field size of the refenced key column, then certain key values cannot be used as foreign key values. It is a serious problem. In OTRS 18.6% (49) and in LedgerSMB 1.1% (3) of foreign key constraints were such.

FK8: A foreign constraint is not redefined in a subtable (see UQ13). In LedgerSMB 4 foreign key constraints of supertables were not enforced in their subtables (in total 18 constraints were missing). The number of not redefined foreign key constraints was even larger (7 – in total 36 missing constraints). However, to mitigate the problem, 18 of the subtable foreign key columns had a check constraint that permitted only one value in the column. In case of one foreign key constraint such check constraint was in five subtables and was missing in one subtable, i.e., it is also an inconsistency.

In case of foreign key constraints one can specify compensating actions (referential actions) that the DBMS undertakes in case of certain violations of the referential integrity rule. The default behavior is NO ACTION, i.e., no compensation takes place. Different actions are suitable in case of different types of relationships.

According to Table 3 in FusionForge and LedgerSMB a variety of different compensating actions were used. In OTRS all the foreign key constraints had the same compensating action properties (NO ACTION) and in Stansoft no foreign keys were declared. Next, we list the problems we found in case of specified compensating actions.

FK9: Cascading update is not needed (based on surrogate keys). If the candidate key that is referenced by a foreign key has an associated sequence generator, i.e., is a surrogate key, then cascading the key value changes to the child table is not needed because surrogate key values will not be changed over time, i.e., the key is immutable. In FusionForge all the foreign key constraints with cascading update had the problem.

Table 3. The percentage of specified foreign key compensating actions.

Compensating action	FusionForge	LedgerSMB	OTRS	Stansoft
ON DELETE CASCADE	75% (129)	8.5% (23)	0% (0)	(0)
ON DELETE SET NULL	0% (0)	1.1% (3)	0% (0)	(0)
ON DELETE SET DEFAULT	1.2% (2)	0.4% (1)	0% (0)	(0)
ON DELETE NO ACTION	23.8% (41)	90% (243)	100% (264)	(0)
ON UPDATE CASCADE	53.5% (92)	0% (0)	0% (0)	(0)
ON UPDATE NO ACTION	46.5% (80)	100% (270)	100% (264)	(0)

FK10: Inconsistency of compensating actions. In FusionForge 27.3% (47) of foreign key constraints had ON UPDATE NO ACTION property while there was at least one foreign key constraint that pointed to the same candidate key of the referenced table and had cascading update property. In this particular case it was a sign that the constraints with the cascading update property should instead have no action property (see FK9). In general, there should not be such inconsistencies.

FK11: Cascading update is probably missing (based on column names, types, or sequence generators). A foreign key constraint should permit cascading updates if it references a natural key, which values can be changed over time. There are different signs that indicate that a foreign key probably references a natural key.

In the following cases the foreign key constraints *did not have* the cascading update property. In LedgerSMB 2.2% (6) and in FusionForge 1.3% (1) of such foreign key constraints covered a column that name had the prefix or suffix *code* (instead of the prefix or suffix *id*). In LedgerSMB 7.8% (21) and in FusionForge 1.3% (1) of such foreign key constraints covered a column that did not have an integer type. In LedgerSMB 4.8% (13) and in FusionForge 3.8% (3) of such foreign key constraints referenced a simple key where the column had an integer type while the column did not have an associated sequence generator and was not covered by a foreign key constraint.

Using the combination of strategies is more efficient than using just one because in LedgerSMB we found in total 34 and in FusionForge 5 different foreign key constraints.

FK12: Cascading delete is probably not needed (based on the relationship type). In FusionForge 32% (55) and in LedgerSMB 3% (8) of foreign key constraints implemented a non-identifying relationship type (foreign key columns are not covered by the primary key) but had the cascading delete property.

FK13: Cascading delete is probably missing (based on the multiplicity of the relationship type). If a foreign key covers the same columns as some key of the table, then conceptually the foreign key implements one-to-one relationship type. It is an identifying relationship type and it could have been created based on a generalization in the conceptual data model. In LedgerSMB 6.3% (17) and in OTRS 0.4% (1) of foreign key constraints implemented such relationship type without specifying the cascading delete.

4.3 Check Constraints

CHK1: Defining a list of permitted values for a column. Factor [15] describes the problem as "Using constraints to restrict values in a column". The name is misleading because all integrity constraints restrict values in columns. In LedgerSMB 0.8% (8) of columns had the list of permitted values specified by using a check constraint. Moreover, FusionForge used an enumerated type to define the list of values for a column.

CHK2: A check constraint covering multiple columns of the same table is missing. Each column with the date or a timestamp type corresponds to an event type. In case of multiple event types of the same entity type there is often a certain order in which these events can happen. For instance, a record is firstly registered and only after that can be updated. In this case there should be check constraints that enforce the order. LedgerSMB had one such constraint in one table but even there it did not cover all the relevant columns. In OTRS 64.7% (75), in Stansoft 20.1% (35), in LedgerSMB 13% (21), and in FusionForge 0.5% (1) of tables had relevant columns without such constraints.

Multiple Boolean columns may be used to record the states of an entity type. There could be dependencies between the states and hence the columns. For instance, an order that is rejected cannot be delivered. There should be check constraints that enforce the dependencies. LedgerSMB had two tables with one such constraint but in one of these not all the Boolean columns were involved. In LedgerSMB 10.5% (17) of tables had relevant columns without such constraints. The column names indicate that there may indeed be dependencies between the columns.

In LedgerSMB 3.1% (5), in OTRS 1.7% (2), and in FusionForge 0.5% (1) of tables implemented adjacency list design for recording hierarchical data. The tables did not have a constraint that prevents an entity to be the parent of itself.

CHK3: A check constraint covering one column is missing. Based on column names and data types we found that in FusionForge 0.5% (5), in Stansoft 0.3% (6), and in OTRS 0.1% (1) of columns contain e-mail addresses without a constraint that prevents the addresses without the @ sign. Based on column names and data types we found that in FusionForge 0.3% (3), in LedgerSMB 0.1% (1), and in OTRS 0.1% (1) of columns contain URLs without a constraint that prevents URLs that do not follow the syntax of URLs – for instance, do not start with a specific scheme. Based on column names and data types we found that in Stansoft 3.1% (60) and in LedgerSMB 1.9% (19) of columns contain data about price or quantity without a constraint that prevents negative values.

Based on column data types we found that in Stansoft 40.1% (775), in OTRS 36% (346), in LedgerSMB 24.8% (243), and in FusionForge 23.1% (253) of columns contain textual values without a constraint that prevents the empty string and strings that consist of only whitespace. Based on the names and types of columns we found non-foreign key columns that were not meant to record large textual values but had TEXT or VARCHAR type without restrictions to the field size (field size in case of VARCHAR or a check constraint). In LedgerSMB the proportion of such columns from all the columns was 17.9% (175), in FusionForge 12.5% (137), and in OTRS 3% (29).

Based on column data types we found that in OTRS 18.4% (177), in LedgerSMB 9.5% (93), in Stansoft 7.3% (141), and in FusionForge 0.3% (3) of columns contain timestamp or date values without a constraint that prevents values that are too far back in history or too far away in the future.

CHK4: Inconsistent check constraints. In LedgerSMB two tables (*Person* and *Robot*) had columns *first_name* and *last_name* but with check constraints that had slightly different regular expression patterns. Another inconsistency is that both tables had column *middle_name* without any check constraint. Discovery of such inconsistencies helps analysts to formulate questions like whether *Robot* has such attributes in the first place and whether the names of the attributes correctly describe the meaning of data.

CHK5: Duplicate check constraints. In LedgerSMB a table had two identical check constraints. Interestingly the same table had two identical unique indexes. Like in case of FK4 the constraint definitions did not specify the constraint name.

4.4 NOT NULL Constraints

NN1: All the columns permit NULLs. In Stansoft 78.2% (136), in FusionForge 14.1% (29), and in LedgerSMB 0.6% (1) of tables permit rows that consist of only NULLs.

NN2: The empty string instead of NULL. A value could be missing from a database for different reasons. To encode the reasons, one may try to use different special values or markers. For instance, in case of textual columns one may use empty strings and NULLs. On the other hand, using these inconsistently may lead to errors. Thus, it may be better to represent missing textual values only with the help of NULL. In FusionForge 50.9% (59) and in LedgerSMB 0.8% (1) of mandatory textual columns had the empty string as the default value. Perhaps the columns should be optional and the default value should be dropped. Many mandatory textual columns in these databases did not have such default value and this is yet another example of inconsistency.

NN3: Surname and given name of a person are both required. In OTRS 1.7% (2) and in LedgerSMB 0.6% (1) of tables had a pair of columns for the surname and given name where both columns were mandatory. This does not consider mononymous persons who have only one name component. A better approach is to make both columns optional and define a check constraint that at least one of the values must be registered.

NN4: Specific columns do not have the NOT NULL constraint. In LedgerSMB 6.5% (64), in FusionForge 5.9% (65), and in Stansoft 0.1% (1) of columns were optional and had a default value that was not the empty string.

Secondly, one can search optional columns that participate in a unique constraint or a unique index. There is no need to unnecessarily distinguish the primary and alternate keys of a table. As the primary key columns are automatically mandatory the same should apply to the alternate key columns. In Stansoft 12.2% (235), in LedgerSMB 1.1% (11), in FusionForge 1% (11), and in OTRS 0.3% (3) of columns were optional and also belonged to a uniqueness constraint or a unique index. Most of the columns were non-foreign key columns. If we excluded foreign key columns, then the number of such columns was reduced to 6 in LedgerSMB and 1 in OTRS. In Stansoft there are no primary key constraints and uniqueness is ensured by using unique indexes. Thus, there can be rows without the value of the key (see NN1). This mistake was also mentioned by [11] (null primary key attributes).

Thirdly, in LedgerSMB 62,2% (46) and in FusionForge 50% (1) of Boolean columns were optional. Optional Boolean columns imply the use of three-valued logic (TRUE, FALSE, UNKNOWN – represented by NULL) instead of two-valued logic. In addition to unnecessary complexity it also means inconsistency because almost the same number of Boolean columns were mandatory. Moreover, in FusionForge 2.3% (25) and in OTRS 1.6% (15) of columns had a non-Boolean type but the column name started with "is_" or "has_", indicating that the column contains representations of truth values. In FusionForge 20% (5) of these columns were optional. There is also an inconsistency in data type usage because some columns in the database have the Boolean type.

5 Final Thoughts

We listed 36 problems with declarative integrity constraints. The number of checked problems was even higher but we only reported those that had at least one occurrence. It is by far the largest number of problems compared with the resources presented in Sect. 2. Together these resources report 12 different problems that are also reported in this paper. In these resources there is no discussion about compensating actions and very little is said about check constraints. The most often reported problems in the resources are missing foreign key constraints (in 11 resources) (see FK1), specifying a list of values for a column (6 resources) (CHK1), missing primary keys (5 resources) (UQ1, UQ6), and permitting NULLs (4 resources) (NN4). In case of missing primary keys, the resources do not differentiate between different situations (the primary key is missing but there is a unique constraint or index or there is no uniqueness at all) whereas our catalog differentiates these reasons. In case of missing constraints or compensating actions we pointed towards different signs that help us to identify these.

Improving the database design is an iterative and incremental process. Declared constraints give information about the meaning of data. The lack of these means the lack of an important source of information for finding design problems. Declaring a constraint may reveal the lack of other constraints. For instance, if a table has no foreign key constraints, then a used query cannot find the missing uniqueness constraints in tables that implement many-to-many relationship types. However, a design problem could be discovered in multiple ways. For instance, if a table with no foreign key constraints does not have any uniqueness constraints as well, then the latter could be discovered before the foreign keys are declared. Fixing a problem may cause new problems. For instance, a new declared key may cover exactly the same columns as a unique index that does not support a declared key, is not partial, and is not function-based. In this case the index becomes redundant.

It is probably easier to implement the search based on system catalog compared to the processing of source code. A downside is that many queries are specific to a DBMS. In PostgreSQL the names of foreign key and check constraints are not necessarily unique within a schema. Thus, for the better precision, we often made queries based on the system catalog tables instead of standardized views. Execution of all the queries from the collection "Find problems about integrity constraints" took about 25 s per database.

A threat to validity is that the research involves a small number of databases. Another threat is that we consider only databases in one DBMS. The technical circumstances in different DBMSs influence the database implementation. These threats are partly mitigated by testing mature databases and using a feature-rich and mature DBMS. Most of the investigated databases are from open-source programs whereas development practices of proprietary software might produce different results in terms of quality.

6 Conclusion

We investigated the use of declarative integrity constraints in the PostgreSQL database of four programs (FusionForge, LedgerSMB, OTRS Community Edition, and Stansoft), three of which were open source. These programs have a long development history and are still actively used and developed. We used system catalog-based queries for identifying the occurrences of design problems (collection "Find problems about integrity constraints" – https://github.com/erki77/database-design-queries). We found that the databases lacked many integrity constraints, had some too restrictive constraints, and had many inconsistencies in case of constraints. For instance, only one of the databases had check constraints and even there many were probably missing. All the databases had a number of tables without any primary key and unique constraints and all the databases had a number of tables that did not participate in any foreign key (referential) constraint – not as the referencing nor as the referenced table. We assumed based on the names and data types of columns that a large number of foreign key and check constraints were missing. One database was almost completely without declarative integrity constraints except not null constraint on 62 columns (3.2% of all the columns). Even if the missing constraints are correctly implemented in the application, which we did not investigate, then the lack of constraints prevents both human users and various programs (including the DBMS) from getting crucial information about the meaning of the data. Moreover, if data is modified by bypassing the application, then none of its potential integrity checks would be applied. We can say that these databases have a lot of technical debt. We are not aware of any other research that presents so many different types of problems with integrity constraints in SQL databases, including problems with check constraints and compensating actions of foreign key constraints.

These results show the need to better educate developers about databases, continuously refactor databases, and the need for a software that makes it easy to discover the occurrences of database design problems right at the time of creating database objects.

The future work could analyze additional databases in terms of constraints as well as analyze databases in terms of other classes of problems. It would also be interesting to find how the use of constraints has evolved over time in the databases.

References

1. Date, C.J.: An Introduction to Database Systems, 8th edn. Pearson, Addison Wesley, Boston (2003)
2. Kapfhammer, G.M., McMinn, P., Wright, C.J.: Search-based testing of relational schema integrity constraints across multiple database management systems. In: Sixth International Conference on Software Testing, Verification and Validation, pp. 31–40. IEEE (2013). https://doi.org/10.1109/ICST.2013.47
3. Bailis, P., Fekete, A., Franklin, M.J., Ghodsi, A., Hellerstein, J.M., Stoica, I.: Feral concurrency control: an empirical investigation of modern application integrity. In: 2015 ACM SIGMOD International Conference on Management of Data, pp. 1327–1342 (2015). https://doi.org/10.1145/2723372.2737784
4. FusionForge. https://github.com/fusionforge/fusionforge. Accessed 26 Oct 2020
5. LedgerSMB. https://github.com/ledgersmb/LedgerSMB. Accessed 26 Nov 2020

6. OTRS/otrs. https://github.com/OTRS/otrs/tree/rel-6_0. Accessed 04 Nov 2020
7. Stansoft. http://www.stansoft.org/download.html. Accessed 21 Nov 2020
8. Eessaar, E.: Automating detection of occurrences of PostgreSQL database design problems. In: Robal, T., Haav, H.-M., Penjam, J., Matulevičius, R. (eds.) DB&IS 2020. CCIS, vol. 1243, pp. 176–189. Springer, Cham (2020). https://doi.org/10.1007/978-3-030-57672-1_14
9. Balogh, G., Gergely, T., Beszédes, Á., Szarka, A., Fábián, Z.: Capturing expert knowledge to guide data flow and structure analysis of large corporate databases. Acta Polytechnica Hungarica **16**(4), 7–26 (2019)
10. Blaha, M.: A retrospective on industrial database reverse engineering projects - part 2. In: Eighth Working Conference on Reverse Engineering, pp. 147–153. IEEE (2001). https://doi.org/10.1109/WCRE.2001.957818
11. Blaha, M.R., Premerlani, W.J.: Observed idiosyncracies of relational database designs. In: 2nd Working Conference on Reverse Engineering, pp. 116–125. IEEE (1995). https://doi.org/10.1109/WCRE.1995.514700
12. Karwin, B.: SQL Antipatterns. Avoiding the Pitfalls of Database Programming. The Pragmatic Bookshelf (2010)
13. Weber, J.H., Cleve, A., Meurice, L., Ruiz, F.J.B.: Managing technical debt in database schemas of critical software. In: Sixth International Workshop on Managing Technical Debt, pp. 43–46. IEEE (2014). https://doi.org/10.1109/MTD.2014.17
14. Al-Barak, M., Bahsoon, R.: Database design debts through examining schema evolution. In: 8th International Workshop on Managing Technical Debt, pp. 17–23. IEEE (2016). https://doi.org/10.1109/MTD.2016.9
15. Factor, P.: SQL Code Smells. Redgate. http://assets.red-gate.com/community/books/sql-code-smells.pdf. Accessed 29 Dec 2019
16. Delplanque, J., Etien, A., Auverlot, O., Mens, T., Anquetil, N., Ducasse, S.: CodeCritics applied to database schema: challenges and first results. In: 24th International Conference on Software Analysis, Evolution and Reengineering, pp. 432–436. IEEE (2017). https://doi.org/10.1109/SANER.2017.7884648
17. Sharma, T., Fragkoulis, M., Rizou, S., Bruntink, M., Spinellis, D.: Smelly relations: measuring and understanding database schema quality. In: 40th International Conference on Software Engineering: Software Engineering in Practice, pp. 55–64. ACM (2018). https://doi.org/10.1145/3183519.3183529
18. Dintyala, P., Narechania, A., Arulraj, J.: SQLCheck: automated detection and diagnosis of SQL anti-patterns. In: 2020 ACM SIGMOD International Conference on Management of Data, pp. 2331–2345 (2020). https://doi.org/10.1145/3318464.3389754
19. Piattini, M., Calero, C., Sahraoui, H.A., Lounis, H.: Object-relational database metrics. L'Objet **7**(4), 477–496 (2001)
20. Foidl, H., Felderer, M., Biffl, S.: Technical debt in data-intensive software systems. In: 45th Euromicro Conference on Software Engineering and Advanced Applications, pp. 338–341. IEEE (2019). https://doi.org/10.1109/SEAA.2019.00058
21. Sein, M.K., Henfridsson, O., Purao, S., Rossi, M., Lindgren, R.: Action design research. MIS Quart. **35**, 37–56 (2011). https://doi.org/10.2307/23043488

Average Consensus with Perron Matrix for Alleviating Inaccurate Sensor Readings Caused by Gaussian Noise in Wireless Sensor Networks

Martin Kenyeres[1(✉)] and Jozef Kenyeres[2]

[1] Institute of Informatics, Slovak Academy of Sciences, Dubravska cesta 9, 845 07 Bratislava, Slovakia
martin.kenyeres@savba.sk

[2] EBCONT proconsult GmbH, Millennium Tower, Handelskai 94-96, 1200 Vienna, Austria
jozef.kenyeres@ebcont.com
http://www.ui.sav.sk/w/en/dep/mcdp/

Abstract. Gaussian noise is a factor that negatively affects almost any signal in real-world systems. In this paper, we consider wireless sensor networks where the sensor measured values differ from the value of the observed unknown parameter due to inaccurate sensor readings caused by Gaussian noise with the zero mean and a varied standard deviation. The goal of our research is to identify whether and to evaluate how efficiently the application of the average consensus algorithm with the Perron matrix bounded by a fully-distributed stopping criterion can alleviate the deviation of noisy measured values from the correct value of the observed parameter. We vary the parameter(s) of the average consensus algorithm, the implemented stopping criterion, and Gaussian noise in order to analyze whether and how the algorithm can compensate for inaccurate sensor readings as well as identifying the best-performing initial configuration of the mentioned algorithm and the implemented stopping criterion.

Keywords: Average consensus · Consensus algorithms · Data aggregation · Distributed computing · Gaussian noise · Information fusion · Sensor fusion · Wireless sensor networks

1 Introduction

Over recent decades, wireless sensor networks (WSNs) have been attracting the attention of both the academy and the industry due to their great potential in various areas such as agriculture, military, smart homes, habitat monitoring, etc. [1–5]. WSNs are typically formed by numerous independent small-size wireless nodes equipped with one or more microcontrollers, an RF transceiver, a

R. Silhavy (Ed.): CSOC 2021, LNNS 230, pp. 391–405, 2021.
https://doi.org/10.1007/978-3-030-77442-4_34

limited energy source, and various sensors such as infrared sensors, sonar sensors, ultrasonic sensors, etc. [6]. The sensor nodes are deployed usually over large-scale geographical areas in order to collect relevant information about the monitored physical quantity (e.g., temperature, pressure, light, etc.) [7]. However, the sensor nodes have to struggle with many negative environmental factors (e.g., radiation, pressure variations, etc.) distorting the sensor readings whereby the measured and the real value of the observed physical quantity may differ from one another [7,8]. Also, the measured data in WSNs can be highly correlated or duplicated, which has a negative impact on QoS as well [8]. Therefore, in many WSN-based applications, mechanisms for data aggregation are implemented in order to suppress these negative factors, thereby optimizing QoS [8]. As can be found in literature, processing data/information into an aggregated form becomes a crucial process not only in WSNs but also in other technical/non-technical industries [9–12]. As stated in [13], WSNs can execute application-layer algorithms in either a centralized (i.e., algorithms are performed by a fusion center) or a distributed manner (i.e., sensor nodes cooperatively solve a problem without any fusion center). Many modern WSN-based applications are typically built-up on distributed approaches primarily due to their significant energy efficiency [13]. Moreover, distributed approaches represent a more scalable and less vulnerable solution to potential threats than centralized approaches [13]. Also, in distributed schemes, a final estimate of the wanted aggregate function can be easily recovered in the case that some sensor nodes are seriously damaged or their energy source is depleted. See Fig. 1 for the general architecture of WSNs.

Fig. 1. General architecture of wireless sensor networks

In many modern WSN-based applications, sensor nodes are required to reach an agreement on a common value (e.g., the average of local measurements, some network parameters, etc.) in a distributed manner [14,15]. The theory of a consensus problem, a subcategory of computer science, has a rich history in various industries such as distributed computing, multi-agent systems, cooperative control, etc. [7,14]. A consensus means that autonomous sensor nodes achieve an agreement on the quantity of interest in spite of no global communication and no central coordination [7]. In literature, many applications of consensus-based algorithms to aggregating data not only in WSNs but also in other industries can be found [16,17]. See Fig. 2 for the architecture of the data aggregation in WSNs.

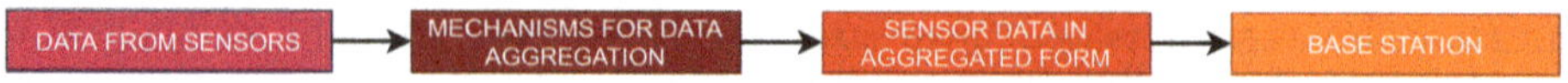

Fig. 2. Architecture of data aggregation over wireless sensor networks

In this paper, we focus our attention on the average consensus algorithm (referred to as AC) with the Perron matrix, which is a distributed algorithm based on neighbor-to-neighbor interactions [18]. Over the past years, AC has been comprehensively studied in various fields such as distributed agreement, synchronization, parallel computing, distributed coordination, distributed data aggregation, etc. [19]. We examine AC with the Perron matrix for distributed averaging, i.e., the algorithm estimates the arithmetic mean from all the initial inner states in a distributed fashion [7]. The adjacent sensor nodes iteratively exchange their current inner states, resulting in the asymptotic convergence of all the initial inner states to the arithmetic mean (or to its approximate value in most cases) [7]. However, the algorithm is multi-functional, meaning that it can estimate not only the arithmetic mean (as its name evokes) but also other aggregation functions (namely, the sum from all the initial values and the graph order) [7]. In Fig. 3, we show the evolution of the inner states over a graph of the line topology formed by five vertices - as seen, the inner state of each sensor node converges to the value of the arithmetic mean.

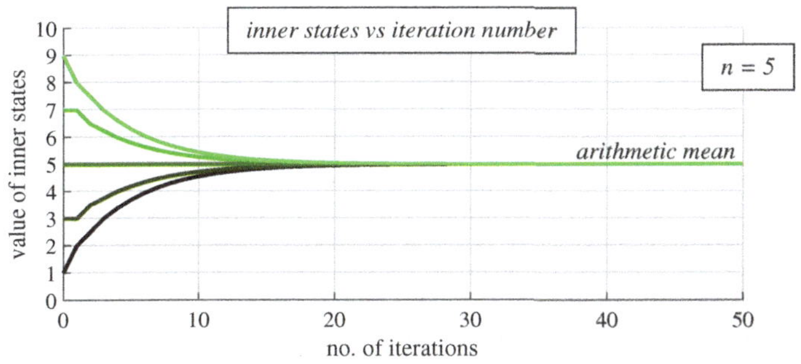

Fig. 3. AC over graph of line topology formed by five vertices - evolution of inner states

The goal of our research presented in this paper is to examine whether and how efficiently the application of AC with the Perron matrix can compensate for inaccurate sensor readings caused by Gaussian noise, i.e., the measured data differ from the value of the observed unknown parameter due to the noise, and AC with a stopping criterion proposed for WSNs is applied to alleviate this deviation. We compare the final estimates of the arithmetic mean under a varied standard deviation of Gaussian noise with the scenario when no data aggregation mechanism is applied. In our experiments, the parameters of the implemented stopping criterion are varied as well as the algorithm with three different initial configurations is examined.

The paper is organized as follows: Sect. 2 is concerned with modeling AC over WSNs, the analyzed AC weights, and the applied stopping criterion. In Sect. 3, we turn readers' attention to the applied research methodology and the used metrics for performance evaluation. Section 4 is focused on the results from numerical experiments and their analysis. The last section (Sect. 5) provides a brief summarization of the research presented in this paper.

2 Theoretical Background into Topic

2.1 Model of Average Consensus Algorithm over Wireless Sensor Networks

We model WSN as a simple finite undirected graph G determined by two sets, namely the vertex set $\mathbf{V}$ and the edge set $\mathbf{E}$ ($G = (\mathbf{V}, \mathbf{E})$) [20]. The vertex set $\mathbf{V}$ contains all the vertices present in a graph G, which represent the sensor nodes in WSNs ($\mathbf{V} = \{v_1, v_2, ..., v_n\}$). The cardinality of this set ($|\mathbf{V}|$) is equal to the graph order n, i.e., $|\mathbf{V}| = n$. The edge set $\mathbf{E} \subset \mathbf{V} \times \mathbf{V}$ consists of all the edges in a graph G, which are representatives of a one-hop connection between two vertices - e_{ij} links v_i and v_j. In AC, a sensor node (represented as v_i) is required to store its inner state $x_i(k)$[1] and to iteratively update it for the next iteration according to a local update rule, which can be globally expressed as the following difference equation [20,21]:

$$\mathbf{x}(k+1) = \mathbf{W} \times \mathbf{x}(k) \tag{1}$$

Here, $\mathbf{x}(k)$ is a vector gathering the inner states of all the sensor nodes at a kth-iteration[2], and $\mathbf{W}$ is the weight matrix of AC affecting numerous aspects of the algorithm such as the convergence rate, the robustness to potential threads, scalability, the difficulty of the real-world implementation, etc. [7]. In this paper, we apply the Perron matrix (i.e., all the edge weights take the same value equal to the mixing parameter) defined as follows [7]:

$$\mathbf{W} = \mathbf{I} - \epsilon.\mathbf{L}(G) \tag{2}$$

Here, $\mathbf{I}$ is the identity matrix, $\mathbf{L}(G)$ is the Laplacian matrix of a graph G, and ϵ is the mixing parameter. As mentioned earlier, we consider scenarios where sensor nodes measure the value of an unknown parameter, but the sensor readings are corrupted by Gaussian noise. Thus, the initial inner states are determined as follows:

$$x_i(0) = \theta + z_i, for\ i = 1,2,\ ...,n \tag{3}$$

Here, θ is the observed unknown parameter - a scalar value to be measured, and z represents IID Gaussian noise:

$$z_i \sim N(\mu, \sigma^2), for\ i = 1,2,\ ...,n \tag{4}$$

Here, $\mu \in \mathbf{R}$ is the mean, and $\sigma > 0$ is the standard deviation. In the case of the correct functioning of AC, the inner states asymptotically converge to the

---

[1] k is a non-negative integer labeling an iteration.

[2] $\mathbf{x}(0)$ contains all the initial inner states.

arithmetic mean calculated from the initial inner states of all the sensor nodes in a network, which can be expressed as follows [21]:

$$\lim_{k\to\infty} \mathbf{x}(k) = \lim_{k\to\infty} \mathbf{W}^k \times \mathbf{x}(0) = \frac{1}{n}.\mathbf{1} \times \mathbf{1}^{\mathrm{T}} \times \mathbf{x}(0) \tag{5}$$

Here, $\mathbf{1}$ is a vector whose all the entries are equal to one, and $\mathbf{1}^{\mathrm{T}}$ is its transpose. As stated in [21], the existence of this limit is necessary for the proper functioning of AC and is ensured, provided that these three convergence conditions are met:

$$\mathbf{1}^{\mathrm{T}} \times \mathbf{W} = \mathbf{1}^{\mathrm{T}}, \mathbf{W} \times \mathbf{1} = \mathbf{1}, \rho(\mathbf{W} - \frac{1}{n} \cdot \mathbf{1} \times \mathbf{1}^{\mathrm{T}}) < 1 \tag{6}$$

Here, ρ is the spectral radius of the underlying matrix and is defined as follows [21]:

$$\rho(\cdot) = \max_{i}\{| \lambda_i(\cdot) |\} \tag{7}$$

Here, λ_i is an ith-largest eigenvalue of the corresponding spectrum. In Fig. 4, we show the evolution of the inner states in the case that the sensor readings are corrupted by Gaussian noise - note that the steady-state of the inner states differs from the value of the estimated unknown parameter θ, which is a consequence of the noisy initial inner states.

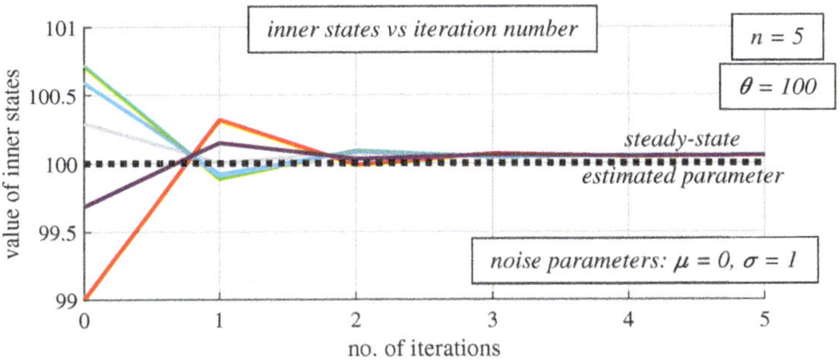

Fig. 4. AC over graph of line topology formed by five vertices - evolution of inner states - sensor readings are corrupted by Gaussian noise

2.2 Analyzed Average Consensus Weights

In this subsection, we introduce the AC weights chosen for evaluation. The first analyzed weights are the Graph Order weights (referred to as GO). In this case, the mixing parameter ϵ is equal to the graph order n, and their weight matrix is doubly-stochastic and defined as follows:

$$[W]_{ij}^{\mathrm{GO}} = \begin{cases} \frac{1}{n}, & \text{if } e_{ij} \in \mathbf{E} \\ 1 - d_i.\frac{1}{n}, & \text{if } i = j \\ 0, & \text{otherwise} \end{cases} \tag{8}$$

Here, d_i represents the degree of v_i. The next analyzed weights are the Maximum Degree weights (referred to as MD), where the mixing parameter ϵ takes the value equaled to the maximum degree of a graph (labeled as Δ). As stated in [22], these weights can be applied, provided that the underlying graph is not bipartite regular. Their weight matrix is doubly-stochastic like in the previous case, and it is defined as [20]:

$$[W]_{ij}^{\mathrm{MD}} = \begin{cases} \frac{1}{\Delta}, & \text{if } e_{ij} \in \mathbf{E} \\ 1 - d_i . \frac{1}{\Delta}, & \text{if } i = j \\ 0, & \text{otherwise} \end{cases} \tag{9}$$

The last examined weights are the Best Constant weights (BC) requiring the exact value about the largest and the second smallest eigenvalue of the corresponding Laplacian spectrum. In contrast to the previous two weights, the weight matrix of BC is not doubly-stochastic for many graphs, and it is defined as follows [21]:

$$[W]_{ij}^{\mathrm{BC}} = \begin{cases} \frac{2}{\lambda_1(\mathbf{L})+\lambda_{n-1}(\mathbf{L})}, & \text{if } e_{ij} \in \mathbf{E} \\ 1 - d_i \cdot \frac{2}{\lambda_1(\mathbf{L})+\lambda_{n-1}(\mathbf{L})}, & \text{if } i = j \\ 0, & \text{otherwise} \end{cases} \tag{10}$$

At the end of this subsection, we compare the initial configuration of the examined AC weights over a graph of the star topology and with the graph order $n = 5$ (see Fig. 5). From the figure, it can be seen that only in the case of BC, weights can take also a negative value.

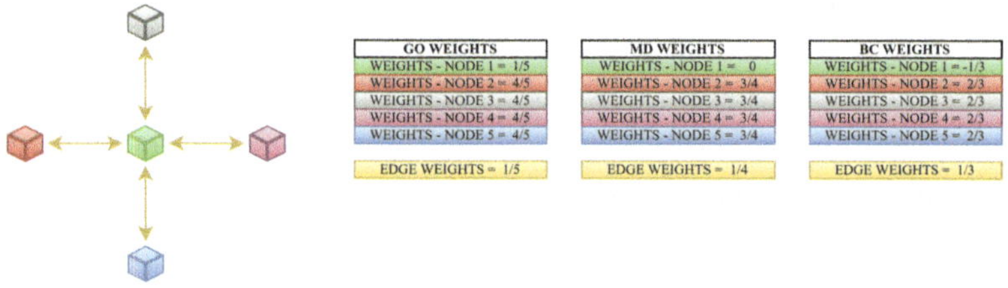

Fig. 5. Average consensus with Graph Order, Maximum Degree, and Best Constant weights over star topology - comparison of initial configurations

2.3 Implemented Stopping Criterion Proposed for Wireless Sensor Networks

This subsection is concerned with the implemented stopping criterion, which is applied in order to bound the execution of AC. As stated in [23], an effective boundary of AC results in energy consumption optimization whereby properly configuring a stopping criterion is a fundamental process especially in technologies with limited energy sources such as WSNs. In this paper, the stopping criterion proposed for WSNs and defined in [24] is applied. This stopping criterion is a fully-distributed approach determined by two time-invariant parameters that are preset and the same at each sensor node and one variable [24]:

- *accuracy* – a positive real constant number usually taking low values
- *counter threshold* – a positive integer with a constant value
- *counter* – a non-negative integer variable initiated by zero and updated independently at each sensor node

The principle of the applied stopping criterion can be described as follows [24]: a sensor node verifies whether the finite difference between its inner states at two subsequent iterations is smaller than the value of *accuracy*. If so, the sensor node increments the current value of *counter* by one, otherwise, the value of *counter* is reset to zero. In the case that the current value of *counter* of a sensor node is equal to *counter threshold*, AC is considered to be locally completed at this sensor node whereby this node does not participate in AC at the next iterations (i.e., it sends and receives no messages any longer).

3 Applied Research Methodology

In this section, we introduce the applied research methodology and the used metric for performance evaluation. As stated earlier, we apply the fully-distributed stopping criterion introduced in [24], which is determined by two preset and constant parameters. In our experiments, their values are set as follows:

- *accuracy* = $\{10^{-1}, 10^{-2}, 10^{-3}, 10^{-4}, 10^{-5}, 10^{-6}\}$
- *counter threshold* = $[1, 100]$

Furthermore, as mentioned earlier, we consider a scenario where the goal of the sensor nodes is to measure the value of an unknown parameter θ, but sensor readings are distorted by Gaussian noise (as mentioned above, it is determined by μ and σ). In our analysis, the mentioned parameters take these values (Fig. 6):

- $\theta = 100$
- $\mu = 0$
- $\sigma = \{1, 3, 5\}$

In our experiments, we generate 100 unique sets of the initial inner states for each setting of Gaussian noise in order to ensure high credibility of the presented research.

In the experimental section, AC is analyzed over the random geometric graph (RGG) with the graph order n equal to 200 whose topology is shown in Fig. 7.

To evaluate and compare the performance of AC with/without Gaussian noise, we apply the mean square error (referred to as MSE) for quantifying the

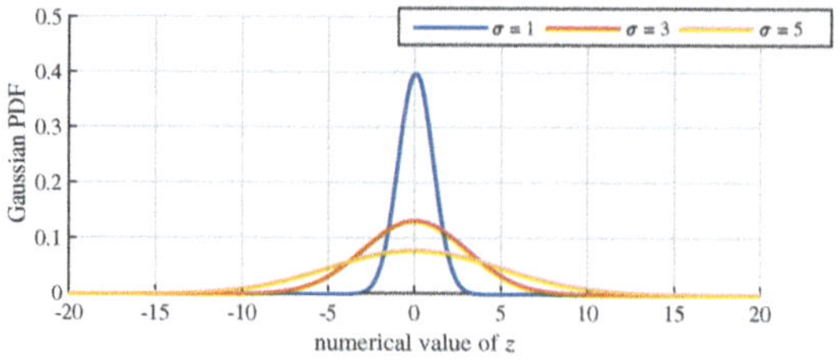

Fig. 6. Gaussian probability density function - three settings of Gaussian noise

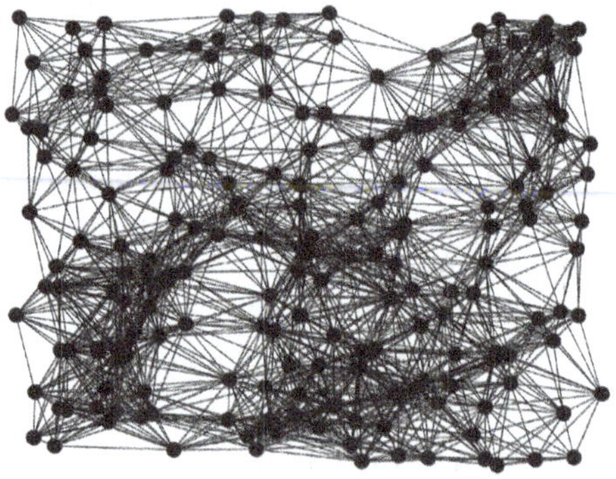

Fig. 7. Topology of random geometric graph formed by 200 vertices

deviation of the final states from the value of θ. It poses a frequently applied metric in various fields and is defined as follows [25,26]:

$$\text{MSE} = \frac{1}{n} . \sum_{i=1}^{n} \left(x_i(k_l) - \theta\right)^2 \tag{11}$$

Here, k_l is the iteration when the consensus among the sensor nodes is achieved. In the figures provided in Sect. 4, we depict MSE averaged over 100 algorithm executions for each setting of Gaussian noise (each execution is carried out with a different set of the initial inner states) and each examined initial configuration of the applied stopping criterion.

Also, in order to evaluate the convergence rate of the analyzed weights, we apply the per-step convergence factor r_{step} defined in (12) [21]. This metric is used for a comparison of AC with/without an implemented stopping criterion and under/not under Gaussian noise.

$$r_{step}(\mathbf{W}) = \left\| \mathbf{W} - \frac{\mathbf{1} \times \mathbf{1}^{\mathrm{T}}}{n} \right\|_2 \tag{12}$$

4 Experimental Part

In this section, we present the results of our numerical experiments executed in Matlab2018b. As mentioned earlier, our goal is to identify whether and how efficiently the application of AC with three different weights can suppress Gaussian noise. Thus, we compare the three mentioned weights with each other and with the scenario when AC for distributed averaging (or more specifically, any mechanism for data aggregation) is not applied. At first, we focus on MSE for $\sigma = 1$ shown in Fig. 8, where each subfigure contains the results for a different value of *accuracy*. From the depicted results, it can be seen that an increase in *counter threshold* results in a decrease in MSE (regardless of the used weights) whereby the precision of the final estimates is increased. However, MSE cannot

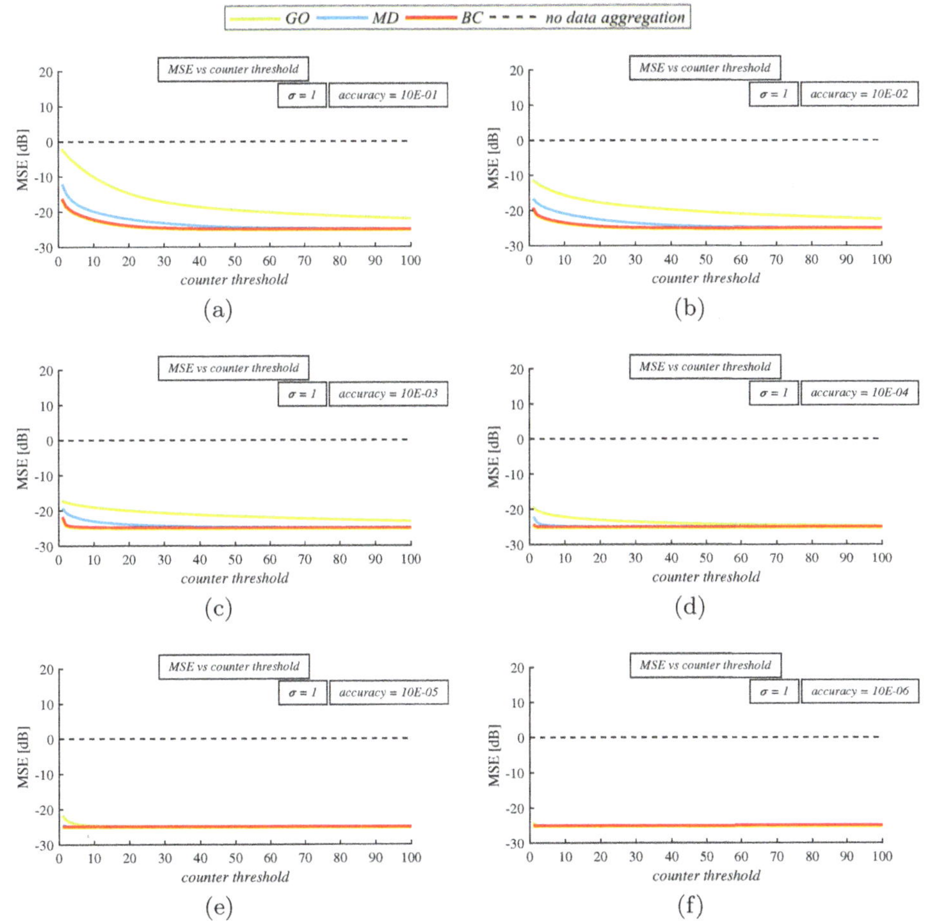

Fig. 8. Mean square error of examined weights of average consensus for varied initial configuration of stopping criterion – $\sigma = 1$

be decreased below −24.72 dB in this scenario due to the presence of Gaussian noise (in the scenario without any data aggregation, MSE is equal to −0.02 dB if $\sigma = 1$). Also, it is seen that BC achieves the highest precision, but the difference in the precisions of the analyzed weights decreases (and is finally equalized) as the value of *counter threshold* is increased. Thus, for higher values of *counter threshold*, the value of the mixing parameter ϵ has no impact on the precision of the final estimates, meaning that all the three examined weights achieve the same precision. Furthermore, we can observe that a decrease in *accuracy* ensures lower values of MSE. For very low values, this parameter does not affect MSE at all like *counter threshold* with high values. Moreover, lower values of *accuracy* cause that the precision of all the three weights is the same for lower values of *counter threshold* than in the case of higher values of this parameter. The

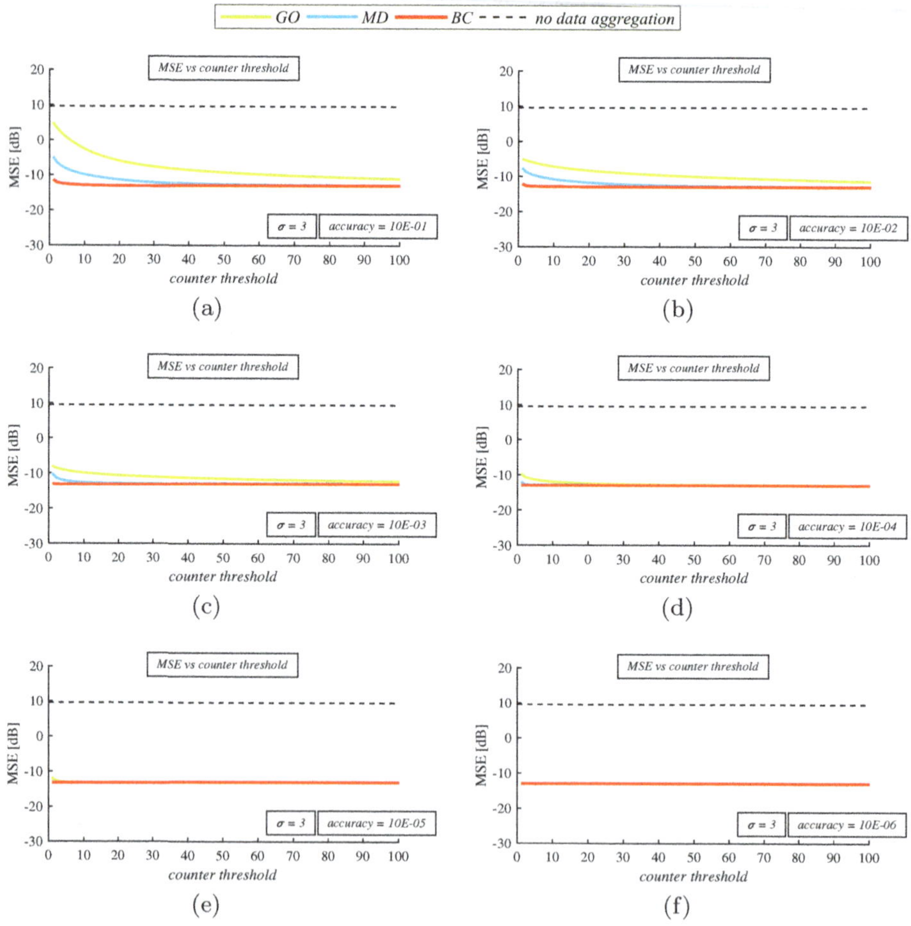

Fig. 9. Mean square error of examined weights of average consensus for varied initial configuration of stopping criterion – $\sigma = 3$

most important observation is that the application of AC (regardless of the used weights and the configuration of the applied stopping criterion) ensures a higher precision of the measured values compared to the scenario when no mechanisms for data aggregation are applied. The precision of the measured values is optimized in the range [1.52, 24.70] dB in the case of GO, in [11.87, 24.70] dB in the case of MD, and in [16.07, 24.70] dB in the case of the application of BC. So, it is seen that for high values of *counter threshold* and low values of *accuracy*, the selection of the weights is irrelevant as it does not affect MSE at all (note that also for *accuracy* $= 10^{-1}$, 10^{-2}, 10^{-3} in the case of the application of GO, the functions have a decreasing character and achieve the same values for high values of *counter threshold* as the functions of MD and BC).

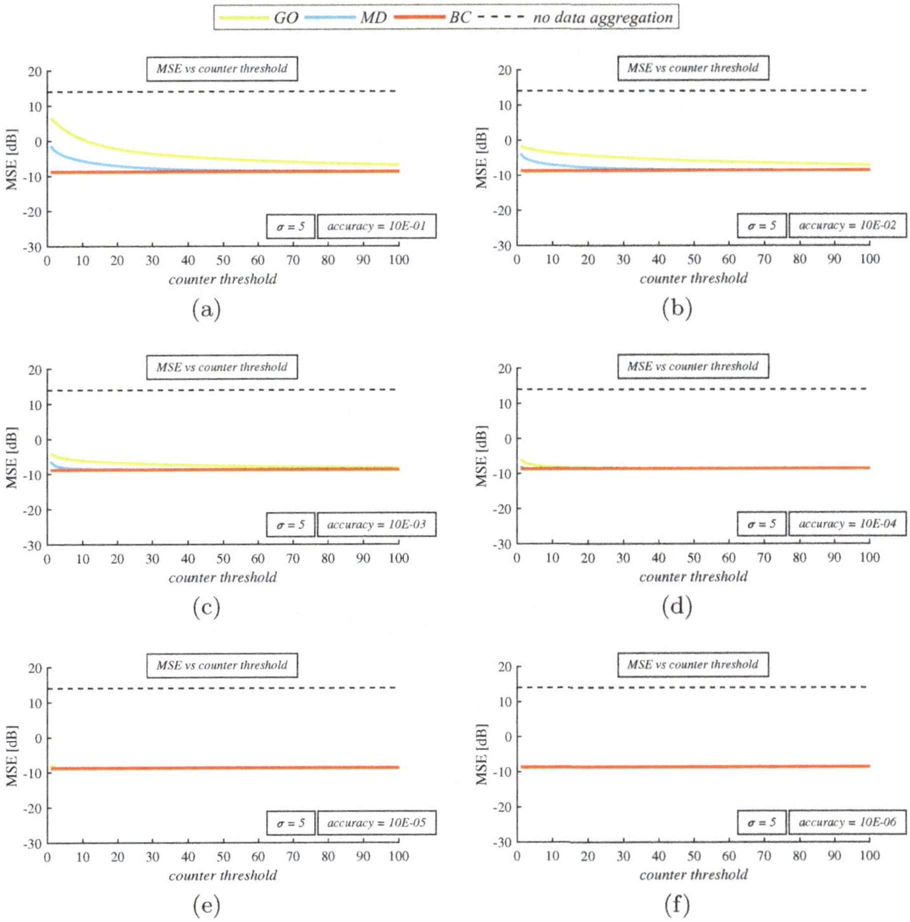

Fig. 10. Mean square error of examined weights of average consensus for varied initial configuration of stopping criterion – $\sigma = 5$

In Fig. 9 and Fig. 10, the values of MSE in the case of $\sigma = 3$ and $\sigma = 5$, respectively, are shown. From the results, we can see that the character of the functions is the same as in the previous analysis. The differences compared to the scenario with $\sigma = 1$ are that MSE is increased as the value of σ increases (including the scenario when data aggregation is not applied - MSE = 9.55 dB if $\sigma = 3$, and MSE = 13.99 dB if $\sigma = 5$). Furthermore, the value of MSE cannot drop below −12.88 dB if $\sigma = 3$ and below −8.61 dB when $\sigma = 5$, and so this threshold takes a greater value for higher σ. Also, the precision of the final states is the same for all the analyzed weights for lower values of *counter threshold* and higher values of *accuracy* if σ is greater. In the case of $\sigma = 3$, the precision is optimized in comparison with the scenario without any data aggregation in the range [4.33, 22.43] dB in the case of GO, in the range [14.28, 22.43] dB in the case of MD, and in the range [20.72, 22.43] dB in the case of BC. When $\sigma = 5$, the application of AC with GO optimizes the precision of the sensor measured values in the range [7.20, 22.60] dB, MD in the range [15.16, 22.60] dB, and BC in [22.60, 22.60] dB. Note that the configuration of the applied stopping criterion has a less intensive impact on MSE as the value of σ increases.

In the last paragraph, we compare the results provided above with the per-step convergence factor r_{step} (12); therefore, with a metric that is affected by neither the presence of Gaussian noise nor the application of the stopping criterion. It is visible from Fig. 8, 9, 10 and Fig. 11 that BC weights, whose r_{step} is the lowest among the analyzed weights (a lower value mean a faster convergence rate [21]), outperform the two other weights (in the cases when the performances are not equal to each other) also when sensor readings are corrupted by Gaussian noise, and AC is bounded by the stopping criterion proposed in [23].

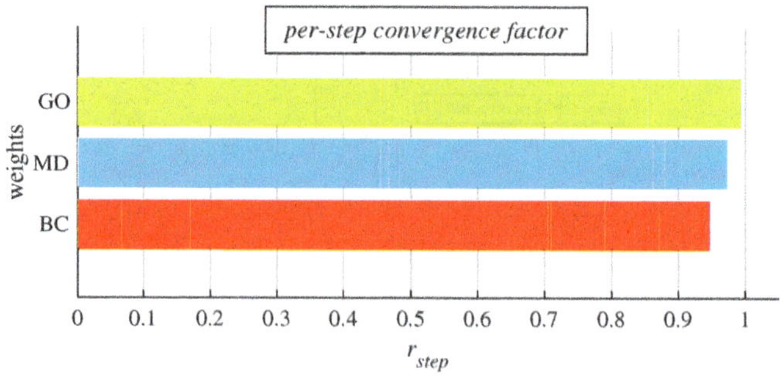

Fig. 11. Per-step convergence factor of examined weights of average consensus

5 Conclusion

In this paper, we analyze whether and how efficiently AC with the Perron matrix with three different weights can compensate for imprecise sensor readings caused by Gaussian noise with a varied standard deviation σ. In the presented research, it is identified that AC for distributed averaging bounded a stopping criterion proposed for WSNs (represented as RGGs) improves the precision of sensor measured values regardless of the applied weights. Furthermore, it is observed that an increase in *counter threshold* results in a decrease in MSE up to a specific threshold determined by the standard deviation of Gaussian noise σ (a greater value of σ causes that both MSE and the value of this threshold are greater). Besides, higher values of σ cause that the mixing parameter ϵ has no impact on MSE for lower values of *counter threshold* and higher values of *accuracy*. Thus, for higher values of σ, the configuration of the applied stopping criterion is less important. Furthermore, it is experimentally identified that the best performance is achieved by the BC weights (i.e., the weight with the lowest per-step convergence factor r_{step}), but for higher values of *counter threshold*, all the three analyzed weights achieve the same precision. Moreover, a decrease in *accuracy* ensures that MSE is smaller (again, it is impossible to exceed the mentioned threshold) as well as causing that the weights achieve the same performance for lower values of *counter threshold*. Also, low values of *accuracy* cause that the value of the mixing parameter ϵ has no impact on MSE. Thus, for many configurations of the applied stopping criterion, it is not relevant which weights are applied.

Thus, the final conclusion of our research presented in this paper is that AC with the Perron matrix can significantly suppress the deviation of the sensor measured values caused by Gaussian noise in WSNs whereby its application with the analyzed stopping criterion is potentially highly beneficial in real-world systems.

Acknowledgment. This work was supported by the VEGA agency under the contract No. 2/0155/19 and by the project CHIST ERA III (SOON) "Social Network of Machines". Since 2019, Martin Kenyeres has been a holder of the Stefan Schwarz Supporting Fund.

References

1. Xiao, L., Boyd, S., Lall, S.: A Scheme for robust distributed sensor fusion based on average consensus. In: 4th International Symposium on Information Processing in Sensor Networks (IPSN 2005), New York, pp. 63–70. IEEE Press (2005). https://doi.org/10.1109/IPSN.2005.1440896
2. Visconti, P., De Fazio, R., Velazquez, C., Del-Valle-soto, C., Giannoccaro, N.I.: Development of sensors-based agri-food traceability system remotely managed by a software platform for optimized farm management. Sensors **20**, 3632 (2020). https://doi.org/10.3390/s20133632
3. Son, Y., Kang, M., Kim, Y., Yoon, I., Noh, D.K.: Energy-efficient cluster management using a mobile charger for solar-powered wireless sensor networks. Sensors **20**, 3668 (2020). https://doi.org/10.3390/s20133668

4. Pranathi, T.Y.S.S., Dhuli, S., Aditya, V.M.V.S., Charisma, B., Jakyakrishna, K.: A hybrid routing protocol for robust wireless sensor networks. In: 12th International Conference on Computational Intelligence and Communication Networks (CICN), New York, pp. 102–106. IEEE Press (2020). https://doi.org/10.1109/CICN49253.2020.9242627
5. Kenyeres, J., Kenyeres, M., Rupp, M., Farkas, P.: Connectivity-based self-localization in WSNs. Radioengineering **22**, 818–827 (2013)
6. Aghaeipoor, F., Mohammadi, M., Sattari Naeini, V.: Target tracking in noisy wireless sensor network using artificial neural network. In: 7th International Symposium on Telecommunications (IST), New York, pp. 720–724. IEEE Press (2014). https://doi.org/10.1109/ISTEL.2014.7000796
7. Kenyeres, M., Kenyeres, J.: Average consensus over mobile wireless sensor networks: weight matrix guaranteeing convergence without reconfiguration of edge weights. Sensors **20**, 3677 (2020). https://doi.org/10.3390/s20133677
8. Izadi, D., Abawajy, J.H., Ghanavati, S., Herawan, T.: A data fusion method in wireless sensor networks. Sensors **15**, 2964–2979 (2015). https://doi.org/10.3390/s150202964
9. Gordon, S., Abawajy, J.H., Ghanavati, S., Herawan, T.: Costs of adjustment, the aggregation problem and investment. Syst. Control Lett. **74**, 422–429 (1992). https://doi.org/10.2307/2109486
10. Krammer, P., Habala, O., Mojzis, J., Hluchy, L., Jurkovic, M.: Anomaly detection method for online discussion. Procedia Comput. Sci. **155**, 311–318 (2019). https://doi.org/10.1016/j.procs.2019.08.045
11. Coroianu, L., Gagolewski, M.: Penalty-based data aggregation in real normed vector spaces. Adv. Intell. Syst. Comput. **981**, 160–171 (2019). https://doi.org/10.1007/978-3-030-19494-9_15
12. Seda, P., Seda, M., Hosek, J.: On mathematical modelling of automated coverage optimization in wireless 5G and beyond deployments. Appl. Sci. **10**, 8853 (2020). https://doi.org/10.3390/app10248853
13. Macua, S.V., Belanovic, P., Zazo, S.: Consensus-based distributed principal component analysis in wireless sensor networks. In: IEEE 11th International Workshop on Signal Processing Advances in Wireless Communications (SPAWC 2010), New York, pp. 1–5. IEEE Press (2010). https://doi.org/10.1109/SPAWC.2010.5671089
14. Braca, P., Marano, S., Matta, V.: Enforcing consensus while monitoring the environment in wireless sensor networks. IEEE Trans. Signal Process. **56**, 3375–3380 (2008). https://doi.org/10.1109/TSP.2008.917855
15. Avrachenkov, K., Chamie, M.E., Neglia, G.: A local average consensus algorithm for wireless sensor networks. In: 7th IEEE International Conference on Distributed Computing in Sensor Systems (DCPSS 2011), New York, pp. 1–6. IEEE Press (2011). https://doi.org/10.1109/DCOSS.2011.5982199
16. He, J., Cai, L., Cheng, P., Pan, J., Shi, L.: Consensus-based data-privacy preserving data aggregation. IEEE Trans. Automat. Contr. **64**, 5222–5229 (2019). https://doi.org/10.1109/TAC.2019.2910171
17. Cocîrlea, D., Dobre, C., Hîrţan, L.-A., Purnichescu-Purtan, R.: Blockchain in intelligent transportation systems. Electronics **9**, 1682 (2020). https://doi.org/10.3390/electronics9101682
18. El Chamie, M., Liu, J., Başar, T.: Design and analysis of distributed averaging with quantized communication. IEEE Trans. Automat. Contr. **61**, 3870–3884 (2016). https://doi.org/10.1109/TAC.2016.2530939

19. Xiao, L., Boyd, S., Kim, S.J.: Distributed average consensus with least-mean-square deviation. J. Parallel. Distrib. Comput. **67**, 33–46 (2007). https://doi.org/10.1016/j.jpdc.2006.08.010
20. Macua, S.V., et al.: How to implement doubly-stochastic matrices for consensus-based distributed algorithms. In: 2014 IEEE 8th Sensor Array and Multichannel Signal Processing Workshop. SAM 2014, New York, pp. 333–336. IEEE Press (2014)
21. Xiao, L., Boyd, S., Kim, S.J.: Fast linear iterations for distributed averaging. Syst. Control Lett. **53**, 65–78 (2004). https://doi.org/10.1016/j.sysconle.2004.02.022
22. Schwarz, V., Hannak, G., Matz, G.: On the convergence of average consensus with generalized metropolis-hasting weights. In: 2014 IEEE International Conference on Acoustics, Speech, and Signal Processing, (ICASSP 2014), New York, pp. 5442–5446. IEEE Press (2014). https://doi.org/10.1109/ICASSP.2014.6854643
23. Al-Nakhala, N., Riley, R., Elfouly, T.: Distributed algorithms in wireless sensor networks: an approach for applying binary consensus in a real testbed. Comput. Netw. **79**, 30–38 (2015). https://doi.org/10.1016/j.comnet.2014.12.011
24. Kenyeres, J., Kenyeres, M., Rupp, M., Farkas, P.: WSN implementation of the average consensus algorithm. In: 17th European Wireless Conference 2011 (EW 2011), Germany, pp. 139–146. VDE (2011)
25. Pereira, S.S., Pages-Zamora, A.: Mean square convergence of consensus algorithms in random WSNs. IEEE Trans. Signal Process. **58**, 2866–2874 (2010). https://doi.org/10.1109/TSP.2010.2043140
26. Schwarz, V., Matz, G.: Mean-square optimal weight design for average consensus. In: 2012 IEEE 13th International Workshop on Signal Processing Advances in Wireless Communications, (SPAWC 2012), New York, pp. 374–378. IEEE Press (2012). https://doi.org/10.1109/SPAWC.2012.6292931

A Conversion of 3D Graphics from Blender to Unreal Engine

Pavel Pokorný(✉) and Miroslav Zapletal

Department of Computer and Communication Systems, Tomas Bata University in Zlín, Faculty of Applied Informatics, Nad Stráněmi 4511, 760 05 Zlín, Czech Republic
{pokorny,m_zapletal}@utb.cz

Abstract. This paper describes the conversion possibilities between Blender and Unreal Engine. Although it may seem at first sight to be a simple issue, the opposite is true. Unlike 2D graphics, where a simple image is transferred from one graphics program to another, the problem is much more complex in 3D graphics. We want to transfer only the simple geometry of 3D objects very rarely. Blender is a tool for the entire 3D graphics pipeline - offers several types of representations, assigning materials and textures to objects, creating simple and more complex animations, various physical simulations, including collision detection (particle systems, fabric objects, flexible objects, liquids), etc. Unreal Engine is a state-of-the-art real-time engine and editor that features photorealistic rendering, dynamic physics and effects, lifelike animation, robust data translation, and much more. Main task of this article is to summarize the results of research aimed at finding out what can be converted between these programs and in what graphic formats.

Keywords: 3D graphics · Graphic formats · Conversion

1 Introduction

Computer graphics is the branch of computer science that deals with generating images with the aid of computers. Almost most computers include screen displays, where the screen image is composed of discrete units called pixels [1], images can have various representation in the memory. Depending on this representation, it is also possible to save them to disk in various graphic formats. In addition to storing all graphic information, these formats are often required to be able to work with other applications than the one that created it.

According to the representation of the image, there are several types of division [2]. The first group is Two-Dimensional Graphics. This includes Raster Graphics as well as Vector Graphics. Raster (Bitmap) Graphics is the more common representation form. A bitmap represents an image via a rectangular grid of pixels, where each individual pixel's spatial location and colour is defined [3]. Vector Graphics represents an image mathematically by means of the use of geometrical primitives like points, lines, curves and polygons. In a sense, one may consider a vector image as stored information regarding the shapes in an image rather than the raw image itself.

R. Silhavy (Ed.): CSOC 2021, LNNS 230, pp. 406–417, 2021.
https://doi.org/10.1007/978-3-030-77442-4_35

Another category is Three-Dimensional Graphics. 3D Graphics, unlike 2D Graphics, use three-dimensional representation of geometric data. This geometric data is often referred to as 3D models. A 3D model is a mathematical representation of any three-dimensional object [4]; a model is not technically a graphic until it is displayed. Apart from a model's geometry, a lot of other information - like material properties, surface textures of animations, is often stored in memory. If we want to save all this information correctly into a file and open and process it correctly in another programme, it is then necessary to choose a suitable graphic format.

This paper describes the conversion possibilities between 3D Blender and Unreal Engine programmes. The most commonly-used graphical formats for the conversion process between these programmes are tested with respect to the amount of graphical information and techniques that Blender and Unreal Engine support. The second section describes the basic features of both programmes. The content of the third section is the characteristics of graphic objects and other features that could be converted between Blender and Unreal Engine. The fourth section contains a short description of the 3D Graphics formats that were tested during the conversion process. The results and experiences learned from the conversion study are summarised in the fifth section.

2 Software Descriptions

In this section, we will present the basic characteristics of the Blender and Unreal Engine programmes. We will focus on the main features and capabilities of both programmes, as well as working with files, since this is important given the content of this paper.

2.1 Blender

Blender [5] is a free and open source 3D creation suite, which is available under GNU GPL License and users can use it on most common platforms – Windows, Linux, Mac OS, etc. It supports the entirety of the 3D pipeline—modelling, animation, rendering, rigging, simulation, compositing and motion tracking, video-editing and a 2D animation pipeline. This software is available under GNU GPL License and users can use it on the most commonly available platforms – Windows, Linux, Mac OS, etc.

Blender opens a world of creativity that has been traditionally exclusive to those able to afford high-end graphics software. Its GUI is an arrangement of windows and panels containing the controls for operating the programme [6]. Blender also uses workspaces that are, essentially, predefined window layouts in order to adapt user interfaces for different tasks like modelling, animating, and scripting.

Since Blender is a very complex programme with extensive possibilities, it keeps everything in its database [7]. This database thus contains all scenes, objects, meshes, textures, etc., and when a user saves their current project, the whole database is saved into one .blend file. Although this format is "open" - (its concept is documented), there are not many applications that can work directly with this format. So, if one ever wants to transfer their work from Blender to another programme one will need a different format. The File - Export menu is used for this purpose.

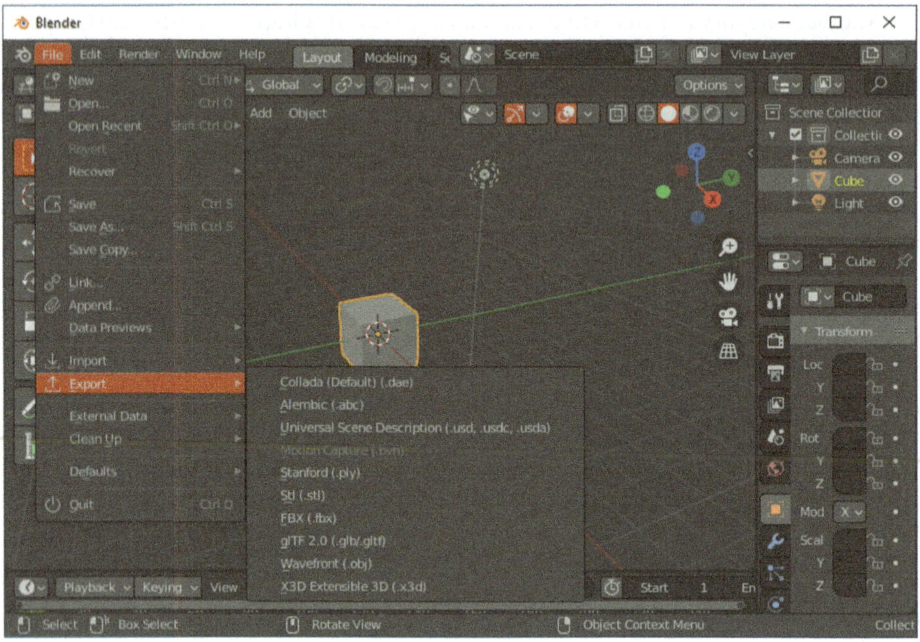

Fig. 1. The blender user interface with file-export menu

Figure 1 shows the default content of this menu. There are 10 different file formats that can keep 3D Graphics (.dae, .abc, .usd, .bhv, .ply, .stl, .fbx, .glb/.gltf, .obj and .x3d). However, this menu does not include all supported formats. Other types of formats can be added to this menu in the form of extension plugins created in Python. Many such plugins can be found on the Internet - or can be created by the user.

2.2 Unreal Engine

Unreal Engine [8] is the world's most open and advanced real-time 3D creation tool. Continuously evolving to serve not only its original purpose as a state-of-the-art game engine, today it gives creators across industries the freedom and control to deliver cutting-edge content, interactive experiences, and immersive virtual worlds.

Unreal Engine is a complete suite of development tools for anyone working with real-time technology. From design visualisations and cinematic experiences to high-quality games across PC, console, mobile, VR, and AR, Unreal Engine provides one with everything they need to create very high-quality applications.

This engine with its own editor, offer complex tools for photo-realistic rendering, dynamic physics and effects, life-like animation, robust data translation and much more [9]. It performs under two licenses. For creators – this license is free to use and 100% royalty-free; and for publishing – this license is free to use and incurs 5% royalties when a user monetises a game/application developed in Unreal Engine and their lifetime gross revenues from that product exceed $1,000,000 USD (Fig. 2).

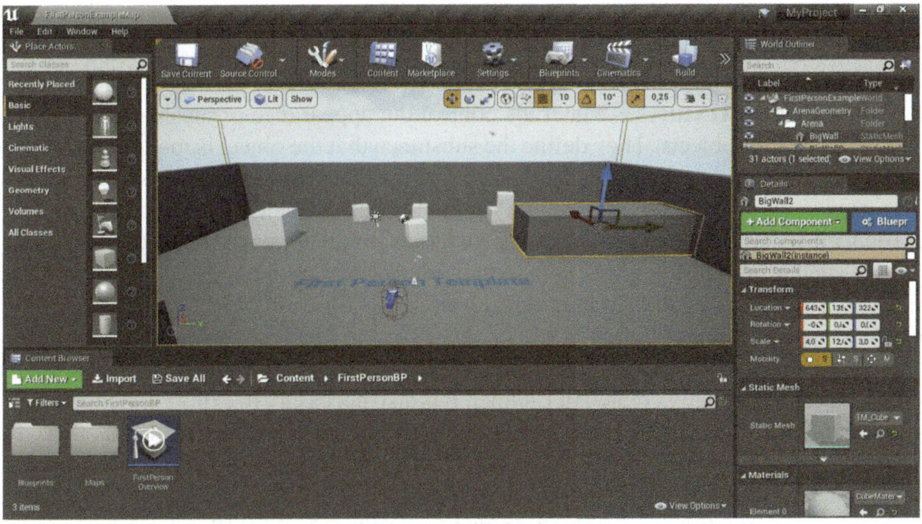

Fig. 2. The unreal engine user interface

3D graphics are added to projects in the asset form in Unreal Engine. An Asset is a piece of content for an Unreal Engine project and can form a variety of resources - 2D images/animations, 3D models, materials, textures, blueprints, sound effects, and visual effects, etc. [10]. In order to include them in a project, Unreal Engine supports a huge number of different file formats. In 3D graphics these formats are mainly supported: .abc, .fbx, and.obj, in basic. Using various plug-ins, Unreal Engine can read other formats like .glb/.gltf or .usd.

3 3D Graphics for Conversion

To determine the correct conversion of various graphical entities between Blender and Unreal Engine, an effort was made to use a wide range of these entities (so-called Data-Blocks), that Blender offers. These are mainly Static Mesh Objects, Particle Systems, Materials and Textures, UV Maps, Character Animations and Physics [7]. These will be described in more detail in this section.

- **Static Mesh Objects** – the most common type of objects. Mesh Objects are composed of vertices, edges and polygonal faces which form a geometric shape and can be edited extensively with Blender's mesh editing tools. Mesh Modelling typically begins with a Mesh Primitive shape (e.g. plane, circle, cube, cylinder, sphere…). From there, users might begin editing and use a large number of modelling tools to create larger, more complex shapes. This type of representation is very suitable for rendering, therefore other types of representations are often converted into mesh objects [11].
- **Particle Systems** – Particles are lots of items emitted from mesh objects - typically in the thousands. Each particle can be a point of light or a mesh, and be static or dynamic. They may also react to many different influences and forces, and have the notion of a

lifespan. Static type particles are joined together and form curves that can represent hair, fur, grass and bristles. Dynamic particles can represent fire, smoke, mist, and other things like dust or magic spells [7].

- **Materials and Textures** – Materials control the appearance of meshes, curves, volumes and other objects. They define the substance that the object is made of, its colour and texture, and how light interacts with it. Physically-based materials are typically created using the Principled BSDF, Principled Hair or Principled Volume shaders in Blender. By using this wide variety of materials, many types of materials can be created - including plastic, glass, skin, metal, cloth, hair, smoke or fire. Textures can be procedurally generated or standard raster images read from disk.
- **UV Maps** – UV maps define UV coordinates in the image textures in order to attain precise mapping of these textures on the object surfaces. Coordinates can be computer generated or manually created in Blender. It is also necessary to export UV coordinates when one wants to open models created with textures in other applications, otherwise the texture will not be applied correctly.
- **Character Animations** – This is a specialised field of the animation process where models of characters have animation that bring them to life. Blender offers a complete animation toolset - specialised Character Animation Pose Editor, Non-linear Animation Editor, Forward/Inverse Kinematics for fast poses and Sound Synchronisation. Riggins Tools include envelope skeleton and automatic skinning, easy weight painting, mirror functionality, bone layers and coloured groups for organisation and B-spline interpolated bones.
- **Physics** – Blender's Physics System allows one to simulate and animate a number of different real-world physical phenomena. These include Rigid Body Animations, Cloth Systems, Soft Body Systems, Fluid Systems, Forces and Collisions. Since we tried to convert Rigid Body Animations and Fluid Systems, they are described in more detail here.
- **Rigid Body** – These animations can be used to simulate the motion of solid objects. They affect the position and orientation of objects but do not deform them. Rigid bodies can be used like regular objects and be part of parent-child relationships, animation constraints and drivers. There are two types of rigid bodies: active and passive. Active bodies are dynamically simulated, while passive bodies remain static. An example of a Rigid Body Animation frame is shown in Fig. 3 - a ball on a chain breaks down stacked cubes.
- **Fluid Systems** – These objects are used to simulate the physical properties of liquids - especially water. While creating a scene in Blender, certain objects can be marked to become a part of the fluid simulation. For a fluid simulation, a user has to have a domain to define the space where the whole simulation proceeds. In the domain settings, users will be able to define the global simulation parameters - like viscosity and gravity. Based on defined objects and set parameters, Blender can calculate the course of the entire simulation.

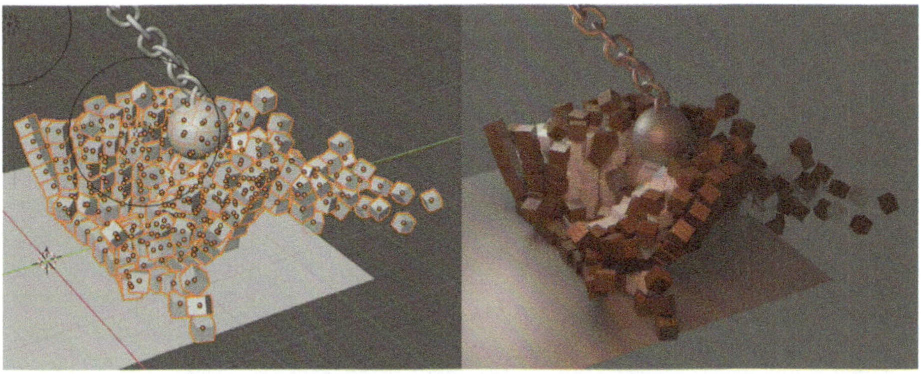

Fig. 3. An example of rigid body animations in blender

4 Conversion Formats

Based on an analysis of the export possibilities from Blender and their import to Unreal Engine, the following formats were selected for conversion: .fbx, .abc, .obj, .glb/gltf and .usd. There is also an extension plug-in for Blender called "Blender for Unreal Engine", which specialises directly in this conversion, so this was also included in the testing process. These are described in detail in this section.

4.1 The .fbx Format

FBX by Autodesk is a proprietary file format designed to facilitate higher-fidelity data exchange between 3ds Max, Maya, MotionBuilder, Mudbox and other propriety and third-party software. The main features it has are single-step inter-operability between the above-mentioned software, easier data exchange, more efficient workflow as well as customisability [12].

Since 2006, an Exporter for this format has been implemented in Blender. Since then, it has undergone intensive development - thanks to which exports to this format are among the most sophisticated [13]. If a user chooses to export into this format in Blender, a window with a number of settings will open - (Fig. 4 - Right). Users can select which objects will be exported - (mesh objects can be exported with Modifiers and normal vectors), scale, orientation, animation, object groups - (Batch Export), etc.

4.2 The .abc Format

Alembic represents an open computer graphics interchange framework [14]. This is focused on efficiently storing the computed results of complex procedural geometric constructions - that are then saved into.abc files. Alembic is very specifically NOT concerned with storing the complex dependency graphs of procedural tools that are used to create the computed results. For example, Alembic will efficiently store the animated vertex positions and animated transformations that result from an arbitrarily

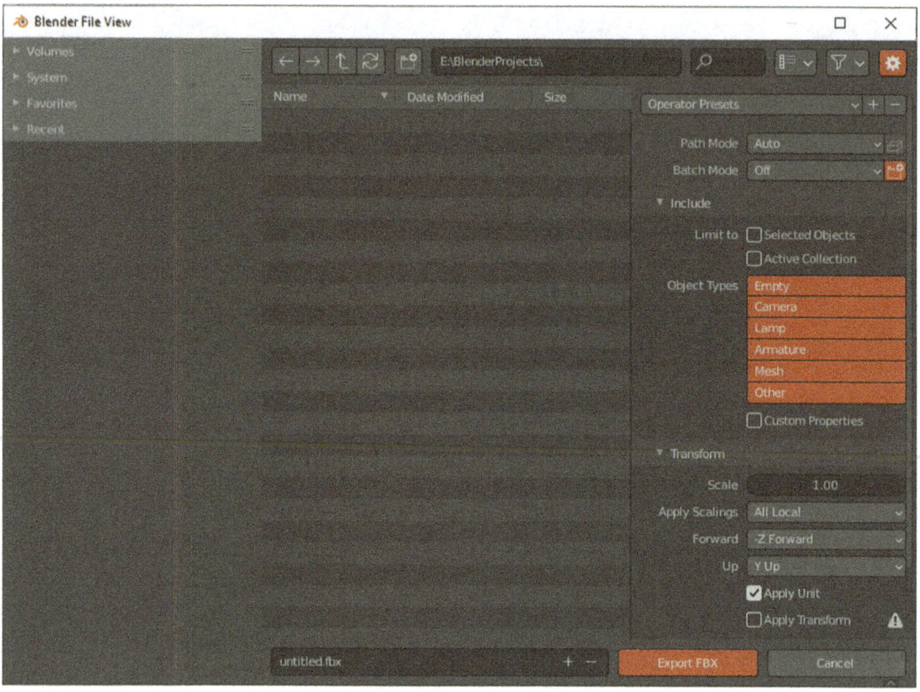

Fig. 4. An export window to .fbx file with adjustable options in blender

complex animation and simulation process that could involve enveloping, corrective shapes, volume-preserving simulations, cloth and flesh simulations, and so on.

When exporting models from scenes to *.abc file in Blender, users can set different export options like scale, frame start and end - (in animations), use instancing, keeping the relationships between objects - (parent/child), export normal, uv maps, etc. [15].

4.3 The .obj Format

Wavefront OBJ (Object) files are formats that contain geometric definitions. It was designed for Wavefront's Advanced Visualizer application to store geometric objects composed of lines, polygons, and free-form curves and surfaces. However, this file format is open and has been adopted by other 3D graphics application vendors [16].

Wavefront is best known for its high-end computer graphics tools, including modelling, animation, and image composition tools. These programmes run on powerful workstations - such as those made by Silicon Graphics, Inc.

The.obj format is a popular plain-text format; however, it has only basic geometry and material support in Blender. There is also support for object groups and Nurbs curves and surfaces. According to the official Blender documentation [17], there is no support for mesh vertex colours, armatures, animation, lights, cameras, empty objects, parenting, or transformations. An export window contains only a few export options in Blender - like global axis orientations, fixed size, or keeping the order of vertices.

4.4 The .glb/.gltf Formats

The glTF™ (GL Transmission Format) is a royalty-free specification for the efficient transmission and loading of 3D scenes and models by engines and applications that minimise the size of 3D assets - and the runtime processing needed to unpack and use them [18]. This format is commonly used on the web, and has support in various 3D engines like Unity3D, Unreal Engine 4, and Godot. There are two format versions – glTF Binary (.glb) - where all data is stored in a single binary file; and glTF Separate (.gltf) - which is text-based and describes the overall structure; this text file follows a .bin file that contains mesh and vector data - and optionally, a number of.png or.jpg files containing image textures referenced by the .gltf file.

The official information [19] says that Blender's importer/exporter supports the following .glb/.gltf features: mesh objects, simple materials and shadeless textures, cameras, punctual lights - (point, spot, and directional), and simple animations - (key-frame, shape key, and skinning). Export options contain properties like merge vertices, pack images, shading type - (flat, smooth) or heuristic for placing bones in Blender.

4.5 The .usd Format

Universal Scene Description (USD) is a framework for the interchange of 3D computer graphics data, created by Pixar [20]. Each of many cooperating applications in the pipeline - (modelling, shading, animation, lighting, fx, rendering), typically these have their own special form of scene description tailored to the specific needs and workflows of the application, and neither readable nor editable by any other application. Universal Scene Description (USD) is the first publicly available software that addresses the need to robustly and scalable interchange and augment arbitrary 3D scenes that may be composed from many elemental assets.

Blender.usd exporter can save static, deforming or arbitrarily animated meshes, curves, meta-balls, cameras, lights - (except area lights), and static particles - (exported as curves). [21] Exporting options include choices for export selection only - animation, static particles, UV maps, normal and materials.

4.6 Blender for Unreal Engine Plug-Ins

This is not an export format, but add-on module that is not implemented in Blender in default. It must be downloaded from [22] and installed in the Blender suite. It simplifies the method of exporting from Blender to Unreal Engine 4 by allowing one to export all the assets of a scene at the same time. It even automatically distributes them in a proper tree structure in correlation with the Unreal Engine 4 pipeline.

This plug-in does not use its own format to save - but uses third–party 3D formats. Its algorithm recognises which format is suitable for each graphic entity in Blender and automatically offers to save them. It prefers the .fbx and .abc file formats, but there is the possibility to save in multiple formats at once. Its export options are very similar to corresponding file format options.

Fig. 5. A 3D scene in blender created for the conversion process

5 Conversion and Results

This section describes the examination of the conversion process between these two programmes. It should be noted that the results below were obtained in Blender version 2.90 and Unreal Engine version 4.25.3. These results may not apply to other versions of these programmes.

It's also a good idea to follow certain rules when converting. The first is that Blender and Unreal Engine use different coordinate systems. Although the z-axis points upwards in both programs, the y-axis has opposite directions in Blender and the Unreal Engine. If a user does not set the correct orientation in the export setting, objects can be imported mirrored in Unreal Engine. It is also good to apply transformations in Blender - (a command in the menu Object-Apply-All Transforms), and to set object origins to meaningful positions. In case a user wants to transform their objects into Unreal Engine after the conversion process, these actions prevent the user from seeing weird transformation numbers. The last thing is to control that all faces' normal vectors point outwards. In Edit mode, it is possible to do that by selecting all faces and then executing the command Mesh-Normals-Recalculate Outside. The reason for doing so is to do with wrong normal orientations, the textures would be mapped on the opposite side of faces. We did all of these for our objects before the converting process started.

In order to find out the conversion possibilities, a comprehensive model of the Blender programme environment was created - this was a 3D street environment scene (Fig. 5). The basis of this scene are static mesh objects - terrain, roads, houses, fences, benches, sidewalks and a car. Some of these models were modelled by hand in Blender, while others were already inserted into the scene finished from the BlendSwap model collection [24]. All of these models have standard material settings with UV mapped textures that were downloaded from [25].

Particle systems were implemented in the scene in two forms. Trees were randomly generated as particles - (the Emitor particle system), and the grass was generated as narrow polygons representing clumps of them (the Hair Particle system) - corresponding materials have been set for them.

In order to have a character animation in the scene, a human 3D model was used. The core of this model are armature objects - (composed from Bone objects), which simulate bones in the human body. These armatures are linked to a human mesh model. Manipulation with these armatures causes the animation of corresponding parts of the mesh model.

As mentioned above, Rigid Body animations are often used to simulate the motion of solid objects including collision among them. To test this in the conversion process, a simple simulation using Rigid Body was created. This included a demolition ball suspended on a chain model. As the ball moves, it hits the wall composed of simple cubes. Upon this impact, cubes fly into space - (Fig. 3).

The last type of object added to the scene is a fluid system. A large container object was created into which fluid flows. There is also one cube-shaped obstacle in this container. The simulation algorithm can calculate the behaviour of this fluid in individual frames of the animation.

Table 1. The conversion process results from blender to unreal engine

Graphical entities	.fbx	.abc	.obj	.glb/.gltf	.usd	Blender for unreal
Static meshes	Yes	Partial	Partial	Partial	Yes	Partial
Materials and textures	Partial	No	Partial	Yes	Partial	No
UV maps	Yes	Partial	Yes	Yes	Yes	No
Character animations	Yes	Partial	Yes	Partial	Yes	Partial
Rigid body	No	No	No	No	No	No
Fluid simulations	No	Yes	No	No	Partial	No

The scene was created in this way, and then the conversion process began. All types of objects mentioned above - as well as 3D graphic formats described in Sect. 4 were used. The final results are listed in Table 1. Particle Systems are not mentioned in this table, because they were converted to mesh objects in the conversion process.

The "Yes" value means that the graphics entity was displayed (in the case of animations or simulations of how they also behaved) in Unreal Engine - exactly in the same way as in Blender. The "No" value says that the entity failed to load in Unreal Engine. The "Partial" value means that the entity was successfully loaded, but was wrongly displayed - or behaved incorrectly in the Unreal Engine environment. In the case of mesh objects, this was most often caused by incorrect transformations, materials being set inaccurately and textures incorrectly mapped. The character animations and fluid simulations were about their incorrect functionalities. Rigid Body animations could not be successfully converted using any of the listed graphic formats.

The test results show that the .fbx format is the most suitable for converting static objects - including UV Maps and character animations. Its biggest advantage is retaining the positions of many objects in the 3D scene. However, the .fbx format has problems with the correct conversion of material properties – their appearance is slightly different between Blender and Unreal Engine. The.glb/.gltf formats are the most suitable for the correct transfer of material settings. The only format that can correctly convert Fluid Simulations is.abc.

The use of the Blender plug-in the "Blender for Unreal Engine 4" has also proved successful since it facilitates large scenes with a large number of objects correctly. This plug-in allows the user to convert all Blender entities into multiple files at once - the entire conversion process can be then significantly accelerated, especially for big 3D graphics projects.

6 Conclusion

The main goal of this paper was to describe the possibilities and ways of converting 3D graphics from Blender to Unreal Engine. These are both world-famous well-recognised programs with free use options. Blender performs the entirety of the 3D pipeline—modelling, rigging, animation, simulation, rendering, compositing and motion tracking, video editing and a 2D animation pipeline. Unreal Engine is a state-of-the-art real-time engine and editor that features photo-realistic rendering, dynamic physics and effects, life-like animation, robust data translation, and much more.

During testing, it was found that the .fbx format is the most suitable for converting static objects due to its ability to properly maintain the distribution of objects in the scene. The .glb/.gltf formats are the most suitable for converting material properties. The .abc format is the most suitable for fluid simulations. At the same time, the "Blender for Unreal Engine" plug-in can be used for more complex environments - which facilitates the conversion of a large number of objects using the .fbx format. The testing process also found that the .usd format is not suitable for converting any of the created models.

This testing was carried out at the beginning of the second half of 2020. Given that both programmes are intensively developed, it can be expected that conversion process possibilities between these programs will continue to improve. Previous developments suggest that the Blender for Unreal plug-in most likely has the greatest potential to improve conversion options between Blender and the Unreal Engine. And there is a new Blender add-on prepared directly by Epic Games - authors of Unreal Engine [23].

References

1. Miano, J.: Compressed Image File Formats: JPEG, PHG, GIF, XBM, 1st edn. Addison-Wesley Professional, Ltd., Texas (1999). ISBN 978-0201604436
2. Murray, J.D.: Encyclopedia of Graphics File Formats, 2nd edn. O'Reilly Media, Sebastopol (1996). ISBN 978-1565921610
3. Tan, L.K.: Image file formats. Biomed. Imaging Interv. J. **2**(1), e6 (2006). https://doi.org/10.2349/biij.2.1.e6
4. Franklin, C.: How 3-D Graphics Work. https://computer.howstuffworks.com/3dgraphics.htm. Accessed 18 Jan 2021

5. Blender.org – Home of the Blender project. https://www.blender.org/. Accessed 18 Jan 2021
6. Blain, J.M.: The Complete Guide to Blender Graphics, 3rd edn. CRC Press (2016). ISBN 978-1-4987-4647-2
7. Blender 2.91 Reference Manual. https://docs.blender.org/manual/en/latest/. Accessed 18 Jan 2021
8. The most powerful real-time 3D creation platform - Unreal Engine. https://www.unrealengine.com/en-US/. Accessed 18 Jan 2021
9. Plowmanm, J.: 3D Game Design with Unreal Engine 4 and Blender, 1st edn. Packt Publishing (2016). ISBN 978-1-78588-146-6
10. Unreal Engine 4 Documentation. https://docs.unrealengine.com/en-US/index.html. Accessed 18 Jan 2021
11. Hughes, J.F., Van Dam, A., Foley, J.D., McGuire, M., Feiner, S.K., Sklar, D.F.: Computer Graphics: Principles and Practice, 3rd edn. Addison-Wesley, Boston (2014). ISBN 978-0-321-39952-6
12. FBX – Adaptable File Formats For 3D Animation Software. https://www.autodesk.com/products/fbx/overview. Accessed 18 Jan 2021
13. Extensions:Py/Scripts/Manual/Export/FBX, https://web.archive.org/web/20090722160843/. http://wiki.blender.org/index.php/Extensions:Py/Scripts/Manual/Export/FBX. Accessed 18 Jan 2021
14. Alembic. http://www.alembic.io/. Accessed 18 Jan 2021
15. Alembic – Blender Manual. https://docs.blender.org/manual/en/latest/files/import_export/alembic.html. Accessed 18 Jan 2021
16. Wavefront OBJ: Summary from the Encyclopedia of Graphics Files. https://www.fileformat.info/format/wavefrontobj/egff.htm. Accessed 18 Jan 2021
17. Wavefront .obj file – Blender Manual. https://docs.blender.org/manual/en/2.80/addons/io_scene_obj.html. Accessed 18 Jan 2021
18. glTF Overview – The Khronos Group Inc. https://www.khronos.org/gltf/. Accessed 18 Jan 2021
19. glTF 2.0 – Blender manual. https://docs.blender.org/manual/en/2.80/addons/io_scene_gltf2.html. Accessed 18 Jan 2021
20. USD Documentation: Introduction to USD. https://graphics.pixar.com/usd/docs/Introduction-to-USD.html. Accessed 18 Jan 2021
21. Universal Scene Desription – Blender Manual. https://docs.blender.org/manual/en/latest/files/import_export/usd.html. Accessed 18 Jan 2021
22. GitHub – xavier150/Blender-For-UnrealEngine-Addons. https://docs.blender.org/manual/en/latest/files/import_export/alembic.html. Accessed 18 Jan 2021
23. Download our new Blender addons – Unreal Engine. https://www.unrealengine.com/en-US/blog/download-our-new-blender-addons. Accessed 18 Jan 2021
24. Blend Swap – Home. https://blendswap.com/. Accessed 18 Jan 2021
25. Textures – Textures for 3D, graphic design and Photoshop! https://www.textures.com/. Accessed 18 Jan 2021

Technology for Dynamic Configuring of the Decision Support System for Personnel Security of the Regional Mining and Chemical Cluster

D. N. Khaliullina[1(✉)], V. V. Bystrov[1], and S. N. Malygina[1,2]

[1] Institute for Informatics and Mathematical Modeling Kola Science Centre, Russian Academy of Sciences, Apatity, Murmansk Region, Russia
{Khaliullina,Bystrov,Malygina}@iimm.ru
[2] Apatity Branch of Murmansk Arctic State University, Apatity, Murmansk Region, Russia

Abstract. The article is devoted to the development of software tools for information and analytical support for personnel security management in a mining and chemical cluster. The authors formulate the concepts of configuration of a network-centric management support system and its dynamic configuring. The implementation of the system configuration in the form of a configuration file is provided. The technology of dynamic configuring of a network-centric decision support system is described. The technology provides automated deployment of a network of software agents and their parametric setting depending on the context of the problem being solved in the subject area.

Keywords: Dynamic configuring · Information system configuration · Network-centric management · Decision support system · Personnel security

1 Introduction

From an economic point of view the Arctic Zone of the Russian Federation is a rather difficult region. The land development of the region is hindered by extreme natural and climatic conditions, and the economy of the territorial units included in the Arctic Zone is focused on mono-cities, which are the main source of budget replenishment. Clustering can become one of the possible solutions for the breakthrough development of such regions. The Murmansk region can be considered as an example of the implementation of the cluster approach at the level of regional management. The government of this constituent entity of the Russian Federation has developed a strategy for economic development until 2025 [1], according to which it is planned to create 7 production clusters in the region. As part of the strategy, special attention is paid to the formation of a mining and chemical cluster, since the extraction and processing of minerals is a key industry in the Murmansk region.

One of the conditions for the formation of a production cluster is the availability of the necessary labor resources in the regional labor market, as well as outside it [2].

R. Silhavy (Ed.): CSOC 2021, LNNS 230, pp. 418–427, 2021.
https://doi.org/10.1007/978-3-030-77442-4_36

Analyzing the staffing of a production cluster is a complex scientific and practical task, for the solution of which it is necessary to use problem-oriented information processing and decision support tools. The Institute for Informatics and Mathematical Modeling of the KSC RAS (IIMM KSC RAS) has been developing in this area of knowledge for several years. The research team of the Institute previously proposed the architecture of the decision support system (DSS) for personnel security of the regional mining and chemical cluster [3]. As part of the current research, a technology for dynamic configuring of such network-centric multi-agent software systems is being developed. This technology makes it possible to form the configuration of the personnel security DSS of a regional mining and chemical cluster, depending on the context of the applied problem being solved.

The use of this technology will allow you to automate the process of deploying a network of software agents and their parametric configuration.

2 Background

The concepts of configuration and dynamic configuring are the key concepts of this study. The definition of the proposed terms, depending on their field of application, is introduced and fixed in various Russian and international standards:

1. GOST R ISO / IEC TO 10032–2007: "Reference model of data management" [4].
2. GOST R ISO / IEC 10746–2–2000: "Information technology. Interconnection of open systems. Data management and open distributed processing. Part 2. The basic model" [5].
3. Federal Standart 1037 [6].
4. ISO/IEC TR 19759:2015 Software Engineering — Guide to the software engineering body of knowledge (SWEBOK) [7].
5. GOST 15971–90 "Information processing systems. Terms and definitions" [8].
6. And others.

Analysis of the given sources of information allows us to conclude that in the field of computer technologies the concept of "configuration" is interpreted as an enumeration of elements of some artificial systems and their adjustment depending on the requirements set by someone. And configuring is the process of forming a configuration.

As part of the current research, a technology for the dynamic configuring of network-centric management systems is being developed. The analysis of published works within the framework of the considered scientific and practical direction allows us to conclude that insufficient attention is paid to the problems of configuring such a class of information systems by the scientific community.

As an example of research in the field of dynamic configuring, one can cite the work that reviews existing solutions for reconfiguring complex systems [9]. Most of the developments described in the article are devoted to changing the system configuration in case of failures and are focused on the aviation industry.

Articles [10, 11] provide an example of software implementation of procedures for configuring a set of microservices depending on the needs of the end user in real time for the Spring cloud platform. This is a practical example of dynamic configuring of service-oriented software systems.

The ONOS (Open Network Operating System) project is another good example of dynamic configuring of complex distributed software systems [12]. Within the framework of this software architecture, a separate module "Dynamic Configuration Subsystem" is implemented. This module provides configuration of ONOS system components (providers, drivers, services, network protocols, etc.) and their modification in a unified mode, depending on the tasks to be solved.

The work that is most closely related to the current research is devoted to the dynamic configuring of the maritime environment simulation system as part of the development of a tactical simulator complex for the Russian Navy [13]. The introduction of dynamic configuring technology provides the possibility of a gradual increase in the number and complexity of models, both during the development of the simulator complex and subsequent modernization. It also provides the independence of the software implementation of simulation models from the other components of the developed complex. Dynamic configuring is reduced to determining the set of parameters of the simulation complex, as well as the list of used computer models, depending on the problem being solved at the moment.

In the framework of the current study, configuring of network-centric management systems is understood as a formalized parametric description of the components of the developed system with an indication of the relationships between them. This description characterizes the current state of the software system as a whole and may vary depending on the problem being solved.

3 Methods

The technology of dynamic configuring of network-centric management systems was developed on the basis of integration of such approaches as:

Theory of Multi-agent Systems. The essence of this theory is a fundamentally new method of solving problems. The multi-agent approach is based on the concept of a mobile software agent, which is implemented and functions as an independent specialized computer program or an element of artificial intelligence [14]. In the classical method a search for a certain clearly defined (deterministic) algorithm that allows finding the best solution to a problem is carried out. In multi-agent technologies in contrast to the classical method the solution is obtained automatically as a result of the interaction of many independent targeted software agents. Software agents have a set of specific characteristics [15]:

- Autonomy - agents are relatively independent in the system.
- Limited representation - the agent's understanding of the system in which it operates is incomplete.
- Decentralized management – there are no agents in the system that manage the entire system.

JADE Platform. Java Agent Development Framework (JADE) is a software development environment for multi-agent systems and applications. Software agents have an architecture in accordance with the FIPA standard for intelligent agents [16]. The platform includes:

- Agent execution environment - agents are registered and run under the control of the environment.
- A library of classes - that are used to develop agent-based systems.
- A set of graphical utilities for administering and monitoring the life-activity of active agents.
- The development of a multi-agent system based on the JADE platform is carried out in the Java language.

Network-Centric Approach to Managing Complex Systems. Within the framework of this approach, methods and technologies for organizing the management of complex systems are collected due to the decentralization of decision-making processes.

Usually, a network-centric management system is understood as a system of managing a distributed system characterized by the principles of openness, self-organization, a weak hierarchy in the decision-making circuit, and the ability to generate goals within itself [17]. The successful solution of management problems within the network-centric approach consists in maintaining the system in as complete and reliable a state as possible, primarily by including in this process the maximum number of available sources of primary, operational information [18]. Traditionally, a network-centric system has the following characteristics:

- a large number of heterogeneous agents and sources of information;
- network architecture with three types of software agents;
- organization of distributed information storage; operating large volumes of information;
- the decision center is decentralized and can change (migrate over the network) depending on the task being solved;
- availability of mechanisms for coordinating goals and for coordinating actions and / or management between nodes of a multi-agent network.

The considered methodological tools allow the development of distributed decision support systems. In the process of functioning of such a class of systems there is a problem of adjustment of the system to solve a specific application problem. The developed technology of dynamic configuring of network-centric management systems is an attempt to solve this problem.

4 Results

The configuration of a multi-level network-centric support system for personnel security management of a mining and chemical cluster is a set of interrelated configuration objects, they are: components of multi-agent software environment, distributed data storage, elements of a distributed simulation environment. The system configuration is supplemented with attribute information that characterizes the technical aspects of ensuring the operability of the software system. Thus, the main purpose of the technology of dynamic configuring of a network-centric system is to regulate the processes of creating and changing its configuration during the functioning of the system, depending on the stage of solving the applied problem.

The technology of dynamic configuring of distributed multi-level decision support systems is based on the results obtained in previous years by the research team of IIMM KSC RAS: the multilevel recurrent hierarchical model of regional security management; method of coordination of network-centric regional security management based on the recurrent model; information technology of coordination of network-centric management of regional security.

The main stages of the developed dynamic configuring technology are (Fig. 1):

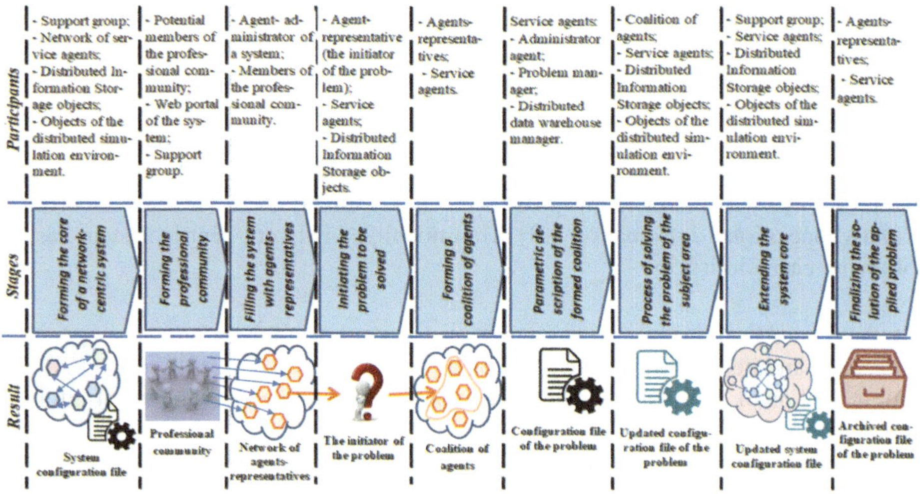

Fig. 1. The main stages of the dynamic configuring technology

1. *Forming the core of a network-centric system.*
   The core of the personnel security management support system is understood as a network of service agents and objects of a distributed information storage, which ensures the implementation of all service functions of the system, in particular, administration of the set of agents; communication of agents; search for agents; management of the catalog of agents; imitation; interaction with the data store; semantic and syntactic analysis; management of the process of solving the set problem of the subject area, etc. Specialists involved in ensuring the operability of the system and its promotion among interested parties (support group) determine the physical location of the main components of the system configuration core (service

agents, software agents, distributed information storage, distributed simulation environment). Software agents are added according to the principle of the JADE agent platform with the recording of the necessary information in a specialized directory and the agent undergoes the procedure of self-registration. The process of forming the system configuration is implemented by the service agent-administrator by collecting parametric descriptions of all core members (in the "request-response" mode) and their automatic adding to the corresponding configuration file.

2. *Forming the professional community.*
   A professional community is a group of people with high qualifications and a set of professional competencies for solving complex applied problems, who does not only offer ideas, but can also translate them into objective reality. When deploying a network-centric system, the initial list of members of the professional community is formed manually by the support group. This list is regulated in accordance with the procedure for preliminary approval of candidates by the support group. The result of this stage is a professional community which is represented in the network by software agents (agents-representatives).
3. *Filling the system with agents-representatives.*
   When installing software that was previously downloaded from the system web portal, the procedure for registering a new software agent starts. This procedure is performed automatically during the interaction of the new agent with the system administrator agent. As a result of questioning a new agent by the agent-administrator, the obtained information about the registered agent is distributed among the catalog of agents, distributed data storage, and a configuration file.
4. *Initiating the problem to be solved.*
   If it is necessary, any member of the professional community, through its software agent-representative, can formulate the problem of ensuring personnel security of the mining and chemical cluster in natural language and initiate the process of its consideration in the system being developed. The problem statement is recorded in the corresponding data structure and converted into a list of keywords (such as the subject of the problem, the object of the problem, and the conditions of the problem) as a result of the communication of specialized agents of the system. Using the generated list of keywords, a search is performed in the distributed information storage of the system for semantically close problems to the specified one.

- If the problem is found in archive, the appropriate system configuration is restored.
- If the problem is relevant, the system invites the initiator of the problem to join the working group that solves this problem.
- If a similar problem is not found, the process of creating a new system configuration for the specified problem is initialized.

If there are semantically close actual problems in the system, a generalized formulation of this group of problems in natural language is formed. This process is accompanied by the selection of the main initiator from among the authors of the initiated problems. The initiator is appointed by analyzing the rank of the agent-representative. The rank is calculated according with assessment of the agent's experience in the system and his competencies.

5. *Forming a coalition of agents involved in solving the initiated problem.*
   This procedure is performed if at the previous stage the initiated problem has not yet been solved in the system, which means that there is no corresponding configuration. When coalition is being formed, the search for agents is carried out on the basis of a certain list of criteria for potential coalition members. The list of criteria is formed in one of the possible ways: based on the description of the problem (in particular, by keywords); based on the criteria proposed by the agent-initiator; hybrid method (a combination of the first and second options).
   A list of potential members of the coalition of agents is formed as a result of the search. In accordance with the list, invitations are sent to the agents to join the coalition. Agent-representatives who respond positively to the request within the specified time period are added to the coalition. This fact is recorded in the corresponding structures of the distributed information storage.
6. *Parametric description of the formed coalition in the form of a configuration file (system configuration).*
   Based on the list of agents (coalition members) formed at the previous stage, their parameters are recorded in the configuration file of the problem under consideration. This procedure is implemented by the agent-administrator as a result of his interaction with specialized agents, such as the problem manager and the manager of the distributed data storage.
7. *The process of solving the problem of the subject area.*
   In accordance with the algorithm for solving the problem of ensuring personnel security [19], developed by the authors earlier, at each stage there is a change in the composition of the program agents included in the coalition. This process involves updating the corresponding configuration file associated with the problem being solved. As an example of actions that lead to configuration updates, the following procedures can be specified: forming a goal tree, synthesizing a function tree, generating process chains, generating a list of performers, etc.
8. *Extending the system core.*
   There is the possibility of increasing the computing power and the volume of stored information of the system being developed, if it is necessary. At a critical level of free space, a request to the support group to increase the information space is sent. The system support group decides how to eliminate the problem of lack of free space by either increasing the disk space on the already existing nodes of the distributed information storage, or adding new nodes. When adding new nodes, the task of updating the configuration of the system core arises, and its solution is carried out by analogy with the algorithm presented at the first stage of this technology. Expansion of the distributed simulation environment when making appropriate changes to the configuration of the system core is carried out in a similar way.
9. *Finalizing the solution of an applied problem.*
   The finalizing of the solution of the problem of the subject area is understood as getting the target results for the problem or concluding that the problem at the moment cannot be solved completely or partially. If the problem remains unresolved, the problem configuration file is archived in the unresolved problems archive. If the problem is solved, this configuration file is added in the archive of the solved problems. All duplicate configurations are removed from all agents who have access

to the configuration file for this problem. Saving the configuration file in the archive is implemented in order to restore it later if it is necessary (for example, initiating a semantically similar problem).

The implementation of the listed stages of the technology of dynamic configuring of the network-centric decision support system can be carried out in a different sequence, since, depending on the method of software implementation some steps can be carried out in parallel.

A feature of the technology is the adaptation of the network-centric system configuration by changing the composition of agents and the services they provide for solving a specific applied problem.

The proposed dynamic configuring technology makes it possible to ensure the functioning of a network-centric system when solving problems of monitoring and identifying potentially dangerous situations for a controlled object, predicting and analyzing possible scenarios of critical situations, as well as automatically generating recommendations for making informed decisions.

From the point of view of the software implementation of the network-centric management support system, the configuration is presented in the form of a configuration file. This file can be presented in various formats, such as XML, JSON, INI, XSL, etc. Figure 2 shows an example of implementing a configuration file in XML format, which is the most common in software systems.

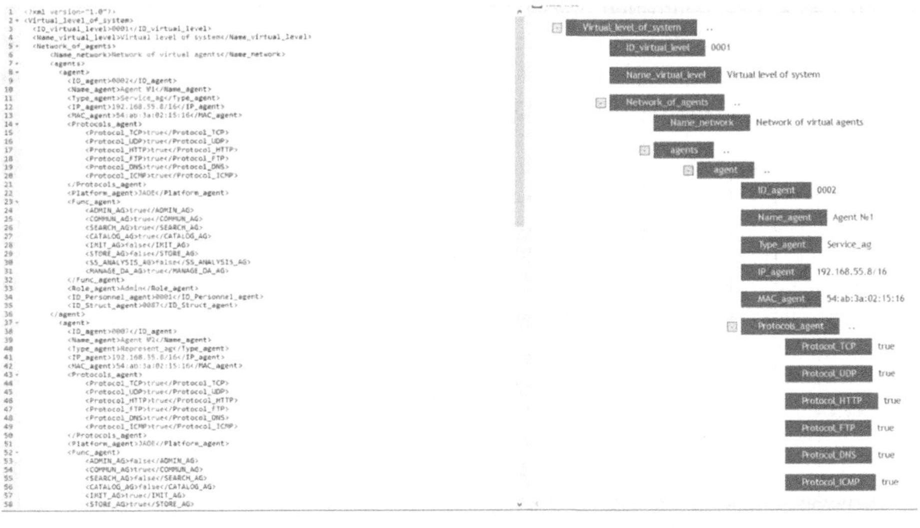

Fig. 2. Example of implementing a configuration of the network-centric system in XML format

Let us note some features of the software implementation of the procedure for operating with the configuration file in the system under development. At the stage of system initialization, the configuration file will contain information about a specific set of software agents and associated data stores. This set of software agents is the core of the system, without the deployment of which it is not possible to operate the entire system under development. The configuration file is updated during any operation that changes the configuration of the system, both its composition and structure, and the parametric description of the configuration unit.

5 Conclusion

This article focuses on the process of dynamic configuring of a network-centric support system for personnel security management in a mining and chemical cluster. Based on analysis of open publications and regulatory documentation in the field of development and maintenance of information systems, the concepts of network-centric system configuration and its dynamic configuring are defined. In accordance with the previously developed general algorithm for the functioning of a network-centric management support system, the technology for dynamic configuring of such system is proposed. The technology is designed for automated deployment of a network of software agents and their parametric configuration depending on the context of the application problem being solved. An example of a software implementation of the decision support system configuration in the form of a configuration file in XML format is given.

The implementation of complex multi-agent network-centric systems allows solving complex tasks for managing personnel security, as well as in other subject areas. The class of such information systems is focused on solving such issues as formalizing and re-using expert knowledge (experience), monitoring and identifying potentially dangerous situations for a controlled object, forecasting and analyzing possible scenarios of critical situations, and also automated generating recommendations for making informed decisions. The proposed dynamic configuring technology makes it possible to ensure the functioning of a network-centric system when solving these tasks.

Acknowledgements. The research was granted by the Russian Foundation for Basic Research (RFBR grant No. 19–07-01193-a «Methods and means of information support of personnel security management of the regional mining and chemical cluster») and supported by the Ministry of Science and Higher Education of the Russian Federation (project No. 0226–2019-0035 «Configuration models and methods of the adaptive multi-level network-centric systems for regional security control in Arctic region of Russian Federation»).

References

1. Strategy of social and economic development of the Murmansk region until 2020 and for the period up to 2025: resolution of the government of the Murmansk region [dated 26.08.2010, no. 383-PP]. https://minec.gov-murman.ru/activities/strat_plan/sub02. Accessed 15 Jan 2021. (in Russia)
2. Grishin, N.N., Grishina, N.A.: Cluster model of industrial development of the Kola region. In: The industrial NORTH, Murmansk: Gelion, no. 10, p. 62 (2009). (in Russia)
3. Bystrov, V.V., Khaliullina, D.N., Malygina, S.N.: Architecture of the decision support system for personnel security of the regional mining and chemical cluster. In: Silhavy, R., Silhavy, P., Prokopova, Z. (eds.) CoMeSySo 2020. AISC, vol. 1294, pp. 442–463. Springer, Cham (2020). https://doi.org/10.1007/978-3-030-63322-6_36

4. GOST R ISO / IEC TO 10032–2007: Reference model of data management. http://protect.gost.ru/document.aspx?control=7&id=173888. Accessed 15 Jan 2021. (in Russia)
5. GOST R ISO / IEC 10746–2–2000: Information technology. Interconnection of open systems. Data management and open distributed processing. Part 2. The basic model. http://protect.gost.ru/document.aspx?control=7&id=138412. Accessed 15 Jan 2021. (in Russia)
6. Federal Standart 1037. https://www.its.bldrdoc.gov/resources/federal-standard-1037c.aspx. Accessed 15 January 2021. (In Russia)
7. ISO/IEC TR 19759:2015 Software Engineering — Guide to the software engineering body of knowledge (SWEBOK). https://www.iso.org/standard/67604.html. Accessed 15 Jan 2021. (in Russia)
8. GOST 15971–90 Information processing systems. Terms and definitions. http://protect.gost.ru/document.aspx?control=7&id=137975. Accessed 15 Jan 2021. (in Russia)
9. Jiang, Z., Zhao, T., Ren, F., Zhao, N.: A Discussion of Dynamic Configuration Technology of Highly Complex System in Information Era. https://www.researchgate.net/publication/317134041_A_Discussion_of_Dynamic_Configuration_Technology_of_Highly_Complex_System_in_Information_Era. Accessed 15 Jan 2021
10. Słota, B.: Dynamic configuration management in microservice architecture with Spring Cloud. http://bartslota.blogspot.com/2018/02/dynamic-configuration-management-in.html. Accessed 15 Jan 2021
11. Kaan, T.: Automating Dynamic Config Management with Spring Cloud Config Bus. https://medium.com/trendyol-tech/automating-dynamic-config-management-with-spring-cloud-config-bus-a34e15397b15. Accessed 15 Jan 2021
12. Sithara Punnassery ONOS. Architecture and Internals Guide. Dynamic Configuration Subsystem. https://wiki.onosproject.org/display/ONOS/Dynamic+Configuration+Subsystem. Accessed 15 Jan 2021
13. Andreev, V.Y., Bazlov, A.F.: Dynamic configuring of the simulation system of the marine environment // Software products and systems – no. 4 – pp.44–46 (2004). (in Russia)
14. Wooldridge, M.: An Introduction to MultiAgent Systems. John Wiley & Sons, p. 366 (2002). ISBN 978–0–471–49691–5.
15. Shoham, Y., Leyton-Brown, K.: Multiagent Systems: Algorithmic, Game-Theoretic, and Logical Foundations, p. 496. Cambridge University Press. (2008). ISBN 978–0–521–89943–7.
16. JAVA Agent DEvelopment Framework. https://jade.tilab.com. Accessed 15 Jan 2021.
17. Efremov, A., Maksimov, D..: Network-centric management system – what is included in this concept? https://docplayer.ru/19261-Setecentricheskaya-sistema-upravleniya-chto-vkladyvaetsya-v-eto-ponyatie.html. Accessed 15 Jan 2021. (in Russia)
18. Ivanyuk V.A., Abdikeev N.M., Pashchenko F.F., Grineva N.V.: Network-centric methods management // Management Sciences. T. 7. no. 1, pp. 26–34 (2017). (in Russia)
19. Bystrov, V.V., Khaliullina, D.N., Malygina, S.N.: Development of the information support system components for personnel security management of the region. In: Silhavy, R. (ed.) CSOC 2020. AISC, vol. 1226, pp. 348–361. Springer, Cham (2020). https://doi.org/10.1007/978-3-030-51974-2_34

Novel Secure Scheme for On-Field Sensors for Data Aggregation in Precision Agriculture

K. Vanishree[1](✉) and G. S. Nagaraja[2]

[1] Department of ISE, R.V College of Engineering, Bengaluru, India
vanishreek@rvce.edu.in
[2] Department of Computer Science and Engineering, R.V College of Engineering, Bengaluru, India
nagarajags@rvce.edu.in

Abstract. With the adoption of various industrial sensors in Precision Agriculture (P.A.), there is an increasing concern of threat during data aggregation by a massive number of on-field sensors to the base station. Owing to the constraint of resources within such farming sensors, it is not feasible to incorporate sophisticated and conventional security approaches. A literature review shows a minimal emphasis on addressing energy and security issues combined relating to P.A. Therefore, the proposed study introduces a novel and simplifies secure authentication mechanism that offers multiple layers of resistance to any adversary for participating in the data aggregation process. The simulation study carried out shows that the proposed system provides better residual energy retention and is high resilience towards a majority of lethal threats reported in P.A.

Keywords: Wireless sensor network · Precision agriculture · Security protocol · Energy efficiency · Data security

1 Introduction

With the modernization of agriculture, there is increasing adoption of various technological advancements giving birth to Precision Agriculture (P.A.) [1, 2]. P.A.'s concept includes usage of industrial sensors, Internet-of-Things, cloud computing, etc., which are quite vulnerable owing to the inclusion of various attacks in it [3, 4]. Frequent usage of on-field sensors promotes Wireless Sensor Network (WSN) usage to perform highly structured farm data aggregation in P.A. With sensors installed in vast acres of farming land, a slight inclusion of any adversary could significantly alter the data, thereby adversely affect the crop cultivation. WSN has already witnessed different attacks and an additional solution towards these attacks [5–8]; however, it is very challenging to ensure the highest security to all the sensors in several thousand to incorporate security protocol. Such sensors are cost-effective, and they have low processing capabilities with local memory. Hence, it is not feasible to execute the conventional encryption algorithm or any other sophisticated algorithm which could affect the data transmission. There are various challenges associated with security threats related to PA [9, 10]; however, less

R. Silhavy (Ed.): CSOC 2021, LNNS 230, pp. 428–436, 2021.
https://doi.org/10.1007/978-3-030-77442-4_37

study is carried out concerning WSN. Therefore, this paper contributes towards harnessing the networking potential of WSN and develop an energy-efficient security protocol in P.A. Energy, which is one of the essential attributes which are potentially affected during a security operation. Therefore, the proposed system contributes towards designing and implementing a unique secure authentication mechanism in P.A. so that no unauthorized node can tamper with the sensitive farming data. The manuscript's organization is as follows: Discussion of existing literature towards security and energy efficiency in WSN in P.A. is done in Sect. 2. At the same time, its associated problems are discussed in Sect. 3. The adopted methodology is discussed in Sect. 4, while algorithm discussion is carried out in Sect. 5. The result is discussed in Sect. 6, while a summary of the proposed work is discussed in Sect. 7.

2 Related Work

At present, various approaches are being carried out towards security and energy in P.A. The work carried out by Ferrag et al. [11] has used blockchain to secure communication in IoT-based agriculture with the main emphasis on achieving privacy. A similar line of security approach adoption was also seen in Wu and Tsai [12]. The idea is to resist distributed denial of service attack in the use case of an intelligent agriculture system. Gupta et al. [13] have discussed the privacy issues associated with smart farming and stated that it is yet an open-end issue in P.A., which is more vulnerable to cyber-physical attacks. A combined study towards security and energy is carried out by Fu et al. [14], where the author has modelled agrometeorology with energy factor to understand the impact of weather conditions. Iqbal et al. [15] have studied security aspects associated with IoT deployment over farming and advice uses of the software-defined network as a security solution. Astillo et al. [16] have developed a model where misbehavior detection is carried out over farming model using Kalman filter. Further adoption of blockchain towards authentication is seen in Arshad et al. [17] to protect unauthorized entries of malicious codes within the application of P.A. Sontowski et al. [18] have developed a prototype for resisting the threat of cyber-attacks in P.A. over on-field sensors. Existing literature has also been carried out towards energy-efficient techniques in WSN associated with farming. Deng et al. [19] have developed an energy harvesting system in WSN from multiple sources using an experimental model. Yu et al. [20] have created a decentralized WSN using single-hop for retaining energy. El-Fouly et al. [21] have designed a fusion mechanism followed by a routing scheme capable of retaining maximum energy. Gulec et al. [22] have developed a harvesting mechanism using solar energy where the network lifetime is extended 6 times. Apart from this, there are various other energy-efficient approaches, e.g., energy harvesting based on 3D printing (Lopez et al. [23]), scheduling using fuzzy logic (Jamroen et al. [24]), bio-inspired algorithm (Zhao et al. [25]), fog computing (Ammad et al. [26]), optimizing using metaheuristic approach (Wang et al. [27]), probability-based routing (Zhang et al. [28]), reinforcement learning (Aoudia et al. [29]), and resource allocation (Azarhava et al. [30]).

3 Problem Description

Review of existing literature towards security and energy efficiency in P.A. has the following research problems: i) there are considerably fewer studies on energy efficiency in P.A. although there are some excellent research attempts on addressing security issues, ii) existing security approaches mainly uses blockchain, which requires dependency of memory, iii) attacks are restricted to few categories and types while solution model is less applicable when exposed to different intrusion scenario, iv) There is combined standard approach where energy and security are addressed together. Therefore, the problem statement is "achieving a balance between robust security and energy efficiency in the presence of multiple environmental parameters in P.A. is quite a challenging task." The next section discusses the solution to address this issue.

4 Proposed Methodology

The proposed system aimed to develop an energy-efficient secure farming data aggregation scheme in P.A. using an analytical model. The study considers a specific set of system parameters to carry out the implementation of a novel and simplified public-key encryption in P.A. The block diagram of the proposed undertaking is as follows:

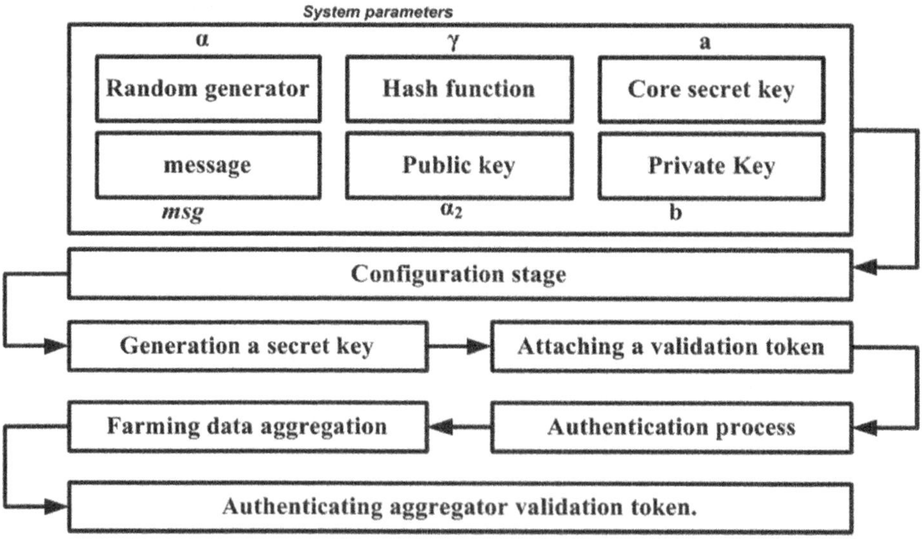

Fig. 1. Block diagram of proposed system

Figure 1 highlights that there are six types of the explicit operation carried out to secure the transmission process. The complete system is based on the fact that a sensor node performs progressive generation of security tokens to ensure multi-tier security in each of the above techniques. Unlike any existing key management approach, the proposed system offers very little secret information storage and focuses on the secret key generation. Connected with a large cloud-based storage unit at the end, the unit accepts the public key and generate further encoded key for public key operation apart from computing private key by the sensors. The study also defines an aggregator node to perform authentication of aggregated data. The sole concept is to retain maximum security with energy efficiency at a target.

5 Algorithm Implementation

The algorithm mainly targets to offer data security in an energy-efficient manner. This algorithm's core target is i) to ensure that only the authenticated sensors in the farming area should participate in the data aggregation process, ii) to ensure that secure communication doesn't demand more resources in the proposed secure P.A. framework. The proposed energy-efficient secure aggregation is as follow:

```
Algorithm for Secure Aggregation in P.A.
Input: h (number of sensors)
Output: ACK (secure data aggregation acknowledgment)
Start
1. For i=0: h
2.       (a, b)∈N  set up
3.        α_0→aα, α_2→bα set up
4.        compute (β_i, n_i)→γ_i(iden_i), aβ_i
5.        generate v_tok→{(r_iα), (n_i+r_iγ), (iden_i), (msg_i)}
6.        For j=0: n
7.                 If λ(n_j+r_jγ, α)= λ(α_o, β_j). λ(r_jα, γ_j)
8.........    flag v_tok as valid
9.                 End
10.      End
11.   generate v_tok for aggregator, agg(v_tok)=(τ_1, τ_2, ...... τ_n)
12.    Fork=0: n
13.           if λ_1= λ_2
14.               Flag ACK→valid data transaction
15.           End
16.      End
17. End
End
```

The proposed algorithm considers the input of all sensor nodes *h* which are installed in the farming area A. The study finds heterogeneous nodes being distributed over different regions (e.g., A_1, A_2...A_m) of the A, i.e., $A = \{A_1, A_2, \ldots, A_m\}$, where m is the number of sub-regions with specific sensors being installed. The study also considers a stationary base station in the corner of a farming region, assuming that it covers all the *h* sensor nodes in *A* (Line-1). The algorithm's first operation is the *configuration stage* where the system opts for (a, b) as randomly generated natural number N (Line-2) followed by computation of random generator α_o and the data center's public key α_2 as shown in Line-3. The variable αis the primary random generator in Line-3. It will also mean that public α_2 and secret key α_3 of the data center will be equivalent to $b\alpha$ and b, respectively. This part of implementation also considers γ_1, γ_2, and γ as hash function while a is considered core secret key, and this complete the first stage of operation. The study considers that the distributed cloud-based storage processing unit receives both public and secret keys. It further generates an encoded public key to avoid misusing the public key.

The next part of implementation is about *generating a secret key*. The sensor computes public key β_i and private key n_i as $\gamma_1(\text{iden}_i)$ and a. γ_1 respectively, where the variable *iden* representsthe identity of the sensor node (Line-4). This algorithm's third operation is to attach *a validation token* with the message to be forwarded. The proposed system generates a validation token v_{tok} as exhibited in Line-5. Four different components are used in obtaining the validation token $v_{tok,}$ i.e., τ_1, τ_2, τ_3, and msg (Line-5). The computation of these are as follows: i) $\tau_1 = n_i + r_i.\gamma$, $\tau_2 = \ r_i\alpha$, $\tau_3 = \gamma_2$ (τ_2, iden_i, *msg*), where *msg* is a message (Line-5). In this process, the sensor node uses a validation token where a private key ni is generated, generating a more random number r_i to compute τ. This completes attaching the validation token with the message.

The next step of this algorithm is to carry out an *authentication process* associated with the sensor node's transmission request. In this process, the sensor node performs authentication by authenticating β and γ as $\gamma_1(\text{iden}_j)$ and $\gamma_2(\tau_2$, iden_j, and $\text{msg}_j)$ respectively. Further, the algorithm computes a condition where the left side of expression in Line-7 is compared with the right-side expression. The variable λ represents a function of cryptography. If both the sides are found to be equal, the validation token is declared to be validated (Line-8). The next process of this algorithm is *farming data aggregation.* In this case, the respective sensor nodes specific to one sub-region within A must forward a validation token $v_{toki} = \{\ \tau_{1,}\ \tau_{2,}\ \text{iden}_i,\ \text{msg}_i\}$. After obtaining the public key α_2, the sensor node computesthe aggregator node validation token (Line-11). The final step of this algorithm is to assess a condition for *authenticating aggregator validation token.* The algorithm formulates two expressions, λ_1 and λ_2 which is equivalent to λ (τ, α) and $[\lambda\ (\alpha_o,\ \beta)^{\chi}.\ \Lambda\ (\tau_i,\ \gamma_i)\ ^{\chi}]$ respectively, where the variable χ is equivalent to the hash of random generator, public key, random generator, the first hash value concerning the variable b (Line-13) if the comparison is found both λ_1 and λ_2 to be equal than it represents an acknowledgment message ACK to highlight a valid data transaction.

From the energy viewpoint, the proposed system offers a specific condition to select an aggregator node. It doesn't let all the farming sensors forward the node to the base station, savingan extensive amount of energy. In this process, an aggregator node is selected based onvast residual energy and with a higher set of connectivity with other nodes in the same sub-region of farming. This ensures a much longer network lifetime of sensors. The complete process doesn't store any information apart from the sensor and public key's identity. Simultaneously, all the rest parameters are computed when there is a necessity of forwarding farming information to the base station. Moreover, retention of sufficient energy ensures higher sustainability of sensor nodes to support extensive rounds of operation.

A closer look into this algorithm will show that public-key cryptography is designed effectively and simplified. Apart from this, the model offers extensive security from any form of attack over the sensor or any communication initiated by it. Therefore, the proposed algorithm provides a potential balance between the sensor's security demands and energy minimization in P.A.

6 Results Discussion

This section discusses the results being obtained by the proposed study. The analytical model consists of 500–1000 sensor nodes being deployed in 1000 × 1000 m^2 with 4000 bytes of packet length and 25 bytes of the control message. The transmittance and receiving energy is fixed to 100 nj considering 1000 simulation rounds. The proposed system is compared with secured LEACH protocol to carry out a comparative analysis, as witnessed in a recent study by Masdari et al. [31]. Secured LEACH offers extensive cryptographic schemes along with energy efficiency and is built exclusively for WSN. Hence, it is an appropriate match for comparingthe proposed system, which also finds security and energy efficiency in WSN deployed for P.A.

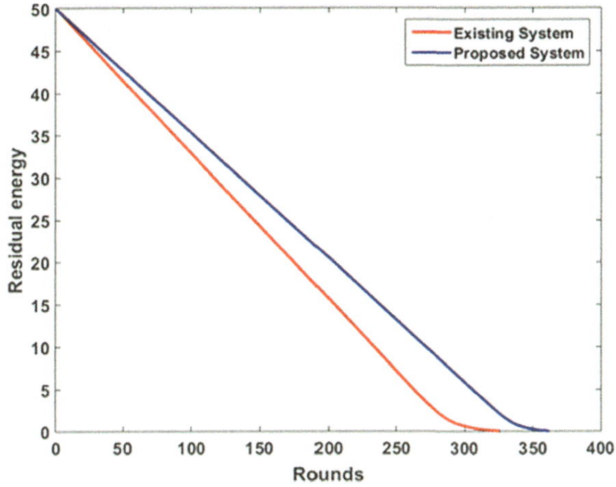

Fig. 2. Comparative analysis of residual energy

Figure 2 highlights the outcome of comparative analysis of residual energy to state that the proposed system offers better energy retention than the existing system. The prime reason behind this is many. The proposed system provides flexibility to place the base station anywhere, restricted in the existing system. Apart from this, the proposed system uses extremely low weight encryption, which is substituted by hashing. Further, generating a secure validation token is carried out in straightforward steps with inclusive parameters that are readily available and require lesser computation. Hence, extensive energy is retained. Apart from this, all nodes are not required to carry out aggregated data transmission to the base station. This task is delegated to a particular node called an aggregator node—this causes potential energy saving.

Figure 3 highlights the comparative analysis of the proposed and existing system concerning alive nodes. The outcome eventually shows that the proposed system can sustain much better rounds compared to the existing protocol.

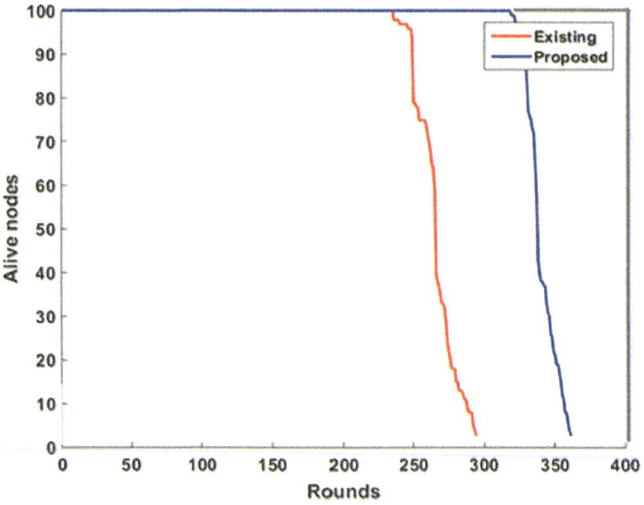

Fig. 3. Comparative analysis of alive nodes

From the security viewpoint, it can be said that the proposed system can potentially resist any key-based attacks, distributed denial-of-service attack, wormhole attack, sinkhole attack, and cyber-physical attack quite quickly. The prime reason behind this is - potential dependencies are being incorporated in formulating security variables in 6 steps of operation as discussed in the algorithm and block diagram. It will mean that attackers will never have any possibility of intruding on the network as their authentication will eventually fail. The existing system cannot facilitate these features.

7 Conclusion

The proposed system discusses a secure authentication mechanism where the on-field sensors in the farming area perform authenticated data transmission to the base station in P.A. Different from any existing security approach, the proposed system doesn't include a sophisticated or complex encryption algorithm, and it carries out entire task of authentication by simplified conditional statements. A hash function is used for encoding the data and other security variables under consideration. Apart from this, a complete operation is executed within 0.3487 s, while the existing system takes approximately 0.65492 s. No memory associated with private keys is stored while the distributed storage unit consistently encodes public keys. Hence, the proposed system offers a simplified framework for energy-efficient and secure authentication in WSN over P.A.

References

1. Kent Shannon, D., David E.: Clay, Precision Agriculture Basics, Wiley Publishers (2020). ISBN: 9780891183662, 0891183663
2. Pedersen, S.M., Lind, K.M.: Precision Agriculture: Technology and Economic Perspectives, Springer International Publishing (2017). ISBN: 9783319687155, 3319687158

3. Mouazen, A., Castrignano, A., Moshou, D., Buttafuoco, G., Naud, O., Khosla, R.: Agricultural Internet of Things and Decision Support for Precision Smart Farming, Elsevier Science, (2020). ISBN: 9780128183748, 0128183748
4. Fagbemi, D.D., Wheeler, D.M., Wheeler, J.C.: The IoT Architect's Guide to Attainable Security and Privacy, CRC Press (2019). ISBN: 9781000762617, 1000762610
5. Xie, H., Yan, Z., Yao, Z., Atiquzzaman, M.: Data collection for security measurement in wireless sensor networks: a survey. IEEE Internet Things J. **6**(2), 2205–2224 (2019). https://doi.org/10.1109/JIOT.2018.2883403
6. Tomić, I., McCann, J.A.: A survey of potential security issues in existing wireless sensor network protocols. IEEE Internet Things J. **4**(6), 1910–1923 (2017). https://doi.org/10.1109/JIOT.2017.2749883
7. Butun, I., Österberg, P., Song, H.: Security of the Internet of Things: vulnerabilities, attacks, and countermeasures. IEEE Commun. Surv. Tutorials **22**(1), 616–644, Firstquarter (2020). https://doi.org/10.1109/COMST.2019.2953364
8. Pundir, S., Wazid, M., Singh, D.P., Das, A.K., Rodrigues, J.J.P.C., Park, Y.: Intrusion detection protocols in wireless sensor networks integrated to the Internet of Things deployment: survey and future challenges. IEEE Access **8**, 3343–3363 (2020). https://doi.org/10.1109/ACCESS.2019.2962829
9. Champion, S.L., et al.: Threats to Precision Agriculture (2018 Public-Private Analytic Exchange Program report) (2020). https://doi.org/10.13140/RG.2.2.20693.37600
10. de Araujo Zanellaab., A.R., Silvac, E., Albini, L.C.P.: Security challenges to smart agriculture: current state, key issues, and future directions, Elsevier, Array, vol. 8 (2020)
11. Ferrag, M.A., Shu, L., Yang, X., Derhab, A., Maglaras, L.: Security and privacy for green IoT-based agriculture: review, blockchain solutions, and challenges. IEEE Access **8**, 32031–32053 (2020). https://doi.org/10.1109/ACCESS.2020.2973178
12. Wu, H., Tsai, C.: An intelligent agriculture network security system based on private blockchains. J. Commun. Netw. **21**(5), 503–508 (2019). https://doi.org/10.1109/JCN.2019.000043
13. Gupta, M., Abdelsalam, M., Khorsandroo, S., Mittal, S.: Security and privacy in smart farming: challenges and opportunities. IEEE Access **8**, 34564–34584 (2020). https://doi.org/10.1109/ACCESS.2020.2975142
14. Fu, X., Yang, D., Guo, Q., Sun, H.: Security analysis of a park-level agricultural energy network considering agrometeorology and energy meteorology. CSEE J. Power Energy Syst. **6**(3), 743–748 (2020). https://doi.org/10.17775/CSEEJPES.2019.03230
15. Iqbal, W., Abbas, H., Daneshmand, M., Rauf, B., Bangash, Y.A.: An in-depth analysis of IoT security requirements, challenges, and their countermeasures via software-defined security. IEEE Internet Things J. **7**(10), 10250–10276 (2020). https://doi.org/10.1109/JIOT.2020.2997651
16. Astillo, P.V., Kim, J., Sharma, V., You, I.: SGF-MD: behavior rule specification-based distributed misbehavior detection of embedded IoT devices in a closed-loop smart greenhouse farming system. IEEE Access **8**, 196235–196252 (2020). https://doi.org/10.1109/ACCESS.2020.3034096
17. Arshad, J., et al.: A novel remote user authentication scheme by using private blockchain-based secure access control for agriculture monitoring. In: 2020 International Conference on Engineering and Emerging Technologies (ICEET), Lahore, Pakistan, pp. 1–9 (2020). https://doi.org/10.1109/ICEET48479.2020.9048218
18. Sontowski, S., et al.: Cyber attacks on smart farming infrastructure. In: 2020 IEEE 6th International Conference on Collaboration and Internet Computing (CIC), Atlanta, GA, USA, pp. 135–143 (2020). https://doi.org/10.1109/CIC50333.2020.00025

19. Deng, F., Yue, X., Fan, X., Guan, S., Xu, Y., Chen, J.: Multisource energy harvesting system for a wireless sensor network node in the field environment. IEEE Internet Things J. **6**(1), 918–927 (2019). https://doi.org/10.1109/JIOT.2018.2865431
20. Yu, C., Yao, D., Yang, L.T., Jin, H.: Energy conservation in progressive decentralized single-hop wireless sensor networks for pervasive computing environment. IEEE Syst. J. **11**(2), 823–834 (2017). https://doi.org/10.1109/JSYST.2014.2339311
21. El-Fouly, F.H., Ramadan, R.A.: E3AF: energy efficient environment-aware fusion based reliable routing in wireless sensor networks. IEEE Access **8**, 112145–112159 (2020). https://doi.org/10.1109/ACCESS.2020.3003155
22. Gulec, O., Haytaoglu, E., Tokat, S.: A Novel distributed CDS algorithm for extending lifetime of WSNs with solar energy harvester nodes for smart agriculture applications. IEEE Access **8**, 58859–58873 (2020). https://doi.org/10.1109/ACCESS.2020.2983112
23. Estrada-López, J.J., Castillo-Atoche, A.A., Sanchez-Sinencio, E.: Design and fabrication of a 3-D printed concentrating solar thermoelectric generator for energy harvesting based wireless sensor nodes. IEEE Sens. Lett. **3**(11), 1–4, Art no. 5500904 (2019). https://doi.org/10.1109/LSENS.2019.2948811
24. Jamroen, C., Komkum, P., Fongkerd, C., Krongpha, W.: An intelligent irrigation scheduling system using low-cost wireless sensor network toward sustainable and precision agriculture. IEEE Access **8**, 172756–172769 (2020). https://doi.org/10.1109/ACCESS.2020.3025590
25. Zhao, X., Cui, Y., Gao, C., Guo, Z., Gao, Q.: Energy-efficient coverage enhancement strategy for 3-D wireless sensor networks based on a vampire bat optimizer. IEEE Internet Things J. **7**(1), 325–338 (2020). https://doi.org/10.1109/JIOT.2019.2952718
26. Ammad, M., et al.: A novel fog-based multi-level energy-efficient framework for IoT-enabled smart environments. IEEE Access **8**, 150010–150026 (2020). https://doi.org/10.1109/ACCESS.2020.3010157
27. Wang, H., Li, K., Pedrycz, W.: An elite hybrid metaheuristic optimization algorithm for maximizing wireless sensor networks lifetime with a sink node. IEEE Sens. J. **20**(10), 5634–5649 (2020). https://doi.org/10.1109/JSEN.2020.2971035
28. Zhang, M., Cai, W.: Energy-efficient depth based probabilistic routing within 2-hop neighborhood for underwater sensor networks. IEEE Sens. Lett. **4**(6), 1–4, 7002304 (2020). https://doi.org/10.1109/LSENS.2020.2995236
29. Ait Aoudia, F., Gautier, M., Berder, O.: RLMan: an energy manager based on reinforcement learning for energy harvesting wireless sensor networks. IEEE Trans. Green Commun. Netw. **2**(2), 408–417 (2018). https://doi.org/10.1109/TGCN.2018.2801725
30. Azarhava, H., Musevi Niya, J.: Energy-efficient resource allocation in wireless energy harvesting sensor networks. IEEE Wirel. Commun. Lett. **9**(7), 1000–1003, (2020). https://doi.org/10.1109/LWC.2020.2978049
31. Masdari, M., Bazarchi, S.M., Bidaki, M.: Analysis of Secure LEACH-Based Clustering Protocols in Wireless Sensor Networks, J. Netw. Comput. Appl. **36**(4), 1243–1260 (2013). http://dx.doi.org/https://doi.org/10.1016/j.jnca.2012.12.017

Conversion Methods of Data Mining Analysis in Algorithms of Statistical and Nowcasting Forecast of Convective Precipitation

David Šaur(✉) and Jaromír Švejda

Faculty of Applied Informatics, Tomas Bata University in Zlin, Nad Stranemi, 4511 Zlín, Czechia
{saur,svejda}@utb.cz

Abstract. This article focuses on the description of computational methods for the conversion of products for the purposes of calculating the statistical and nowcasting prediction of convective precipitation. Convective precipitation can reach very high intensities and cause flash floods, which can endanger human lives and health and cause material damage on a local scale. The methodological part describes computational methods designed for the conversion of input data, which are radar products CAPPI and VIL, including the combined product SUM MERGE. The resulting part is focused on the evaluation of these computational methods, where the outputs of this evaluation are applied to the conversion algorithm.

Keywords: Weather forecast · Convective precipitation · Algorithm · Conversion methods · Data mining · Radar measurement · Crisis management

1 Introduction

Nowadays, the effects and impacts of global climate change have been affecting our entire society. Natural disasters are one of the impacts of climate change, which negatively affect the extent of an emergency or crisis situation a certain size of the area [2, 3]. In the Czech Republic, they most often occur as floods, droughts, wind or snow disasters, or fires [1]. Flash floods are the most common natural disaster with local negative effects and high intensity caused by a combination of factors such as heavy convective (torrential) precipitation, strong soil saturation and slow convective storm speed. Heavy convective precipitation is one of the main causes of flash floods, the local and very short occurrence of which is the main subject of this research [1–3].

The forecast of convective precipitation is realized by numerical models of weather prediction (NWP models) and nowcasting systems, or a combination of both forecasting systems in practice. The main disadvantage of these systems is that it does not allow affecting all factors and conditions influencing the occurrence of convective (torrential) precipitation. For example, many numerical models use the hydrostatic core of the convection parameterization model, where a major problem is the lack of data from ground meteorological stations and aerological soundings due to incomplete initial conditions

R. Silhavy (Ed.): CSOC 2021, LNNS 230, pp. 437–450, 2021.
https://doi.org/10.1007/978-3-030-77442-4_38

for computing the forecast [4–6]. On the other hand, nowcasting systems operate with radar measurement outputs that do not consider local and orographic conditions for the initiation of convection in the ground layer of the atmosphere. Based on these shortcomings, the Algorithm of Storm Prediction has been developed, which uses input data from NWP models, not directly precipitation forecast. For this reason, input data are data representing temperature, humidity, wind and orographic conditions for the initial formation of convection, such as special indices of instability and relief characteristics such as terrain slope and orientation, ridge width and valley and other characteristics. These data significantly better represent the conditions of short-term dynamic development of convective clouds [7–9].

Currently, the Algorithm of Storm Prediction evaluates only data from NWP models. The second proposed approach, which will be implemented in this algorithm, is a forecast based on a statistical analysis of data on historical situations associated with the occurrence of convective precipitation in the Czech Republic in the years 2003 to 2020. The crucial goal of this statistical analysis is to find one or more historical situations that are most similar to the predicted situation based on the defined comparative-selective attributes. This analysis is feasible only under the assumption of data obtained from the algorithm of conversion data mining analysis [10–12] which will be shown in this article. The input data are data from the CAPPI, VIL radar products and the SUM MERGE combined product in the PNG graphic format converted by this algorithm into values of coefficients in the interval from 0 to 3 for 206 municipalities with extended powers in the Czech Republic [1, 13–15].

The aim of this article is to provide information on computational methods of data mining processing of radar and station measurement data, which will be used to calculate predictions of convective precipitation and the risk of flash floods for all 206 municipalities with extended powers in the Czech Republic, including the Zlín region. These outputs will be intended to support the decision-making of the crisis management authorities of the region as part of the early warning system against these dangerous phenomena.

2 Methods

This chapter focuses on the description of computational methods for the purpose of conversion of data of radar and station measurements of convective precipitation to values of coefficients from 0 to 3 in order to unify the values of parameters with different units of their quantities.

The prediction of convective precipitation will be realized by the following two algorithms:

1. Algorithm for statistical prediction of convective precipitation - forecasts calculated from the outputs of statistical analysis of historical situations and selected triggering predictors of convection and storm intensity.
2. Algorithm for nowcasting convective precipitation forecast - forecasts calculated on the basis of the same outputs as for the statistical forecast, including outputs from nowcasting (very short-term convective precipitation forecasts).

Statistical analysis of historical situations includes, to determine one or more historical situations for the predicted situation, the following comparatively selective attributes listed in Table 1:

Table 1. Compared-selected atributes of statistical analyses

Atributes	Forecast parameter (NWP models)
1. Direction and speed precipitation movement	Wind direction of 600 hPa Corfidi vector of speed movement of convective storm
2. Storm tracks	Penetration prediction of precipitation
3. Temporal occurrence	Penetration prediction of temporal occurrence
4. Type of convection clouds	Prediction parameters of fronta lor nonfrontal convection clouds

Table 1 present the four attributes for selecting one or more historical situations in the left column and the associated prediction parameters in the right column. The first attribute contains information on the selection of a group of one direction of precipitation from a total of eight directions movement and at the same time information on the speed of precipitation movement calculated by the Corfidi vector of convective storm motion. Prediction of the speed of precipitation movement is a very important factor influencing the occurrence of flash floods. It is true that the slower the convective storm moves, the higher the risk of flash floods while other conditions are met.

The second attribute is focused on the selection of a more accurate prediction of convective precipitation using storm tracks, which were determined on the basis of the first attribute (the direction of precipitation movement).

The third attribute addresses the temporal occurrence of precipitation to determine the specifics of day and night convection for the purposes of applying a given set of weights of individual relief characteristics.

The last attribute is aimed on determining the type of convective cloud, which will be used in the calculation with the weights of the relief characteristics as in a similar case as for the third attribute.

Selecting one group or historical situations are then calculated based on the input station and radar data, these two outputs:

- Statistical estimation of the probability of convective precipitation occurrence
- Statistical estimation of risk of flash floods.

The following sections describe the input data and methods for calculating data conversion in more detail.

2.1 Input Data

The input data for the data conversion algorithm are:

- **Radar product CAPPI** (Constant Altitude Plan Position Indicator) is a radar reflectivity field computed for a given altitude. Radar reflectivity is calculated by Marshall-Palmer relation:

$$Z = aI^b \tag{1}$$

$$I = 10(Z - 10log(a))/10b, \tag{2}$$

where a and b are the experimentally determined constants (a = 200, b = 1,6).

- **Radar product VIL** is the vertically integrated liquid water content determined assuming a Marshall-Palmer distribution; the VIL value is determined according to the following formula:

$$VIL = 3,44.10^{-6} \int_{hz}^{ht} Z^{4/7} dh, \tag{3}$$

where Z [mm6.m-3] is radar reflectivity, hz [m] the height of the cloud level and ht [m] is the height of the upper limit of clouds.

- **Combined product SUM MERGE** is a product that includes outputs from the CAPPI radar product and station precipitation measurements from ground meteorological stations. The amount of precipitation is calculated using interpolation methods, which are the inverse distance method and the crigging method [1].

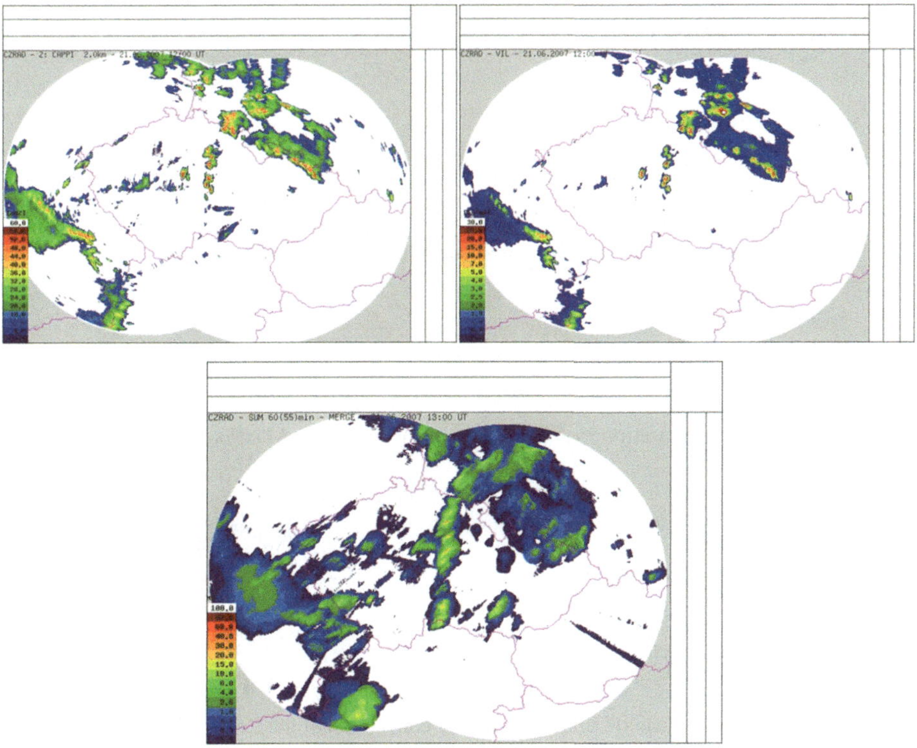

Fig. 1. Radar and combined products as input data [16]

2.2 Conoversion Methods

Data conversion is performed using the following computational methods:

- MaxPixels.
- MAQ (MaxPixels & Average & Quartile).
- Majority.
- Average.

MaxPixels Conversion Method

The aim of this method is to determine the value of the coefficient based on the maximum value of one or more pixels with the classification scale of the given radar product (CAPPI or VIL) for the territory of municipalities with extended powers according to the formula:

$$C_{max} = (value) \tag{4}$$

Figure 1 illustrate the method settings, where the MaxPixels method is 1. The number of pixels is two due to possible display errors (Figs. 2 and 3).

```json
{
  "Setting": {
    "method": [
      "max",
      "average",
      "majority",
      "maq"
    ],
    "colorOutput": [ "#FFF", "#66FF33", "#FFFF00", "#FFC000", "#FF0000" ],
    "methodActive": 1,
    "maxPixels": 2,
    "spectrum": "config/scales/Model_Radarové_snímky&CAPPI_2_km.csv",
    "data": "data/test/",
    "export": "export/",
    "exportType": [ "text"],

  }
}
```

Fig. 2. Configuration json file with parameters of the MaxPixels method

Fig. 3. Map outputs of the MaxPixels method after conversion of the CAPPI (left) and VIL (right) radar products

MAQ Conversion Method

The MAQ (Maximum Average Quartile) is a new conversion method which is determined to more accurately calculation of the converted data for the SUM MERGE combined product according to the following formula:

$$C_{MAQ} = avg(max(values), avg(quantil(0.95), values)) \tag{5}$$

This algorithm counts the value of required parameter according to the following steps:

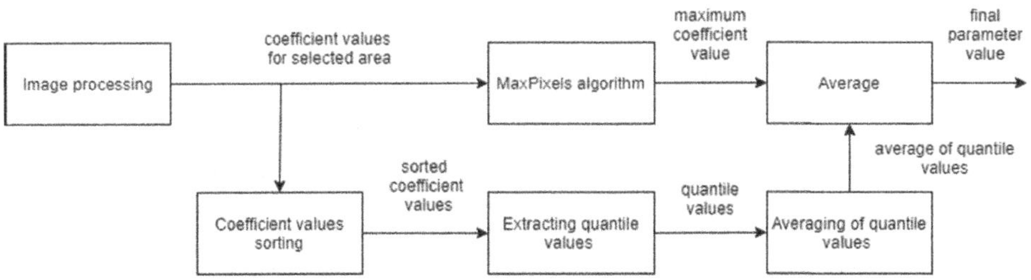

Fig. 4. Scheme of MAQ method

- During image processing, coefficient values are assigned to individual pixels of processed area (this area is clearly defined by real borders of territorial unit obtained from processed image map). Then, these coefficient values are used in two computing branches.
- The first computing branch includes MaxPixels algorithm, whose task is to find the maximum value among all processed pixels.
- The second branch uses quantile characteristic for data processing. First, it is necessary to sort the values in ascending order. Then it is possible to extract the part of coefficient values according to used quantile function. Current computation settings use 0.95 quantile. Finally, the average of extracted part of coefficient values is counted; thus for 0.95 quantile, this step processes only last 5% of all input values. The output of this branch is the average of quantile values.
- Outputs of both computing branches are then combined using average function to obtain final parameter value at the end of processing.

The main task of the second computing branch is to refine the result obtained from the first computing branch (MaxPixels algorithm) according to Fig. 5:

Figure 4 documents the data conversion procedure according to the MAQ method, which is shown in more detail in four steps.

```
{
  "Setting": {
    "method": [
      "max",
      "average",
      "majority",
      "maq"
    ],
    "colorOutput": [ "#FFF", "#66FF33", "#FFFF00", "#FFC000", "#FF0000" ],
    "methodActive": 3,
    "maxPixels": 2,
    "quartile": 0.95,
    "spectrum": "config/scales/Model_Sumarizace_srazek&Radary_srážkoměry.csv",
    "data": "data/test/",
    "export": "export/"
  }
}
```

Fig. 5. Configuration json file with parameters of the MAQ method

Figure 5 represents the setting of the MAQ method with a percentile value of 0.95, which was determined based on testing of this method. The test results are given in the third chapter.

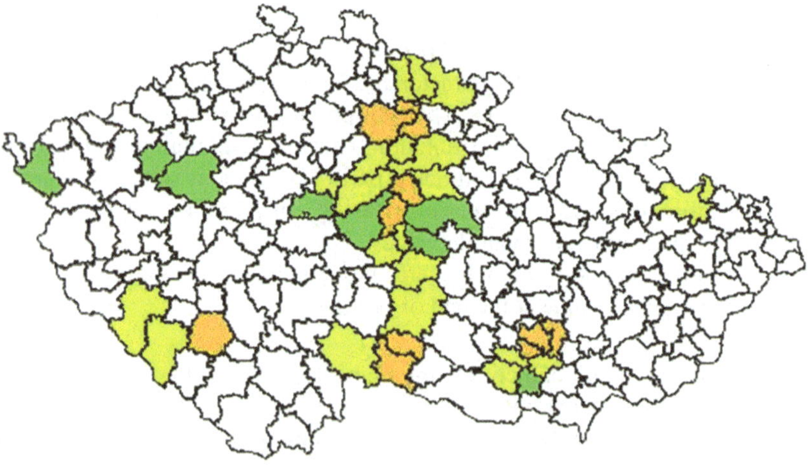

Fig. 6. Map output of the MAQ method after conversion of the SUM MERGE combine product

Figure 6 illustrate the map output obtained by the MAQ conversion method. The major goal of this method is to obtain more accurate data about the measured situation. The rationale for using this method is described in the following chapter.

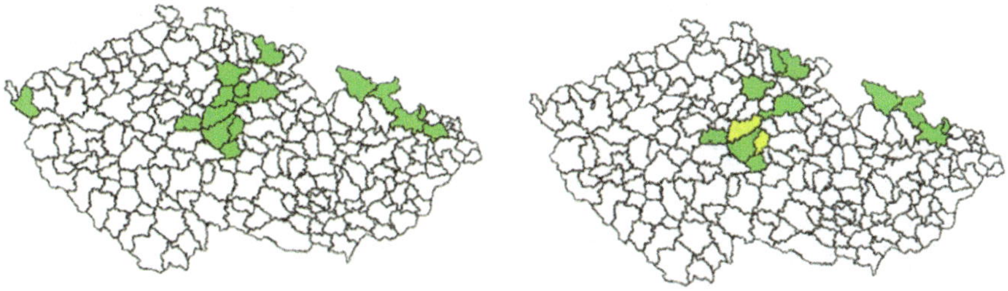

Fig. 7. Map outputs of the Majority method after conversion of the CAPPI (left) and VIL (right) radar products

Majority Method

The aim of the Majority method is to determine the values of coefficients on the basis of the most frequently occurring value of pixels from the classification scale of a given radar product for the territory of municipalities with extended powers. The outputs of this method are calculated as the most repetitive pixel value in the set of all pixels:

$$C_{MJ} = countif(values) \quad (6)$$

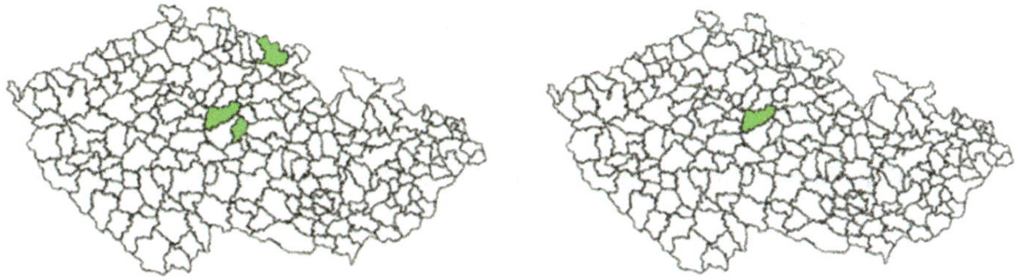

Fig. 8. Map outputs of the Average method after conversion of the CAPPI (left) and VIL (right) radar products

Figure 7 show the outputs of the Majority method on the conversion analysis of CAPPI and VIL radar products. The map outputs show that this method does not provide an ideal output in terms of capturing the most intense precipitation.

Average Method

This method is focused on determining the value of the coefficient based on the average value of all pixels from the classification scale of the radar product for the territory of municipalities with extended powers. The outputs of this method are calculated as the average value of a pixel in the set of all pixels (Fig. 8):

$$C_{AV} = avg(values) \quad (7)$$

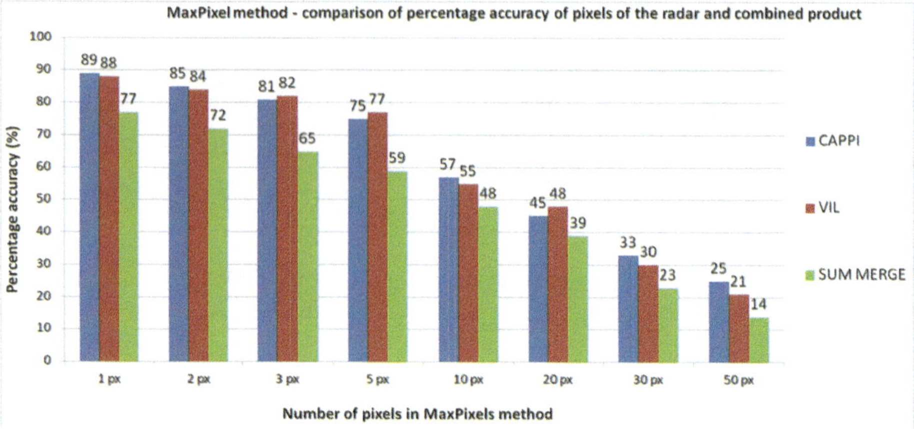

Fig. 9. Percentage accuracy of the MaxPixels method for input data

3 Results

This chapter summarizes the results of evaluating the accuracy of these computational methods, which were then used to configure the conversion algorithm:

- MaxPixels for CAPPI, VIL a SUM MERGE product.
- MAQ method.
- Summarization of all conversion methods.

Values of accuracy computational methods were calculated on this formula:

$$\bar{X} = \frac{1}{n\,i} \sum\nolimits_{i=1}^{n} x \tag{8}$$

where $\sum$xi is the number of accuracy evaluation points for the given conversion method.

First, the accuracy of the MaxPixels method for the input data is evaluated:

Figure 9 document the average accuracy of the number of pixels for the conversion of the CAPPI and VIL radar products, and the SUM MERGE combined product. The graph shows that with increasing number of pixels, the accuracy of the MaxPixels conversion method decreases linearly. The number of pixels of the CAPPI and VIL radar products was first set to 1. Further testing revealed that a value of 1 pixel sometimes provided skewed results due to the thicker thickness of the polygon boundaries of municipalities with extended powers. Therefore, the number of pixels was set to 2. At the same time, the accuracy of SUM MERGE did not produce very satisfactory results, and therefore the MAQ method had to be used.

Second, an evaluation of the accuracy of the MAQ method was performed on the analysis of historical situations associated with the occurrence of:

- Flat convective precipitation.
- Local convective precipitation.

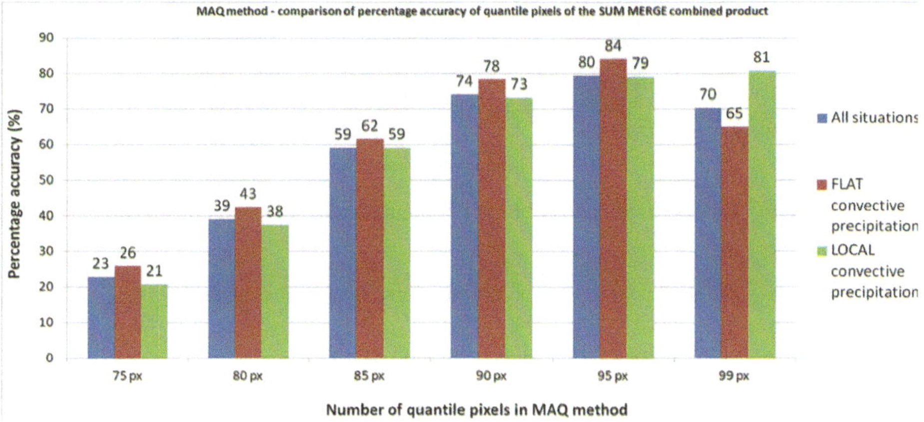

Fig. 10. Percentage accuracy of the MAQ method for surface and local convective precipitation

Figure 10 show that with increasing number of pixels, the accuracy of the MAQ method also increased during the conversion analysis of the SUM MERGE combined product. The highest values of accuracy were achieved for surface convective precipitation for the quantile with a value of 0.95. The average accuracy of this quantile gave the best results and was therefore used as a reference quantile for this product only. On the other hand, better results were obtained for the highest number of pixels in situations associated with local heavy convective precipitation, where it would probably be better to use the MaxPixels method.

Figure 11 summarize the results of all computational conversion methods. The Majority and Average methods generally achieved lower accuracy values than the MaxPixels and MAQ methods for SUM MERGE product.

Table 2 present the methods and their parameter settings for the conversion of CAPPI and VIL radar products, including SUM MERGE product. This configuration is based on the evaluation of calculation methods in Chapter 3.

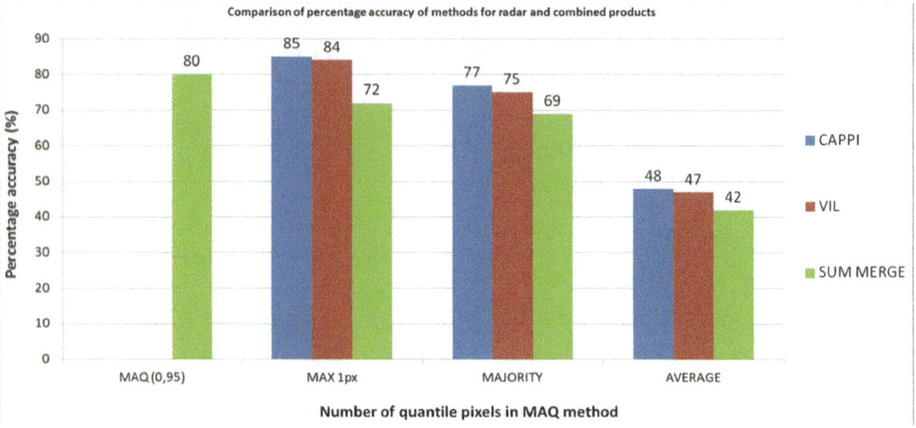

Fig. 11. Results summarization of computational conversion methods

Table 2. Summary of computational conversion method

Input data (product)	Method parameters
CAPPI	MaxPixels = 2
VIL	MaxPixels = 2
SUM MERGE	MAQ (quantile) = 0,95

4 Conclusion

The aim of this article was to provide information on computational methods aimed at the conversion of radar and station data for the purpose of calculating the outputs of statistical and nowcasting prediction algorithms. This article also included a summary of the results of the accuracy evaluation of the individual methods that were used for the conversion algorithm. The accuracy of calculation methods was calculated for area and local convective precipitation. A total of four methods were evaluated - MaxPixels, MAQ, Majority and Average. The best results were obtained with the MaxPixels method for CAPPI and VIL radar products for values of one or two pixels. In practice, two pixels were used due to skewed calculation results. The MAQ method obtained the best results especially for the SUM MERGE product. This product is very important because it best captures the real state of measured hourly precipitation, especially for the occurrence of torrential convective precipitation. The quantile with a value of 0.95 (95 pixels) achieved the best results out of a total of six-pixel values. In addition to these methods, the Majority and Average methods were also tested. These methods did not give very satisfactory results and were therefore not used for data conversion calculations.

The aim of this evaluation was to obtain accurate information about the parameters of calculation methods, on the basis of which the converted values of coefficients from 0 to 3 are calculated for partial outputs of the statistical forecast of convective precipitation. The outputs from the algorithms of statistical and nowcasting prediction of convective precipitation are intended as information support for crisis management authorities in the case of implementation of preventive and preparatory flood control measures.

Acknowledgments. This work was supported by the project No. VI20192022134 - System of more accurate prediction of convective precipitation over the regional territorial unit.

References

1. Šaur, D.: Support for crisis management of the region in terms of evaluation of flood events. Zlín: Academia Centrum TBU in Zlín (2017). ISBN 978–80–7454–712–6.
2. Rulfová, Z., Beranová, R., Kyselý, J.: Climate change scenarios of convective and large-scale precipitation in the Czech Republic based on EURO-CORDEX data. Int. J. Climatol. **37**(5), 2451–2465 (2017). https://doi.org/10.1002/joc.4857
3. Zdenek, S., Dusan, V., Jan, S., Ivan, M., Miroslav, M.: Protection from flash floods. In: (2015) Proceedings of the 26th International Business Information Management Association Conference - Innovation Management and Sustainable Economic Competitive Advantage: From Regional Development to Global Growth, IBIMA, pp. 1359–1363 (2015). https://www.scopus.com/inward/record.uri?eid=2-s2.0-84976391745&partnerID=40&md5=923aa2f309578593d8b5e2cc503d02de
4. Flora, M.L., Potvin, C.K., Wicker, L.J.: Practical predictability of supercells: exploring ensemble forecast sensitivity to initial condition spread. Mon. Weather Rev. **146**(8), 2361–2379 (2018). https://doi.org/10.1175/MWR-D-17-0374.1.ISSN0027-0644
5. Ravazzani, G., et al.: Potentialities of ensemble strategies for flood forecasting over the milano urban area. J. Hydrol. **539**, 237–253 (2016). https://doi.org/10.1016/j.jhydrol.2016.05.023
6. Wang, Y., Belluš, M., Ehrlich, A., et al.: 27 Years of regional cooperation for limited area modelling in Central Europe. Bull. Am. Meteor. Soc. **99**(7), 1415–1432 (2018). https://doi.org/10.1175/BAMS-D-16-0321.1.ISSN0003-0007
7. Novák, P.: The Czech Hydrometeorological Institute's Severe Storm Nowcasting System. https://doi.org/10.1016/j.atmosres.2005.09.014
8. Mejsnar, J., Sokol, Z., Minářová, J.: Limits of precipitation nowcasting by extrapolation of radar reflectivity for warm season in Central Europe. Atmos. Res. **213**, 288–301 (2018). https://doi.org/10.1016/j.atmosres.2018.06.005.ISSN01698095
9. James, P.M., Reichert, B.K., Heizenreder, D.: NowCastMIX: automatic integrated warnings for severe convection on nowcasting time scales at the german weather service. Weather Forecast. **33**(5), 1413–1433 (2018). https://doi.org/10.1175/WAF-D-18-0038.1.ISSN0882-8156
10. Ren, X., Li, X., Ren K., Song J., Xu Z., Deng K., Wang X. Deep learning-based weather prediction: a survey. Big Data Res. **23** (2021). ISSN 22145796. https://doi.org/10.1016/j.bdr.2020.100178
11. Qamar, S., Khalique, A., Grzegorczyk, M.A.: On the Bayesian network based data mining framework for the choice of appropriate time scale for regional analysis of drought Hazard. Theoret. Appl. Climatol. **143**(3–4), 1677–1695 (2021). https://doi.org/10.1007/s00704-021-03530-2

12. Rahmat, F., Zulkafli Z., Juraiza Ishak, A., Mohd Noor, S.B., Yahaya, H. Masrani, A.: Exploratory data analysis and artificial neural network for prediction of leptospirosis occurrence in Seremban, Malaysia based on meteorological data. Front. Earth Sci. **8** (2020). ISSN 2296–6463. DOI:https://doi.org/10.3389/feart.2020.00377
13. Ahijevych, D., Pinto, J.O., Williams, J.K., Steiner, M.: Probabilistic forecasts of mesoscale convective system initiation using the random forest data mining technique. Weather Forecast. **31**(2), 581–599 (2016). https://doi.org/10.1175/WAF-D-15-0113.1
14. Šaur, D., Švejda, J., Žák, R. Conversion of meteorological input data implemented in the algorithm of storm prediction. In Advances in Intelligent Systems and Computing. Berlín : Springer Verlag, vol. 986, pp. 357–367 (2019). ISSN 21945357. ISBN 978–3–030–19812–1
15. Šaur, D.: Statistical, quantitative probability and nowcasting forecasting methods of severe convective storms. WSEAS Trans. Environ. Dev., roč. 14, č. neuveden, s. 607–618 (2018). ISSN 1790–5079. Biological Algorithms (5) – Neural Networks: Learning – Backpropagation, https://www.root.cz/clanky/biologicke-algoritmy-5-neuronove-site/
16. Czech Hydrometeorological Institute. Radar and station data (2020)

Interval of Spatial Correlation of Fading in a Single-Beam Decameter Radio Line

V. P. Pashintsev[1], S. A. Koval'[2], A. F. Chipiga[1], and Alexander D. Skorik[3](✉)

[1] North Caucasus Federal University, Stavropol, Russian Federation
[2] Military Academy of Communications named after Marshal of the Soviet Union, S. M. Budyonny, Saint Petersburg, Russian Federation
[3] JSC "Russian Institute of Powerful Radio Engineering", Saint Petersburg, Russian Federation

Abstract. A method for determining the interval of spatial correlation of fading in a decameter radio line with a single discrete beam (mode), which are caused by wave diffraction on small-scale inhomogeneities of the ionosphere, was developed. The dependence of the interval of the spatial correlation of the fading in a single-beam of decameter radio line from parameters of small-scale ionospheric inhomogeneities that are equivalent to the length of the radio line and from selection of the operating frequency signal through the value of standard deviation of fluctuations of the phase of the wave front at the outlet of the inhomogeneous ionosphere was obtained. It is shown that under conditions when the fluctuations of the phase front of the wave at the ionospheric outlet exceed 1.25 radians, the known simplified expression for estimating the interval of spatial correlation of fading can be used in a single-beam decameter radio line with an error of no more than 5% .

Keywords: Decameter radio line · Diffusivity · Small-scale inhomogeneities of the ionosphere · Phase front fluctuations · Fading · Normalized spatial correlation function · Spatial correlation interval

1 Introduction

It is known that in decameter (DCM) radio lines with a single path (beam) of wave propagation to the receiving point, interference fading of signals occurs due to scattering (diffraction) of the wave on small-scale inhomogeneities of the ionosphere, which obey the Rayleigh or Rice distribution laws [1, 2].

The presence of fading causes a significant deterioration in the noise immunity of signal reception in the DCM radio line. The most widely known [2–8] method of combating fading and improving the noise immunity of signal reception is the use of spatially spaced signal reception. The greatest noise immunity is achieved when the condition of uncorrelated fading is met at the outputs of antennas spaced by a distance $\Delta\rho$ in the space $\Delta\rho \gg \Delta\rho_c$, where $\Delta\rho_c$ is the interval of spatial correlation of fading [4].

R. Silhavy (Ed.): CSOC 2021, LNNS 230, pp. 451–468, 2021.
https://doi.org/10.1007/978-3-030-77442-4_39

The ratio $\Delta\rho/\Delta\rho_c$ determines a value of spatial correlation coefficient (or normalized spatial correlation function) of fading in the DCM channel (radio line) of communication [6, 7] $K_n(\Delta\rho) = \exp\left(-(\Delta\rho/\Delta\rho_c)^2\right)$. When the condition of uncorrelation $\Delta\rho \gg \Delta\rho_c$ is met, $K_n(\Delta\rho) = \exp\left(-(\Delta\rho/\Delta\rho_c)^2\right) \approx 0$. With a small antenna spacing and the condition of complete correlation $\Delta\rho \ll \Delta\rho_c$ of fading at their outputs $K_n(\Delta\rho) = \exp\left(-(\Delta\rho/\Delta\rho_c)^2\right) \approx 1$. In this case, the noise immunity of spatially spaced signal reception is significantly worse than in the case of uncorrelated fading, and the energy loss can reach several tens of decibels [5].

It is obvious that the solution of the problem of ensuring the required noise immunity of spatially spaced signal reception in the DCM radio line should be based on determining the interval of spatial correlation of fading $\Delta\rho_c$.

Currently, the value of the interval of spatial correlation of fading $\Delta\rho_c$ is determined experimentally ambiguously [3, 4, 6, 7]. It varies in the range from $\Delta\rho_c = (10\ldots25)\lambda_0$ [3] to $\Delta\rho_c = (10\ldots100)\lambda_0$ [4], which makes it difficult to design antenna systems for receiving signals with a given wavelength $\lambda_0 = c/f_0$. In practice, for DCM communication systems, the amount of spacing is usually chosen as either $\Delta\rho_c = 10\lambda_0$ [3, 7] or $\Delta\rho_c \approx 200$ m [4, 8].

According to [7], in a DCM radio line, the interval of spatial correlation $\Delta\rho_c$ of fading depends on the size (l) of inhomogeneities in the ionosphere, the length (L), of the radio line, the carrier (operating) frequency (f_0) of the wave, and the direction of spacing.

The analytical dependence $\Delta\rho_c = \psi(l, L, f_0)$ or a single-beam DCM radio line is found in [9] based on the definition of: 1) normalized spatial correlation function in the traditional form $K_n(\Delta\rho) \approx \exp\left(-(\Delta\rho/\Delta\rho_c)^2\right)$, 2) the analytic dependence $\Delta\rho_c \approx l_s/\sigma_\varphi$ of the interval of spatial correlation of fading of the received signals from fluctuations of the phase of the wave front at the outlet of the inhomogeneous ionosphere $\sigma_\varphi = \psi(l, L, f_0)$, depending on the size of inhomogeneities, length of the radio line and operating frequency of signals.

However, the expressions obtained in [9] for the normalized spatial correlation function of a single-beam DCM radio line and the interval of spatial correlation of fading of received signals.

$$K_n(\Delta\rho) \approx \exp\left(-(\Delta\rho/\Delta\rho_c)^2\right); \tag{1}$$

$$\Delta\rho_c \approx l_s/\sigma_\varphi \tag{2}$$

are approximate and require refinement.

It should be noted that in [8] for a single-beam DCM radio line, a complex dependence $K_n(\Delta\rho) = \psi(l, \sigma_\varphi)$ of the spatial correlation coefficient of fading on the size of inhomogeneities and fluctuations of the phase front of the wave at the outlet of the inhomogeneous ionosphere was obtained. However, the value σ_φ for a given DCM communication range is determined experimentally and therefore the desired dependence $\Delta\rho_c = \psi(\sigma_\varphi) = \psi(l, L, f_0)$ is not established in [8].

The goal of the article is to develop a refined method for determining the dependence $\Delta\rho_c = \psi(\sigma_\varphi) = \psi(l, L, f_0)$ of the spatial correlation interval of fading in a single-beam decameter radio line on the parameters of ionospheric inhomogeneities, the length of the radio line and the operating frequency (f_0) of the signal.

2 Analysis of DCM Wave Propagation in the Ionosphere with Small-Scale Inhomogeneities

To achieve this goal, we first analyze the process of propagation (Fig. 1) of a monochromatic wave with a carrier (operating) frequency f_0 n a single-beam DCM radio line in two stages: 1) from the transmitter (TRM) to the outlet from the reflecting layer F of the ionosphere, taking into account the influence of small-scale ($10\ldots10^3$ m) inhomogeneities of the ionosphere (Fig. 1a); 2) from the outlet from the ionosphere to the location of the receiver (RCV) (Fig. 1b).

Let the TRM emit a complex monochromatic signal with frequency $\omega_0 = 2\pi f_0$, amplitude $S_t = \sqrt{P_t}$, power P_t and initial phase φ_t:

$$\dot{s}_t(t) = S_t \exp(j(\omega_0 t - \varphi_t)) = \sqrt{P_t} \exp(j(\omega_0 t - \varphi_t)). \tag{3}$$

The ionospheric reflecting layer F is characterized with the following parameters [7, 10–14]:

1) change in the height h of the average electron concentration (EC) $\bar{N}(h)$, el/m^3, from the lower boundary h_0 of the layer to the height of the maximum ionization of the ionosphere with a critical frequency $f_{cr} = \left(80,8\bar{N}(h_m)\right)^{0,5}$ Hz;
2) spatial fluctuations of the EC $\Delta N(x, y, h) = \Delta N(\rho, h)$ in small-scale inhomogeneities of the ionosphere relative to the average (background) value $\bar{N}(h)$, characterized with the characteristic (average) scale $l = l_s \sim 10\ldots10^3$ m, the standard deviation (SD) $\sigma_{\Delta N}(h) = \left(\overline{\Delta N^2(\rho, h)}\right)^{0,5}$, el/m^3, and the intensity of inhomogeneities (approximately the same for any heights h):

$$\beta_i = \sigma_{\Delta N}(h) / \bar{N}(h) \approx const \tag{4}$$

The front of the emitted wave (3) which is falling the lower boundary (h_0) of the reflecting layer F of the ionosphere at an angle θ_0 will be flat (line AB in Fig. 1a) with an amplitude $A_0 = S_t\sqrt{K_{at}} = \sqrt{{}_tK_{at}}$, where K_{at} is the attenuation coefficient of the wave in power, and the initial phase, the value of which can be assumed to be $\varphi_t = 0$.

During the propagation of a DCM wave in an inhomogeneous ionosphere with an electron concentration $N(\rho, h) = \bar{N}(h) + \Delta N(\rho, h)$ and reflection at height $h = h_r$ its amplitude-phase front will be distorted in space (ρ) due to diffraction effects (line CD in Fig. 1a).

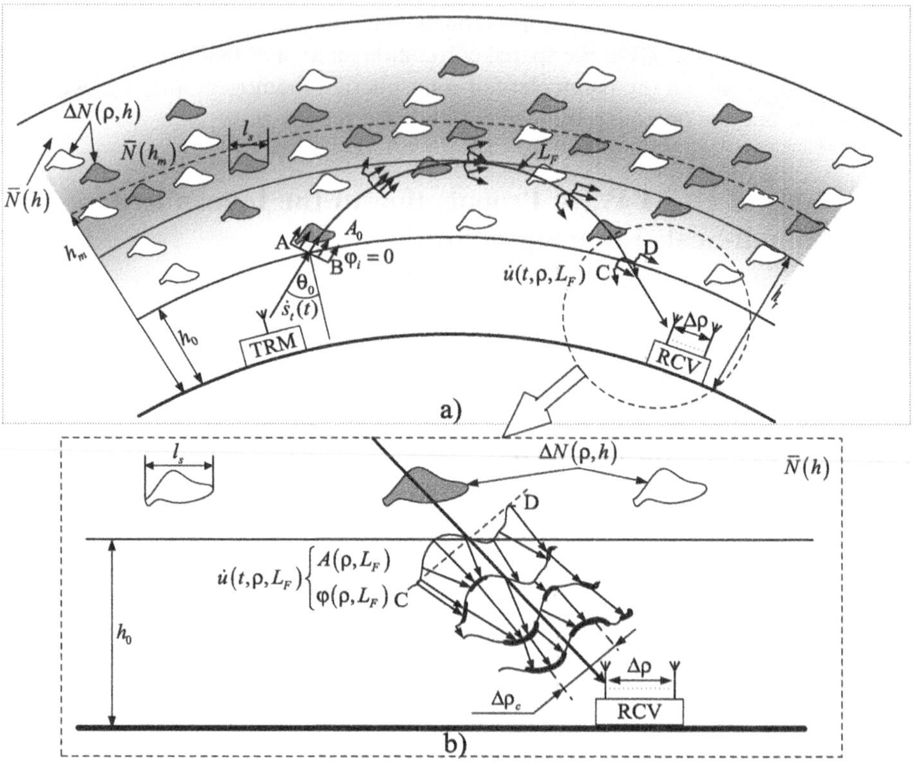

Fig. 1. The process of wave propagation in a single-beam DCM radio line inside the inhomogeneous ionosphere (a) and beyond it to the receiving point (b)

The complex field of the output wave (Fig. 1a) that has passed the real path $L = L_F$ in the inhomogeneous reflecting layer F of the ionosphere can be written as [9, 11–13]

$$\dot{u}(t, \rho, L_F) = A(\rho, L_F)\exp(j(\omega_0 t - \varphi(\rho, L_F))). \tag{5}$$

Amplitude front of the wave (5) at the outlet of an inhomogeneous ionospheric layer

$$A(\rho, L_F) = A_0 \exp(\chi(\rho, L_F)) = \sqrt{P_t K_{at}} \exp(\chi(\rho, L_F)) \tag{6}$$

is determined by the amplitude $A_0 = \sqrt{P_t K_{at}}$ of the field wave in the ionosphere without considering inhomogeneities of EC (when $\Delta N(\rho, h) = 0$) and by fluctuation of level $\chi(\rho, L_F) = \ln((\rho, L_F)/A_0)$ in the amplitude front of the wave relative to A_0.

The phase front of the output wave (5) is defined as the sum

$$\varphi(\rho, L_F) = \overline{\varphi}(L_F) + \Delta\varphi(\rho, L_F) \tag{7}$$

of the average value of the wave phase run $\overline{\varphi}(L_F)$ in the ionosphere, determined by the change in height of the average value of the EC $\bar{N}(h)$, and the fluctuations of the phase front $\Delta\varphi(\rho, L_F)$, determined by the inhomogeneities of the EC $\Delta N(\rho, h)$.

The average value of the phase of the wave that passed through the real path L_F in the reflecting F-layer of the ionosphere consists of two components [9, 12, 13]:

$$\overline{\varphi}(L_F) = k_0 L_F + \overline{\varphi} = k_0 L_F - g\pi \int_{L_F} \overline{N}(h)dl \Big/ cf_0 \approx \omega_0 L_F/c - 80,8\pi\overline{N_T}/cf_0, \quad (8)$$

where $k_0 L_F$ is the wave phase run in free space on the path L_F; $k_0 = \omega_0/c = 2\pi f_0/c = 2\pi/\lambda_0$ is the wave number;

$$\overline{\varphi} = -g\pi \int_{L_F} \overline{N}(h)dl \Big/ cf_0 \approx -80,8\pi\overline{N_T}/cf_0, \quad (9)$$

is a correction to the average phase value in the output wave front. Here, the coefficient g in the SI has a dimension [m^3/s^2] and a value of 80,3…80,8 (in the future, we will assume that $g \approx 80,8$); dl is an element of the path along the wave propagation path in the F layer of the ionosphere;

$$\overline{N_T} = \int_{L_F} \overline{N}(h)dl \quad (10)$$

is the average value of the total electron content (TEC) of the ionosphere (in other words, its integral electron concentration) along the path of wave propagation with the length L_F.

Fluctuations of the phase front of the wave $\Delta\varphi(\rho, L_F)$ at the outlet of the inhomogeneous ionosphere are determined by the operating frequency f_0 and small-scale ($l = (,,h) = (\rho, h) \sim 10...10^3$ m) fluctuations of the EC $\Delta N(\rho, h)$ on the real path L_F along the wave propagation path in the F layer [12, 14]

$$\begin{aligned}\Delta\varphi(\rho, L_F) &= -(\lambda_0 r_e) \int_{L_F} \Delta N(\rho, h)dl \\ &= (g\pi/f_0) \int_{L_F} \Delta N(\rho, h)dl \approx -(80,8\pi/f_0)\Delta N_T(\rho, L'_F) \quad (11)\end{aligned}$$

where r_e is the classical electron radius;

$$\Delta N_T(\rho, L'_F) = \int_{L_F} \Delta N(\rho, h)dl \quad (12)$$

are TEC fluctuations caused by small-scale EC inhomogeneities $\Delta N(\rho, h)$ along the wave propagation path in the ionosphere with a length L_F.

It is difficult to determine the statistical characteristics of the ionospheric TEC fluctuations $\Delta N_T(\rho, L'_F)$ since small-scale EC inhomogeneities $\Delta N(\rho, h)$ along the real path L_F of the DCM wave propagation in the ionosphere cannot be considered statistically homogeneous. Indeed, according to expression (4) $\beta_i = \sigma_{\Delta N}(h)/\bar{N}(h) \approx onst$ the highest value of the SD of fluctuations of EC will take place at the reflection height of the DCM wave, where the average value of EC will be the greatest.

(i.e. $\bar{N}(h) > \bar{N}(h)$) and the following inequality will be satisfied:

$$\sigma_{\Delta N}(h_r) = \left(\overline{\Delta N^2(\rho, h_r)}\right)^{0,5} = \beta_i \bar{N}(h_r) > \sigma_{\Delta N}(h) = \left(\overline{\Delta N^2(\rho, h)}\right)^{0,5} = \beta_i \bar{N}(h). \tag{13}$$

According to (13), the greatest contribution to the magnitude of the fluctuations of the TEC (10) $\Delta N_T(\rho, L\prime_F)$ along the real path L_F of propagation of the DCM wave in the ionosphere will be made by fluctuations of the EC at the reflection height $\Delta N(\rho, h = h_r)$. Therefore, in the future, it is advisable to replace the consideration of the real path L_F of propagation of the DCM wave in the ionosphere with an inhomogeneous height of the average EC $\bar{N}(h)$ with a smaller equivalent path $L_e < L_F$ in the ionosphere with a higher homogeneous average EC $\bar{N}(h_r) > \bar{N}(h)$.

The equivalent homogeneous path L_e s determined from formula (10) for the total electron content (TEC) of the ionosphere ($\overline{N}$) in a "tube" with the cross section of 1 m^2 along the real path of propagation of the DCM wave in the ionospheric layer F with an inhomogeneous height (h) of average electron concentration $\bar{N}(h)$ according to the expression [12, 18, 19]

$$N_T = \int_{L_F} \overline{N}(h)dl = L_F\overline{N}(L_F) = L_e\overline{N}(h_r), \tag{14}$$

where $\bar{N}(L_F)$ is the value of the average electron concentration in the "tube" averaged along the real path L_F (i.e., homogeneous value); L_e is the equivalent length of the "tube" ($L_e < L_F$) with a homogenous average electron concentration corresponding to the reflection height $\bar{N}(h_r) > \bar{N}(L_F)$.

In accordance with (14), the expression (12) can be written in terms of the equivalent wave path L_e in the ionosphere as [12]

$$\Delta N_T(\rho, L\prime_F) = \int_{L_F} \Delta N(\rho, h)dl = \int_{L_e} \Delta N(\rho, h = h_r)dl$$
$$= L_e \Delta N(\rho, h = h_r) = \Delta N_T(\rho, h_r, L_e). \tag{15}$$

Here $\Delta N_T(\rho, h_r, L\prime_e)$ has a physical meaning of the fluctuations of the ionospheric TEC during the propagation of a DCM wave in a "tube" with a cross-section of 1 m^2 with statistically homogenously placed inhomogeneities with the largest fluctuations of the EC corresponding to the height of the wave reflection as $\Delta N(\rho, h_r) > \Delta N(\rho, h)$. In this case, the length of the "tube" corresponds to the equivalent path $L\prime_e$ of propagation of the DCM wave, the length of which is less than the real path $L_e < L\prime_F$ so that the equality (15) $\Delta N_T(\rho, L\prime_F) = \Delta N_T(\rho, h_r, L\prime_e)$ is satisfied.

The correlation function of TEC fluctuations (15) $\Delta N_T(\rho, h_r, L_e)$ along two trajectories of propagation of a DCM wave with a length $L_{/e}$ separated by a distance $\Delta\rho = \rho_1 - \rho_2$ is expressed in terms of the correlation function of EC fluctuations $\Delta N(\rho, h_r)$ as [14]

$$B_{\Delta N_T}(\Delta\rho) = \overline{\Delta N_T(\rho_1, h_r, L_e)\Delta N_T(\rho_2, h_r, L_e)} = L_e \int_{-\infty}^{\infty} B_{\Delta N}(\Delta\rho, h_r) dh_r = L_e A_{\Delta N}(\Delta\rho), \tag{16}$$

where $B_{\Delta N}(\Delta\rho, h_r) = \overline{\Delta N(\rho_1, h_r)\Delta N(\rho_2, h_r)}$. The limits of integration in (16) are extended from $-\infty$ to $+\infty$, since the equivalent wave path in the ionosphere (L_e) significantly exceeds the radius of correlation (l_s) of inhomogeneities over space.

Let's determine the correlation function of TEC fluctuations (16) for the Gaussian correlation function of EC fluctuations $\Delta N(\rho, h)$, which has the form [15–17]

$$B_{\Delta N}(\Delta\rho, h_r) = \sigma^2_{\Delta N}(h_r) \exp\left(-(\Delta\rho^2 + h_r^2)/l_s^2\right). \tag{17}$$

Substituting (17) into expression (16) gives us

$$B_{\Delta N_T}(\Delta\rho) = L_e \sigma^2_{\Delta N}(h_r) \exp\left(-(\Delta\rho/l_s)^2\right) \int_{-\infty}^{\infty} \exp\left(-(h_r/l_s)^2\right) dh_r.$$

Then, using the tabular integral

$$\int_{-\infty}^{\infty} \exp\left(-p^2x^2 \pm qx\right)dx = \sqrt{\pi/p^2}\exp\left(q^2/4p^2\right), \text{ when } p > 0,$$

an expression is obtained for the correlation function of small-scale fluctuations of the TEC during the propagation of the DCM wave in the form

$$B_{\Delta N_T}(\Delta\rho) = \sqrt{\pi}\sigma^2_{\Delta N}(h_r) L_e l_s \exp\left(-(\Delta\rho/l_s)^2\right). \tag{18}$$

From (18) with $\Delta\rho = 0$ taking into account expression (13) we can obtain an expression for the dispersion of small-scale fluctuations of TEC in the propagation of the DCM wave in the form

$$\sigma^2_{\Delta N_T} = _{\Delta N_T}(0) = \sqrt{\pi} L_e l_s \sigma^2_{\Delta N}(h_r) = \sqrt{\pi} L_e l_s \left(\beta_i \overline{N}(h_r)\right)^2. \tag{19}$$

In accordance with expression (11), the SD of fluctuations in the phase front $\Delta\varphi(\rho, L_F)$ of the wave at the outlet of the inhomogeneous ionosphere is defined as

$$\sigma_\varphi = (\lambda_0 r_e)\sigma_{\Delta N_T} = (g\pi/f_0)\sigma_{\Delta N_T} \approx (80, 8\pi/f_0)\sigma_{\Delta N_T}. \tag{20}$$

Substitution of expression (19) into expression (20) allows to write it as the dependence of the SD of fluctuations of the phase front of DCM wave at the outlet of the inhomogeneous ionosphere on the intensity of inhomogeneities β_i and average EC $\overline{N}(h_r)$ at the reflection height of the wave:

$$\sigma_\varphi \approx (80,8\pi/f_0)\beta_i\overline{N}(h_r)\left(\sqrt{\pi}L_e l_s\right)^{0,5}. \quad (21)$$

Recall that in the DCM radio line, the average value of the EC at the reflection height $\overline{N}(h_r)$ determines the frequency f_v of the vertically directed reflected wave and the choice of the operating frequency f_0 of the wave falling at an angle θ_0 on the lower boundary $h = h_0$ of the spherical reflecting layer F of the ionosphere (Fig. 1a) [3, 7, 12]

$$f_0 = f_v K_s \sec\theta_0 = \left(80,8\overline{N(h_r)}\right)^{0,5} K_s \sec\theta_0, \quad (22)$$

where $K_s \leq 1$ is the sphericity coefficient of the Earth and ionosphere. From here, we express $\overline{N}(h_r)$ in terms of f_0 and substitute it in (21). As a result, the expression for the SD of phase fluctuations in the DCM wave front at the outlet of the spherical reflecting layer F of the ionosphere with small-scale inhomogeneities takes the most obvious form

$$\sigma_\varphi \approx f_0 \pi\beta_i\left(\sqrt{\pi}L_e l_s\right)^{0,5}/K_s^2 \sec^2\theta_0. \quad (23)$$

This expression differs from the known one [12, 18, 19] $\sigma_\varphi \approx f_0\pi\beta_i(2L_e l_s)^{0,5}/K_s^2 \sec^2\theta_0$ by the presence of a cofactor $(\pi)^{0,25} \approx 1,33$ instead of $(2)^{0,5} \approx 1,41$.

The analysis of expression (23) shows that as the operating frequency f_0 of the single-beam DCM radio line increases and approaches the maximum applicable frequency (MAF)

$$f_m = f_{cr} K_s \sec\theta_0 = \left(80,8\overline{N(h_m)}\right)^{0,5} K_s \sec\theta_0 \quad (24)$$

a proportional increase in the SD of phase fluctuations $\sigma_\varphi \sim f_0\beta_i(L_e l_s)^{0,5}/K_s^2 \sec^2\theta_0$ in the front of the DCM wave at the ionosphere's outlet occurs. In addition, the value of σ_φ increases in direct proportion to the increase in the intensity of inhomogeneities β_i. The latter characterizes the degree of ionospheric diffusivity [20] and can increase from $\beta_i \approx 10^{-3}...10^{-2}$ in the normal mid-latitude ionosphere to $\beta_i \approx 10^{-1}$ in polar and equatorial latitudes [11, 13]. To a lesser extent, the increase in $\sigma_\varphi \sim (L_e l_s)^{0,5}$ depends on the length of the equivalent path L_e in the inhomogeneous ionosphere and the characteristic scale l_s of inhomogeneities. In addition, the growth of $\sigma_\varphi \sim \left(K_s^2 \sec^2\theta_0\right)^{-1}$ is significantly affected by a decrease in the angle θ_0 of the wave's falling on the lower boundary of the spherical reflecting layer of the ionosphere (at a given distance of communication) and the sphericity of the ionosphere (decrease in the coefficient $K_s < 1$).

Thus, a refined analytical dependence (23) $\sigma_\varphi = \psi(\beta_i, l_s, L_e, K_s, f_0)$ of the SD of phase fluctuations in the front of the DCM wave at the outlet of the reflecting layer F on the parameters of small-scale inhomogeneities (their intensity β_i and characteristic scale l_s), equivalent length of the radio line (L_e), its geometric parameters (θ_0,K_s) and the choice of the operating frequency (f_0) is obtained. This dependence is the basis for obtaining the desired dependence of the interval of spatial correlation of fading ($\Delta\rho_c$) on the above parameters in terms of the value $\sigma_\varphi = \psi(\beta_i, l_s, L_e, K_s, f_0)$.

3 Spatial Correlation Function of a Single-Beam Decameter Radio Line

It is known [17] that when the standard deviation of the fluctuations of the phase front of the wave at the ionospheric outlet is $\sigma_\varphi < 1$ radian, the diffraction pattern on the Earth's surface is such that the radius (interval) of the phase correlation is equal to the radius of the correlation of the fluctuations of the electron concentration of the ionosphere. As the phase fluctuation increases as $\sigma_\varphi > 1$, the phase correlation radius observed on the Earth's surface will decrease in comparison with the correlation radius of electron concentration fluctuations determined by the characteristic scale $l_s = 200$–500 m [7, 8, 17, 21] of inhomogeneities of the ionospheric electron concentration.

The approximate expressions (1) and (2) obtained in [9] for the normalized spatial correlation function of a single-beam DCM radio line $K_n(\Delta\rho) = \exp\left(-(\Delta\rho/\Delta\rho_c)^2\right)$ and the interval of spatial correlation of fading $\Delta\rho_c \approx l_s/\sigma_\varphi$ do not fully correspond to the given experimental data.

In a single-beam DCM radio line (channel) of communication in the normal (undisturbed) state of the ionosphere, the SD of fluctuations in the phase front of the wave at the ionosphere outlet is small and is $\sigma_\varphi = 0{,}1\ldots1{,}2$ radians [21]. Therefore, with $l_s = 200$–500 m the spatial correlation interval according to formula (2) should be within $\Delta\rho_c \approx l_s/\sigma_\varphi = 170\ldots5000$ m that does not correspond to known [21] values (200–500 m). Thus, the estimation of the spatial correlation function and the spatial correlation interval of the DCM radio line using expressions (1) and (2) will have a significant error (especially with small values of $\sigma_\varphi \sim 0,1$ radians, when $\Delta\rho_c \approx l_s/\sigma_\varphi = 200..500/0,1 = 2000...5000$ m).

Therefore, for DCM radio channel it is necessary to more accurately describe the distribution of DCM waves beyond the ionospheric layer and the spatial correlation function $K_n(\Delta\rho)$ of the channel defined by an interval of spatial correlation $\Delta\rho_c$ of fading.

The process of propagation of the DCM wave beyond the ionospheric layer to the location of the receiver (RCV) is accompanied by diffraction effects in free space (see Fig. 1b), the essence of which is as follows.

The presence of fluctuations (distortions) of the amplitude front $A(\rho, L_e)$ and phase front $\Delta\varphi(\rho, L_e)$ of the wave at the outlet of an inhomogeneous ionospheric layer leads to the fact that behind the ionosphere, different sections (ρ_i) of this front will propagate perpendicular to the surface ρ_i in different directions. As a result of their interference, regions with increased and reduced field density arise, i.e., the amplitude is redistributed along the wave front (this is illustrated in Fig. 1b as a random thickness of the wavefront line). As you move away from the ionospheric layer, amplitude fluctuations of the wave's front will increase, and at the receiving point they will appear as fading.

The fading characteristics (including the spatial correlation coefficient $K_n(\Delta\rho)$) are mainly influenced by fluctuations in the phase front $\Delta\varphi(\rho, L_e)$ of the wave at the ionospheric outlet.

In [8], the dependence $K_n(\Delta\rho) = \psi(\Delta\rho, l_s, \sigma_\varphi)$ of the spatial correlation coefficient of fading of the received wave in a single-beam DCM radio line on the scale of inhomogeneities and the dispersion of fluctuations of the phase front of the wave at the outlet the ionosphere is given in the following form:

$$K_n(\Delta\rho) = \frac{\exp\left(-\sigma_\varphi^2\left(1 - \exp(-\Delta\rho^2/l_s^2)\right)\right) - \exp\left(-\sigma_\varphi^2\right)}{1 - \exp\left(-\sigma_\varphi^2\right)}. \tag{25}$$

However, the dependence graphs $K_n(\Delta\rho) = \psi(\Delta\rho, l_s, \sigma_\varphi)$ constructed in [8] according to formula (25) are limited by the constant values of the characteristic inhomogeneity scale $l_s = 200$ m and the dispersion of fluctuations of the phase front of the DCM wave at the outlet of the inhomogeneous ionosphere $\sigma_\varphi^2 = \pi^2/3 \approx 3,3$ (i.e., $\sigma_\varphi \approx 1,8$).

In addition, in [8], the dependence $\sigma_\varphi^2 = \psi(\gamma^2(\tau_c))$ of the dispersion of fluctuations of the phase front of the DCM wave at the outlet of the inhomogeneous ionosphere on the depth of fading (Rice parameter $\gamma^2 = \psi(\tau_c)$) is established based on experimental measurements of the time correlation interval τ_c of fading of received signals in a single-beam DCM radio line.

It follows that [8] presents a method that allows to determine the dependence $K_n(\Delta\rho) = \psi(\Delta\rho, l_s, \sigma_\varphi) = \psi(\Delta\rho, l_s, \tau_c)$ of the normalized spatial correlation function of fading in the DCM radio line on the spacing distance $\Delta\rho$, the characteristic scale of small-scale ionospheric inhomogeneities l_s and the dispersion of fluctuations in the phase front $\sigma_\varphi^2 = \psi(\tau)$ of the DCM wave at the outlet of the inhomogeneous ionosphere, which depends on the time correlation interval τ_c of fading of received signals in a single-beam DCM radio line. According to [8], the main downside of the above dependence $K_n(\Delta\rho) = \psi(\Delta\rho, l_s, \sigma_\varphi) = \psi(\Delta\rho, l_s, \tau_c)$ is that it does not depend on the choice of the operating frequency relative to the MAF (f_0/f_m) and the state of the ionosphere. Therefore, it does not make it possible to determine the desired dependence $\Delta\rho_c = \psi(l_s, \sigma_\varphi) = \psi(\beta_i, l_s, L_e, f_0)$ of the interval of spatial correlation of fading with a given value $K_n(\Delta\rho)$.

It can be shown that the elimination of these downsides is provided by the use of the expression for the dispersion σ_φ^2 of fluctuations of the phase front of the DCM wave at the outlet of an inhomogeneous ionosphere in formula (25). According to (23), it is defined as

$$\sigma_\varphi^2 \approx \sqrt{\pi} L_e l_s \left(f_0 \pi \beta_i / K_s^2 \sec^2\theta_0\right)^2. \tag{26}$$

The validity of the application of expression (26) in (25) can be confirmed the following way.

In the special case of using receiving antennas with a small spacing interval $\Delta\rho \ll l_s$ we have $\exp(-\Delta r^2/l_s^2) \approx 1 - \Delta r^2/l_s^2$. Therefore, the expression (25) is reduced to the form

$$K_n(\Delta\rho) \approx \frac{\exp\left(-\sigma_\varphi^2 \Delta\rho^2 / l_s^2\right) - \exp\left(-\sigma_\varphi^2\right)}{1 - \exp\left(-\sigma_\varphi^2\right)}. \tag{27}$$

The numerator of (27) with $\Delta\rho << l_s$ can be written in approximate form as

$$\exp\left(-\frac{\sigma_\varphi^2\Delta\rho^2}{l_s^2}\right)\left(1-\exp\left(-\sigma_\varphi^2\left(1-\frac{\Delta\rho^2}{l_s^2}\right)\right)\right)\approx\exp\left(-\frac{\sigma_\varphi^2\Delta\rho^2}{l_s^2}\right)\left(1-\exp\left(-\sigma_\varphi^2\right)\right).$$

Substituting this expression into the numerator of (27) allows us to write it in the well-known [9] form (1, 2)

$$K_{\text{n}}(\Delta\rho)\approx\exp\left(-\sigma_\varphi^2\Delta\rho^2/l_s^2\right)=\exp\left(-\Delta\rho^2/\left(l_s/\sigma_\varphi\right)^2\right)=\exp\left(-\Delta\rho^2/\Delta\rho_c^2\right),$$

where $\Delta\rho_c \approx l_s/\sigma_\varphi$ is the interval of spatial correlation of fading, and σ_φ is described with expression (23) as $\sigma_\varphi \approx f_0\pi\beta_{\text{i}}\left(\sqrt{\pi}L_e l_s\right)^{0,5}/K_s^2\sec^2\theta_0$.

Thus, the obtained expressions (25) and (26) allow us to establish a refined analytical dependence

$$K_{\text{n}}(\Delta\rho)=\psi(\Delta\rho, l_s, \sigma_\varphi)=\psi(\Delta\rho, \beta_{\text{i}}, l_s, L_e, f_0)$$

of the normalized spatial correlation function of the fading in the DCM radio line on a distance of spacing $\Delta\rho$, the characteristic scale of small-scale ionospheric inhomogeneities l_s and dispersion of fluctuations of the phase front of the DCM wave at the outlet of the inhomogeneous ionosphere $\sigma_\varphi = \psi(\beta_{\text{i}}, l_s, L_e, f_0)$, which depends on the parameters of small-scale inhomogeneities (β_{i}, l_s), the equivalent length of the DCM radio line (L_e) and the choice of the operating frequency (f_0).

4 Analysis of the Refined Spatial Correlation Function of A Single-Beam Decameter Radio Line

In contrast to the simplified formula (1, 2), the obtained refined expression (25, 26) for the normalized spatial correlation function (CF) of the DCM radio line $K_{\text{n}}(\Delta\rho)$ does not allow to determine the spatial correlation interval $\Delta\rho_c$ analytically. However, its value can be found by the graphoanalytic method from the functional dependence (25) $K_{\text{n}}(\Delta\rho)$ at a given level, for example $K_{\text{n}} = 1/e \approx 0,37$. At the same time, when performing calculations of $\Delta\rho_c$, we determine the limits of applicability of simplified expressions (1) and (2) for a single-beam DCM radio line.

Let's calculate $K_{\text{n}}(\Delta\rho)$ using the known (1, 2) and obtained (25, 26) formulas for the spatial CF of a single-beam DCM radio line with the parameters given in [19]: the range of communication is $R = 2000$ KM, the critical frequency of reflection of a vertically directed wave from the ionosphere is $f_{\text{cr}} = \left(80,8\overline{N}(h_m)\right)^{0,5} = 7\,\text{MHz}$, the height of the lower boundary of the reflecting layer of the ionosphere is $h_0 = 250$ km, the half-thickness of the reflective layer is $z_m = h_m - h_0 = 100$ km, the characteristic scale of inhomogeneities is $l_s = 500$ m.

In both cases, σ_φ will be calculated using the formula (23), taking into account the weak influence of the sphericity of the ionosphere ($K_s \approx 1$):

$$\sigma_\varphi \approx f_0(h_{ef})\pi\beta_{\text{i}}\left(\sqrt{\pi}L_e l_s\right)^{0,5}/\sec^2\theta_0, \tag{28}$$

where the operating frequency is defined according to (22) as

$$f_0 \approx f_w \sec\theta_0 = \left(80,8\overline{N(h_r)}\right)^{0,5} \sec\theta_0. \quad (29)$$

Angle of incidence of a wave on a flat ionosphere, taking into account the sphericity of the Earth (with a radius $R_E = 6370\,\text{km}$) is determined with the given range of communication R and the results of measurements of the effective height $h_{ef}(f_w)$ of the wave reflection with the frequency f_w in expression [7, 19, 22]:

$$\sec\theta_0 = \sqrt{1 + \text{tg}^2\theta_0} = \sqrt{1 + \left(\frac{\sin(R/2R_E)}{1 + (h_{ef}(f_w)/R_E) - \cos(R/2R_E)}\right)^2}. \quad (30)$$

The effective height of the wave reflection from the ionosphere is calculated in accordance with the well-known expression [7, 22]

$$h_{ef}(f_w) = h_0 + \frac{z_m}{2}\frac{f_w}{f_{cr}} \ln\frac{1 + (f_w/f_{cr})}{1 - (f_w/f_{cr})}. \quad (31)$$

The value of the equivalent path L_e of wave propagation in the ionosphere, which is included in the expression (28), is determined according to the expression [19]

$$L_e = (h_{ef}(f_w) - h_0)\left(1 + \frac{f_{cr}^2}{f_w^2} - \frac{z_m}{h_{ef}(f_w) - h_0}\right)\sqrt{\sec^2\theta_0 - \frac{1}{2}\left(1 + \frac{f_{cr}^2}{f_w^2} - \frac{z_m}{h_{ef}(f_w) - h_0}\right)}. \quad (32)$$

Table 1 shows, as an example, the results of calculations of $K_n(\Delta\rho)$ using expressions (1) and (25) for typical parameters of small-scale inhomogeneities of the mid-latitude ionosphere (the intensity of inhomogeneities is $\beta_i = 5\cdot 10^{-3}$ and the characteristic scale is $l_s = 500$ m) for three values of antennas spacing: $\Delta\rho = 100$ m, 200 m and 500 m.

Table 1. Calculation of the normalized spatial correlation function of a single-beam DCM radio line $K_n(\Delta\rho)$

f_w, MHz	$\frac{f_w}{f_{cr}}$	θ_0, deg	f_0, MHz	σ_φ, rad	$K_n(\Delta\rho)$ using (25)	$K_n(\Delta\rho)$ using (1)	$K_n(\Delta\rho)$ using (25)	$K_n(\Delta\rho)$ using (1)	$K_n(\Delta\rho)$ using (25)	$K_n(\Delta\rho)$ using (1)
					with $\Delta\rho = 100$ m		with $\Delta\rho = 200$ m		with $\Delta\rho = 500$ m	
1,4	0,2	71,5	4,4	0,09	0,961	1	0,852	0,999	0,367	0,991
2,1	0,3	71,3	6,5	0,22	0,96	0,998	0,849	0,993	0,362	0,955
2,8	0,4	70,9	8,5	0,39	0,958	0,994	0,842	0,976	0,35	0,857
3,5	0,5	70,3	10,4	0,64	0,953	0,984	0,826	0,937	0,322	0,668

(continued)

Table 1. *(continued)*

f_w, MHz	$\frac{f_w}{f_{cr}}$	θ_0, deg	f_0, MHz	σ_φ, rad	$K_n(\Delta\rho)$ using (25)	$K_n(\Delta\rho)$ using (1)	$K_n(\Delta\rho)$ using (25)	$K_n(\Delta\rho)$ using (1)	$K_n(\Delta\rho)$ using (25)	$K_n(\Delta\rho)$ using (1)
					with $\Delta\rho = 100$ m		with $\Delta\rho = 200$ m		with $\Delta\rho = 500$ m	
4,2	0,6	69,6	12,1	0,96	0,941	0,964	0,789	0,863	0,267	0,399
4,9	0,7	68,6	13,5	1,39	0,915	0,926	0,71	0,735	0,176	0,146
5,6	0,8	67,3	14,5	1,98	0,855	0,855	0,552	0,535	0,066	0,02
6,3	0,9	65,1	15,0	2,89	0,721	0,717	0,292	0,264	5·10–3	2,5·10–4

In Figs. 2, 3, and 4, a solid line shows the dependence graph $K_n(\Delta\rho)$, constructed according to expression (25), and a dotted line – according to formulas (1, 2), with the same parameters of small-scale inhomogeneities in the mid-latitude undisturbed ionosphere (when $\beta_i = 5 \cdot 10^{-3}$, $l_s = 500$ m) and with three values of operating frequencies: $f_0 = 4{,}4$ MHz (when $f_w = 1, 4$ MHz); $f_0 = 10{,}4$ MHz (when $f_w = 3{,}5$ MHz); $f_0 = 14{,}5$ MHz (when $f_w = 5{,}6$ MHz).

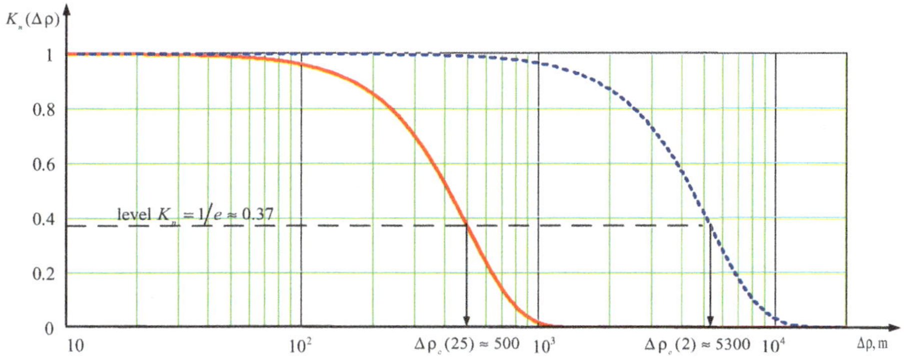

Fig. 2. Dependences of the normalized correlation function of a single-beam DCM radio line on the spatial separation with the operating frequency $f_0 = 4{,}4$ MHz, undisturbed ionosphere ($\beta_i = 5 \cdot 10^{-3}$, $l_s = 500$ m) and $\sigma_\varphi = 0{,}09$ radians.

The analysis of graphs in Fig. 2 shows that in the conditions of the undisturbed mid-latitude ionosphere ($\beta_i = 5 \cdot 10^{-3}$, $l_s = 500$ m), the operating frequency $f_0 = 4{,}4$ MHz and very small values of the SD of phase fluctuations ($\sigma_\varphi = 0, 09$ radians), interval of the spatial correlation of the fading calculated with the known expression (1) is $\Delta\rho_c = \Delta\rho_c(1) = 5300$ m, i.e. is significantly higher than the value of the interval of the spatial correlation, defined graphically according to the received updated according to (25) ($\Delta\rho_c = \Delta\rho_c(25) = 500$ m).

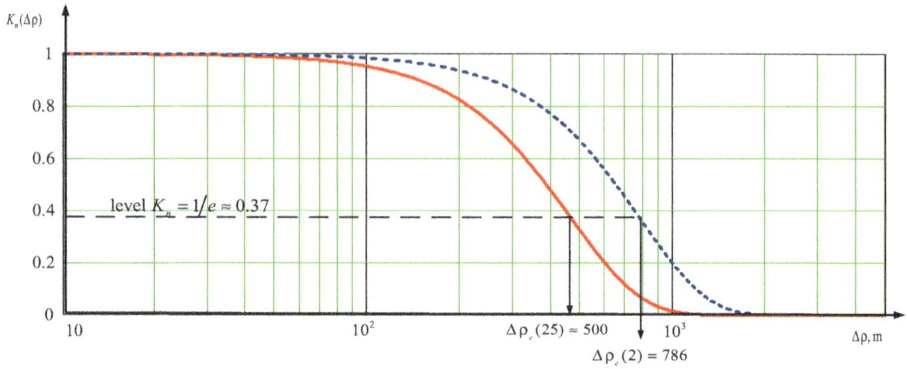

Fig. 3. Dependences of the normalized correlation function of a single-beam DCM radio line on the spatial separation with the operating frequency $f_0 = 10{,}4$ MHz, undisturbed ionosphere ($\beta_i = 5 \cdot 10^{-3}, l_s = 500$ m) and $\sigma_\varphi = 0{,}64$ radians.

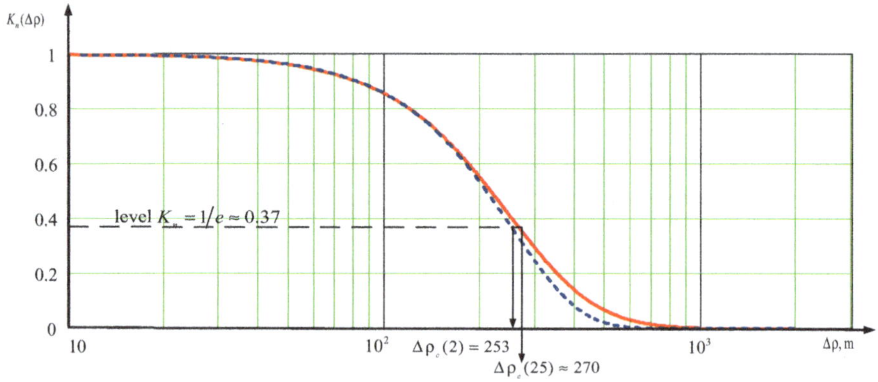

Fig. 4. Dependences of the normalized correlation function of a single-beam DCM radio line on the spatial separation with the operating frequency $f_0 = 14{,}5$ MHz, undisturbed ionosphere ($\beta_i = 5 \cdot 10^{-3}, l_s = 500$ m) and $\sigma_\varphi = 1{,}98$ radians.

The analysis of graphs in Fig. 3 shows that in the normal ionosphere ($\beta_i = 5 \cdot 10^{-3}$, $l_s = 500$ m) the interval of spatial correlation of fading calculated with expression (1) when the operating frequency $f_0 = 10{,}4$ MHz is selected and the SD of phase fluctuations is relatively small ($\sigma_\varphi = 0, 64$ radians) is $\Delta\rho_c = \Delta\rho_c(1) = 786$ m, i.e. it turns out to be somewhat overestimated in comparison with the value of the spatial correlation interval determined graphically from the dependence (25) ($\Delta\rho_c = \Delta\rho_c(46) = 500$ m).

The analysis of graphs in Fig. 3 shows that in the normal ionosphere ($\beta_i = 5 \cdot 10^{-3}$, $l_s = 500$ m) the interval of spatial correlation of fading calculated with expression (1) when the operating frequency $f_0 = 14{,}5$ MHz is selected and the SD of phase fluctuations is large ($\sigma_\varphi = 1, 98$ radians) is $\Delta\rho_c = \Delta\rho_c(1) = 253$ m, i.e. it coincides with an accuracy of up to 5% with the value of the spatial correlation interval determined graphically from the dependence (25) ($\Delta\rho_c = \Delta\rho_c(25) = 270$ m).

The reliability of the obtained results of calculating the spatial correlation interval determined from the graphs of Fig. 2, 3, and 4 in accordance with the refined dependence (25) $\Delta\rho_c = \Delta\rho_c(25)$ is confirmed with experimental data [17]. According to the latter, if the wave frequency is such that $\sigma_\varphi < 1$ radians, then the diffraction pattern on the Earth's surface is such that the interval of spatial correlation ($\Delta\rho_c$) of fading is equal to the characteristic scale of EC fluctuations in the inhomogeneities of the ionosphere (l_s). According to Fig. 2 and 3 with $l_s = 500$ m, $\sigma_\varphi = 0{,}09$ radians and 0,64 radians, the value of $\Delta\rho_c(25) = l_s = 500$ m (while calculations using the simplified formula (1) give an overestimated result, for example $\Delta\rho_c(1) = 5300$ m $> > l_s = 500$ m). According to [17], if the wave frequency is such that $\sigma_\varphi > 1$ radians, then the interval of spatial correlation ($\Delta\rho_c$) of fading is less than the characteristic scale of ionospheric inhomogeneities (l_s). According to Fig. 4, with $\sigma_\varphi = 1{,}98$ radians we have $\Delta\rho_c(25) = 270$ m $< l_s = 500$ m.

We analyze the limits of applicability of the simplified expression (2) for estimating the interval of spatial correlation of fading in a single-beam DCM radio line. Table 2 shows the values $\Delta\rho_c = \psi(\sigma_\varphi)$ of the interval of spatial correlation of fading as a function of the SD of fluctuations of the phase front σ_φ of the wave at the ionosphere outlet, obtained by the analytical formula (2) and graphically according to the dependence (25) $K_n(\Delta\rho) = \psi(\Delta\rho, l_s, \sigma_\varphi)$ with the characteristic scale of inhomogeneities $l_s = 500$ m.

Table 2. Dependence of the interval of spatial correlation of fading on the SD of fluctuations of the phase front of the wave at the ionosphere outlet

σ_φ, radians	0,1	0,5	1	1,25	1,5	2	3	4	5	10
$\Delta\rho_c(25)$, m	495	480	425	381	340	263	172	127	101	50,3
$\Delta\rho_c(2)$, m	5000	1000	500	400	334	250	166	125	100	50

It is possible to construct a table of the dependence of the intervals of spatial correlation of fading on the SD of fluctuations of the phase front of the wave at the ionosphere outlet the similar way with a smaller characteristic scale of inhomogeneities ($l_s = 200$ m).

In Fig. 5, solid lines show the dependences $\Delta\rho_c = \Delta\rho_c(25) = \psi(\sigma_\varphi, l_s)$, obtained from the graphs according to expression (25) for various characteristic scales of ionospheric inhomogeneities ($l_s = 500$ m and $l_s = 200$ m), dashed lines – according to expression (2) $\Delta\rho_c = \Delta\rho_c(2) \approx l_s/\sigma_\varphi$ for $l_s = 500$ m and $l_s = 200$ m.

The analysis of graphs (Fig. 5) shows that when the operating frequency f_0 approaches the maximum applicable frequency f_m or perturbation (diffusivity) of the ionosphere (when the intensity inhomogeneities may increase from values $\beta_i = 5 \cdot 10^{-3}$ to $\beta_i \approx 10^{-1}$), causing an increase in the SD of phase fluctuations to values of $\sigma_\varphi > 1,25$ radians, the well-known expression (2) can be used to estimate the spatial correlation interval single-beam DCM radio line with a relative error $\Delta = 100\%(\Delta\rho_c(2) - \Delta\rho_c(25))/\Delta\rho_c(25) \leq 5\%$. It can be shown that the conclusion that the simplified expression (1) can be used to estimate the spatial correlation interval of a single-beam DCM radio line with a relative error of no more than 5% remains valid (see

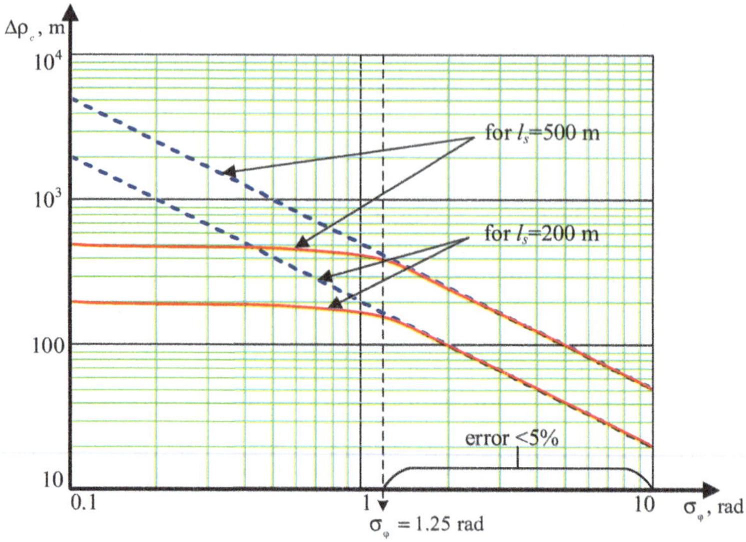

Fig. 5. Dependences of the interval of spatial correlation of fading on the SD of fluctuations of the phase front of the wave at the ionosphere outlet with the characteristic scale of inhomogeneities $l_s = 500$ and $l_s = 200$ m.

Fig. 5) for other scales of small-scale inhomogeneities of the ionosphere (for example, $l_s = 200$ m).

5 Conclusion

Based on the analysis of the propagation of a DCM wave in an inhomogeneous ionosphere (Fig. 1), a refined analytical dependence (23) $\sigma_\varphi = \psi(\beta_i, l_s, L_e, K_s, f_0)$ of the SD of phase fluctuations in the DCM wave front at the outlet of the reflecting layer F on the parameters of small-scale inhomogeneities (their intensity β_i and characteristic scale l_s), equivalent the length of the radio line (L_e), its geometric parameters (θ_0, K_s) and the choice of the operating frequency (f_0) is obtained.

This dependence is the basis for determining the refined analytical dependence (25) $K_n(\Delta\rho) = \psi(\Delta\rho, l_s, \sigma_\varphi)$ of the normalized spatial correlation function of the fading of the DCM wave on the characteristic scale of small-scale ionospheric inhomogeneities l_s and the SD of fluctuations of the phase front of the DCM wave at the outlet of the inhomogeneous ionosphere (23) $\sigma_\varphi = \psi(\beta_i, l_s, L_e, K_s, f_0)$.

In turn, the dependence (25) is the basis for determining a graphical way (Fig. 2, 3, and 4) of the desired refined dependence $\Delta\rho_c = \psi(\sigma_\varphi) = \psi(\beta_i, l_s, L_e, f_0)$ of the interval of spatial correlation of fading on the above parameters using the value of the SD of fluctuations of the phase front of the output wave (23) σ_φ in the single-beam decameter radio line parameters of the ionospheric inhomogeneities (β_i, l_s,), equivalent length of the radio line (L_e) and the choice of the operating frequency (f_0) of signal.

In addition, it is shown (Fig. 5) that when the operating frequency f_0 of the radio line approaches the maximum applicable frequency or under conditions of ionospheric diffusivity, when $\sigma_\varphi > 1,25$ radians, известное выражение (2) the known expression (2) can also be used to estimate the spatial correlation interval of a single-beam DCM radio line with an error of no more than 5%.

References

1. Sarwate, V.V.: Electromagnetic Fields and Waves. Bohem Press, Zurich (1993)
2. Yau, K.S.B., Coleman, C.J., Cervera, M.A.: Investigation on fading of High Frequency radio signals propagating in the ionosphere - Results from a Jindalee radar experiment. In: 10th IET International Conference on Ionospheric Radio Systems and Techniques (IRST 2006). IET, London (2006)
3. Grudinskaya, G.P.: Radio Wave Propagation. Vysshaya Shkola, Moscow (1975)
4. Stein, S., Jones, J.: Modern Communication Principles. McGraw-Hill Telecommunications, New York (1967)
5. Andronov, I.S., Fink, L.M.: Transmission of Discrete Messages Over Parallel Channels. Sovetskoe Radio, Moscow (1971)
6. Buga, N.N.: Fundamentals of the theory of communication and data transmission. Part 2. LVIKA, Leningrad (1970)
7. Kalinin, A.I., Cherenkova, E.L.: Radio Wave Propagation and Operation of Radio Lines. Svyaz', Moscow (1971)
8. Chernov, Yu.A.: Special issues of radio wave propagation in communication and broadcasting networks. TECHNOSPHERE, Moscow (2018)
9. Pashintsev, V.P., Porsev, A.V., Koval', S.A., Alekseev, D.V., Senokosova, A.V.: Spatial correlation function of a short-wave communication channel. Proc. Inst. Eng. Phys. **2**(12), 81–81 (2009)
10. Fabrizio, G.A.: High Frequency Over-the-Horizon Radar: Fundamental Principles, Signal Processing, and Practical Applications. McGraw-Hill Education, New York (2013)
11. Maslov, O.N., Pashintsev, V.P.: Models of transionospheric radio channels and noise immunity of space communication systems. Inf. Technol. (4) (2006)
12. Pashintsev, V.P., Kolosov, L.V., Tishkin, S.A., Antonov, V.V.: Application of the phase-screen theory for developing a model of a one-hop decameter communication link. J. Commun. Technol. Electron. **1**(41), 16–21 (1996)
13. Pashintsev, V.P., Kolosov, L.V., Tishkin, S.A., Smirnov, A.A.: Influence of the Ionosphere on signal detection in space communications systems. J. Commun. Technol. Electron. **2**(44), 132–139 (1999)
14. Yeh, K.C., Liu, C.H.: Radio wave scintillations in the ionosphere. Proc. Inst. Electr. Electron. Eng. **4**(70), 324–360 (1982)
15. Rytov, S.M., Kravtsov, Yu.N., Tatarskiy, V.I.: Introduction to Statistical Radiophysics, Part 2. Science, Moscow (1978)
16. Isimaru, A.: Propagation and Scattering of Waves in Randomly Inhomogeneous Environment, vol. 2. Mir, Moscow (1981)
17. Kolosov, M.A., Armand, N.A., Yakovlev, O.I.: Radio Wave Propagation in Space Communications. Svyaz', Moscow (1969)
18. Pashintsev, V.P., Skorik, A.D., Koval', S.A., Alekseev, D.V., Senokosov, M.A.: Algorithm for calculating the frequency correlation interval of a short-wave radio line taking into account the sphericity and small-scale inhomogeneities of the ionosphere. Syst. Control Commun. Secur. **2**, 49–72 (2020)

19. Pashintsev, V.P., Tishkin, S.A., Ivannikov, A.I., Borovlev, I.I.: Calculation of the fading depth parameter in a single-beam decameter radio line. Radioelectron. Commun. Syst. **12**(44), 57–65 (2001)
20. Pashintsev, V.P., Omel'chyuk, A.V., Koval', S.A., Galushko, Y.: Method for determining the value of intensity of inhomogeneities in the ionospheric sounding data. Dual Technol. **1**(46), 38–41 (2009)
21. Kirillov, N.E.: Noise-Proof Signal Transmission Over Linear Channels With Randomly Changing Parameters. Sovetskoe Radio, Moscow (1971)
22. Davies, K.: Ionospheric Radio Waves. Blaisdell Publishing, Waltham (1969)

Virtual Reality Model Assessment Platform: Transfer Models to a Standalone Virtual Reality Headset

Jakub Tomeš(✉), Naďa Tylová, Jan Kohout, and Jan Mareš

Department of Computing and Control Engineering Technická, University of Chemistry and Technology in Prague, 1905/5, 166 28 Praha 6, Prague, Czech Republic
jakub.tomes@vscht.cz

Abstract. The goal of the paper is to introduce a Virtual Reality Model Assessment Platform. VRMAP (for short) is to be a lightweight, simple to use platform that allows users from various fields to easily check out models in virtual reality, as there are few options to do so natively on the headset. The platform consists of a Model Repository and VR client app. Models can be uploaded to the Model Repository and then downloaded and spawned in the client app for assessment or observation. The VR client app is created in Unity Engine using the new Unity XR framework and supports the GL Transmission Format (glTF), a royalty-free, extensible, common publishing format for 3D models created by Khronos Group. Many of the mainstream formats used in 3D printing and computer graphics can be easily converted to glTF and thus imported into the VRMAP. The Model Repository is built using Express, Node, Typescript and MongoDB. It allows for upload, storage, and download of models. Although this implementation is fully functional, it acts as an example and as long as other, more robust implementation adheres to the defined API standard, it will work as a drop-in replacement.

Keywords: Virtual reality · 3D modeling · Integration · Unity

1 Introduction

Virtual reality (VR) is a highly interactive human-computer interface that simulates a realistic environment with a strong sense of 3D space [5,18]. VR headsets as we know them today were developed in the 1980s [27]. The reduced price/performance ratio has positive implications for hospitals, educational institutions, museums, and other organizations where funding of new technologies is often limited [18]. Many VR applications have been developed not only for gaming, but also for the teaching, learning, and practice of medicine [10,19], physical sciences, and engineering among others [25].

R. Silhavy (Ed.): CSOC 2021, LNNS 230, pp. 469–479, 2021.
https://doi.org/10.1007/978-3-030-77442-4_40

The main purpose of VR headsets is to visualize 3D scenes with a high level of interaction. Virtual reality allows the observation and examination of areas and events unavailable or impossible by other technologies. Several studies have discussed visualization in VR compared with traditional planar 2D visualizations. Participants were asked to complete the tasks using both displaying methods, i. e. exploring data [14], a therapeutic game [13], or Depth Perception [2] Because of the subjective property of the VR experience, all studies were included in the finished POST-QUESTIONNAIRE survey analyzing the satisfaction level. We found that there is no difference in the overall data exploration workload between mediums. A promising result for the VR visualization tool is that [14]. Interacting with a VR model contributes to raising motivation and interest, conditions that are recognized as crucial in the learning process [1].

First, headsets were made as mounted on a jointed mechanical arm with tracking sensors located at the joints. High-performance systems that were previously exclusive to research institutions with a well-funded budget can now be constructed relatively cheaper [18]. Nowadays, the most widely used VR sets are Oculus Rift and HTC Vive [6]. Both technologies still require hardware components such as computer workstations with advanced graphics card and tracking system. That is why those systems are replaced with complete wireless systems such as Oculus Quest [24].

Using a wireless headset without requirements for a well-performing computer is a suitable solution for users without advanced software skills. To explore 3D models, the user is usually dependent on existing VR applications and their databases, as was presented in [9]. In this paper, we are presenting a platform that imports a given 3D model into the VR viewer in a standalone headset without using additional programming.

Virtual reality allows for close observation and interaction with an object that could otherwise be only seen on a 2D computer screen [26]. Our goal is to create a tool that would allow comfortable viewing of models from various fields like CAD, 3D printing, modeling, or medical data visualization. Furthermore, virtual reality has a positive impact on data evaluation, providing a better understanding of the data [15].

2 Materials and Methods

Equipment needed for a working VR setup consists of a Headset, a unit that goes to the users head. Controller to interact with the virtual world and a way of tracking the movement of headset or controllers. The Oculus Quest is a completely standalone virtual reality headset or VR headset for short. It no longer needs the computational power of a computer to handle the rendering and logic of a VR application, as it is equipped with Snapdragon 835 processor, 4 GB of RAM, and 64 or 128 GB of storage [17].

The headset utilizes an inside-out tracking system, which ensures ease of setup and transport, while maintaining similar tracking performance to the older outside-in system employed in the Oculus Rift. The headset comes with a set of controllers. The controller is equipped with one analog hat joystick, two digital buttons, and two analog buttons. Game engine is the foundation on which a game is later built. It serves as a layer of abstraction between our game logic and, e.g., the code that handles rendering three-dimensional models on the screen on multiple platforms. It can also handle tasks like loading models, managing resources, etc.

Unity was chosen because of its built-in support for numerous virtual reality headsets, a fully customizable and highly modular rendering system called the Rendering Pipeline, intuitive editor, good performance, C# scripting back-end and up to date documentation [21]. Unity XR is a plugin developed by the Unity team which handles communication with a wide variety of VR and AR devices and provides a common API [23].

Node.js provides a way to build applications using Javascript or other languages that compile to it [8]. The Express framework can be leveraged to quickly build web application, APIs, etc. on the Node.js platform [7]. MongoDB is a scalable and flexible document based database solution. That is used to store file and file provider data [16].

The glTF standard for model format was chosen due to the great diversity of file formats across the fields we would like to support. By supporting the GLB file format, we ensure that most formats can be loaded into the app. glTF is a fairly new standard created by Khronos Group, which should act as a common way of transferring 3D data for most computer graphics and data visualization fields [11]. Since glTf is still a novel format, Blender and Blackthread.io were used for stl to glTF conversions [3,4]. This is only a small slice of what is available, however it covers our needs while working on this project.

3 Software Development

Developed VR application consists of multiple systems which interact together to provide a fully fleshed experience. The architecture is shown in Fig. 1. The input system processes the input data provided by the Unity XR plugin [23] and it exposes the current state of control elements and fires events when the state changes, this eliminates the need to check the state change in multiple systems every frame.

The movement system utilizes the input system and allows the user to teleport, snap, turn, and walk either in a roomscale VR setup or using the thumbstick on a controller. The teleportation is controlled by headset orientation as it provides a more natural way to teleport and it reduces motion sickness. The teleportation marker is hidden by default and activates only when a finger is present on the left thumbstick. Teleport is then initiated by pushing the thumbstick forward.

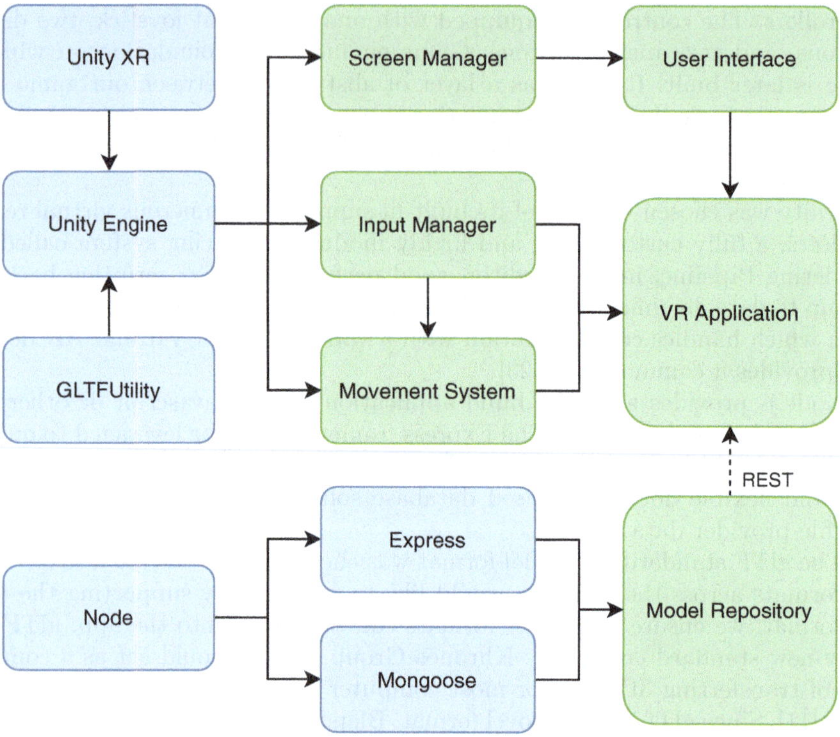

Fig. 1. Architecture of the VR application and the model repository. (Blue = external libraries, Green = our systems) (Color figure online)

The left thumbstick also controls snap turning. A user can snap-turn by 45 degrees by pushing the thumbstick either left or right. The right thumbstick is used to control walking. The manual controls were implemented to provide a way to use the application from a chair or a confined space.

3.1 GUI and Screen Manager

A user interface can be summoned at any time in the scene in the form of an information orb. This was an important design decision since the UI is hidden at most times and does not obstruct a part of the user's view. The info orb is a sphere model with a canvas floating above it. When a user summons the UI, they interact with it by pointing at it with a controller, similarly to a laser pointer and clicking the trigger button on the controller.

The screen manager has a reference to the canvas. It stores a screen stack and facilitates the opening and closing of new screens and other screen operations. The GUI is separated into multiple screens with a menu that gives access to other screens as the initial screen.

3.2 File Management

File management is done via a set of screens. The user first connects to a repository and chooses a model that they want to download. Once the model is downloaded and stored locally, it can be loaded into the scene from another screen.

The model repository is split into three subscreens that show file providers, files, and file detail, respectively. Each subscreen communicates with the model repository using a the new UnityWebRequest [22] inside a coroutine. This ensures that the user interface remains responsive even with a slow connection.

All downloaded models are stored in the glTF binary format on the headset. glTFUtility was used to load the models into unity from local storage [20]. Loaded models were then prepared by adding a MeshCollider, Rigidbody and XRInteractable where applicable.

3.3 Model Repository

The Model repository acts as a cloud storage platform for the VR application. It is split into a mandatory and optional section. The VR application communicates with the mandatory section to access information about file providers and files. The optional section is not standardized and it facilitates registration and authorization of file providers, file creation, update and removal, and any other additional features.

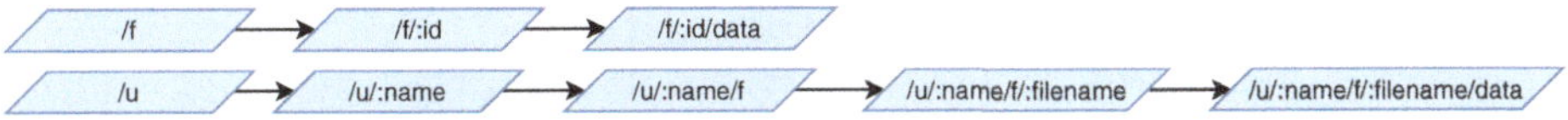

Fig. 2. Model repository endpoint architecture scheme

The model repository was implemented in Node using the Express framework and Mongoose, the architecture can be seen in Fig. 1. The VR application communicates with the model repository via http requests to the individual endpoints which return data in JSON format, so it can be replaced entirely with a different implementation that adheres to the endpoint structure in Fig. 2 and Table 1.

Table 1. Model repository endpoint architecture

Endpoint	Description
/f	Array of available file ids
/f/:id	Specified file
/f/:id/data	Specified file including its data in Base64 or a link to file download
/u	Array of file providers (users)
/u/:name	File provider data
/u/:name/f	Files that the provider uploaded
/u/:name/f/:filename	Specific file data
/u/:name/f/:filename/data	Specified file including its data in Base64 or a link to file download

4 Results and Discussion

The VR application provides a smooth motion around the virtual world that feels intuitive after a few minutes of use. The app wirelessly connects to the repository and allows the user to browse and download stored models via the UI (Fig. 4). Downloaded models can then be loaded into the scene with a click of a button and the user can interact with the loaded models if they are flagged as interactable.

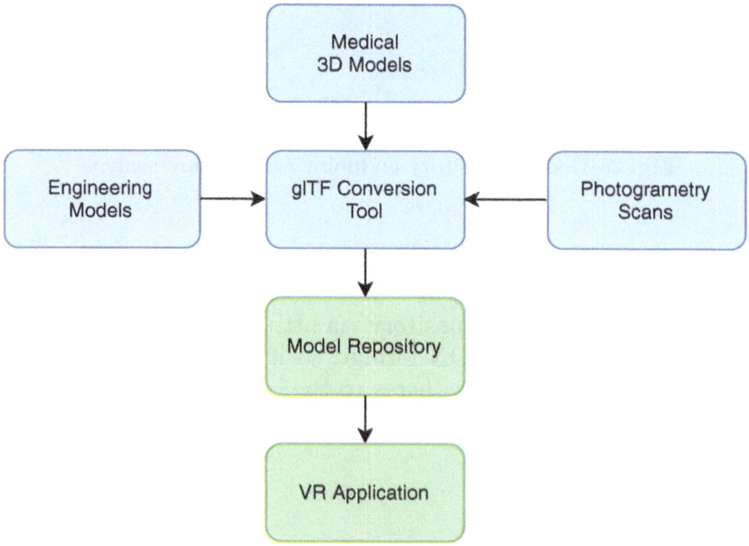

Fig. 3. Diagram with targeted use case of the platform. GLTF files can converted from many formats are uploaded to the model repository and the downloaded and viewed in the application.

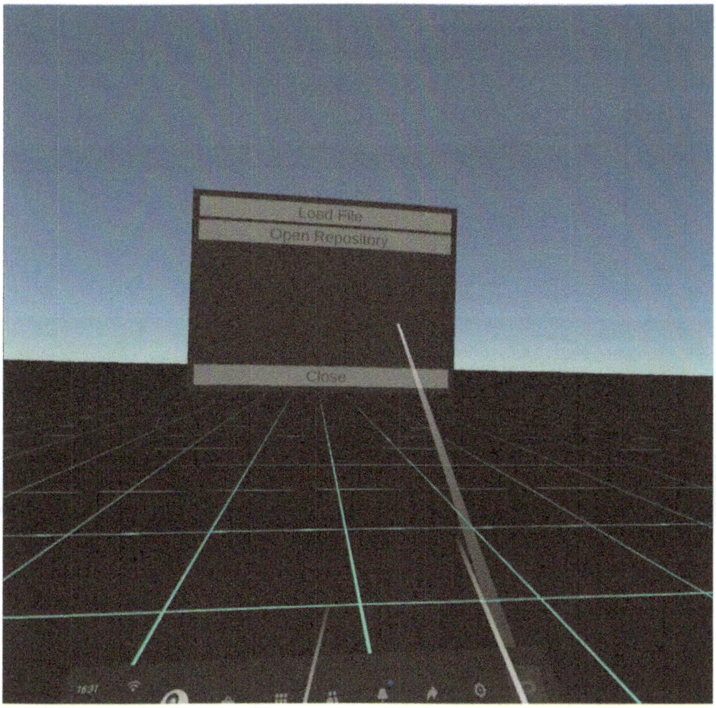

Fig. 4. Loaded application, with UI open on the main screen. Screenshot was captured in the app on an Oculus Quest v1.

The workflow presented in Fig. 3 was tested by loading models from the glTF sample model repository maintained by Khronos Group [12] (Fig. 5). Models were created with photogrammetry (Fig. 6) and medical data that was extracted from a CT scan (Fig. 7).

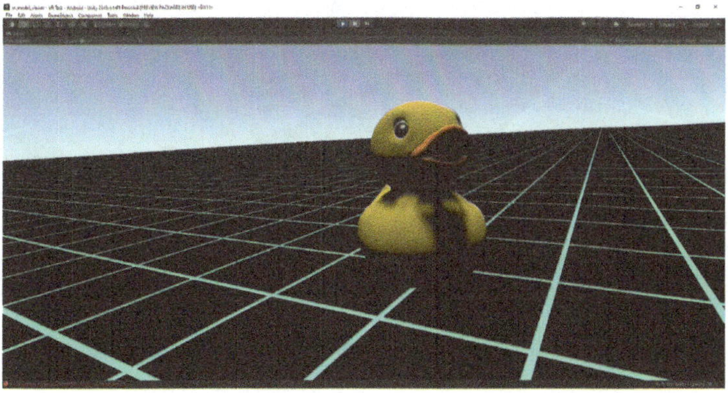

Fig. 5. Duck model from the Khronos Group glTF sample model repository v2.0 loaded in the application, captured from Unity Editor 2018.4 with the Oculus Quest v1

The Duck model in Fig. 5 was used as a main test model, due to its relatively small size and manageable polygon count. It was loaded without any issues.

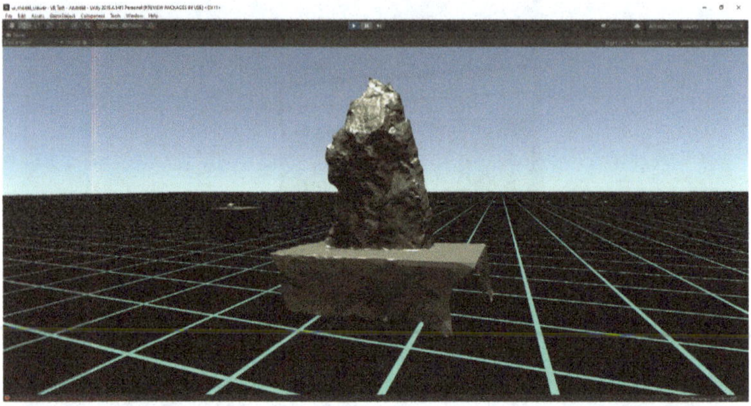

Fig. 6. Rock model constructed with Meshroom from a photo set captured with the Sony a7 II and Tamron 28–75 f2.8. The resulting model was then simplified in Blender. Screenshot was captured from Unity Engine 2018.4 with the Oculus Quest v1 headset inside of the application. The model was loaded via the Model Repository.

The Rock model in Fig. 6 was successfully loaded after some postprocessing in Blender. The number of polygons had to be reduced and the artifacts from photogrametry had to be removed. This model demonstrates the level of detail you can get in VR quite well.

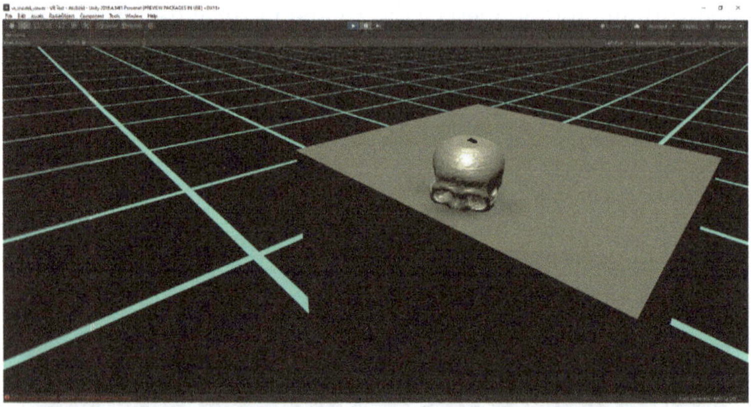

Fig. 7. Model of a human skull reconstructed from CT scan data via our compute shader slice to volume pipeline. The model was simplified and smoothed in Blender and then uploaded to the model repository. The screenshot was taken in Unity Editor 2018.4 with the Oculus Quest v1 and the model was loaded from the model repository.

The skull model from Fig. 7 was reconstructed with the experimental compute shader pipeline, however, the produced models have very high polygon counts, so it had to be greatly simplified to be viewable in virtual reality without inducing motion sickness due to the low framerate, similarly to the Rock model. Limits of the device that runs the application have to be taken into consideration when uploading models to the repository and support for multiple levels of detail will be a useful addition in a future version.

The novelty lies primarily in the fact that this (user friendly) application will be available to clinicians, especially specialists in onco-surgery, traumatology, and radiology. In these disciplines, the great advantage is that the MDs have the opportunity to analyze the problem in detail using a model in virtual reality. At the same time (especially in traumatology), speed is also necessary. The clinician must (i) have the data immediately available, (ii) be able to generate the model himself, and (iii) be able to control the application himself. Our approach meets all these criteria.

5 Conclusion

In conclusion, a combination of a remote model storage system and a local viewing software running on a standalone VR headset seems like a robust way to improve user experience inside of virtual reality. You gain the ability to use a standalone virtual reality headset, which is not only easier to use while at an established working place, but also deployed in the field. A successful adoption of such technology requires improvement in key areas. We expect the application will have a wide use in biomedicine as a viable tool for MD, after the improvements are implemented.

Firstly, there needs to be an application that runs on the users computer and allows them to effortlessly load models into a remote model repository or even start hosting a model repository in the local network from their computer with a click of a button.

Secondly, the model repository app needs to be improved with the end user in mind. There needs to be a way to load an environment that suits the user best. A way to load interactive models in the form of Asset Bundles would also greatly improve the usability and widen the target audience.

Furthermore, model conversion should be handled on the backend, so the user does not have to think about the file format they are using, as long as it is supported by the model repository. The model repository should also handle multiple levels of detail of the models to accommodate various performance or internet bandwidth restrictions in different use cases.

Another area where the application can be improved is the user interface interaction. We use a common way of interaction with the UI, however this can be difficult in some situations, e.g., when users hands are shaky for some reason. We decided to improve the interaction with the UI in future versions by utilizing the joystick to move the focus between UI elements instead of pointing.

Acknowledgements. Financial support from specific university research (A1_FCHI_2020_002) and LTAIN19007 Development of Advanced Computational Algorithms for evaluating post-surgery rehabilitation. The financial support is gratefully acknowledged.

References

1. Virtual reality training for health-care professionals: Cyberpsychology and Behavior **6**(4), 389–395 (2003)
2. Armbrüster, C., Wolter, M., Kuhlen, T., Spijkers, W., Fimm, B.: Depth perception in virtual reality: distance estimations in peri- and extrapersonal space. Cyberpsychol. Behav. Impact Internet, Multimedia Virtual Reality Behav. Soc. **11**, 9–15 (2008)
3. Blender: About. https://www.blender.org. Accessed 23 Nov 2020
4. Blue, L.: gltf converter. https://blackthread.io/gltf-converter/. Accessed 23 Nov 2020
5. Bryson, S.: Virtual reality in scientific visualization. Commun. ACM **39**(5), 62–71 (1996). https://doi.org/10.1145/229459.229467
6. Checa, D., Bustillo, A.: A review of immersive virtual reality serious games to enhance learning and training. Multimedia Tools Appl. 5501–5527 (2019). https://doi.org/10.1007/s11042-019-08348-9
7. Express.js: About. http://expressjs.com. Accessed 23 Nov 2020
8. Foundation, O.: About. https://nodejs.org/en/about/. Accessed 23 Nov 2020
9. García Hernandez, R., Kranzlmüller, D.: Nomad VR: multiplatform virtual reality viewer for chemistry simulations. Comput. Phys. Commun. 237 (2018)
10. González Izard, S., Juanes Méndez, J.A., Ruisoto Palomera, P., García-Peñalvo, F.J.: Applications of virtual and augmented reality in biomedical imaging. J. Med. Syst. **43**(4), 1–5 (2019). https://doi.org/10.1007/s10916-019-1239-z
11. Group, K.: gltf overview. https://www.khronos.org/gltf/. Accessed 23 Nov 2020
12. Group, K.: gltf sample models. https://github.com/KhronosGroup/glTF-Sample-Models. Accessed 23 Nov 2020
13. Lledó, L.D., et al.: A comparative analysis of 2d and 3d tasks for virtual reality therapies based on robotic-assisted neurorehabilitation for post-stroke patients. Front. Aging Neurosci. **8**, 205 (2016). https://www.frontiersin.org/article/10.3389/fnagi.2016.00205
14. Millais, P., Jones, S., Kelly, R.: Exploring data in virtual reality: comparisons with 2d data visualizations. In: CHI 2018 - Extended Abstracts of the 2018 CHI Conference on Human Factors in Computing Systems. vol. 2018-April. Association for Computing Machinery, USA United States, April 2018
15. Millais, P., Jones, S.L., Kelly, R.: Exploring data in virtual reality: Comparisons with 2d data visualizations. In: Extended Abstracts of the 2018 CHI Conference on Human Factors in Computing Systems. CHI EA 2018, New York, NY, USA, pp. 1–6. Association for Computing Machinery (2018). https://doi.org/10.1145/3170427.3188537
16. MongoDB, I.: What is mongodb. https://www.mongodb.com/what-is-mongodb. Accessed 23 Nov 2020
17. Oculus: Oculus device specifications. https://developer.oculus.com/learn/oculus-device-specs. Accessed 23 Nov 2020
18. Onyesolu, M.O., Eze, F.U.: Understanding virtual reality technology: advances and applications. Adv. Comput. Sci. Eng. 53–70 (2011)

19. Robb, R.: Medical imaging and virtual reality: A personal perspective. Virtual Reality **12**(4), 235–257 (2008)
20. Siccity: Gltfutility. https://github.com/Siccity/GLTFUtility. Accessed 13 Sept 2020
21. Technologies, U.: Unity documentation. https://docs.unity3d.com/Manual/. Accessed 20 Nov 2020
22. Technologies, U.: Unity documentation: Unitywebrequest. https://docs.unity3d.com/Manual/UnityWebRequest.html. Accessed 20 Nov 2020
23. Technologies, U.: Unity documentation: Xr. https://docs.unity3d.com/Manual/XR.html. Accessed 20 Nov 2020
24. Todd, C.: https://www.cs.csustan.edu/~dkim/files/college_poster_session/2020/game_development_presentation.pdf (2021)
25. Turner, C.J., Hutabarat, W., Oyekan, J., Tiwari, A.: Discrete event simulation and virtual reality use in industry: new opportunities and future trends. IEEE Trans. Hum.-Mach. Syste. **46**(6), 882–894 (2016)
26. Hageman, A.: Virtual reality. Nursing **24**(3), 3–3 (2018). https://doi.org/10.1007/s41193-018-0032-6
27. Zheng, J., Chan, K., Gibson, I.: Virtual reality. IEEE Potentials **17**(2), 20–23 (1998)

Implementing Dynamic Mathematics Software in Calculus II for Engineering Students: Quadratic Surfaces

Tommy Tanu Wijaya, Jianlan Tang(✉), Li Li, and Aditya Purnama

Guangxi Normal University, Guilin, China
tjlwxt@mailbox.gxnu.edu.cn

Abstract. This research aims to analyze the learning effectiveness of using technology for engineering students when learning parabolic model. For engineering students, Parabolic can be learned in calculus II. To solve parabolic questions, a deep and good understanding of the concept is needed. An active learning condition using technology can help engineering students understand the parabolic concept and increase their learning interest. The learning concept that was used in this research is learning by doing. engineering students will not only listen to the teachers' explanation, but they will also try to solve the problem using Hawgent. This research is done at one of university in Guilin, China. There is a total of 63 sample engineering students. This research result showed that engineering students' learning interest and confidence in designing parabolic models increased and guided engineering students to achieve deep learning.

Keywords: Calculus · Hawgent · Engineering student · Parabolic model

1 Introduction

Quadratics surface concept is one of the subjects that should be mastered by engineering students [1]. A quadratic surface known as quadratics has 17 different types (see in Table 1) with different difficulties [2]. Every quadratic surface that they learn in calculus is one of the basic courses for engineering students as they will not succeed without mastering these courses. The question development for the quadratic surface will not only focus on finding the volume but on various variations and difficulty levels. Not every engineering students' ability is equal, so some engineering students find quadratic surface difficult [3]. When students find a topic difficult to learn, they will not achieve a maximal learning outcome and their learning interest is also low. Also, the minimal class hours and an abstract topic require good visualization skills and a strong basic concept understanding. Researchers see that the use of technology in the learning process can help students to overcome their problems.

There are many researchers that use technology to help students understand an abstract mathematics topic. Adnan A. et al. [4] showed how to use GeoGebra software on geometry. Zengin Y. et al. [5] researched the effect of learning GeoGebra on

R. Silhavy (Ed.): CSOC 2021, LNNS 230, pp. 480–491, 2021.
https://doi.org/10.1007/978-3-030-77442-4_41

Table 1. 17 quadratic surfaces and it's characteristics

Surface	Equation	$\rho 3$	$\rho 4$	k
planes	$x^2 = 0$	1	1	
Elliptic cone	$\frac{x^2}{a^2} + \frac{y^2}{b^2} + \frac{z^2}{c^2} = 0$	3	3	0
Ellipsoid	$\frac{x^2}{a^2} + \frac{y^2}{b^2} + \frac{z^2}{c^2} = 1$	3	4	1
Elliptic cylinder	$\frac{x^2}{a^2} + \frac{y^2}{b^2} = 1$	2	3	1
Elliptic paraboloid	$\frac{x^2}{a^2} + \frac{y^2}{b^2} = z$	2	4	1
Hyperbolic cylinder	$\frac{x^2}{a^2} - \frac{y^2}{b^2} = -1$	2	3	0
Hyperboloid paraboloid	$\frac{y^2}{b^2} - \frac{x^2}{a^2} = z$	2	4	0
Hyperboloid of one sheet	$\frac{x^2}{a^2} + \frac{y^2}{b^2} - \frac{z^2}{c^2} = 1$	3	4	0
Hyperboloid of two sheet	$\frac{x^2}{a^2} + \frac{y^2}{b^2} - \frac{z^2}{c^2} = -1$	3	4	0
Parabolic cylinder	$x^2 + 2rz = 0$	1	3	
Parallel planes	$x^2 = a^2$	1	2	

students' learning achievement and found out that the class that uses GeoGebra software has a better learning outcome than the class that did not. Tan S. et al. [6] used Hawgent software to explain the sides of a circle and also found out that using technology has a better outcome than the traditional method. Using technology can also improve students' problem-solving [7–10], mathematical ability [11], and mathematics basic concept understanding [12–14]. Only a few papers discussed the development and effect of using software for university students specifically for engineering students. However, not all university students understand the topic taught and still need technology to help them understand as many university-level materials require the software. With this in mind, researchers believe that the use of technology positively impacts students and can be researched further and deeper.

These days, there is a lot of mathematics software that can be downloaded freely such as GeoGebra [15], geometer sketchpad [16], maple [17], and etc. This software has the same function which is to draw a function or other things to visualize the shape. Hawgent is one of the mathematics software similar to other dynamic mathematics software [18, 19]. Hawgent still uses Chinese because it still focuses on domestic usage or people who can understand Chinese.

Engineering students and lecturers can use that software when explaining calculus II, especially on the quadratic surface. By using Hawgent, the quadratic surface can be visualized in a very interesting way. In other words, it can help engineering students to learn quadratic surfaces.

The focus of this research is the implementation of Hawgent on engineering students on the quadratic surface. Below is the purpose of this research:

- See the difficulties faced by students when solving quadratic surface problems
- Proof that Hawgent can help engineering students to visualize the shapes of quadratic surfaces.
- engineering students' response on Hawgent as a learning media

2 Method

This research was done in 2019–2020. Researchers gave 6 pre-test questions to 63 1st year engineering students of which 47 among them are female and 16 of them are male in one of University, China. The pre-test was given to see the difficulties faced by engineering students on the quadratic surface. A purposive sample is used in this research.

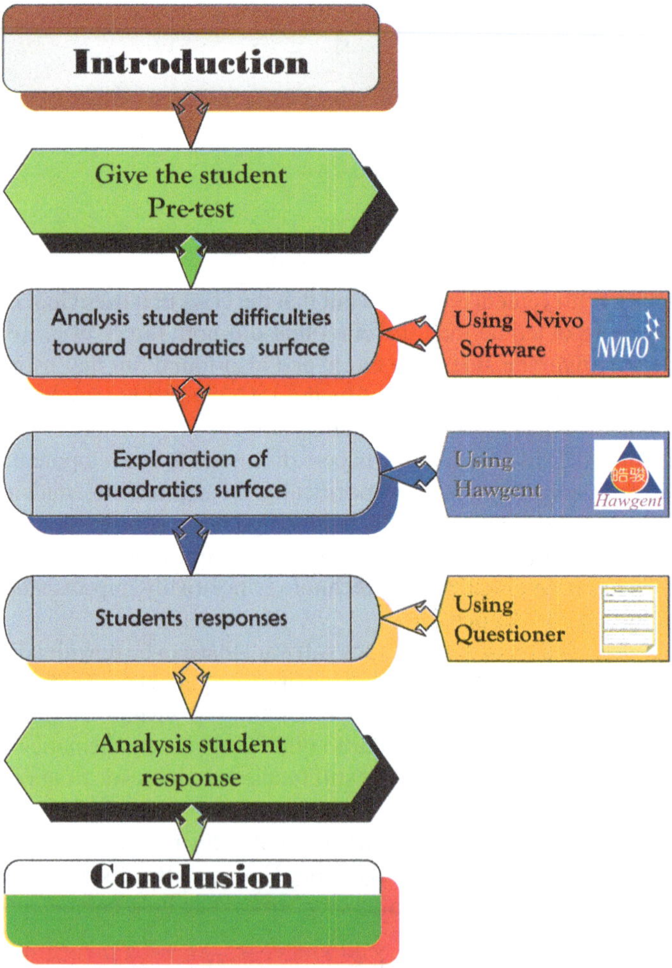

Fig. 1. Research steps

The research steps can be seen in Fig. 1. Firstly, researchers explain the introduction on calculus II and quadratic surface, learning aim, types, characteristics, and examples of the quadratic surface. After 30 min of introduction, given 6 pre-test questions to know engineering students' initial ability on the quadratic surface. 60 min given to students to do the pre-test questions. engineering students' answers analyzed by researchers to know the difficulties faced by engineering students in answering quadratic surface questions. In this step, the analysis use the qualitative method with the help of Nvivo software. The analysis of engineering students' difficulties serve as a reference for researchers to teach using Hawgent software in the next meeting.

In the second meeting, researchers point out the mistakes that students made in their pre-test. Researchers then explain quadratic surfaces using Hawgent. Throughout the teaching-learning activity, researchers analyze engineering students' responses during class. At the end of the teaching, researchers gave out questionnaires to students. The last step is analyzing the questionnaire result and conclude.

2.1 Instruments

The research instrument is divided into 2 which are pre-test questions that consist of 6 questions to measure students' difficulties in solving calculus II questions (quadratic surface) and a questionnaire. The questionnaire is given after the meeting to see the advantage of using Hawgent on calculus II and engineering students' opinion on Hawgent. 5 points will be given if the student agrees to the statement and 1 point will be given if they strongly disagree. The questionnaire can be seen in Table 2.

Table 2. Questionnaire on engineering students' opinion towards Hawgent on calculus II

Items	On a scale of 1–5, what is your opinion on Hawgent	Scale				
		5	4	3	2	1
Q1	Help students to understand the quadratic surface					
Q2	Improve analysis and problem-solving skills					
Q3	Improve creative thinking ability					
Q4	Can visualize the shapes of quadratic surface					
Q5	It made the class more interesting and fun					
Items	On a scale of 1–5, what is your opinion of using Hawgent on calculus II	Skala Scale				
		5	4	3	2	1
Q6	Improve students' learning interest and motivation					
Q7	Improve skills of using technology in class					
Q8	Help understand the quadratic surface concept					
Q9	A creative way to visualize quadratic surface					
Q10	Improve students' critical thinking and help achieve deep learning					

Table 3. Pre-test question on the quadratic surface

Match the figure in the right with the equation. If the figure is a quadratic surface, name the figure. The answer should come with the solution.

	Equation		Graph
1	$2x^2 + y^2 + 3z^2 = 1$	A	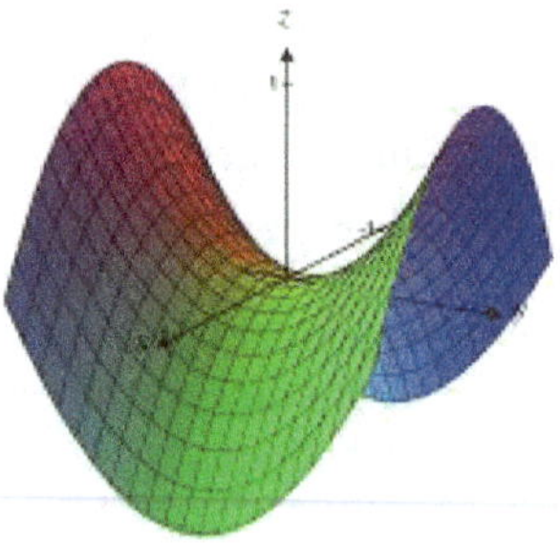
2	$y^2 = 4x$	B	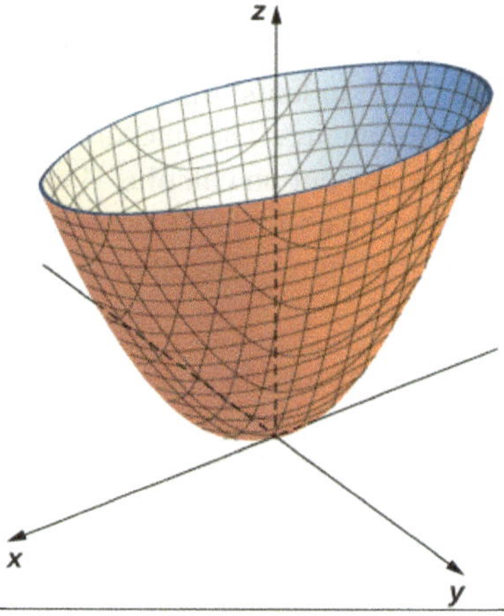
3	$x^2 + z^2 = y$	C	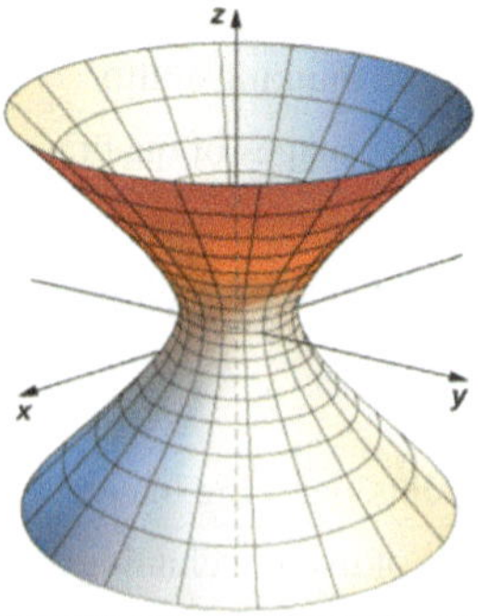

(continued)

Table 3. (*continued*)

4	$x^2 + 4y^2 = 2z$	D	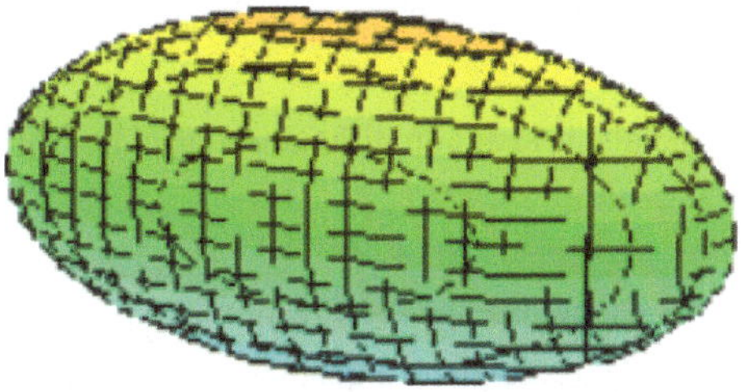
5	$2x^2 + y^2 - z^2 = 1$	E	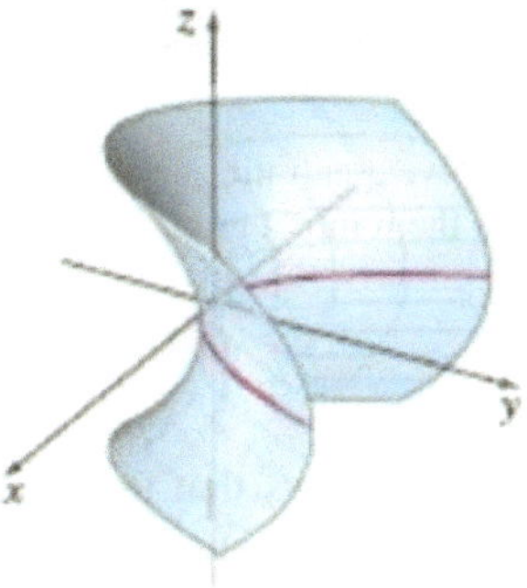
6	$4y^2 - x^2 = z$	F	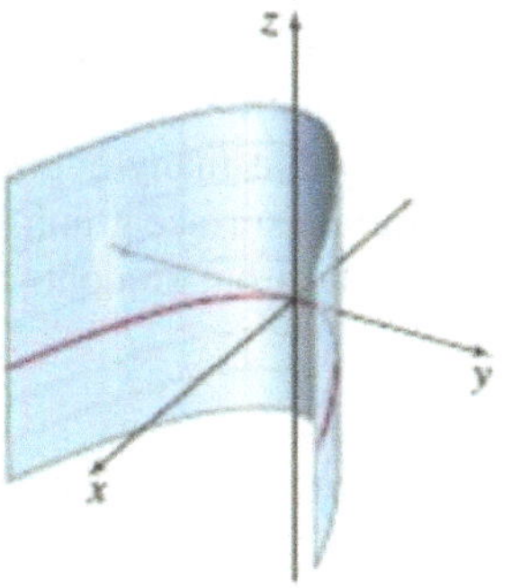

The second instrument is the pre-test question that focuses on finding the volume and surface area of quadratic surfaces (Ellipsoid, Cone, Elliptic Paraboloid, Hyperboloid of one sheet, Hyperboloid of two sheets). Students are asked to draw a quadratic surface, determine the formula from the picture and find the quadratic surface volume. This way, researchers can analyze students' skills in calculus II. The pre-test question can be seen in Table 3.

3 Difficulties Faced by Students on Quadratic Surface

The pre-test result shows that when teachers explain quadratic surface without the help of software is not very good. Some of the difficulties faced by students that researchers found from the pre-test can be seen in Table 4. From a total of 63 engineering students, there were only 12 (19.05%) students that answered all the questions right. While most students have various difficulties.

Table 4. Analysis result on difficulties faced by engineering students during the pre-test

No.	Difficulties faced by engineering students	Number of students	Percentage
1	Students cannot change the equation into a graph	51	80.95%
2	When students are given a quadratic surface figure, they cannot write the general equation	44	69.84%
3	Students do not know the characteristic of each quadratic surface	55	87.30%
4	Students can solve quadratic surface problems but cannot change them into a graph	48	76.19%
5	Operation error	23	36.51%

The analysis result shows that the main factors that affect students' mistakes are because students have not mastered the basic concept and the characteristic of each quadratic surface. The lack of visualization when the teacher explains the quadratic surface causes students to understand each quadratic surface's concept and characteristic. One of the examples is that when students solve problems on Hyperboloid of one sheet, students assume that the equation is Hyperboloid of two sheets.

Researchers found out that many engineering students can solve the equation but they can't draw the quadratic surface graph. This is because students can only solve and memorize the solving pattern but students are still not good at changing the equation into a graph.

4 Students' Behavior Initial Analysis During the Calculus II Class

Students' attitudes towards a new topic should be positive. A positive attitude and learning interest can improve students' absorption potential on what is taught by teachers. A good learning interest can make students achieve deep learning and improve their critical and creative thinking ability. However, researchers found out that students' learning interest during the first class is not very good. This can be seen as some students play with their phones, talk to other students and sleep during class.

The teacher teaches by using PowerPoint and explaining some questions on the board, making students feel that calculus II is boring and hard to understand. Students were not active in answering questions given by the teacher. One a few students who have a good understanding of calculus can follow the teacher's lesson.

The lack of students' learning interest and the class atmosphere that is not supportive, researchers consider these to experiment by using Hawgent to help the student visualize the quadratic surface and explain each quadratic surface type's characteristic.

5 Using Hawgent to Help Visualize Quadratic Surface

In the second meeting, teachers elaborate on students' mistakes and difficulties that can be seen in Table 3. Researchers then start to teach using Hawgent and show step-by-step how to draw a quadratic surface graph from an equation (Fig. 2). After each quadratic surface graph is made, researchers then explain the basic concept and characteristics of each quadratic surface (Fig. 3).

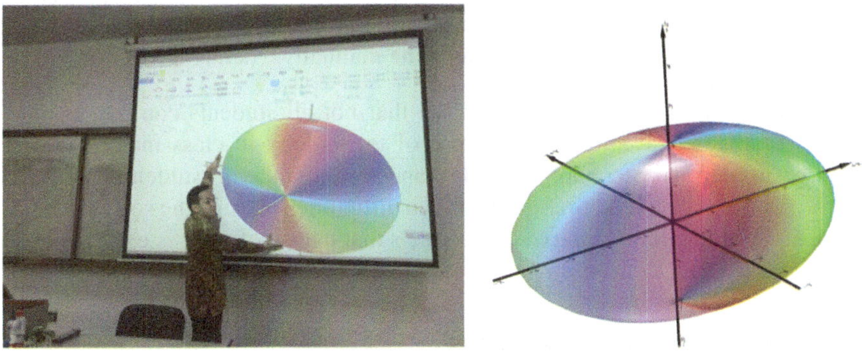

Fig. 2. Ellipsoid

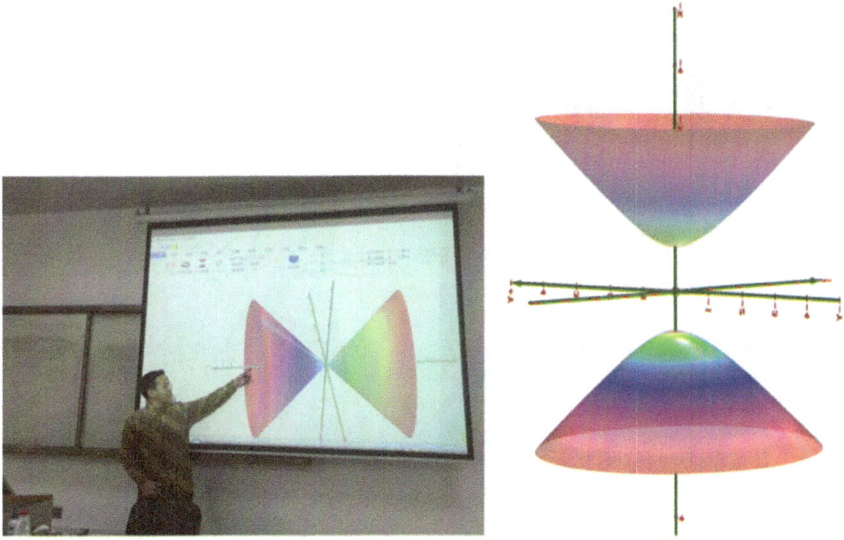

Fig. 3. Hyperboloid of two sheets

Researchers know that learning by doing has a better result than just listening and seeing the teacher explain a topic. That is why, after researchers explain everything in class, the teachers ask a random engineering students to come forward and try to draw a hyperboloid of one sheet from its equation using Hawgent. Teachers guide the engineering students step-by-steps when they are drawing the hyperboloid of one sheet graph using Hawgent. At first, the student looked confused and have trouble using Hawgent but after the student successfully made the hyperboloid of one sheet graph that looks like Guangzhou Tower (see Fig. 4), the student feels that Hawgent is interesting, has a clear graphic, easy to use.

> "When the teacher asked me to come forward and draw quadratic surface, I feel foreign and difficult maybe it's because I can't use mathematical software. However, after I'm able to convert the equation into a graph, I feel that using technology to learn quadratic surface is easier than done manually." (student 1)

From the student statement, we can see that not all students can use technology especially mathematical software. The use of technology in class make engineering students get used to using technology in class and help engineering students to understand an abstract concept better. At first, students had difficulties in using the software but from the students' statement, we can see that he feels that using software is easier than doing it manually.

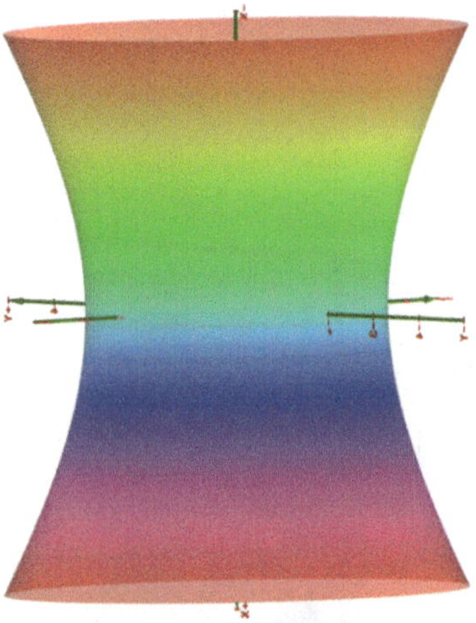

Fig. 4. Engineering Students' work result in making a Hyperboloid of one sheet graph that looks like Guangzhou tower

At this stage, students have mastered using technology. This skill can be utilized and directed at mastering software in schools. When students first use the software, they will feel foreign and difficult. However, after they get used to the technology, students will feel that it's easier to use technology and more accurate in solving problems compared to the manual method. Just like the use of a calculator that is commonly used to help people to count big numbers rather than counting manually.

6 Engineering Students' Response on Learning Calculus II Using Hawgent

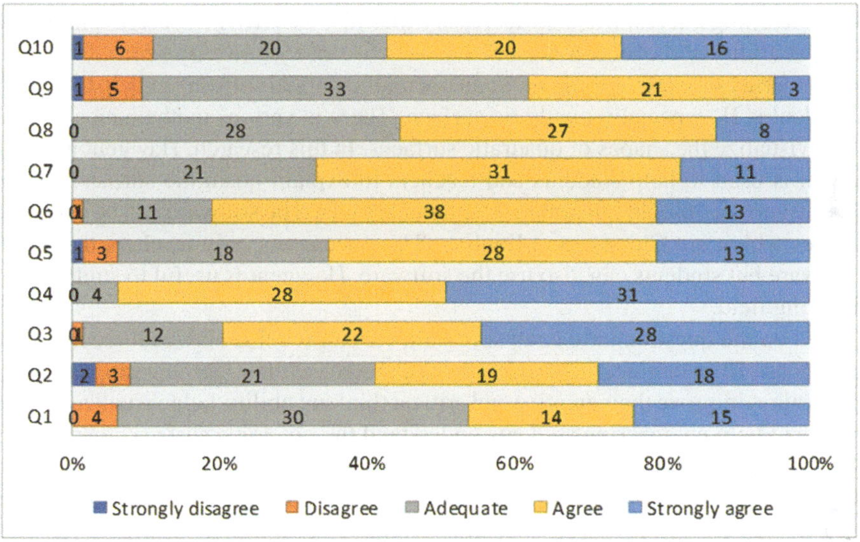

Fig. 5. Engineering Students' response on learning calculus II using Hawgent

After researchers teach students how to use Hawgent on the quadratic surface, researchers gave out questionnaires to know engineering students' responses towards Hawgent on the quadratic surface. Based on the questionnaire result in Fig. 5, most students agree with the statements as it has the highest average of 39.37% while adequate, strongly agree, disagree, strongly disagree has an average of 31.43%, 24.76%, 3.65%, and 0.79% respectively.

From the result above, we can conclude that students have a positive response towards technology. This is in line with previous researchers that showed that Hawgent receives positive response from students. The integration of technology in class can always improve engineering students' soft and hard skills.

One of the interesting things that can be seen from this questionnaire result is that using Hawgent increases students' learning interest. The interesting and colorful design made students not bored in learning quadratic surfaces. engineering students' can also easily input numbers in Hawgent to change and find the area of the quadratic surface. When students' learning interest increases, their learning outcome, and engineering students' ability will also improve. This can also suggest teachers use learning media to increase students' learning interest so that class can be more optimal dan material absorption to be easier.

7 Conclusion and Limitation of the Study

Engineering students still face difficulties in solving quadratic surface problems. One of the difficulties is they are unable to master the characteristic and basic concept of the quadratic surface. A weak understanding of the basic concept can result in students having difficulties in solving various quadratic surface problems. On the other side, students' learning interest will also decrease which also resulted in a decrease in students' absorption ability and concentration.

Engineering students and teachers can use mathematical software to help them understand calculus II, especially quadratic surface. There is various mathematical software that can visualize the shapes of quadratic surfaces. In this research, Hawgent is the software that is used to help students and teachers to explain quadratic surface. Hawgent v3.0 is easy to use, has an interesting layout and colors. The software can be used in two languages which are Chinese and English. This means not only teachers that can use the software but students can also use the software. Hawgent is useful to students in the engineering field.

Based on the questionnaire analysis result, we can see that Hawgent receives a positive response from engineering students. It can increase their learning interest, achieve deep learning, improve their creative and critical thinking ability, help visualize quadratic surface, and help engineering students understand the quadratic surface concept.

Seeing the positive response, the use of technology in the education field can be further developed especially in the engineering field as most of their theory is abstract. The use of technology in class can also improve students' creativity. Hawgent can always be used to visualize and make the class more interesting for every concept. The use of technology can improve education quality in the future.

This research's limitation is that it can only see engineering students' difficulties when learning quadratic surface and engineering students' attitude towards Hawgent. Deeper research on the post-test can be done in the next research. Suggestions for future research to applicate Hawgent to other topics and comparison research between Hawgent and other dynamic mathematics software can also be done.

Acknowledgments. The development of learning media using hawgent was supported by the province of Guangxi, China (2020JGA129).

References

1. Hahn, A.J.: Basic Calculus of Planetary Orbits and Interplanetary Flight: The Missions of the Voyagers, Cassini, and Juno. Springer, Cham (2020). https://doi.org/10.1007/978-3-030-24868-0
2. Maekawa, T.: Self-intersections of offsets of quadratic surfaces: Part I, Explicit surfaces. Eng. Comput. **14**(1), 1–13 (1998)

3. Hong, K.J., Kim, M.S., Choi, D.H.: Efficient approximation method for constructing quadratic response surface model. KSME Int. J. **15**(7), 876–888 (2001)
4. Akkaya, A., Tatar, E., Kağizmanli, T.B.: Using dynamic software in teaching of the symmetry in analytic geometry: the case of GeoGebra. Procedia - Soc. Behav. Sci. **15**, 2540–2544 (2011)
5. Zengin, Y., Furkan, H., Kutluca, T.: The effect of dynamic mathematics software geogebra on student achievement in teaching of trigonometry. Procedia - Soc. Behav. Sci. **31**(2011), 183–187 (2012)
6. Tan, S., Zou, L., Wijaya, T.T., Suci, N., Dewi, S.: Improving student creative thinking ability with problem based learning approach using hawgent. J. Educ. **02**(04), 303–312 (2020)
7. Zhang, L., Zhou, Y., Wijaya, T.T.: Hawgent dynamic mathematics software to improve problem-solving ability in teaching triangles. J. Phys. Conf. Ser. **1663**(1) (2020)
8. Alhazzani, N.: MOOC's impact on higher education. Soc. Sci. Humanit. Open **2**(1), 100030 (2020)
9. Yadav, S., Chakraborty, P., Kochar, G., Ansari, D.: Interaction of children with an augmented reality smartphone app. Int. J. Inf. Technol. **12**(3), 711–716 (2020). https://doi.org/10.1007/s41870-020-00460-6
10. Sahin, A., Gulacar, O., Stuessy, C.: High school students' perceptions of the effects of international science Olympiad on their STEM career aspirations and twenty-first century skill development. Res. Sci. Educ. **45**(6), 785–805 (2014). https://doi.org/10.1007/s11165-014-9439-5
11. Mudaly, V., Fletcher, T.: The effectiveness of geogebra when teaching linear functions using the IPad. In: Problem Education 21st Century, vol. 77, no. 1, pp. 55–81 (2019)
12. Fani, R.A., Sukoco, P.: Volleyball learning media using method of teaching games for understanding adobe flash-based. Psychol. Eval. Technol. Educ. Res. **2**(1), 34–50 (2019)
13. Suan, L., Ying, Z., Wijaya, T.T.: Using hawgent dynamic mathematics software in teaching arithmetic operation. Int. J. Educ. Learn. **2**(1), 25–31 (2020)
14. Reeves, T., Caglayan, E., Torr, R.: Don't shoot ! understanding students' experiences of video-based learning and assessment in the arts. Video J. Educ. Pedagog. 1–13 (2017)
15. Daniela, L.: Smart pedagogy for technology-enhanced learning. In: Daniela, L. (ed.) Didactics of Smart Pedagogy, pp. 3–21. Springer, Cham (2019). https://doi.org/10.1007/978-3-030-01551-0_1
16. Danday, B.A., Monterola, S.L.C.: Effects of microteaching physics lesson study on pre-service teachers' critical thinking. J. Balt. Sci. Educ. **18**(5), 692–707 (2019)
17. Saha, R.A., Ayub, A.F.M., Tarmizi, R.A.: The effects of GeoGebra on mathematics achievement: Enlightening Coordinate Geometry learning. Procedia - Soc. Behav. Sci. **8**(5), 686–693 (2010)
18. Wijaya, T.T., et al.: Hawgent dynamic mathematic software as mathematics learning media for teaching quadratic functions. In: Journal of Physics: Conference Series, vol. 1592, no. 1 (2020)
19. Wijaya, T.T., Ying, Z., Purnama, A.: Using hawgent dynamic mathematics software in teaching trigonometry. Int. J. Emerg. Technol. Learn. **15**(10), 215–222 (2020)

A Machine Learning Framework for Inter-frame Prediction for Effective Motion Estimation

S. K. Veena(✉) and K. Mahesh Rao

Department of ECE, MITM, Mysuru, Karnataka, India
veenask_ece@mitmysore.in

Abstract. Motion estimation is an essential technique to be carried out during the video encoding, characterized by the complex inter-frame prediction challenges. A review of existing approaches exhibits that there is still enormous scope for improvement. Therefore, the proposed system introduces a unique and straightforward process for performing inter-frame prediction by applying an enhanced exhaustive search mechanism followed by a simplified machine learning scheme. The further encoding process is carried out with respect to quantization parameters, frame size, and intra block size. The simulation carried out in MATLAB shows that the proposed system offers a good signal quality and response time compared to H.265 and H.264.

Keywords: Video encoding · H.265 · H.264 · Motion estimation · Motion compensation · Inter-frame prediction

1 Introduction

Motion estimation is an essential process to extract important information from the given video frames [1]. This process is responsible for assessing an object's mobility for a given sequence of an image to extract potential vectors representing estimated motion [2]. This process also contributes to data compression by utilizing the mobile object's information [3]. The issues associated with temporal redundancies are addressed using motion estimation, followed by motion compensation [4]. Some significant challenges are associated with this process regarding real video scenes due to the combination of rotation and translation [5]. Simultaneously, this process is quite expensive from a computational viewpoint as the search is carried out at the position of pixel considering multiple variants of reference frames [6]. In this process, H.265 is considered to offer better performance than H.264 concerning the file's reduced size, hence reduced bandwidth [7]. At present, there have been various works carried out towards motion estimation and compensation techniques [8–10], where all the techniques have their benefits and limitation. The prime issue still resides in inter-frame prediction, where achieving better signal quality and faster response time is an open-end challenge.

Therefore, this paper presents a discussion about a simplified and novel inter-frame prediction for offering better motion estimation using an enhanced exhaustive search mechanism. The paper's organization is as follows: Sect. 2 discusses the current

R. Silhavy (Ed.): CSOC 2021, LNNS 230, pp. 492–500, 2021.
https://doi.org/10.1007/978-3-030-77442-4_42

research work followed by problem identification in Sect. 3. Section 4 discusses the proposed methodology, followed by an elaborate discussion of algorithm implementation in Sect. 5. Comparative analysis of the accomplished result is discussed under Sect. 6, followed by the conclusion in Sect. 7.

2 Related Work

The different variants of motion estimation models are being investigated in present times. The recent work carried out by Park, and Kang [11] have used the affine motion estimation model to control the temporal redundancies. Such an approach and compressive sensing also seem to offer better motion estimation schemes, as witnessed in Zhang et al. [12]. Ray et al. [13] have used recent encoding HEVC and evolved up with a progressive and adaptive motion vector scheme for faster access and lower delay. Luo et al. [14] have introduced a unique study where a three-dimensional geometry-based policy is applied for flow estimation. Compensation using Wang et al. [15] has used a segmentation-based approach over reference frames to identify the zones with motion compensation using versatile video coding. Saliency based on space and time is used by Li et al. [16] to resist unwanted computation. Liu et al. [17] have carried out a work towards intra prediction using convolution neural network claimed to offer better model capacity. The adoption of HEVC by Wang et al. [18] has considered moving the camera to perform background modeling with reduced complexity. There are also studies associated with transcoding towards HEVC for improving decision in the fast mode, as seen in Franche et al. [19]. Flow estimation is another unique approach where the adoption of cost control will give better results, as seen in the study of Yang et al. [20]. Padin et al. [21] have used the prediction of footprints for improving bi-directional frames to offer better compression performance. The current study shows that the adoption of machine learning, e.g., convolution network contributes to pose prediction, as seen in Zhao et al. [22]. Khan et al. [23] have emphasized developing a tracking mechanism using the Kalman filter, which could also be used in motion estimation.

Similarly, various other approaches have contributed to motion estimation. e.g., prediction using multi-hypothesis in Chen et al. [24], parametric video prediction in Ho et al. [25], derivation of the motion vector in HEVC in Wang et al. [26], convolution network-based search for the motion vector in Choi et al. [27], video coding using the difference of edge position in Asikuzzaman et al. [28], interpolation using an adaptive method in Lin et al. [29], and motion estimation of the object in Asikuzzaman et al. [30]. The prior work of Veena SK and Mahesh Rao K [31, 32] discusses video compression technique and performance analysis of the video codec. The next section outlines the research problems associated with it.

3 Problem Description

A review of different existing approaches towards motion estimation shows that they are still associated with the following challenges viz. i) adoption of complex approaches which offers better parametric performance score but at the cost of computational efficiency, ii) majority of the studies have considered existing H.265 standards which have issues associated with transcoding with the higher dependency of processors, iii) a simplified architectural model for encoding is still missing in the literature, and iv) adoption of machine learning towards optimizing the performance of encoding still requires more simplification. Therefore, the problem can be stated as "*designing an optimal encoding process which can balance computational efficiency with signal quality while performing motion estimation.*"

4 Proposed Methodology

The proposed study discusses a simplified motion estimation model to perform inter-frame prediction using analytical modelling. The core idea is mainly focused on performing motion estimation and a machine learning approach to carry out this prediction. The adopted architecture of the proposed system is as follows:

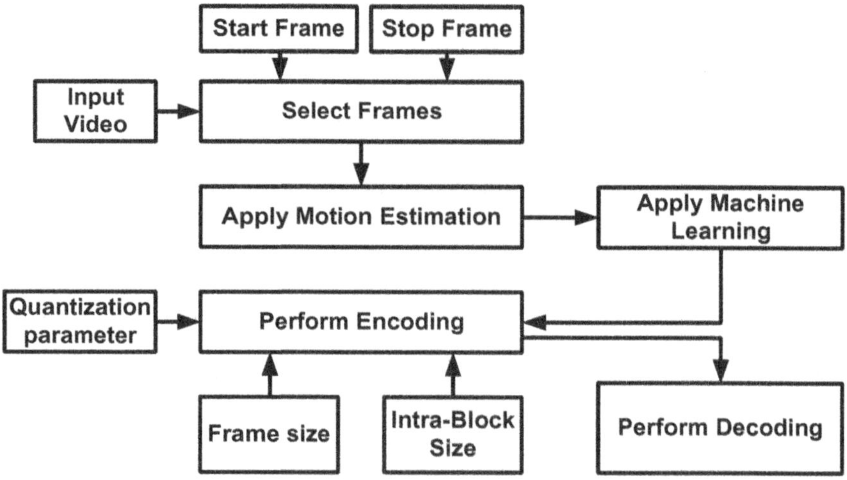

Fig. 1. Adopted architecture of proposed system

According to this scheme shown in Fig. 1, the input frames are extracted from its test video to apply motion estimation. The proposed system uses an enhanced exhaustive search algorithm for performing motion estimation. Further, the proposed system uses a simplified machine learning approach based on clustering pixels. This operation is further followed by selecting quantization parameters, frame size, and intra-block size to encode the frames saved in bitstreams. The receiver performs a decoding operation.

5 Algorithm Implementation

The proposed system's implementation is carried out considering two different variants of inter-frame prediction associated with the motion compensation, i.e., the search algorithm and learning algorithm. The steps of this algorithm are as follows:

Algorithm for Exhaustive Frame Prediction
Input: $frame_{max}$ (maximum number of frames)
Output: $frame_{pred}$ (predicted frame)
Start
1. **For** i=1: $frame_{max}$
2. $I_m = f_1(I_{mrec})$
3. $frame_{ref}(m, n) = I_{mrec}(m, n)$
4. **For** j=1: ($frame_{max}$-1)
5. $frame_2(m, n) = I_{m2}(m, n)$
6. **For** $k=m: n$
7. **For** p=1:8
8. **For** q=1:8
9. $I_{abs}(p. q) = \Sigma(\Sigma(I_{m16}(p_1, q_1) - I_{m8})^2)$
10. $frame_{pred} = I_{m16}(p_2, q_2)$
11. **End**
12. **End**
13. $Res = |frame_2 - frame_{pred}|$
14. **End**
15. **End**
16. **End**
End

The algorithm considers a minimum of 1 to a maximum number of frames (Line-1). A matrix is constructed which stores reconstructed frames I_{mrec} considering the selected frames corresponding to I frame, P frame, and B frame. A function f_1(x) is applied to I_{mrec}, which extracts the grayscale of it followed by resizing the image, and the resultant frame is stored in a matrix I_m (Line-2). The proposed system selects a dimension m and n as rows and columns for the new matrix extracted from I_m and stores it back as reference frame $frame_{ref}$ (Line-3). Considering the first to the last frame (Line-4), the algorithm obtains the next frame I_{m2} using similar function f_1(x), which is further mapped with the matrix of dimension m and n to obtain the next frame $frame_2$ (Line-5).

The algorithm obtains an absolute image L_{abs} for all m and n dimensions (Line-6), and the search window is generated with prows and q columns where its range is fixed between [1–8] as shown in Line-7 and Line-8. An absolute image I_{abs} is obtained considering reference frame I_{m16} and next frame I_{m8} using a comparative method (Line-9). Predicted frame $frame_{pred}$ is obtained by extracting correlated values in both images' frames (Line-10). The residual image *Res* is obtained by the absolute difference between the next frame $frame_2$ and the predicted frame $frame_{pred}$ (Line-13). The sum of absolute difference can be calculated from the obtained residual image *Res*. This algorithm is applied for performing motion compensation over the block. The next part of the implementation is about applying machine learning to the selected frame. The steps of the algorithm are as shown below:

Algorithm for Machine Learning
Input: I_m (selected frames)
Output: h1_comp (compensated frame)
Start
1. I_{size}=size(I_m)
2. obtain centroid $c=f_2(h_1, h_2)$
3. $[c_1, idx]=f_3(h_1, c, iter)$
4. h1_comp=$f_4\{c_1(idx)\}$

End

This algorithm takes the selected frame's input (I, P, B) and obtain its size, stored back in matrix I_{size} (Line-1). The algorithm applies a function $f_2(x)$ considering two input arguments h_1 and h_2, which corresponds to the resized image of I_m and the number of clusters, respectively, to obtain the primary centroid (Line-2). The algorithm also implements a function $f_3(x)$ that iterates this clustering process considering the input argument of h_1, centroid *c*, and iteration *iter* (Line-3). This operation will lead to the generation of updated centroid c_1 and index *idx* (Line-3). The function $f_4(x)$ is applied for finding the centroid in proximity using corresponding index *idx* to generate a distinct matrix h1_comp for all the centroid corresponding to the images (Line-4). Hence the outcome of this matrix is a compensated frame, which is also compressed with limited colors. The proposed system uses a machine-learning mechanism that initiates with the primary group of arbitrarily chosen centroid used as the preliminary points for all the clusters, followed by exercising the iterative operation for evaluating the optimal centroidal position. This is the mechanism adopted for learning data. The learning process is stopped when i) the system accomplishes its configured iteration number or ii) stabilization of centroid.

6 Results Discussion

This section discusses the outcomes obtained from the proposed study. The proposed system considers testing with a video file of mp4 extension with 480 frame width and 360 frame height to assess the outcomes. The data rate is 631 kbps, while the total bitrate is 739 kbps. The frame rate is 25 frames per second with the same audio rate of 44.100 kHz. The size of the overall file is 272 kilobytes. Any equivalent form of video can be used to assess the proposed framework. The proposed system's outcome is also compared with the existing standards of H.265 and H.264 concerning Peak Signal-To-Noise Ratio (PSNR) and processing time. The outcome discussed in this section is obtained by considering the same computing platform with consideration of the same input video and resources as CPU processing speed (Fig. 2).

PSNR performance of the proposed system showcase that it offers approximately 80% better performance than H.264 and approximately 20% better performance concerning H.265. The proposed algorithm constructs a relationship using referential pixels associated with the reference block to eliminate the non-matching candidate in the faster track. Hence the proposed algorithm offers a better adaptive matching scheme despite using conventional search schemes deployed in motion compensation schemes in H.265 and H.264. This results in enhanced signal quality in the proposed exhaustive search

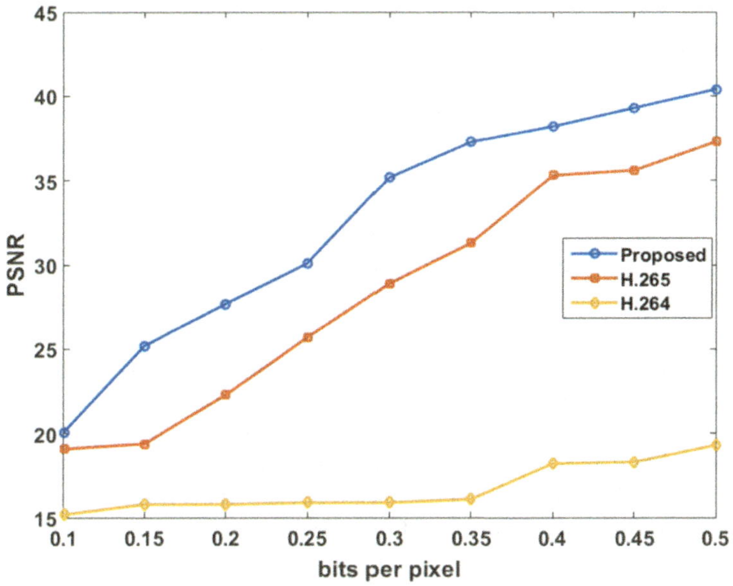

Fig. 2. Comparative analysis of PSNR

algorithm for inter-frame prediction. Hence, the PSNR performance of the proposed system is quite justified over an increasing bits-per-pixel value as test parameters. Similar performance is also noticed for any equivalent test video.

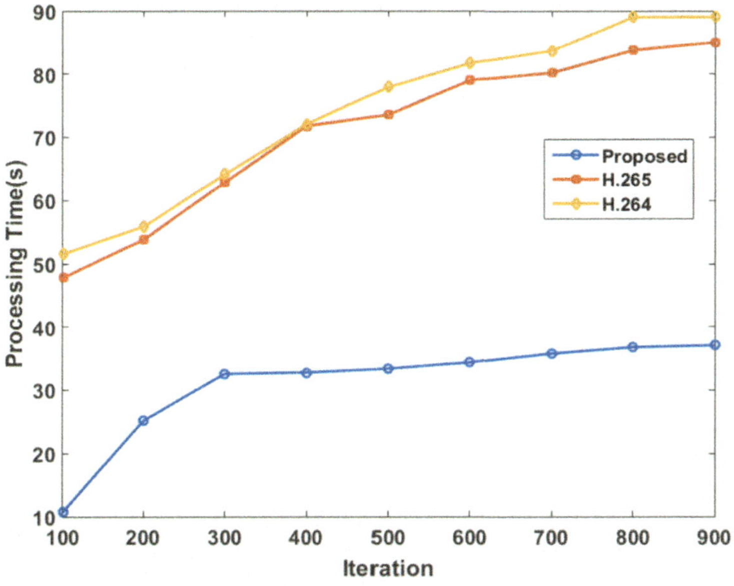

Fig. 3. Comparative analysis of processing time

A closer look into Fig. 3 highlights that the proposed system offers approximately 62% reduced processing time in seconds than H.265 and H.264. It is now known that the proposed system targets to truncate the massive amount of redundancy associated with spatial and temporal forms, thereby assisting in compression performance. The prime reason for better performance is comparing the current block with a search window to obtain the optimal match using the machine learning approach. A closer look into the proposed machine learning approach will show that it is less iterative and less complicated than other conventional machine learning approaches. A better form of convergence performance is ensured in the proposed study, which offers better scalability towards a larger dataset. Apart from this, it can also be found that the proposed system performs unique indexing of frames where the size is modified as per the block size leading to much faster operation. However, the existing approach, e.g., H.264, is characterized by many dependencies on processing factors, demanding more hardware usage. There is also a potential reason for H.265 to exhibit nearly similar performance concerning processing time. Although H.265 is a better version of H.264concerning its compression efficiency and reduced channel capacity consumption, there is still an enormous scope of using more simplified architecture for H.265. With increasing iteration, the processing time of H.264 and H.265 keeps on increasing, whereas the proposed system performance time is found to be reduced. This claims for better cost-effectiveness from a computational operation viewpoint. One of the significant advantages of the proposed machine learning approach is that it is highly applicable to unlabeled numerical data while it is also efficient and fast.

7 Conclusion

This paper discusses the model of motion estimation. The proposed system contributes to designing a novel motion estimation approach where the motion codes are subjected to encoding using an enhanced exhaustive search mechanism. The proposed system's second contribution is to extract potential information of complex motion forms, including occlusion, rotation, and warping. The third contribution of the proposed system is that it uses a machine learning approach towards the motion codes, directly impacting video compression followed by decoding. The study outcome shows that the proposed scheme offers a better signal to noise quality and response time than existing video coding standards.

Acknowledgment. Authors gratefully thank the Maharaja Research foundation and Visvesvaraya Technological University, Jnana Sangama, Belagavi for the support extended to this research work.

References

1. Chakrabarti, I., Batta, K.N.S., Chatterjee, S.K.: Motion Estimation for Video Coding, p. 157. Springer, Cham (2015). https://doi.org/10.1007/978-3-319-14376-7
2. Metkar, S., Talbar, S.: Motion Estimation Techniques for Digital Video Coding. Springer (2013). https://doi.org/10.1007/978-81-322-1097-9

3. Ahmed, Z., et al.: Lossy and lossless video frame compression: a novel approach for high-temporal video data analytics. Remote Sens. **12**(6), 1004 (2020)
4. Nicolas, H., Moscheni, F.: Temporal redundancy reduction using a motion model hierarchy and tracking for image sequence coding. In: Visual Communications and Image Processing 1993, vol. 2094, pp. 1548–1557. International Society for Optics and Photonics (1993)
5. Tian, M., Guan, B., Xing, Z., Fraundorfer, F.: Efficient ego-motion estimation for multi-camera systems with decoupled rotation and translation. IEEE Access **8**, 153804–153814 (2020)
6. Pakdaman, F., Hashemi, M.R., Ghanbari, M.: A low complexity and computationally scalable fast motion estimation algorithm for HEVC. Multimedia Tools Appl. **79**(17–18), 11639–11666 (2020). https://doi.org/10.1007/s11042-019-08593-y
7. Hsieh, J.-H., Syu, J.-C., Zhang, Z.-Y.: Coding efficient motion estimation rate control for H.265/HEVC. In: 2020 IEEE 9th Global Conference on Consumer Electronics (GCCE), Kobe, pp. 442–443 (2020). https://doi.org/10.1109/GCCE50665.2020.9291725
8. Paramkusam, A.V., Arun, V.: A survey on block matching algorithms for video coding. Int. J. Electr. Comput. Eng. (2088–8708) **7**(1) (2017)
9. Huang, Y.-W., Chen, C.-Y., Tsai, C.-H., Shen, C.-F., Chen, L.-G.: Survey on block matching motion estimation algorithms and architectures with new results. J. VLSI Sig. Process. Syst. Sig. Image Video Technol. **42**(3), 297–320 (2006)
10. Zhang, D., Huang, Q., Wang, Y., Ye, F., Wang, Y.: Survey on block-based motion estimation for frame rate conversion. In: Eleventh International Conference on Digital Image Processing (ICDIP 2019), vol. 11179, p. 111794N. International Society for Optics and Photonics (2019)
11. Park, S., Kang, J.: Fast affine motion estimation for Versatile Video Coding (VVC) encoding. IEEE Access **7**, 158075–158084 (2019). https://doi.org/10.1109/ACCESS.2019.2950388
12. Zhang, R., Wu, S., Wang, Y., Jiao, J.: High-Performance distributed compressive video sensing: jointly exploiting the HEVC motion estimation and the $\ell 1 - \ell 1$ reconstruction. IEEE Access **8**, 31306–31316 (2020). https://doi.org/10.1109/ACCESS.2020.2973392
13. Ray, B., Larabi, M., Jung, J.: Flexible motion vector resolution prediction for video coding. IEEE Trans. Circuits Syst. Video Technol. **29**(11), 3302–3316 (2019). https://doi.org/10.1109/TCSVT.2018.2878401
14. Luo, C., et al.: Every pixel counts++: joint learning of geometry and motion with 3D holistic understanding. IEEE Trans. Pattern Anal. Mach. Intell. 42(10), 2624–2641 (2020). https://doi.org/10.1109/TPAMI.2019.2930258.
15. Wang, Z., Wang, S., Zhang, X., Wang, S., Ma, S.: Three-zone segmentation-based motion compensation for video compression. IEEE Trans. Image Process. **28**(10), 5091–5104 (2019). https://doi.org/10.1109/TIP.2019.2910382
16. Li, R., Lv, Y., Liu, Z.: Multi-scheme frame rate up-conversion using space-time saliency. IEEE Access **6**, 1905–1915 (2018). https://doi.org/10.1109/ACCESS.2017.2780822
17. Liu, J., Xia, S., Yang, W.: Deep reference generation with multi-domain hierarchical constraints for inter prediction. IEEE Trans. Multimedia **22**(10), 2497–2510 (2020). https://doi.org/10.1109/TMM.2019.2961504
18. Wang, G., Li, B., Zhang, Y., Yang, J.: Background modeling and referencing for moving cameras-captured surveillance video coding in HEVC. IEEE Trans. Multimedia **20**(11), 2921–2934 (2018). https://doi.org/10.1109/TMM.2018.2829163
19. Franche, J., Coulombe, S.: Efficient H.264-to-HEVC transcoding based on motion propagation and post-order traversal of coding tree units. IEEE Trans. Circuits Syst. Video Technol. **28**(12), 3452–3466 (2018). https://doi.org/10.1109/TCSVT.2017.2754491
20. Yang, F., Cheng, Y., Weijer, J.V.D., Mozerov, M.G.: Improved discrete optical flow estimation with triple image matching cost. IEEE Access **8**, 17093–17102 (2020). https://doi.org/10.1109/ACCESS.2020.2968180

21. Vázquez-Padín, D., Fontani, M., Shullani, D., Pérez-González, F., Piva, A., Barni, M.: Video integrity verification and GOP size estimation via generalized variation of prediction footprint. IEEE Trans. Inf. Forensics Secur. **15**, 1815–1830 (2020). https://doi.org/10.1109/TIFS.2019.2951313
22. Zhao, Y., Dou, Y.: Pose-forecasting aided human video prediction with graph convolutional networks. IEEE Access **8**, 147256–147264 (2020). https://doi.org/10.1109/ACCESS.2020.2995383
23. Khan, B., Ali, A., Jalil, A., Mehmood, K., Murad, M., Awan, H.: AFAM-PEC: adaptive failure avoidance tracking mechanism using prediction-estimation collaboration. IEEE Access **8**, 149077–149092 (2020). https://doi.org/10.1109/ACCESS.2020.3015580
24. Chen, C., Zhou, C., Liu, P., Zhang, D.: Iterative reweighted Tikhonov-Regularized multihypothesis prediction scheme for distributed compressive video sensing. IEEE Trans. Circuits Syst. Video Technol. **30**(1), 1 (2020). https://doi.org/10.1109/TCSVT.2018.2886310
25. Ho, Y., Chan, C., Alexandre, D., Peng, W., Chang, C.: P-frame coding proposal by NCTU: parametric video prediction through backprop-based motion estimation. In: 2020 IEEE/CVF Conference on Computer Vision and Pattern Recognition Workshops (CVPRW), Seattle, WA, USA, pp. 598–601 (2020). https://doi.org/10.1109/CVPRW50498.2020.00083
26. Wang, Z., Wang, S., Zhang, X., Wang, S., Ma, S.: Multi-hypothesis prediction based on implicit motion vector derivation for video coding. In: 2018 25th IEEE International Conference on Image Processing (ICIP), Athens, pp. 2541–2545 (2018). https://doi.org/10.1109/ICIP.2018.8451264
27. Choi, G., Heo, P., Oh, S.R., Park, H.: A new motion estimation method for motion-compensated frame interpolation using a convolutional neural network. In: 2017 IEEE International Conference on Image Processing (ICIP), Beijing, pp. 800–804 (2017). https://doi.org/10.1109/ICIP.2017.8296391
28. Asikuzzaman, M., Pickering, M.R.: Object-based motion estimation using the EPD similarity measure. In: 2018 Picture Coding Symposium (PCS), San Fran-cisco, CA, pp. 228–232 (2018). https://doi.org/10.1109/PCS.2018.8456287
29. Lin, W., Nanjundaswamy, T., Rose, K.: Adaptive interpolated motion-compensated prediction. In: 2017 IEEE International Conference on Image Processing (ICIP), Beijing, pp. 943–947 (2017). https://doi.org/10.1109/ICIP.2017.8296420
30. Asikuzzaman, M., Suman, A.A., Pickering, M.R.: EPD similarity measure and demons algorithm for object-based motion estimation. In: 2018 Digital Image Computing: Techniques and Applications (DICTA), Canberra, Australia, pp. 1–7 (2018). https://doi.org/10.1109/DICTA.2018.8615826
31. Veena, S.K., Mahesh Rao, K.: Insights on video compression strategies using machine learning. Int. J. Eng. Adv. Technol. (IJEAT) **9**(1) (2019). ISSN 2249 – 8958
32. Veena, S.K., Mahesh Rao, K.: Performance evaluation of joint rate-distortion model of video codec. In: Silhavy, R. (ed.) CSOC 2020. AISC, vol. 1224, pp. 63–72. Springer, Cham (2020). https://doi.org/10.1007/978-3-030-51965-0_6

Distributed Library Model Based on Distributed Ledger Technology for Monitoring and Diagnostics System

E. V. Melnik(✉), M. V. Orda-Zhigulina, and D. V. Orda-Zhigulina

Southern Scientific Center of the Russian Academy of Sciences, st. Chehova, 41, Rostov-on-Don 344006, Russia

Abstract. The authors suggest a new model of distributed data library which is implemented on distributed ledger. The suggested model consists of the set of distributed ledger programs; the set of scripts for organizing access and storage of data in distributed ledger; local dispatcher for organizing distributed computing, launching and dispatching services of external subsystems for decentralized heterogeneous reconfigurable monitoring and diagnostics systems for different applications. Also, the decentralized distributed library structure is proposed for modern monitoring and diagnostics systems. Distributed ledger technology allows to efficiently combine different types of data sources, heterogeneous computing resources and communication channels with different bandwidth and reliability. The distributed ledger is backbone element to build the system and to connect other systems components about itself. The distributed library is intended for storing the original and processed data. Also, the distributed library includes local dispatchers and functional applications for data access.

Keywords: Data collection · Data mining · Processing and storage systems · Fog computing · Internet of things · Decentralized decision support system · Blockchain

1 Introduction

In order to create modern monitoring and diagnostics systems for different fields of application and also based of previous studies [1–9], the authors have developed a monitoring and diagnostics system component - the distributed library based on distributed ledger technology.

The developed component is a set of distributed ledger programs; a set of scripts for organizing access and storing data in a distributed ledger and a local dispatcher for organizing distributed computing and launching and dispatching services of external subsystems. This complex of programs can be the basis for the development of functional applications from external subsystems. These subsystems can carry out the necessary data processing in addition to reading and writing data to the ledger.

Previously, elements of methods for processing and transmitting information for monitoring and diagnostics systems for different fields of application, methods and

R. Silhavy (Ed.): CSOC 2021, LNNS 230, pp. 501–509, 2021.
https://doi.org/10.1007/978-3-030-77442-4_43

algorithms for increasing reliability due to the reconfiguration of resources (redistribution of functions) based on such technologies of the digital economy as "fog" and "edge" computing and distributed ledger [1, 7].

Also, on the basis of previously developed methods [1–5], algorithms for the operation of services of external plug-in subsystems were developed. If necessary the external developers using these subsystems can create their own functional applications that ensure the interface of plug-in external systems with the developed components of the monitoring and diagnostics system which Fig. 1 demonstrates.

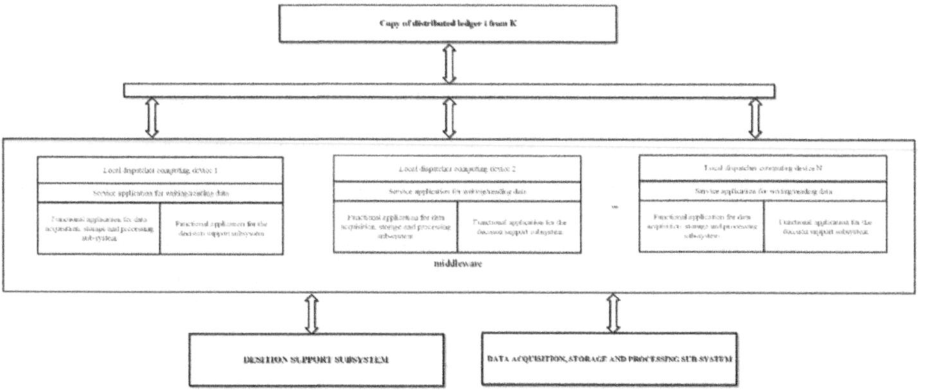

Fig. 1. To the development of the structure of the monitoring and diagnostics system - the structure of the copy of the distributed ledger.

In addition to reading and writing data to the ledger such functional applications should ensure the necessary data processing. Also an algorithm was developed for the local dispatcher which ensures the interaction of functional applications in a distributed environment.

The distributed ledger is a backbone element around which the entire system is built and to which all other components are connected. Individual copies of the distributed ledger can be "brought closer" to consumers through the use of fog layer devices. Therefore it is advisable to associate functional applications that handle data processing with each such copy of the distributed ledger.

The distributed library is intended for storing the original and processed data. At the same time it includes not only a distributed ledger but also local dispatchers and functional applications for organizing data access.

Local dispatchers are included in the distributed library for organizing the interaction of subsystems, reducing the load on the network, reducing latency, and also for using the resources of the fog layer devices as backup.

At the same time local dispatchers ensure the distribution of the computational load between all devices in the system, transferring the computational load of the between the layers of the edge and fog computing [1].

2 The Principles of Building a Distributed Library Based on a Distributed Ledger

Figure 2 demonstrates an example of using the model and the circulation of data in it is the connection to a distributed library of a subsystem for collecting, storing and primary processing of data and a subsystem for decision support.

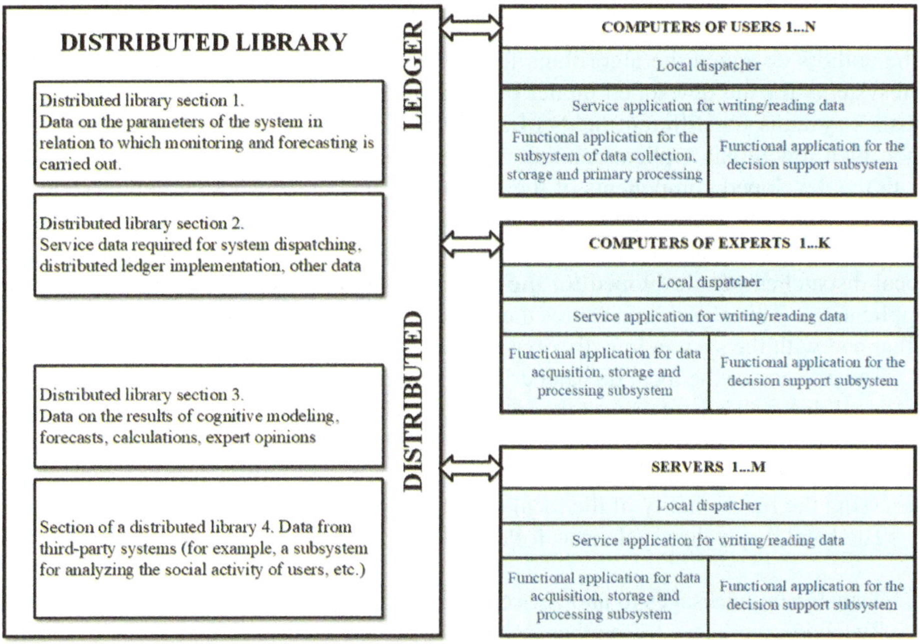

Fig. 2. Distributed library of monitoring and diagnostics system for different fields of application.

As can be seen from Fig. 2 data collection, its description, the results of its processing receive from the subsystem to the distributed library. Later this data will be available to users and experts. The structure of the distributed library has four sections as shown in Fig. 2. If necessary the services provided by the subsystem can be executed on the layer of cloud or fog computing. The decision support subsystem also has access to the distributed ledger, and therefore to the data received from the collection subsystem.

The data on the results of cognitive modeling, the expert opinion is gone the distributed library from this subsystem. If necessary modeling tasks can be performed on the computational nodes on the layer of cloud and/or fog computing. An algorithm was developed for the main loop of the local dispatcher to implement the distributed library. The developed algorithm provides decentralized interaction of functional applications in a distributed environment without a clear leader and with fault-tolerance for whole system.

In this case a software library is provided to each executable functional application to access the functionality of the local dispatcher prototype. The algorithm has two modes of operation: synchronous and asynchronous. Algorithms for organizing access to distributed ledger data were also developed.

There is only computing verification of the proposed model, which should have given more weight to the paper, so it the model is still in developing and it will be in future to improve the quality and significance of research.

2.1 Algorithms of the Model of a Component of the Monitoring and Diagnostics System - A Distributed Library Based on Distributed Ledger Technology

The authors designed the algorithms for the operation of services of external plug-in subsystems for the developed model of a distributed library for monitoring and diagnostics systems for different fields of application. The external developers can create their own functional applications that ensure the interface of plug-in external systems with the developed components of the monitoring and diagnostics system as shown at Fig. 1. These functional applications should ensure the necessary data processing in addition to reading and writing data to the ledger. An algorithm for the operation of a local dispatcher was developed for the subsequent implementation of these functional applications. This algorithm ensures the interaction of functional applications with each other and with the system in a distributed environment.

The essence of the local dispatcher's work algorithm is as follows: the main task of the local dispatcher is to ensure the interaction of service applications in a distributed environment, including a copy of the ledger and fault-tolerance of the whole system. In this case each running service application is provided with a software library for accessing the functionality of the local dispatcher prototype.

The local dispatcher cycle is as follows:

1. Transferring messages to another service.
2. Receiving messages from other tasks.
3. Remote call of the functions of applications located on the nodes of the system is carried out (access to the functions of one application from the side of another as to usual own functions).
4. The publisher application remotely calls the event handlers located in the subscribing applications.
5. The task status is checked by regularly calling the "pulse" function of the task.
6. The status of the main process of each task is checked.
7. Provide each task with information about the current status of other tasks.
8. If the sign is set of the end of the work of the local dispatcher program, then go to step 9, otherwise go to step 1.
9. Exit.

2.2 Connection Point

The main mechanism for exchanging data between service applications is the connection point.

A connection point is a named group of events and remotely called procedures (functions) that operates within the entire set of functional software components. In the context of the event model the application that registers the connection point (triggering events) is called the "publisher", and the applications that receive the connection point event notification are called "subscribers".

Subscription exchange assumes that some of the tasks are sources of information of a certain type, about which they "report" to "their" local dispatcher at the beginning of work. Other tasks should "subscribe" to it by contacting "their" local dispatcher in the case interested in obtaining this information. In fact in this case the collective of local agents implements a "connection point" to data of a certain type. Further when the source-task is ready to send another "portion" of data it transfers it to "its" local dispatcher which according to the available list "Subscribers" carries out the mailing. This mechanism ensures flexibility in managing the structure of a task; a task is bound not to other tasks, but to data of a certain type. This flexibility is important during task development as well as when performing software modifications to a distributed control system.

There are two modes of operation - synchronous and asynchronous. In the case with a synchronous call the source at the call site waits for the result after which it continues to function. In an asynchronous call after the request is sent the task continues to run. The special handler is assigned to receive the response to which the result is passed upon receipt. There is a standard handler for receiving the results of an asynchronous remote procedure or event call which in the case of sending messages and sending data puts the result in the message queue.

Also the algorithms for the operation of service programs were developed for the model of a component of the monitoring and diagnostics system - a distributed data library based on the distributed ledger technology.

2.3 Service Programs

Service programs called services provided by external subsystems for collecting, storing, transferring and analyzing data. Two sets (templates) of service programs have been developed in addition to the local dispatcher as part of the implementation of the system prototype: "subtask of data reception" and "subtask of data processing". The "data reception subtask" receives data from smart sensors and submits a request to process this data on the "bulletin board". The "data processing subtask" analyzes the queries in the system and processes them.

Also the "data reception subtask" monitors the activity of the node that took over the task and if necessary relays a new request for execution.

Further the "data processing subtask" writes the final data one-time or at a predetermined frequency to the distributed ledger. An example of a general scheme for the operation of service programs is shown in Fig. 3.

The possible configurations of the distribution of data streams are shown in the Fig. 3 with by a dashed line. In the configuration of data reception subtasks a list of smart sensors is specified from which the program receives data.

In the above example the "data reception subtask 1" receives data from the first and second. The "data reception subtask 2" receives data from the second and fourth smart sensor.

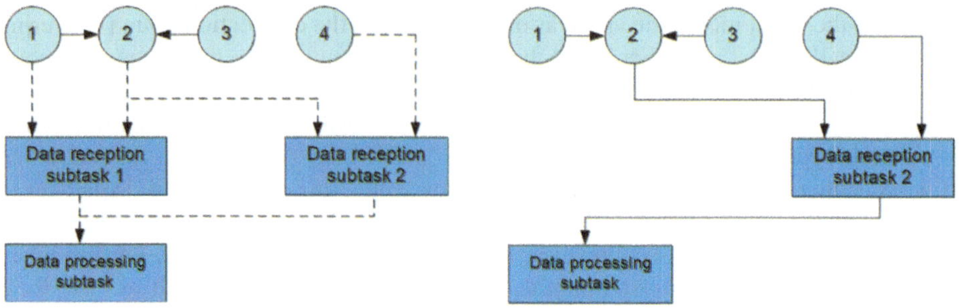

Fig. 3. Configuration of the set of tasks and the final selection.

At the start if data from applications is not yet processed both applications place requests for processing in a distributed ledger based on Ethereum from where the processing subproblem processing application selects one of them.

In the example "data processing subtask" confirms (also through the "bulletin board" of the distributed ledger) the work with the application-publisher "data reception task 2" and the resulting data streams are already shown by a solid line.

2.4 The Algorithm of the Service Program "Data Reception Subtask"

The algorithm of the service program "data reception subtask" is as follows:

The service (utility) program "data reception subtask" is a simple deterministic finite state machine (DFSM) with three states:

- search for a performer;
- data processing;
- stop.

The program can switch to one of these three states depending on certain conditions. The algorithm of the "data reception subtask" service program is as follows:

1. initialization of functions of interaction with the local dispatcher;
2. determination of the configuration of the task - the list of intelligent sensors of the distributed network of sensors with which the work takes place;
3. creation of a named object of interaction with other functional programs - connection points;
4. the initial state of the DFSM is the search for a data processor;
5. if a stop command is received from the local dispatcher - end of work;
6. if we are in a state of searching for a data processor, then point 7, otherwise go to point 9;
7. if there is no record about the processing request in the distributed ledger or no record about the data processing itself; then place a request record in the ledger;
8. if there is a record in the distributed ledger about the readiness of data processing; then set the state - data processing and go to step 9, otherwise go to step 6;

9. the entry in the bulletin board does not correspond to the data processing - go to point 12
10. receive information from the local dispatcher about the activity of the computing node that processes the data. If the node is active, then step 11, otherwise set the state to stop and go to step 12;
11. form a single data packet from the sensors and send it to the connection point, go to point 9;
12. if the entry in the distributed ledger does not correspond to data processing, then go to the state - search for a data processor and go to step 5, otherwise wait.

2.5 The Algorithm of the Service Program "Data Processing Subtask"

The algorithm of the service program "data processing subtask" is as follows: the service (utility) program of the data handler "data processing subtask" is also a deterministic finite state machine with three states - task search, task processing and stop.

In more detail the operation algorithm is the following sequence of operations:

1. initialization of functions of interaction with the local dispatcher;
2. task configuration definitions - a list of possible data sources;
3. initial state - task search;
4. if a stop command is received from the local dispatcher - end of work;
5. if the state is the search for a task, go to step 6, otherwise go to step 10;
6. get a list of requests available for processing from the "bulletin board" of the distributed ledger;
7. if in the list of available requests there is a request suitable for us, then set the sign of readiness for data processing in the distributed ledger and go to step 8, otherwise go to step 4;
8. if it was not possible to connect to the connection point, then go to the stop state and go to step 11;
9. set the state - data processing, set the computational load for the execution of the selected task to the local dispatcher;
10. if the state is data processing, then go to point 11, otherwise go to point 12;
11. if the record about the current state of the task execution changes in the distributed ledger, then complete the connection point processing, switch to the task search state and switch to state 4;
12. If the state is stop, then waiting for the timeout to work - at the end, the transition to the state - search for a request, otherwise the transition to state 4.

The algorithm description of the data handler program does not explicitly mention the data processing operation. This is because the data handler is a connection point callback function. The function is called automatically assuming that the application is connected to the connection point and a data packet has arrived.

The data processing function looks like this:

1. if the data format is as expected - accept the data packet, otherwise skip;
2. place a packet with data into a dynamic buffer;

3. if one of two events has occurred - a specified number of packets have arrived, or a processing timeout has come - call the data processing function for the accumulated data array, otherwise go to step 2;
4. put the final result into a distributed ledger.

3 Conclusion

The authors have developed methods and algorithms for building distributed library components: middleware (service) software including a local dispatcher; a service application template for writing/reading data. The functional application for a subsystem for collecting, storing and primary data processing; the functional application for decision support subsystem and distributed ledger can be developed on the basis developed methods and algorithms.

The above methods and algorithms allow organizing the decentralized interaction of all system components and increasing their reliability, efficiency and speed of operation through the simultaneous use of digital economy technologies such as a distributed ledger, data mining technologies, fog and cloud computing and local agents (dispatchers). The model of the monitoring system component was designed to check the performance of the developed methods and algorithms. This model is distributed library based on the distributed ledger technology.

The developed model is a complex of distributed ledger programs; a set of scripts for organizing access and storage of data in a distributed ledger; a local dispatcher for organizing distributed computing, launching and dispatching services of external subsystems in a decentralized heterogeneous reconfigurable monitoring and diagnostics system. At the same time a decentralized structure is the most effective in the construction of modern monitoring and diagnostics systems. The distributed ledger technology allows you to effectively combine disparate data sources, heterogeneous computing resources and communication channels with different bandwidth and reliability. The publication was prepared as part of GZ SSC RAS N GR project №AAAA-A19-119011190173–6 and RFBR projects 18–05-80092 and 18–29-22086.

References

1. Orda-Zhigulina, M.V., Melnik, E.V., Ivanov, D.Y., Rodina, A.A., Orda-Zhigulina, D.V.: Combined method of monitoring and predicting of hazardous phenomena. In: Silhavy, R. (ed.) CSOC 2019. AISC, vol. 984, pp. 55–61. Springer, Cham (2019). https://doi.org/10.1007/978-3-030-19807-7_6
2. Orda-Zhigulina, D.V., Orda-Zhigulina, M.V., Rodina, A.A.: Cognitive model for monitoring and predicting dangerous natural processes for hydro ecosystem analysis. In: Silhavy, R., Silhavy, P., Prokopova, Z. (eds.) CoMeSySo. AISC, vol. 1294, pp. 688–695. Springer, Cham (2020). https://doi.org/10.1007/978-3-030-63322-6_58
3. Melnik, E.V., Bulysheva, N.I., Orda-Zhigulina, M.V., Orda-Zhigulina, D.V.: Component of decision support subsystem for monitoring and predicting of hazardous processes at the base of analysis of macro zoobenthos communities of azov sea. In: Proceedings of the Computational Methods in Systems and Software, pp. 676–687. Springer, Cham (2020)

4. Melnik, E.V., Orda-Zhigulina, M.V., Orda-Zhigulina, D.V., Ivanov, D.Y., Rodina, A.A.: Fog computing in new approach for monitoring of hazardous phenomena. In: Journal of Physics: Conference Series, vol. 1333, No. 7, p. 072016. IOP Publishing, Wroclaw(2019)
5. Melnik, E.V., Orda-Zhigulina, M.V., Orda-Zhigulina, D.V., Ivanov, D.Y., Rodina, A.A.: Primenenie tehnologij cifrovoj ekonomiki pri razrabotke sredstv monitoringa i prognozirovanija opasnyh processov i obespechenija bezopasnosti naselenija i beregovoj infrastruktury. In: Zakonomernosti formirovanija i vozdejstvija morskih, atmosfernyh opasnyh javlenij i katastrof na pribrezhnuju zonu RF v uslovijah global'nyh klimaticheskih i industrial'nyh vyzovov ("Opasnye javlenija"), pp. 289–291. Izdatelstvo SSC RAS, Rostov-on-Don (2019)
6. A New Approach to Consensus: Swirlds HashGraph. http://sammantics.com/blog/2016/7/27/hashgraph- consensus. Accessed 02 Sept 2020
7. Hedera Hashgraph. Explained. https://medium.com/datadriveninvestor/hedera-hashgraph-explained-c5d8ce4730a6. Accessed 02 Sept 2020
8. Hedera Hashgraph, Consensus, and Scalability. https://medium.com/@saratechnologiesinc/hedera-hashgraph-consensus-and-scalability-2315133a3e33. Accessed 02 Sept 2020
9. Colin LeMahieu. Nano: A feeless distributed cryptocurrency network. https://content.nano.org/whitepaper/Nano_Whitepaper_en.pdf. Accessed 25 Jan 2021

Business Process Reengineering for Health Care Management and Medical Insurance: The Case of Russia

Yuri Antokhin[1](✉), Irina Sokolova[2], and Anton Suprun[3]

[1] City Polyclinic No. 86, 195299 St. Petersburg, Russia
[2] St. Petersburg State University, 199034 St. Petersburg, Russia
[3] ITMO University, 197101 St. Petersburg, Russia

Abstract. The purpose of the study was to determine the state of the market for information systems and services in health care, and to develop requirements to software suitable for 85 territorial compulsory healthcare insurance (CHI) funds as an important component of national e-health in Russian Federation . The authors have found out that reliable solutions for the digitalization of CHI are not yet available and revealed the reasons and disadvantages of the current situation in the healthcare-related IT market, and the means to improve it. To develop proposals for the creation of a unified software package for data processing and digitalization of business processes at territorial CHI funds, the authors fulfilled reengineering of CHI business processes. Particularly, they analyzed the information system currently operating in the fund of the Leningrad Region and estimated the time required for manual operations that can be sufficiently reduced. As a result, specific measures for the digitalization of the activities of the territorial CHI fund were worked out, along with the timing of these measures to be fulfilled, financing, and evaluated investment attractiveness.

Keywords: Health care · Digital platform · Digital solutions · Medical insurance

1 Introduction

E-health has a significant impact on how healthcare is organized around the world, making healthcare systems more efficient and better meeting people's needs.

In Russia the unified state information system in the field of healthcare should cover 100% of medical organizations by 2024. The federal project "Creation of a unified digital circuit in healthcare" [12] has been operating in Russia since 2019 and currently regulates the development of e-health.

The health care payment model includes compulsory health insurance (CHI) as the main component. Employers and the government reserve cash for medical care in territorial CHI funds. Every region in Russia has got its own territorial CHI fund. CHI funds transfer money to be paid for medical care to insurance organizations, which, in their turn, pay to medical organizations that treated patients. The amounts to be paid are

R. Silhavy (Ed.): CSOC 2021, LNNS 230, pp. 510–521, 2021.
https://doi.org/10.1007/978-3-030-77442-4_44

in proportion to the number of inhabitants for ambulance and outpatient medical care or can be calculated for each case of inpatient medical care.

The health care payment model is valid for paying not only for first aid and general practitioners, but also for many types of expensive high-tech medical care. In addition, CHI funds and insurance organizations control the quality of medical care and the correctness and timeliness of the funds transfer. The model uses information about a significant number of complex business processes that determine the order of delivery, payment and quality control of medical care. We can observe such business processes in compulsory medical insurance funds, insurance and medical organizations. A single digital health care circuit should combine information flows that describe these business processes [12]. Such a combination of information on the procedure for the provision, quality control and payment of medical care will help increase the availability and quality of medical care for citizens.

Now in Russia, many organizations are developing information systems for health care. CHI funds, insurance and medical organizations independently determine the requirements for information systems for their needs and place orders in the Unified Information System in the field of procurement (https://zakupki.gov.ru/epz/main/public/home.html).

The purpose of our study was to determine the state of the market for information systems and services in health care, in particular for territorial CHI funds, and to develop requirements for complete software package for data processing and digitalization of business processes in the territorial CHI fund.

Digitalization of healthcare provides new opportunities for citizens to become central elements in taking care of their health and lifestyle and will ensure universal coverage of the population with medical services [1]. At the same time, according to Hermes, S., Riasanow, T., Clemons, E.K. et al. [4], healthcare lags behind other sectors of the economy regarding an advance in digital ecosystems. Researches of the healthcare digitalization do not take into account the prospects for interaction between healthcare organizations, the emergence of new markets in the industry, and the emergence of new opportunities for patients. The authors of [1] describe the priorities for the development of digital health care. First, the organizations that own and manage the data must ensure that it is stored, integrated and processed efficiently. Second, digital technology must actively support disease prevention. Third, digital technologies should make it more accessible for patients to receive high-tech medical care. Fourth, digital technologies should provide people with access to personalized medicine.

Russian health care specialists write [5, 7] that in Russia there are conditions for active digitalization of health care, in particular, a high level of education of the population, high availability of Internet communications for the population, many small geographically distributed settlements. In Russia, the strategy for the development of the information society and the national program for digitalization of the economy [13] provide regulatory framework for the development of information technologies in all sectors of the economy.

Yusif, S., Hafeez-Baig, A. & Soar, J. [22] analyzed the factors to be used in assessing conditions for developing medical information technologies: legislation and political institutions; the level of information and communication technologies used in health care;

financial security; staff readiness to use information technology; leadership readiness to manage changes. The authors conducted in-depth interviews with healthcare professionals and identified 23 factors that can drive the adoption of health information technology, and 29 factors that can hinder their adoption in Ghana. The results of that study provide information on improving conditions for digital health development. Taking into account the selected factors can help to compose a scenario for establishing digital technologies in healthcare.

Interoperability between disparate health information systems is a major challenge in digital health development. Empirical studies by Weber, J.H., Ho, J. [21] have shown that very often the existing interfaces of interaction between various health information systems work poorly. An analysis of Russian publications on the problems of digitalization of healthcare [3, 5–7, 17, 20] showed that a transition from a centralized model of data processing and control to distributed systems is necessary. The problem of coordinating such distributed systems can be solved at the level of territorial CHI funds [5, 17]. Territorial CHI funds are a source of information for many departments, since they actually accumulate information about the health of the population, medical care for patients, and its financing.

Health information systems must become more and more interoperable. However, now problems arise with the compatibility of information systems developed in different periods for different organizations (health authorities, clinics, hospitals, insurance organizations). The cost and quality of information systems are important factors for the successful development of e-health [15]. The cost of software for healthcare can be estimated both in traditional ways and using new tools, such as described in works [8, 16]. In any case, we should consider the fact that pricing strategies for information products or services are aimed at increasing profits. In addition, research into the causes and consequences of software implementation in the healthcare industry must be carried out [8].

2 Methods

We used data from the software register of the Ministry of Telecom and Mass Communications [18] and data from the unified information system in the field of procurement on the procurement of goods and services for informatization of healthcare organizations [19]. The data was quite complete because all procurements for public organization are to be registered, and it is obligatory to register software in order to attend open tenders on providing IT products and services to CHI funds. That is why we could trace down all contracts in that field.

We processed those data to assess the state of the market for information systems and services for territorial CHI funds. For that we used descriptive statistics methods.

Since we worked out that most of cash was spent on reworking software products, rather than their development, the bulk of the study was aimed at developing requirements for typical software components to be suitable for CHI funds.

To develop requirements to the software package, we performed reengineering of CHI business processes. We analyzed the information system currently operating in the territorial CHI fund of the Leningrad region, identified the processes in this system and highlighted processes that are implemented manually by employees to estimate the time required for manual operations.

To develop approaches to the design of a typical information system for territorial CHI funds, we examined the existing information flows and data processing procedures for the particular territorial CHI fund and substantiated the requirements for the new software system.

In addition, we analyzed the legal documents that determine the existing and future requirements for the interaction between organizations in the compulsory health insurance system and the rules for the provision of medical care [2, 10–14].

3 Results

3.1 Conditions for Digitalization of Territorial Compulsory Health Insurance Funds

In Russia, the creation of a single digital circuit [12] is the core of digital health care. The digital circuit provides citizens with access to services in the field of health care and drug provision.

The digital health care circuit is focused on the unification of information processing and exchange processes, the regulation of data formats, the standardization of data collection procedures, the creation of standard software solutions for subjects of the health care system. One of the priority tasks of the National Healthcare Project is the creation of such a digital circuit. The cost of developing the digital circuit is 177.7 billion rubles. for the period from 2019 to 2024 [14].

The development of information systems for compulsory health insurance is taking place within the framework of creating the digital circuit. General principles for the construction and functioning of information systems, the procedure for information interaction in the field of compulsory health insurance were approved in Order No. 79 of the Federal Fund for Compulsory Health Insurance [11].

In accordance with the order, the components of the information systems being created must implement the following five main processes:

- maintaining a regional register of insured citizens;
- registration of each case of medical assistance;
- updating the site of the territorial CHI fund;
- updating the normative and reference information of the healthcare system and maintaining classifiers, registers and lists in the field of compulsory medical insurance;
- keeping a log of the citizens' appeals.

In addition to those processes, a mandatory requirement for the information system is the implementation of protocols for interaction of the territorial CHI fund with medical and insurance organizations.

Typical solutions for the implementation of the listed processes are presented in the software market in small quantities.

Only seven software systems for territorial CHI funds are indicated in the register of Russian software [18]. Of those, four software systems provide a comprehensive implementation of the listed processes, and only two of them are digital platforms for organizing interaction between territorial funds, medical and insurance organizations. Over 250 software systems are intended for medical organizations [18]. We explain this fact by a greater number of medical and insurance organizations that are the potential consumers of information systems. The number of medical organizations registered by the Federal CHI Fund is more than 14 thousand to be added with more than 250 insurance organizations [9], while there are only 85 territorial CHI funds.

We analyzed 1,596 contracts for the design, modification and maintenance of software systems executed from January 2014 to December 2019 to ensure information exchange in the compulsory health insurance system. The territorial funds of the CHI were the customers of only 148 such contracts for a total amount of 182.8 million rubles. Five organizations were executors of these contracts. Three of these organizations have 99% of the market volume in monetary terms, 59%, 29% and 11%, respectively. Most of the contracts in monetary terms were concluded for maintenance (88 contracts for 81 million rubles). Contracts for creation and adaptation are practically the same in terms of market size (see Fig. **1**).

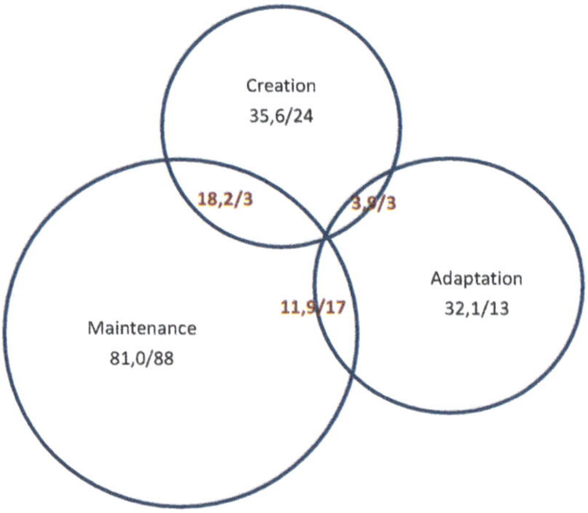

Fig. 1. Volumes of procurement of services for information support of the activities of territorial CHI funds for the period 2014–2019. The numbers indicate the volume of purchases in million rubles/number of contracts (worked out by the authors for the data of procurement system https://zakupki.gov.ru/).

In addition, we found out that those three main players in the analyzed market provide information services for territorial CHI funds in 30 regions out of overall 85 regions of Russia, while the spheres of territorial influence of those companies practically do not overlap. The companies sold 60% to 94% services in monetary terms in 3–5 regions of their influence.

There are two peculiarities we need to comment.

First, the majority of territorial CHI funds did not purchase information systems for compulsory health insurance. Our own experience confirm that such funds inherit old, usually self-made products, being adapted by staff programmers. However, that practice results in a large number of manual operations and risks systems' discontinuity as such self-made products are usually poorly documented and are sensitive to alteration of input/output data formats.

Second, there are a lot of contracts per one CHI fund, because there is more than one software product needed for exchanges with patients, insurance organizations, clinics, hospitals, and federal CHI fund. Frequent changes in exchange algorithms and protocols result in additional adaptations of software products.

We analyzed the dynamics of contracts fulfilled in 19 regions by the most active company with 60% of the market volume over the observation period. We found out that the company was focused on maintenance contracts (58% in monetary form and 63% of the total number of contracts). Creation contracts are in the second place: 28% of the total amount of costs and 15% of the total number of contracts. The interval estimates of average costs per contract was 1.3 ± 0.3 million rubles for maintenance contracts and 1.8 ± 1.1 million rubles for creation contracts. The frequency of contracting was 9.4 and 2.3 contracts per year, respectively, for maintenance and creation (see Fig. 2).

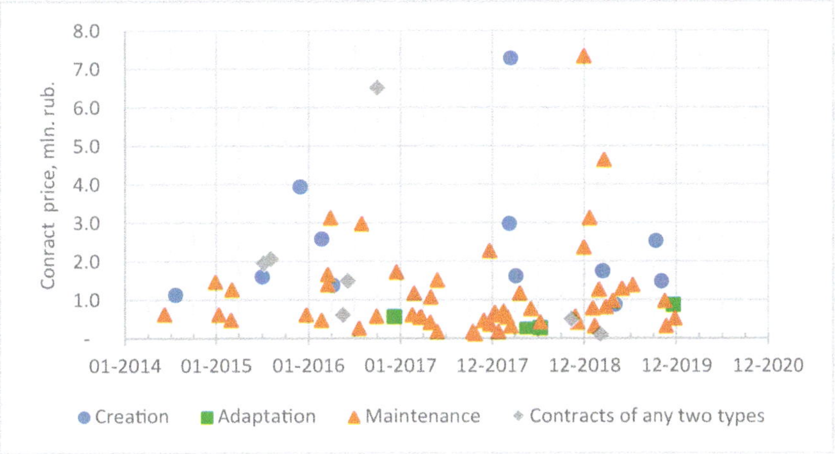

Fig. 2. A figure caption is always placed below the illustration. Short captions are centered, while long ones are justified. The macro button chooses the correct format automatically.

Looking at the results of the market analysis performed, we can conclude that competition in the information services market for territorial CHI funds is weak, there are no standard solutions. Therefore, it is necessary to develop approaches to designing a typical (model) information system for territorial CHI funds.

A typical solution should be adapted to the existing circuit of business process management of territorial CHI funds. It is possible that modifications of this control circuit, taking into account modern requirements for information interaction of participants in the CHI, would be necessary. The new information system should be implemented in stages without prejudice to current activities. In addition, staff training should be provided.

3.2 Digitalization of Interaction Between Participants of the CHI System

To develop approaches to the design of a typical information system for territorial CHI funds, we examined the existing information flows and data processing procedures for the particular territorial CHI fund and substantiated the requirements for the new software system.

The territorial CHI funds perform the functions of controlling the volume, timing, quality and conditions for the provision of medical care for compulsory health insurance [10]. The MHI Fund of the Leningrad Region has been collecting and consolidating information from participants of compulsory health insurance (citizens, medical organizations, insurance organizations, other territorial funds and the federal fund) since it was established in 1995. Later, information from the pension fund, tax inspection, bureau of medical and social expertise, and other sources was added to these data. Therefore, new types and forms of reporting are now necessary for work.

Information systems of the territorial CHI fund were formed gradually, relying on technological platforms and the regulatory framework that were available at the time of development. On the one hand, this gradual development was a guarantee of the continuous operation of information systems. On the other hand, the gradual development of information systems has led to the obsolescence of their individual blocks, disruption of connections between individual blocks, as well as to "information layers" in the form of tables filled in by the fund's employees manually or semi-automatically. The described negative aspects lead to ineffective use of working time and affect the work of the territorial fund.

Therefore, further we analyze the current processes and existing information systems in the CHI fund of the Leningrad Region and develop proposals for improving information systems and digitalizing employees' work processes. We consider the CHI fund of the Leningrad region as a typical territorial CHI fund that can be used for working out standard replicable software solutions.

The existing data management system on compulsory health insurance in the Leningrad Region is divided into two circuits: the first circuit served by the territorial fund, the second circuit served by third-party departments and organizations.

With the existing data management system, a number of problems arise that must be solved as a result of creating a software package for data processing:

- fragmentation of information systems in the territorial fund;
- large time expenditures for obtaining information;
- a large number of fund employees that are busy obtaining information from various sources;
- incomplete automation of data processing;
- incomplete compliance with the goals of the implementation of national projects of the Russian Federation.

The main stakeholders of the project to create a unified data processing software complex in the compulsory health insurance fund of the Leningrad Region are: the Ministry of Health, the Ministry of Labor and Social Protection of the Population, the Federal Fund for Mandatory Health Insurance, territorial CHI fund of the Leningrad

Region, medical organizations included in the register of compulsory health insurance of the Leningrad Region, insurance organizations operating in the Leningrad Region.

We propose to implement the creation of a unified software package for data processing in the CHI fund as a unified system of three blocks: 1) account management, 2) information support for insured persons, 3) planning and control of the territorial compulsory health insurance program.

In order to assess the economic and organizational efficiency of the project, we analyzed the work processes of the fund's employees. The analysis showed which time spending should be reduced through digitalization.

We analyzed the work processes of the employees of the financial and economic department, the department for the organization of compulsory medical insurance and the department of information technology, identified groups of work processes subject to digitalization (automated workstation (AWS), automated information subsystem (AIS), etc.), assessed temporary costs for each of the processes in each department. Below we present the total (in total for all departments) indicators of time consumption by process groups for the key elements of the information system.

1. Processing of information received from state departments of registration of births and deaths of citizens, preparation of data for inclusion in a return message - 8 h/month.
2. Planning and accounting of preventive measures, including: accounting citizens' appeals to the fund and insurance organizations; filling out a registration and control card (manual entry of previously unreported citizens); control of the processing time of the appeal; viewing the history of calls for a specific completed treatment; creation of reports in free form or according to form templates 254 h/month.
3. Daily (no later than 09.00 local time) information update at the level of a medical organization for the past day: the number of vacant places for hospitalization in the context of profiles (departments) of medical care for the current day and for the next ten working days with an indication of the planned date of hospital places to be released; insured persons for whom the planned hospitalization did not take place, 263 h/month.
4. Economic justification and monitoring of the territorial compulsory health insurance program for the next financial year and planning period, and next month, 118 h/month. and additionally, annually - 271 h/year.
5. Providing automatic support for format-logical control and medical and economic control of registers of accounts for medical care from medical organizations and from CHI funds 33 h/month.
6. Ensuring the exchange of information on cases of medical care with other subjects of the CHI system 116 h/month.
7. Provision of document circulation 256 h/month.

Analysis of the workflows shows that the total time spent on the processes to be reduced in the project is more than 100 h for monthly work, and 271 h for annual work.

The total duration of the implementation of the information platform using ready-made standard software solutions will take 83 working days, including the parallel execution of work in three blocks: account management 83 days; planning of the territorial compulsory medical insurance program 61 days, information support of the insured 69 days. We assessed the investment attractiveness of the project taking into account the discount rate of 12% (Table 1).

Table 1. Investment indicators for the development of a platform for digitalizing the work processes of the compulsory health insurance fund.

Indicator	Value
Investment volume, rubles	13 891 800
Cost savings amount, rubles a year	5 414 284
Years for payback	4,5
The amount of cost savings over 4.5 years, rubles	24 364 278
Discount rate	12%
Coefficient	3,32910
Net present value of the project, rubles	4 132 886,59
Return on investment (ROI)	1,30
Internal ROI	13%
The ratio of the amount of cost savings and the amount of initial investment (minimum)	0,39

We distinguished external and internal risks of the project.

External risks:

- changes in legislation in the field of compulsory health insurance;
- risks associated with the f competitive methods for identifying suppliers (contractors, performers) and the execution of contracts in accordance with applicable laws, for example, these can be risks associated with the failure of the contract, working with an unscrupulous supplier, etc.

Internal risks:

- the working staff will not be able to work in the new conditions; it will be necessary to spend time looking for new qualified workers;
- the existing funding of the fund will not be enough for the implementation of the project, it is possible that non-project work processes that require digitalization may appear, and as a result, additional costs may arise;
- resistance of personnel to changes in the work process;
- the need for partial or complete reorganization of the territorial CHI fund of the Leningrad region.
- a temporary increase in the workload on employees during the implementation of the software package.

It is possible to mitigate most of the internal risks by training employees and increasing their digital literacy.

4 Discussions

Healthcare lags behind other sectors of the economy in terms of the implementation of digital ecosystems. Health information systems need to become more and more interoperable. However, now problems arise with the compatibility of information systems developed in different periods for different organizations (health authorities, clinics, hospitals, insurance organizations). The cost and quality of information systems are important factors for the successful development of e-health.

Only seven software systems intended for territorial CHI funds are registered by the Ministry of Telecom and Mass Communications. Only two of them can be regarded as digital platforms for organizing interaction between territorial funds, medical and insurance organizations.

For the period from January 2014 to December 2019, the territorial CHI funds placed a total of 148 contracts for the development, modification and maintenance of information systems at the unified information system in the field of procurement. The total amount of these contracts was 182.8 million rubles. Five organizations were executors of these contracts. Three of these organizations got 99% of the market volume in monetary terms. More than half of the contracts in monetary terms were for maintenance. Contracts for creation and adaptation are practically the same in terms of market size. Those three main market players provided information services for territorial CHI funds in 30 out of 85 regions of Russia, while the spheres of territorial influence of companies practically do not overlap. The companies sold 60% to 94% of orders in monetary terms in each of 3–5 regions of their influence. The main market player (60% of the total market volume) was represented in 19 regions of Russia. The company is focused on maintenance contracts. Creation contracts are in second place. The interval estimates of average cost per contract is 1.3 ± 0.3 million rubles for maintenance contracts and 1.8 ± 1.1 million rubles for creation contracts. On average, the company fulfills 9.4 and 2.3 contracts per year, respectively, for maintenance and creation.

The CHI fund of the Leningrad Region is a typical territorial CHI fund. The existing data management system on compulsory health insurance in the Leningrad Region is divided into two circuits: the first circuit served by the territorial fund, the second circuit served by third-party departments and organizations.

With the existing data management system, a number of problems arise that must be solved as a result of creating a software package for data processing:

- fragmentation of information systems in the territorial fund;
- large time expenditures for obtaining information;
- a large number of fund employees that are busy obtaining information from various sources;
- incomplete automation of data processing;
- incomplete compliance with the goals of the implementation of national projects of the Russian Federation.

We propose to implement the creation of a unified software package for data processing in the CHI fund as a unified system of three blocks: 1) account management, 2) information support for insured persons, 3) planning and control of the territorial compulsory health insurance program.

The implementation of the project can become a pilot system for territorial CHI funds. The project will be completed in 83 working days. Digitalization will affect 40 main work processes (about 30 employees of the territorial fund and about 10 employees of medical insurance organizations).

The project has a high social significance. The project is aimed not only at the processes taking place within the fund, but also concerns the entire insured population of the Leningrad region (about 1.6 million people).

The project has a relatively low financial investment of 13.9 million rubles. The project has a payback period of 4.5 years. Savings from digitalization of work processes will amount to 5,414,284 rubles a year.

The implementation of the project will significantly improve the quality of the work performed by the territorial fund of the Leningrad Region. In addition, the standard scheme for implementing the project can be replicated to other compulsory health insurance funds.

References

1. André, A. (ed.): Digital Medicine. HI, Springer, Cham (2019). https://doi.org/10.1007/978-3-319-98216-8
2. Federal Law of November 29, 2010 No. 326-FZ (as amended on 26 July 2019) "On Compulsory Health Insurance in the Russian Federation". http://www.consultant.ru/document/cons_doc_LAW_107289/
3. Gomaleev, A.O.: Creating mechanisms for the interaction of medical organizations on the basis of a unified state information system in the field of healthcare (EGISP). J. Econ. Bus. **51**, 79–84 (2017). https://doi.org/10.24411/2411-0450-2019-10824
4. Hermes, S., Riasanow, T., Clemons, E.K., Böhm, M., Krcmar, H.: The digital transformation of the healthcare industry: exploring the rise of emerging platform ecosystems and their influence on the role of patients. Bus. Res. **13**(3), 1033–1069 (2020). https://doi.org/10.1007/s40685-020-00125-x
5. Karpov, O.E., Subbotin, S.A., Shishkanov, D.V., Zamyatin, M.N.: Digital healthcare. Necessity and background Doct. Inf. Technol. **3**, 5–22 (2017)
6. Khmelenko, A., Maksimova, T.G.: System analysis and optimization of business processes in a private medical clinic Econ. Right. Innov. **2**, 37–41 (2019)
7. Lipatov, V.A., Kaitsev, I.G., Severinov, D.A.: On the problems of implementing IT systems in practical healthcare. Bull. Sib. Med. **17**(1), 177–190 (2018). https://doi.org/10.20538/1682-0363-2018-1-177-190
8. Manjula, P., Balachandra, P.: An analysis on pricing strategies of software 'I-Med' in healthcare industry. In: Mishra, D., Nayak, M., Joshi, A. (eds.) Information and Communication Technology for Sustainable Development. Lecture Notes in Networks and Systems, vol 9. Springer, Singapore (2018). https://doi.org/10.1007/978-981-10-3932-4_25
9. Medical insurance organizations and medical organizations. FFOMS. http://ffoms.ru/system-oms/registry/

10. Order of the Federal Compulsory Medical Insurance Fund dated February 28, 2019 No. 36 "On approval of the Procedure for organizing and monitoring the volumes, terms, quality and conditions of providing medical assistance for compulsory medical insurance". https://www.garant.ru/products/ipo/prime/doc/72171640/
11. Order of the FFOMS dated 04.04.2011 N 79 (as amended on 01/15/2020) "On the approval of the General principles for the construction and functioning of information systems and the procedure for information interaction in the field of compulsory health insurance". http://www.consultant.ru/document/cons_doc_LAW_113397/
12. Passport of the federal project "Creation of a unified digital circuit in health care based on a unified state information system in the health sector". https://minzdrav.gov.ru/poleznye-resursy/natsproektzdravoohranenie/tsifra
13. Passport of the national program "Digital Economy of the Russian Federation" (approved by the Presidium of the Presidential Council for Strategic Development and National Projects on December 24, 2018 N 16). https://base.garant.ru/72190282/
14. Passport of the national project "Healthcare" (approved by the Presidium of the Presidential Council for Strategic Development and National Projects, minutes of 24 December 2018 N 16). http://www.consultant.ru/document/cons_doc_LAW_319209/
15. Report on the WHO Symposium on the Future of Digital Health Systems in the European Region, Copenhagen, Denmark, 6–8 February 2019. Copenhagen: WHO Regional O ce for Europe; 2019. License: CC BY-NC-SA 3.0 IGO. https://apps.who.int/iris/bitstream/handle/10665/329032/9789289059992-eng.pdf
16. Bansal, J.C., Gupta, M.K., Sharma, H., Agarwal, B. (eds.): ICCIS 2019. LNNS, vol. 120. Springer, Singapore (2020). https://doi.org/10.1007/978-981-15-3325-9
17. Stefanova, N.A., Andronova, I.V.: Problems of digitalization of the healthcare sector: Russian and foreign experience. Bull. Samara Univ. Econ. Manage. **9**(3), 31–35 (2018)
18. The official site of the operator of the unified registry of Russian programs for electronic computers and databases on the Internet. https://reestr.minsvyaz.ru/search/?how=d&q=foms&where=reestr
19. Unified information system in the field of procurement. https://zakupki.gov.ru/epz/main/public/home.html
20. Verzilin, D., Gorovykh, E., Maximova, T., Sokolova, I., Gokinaeva, I.: Information technologies for public policy measures adjustments in the social sphere and environmental protection. In: Proceedings of the 32nd International Business Information Management Association Conference, IBIMA 2018 - Vision 2020: Sustainable Economic Development and Application of Innovation Management from Regional expansion to Global Growth (2018)
21. Weber, J.H., Ho, J.: Applying bidirectional transformations to the design of interoperable EMR systems. J. Healthc. Inf. Res. **4**(2), 138–150 (2020). https://doi.org/10.1007/s41666-019-00065-0
22. Yusif, S., Hafeez-, A., Soar, J.: An exploratory study of the readiness of public healthcare facilities in developing countries to adopt health information technology (HIT)/e-Health: the case of Ghana. J. Healthc. Inf. Res. **4**(2), 189–214 (2020). https://doi.org/10.1007/s41666-020-00070-8

An Approach to Adjust Effort Estimation of Function Point Analysis

Huynh Thai Hoc(✉), Vo Van Hai, and Ho Le Thi Kim Nhung

Faculty of Applied Informatics, Tomas Bata University in Zlin, Nad Stranemi 4511, 76001 Zlin, Czech Republic
{huynh_thai,vo_van,lho}@utb.cz

Abstract. This study presents a modified approach to adjust a software development effort estimation. The AdamOptimizer-based regression model is adopted to adjust and enhance the accuracy of effort estimation. This approach is derived into three phases. The first step deals with the logarithmized formula of effort estimation computed by Function Point Analysis and Productivity Delivery Rate. The Adam-Optimizer-based regression model is examined in the second phase, and the ISBSG repository 2020 release R1 is considered as a historical dataset in this paper. Moreover, the K-Fold cross-validation technique is adopted to tunning the training model. In the following phase, all results are evaluated by statistical significance and the goodness of fit measure. Finally, a proposed approach is compared with others: Capers Jones, and the Mean Effort.

Keywords: Software effort estimation · FPA · Capers Jones · Adj-Effort · AdamOptimizer · K-Fold

1 Introduction

Effort estimation of software development is a significant challenge in the process of software project building [1–4]. This process is likely the first phase to estimate the development resources (i.e., cost, time). The accurate estimation might lead to the proper contribution to resources for a project. It is probably an important initial-step of project planning. Underestimating or overestimating might cause less accuracy in allocating the budgets, and then might be probably a significant cause of project failure [5]. This problem might boost researchers to study it, and many techniques have been proposed for the last decade regarding how to improve the effort estimation.

In general, there are two types of techniques used in software project effort estimation, that are, non-algorithmic and algorithmic [1, 3, 6]. Non-algorithmic is a technique that predicting the Effort might be based on either the expert's opinion or analytical comparisons with the historical projects. For example, Expert Judgment, Analogy, Price-to-win, Bottom-up, Top-up, Wideband Delphi, Planning Poker [1, 3, 7]. The algorithmic technique is a mathematical-based approach that mainly focuses on algorithms such as COCOMO, Use Case Points (UCP), Function Point Analysis (FPA).

R. Silhavy (Ed.): CSOC 2021, LNNS 230, pp. 522–537, 2021.
https://doi.org/10.1007/978-3-030-77442-4_45

Most algorithmic methods might mainly focus on optimizing factors such as the proportion of effort estimation known as Productivity Delivery Rate (PDR) or environmental factors that impact the development effort [5, 7–9]. These indicators impact the measurement of software effort. The purpose of these optimations is to make them fitter with the actual data. The better value of these indicators might lead to more accuracy of the estimation. In order to obtain those targets, scientists proposed many approaches to compute the predicted results and try to fit with the actual value. In their researches, some applied regression models, such as least squares regression, multiple linear regression [5, 7–9]. The other integrated regression methods work with fuzzy models, deep learning techniques [9, 10], etc.

Some studies presented other optimizable measures to compute the Effort of the application development. Prokopová et al. [11] worked with the VAF factor to identify its influence on the estimation of the effort size. Šilhavý et al. [12] improved this Effort by proposing a new CVS model, and dataset segmentation was applied to estimate it. Moreover, Artificial neural networks (ANN) were also adopted by Wena et al. [13] to do this estimation.

On the other hand, the term PDR is considered as the proportion of effort size. This term has always been adopted to calculate the required software project effort by multiplying with UCP size or FPA size.

$$\text{Effort} = \text{Size} \times PDR \tag{1}$$

Additionally, Nassif et al. applied a fuzzy-based methodology to propose four levels of Productivity based on a summation of environmental factors [14]. Moreover, the relationship between project Productivity and environmental factors was also studied by Azzeh et al. [15, 16]. They concluded that such indicators might be good factors for Productivity prediction when observational projects are available. Besides, Hai et al. [17] have proposed Productivity as a Multiple Regression model with independent variables (VAF, EI, EO, EQ, ILF, EIF) in the measured Effort of function point analysis based on the International Software Benchmarking Standards Group (ISBSG) 2018/release R2. The authors report that the result might be better than the existing method, compared with the Effort measured by Mean Productivity based on Eq. 1, and Capers Jones.

The purpose of this paper is to propose an adjusted technique to predict effort estimation based on the original formula of FPA size (Eq. 1). The formula will be transformed into a logarithm-based equation, and then apply the Regression Model to find a good (best) fitness model for estimating Effort. The proposed technique evaluation will be based on datasets extracted from the ISBSG repository released in August 2020. This technique might be called an Adjust Effort Estimation of Function Point Analysis (Adj-Effort).

The rest of the paper is structured as follows: Sect. 2 shows the Research Questions; Sect. 3 presents the background of Function Point Analysis as well as the variant of Effort Estimation; Sect. 4 illustrates the data pre-processing; Sect. 5 proposes the new approach - the Adjust Effort Estimation by Adam-optimizer method; Sect. 6 discusses the criteria validation; Sect. 7 presents the result and discussion; and conclusion – future work is presented in the last section.

2 Research Questions

In order to conduct the proposed model is the "good-fitness-model" for adjusting effort estimation, two research questions should be answered:

(1) RQ1: how close the real efforts are to the predicted regression model? Answering this question determine a statistical measure, including the Coefficient of determination (R^2), and adjusted R^2.
(2) RQ2: Is this a better-fitness model than Capers Jones and Original Effort method? Answering this question is to determine the SSE, MAE, PRED(0.25), the minimised SSE and MAE, and the maximised $PRED(0.25)$. In addition, the paired t-test based on the MAE measurement is examined [18].

$H_o : MAE_{Adj-Effort} = MAE_{CapersJones/OrginalEffort}$: There is no estimation error difference between these methods.

$H_1 : MAE_{CapersJones/OriginalEffort} > MAE_{Adj-Effort}$: Estimation capability of project size by the Adj-Effort might be more feasible than by Capers Jones and Original Effort method.

3 Background

3.1 Function Point Analysis

Function Point Analysis (FPA) was first introduced in 1979 by Alan J. Albrecht when he was a Program Manager of the Application and Maintenance Measurement Program for IBM [19, 20]. It is a Function Point to count the functional size and complexity of a software-based on user requirements [21]. His approach is to quantify the number of External Inputs (EI), External Inquiries (EQ), External Outputs (EO), Internal Logic Files (ILF), and External Interface Files (EIF) that are delivered by software projects. In 1984, the International Function Point Users Group (IFPUG) was established to define the set of rules for the basic components (BCs), including transaction function types (EI, EO, EQ) and data function types (EIF, ILF) [22]. The complexity weights of each component are shown in Table 1.

Table 1. Complexity weights of components [19, 23]

Basic Components (BCs)	Complexity Weight (CWs)		
	Low	Medium	Large
EI	3	4	6
EO	4	5	7
EQ	3	4	6
EIF	5	7	10
ILF	7	10	15

In the Counting Practices Manual, version 4.3.1 (2010) [24], the FPA is organized by the IFPUG, which is accountable for the development as well as the improvement of its rules. The IFPUG's FPA is known as the original function point analysis and is standardized by ISO/IEC 20926:2010. Additionally, there are four other Functional Software Measurements (FSM) of ISO/IEC, including MarkII, MESMA, COSMIC, FISMA. These methods are out of the scope of this paper; they might be found in [25].

The FPA has a majority of characteristics that it is possible to employ to estimate software development in the early phases of the project [26]. Firstly, function points might be counted based solely on either requirements specifications or design specifications. It seems the initial phase of the projects. Secondly, they are unrelated to language programming, specific technologies for developing, or any other data processing [23]. Moreover, non-technical users of software might be easier to understand the function points because they are built based on the user's external view of the system [27].

Table 2. General systems characteristics [19, 23]

GSC factors	Characteristic	Description
F1	Data communications	Does the system require backup and recovery?
F2	Distributed functions	Are data required for communication?
F3	Performance	Does the system include a distributed processing function?
F4	Heavily used configuration	Is critical performance required?
F5	Transaction rate	Will the system work during heavy loads?
F6	Online data entry	Does the system require direct data input?
F7	End-user efficiency	Do data inputs require multiple screens or operations?
F8	Online update	Are the main files up-to-date?
F9	Complex processing	Are inputs, outputs, files, and queries intricate?
F10	Reusability	Is internal processing complicated and complex?
F11	Installation ease	Is the code designed for reuse?
F12	Operational ease	Are conversions and installation included in design?
F13	Multiple sites	Is the system designed for multiple installations in different locations?
F14	Facilitate change	Is the application designed to make it easy for users to make changes?

To count function points, a linear combination between function types (EI, EO, EO, EIF, ILF) with appropriate three levels of complexity weights are established. This function count is also known as Unadjusted Function Points (UFP). The UFP formula is shown in Eq. 2.

$$UFP = \sum\nolimits_{i=1}^{5} \sum\nolimits_{j=1}^{3} BCs_{ij} \times CWs_{ij} \tag{2}$$

where BCs_{ij} is the count of component i at level j, and CWs_{ij} is an appropriate weight given in Table 1.

The final counting of function points is driven by multiplying UFP with adjustment influent factors, General Systems Characteristics (GSC). These might help the counting of UFP to be more accurate [27]. Additionally, VAF is considered as Value Adjustment Factor described by the formula:

$$\mathrm{VAF} = 0.65 + 0.01 \times \sum\nolimits_{i=1}^{14} F_i \times Rating_{Influence} \quad (3)$$

Where F_i is the influence of the GSC factor. There is a total of 14 factors impacting the system. These factors are illustrated in Table 2, and the influence factors rating is shown in Table 3.

Table 3. Influent factors rating [19, 23]

Influence	Rating
None	0
Insignificant	1
Moderate	2
Average	3
Significant	4
Strong significant	5

The Adjusted Function Points (AFP) is then the following:

$$AFP = \mathrm{UFP} \times VAF \quad (4)$$

As mentioned in Sect. 1, based on the AFP, the measurement of Software Effort Estimation is shown in Eq. 5, where Effort (called Original Effort) is measured in person-hours and AFP is considered as a software size.

$$\mathrm{Effort} = \mathrm{AFP} \times \mathrm{PDR} \quad (5)$$

3.2 The Variant of Sofware Effort Estimation

Capers Jones' Approach

From 1997 to 2007, Capers Jones proposed thumb rules to measure the sizing of software project [28]. The first version was published in 1997, and then it was updated in 2003. In 2007, it continued to edit and become the IFPUG counting rules version 4.1. According to the author [28], there are ten simple rules for software sizing estimation, which the last four rules regarding schedules, resources, and costs. As mentioned in the 10th rule, the estimating software development effort is employed by compiling the 7th and 8th rules. As a result, the work effort estimation is presented as the following [17, 28, 29]:

$$\text{Effort}_{\textit{Capers Jones}} = \frac{AFP}{150} \times AFP^{0.4} \quad (6)$$

where $\text{Effort}_{\textit{Capers Jones}}$ is the Effort measured in person-month; AFP divides 150 that illustrates the staff size (the 7th rule) [28] and AFP raises to power 0.4 present the approximate schedule (the 8th rule) [28].

Original Effort with Mean Productivity – the Mean Effort

The Mean Effort is a predicted size to measure an estimated software development effort based on Mean Productivity. It is computed by multiplying Adjusted Function Points with mean Product Delivery Rate value - PDR_{Mean} (Eq. 8). The PDR_{Mean} is expressed in Eq. 7, where n is the number of observed projects. This is a kind of Effort used as a sample method to compare with other predicted efforts. Prokopova et al. (2018), or Hai et al. (2020) have adopted the Mean Effort to evaluate the accuracy of their approaches compared with it.

$$\text{PDR}_{Mean} = \frac{1}{n}\sum\nolimits_{n}^{1} PDR \quad (7)$$

$$\text{Effort}_{Mean} = \text{AFP} \times \text{PDR}_{Mean} \quad (8)$$

4 Data Pre-processing

In this research, the ISBSG project repository Release 2020 will be adopted as the historical dataset. The ISBSG has 251 recorded attribute values, 9592 projects. These attributes are divided into many groups. The first group of these datasets is Rating. This group contains an ISBSG rating code that is in order from high-quality reviewers to low-quality reviewers, denoted A, B, C, or D. The second group is software age, it records the year of the project, ranges from 1989 to 2019. The information of Development type is also presented, it is categorized into three types, such as new development, enhancement, and re-development. Furthermore, the group of attributes regarding Sizing (UFP, AFP, VAF), Effort (Normalised Work Effort Level 1, Normalised Work Effort, Summary Work Effort), Productivity (Normalised Level 1 PDR - unadjusted function points, Normalised PDR, Pre 2002 PDR), Size Attributes (EI, EO, EQ, EIF, ILF) are recorded in these datasets.

These attributes are considered as a majority of datasets. It is noted that Normalised Level 1 Productivity Delivery Rate is the Productivity currently recommended by ISBSG [30, 31]. Additionally, the other project-relevant information such as Effort Attributes (Team Size Group, Resource Level), Tool Data, Project team (experience BA, IT, project manager), and so on is also recorded in this dataset.

In this study, Summary Work Effort, a total effort in hours recorded against the project, and Normalised Level 1 PDR will be used as the "Real-Effort" and PDR for the training dataset. Moreover, due to Normalised Level 1 PDR is selected as a PDR in Eq. 5 and 7, the value of AFP in the training dataset is also re-calculated by Eq. 4 based on the value of UFP and VAF.

There are many projects in the ISBSG repository released in 2020, it contains all 9592 projects. However, this proposal solely chooses projects that have sufficient good quality factors with the IFPUG-based development type. The following guidelines are adopted to filter the datasets.

(1) As recommended by ISBSG as well as the selection in the publish [10, 17], the research only chooses projects with data quality A and B. The other might be insufficient credibility due to either significant data not being provided or one factor or a combination of factors [30, 31], reducing the size of the dataset to 8619 projects.
(2) This paper mainly focuses on FPA based counting approach as the IFPUG, the size is measured by MarkII, MESMA, COSMIC, FISMA that is ignored, leaving only 6365 records.
(3) New development projects used solely [10], other types of project are considered removed, resulting in the total projects to 1460
(4) The missing data on Summary Work Effort (Effort), Normalised Level 1 Productivity Delivery Rate (PDR), UFP, VAF are removed, and the missed AFP might be recovered by Eq. 4, bring the remaining projects to 583.

5 Regression-Based Adjusting Effort

A Regression Model is the most common method for identifying relationships between factors of effort software estimation [3]. One of the factors is the dependent (response) variable, and the others are independent (preditor) variables [32]. This relationship may be used for predicting the values of the response variable for a given set of independent variables. The outcome in this model can be examined based on the historical data. This process is equivalent to optimizing the model's suitable coefficients, including its functional form and coefficients based on sample observational data.

A genre of a regression model is presented as Eq. 9, a linear regression model, where the dependent variable (Effort) is represented by Y and the predictor variable (AFP × PDR) by X; β_0 is an Intercept Parameter, and β_1 is called Regression Coefficient, ε is depicted as the error residuals.

$$Y \approx \beta_0 + \beta_1 \mathrm{X} + \varepsilon \tag{9}$$

Figure 1 and Fig. 2, the histograms of software effort and adjusted function points, illustrate that the selected dataset is not a normal distribution. In pact, this sub-dataset might be adopted to attain a predicted effort. However, it might result in an inaccurate outcome [6]. Nevertheless, after the Effort and the AFP are normalized by logarithmic technique, their histograms are slightly normally distributed. These figures are shown in Fig. 3 and Fig. 4. As a result, the prediction effort based on the historical dataset might be obtained by a log-based regression model instead of Eq. 9.

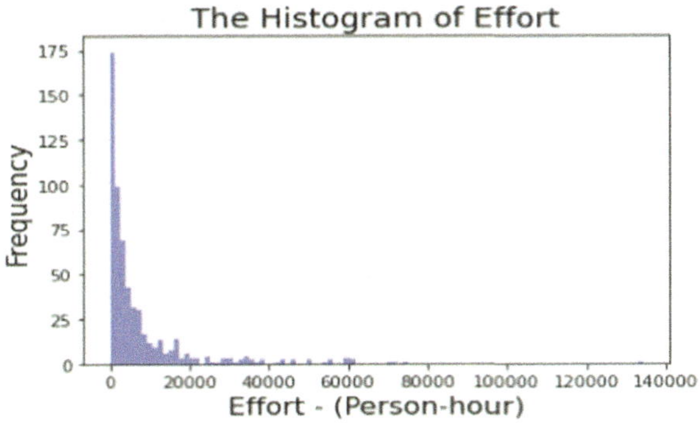

Fig. 1. The histogram of effort

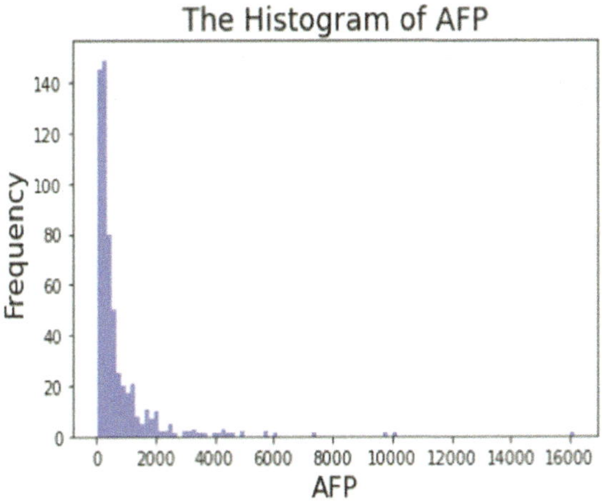

Fig. 2. The histogram of AFP

There are several steps to gain the log-based regression model. First, Eq. 5 is transformed by the logarithmic method, and then the regression model is adopted on it. These equations are given as Eq. 10, and 11, where ln(Effort) as a dependent variable, and $\ln(AFP)$, $\ln(PDR)$ as independent variables. Second, the excellent fitness model in Eq. 11 might be obtained by minimizing the error residual (ε).

$$\ln(\text{Effort}) = \ln(\text{AFP} \times \text{PDR}) = \ln(AFP) + \ln(AFP) \tag{10}$$

$$\ln(\text{Effort}) = \beta_0 + \beta_1 \ln(AFP) + \beta_2 \ln(PDR) + \varepsilon \tag{11}$$

There are some approaches to minimize the error residual. Radek Silhavy et al. (2015) have adopted the Multiple Least Square Regression method, namely AOM, to improve UCP. However, Hoc et al. (2020) have used the Adam-Optimizer module [3] to attain the predictive UCP. It is a kind of open-source, available in the TensorFlow

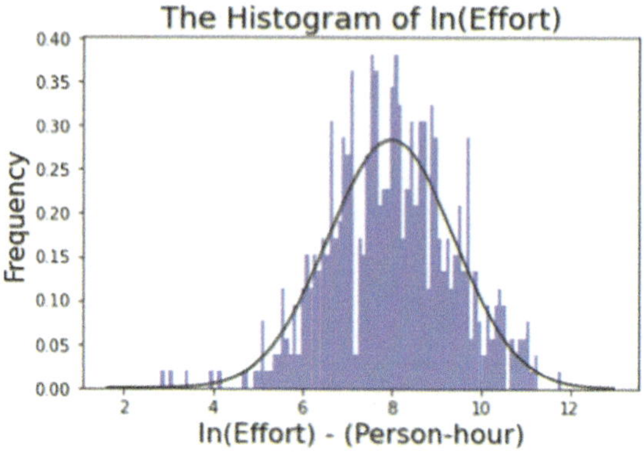

Fig. 3. The histogram of ln(Effort)

package [33], developing by the Google Brain team [33, 35]. Comparing with the AOM on the same historical dataset, they concluded that their approach was better than the AOM.

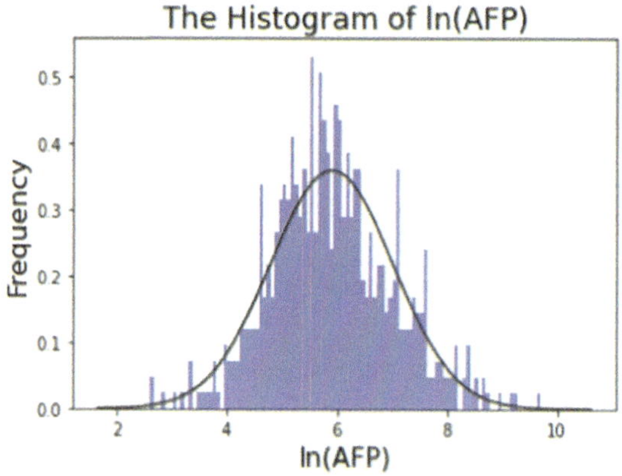

Fig. 4. The histogram of ln(AFP)

Finally, due to the outperformance of the Adam-Optimizer [34, 35], this module is applied to find suitable unknown values (β_0, β_1, β_2) in Eq. 11. It means that the final adjust Effort (Adj-Effort) is computed according to the formula:

$$\ln(\text{Effort}) \approx \beta_0 + \ln\left(AFP^{\beta_1}\right) + \ln\left(PDR^{\beta_2}\right) \tag{12}$$

$$\text{Effort} \approx e^{\beta_0 + \ln\left(AFP^{\beta_1}\right) + \ln\left(PDR^{\beta_2}\right)} \tag{13}$$

$$\text{Effort} \approx e^{\beta_0} \times AFP^{\beta_1} \times PDR^{\beta_2} \tag{14}$$

5.1 Dataset Description

As mentioned in Sect. 4, the Adj-Effort technique will use 583 filtered projects of the ISBSG repository released in 2020 as observational data. Figure 5 presents a boxplot of the log-based dataset before (Fig. 5.a) and after (Fig. 5.b) removing outliers. As can be seen, there are several noticeable outliers in that dataset. According to many researchers [17], removing outliers is suitable for accurate and reliable software effort estimation based on historical datasets. In this study, the interquartile range (IQR) method is adopted to eliminate the outliers. If the value of ln(Effort), ln(PDR), and ln(AFP) out of the range from $Q1 - 1.5 \times IRQ$ to $Q3 + 1.5 \times IRQ$, it might be considered to ignore, Q1 is the lower quartile, Q3 is the upper quartile, and IRQ is expressed as the following:

$$IQR = Q3 - Q1 \tag{15}$$

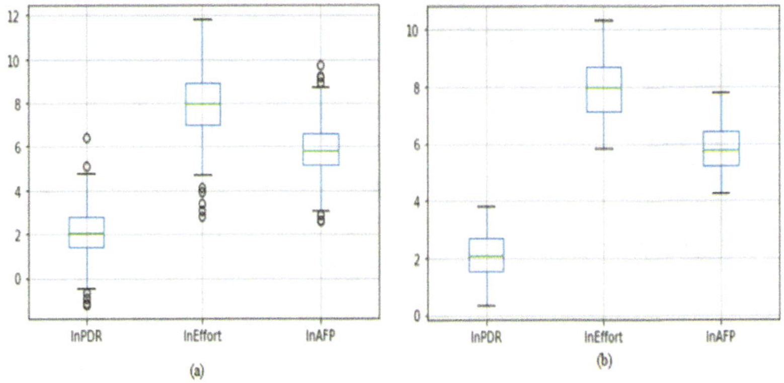

Fig. 5. The Boxplot of the observational dataset before (a) and after (b) remove outliers

Table 4 shows the descriptive characteristics of the logarithm of Summary Work Effort before removing outliers (initial) and after removing outliers, where N stands for the number of observed projects, 583 and 450, respectively. It is noted that the Mean value is slightly equal on both datasets, 7.9793, after adopting outliers. Additionally, the Standard Deviation, the value at 25%, 50%, and 75% are insignificantly different. However, the Minimum and Maximum at Remove Outliers Dataset might be better than the Initial dataset, compared with the others. As a result, the removed outliers-based dataset is used in this research.

Table 4. Descriptive observed datasets – ln(Effort)

Dataset	N	Mean	Standard deviation	Minimum	25%	50%	75%	Maximum
Initial	583	7.9953	1.4042	2.833	7.0148	8.0044	8.9344	11.8072
Remove outliers	450	7.9793	1.0344	5.8200	7.1323	8.0019	8.7073	10.2955

5.2 Adj-Effort Setting

The Adj-Effort technique is processed in three steps - shown in Fig. 6. The first step filters the raw dataset (2020) by adopting Data processing criteria – presented in Sect. 4. Second, the selected sub-dataset, including Effort, PDR, and AFP are logarithmized to make them a normal distribution. In the following phase of the process, the pre-processed projects are randomly divided into two parts. 80% of the projects are used to train the Adj-Effort model. The tested part – 20% of the projects are considered as unknown projects, where the data on Effort, AFP are available in this dataset, they are considered as real values; and the mean of the PDR ($PDR_{Mean} = 11.2569$), gained from the training dataset, is used as a value of the PDR in the testing phase.

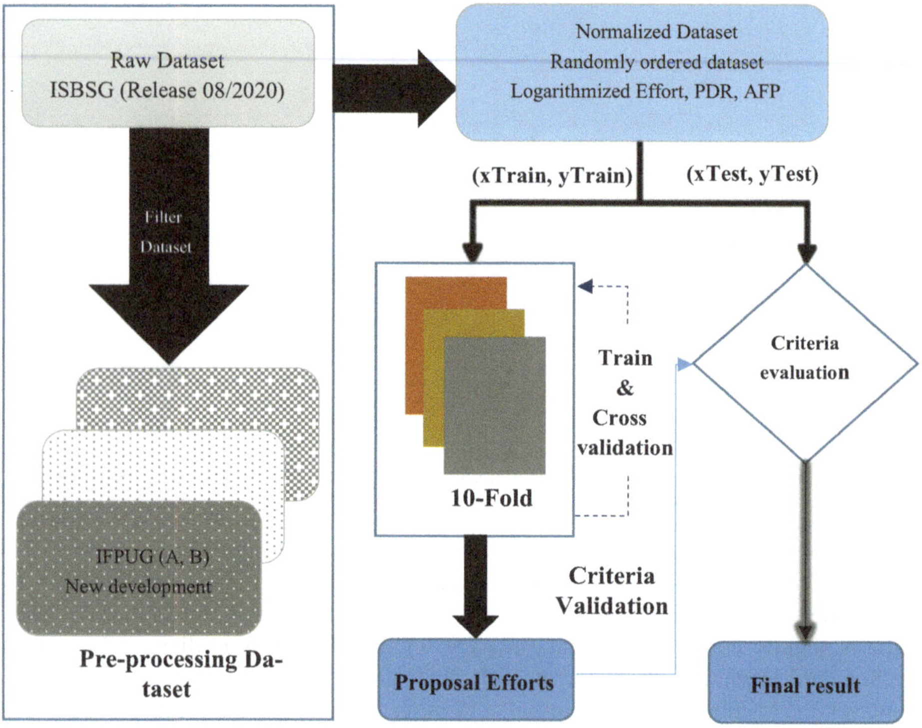

Fig. 6. The Adj-Effort with 10-Fold cross-validation flow diagram

Additionally, the non-cross validation and cross-validation methods are examined in optimizing the Effort. 10-Fold cross-validation on the training set is adopted to examine the performance and reliability of the proposed technique. The average 10-Fold-based Effort is obtained in this phase. Finally, the effort models obtained from non-cross validation and cross-validation are evaluated based on the criteria presented in Sect. 6 and compared with Capers Jones and the Mean Effort method to achieve the final Adj-Effort.

6 Criteria Validation

This section presents the standard criteria to validate the accuracy of the effort estimation of each method. MMRE is the Mean Magnitude of Relative Error, and PRED(x) is the Prediction level. Both are computed based on the Magnitude of Relative Error (MRE). SSE (Sum of Squared Errors) is a metric to compute modeling error variation. MMRE, PRED(x), or SSE are the most well-known measurement of accurate effort estimation [3, 36, 37]. Furthermore, MAE stands for the Mean Absolute Error; it represents the difference between the actual and predicted values. Finally, R^2 *or Adjusted* R^2 are the statistical indicators to present how close the data are to the fitted regression model. Their values might from 0% to 100%, the higher $R^2/Adjusted\ R^2$ means the more accuracy of the proposed approach [38]. The mentioned criteria equations are given as follows:

$$MAE = \frac{\sum_i^N |\hat{y}_i - y_i|}{N} \tag{16}$$

$$MRE = \frac{|\hat{y}_i - y_i|}{y_i} \tag{17}$$

$$MMRE = \frac{\sum_i^N MRE}{N} \tag{18}$$

$$\text{PRED}(x) = \frac{1}{N}\sum_i^N \begin{cases} 1\ if\ MRE_i\ \leq x \\ 0\ otherwise \end{cases} \tag{19}$$

$$SSE = \sum_i^N (y_i - \hat{y}_i)^2 \tag{20}$$

$$R^2 = 1 - \left[\frac{\sum_i^N (y_i - \hat{y}_i)^2}{\sum_i^N (y_i - \bar{y})^2}\right] \tag{21}$$

$$Adjusted\ R^2 = 1 - \frac{N-1}{N-k-1}\left(1 - R^2\right) \tag{22}$$

Where N is the number of observations, k is the number of preditors, y_i the actual value, $\hat{y}_i$ the predicted value, $\bar{y}$ the mean of actual value and the x value is considered to be 0.25 - as recommended in many studies [7, 36, 39]. Some authors also use PRED(0.20) or PRED(0.30) with little differing results [7].

7 Results and Discussion

In this section, the results of proposal efforts are discussed based on the criteria validation in Sect. 6. It is noted that two predictive models are shown in Table 5. The first one is obtained without applying cross-validation (non-cross validation). The second one is attained by dividing the training dataset into 10-folds, one of them for validating and the rest for training. As can be seen, the obtained models might be good fitness models compared with the actual Effort due to both R^2 and *Adjusted* R^2 are strongly high [34]. These values might answer the RQ1 that predicted models might slightly close with the real one.

Table 5. The predictive effort (Adj-Effort)

	Regression models	R-squared	Adj-R-squared
Adj-Effort	Effort $\approx$ 2.45994 $\times$ $AFP^{0.8999}$ $\times$ $PDR^{0.8506}$	0.8357	0.8349
Adj-Effort/10-Fold	Effort $\approx$ 2.4639 $\times$ $AFP^{0.8813}$ $\times$ $PDR^{0.8433}$	0.8866	0.8859

Table 6 shows a summary of the predictive statistics in terms of the effort that was adopted to validate the accuracy between Adj-Efforts (non-cross validation and 10-Fold cross-validation) and other methods (Capers Jones, the Mean Effort). As can be seen, both the regression-based Adj-Efforts perform well and give relatively accurate results with values of PRED(0.25) $\geq$ 21% and MAE $\leq$ 2530. It is noticeable that both Capers Jones and the Mean Effort yields worse results than Adj-Effort models. Although the MMRE of the Adj-Effort is slightly higher than Capers Jones, the results gained from the Adj-Effort are likely more accurate than the ones estimated by Capers Jones and the Mean Effort (i.e. MAE-Adj-Effort < MAE-Capers Jones/ Mean Effort, PRED(0.25)-Adj-Effort > PRED(0.25)-Capers Jones/Mean Effort), and SSE-Adj-Effort < SSE-Capers Jones/Mean Effort as well - matching the RQ2 requirement in Sect. 5.

Table 6. The performance effort estimation comparison methods

	Adj-Effort	Adj-Effort 10-Fold	Capers Jones	Original effort
MMRE	1.17	1.01	0.99	1.26
MAE	2527.89	2447.72	4509.96	2815.93
PRED(0.25)	0.21	0.26	0.0	0.20
SSE	36355.40	36020.93	65488.69	39939.61
N	90	90	90	90

On the other hand, based on MAE, PRED(0.25), and SSE, it is possible to conduct that Adj-Effort with/without 10-Fold might be more accurate than the others. It means that two models might be used as the best-predicted estimation. However, all criteria obtained from Adj-Effort/10-Fold are all more significant than those obtained from Adj-Effort without cross-validation. For example, PRED(0.25) attains the maximum, while MMRE, MAE and SSE reach the minimum when compared with Adj-Effort/without cross-validation. It expresses the meaning that Adj-Effort/10-Fold might be more accurate than the other one.

In addition, the paired t-test result of the hypothesis is given in Table 7. There is a significant difference in MAE between Adj-Effort/10-Fold and Capers Jones −(p value = 0.000004 < 0.05); the Mean Effort (p value = 0.0271210 < 0.05). Based on the paired t-test results, one might reject the null hypothesis and accept the alternative hypothesis. This might reveal that Adj-Effort/10-Fold is the best in the Estimation Accuracy Field.

Table 7. Adj-Effort/10-Fold hypothesis t-test results

	Degree of Freedom	t-value	p-value
Capers Jones	89	4.89	0. 000004
Mean effort	89	2.25	0.0271210

8 Conclusion and Future Work

In this study, FPA-based effort estimation is proposed, namely Adj-Effort, and the AdamOptimizer-based regression model is used to adjust and improve that estimation. The model was trained and tested, employing 583 filtered projects of the ISBSG repository 2020. The training dataset was divided into 10-Fold, and then applied cross-validation to tunning the precise of the accurate predictive model.

As discussed in Sect. 7, the Adj-Effort is more consistent than Capers Jones and the Mean Effort. Moreover, based on the analysed results in that section, we conclude that with $R^2 = 0.8866$, and $Adjusted\ R^2 = 0.8859$, it is noticeable that the adjusted Effort of projects gained from the Adj-Effort/10-Fold, overall, is close to the actual Effort of projects. In other words, the real Effort of the project is considered to slightly fit with the estimated Effort from the Adj-Effort/10-Fold.

However, as presented in Sect. 4, there are many factors in the ISBSG repository. It might influence the quality of the estimation. Therefore, in our future research, a statistical analysis of the different factors affecting the accuracy of effort estimation software will be considered.

Acknowledgment. This work was supported by the Faculty of Applied Informatics, Tomas Bata University in Zlín, under Project IGA/CebiaTech/2021/001.

References

1. Muketha, G.: A review of Agile software effort estimation methods. Int. J. Comput. Appl. Technol. Res. **5**(9), 612–618 (2016)
2. Trendowicz, A., Jeffery, R.: Software Project Effort Estimation. Springer, Cham (2014). https://doi.org/10.1007/978-3-319-03629-8

3. Hoc, H.T., Van Hai, V., Le Thi Kim Nhung, H.: A review of the regression models applicable to software project effort estimation. In: Silhavy, R., Silhavy, P., Prokopova, Z. (eds.) CoMeSySo 2019 2019. AISC, vol. 1047, pp. 399–407. Springer, Cham (2019). https://doi.org/10.1007/978-3-030-31362-3_39
4. Van Hai, V., Le Thi Kim Nhung, H., Hoc, H.T.: A review of software effort estimation by using functional points analysis. In: Silhavy, R., Silhavy, P., Prokopova, Z. (eds.) CoMeSySo 2019 2019. AISC, vol. 1047, pp. 408–422. Springer, Cham (2019). https://doi.org/10.1007/978-3-030-31362-3_40
5. Seo, Y.-S., Bae, D.-H., Jeffery, R.: AREION: software effort estimation based on multiple regressions with adaptive recursive data partitioning. Inf. Softw. Technol. **55**(10), 1710–1725 (2013)
6. Nassif, A.B., Ho, D., Capretz, L.F.: Towards an early software estimation using log-linear regression and a multilayer perceptron model. J. Syst. Softw. **86**(1), 144–160 (2013)
7. Fedotova, O., Teixeira, L., Alvelos, H.: Software effort estimation with multiple linear regression: review and practical application. J. Inf. Sci. Eng. **29**(5), 925–945 (2013)
8. Silhavy, P., Silhavy, R., Prokopova, Z.: Analysis and selection of a regression model for the use case points method using a stepwise approach. J. Syst. Softw. **125**, 1–14 (2017)
9. Hoc, H.T., Van Hai, V., Le Thi Kim Nhung, H.: AdamOptimizer for the optimisation of use case points estimation. In: Silhavy, R., Silhavy, P., Prokopova, Z. (eds.) CoMeSySo 2020. AISC, vol. 1294, pp. 747–756. Springer, Cham (2020). https://doi.org/10.1007/978-3-030-63322-6_63
10. Nassif, A.B., Azzeh, M., Idri, A., Abran, A.: Software development effort estimation using regression fuzzy models. Comput. Intell. Neurosci. **2019**, 8367214 (2019)
11. Prokopova, Z., Silhavy, P., Silhavy, R.: VAF factor influence on the accuracy of the effort estimation provided by modified function points methods. In: Annals of DAAAM & Proceedings (2018)
12. Silhavy, P., Silhavy, R., Prokopova, Z.: Categorical variable segmentation model for software development effort estimation. IEEE Access **7**, 9618–9626 (2019)
13. Wena, J., Lia, S., Linb, Z., Huc, Y., Huangd, C.: Systematic literature review of machine learning based software development effort estimation models. Inf. Softw. Technol. **54**(1), 41–59 (2012)
14. Azzeh, M., Nassif, A.B.: Project productivity evaluation in early software effort estimation. J. Softw. Evol. Process **30**, e2110 (2018)
15. Symons, C.R.: Function point analysis: difficulties and improvements. IEEE Trans. Softw. Eng. **14**(1), 2–11 (1988)
16. Heiat, A.: Comparison of artificial neural network and regression models for estimating software development effort. Inf. Softw. Technol. **44**(15), 911–922 (2002)
17. Hai V.V., Nhung H.L.T.K., Hoc H.T.: Productivity optimizing model for improving software effort estimation (2020)
18. Ross, A., Willson, V.L.: Paired samples T-test. In: Basic and Advanced Statistical Tests. SensePublishers, Rotterdam (2017)
19. Albrecht, A.J.: Measuring application development productivity. In: Proceedings of the Joint Share/Guide/IBM Application Development Symposium, pp. 83–92 (1979)
20. Behrens, C.A.: Measuring the productivity of computer systems development activities with function points. IEEE Trans. Softw. Eng. **SE-9**(6) (1983)
21. IFPUG: International Function Point Users Group. http://www.ifpug.org/
22. Abran, A., Robillard, P.N.: Function points: a study of their measurement processes and scale transformations. J. Syst. Softw. **25**(2), 171–184 (1994)
23. Albrecht, A.J., Gaffney, J.E.: Software function, source lines of code, and development effort prediction: a software science validation. IEEE Trans. Softw. Eng. **SE-9**(6), 639–648 (1983)

24. IFPUG: Function Point Counting Practices Manual, release 4.3.1. Int. Function. International Function Point Users Group, Westerville, Ohio (2010)
25. Gencel, Ç., Demirörs, O.: Conceptual differences among functional size measurement methods. In: First International Symposium on Empirical Software Engineering and Measurement (ESEM 2007), Madrid, pp. 305–313 (2007)
26. Low, G.C., Jeffery, D.R.: Function points in the estimation and evaluation of the software process. IEEE Trans. Softw. Eng. **16**, 64–71 (1990)
27. Matson, J.E., Barrett, B.E., Mellichamp, J.M.: Software development cost estimation using function points. IEEE Press **20**(4), 275–287 (1994)
28. Jones, T.: Software Estimating Rules of Thumb, Version 3. Software Productivity Research, March 2007
29. Czarnacka-Chrobot, B.: The role of benchmarking data in the software development and enhancement projects effort planning. Front. Artif. Intell. Appl. **199**, 106–127 (2009)
30. ISBSG Release R2: ISBSG (2018)
31. ISBSG Release R1: ISBSG (2020)
32. Chatterjee, S., Hadi, A.S.: Regression Analysis by Example, 5th edn. Wiley, Hoboken (2012)
33. Ramsundar, B., Zadeh, R.B.: TensorFlow for Deep Learning: From Linear Regression to Reinforcement Learning, pp. 43–73. O'Reilly Media (2018)
34. Sawant, A., Bhandari, M., Yadav, R., Yele, R., Bendale, M.S.: Brain cancer detection from MRI: a machine learning approach (TensorFlow). Int. Res. J. Eng. Technol. (IRJET) **05**(April), 2092 (2018)
35. Kingma, D.P., Ba, J.: Adam: a method for stochastic. In: Proceedings of the 3rd International Conference on Learning Representations (2014)
36. Silhavy, R., Silhavy, P., Prokopova, Z.: Algorithmic optimisation method for improving use case points estimation. PLoS ONE **10**(11), e0141887 (2015)
37. Nhung, H.L.T.K., Hoc, H.T., Hai, V.V.: A review of use case-based development effort estimation methods in the system development context. In: Silhavy, R., Silhavy, P., Prokopova, Z. (eds.) Intelligent Systems Applications in Software Engineering. CoMeSySo 2019 2019. Advances in Intelligent Systems and Computing, vol. 1046. Springer, Cham (2019). https://doi.org/10.1007/978-3-030-30329-7_44
38. Moore, D.S., Notz, W.I, Flinger, M. A.: The Basic Practice of Statistics, p. 138, 6th edn. W. H. Freeman and Company, New York (2013)
39. Silhavy, P., Silhavy, R., Prokopova, Z.: Stepwise regression clustering method in function points estimation. In: Silhavy, R., Silhavy, P., Prokopova, Z. (eds.) CoMeSySo 2018. AISC, vol. 859, pp. 333–340. Springer, Cham (2019). https://doi.org/10.1007/978-3-030-00211-4_29

M-Learning Autonomous Model for Learning Mathematics with MatiMath Application in Students of a Confessional

Sol A. Mamani(✉), Carlos R. Mamani, and Juan J. Soria

Professional School of Systems Engineering, Universidad Peruana Unión, Lima, Perú
solmamani@upeu.edu.pe

Abstract. Teaching-learning in mobile virtual environments is having a greater presence and impact on students and teachers, giving them more tools to learn, so an appropriate model for learning mathematics through mobile devices is sought. The study aims to propose an application based on the m-learning requirements that provides support to students and teachers in the area of mathematics fulfilling criteria such as technological requirements, pedagogical principles and UX elements. Which was carried out through 4 phases design, development, implementation and evaluation, where it was obtained as a result through the T test for independent samples that the students who used the application reached an average of 14,1 in contrast to 8.46 of the students who did not use the application so it is concluded that the MatiMath application had a positive influence and provided support in the learning process in the area of mathematics of university students.

Keywords: M-Learning · Mobile app · Learning · Math

1 Introduction

One of the frequent difficulties that a percentage of students face when starting their studies in Peruvian universities is that they do not understand the topics of Basic Mathematics, and therefore disapprove of the subject of mathematics that universities program in their respective curricula [1]. One of the causes of this problem could be the low levels of achievement in the area of mathematics among graduates of basic education. The Program for International Student Assessment (PISA) [2] applies a standard test to measure the competences in Science, Mathematics and Reading Comprehension in students aged 15 to 16, that is to say in students who are finishing their basic education. According to the report of the Program for the International Evaluation of Students (2018) Peru ranks 64 out of 77 countries, on the other hand 60.3% of students are at level 1 out of 6, which means that 6 out of 10 students who complete the basic training level (secondary) would not be in a position to meet the challenges of university or technical vocational training requiring basic mathematical skills and competences [3].

The results of the entry assessments of mathematics at the Universidad Peruana Union in the year 2020 of first-year students reveal that students arrive at university with a low level of basic math skills and competencies, 81% are disapproved. Mathematics

R. Silhavy (Ed.): CSOC 2021, LNNS 230, pp. 538–551, 2021.
https://doi.org/10.1007/978-3-030-77442-4_46

teachers claim that students who have an adequate command of basic mathematical procedures and algorithms, show better chances of achieving the learning results of units of mathematics courses, while students who "have no basis"; show insecurity, do not understand what is explained, show fear, do not fulfill tasks, show discouragement, and look for different ways to evade math topics and activities [1, 3].

Teachers with years of experience indicate that a student who has difficulties in the basics of mathematics urgently needs to regain confidence, needs to consult a tutor or teacher to give him confidence, needs a repeat explanation, need to practice with exercises prepared specifically for the developed theme, and need to see their state of advancement [4].

In recent years, the use of mobile technologies has increased in several fields such as banking, economics, tourism, entertainment, information management, as well as in education; in the latter it has shown a constant evolution and expansion, generating an opportunity to make better tools and resources available to students for their learning. Mobile learning, known as m-learning, is an emerging tool for the education system because it provides opportunities for innovation in favor of the teaching process - learning. Considering that users are increasingly demanding accessibility and availability, m-learning is presented as the next generation of e-learning [5, 6]. Currently in the market is m-learning in a generic format including for learning in the area of mathematics, providing students with resources, tools, teachers and tutors with some limitations in their scope. The effectiveness and expansion of m-learning is based on the strategic integration of the teaching and learning process semi-personalized and responsive to the needs presented in the learning process. The virtual student requires a well-structured content, the application and continuous practice of the knowledge they are acquiring and also see the progress of their learning. Teachers encounter limitations when it comes to demonstrating the learning outcomes of students in remote mode, and even more difficult if they apply the same assessment to the whole group of students, because a plagiarism using the internet facilities, would significantly affect data on achievement levels in mathematics learning. An asynchronous learning model has a number of advantages over traditional learning by allowing students to learn from anywhere in the world, if they have access to the internet, in the time that both the teacher and the student have available, the costs are low. In some cases it does not have a planned and targeted development structure or method [7]. In order to increase the quality of the m-learning tools aimed at teaching-learning in the subject of mathematics, the present research implemented an M-Learning Model Learning based on the basic elements of learning in virtual and mathematical environments, aiming to go beyond the standardization of current m-learning models that have different specifications and standards [7, 8].

Students with difficulties in math courses require a learning model that puts at their disposal the tools, content and instructional design to ensure their math learning. The Peruvian government issued Supreme Decree N° 044-2020-PCM on March 16, 2020, officially announcing a state of emergency and mandatory preventive isolation [19]. So students and teachers from different institutions had to develop their classes in a virtual way [20]. This situation forced universities to enable platforms and adopt learning models to develop classes through the internet, and teachers had to learn how to manage resources for this mode of teaching [21]. Each area of a curriculum has its own characteristics, and therefore requires teaching resources and process - particular learning, and having an M - Learning model created specifically for the learning of mathematics; is an important tool for learning mathematics in remote mode [22].

2 Literature Review

2.1 E-Learning

The term e-learning qualifies as a form of learning based on the use of new technologies that allows access to an online, personalized training through the Internet and other electronic means for the development of competences in the teaching process-learning regardless of time and place [3].

E-learning not only refers to the internet as a technological support for teaching-learning, but also to the set of resources such as videos, audios, live broadcasts, wireless and mobile devices; it is an online learning. One of the main objectives of e-learning is the development of personalized, understandable and dynamic learning content in real time. E-learning can be asynchronous, that is, at the pace of the student or synchronous led by a tutor, in a general concept it encompasses flexible distance learning, or also as a complementary and combined learning together with traditional face-to-face learning. As technologies develop and encompass more environments, e-learning is taking on a greater position and delivering better results [3, 4, 13].

2.2 M-Learning

The growth of computing has had a major impact on learning, with devices suitable for mobile learning including digital media players, smartphones and personal digital assistants. Ktoridou and Eteokleous defined m-learning as learning using mobile devices and portable IT devices, one of the key benefits of these devices is that they allow students to vary their study location, the use of mobile devices in a learning context allows a learner to learn anywhere and at any time, thus eradicating geographical boundaries, enabling cooperative learning environments that encounter group and individual interaction, having instant communication with other users [1, 14].

Although the primary objectives that motivate the implementation of m-learning are cost, adaptability and scalability should be remembered that the use of technology must be driven by pedagogical considerations. The main pedagogical aspect related to m-learning is finding ways to integrate mobile tools into teaching and learning activities [14].

2.3 M-Learning Standard and Specification Objects

The M - Learning models are based on principles or prescriptions whose validity has been proven by their results, some are mentioned below:

Constructivism. Construct knowledge by interpreting new knowledge based on prior knowledge, where the student must have social interactions with his peers actively participating in the construction of information [14, 15].

Combined. It combines classroom instruction with virtual learning and maximizes the benefits of both face-to-face and online methods. This model combines different advantages of face-to-face education and m-learning to ensure a stable learning environment for students. Where three features stand out: shift from teacher-centered instruction to student-centered instruction that becomes an active apprentice, increase of student content to tutor/teacher and integrated mechanisms of formative and summative evaluation for students and instructors, this makes this learning effective [15, 16].

Collaborative. This learning environment has been more effective at the time of collaboration, defined as good management of people participating in an assignment using technologies as mobile tools. And also, these types of learning environments are generally environments that provide information and exchange of views between group members and experts through the technology that supports collaborative learning tools [17].

Active. It is a process where students engage in higher-order thinking tasks, such as analysis, synthesis, and evaluation. Mobile devices and tools enrich the learning process of students as they are versatile and active [4, 18].

2.4 M-Learning Standards and Specifications

Organizations such as: IEEE, AICC, IMS and ADL developed standards and specifications as guidelines and best practices on the description and use of m-learning content, which are divided into three general groups: metadata, content package and educational design [15].

2.5 Metadata

It consists of one or more educational resources that are generated from the development of processes and interaction with users, one of the most used is the LOM (Learning Object Metadata) which aims to facilitate the interoperability of Learning Objects, support their reuse and assist in their discovery [4, 16].

2.6 Content Packs

It refers to the methodology for storing m-learning material and making it available on different platforms (operating systems). One of the most widely used formats is IMS Content Packaging (IMS CP) [15].

2.7 Educational Design

Includes pedagogical activities and instructional sequences with structure of content distribution [1, 8].

3 Methodology

3.1 Criteria of the Model

The following table shows the fundamental criteria of the m-learning model proposed, divided into 3 types: Technological, pedagogical and elements Ux, criteria contemplated in the development and implementation. Table 1 shows the 3 principles on which the MatiMath model is based.

Table 1. Types and criteria of the M-learning model.

Type	Criteria	Description
Technological resources	Portability	Ability of the application to run on different operating systems [17]
	Accessibility	Application accessible to different operating systems [18]
	Security	Availability of mechanisms that control and protect data generated as a result of application use [18]
	Facility	Ease with which users manage to interact in the app the first time they use it [17, 18]
	Response time	Files uploaded to the server have to support network changes and not exceed one minute [17, 18]
	Responsive	The application design has to support changes in size, screen resolutions and orientation [11, 17, 18]
	Interface	Interactions with user-friendly buttons and tools [18]
Pedagogical grounds	Content	Distributed information on the content of each course should be up-to-date and should facilitate updating [1, 8]
	Information distributed	Course content should be fragmented and organized into sessions, topics and units, so that students can easily locate them [8, 14]
	Personification	The student should be given the facility to repeat an explanation and to access practice exercises as often as his or her learning pace requires it [1, 8]

(continued)

Table 1. (*continued*)

Type	Criteria	Description
Elements UX	Monitoring	The student should be given the facility to repeat an explanation and to access practice exercises as often as his or her learning pace requires it [14, 18]
	Motivation	The application must be friendly, intuitive, must make resources available quickly and simply [1, 18]

3.2 M-Learning Model Requirements for Math Learning

Figure 1 shows the schema of the model raised with their respective characteristics (Table 2).

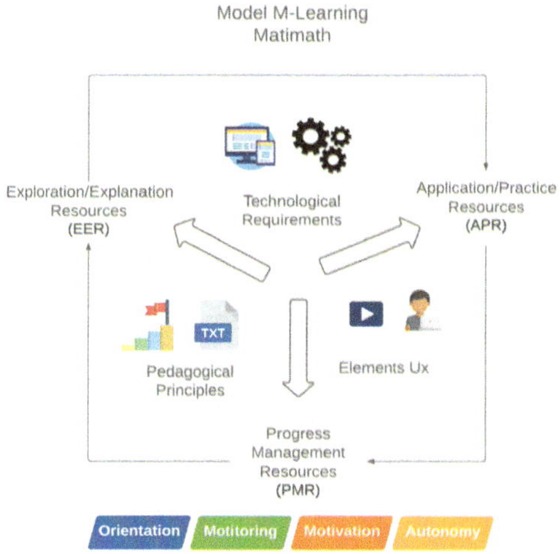

Fig. 1. Graphic of the M-learning MatiMath model.

Table 2. Fundamentals for mathematical learning.

Fundamentals	Description
Exploration resources/explanation	Every student needs a content path organized in a natural sequence, and each unit of content has a sufficient explanation [1, 8, 14]

(*continued*)

Table 2. (*continued*)

Fundamentals	Description
Resources for implementation/practice	The repetition of an action, of a procedure, of a process is the core element of learning because it promotes assimilation [1, 14, 18]
Resources for progress management	Every student needs objective information about their progress; the more genuine the report of their progress, the more motivating it will be [8, 18]

3.3 Development of the Application

A hybrid mobile application was developed in the dart language, with reach to the android and iOS operating systems, which are used by students selected for this research. For the server-side application NodeJS and MongoDB database were used, the server is published in the Amazon Web Service EC2 instance, and the resources were stored in the S3 service also from Amazon Web Service. In Fig. 2 and 3 we can see the main features developed in the MatiMath application. As also in Table 3 is pseudocode that refers to programming for students evaluations.

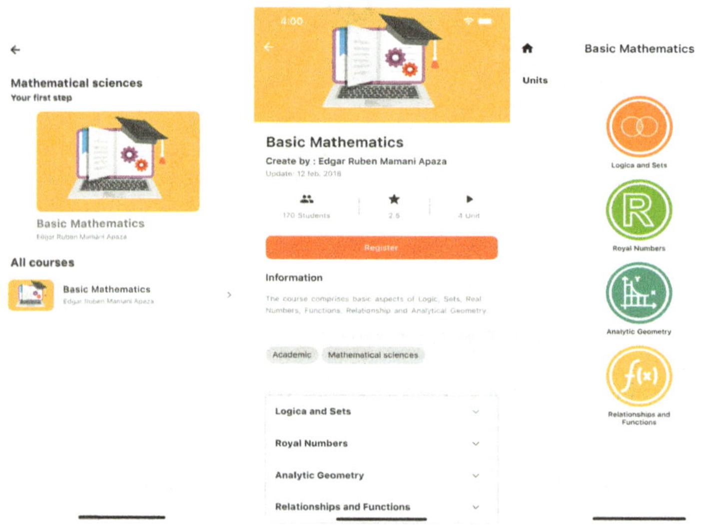

Fig. 2. Course and application units MatiMath.

3.4 Implementation of the Application

This research had as population 350 students, with a non-probabilistic sample and convenience of 65 students. For the implementation of the MatiMath application the following sequence was followed:

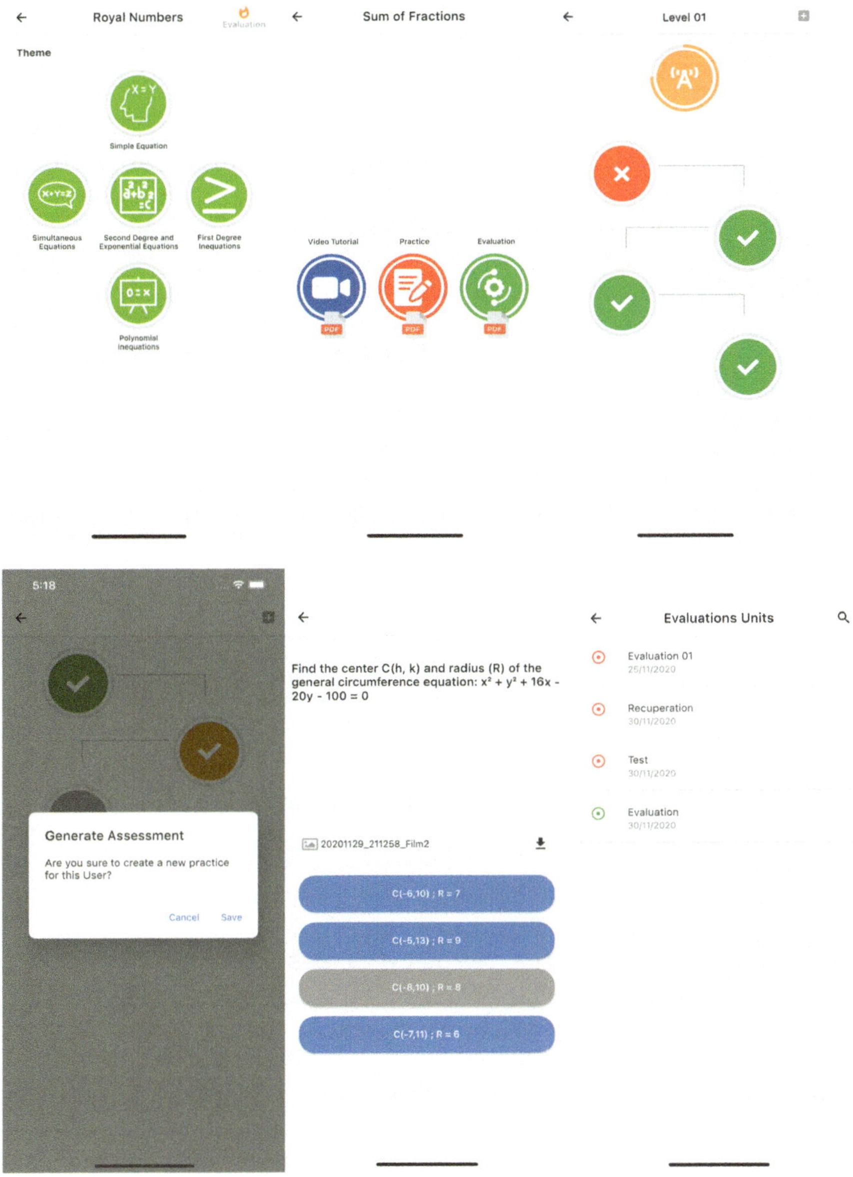

Fig. 3. Topics, session, level and evaluation of the MatiMath application.

Production and Empowerment of Resources. The content for the Basic Mathematics course was prepared under the criteria of the m-learning model proposed, as a first instance an orientation map was prepared with the units, themes, sessions and levels that were developed, as well as the videos, documents and exercises. Then all the resources were loaded to the server by the postman tool, coming having 3 units, 11 themes, 37 sessions, 111 resources (videos, pdf and ppt) and 1309 exercises.

Table 3. Pseudocode, student assessment verification.

```
for(evaluation of evaluationLevel){
 var stateEvlauation = 'Q';//STATES EVALUATION
Q:No answer, F:Anwered Error, V: Answered Cor-
retly, I: Error
if(evaluation.res.length>1){

if(evaluation.items.length>0)stateEvaluation =
'F';
                        for(evaluationItems of
evaluation.res){
 var status = false;
evaluationItems.resUser = false;
for(var i=0; i<evaluation.items.length; i++){
if((evaluationItems._id).equals(evaluation.items
[i]))status = true;
                                   }
if(status){
evaluationItems.repUsuario = true;
if(evaluationItems.state)stateEvaluation =
'V';//
                               }
                               }
evaluation.stateEvaluation = stateEvaluation;
                           }
                       }
result.stateResp = true;
result.message = 'Successfull';
result.evaluations = evaluationsLevel;
```

Capacitation. The app was available to students from the Play Store app store. In the first instance, all students were trained in how to use the application, and a demo of the evaluations, practices and progress of each completed session was also made. It made support to all users in a personalized way with some drawbacks that they had such as password change, user recovery, change of uploaded files, generation of new evaluations, evaluation enabled, as well as some that had regarding the characteristics of your mobile device. It was present in each class done, 4 times a week and also in different schedules so that the student can follow the learning process without any problems.

Monitoring. The MatiMath application has the facility to visualize the progress report of the practice exercises and evaluations in real time for each student, so a more personalized follow-up was made on the sessions that present more difficulty providing tutorials and review of the sessions.

Evaluation. The m-learning model allows for customized configuration of evaluations and thanks to the number of exercises per session found in the database, the structure of the application and content allows you to configure evaluations according to the advanced level of the students, as well as gives each student a completely different evaluation since they take exercises randomly. The teacher can visualize the completed student exercises, generate another assessment, enable, disable, publish, and correct the assessments in real time. So the evaluation was done asynchronously. In the Fig. 4 shows the functionalities of the teaching role in the MatiMath application.

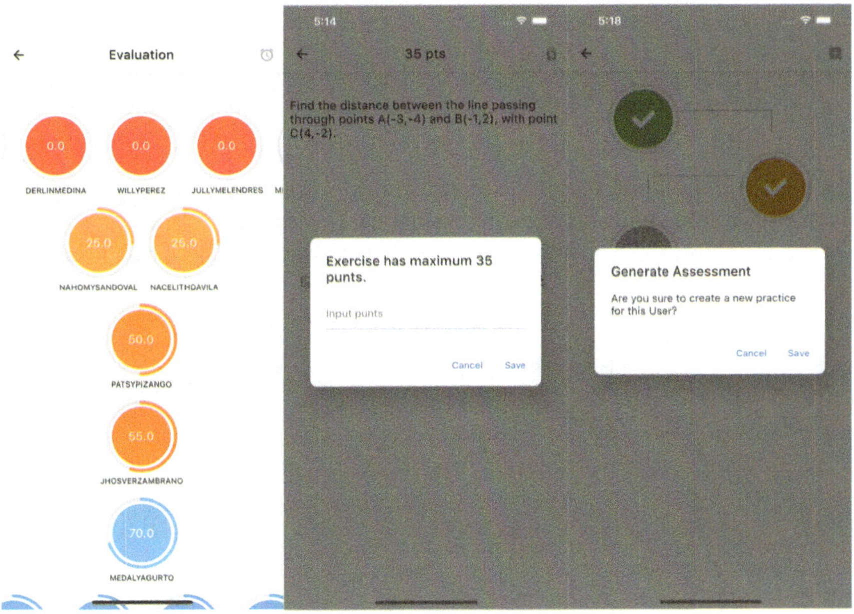

Fig. 4. Evaluations of the student beggar application MatiMath.

4 Results

The results obtained by the statistics of play store that is linked to our MatiMath application are the following:

The MatiMath application deploys in two iOS and Android operating systems, and within its execution 16 compatible devices, where they represented 95% of active users during the period of execution of the project as seen in the Fig. 5.

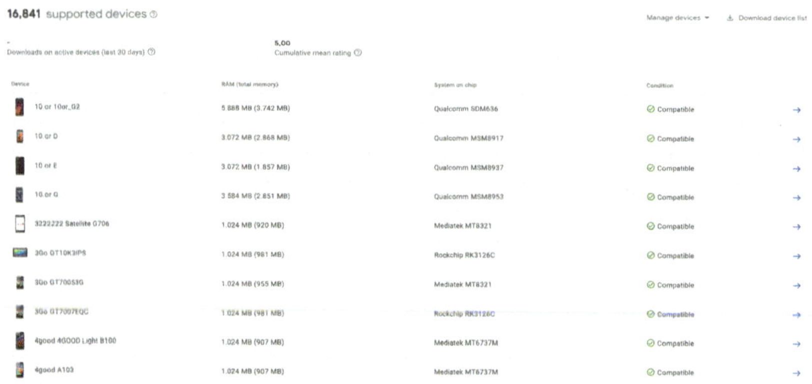

Device	RAM (total memory)	System on chip	Condition
10 or 10or_G2	5.888 MB (3.742 MB)	Qualcomm SDM636	Compatible
10.or D	3.072 MB (2.868 MB)	Qualcomm MSM8917	Compatible
10 or E	3.072 MB (1.857 MB)	Qualcomm MSM8937	Compatible
10.or G	3.584 MB (2.851 MB)	Qualcomm MSM8953	Compatible
3222222 Satellite G706	1.024 MB (920 MB)	Mediatek MT8321	Compatible
3Go GT10K3IPS	1.024 MB (981 MB)	Rockchip RK3126C	Compatible
3Go GT70053G	1.024 MB (955 MB)	Mediatek MT8321	Compatible
3Go GT7007EQC	1.024 MB (981 MB)	Rockchip RK3126C	Compatible
4good 4GOOD Light B100	1.024 MB (907 MB)	Mediatek MT6737M	Compatible
4good A103	1.024 MB (907 MB)	Mediatek MT6737M	Compatible

Fig. 5. Graphic of MatiMath compatibles devices. Ft. Play Store

We can observe in the Fig. 6 the report of errors ANRs that is the total number of errors that have stopped the application during the deployment period, it is seen that in the course of the project no RNAs error was detected so it follows that the application was developed in a stable way.

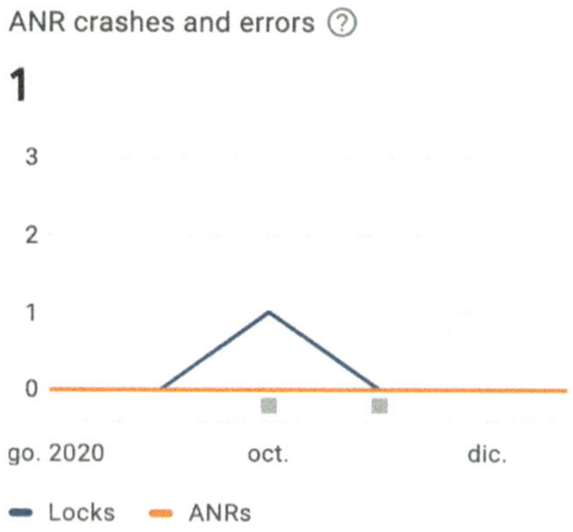

Fig. 6. Graph of the ANRs errors of the MatiMath application. Ft. Play Store

We also see the growth of the application according to the users in the category of education, where it shows the reception that MatiMath had in the app store, as the course was being implemented in the application generated a greater interest in users as seen in the Fig. 7.

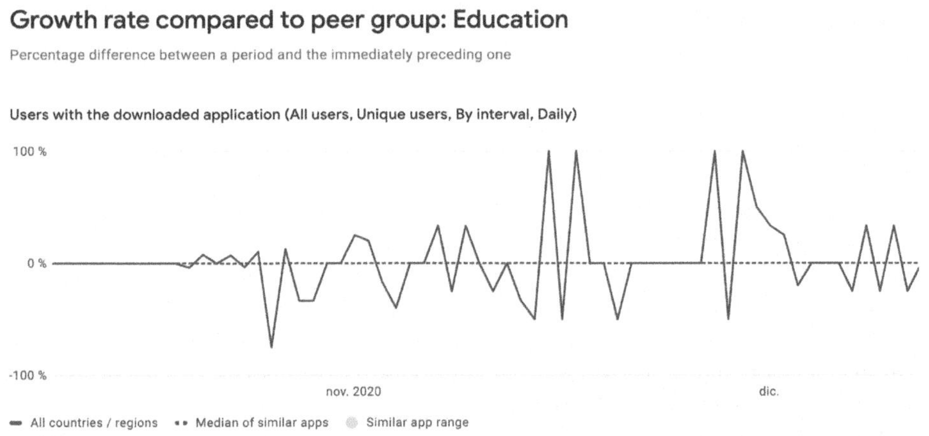

Fig. 7. Graphic of the user adaptation to the MatiMath application. Ft. Play Store

The size of the apk, when installed on each mobile device is considered acceptable (22.2 MB) by Play Store, since also the functionalities in the application are mostly remote not storing data in the device if not consulted the information to the server as seen in the Fig. 8.

Means

Archive	Download size
Original file	23.2 MB
Distribution APK	23.2 MB

Fig. 8. Size of apk to install on devices. Ft. Play Store

Students who needed help and used the app for their consultations and internships achieved an average score of 14.1 with a standard deviation of 2,682, and students who needed support but did not use the app scored 8,46 with a standard deviation of 2,375. The bilateral significance is 0.010, with $0.010 < 0.05$, it can be said that for a significance level of 5%; there is a significant difference between the learning outcomes of students who needed support and who used the app and students who needed support who did not use the Matimath app (Table 4).

Table 4. Student T test for independent samples

Status	Abbreviation	Average	Standard deviation	Bilateral significance
Used app	APP	14,1	2,682	0,010
No used app	NAPP	8,46	2,375	

5 Conclusions

MatiMath application meets technological requirements, pedagogical principles and UX elements raised in the M-Learning Model for mathematical learning, having 95% of recurring users, so it is positioned in one of the effective applications in the education category in the Play Store store reaching 4 times the 100% daily interval.

The development of the Model M-Learning and later the implementation of the MatiMath application allows in the development environment a better traceability of the requirements since based on the needs and objectives of the project allows to capture in a better way the orientation of the project having thus best results.

The MatiMath application positively influenced students who required mathematical reinforcement, obtaining an average of 14.1 in contrast to an 8,46 of the students who did not use the application and who also needed to reinforce basic mathematical knowledge as it was also a help tool in remote classes due to the circumstance presented. It is finally concluded that the application was a support tool in learning and teaching in end users thanks to its modeling taking into account the three pillars of the model technological requirements, pedagogical principles and UX elements.

References

1. Ozdamli, F.: Pedagogical framework of m-learning. Procedia S oc. Behav. Sci. **31**, 927–931 (2012). https://doi.org/10.1016/j.sbspro.2011.12.171
2. Educación Financiera Pisa 2018. https://es.calameo.com/read/0062866257073a6bcb307?page=1&view=slide. Accessed 28 Jan 2021
3. Powell, S.R., et al.: Data-based individualization in mathematics to support middleschool teachers and their students with mathematics learning difficulty. Stud. Educ. Eval. 2020. https://doi.org/10.1016/j.stueduc.2020.100897
4. Hillmayr, D., Ziernwald, L., Reinhold, F., Hofer, S.I., Reiss, K.M.: The potential of digital tools to enhance mathematics and science learning in secondary schools: a context-specific meta-analysis. Comput. Educ. **153**(September), 103897 (2020). https://doi.org/10.1016/j.compedu.2020.103897
5. Sun, G., Gan, J., Liu, S., Lang, F., Lu, Z. (eds.): eLEOT 2019. LNICSSITE, vol. 299. Springer, Cham (2019). https://doi.org/10.1007/978-3-030-35095-6
6. Bass, H., Evia, G.B., Seiler, R.: e-learning mathematics (2006)
7. Anaya Rivera, K.: Un modelo de enseñanza-aprendizaje virtual: análisis, diseño y aplicación en un sistema universitario Mexicano. Univ. Granada 172 (2004). http://decsai.ugr.es/Documentos/tesis_dpto/78.pdf/
8. Morales, E.: Gestión del conocimiento en sistemas e-learning, basado en objetos de aprendizaje, p. 434 (2007). https://gredos.usal.es/jspui/handle/10366/21864
9. Najar, P., Ledesma, E., Rocabado, S., Herrera, S., Palavecino, R.: Eficiencia de aplicaciones móviles según su arquitectura (2014)
10. Benavides-Varela, S., Zandonella Callegher, C., Fagiolini, B., Leo, I., Altoè, G., Lucangeli, D.: Effectiveness of digital-based interventions for children with mathematical learning difficulties: a meta-analysis. Comput. Educ. **157**(July), 103953 (2020). https://doi.org/10.1016/j.compedu.2020.103953
11. Heidkamp-kergel, D.K.B.: und digitales Lernen

12. Rivera, C.P.: Modelo de Sistema e-learning adaptativo para el nivel superior, utilizando aprendizaje colaborativo basado en proyectos, considerando estilos de aprendizaje y estilos de pensamiento (2018)
13. José, P., Mũ, M., Merino, M.: Tería de Modelado del E-learning y Aplicación a un Sistema de pistas Adaptativo en Tutoría Inteligente Utilizando Web Sistemática (2009). https://e-archivo.uc3m.es/bitstream/handle/10016/5568/Tesis_Pedro_Munoz_Merino.pdf
14. Zwart, D.P., Noroozi, O., Van Luit, J.E.H., Goei, S.L., Nieuwenhuis, A.: Effects of digital learning materials on nursing students' mathematics learning, self-efficacy, and task value in vocational education. Nurse Educ. Pract. **44**(February), 102755 (2020). https://doi.org/10.1016/j.nepr.2020.102755
15. Xhafa, F., Caballé, S., Abraham, A., Daradoumis, T., Perez, A.A.J.: Studies in computational intelligence: preface. Stud. Comput. Intell. **273**(May), 2010 (2014). https://doi.org/10.1007/978-3-642-11224-9
16. Schinzel, B., Taeger, J., Gorny, P., Dreier, T., Hrsg, B.H.: E-Learning im Hochschulverbund (2004)
17. Sarrab, M., Elgamel, L.: Mobile learning (M-Learning) and. Int. J. Distrib. Parallel Syst. **3**(4), 31–39 (2013)
18. Enriquez, J.G., Casas, S.I.: Usabilidad en aplicaciones móviles. Inf. Científicos Técnicos - UNPA **5**(2), 25–47 (2014). https://doi.org/10.22305/ict-unpa.v5i2.71
19. Emergency Decree No. 036-2020: Decreto de urgencia. El Peru. (2020). https://cdn.www.gob.pe/uploads/document/file/581842/DU036_2020.pdf
20. Feria-Cuevas,Y., Rodríguez-Morán, M., Torres-Morán, M.I., Pimienta-Barrios, E.: Panorama De Conexión Durante Las Clases Virtuales En Una Muestra De Estudiantes Universitarios. e-CUCBA 14, 25–33 (2020). https://doi.org/10.32870/e-cucba.v0i14.160
21. Turchi, L.B., Bondar, N.A., Aguilar, L.L.: What really changed? Environments, instruction, and 21st century tools in emergency online English language arts teaching in United States schools during the first pandemic response. Front. Educ. **5**(November), 1–13 (2020). https://doi.org/10.3389/feduc.2020.583963
22. Ruiz-Guerrero, A.: Our self-access experience in times of COVID. SiSal J. **11**(3), 250–262 (2020). https://doi.org/10.37237/110311

Methods and Models of Resource Allocation in Load Balancing Clusters

Valery Petrovich Mochalov, Gennadiy Ivanovich Linets, and Ilya Sergeevich Palkanov(✉)

North-Caucasus Federal University, Stavropol, Russia

Abstract. Modern data centers are complex and effective solutions for managing enterprises and corporations, organizing data processing and storage systems, and distributing software applications among available resources. This paper considers load balancing clusters of data centers that contain a definite set of application servers, file servers, data storage systems, and an input-output system, all interconnected by a switching system and communication channels. The goal of research is to increase the efficiency of virtualized data centers using a new methodological approach, as well as rational methods and models, to distribute their load among the available software and hardware means. Interrelated mathematical methods and models, necessary to develop specialized software for the rational planning and allocation of the data center's physical resources on a given time interval, are proposed. These methods and models are based on the system of correct mappings of the set of request parameters into the known characteristics of the data center's physical resources. As an optimization criterion, the maximum system performance for a certain time interval is used.

Keywords: Data centers · Computing clusters · Load balancing service · Efficient resource allocation · System of mappings

1 Introduction

The information structure of a data center includes the following components: a server complex, represented by load balancing clusters, which uses multi-platform solutions for equalizing the load distribution between servers; memory facilities, which include the set of storage devices in the form of disk arrays, a data storage system, and copying and restoring facilities; communication and switching facilities; telecommunication facilities, which provide interaction between the operators and users of the data center. A formal model of a data center cluster is shown in Fig. 1. In the case of a load balancing cluster, a set of resource requests comes from different users to a load balancer, which distributes them between the cluster servers according to definite rules. An incoming request can be directed to the first available resource of the cluster, or a resource can be randomly selected from the set of available ones sufficient to implement this request. Such solutions can be based on well-known heuristic greedy algorithms. For example, load balancing methods embedded in the cyclic logic of the Round Robin algorithm

R. Silhavy (Ed.): CSOC 2021, LNNS 230, pp. 552–563, 2021.
https://doi.org/10.1007/978-3-030-77442-4_47

generate an appropriate IP address for any request, thereby ensuring a circular loop for all requests and the same number of requests directed to each server.

Although the running time of this algorithm is acceptable, it is not quite suitable for practical use. Round Robin ignores the available volumes of server resources, their state, performance, current load, and the number of supported connections. Advanced modifications of this algorithm, designed for improving the characteristics of clusters, such as Least Connections, Weighted Least Connections, Locality-Based Least Connections Scheduling, Destination Hash Scheduling, Source Hash Scheduling, and Sticky Sessions, find servers with the lowest load and redirects a new request to them. However, like Round Robin, they neither fully implement an optimal technology of load balancing nor ensure efficient clustering and optimal server loading with a minimum application execution time. The load distribution systems of widespread modern hypervisors (VMware Infrastructure, VMware ESXi Server, Microsoft Hyper-V) have direct application distribution methods as well as experimental settings selected empirically, which are often based on statistics and resource utilization forecasting [1–3, 7]. For example, a resource scheduler based on the OpenStack platform algorithms neither properly matches the dynamically changing and unpredictable load rate nor fully reflects the real-time dynamics of all ongoing processes. These algorithms do not provide a complete solution to balance the load of data centers and to allocate physical server resources among virtual machines and network applications in real time [4–6, 9]. Therefore, the quality of service is not properly guaranteed. For a higher efficiency of the hardware-software platform of data centers, it is necessary to develop rational methods and models implementing, first, resource allocation service for load balancing clusters and, second, optimization algorithms for placing requests and software applications on physical servers in real time.

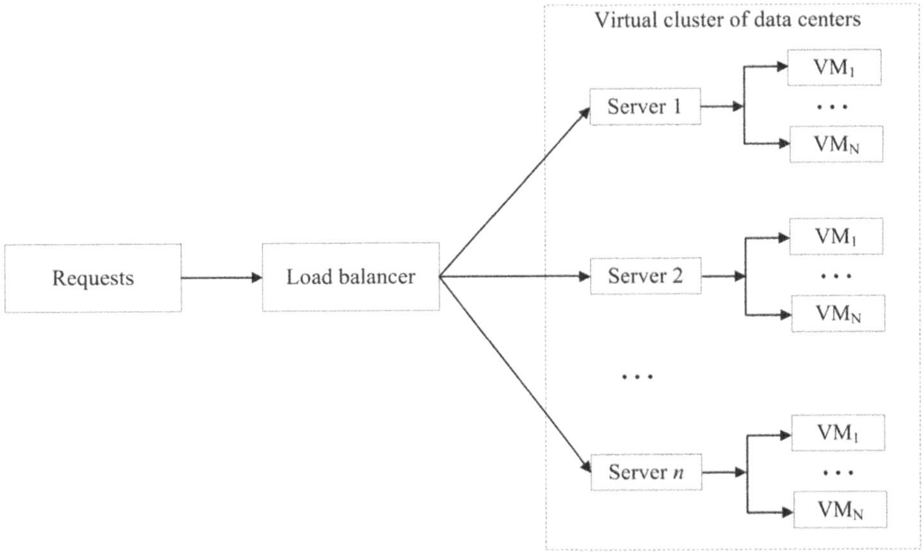

Fig. 1. Formal model of data center cluster.

The performance of a virtual data center, the RAM of computing servers and the throughputs of a switching system and data transmission channels are all limited. To eliminate the problem of overloading the resources of a virtual data center, as well as to match the type of requests with the technical parameters of the requested resources, a series of constraints have to be imposed on the relationship between the parameters of a request system and the corresponding resource characteristics. As an optimality criterion, the maximum system performance on a fixed time interval can be used.

2 Resource Allocation in Load Balancing Clusters: Problem Statement

Each request for implementing application functions can be executed on one of the cluster servers, subject to the correct relationship between the parameters of the request implemented by software applications of different complexity levels and the corresponding characteristics of the physical resources of the data center. The structural diagram of the management system for the hardware-software platform of a data center is shown in Fig. 2.

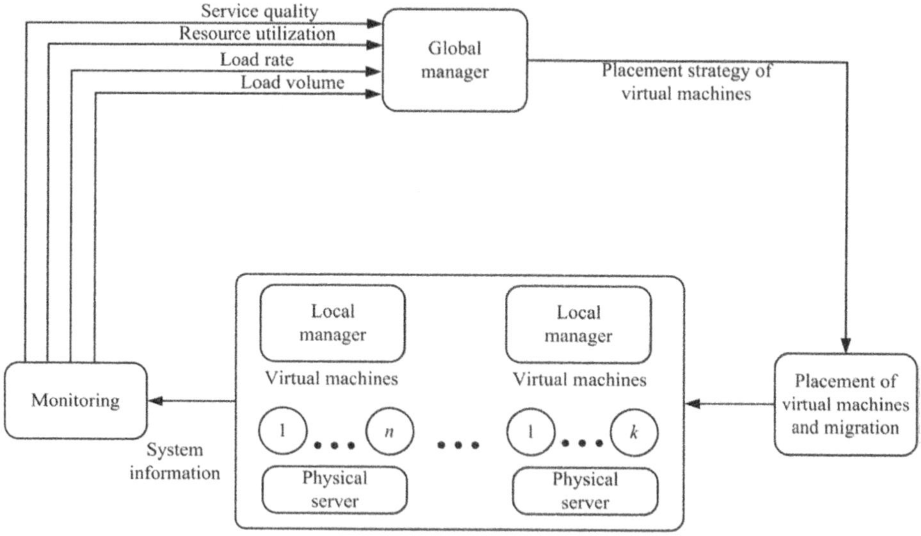

Fig. 2. Management system for hardware-software platform of data center.

At the request of the data center users, this system creates a VM with definite resource parameters. The local manager controls the load of physical servers and the placement of requests for the resources of VMs. The data center monitoring system transfers the information received to the global manager; the latter migrates VMs and manages the redistribution of physical servers. The following tasks are sequentially implemented:

– receiving requests for services from users;
– distributing the incoming requests among cluster servers;
– selecting a cluster server capable of implementing user requests;
– redirecting requests to the cluster server selected;
– placing the software applications under implementation on the cluster servers;
– forming a set of virtual machines for implementing software applications;
– receiving the results of solving all these tasks;
– sending the results to users.

The composition and structure of the server complex of a data center [5, 8] can be represented by a quadruple:

$$H = (P \cup M \cup K, L), \tag{1}$$

where P denotes the set of cluster servers;
M is the set of data storage devices;
K, L is the set of telecommunications equipment resources.
The requests for implementing applications can be formalized by:

$$G = (W \cup S, E), \tag{2}$$

where W denotes the set of request attributes;
S and E are the characteristics of cluster resources requested by users.
The relationship between the characteristics of available physical resources and the parameters of requests can be described by a mapping:

$$G \to H = \{W \to P, S \to M, E \to \{K, L\}\}. \tag{3}$$

This mapping is correct under the following constraints on the requests and resources.

1. The model of constraints on the distribution of requests among cluster servers.

– Each software application i, $i = \overline{1, n}$, is executed on a single jth server, $j = \overline{1, m}$, i.e.,

$$\sum_{j=1}^{m} x_{ij} = 1, \quad x_{ij} \in \{0, 1\} \tag{4}$$

– The execution time of the software applications distributed to the j th server must not exceed a given threshold $t_{\max}$.

$$\sum_{i=1}^{m} x_{ij}\tau_{ij} \le t_{\max}, \quad \tau_{ij} = Q_i / S_j, \tag{5}$$

where Q_i denotes the complexity of the i th application;
S_j is the performance of the j th server.
The request distribution procedure presented below guarantees quite acceptable results in the case of short-term load planning for cluster servers.

- Rearrange all software applications in the ascending order of their execution times.
- Find a number p of the software application for which

$$t_1 = \sum_{i=1}^{p} \tau_i, \tag{6}$$

where τ_i is the execution time of the i th software application with a minimum deviation from $t_{\max}$,

$$t_1 - t_{\max} \rightarrow \min . \tag{7}$$

- Pass to the $(p + 1)$ th software application and repeat the same operations for the remaining software applications, obtaining $t_2,\ t_3$, and so on.
- Distribute the groups of requests to the corresponding servers of the cluster.
- The set of software applications $F = \{f_1, f_2, \ldots, f_n\}$ is distributed among $k \in K$ VMs taking into account the limited resources of data centers. A rational number of VMs involved and the software applications distributed among them are determined under the following constraints:
- An application can be executed on a single VM only, i.e.,

$$\sum_{n} a_{jn} \leq 1, \tag{8}$$

where $a_{jn} = \begin{cases} 1 \text{ if application is executed on } n \text{ VMs,} \\ 0 \text{ otherwise,} \end{cases}$

$j = \overline{1, n},\quad n = \overline{1, k}.$

- The applications distributed to VMs request the total volumes of resources not exceeding the available ones, i.e.,

$$\sum_{n} a_{jn} C_j \leq C_n, \quad \sum_{n} a_{jn} m_j \leq M_n, \tag{9}$$

where C_n and M_n are the performance and memory capacity of the j th VM, respectively.

- The applications are indivisible for all time intervals t_n of their execution, i.e.,

$$\sum_{t=0}^{t_n} \sum_{n} a_{jn} = t_j \sum_{n} a_{jn}. \tag{10}$$

This problem belongs to the class of NP-complete ones. It can be solved using the heuristic algorithm presented below; also, see [10, 12].

- Rearrange all software applications $j \in r$ in the descending order of their requests for the resources of VMs.
- Rank all VMs $k \in K$ by their performance, top to bottom.
- For a next application $j(k)$ on each time interval t_n, select a VM being able to execute this application.
- If such a VM is not selected, the application will not be executed. Otherwise the required volumes of resources are reserved on the VM.

This algorithm has a polynomial complexity and yields an almost optimal distribution; see [11, 14].

3. The model of constraints on the distribution of virtual machines among cluster servers.

A data center contains a definite set of application servers and a definite set of file servers, $S_i, i = \left(\overline{1, L}\right)$, and $H_j, j = \left(\overline{1, M}\right)$, respectively. It is required to distribute among them the set of all VM_k, $k = \left(\overline{1, N}\right)$, under the following constraints:

- Each VM can be placed on a single server only, i.e.,

$$\sum_{i=1}^{L}\sum_{j=1}^{M}\sum_{k=1}^{N} x_{ijk} = 1, \tag{11}$$

where $x_{ijk} = \begin{cases} 1 \text{ if } VM_k \text{ is placed on } S_i \text{ or } H_j, \\ 0 \text{ otherwise.} \end{cases}$

- The memory resources (RAM) requested by VMs from physical servers satisfy the inequality.

$$\sum_{i=1}^{L}\sum_{j=1}^{M}\sum_{k=1}^{N} RAM_k x_{ijk} < RAM_{ij}. \tag{12}$$

The resources utilized by all VMs do not exceed the resources of servers.

- The server performance constraint, which is written as

$$\sum_{i=1}^{L}\sum_{j=1}^{M}\sum_{k=1}^{N} CPU_k x_{ijk} < CPU_j. \tag{13}$$

- The disk storage capacity constraint, which is written as

$$\sum_{i=1}^{L}\sum_{j=1}^{M}\sum_{k=1}^{N} S_k x_{ijk} < S_{ij}. \tag{14}$$

– The performance constraint on the data input-output system, which is written as

$$\sum_{i=1}^{L}\sum_{j=1}^{M}\sum_{k=1}^{N} IOPS_k x_{ijk} < IOPS_{ij}. \tag{15}$$

Physical servers and storage systems are interconnected by a switching system. Therefore, when allocating the resources of physical servers, it is necessary to consider the main parameters of this system and virtual communication channels between the VM and the data storage system. The total throughput of all communication channels and switching systems must satisfy the conditions

$$\sum_{i=1}^{n}\sum_{l=1}^{L} r_i l < \Pi, \quad \sum_{i=1}^{n}\sum_{l=1}^{L} r_i l < K \tag{16}$$

where $r_i l$ is the required throughput of the ith VM under the utilization of l virtual channels;

Π and K are the total throughputs of physical channels and the switching system.

Choose the following criteria of optimality:

– the maximum load of data center equipment,

$$K_1 = \sum_{i=1}^{L}\sum_{j=1}^{M}\sum_{k=1}^{N} CPU_k x_{ijk} \cdot RAM_k x_{ijk} \tag{17}$$

– the maximum performance of the input-output system,

$$K_2 = \sum_{i=1}^{L}\sum_{j=1}^{M}\sum_{k=1}^{N} IOPS_k x_{ijk} \tag{18}$$

– the maximum performance of the switching system and transfer system,

$$K_3 = \sum_{i=1}^{n}\sum_{l=1}^{L} r_i l \tag{19}$$

The optimal allocation problem of data center resources is NP-hard; therefore, it should be solved using non-polynomial algorithms for offline or online placement [13, 16].

The algorithm proposed in this paper performs the sequential processing of all requests for VMs placement. They are ranked in the descending order of the requested resources. The external data center scheduler allocates the maximum volume of all resources (CPU, RAM, disk, IOPS, etc.) to each VM. A physical resource with a minimum residual sum of its parameters is determined, and a VM is placed on it using the following scheme:

1. All VMs [4] are ranked by the value of the characteristic $V_i = \sum_{j=1}^{K} C_j V_j^i$, where V_j^i is the request of VM_i for resource j, $j = \overline{1, K}$, and C_j denotes the significance of this resource.
2. In accordance with the initial scheme, a VM with the highest resource requirements is selected and then placed on a physical element with a minimum resource. Here a most reasonable method is to use an iterative algorithm based on dynamic programming.

For example, the iterative greedy algorithm runs as follows [3, 15]. As the input processes a_i of the set $S = \{a_1, a_2, \ldots, a_n\}$ the final times f_i of their implementation are used. The sequentially selected processes are combined by the Greedy Activity Selector procedure (s, f) into a set A in which f_i is the maximum final time of all processes: $f_i = \max\{f_k : a_k \in A\}$. Greedy Activity Selector (s, f) can be described by the code:

- $n \leftarrow length[s]$;
- $A \leftarrow \{a_1\}$;
- $i \leftarrow 1$;
- *for* $m \leftarrow 2$ *to* n;
- *do if* $S_m \geq f_i$;
- *then* $A \leftarrow A \cup \{a_m\}$;
- $i \leftarrow m$;
- *return* A.

3. The external data center scheduler estimates the performance of the VM placed to the server.
4. If the VM's performance is below the specified level, the external scheduler moves it to a physical element with a higher level of resources and goes back to Step 3.
5. If the VM's performance is higher than the specified level, the external scheduler moves it to a physical element with a lower level of resources and goes back to Step 3.
6. If it is impossible to distribute VM, the transition to the limited enumeration procedure is performed. The enumeration depth, which determines the maximum number of servers for which the distribution is assigned, must guarantee the required balance between the quality of service and the VM distribution time (find an acceptable solution in an allowed time). The quality of service can be estimated by the probability of blocking for service requests; otherwise, by the probability of no admissible servers per unit time. This can be done using the Erlang formula

$$P_\delta = \frac{a^n/n!}{\sum_{k=0}^{n} a^k/k!}, \tag{20}$$

where P_δ denotes the probability of blocking;

n is the number of servers;

a is the rate of requests.

This probability can be calculated iteratively [17]:

$$P_{\delta} = B(n,\ a) = \frac{B(n-1,\ a)}{B(n-a) + n/a}, \tag{21}$$

where $n = 1, 2, \ldots,$ and $B(0,\ a) = 1$.

In the case of large values of n and a, the computing time of P_{δ} can be reduced by choosing a required deviation $P_{\delta}(i) - P_{\delta}(i-1) < \delta$ and terminating the enumeration procedure as soon as the value δ is achieved [10]. Here δ actually specifies the accuracy of calculations.

7. If the resource is not found, then the next VM is placed, the resource requested by the previous VM is decreased, and it is included in the common ordered queue.
8. If the resource is found, then the VM is eliminated from the queue.
9. The time-optimal placement is selected under the existing constraints.

3 Numerical Experiment

The formal model of a load balancing cluster (see Fig. 3) was developed in AnyLogic 7 [18, 19] for a cluster based on the Huawei 2488 platform, Intel (R) Xeon (R) Gold 6154 processor. The model itself and the request distribution system were implemented using elements of Enterprise Library.

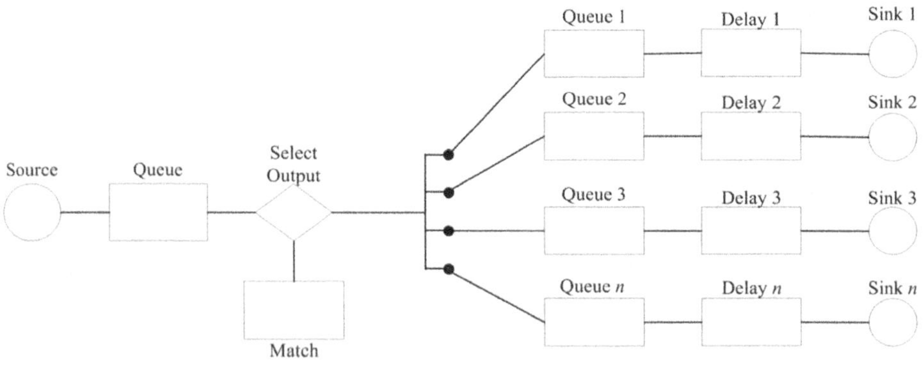

Fig. 3. Model of server cluster.

The following blocks are shown in Fig. 3:

- Source, the generation block;
- Queue, the queue formation block;
- Select;
- Delay, the delay formation block;

- Match, the matching block;
- Sink, the transaction removal block.

The volume of requested VM resources was uniformly distributed in the range of 20–80% of the server resources. The ratio of CPU time and RAM was uniformly distributed. The switching system throughput was limited to 1 Gb/s. The duration of the VM migration was defined by the Gaussian distribution law. The system characteristics were studied in three stages, via the sequential use of the proposed methods for distributing (1) the requests, (2) the requests and VMs, and (3) the requests, VMs and software applications. The new resource allocation algorithms proposed in this paper were compared with the classical Round Robin algorithm by the execution times of requests; see Table 1.

Table 1. Execution time of requests by data center cluster.

Number of software applications	Number of servers	T, ms			
		T_0	T_1	T_2	T_3
50	5	4576	4112	3742	2976
100	7	5844	5231	4806	3217
150	10	6733	5820	4173	3580
200	12	6915	6113	5633	4512
300	15	7062	6549	5321	4893

The notations are the following:

- T_0 the execution time of requests in the Round Robin algorithm;
- T_1 as the execution time of requests in the new resource allocation algorithm used for the requests;
- T_2 as the execution time of requests in the new resource allocation algorithm used for the requests and VMs;
- T_3 as the execution time of requests in the new resource allocation algorithm used for the requests, VMs and applications.

The new load balancing methods proposed in this paper have an acceptable execution time of requests when the number of cluster servers does not exceed 15. With an increase in the number of servers, the time grows exponentially. According to the simulation results presented in Table 1, the highest performance is provided by the methods used for the rational distribution of the requests, VMs and software applications. In comparison with the classical Round Robin algorithm, the execution time of requests in this case decreases on average by 15%.

4 Conclusion

This paper has considered the problem of constructing an effective mechanism for implementing user requests for data center services via rational resource allocation methods. The methods and models of resource allocation in load balancing clusters presented above involve a system of correct mappings of the set of request parameters into the known characteristics of data center physical resources and an iterative greedy algorithm. The computation time has been reduced by imposing constraints on the admissible depth of search. Methods for the efficient distribution of requests among cluster servers, a model for distributing software applications among the set of virtual machines in a cluster, a method for distributing virtual machines among cluster servers, and a corresponding model have been considered. Simulation models for the management system of data center clusters virtualization have been developed (State registration certificates of computer programs No. 2019664647 dated November 11, 2019, and No. 2020617795 dated July 17, 2020). These models have been studied experimentally. The new methods suggested in this paper can be used to distribute software applications among the VMs of a corporate data center, to select the composition of virtual machines, and to place virtual machines on the physical servers of a data center, all done in a rational way.

Acknowledgments. This work was supported by the Russian Foundation for Basic Research, project no. 19-07-00856/20.

References

1. Mochalov, V.P., Linets, G.I., Bratchenko, N.Y., Govorova, S.V.: An analytical model of a corporate software-controlled network switch. Scalable Comput. **21**(2), 337–346 (2020)
2. Boev, V.: Komp'yuternoe modelirovanie: Posobie dlya prakticheskikh zanyatii, kursovogo i diplomnogo proektirovaniya v AnyLogic7 [Computer Modeling: A Manual for Practical Classes, Coursework and Diploma Projects in AnyLogic7], 432p. VAS Publications, St. Petersburg (2014). (In Russian)
3. Kim, T.-H., Kim, S., Park, H.-U., Kim, M.-S.: Analysis of security session reusing in distribution server system. In: Gavrilova, M.L., Gervasi, O., Kumar, V., Tan, C.J.K., Taniar, D., Laganá, A., Mun, Y., Choo, H. (eds.) ICCSA 2006. LNCS, vol. 3984, pp. 700–707. Springer, Heidelberg (2006). https://doi.org/10.1007/11751649_77
4. Khritankov, A.: Modeli i algoritmy raspredeleniya nagruzki. Algoritmy na osnove setei SMO [Models and Algorithms for Load Balancing. Algorithms Based on Networks of Queuing Systems]. Informatsionnye tekhnologii i vychislitel'nye seti – Information Technologies and Computer Networks, vol. 3 (2009). (In Russian)
5. Ivanisenko, I., Kirichenko, L., Radivilova, T.: Metody balansirovki s uchetom multifraktal'nyh svoistv nagruzki [Balancing Methods Considering Multifractal Properties of Load]. Int. J. Inf. Content Process. **2**(4), 345–368 (2015). (In Russian)
6. Panchenko T.V.: Genetic Algorithms [Text]: A Manual/Yu.Yu. Tarasevich. Astrakhan University, Astrakhan, 87 p. (2007). (In Russian)
7. Holland, J.H.: Adaptation in Natural and Artificial Systems: An Introductory Analysis with Applications to Biology, Control, and Artificial Intelligence, 211 p. J.H. Holland, The MIT Press, Cambridge (1992)

8. Michalewicz, Z.: Genetic algorithms + Data Structures = Evolution Programs, 387 p. Springer, New York (1996). https://doi.org/10.1007/978-3-662-03315-9
9. Tsoi Yu.R., Spitsyn, V.G.: Genetic Algorithm. In: Knowledge Representation in Information Systems: A Manual, 146 p. Tomsk Polytechnic University, Tomsk (2006). (In Russian)
10. Mitchell, M.: An Introduction to Genetic Algorithms, 158 p. MIT Press, Cambridge (1999)
11. Periaux, J., Sefrioui, M.: Evolutionary computational methods for complex design in aerodynamics. In: AIAA-98-0222, 15 p., Reno (1998)
12. Periaux, J., Chen, H.Q., Mantel, B., Sefrioui, M., Sui, H.T.: Combining game theory and genetic algorithms with application to DDM-nozzle optimization problems. Finite Elem. Anal. Des. **37**(5), 417–429 (2001)
13. Zhirkov, A.: Supercomputers: Development, trends, usage. Eurotech HPC solutions review [Superkomp'yutery: razvitie, tendentsii, primenenie. Obzor HPC-reshenii Eurotech]. Present Automation Solutions, no. 2, p. 201 (2014)
14. Beloglazov, A., Buyya, R.: Optimal online deterministic algorithms and adaptive heuristics for energy and performance efficient dynamic consolidation of virtual machines in cloud data centers. Concurr. Comput. Pract. Exp. (CCPE) **24**(13), 1397–1420 (2012)
15. Mochalov, V.P., Bratchenko, N.Y., Yakovlev, S.V.: Analytical model of object request broker based on CORBA standard. J. Phys. Conf. Ser. **1015**(2), 022012 (2018). https://doi.org/10.1088/1742-6596/1015/2/022012
16. Mochalov, V.P., Bratchenko, N.Y., Yakovlev, S.V.: Analytical model of integration system for program components of distributed object applications. In: International Russian Automation Conference (RusAutoCon 2018), no. 8501806 (2018). https://doi.org/10.1109/RUSAUTOCON.2018.8501806
17. Mochalov, V., Bratchenko, N., Linets, G., Yakovlev, S.: Distributed management systems for infocommunication networks: a model based on TM Forum Frameworx. Computers **8**(2), 45 (2019). https://doi.org/10.3390/computers8020045
18. Mochalov, V.P., Bratchenko, N.Y., Yakovlev, S.V.: Process-oriented management system for infocommunication networks and services based on TM forum Frameworx. In: Proceedings of 2019 International Russian Automation Conference (RusAutoCon 2019), no. 8867619 (2019).https://doi.org/10.1109/RUSAUTOCON.2019.8867619
19. Vdovin, P.M., Kostenko, V.A.: Algorithm for resource allocation in data centers with independent schedulers for different types of resources. J. Comput. Syst. Sci. Int. **53**(6.1), 854–867 (2014)

Requirements Engineering in Agile Software Startups - Insights from Multiple Case Studies

Varun Gupta, Thomas Hanne(✉), and Rainer Telesko

University of Applied Sciences and Arts Northwestern Switzerland, Olten, Switzerland
thomas.hanne@fhnw.ch

Abstract. This exploratory case study was conducted with five IT startups in order to investigate how requirement engineering-related activities are performed and what is the state of maturity with the practices & tools used. Another objective was to study that how the startups managed their practices during Corona Virus (COVID19) pandemic time. The results indicate that flexibility and access to the online tools were the main strengths of the startups to cope up with the pandemic situation while fluctuating market demands affected them marginally. The startups do vary in domain, team size, practices and selection of the tools, with matured startups having more structured (but flexible) processes compared to younger startups. The young startups have the opportunity to learn from the practices of the matured startups, to adopt the learning in their working context. The previous software development experience of the startups and its founders does affect the maturity of the practices and selection of the tools. The flexibility and agility as evident in the working context of the startups helped them to turn pandemic situation into their business opportunities.

Keywords: Software startups · Requirement engineering practices · COVID-19

1 Introduction

Software Engineering is a field of computer science investigated for a long time. No matter which specific procedure model in software engineering is followed, requirements engineering (i.e. elicitation, analysis, and documentation of requirements) is usually considered as one of the first and possibly most important steps. The question whether there are particular requirement engineering practices for startups has so far not much been considered in the literature. As such companies differ in various aspects from longer established ones, e.g. regarding the maturity of development processes in general, one might expect that requirement engineering is also done differently. On the one hand, such processes may be less structured, less mature but more agile and flexible than in older and larger companies. On the other hand, some startup companies, especially in the tech sector, may have a strong IT background of their founders and might therefore show the usage of up-to-date approaches in requirement engineering. However, because of financial limitations and difficulties in setting up a team of highly qualified software developers, startups may have problems to exploit their potential of technical expertise.

R. Silhavy (Ed.): CSOC 2021, LNNS 230, pp. 564–577, 2021.
https://doi.org/10.1007/978-3-030-77442-4_48

One way to cope with this may be a stronger involvement of freelancers, outsourcing or offshoring of software development or other forms of flexible and affordable integration of human resources in software development.

The purpose of our study is to consider a few cases of startup companies with a strong part of software development in their business models. These companies are quite different regarding the product or service being offered, the approached markets or customers, the age and size of the company, and other internal aspects. Based on interviews and earlier contacts with the companies we compile information specific to requirement engineering and compare it. Reasons for similarities and differences in requirement engineering are discussed.

The paper is structured as follows: Sect. 2 discusses the background of our study related to agile methods. In Sect. 3, related work is discussed. Section 4 discusses the case study methodology used for our research. In Sect. 5, the criteria for assessing requirement engineering in the considered companies are discussed. In Sect. 6, the "agile" startups considered for our study are characterized. In Sect. 7, the main findings of our study are presented. Section 8 includes a discussion of flexibility aspects in startups and Sect. 9 concludes the paper.

2 Background on Agile Procedure Models and Agile Requirements Engineering

2.1 Agile Procedure Models

Over the last years, it turned out that the classical waterfall model in software engineering has severe disadvantages, like not incorporating changing requirements, inadequately considering the "voice of the customer" and not providing any intermediate results to the customer. Agile procedure models present a possible solution for minimizing these disadvantages. In the meantime, these models have become mainstream, even in larger organizations. In order to understand the very nature of agile models, the best source is the Agile Manifesto (http://agilemanifesto.org/):

- Individuals and interactions over processes and tools: The knowledge is in the head of the developers, not in the documented processes.
- Working software over comprehensive documentation: Agile is about showing early results to the customer, not just design documents.
- Customer collaboration over contract negotiation: Regular exchange with the customer is vital for the project.
- Responding to change over following a plan: Changes in requirements will happen and the procedure model has to take that into account.

Classical development models execute the development phases analysis, specification and design, implementation and testing in a sequential manner. Agile models propose a different approach with two important cornerstones: Not all requirements are considered (i.e. analyzed, implemented, ...) at the same time, but in sets based on descending business importance and the development of sets of requirements is conducted in iterations (called Sprints in Scrum), where the results are immediately shown to the customer.

It is clear that this procedure also requires re-thinking requirements engineering. In the literature respective approaches are called "agile requirements engineering" (ARE).

2.2 Agile Requirements Engineering

The techniques used in classical requirements engineering (Pohl and Rupp 2015) are the same that can be used in ARE, however the way to perform requirements engineering as well as the possible pitfalls are totally different. A good overview about the ARE discipline with approaches, project examples and challenges can be found in (Telesko 2018). There is no standardized definition of ARE, a good definition reflecting the main idea of ARE is provided by Heikkilä et al. (2015, p. 205):

> "In agile RE, the requirements are elicited, analyzed and specified in an ongoing and close collaboration with a customer or customer representative in order to achieve high reactivity to changes in the requirements and in the environment. Continuous requirements re-evaluation is vital for the success of the solution system, and the close collaboration with the customer or customer representative is the essential method of requirements and system validation."

On a continuum of different ARE approaches the following two extreme positions can be identified:

- Minimal documentation: For the selected requirements it is sufficient to document them for example with one user story per feature.
- Maximal documentation: For the selected requirements it is necessary to extensively document them with UML diagrams and a fully-fledged text template (i.e. a text template for requirements containing a dozen of different attributes).

The main challenge is now how to apply ARE adequately for a development project in order to end up with a superior cost-benefit ratio (Oesterreich 2010). On the one side ARE demands a lightweight documentation which can be changed rapidly and is only supporting but not replacing the developers (see principle 2 of the Agile Manifesto - "Working software over comprehensive documentation"). On the other side, also in the agile world we still have issues like fluctuation of developers, handling large projects, teams with inexperienced members, dealing with compliance issues, managing DevOps etc. Some guiding questions which assist in finding the right way of documentation can be found in (Oesterreich 2010).

- How familiar are the stakeholders (i.e. product owners, domain experts, developers, testers, etc.) with the domain?
- How homogeneously are the business processes and use cases executed by different users?
- How many different variants and exceptions have to be considered?

- How new are the procedures in the software compared to the old solution?
- How many different persons are providing or influencing requirements?
- How big is the risk that using very little requirements engineering may lead to overlooked dependencies or details that may later cause extra costs or delays?
- How serious are the consequences if important functional variants or details are forgotten?

In order to cover some new trends, one may add here further questions like:

- Knowledge management/fluctuation: How strong is the (expected) fluctuation within the development team?
- Compliance: Do compliance issues (like the Sarbanes-Oxley Act, Basel II etc.) have to be handled which makes a more extensive documentation necessary?
- DevOps: Is a development approach planned which envisages additional work for the developers (e.g. DevOps)?

ARE is also covered by business analysis frameworks like IIBA (International Institute of Business Analysis International) (https://www.iiba.org/, (and IREB® (International Requirements Engineering Board, https://www.ireb.org/en). The IIBA defines six knowledge management areas for business analysis-related tasks in the BABOK Guide. The Agile Extension to the BABOK® Guide defines agile techniques (e.g. related to Scrum, Kanban like backlog management, prototyping, personas etc.) which can be used when executing tasks in the knowledge areas. The IREB follows a similar approach. The core requirement engineering processes (eliciting, documenting, checking and managing requirements) are defined in the CPRE-FL (Certified Professional for Requirements Engineering - Foundation Level). The RE@Agile Primer specifies agile techniques to be used within the IREB framework.

3 Related Work on Requirement Engineering in Startups

Gupta et al. (2020a) conducted a mapping study to identify the state of art of requirement engineering in the startup context. The overall aim is to improve the requirement engineering practices helping startups to improve their value propositions to attain higher success levels. To accomplish the objectives, the systematic mapping protocol was executed by executing the search string against four bibliographic databases, i.e. IEEE Xplore, ACM digital library, SpringerLink and ScienceDirect for the time period 2015 to 2020 (both end inclusive). The search string was formulated as (Startup*) AND ((“Requirement Engineering”) OR (“Requirement Management”) OR (“Requirements Engineering”) OR (“Requirements Management”)).

A total of 461 papers were identified from the search, which after being subjected to inclusion and exclusion criteria were reduced to 16. Forward snowballing (through Google citations) and analysis (as per inclusion and exclusion criteria) was applied to these 16 studies which resulted in another 4 papers, leading to final 20 studies being included for the research study.

The results indicated that researchers focused on the generic requirement engineering process (40%) (rather than on individual activities) followed by product validation (20%), requirement elicitation (17.5%) and requirement validation (12.5%). Requirement prioritization and documentation are the least focused activities (5% each). Major studies are empirical research studies (85%) followed by validation approaches (solutions with laboratory validations) (10%) and experience (5%). Solutions in general, validated solutions and solutions validated in real settings are very limited. Major focus of studies is to attain a good product/market fit (62.5%) followed by process improvement (27.5%) and least focus on studying ongoing processes (10%). The studies analyzed in the systematic mapping study do not report any freelancer involvement in requirement engineering activities of the software startups. Further the research focus in this area in form of validated solutions is not enough to support startups to overcome the failure related issues related to requirement engineering processes.

The main findings of the literature survey were that Requirement Engineering in startups is still in its infancy stage. To facilitate knowledge transfer amongst the startups and between literature and startups, a lot of research needs to be conducted to enhance the body of knowledge. This study is conducted to explore the requirement engineering related aspects in the startup contexts through case studies with the five startups. The results will enhance the body of knowledge that will support other startups to improve their practices and enhance their success rates.

4 Case Study Design

The exploratory case study is conducted with five startups located in Switzerland, Germany and India. Three startups i.e. AbiliCor, Kiwih and Swip gave their consent for remaining non anonymous in the report while two startups named A and B participated in the case study with the condition that their identity should be kept anonymous. The selection of startups was based on the goal to reach sufficient heterogeneity with respect to the region, the business model, and the age and maturity degree of the startups. In addition, these were companies where we had closer cooperation in the past which is often a precondition to gain deeper insights from their business models and software development practices.

The objective of our study is to investigate the maturity of the requirement engineering practices in the studied startups from the perspectives of the business models, requirement management, documentation practices and organization maturity in terms of software engineering tools used & competencies in the team of software engineers. This research employs the case study protocol as given in (Runeson and Höst 2009).

Runeson and Höst (2009) presented the guidelines to conduct and report the case studies in software engineering which suggest executing five steps, i.e. case study design, data collection procedures, collecting evidence, analysis of collected data and reporting. Case study design is the identification of the case study protocol, which specifies case selection procedures, data collection procedures, analysis procedures, validity and ethical considerations etc. The data collection methods are executed to collect data from various data sources, which then are analyzed using various analysis techniques leading to generation and/or testing of the hypothesis. The data is usually a mix of qualitative (major) and quantitative (minor) data. The results of the case study are then reported to its intended audience using suitable reporting formats.

The representatives of the startups were interviewed using video conferencing tools (e.g. Skype) in mid-February 2021 in repetitive and iterative manner. Data was collected by a single researcher with individual startups. To address the threat to internal validity, the data collected by the researcher was shared in biweekly meetings between the researchers. The outcome of the meeting results in agreements over the interpretation of the aspects related to the maturity of requirements engineering practices and/or agreements over the missing perspectives and need for more clarifications. In either of two cases, the next round of interviews was conducted with the purpose of getting validation of the interpretations of the researchers and further insights to address doubts or missing perspectives. The series of interviews and biweekly meetings continued until last week of February 2021 when the researchers decided to end the interview sessions as no new information was being mined from the startup representatives. The field notes were taken during the interviews and were validated by the startup representatives after the end of the interviews. The interview transcript was subjected to grounded theory for quantitative data analysis.

By conducting the case study this paper tries to answer the following research question: "What is the state of maturity of requirement engineering practices in software startups?".

5 Criteria for Assessment

We elaborated some criteria which allow us to assess companies with regards to their requirements engineering maturity. In the following these criteria are described in detail. The criteria selected focus on the organizational, human and technical dimensions and enable a holistic insight.

- Business model and customer situation: This criterion defines the overall situation of the IT-company with regards to the business model, customers and revenues.
- Maturity of software organization: Here the skill level of the team for all tasks of the software life cycle has to be investigated, not only requirements engineering (i.e. analysis, design, implementation, testing, maintenance, etc.). Also, the quality of the tools used as well as their degree of interconnection has to be discussed here (i.e. tools for elicitation, documentation, development, testing, etc.).
- Documenting requirements: The documentation chosen has to be aligned with customer and project needs. Finding the right level of documentation is an ongoing research topic and extensively discussed in the scientific community.
- Checking requirements: Here it should be discussed what techniques are foreseen in detail for ensuring verification and validation (i.e. walkthrough, technical review, prototyping, simulation, consistency checks, checklists, etc.).
- Managing requirements: This criterion discusses what techniques are used to create transparency for the customers, the project team and the management.

6 Studied Startups

For our empirical analysis concerning requirements engineering we selected five startups from Europe (Switzerland, Germany) and India. All startups deal either with consulting services (project management, strategic planning etc.) or software development (mainly Web development). One startup has a strong research background. The number of employees varies between 1 and 10. The authors compare different startups during the whole lifecycle, from a very early to a more mature stage. The interviews were conducted with the founders and senior managers of the startups to elicit the relevant information. In the following subsection details of the three startups are provided which do not require anonymous treatment.

6.1 AbiliCor

AbiliCor (https://www.abilicor.ch/) was founded by three people in 2017 (S. Gatziu, C. Giovanoli, M. Peter) as a startup within the Institute for Information Systems at the University of Applied Sciences Switzerland. AbiliCor's mission is to support companies in the area of digital transformation. The business model consists of a procedure encompassing three steps:

- Step 1 "Discover": Internal and external analysis of the technological, economic and social trends with the Digital Backpack Assessment to focus on new or previously unknown. Output is to determine the case for change as a range of potential transformation areas.
- Step 2 "Orientate": Determination of the strategic thrust based on a location determination with the Transformation Compass. It unites all aspects of a transformation. Output is to develop a digital strategy and the definition of a strategy map.
- Step 3 "Navigate": Control of the implementation with the digital roadmap depending on the resources, skills, processes and systems. Output is the compilation of the strategic project portfolio into a digital agenda.

Customers can use the AbiliCor IT services (on-line tools) based on a subscription model. Additionally they may take advantage of consulting services (e.g. workshops) offered by AbiliCor guides. AbiliCor guides support companies in transformation tasks (training in the AbiliCor procedure for the transformation, comprehensive package of documents for designing customer workshops including access to online tools, constant updates of the approach to transforming companies). There are also joint efforts with so-called AbiliCor partners in order to develop new tools, using synergies and to enable the knowledge transfer close to research.

Software products are based on an "ad-hoc" approach of Scrum. For managing features and bugs the visual collaboration tool Trello (trello.com) is used. Currently there exist four in-house programmers (one senior and three junior programmers) and three external developers. The senior programmer acts as Scrum Master, the foundation members take - alternatively - the role as Product Owner.

With regards to the software stack used there were some changes in the last years. At the very beginning the MEAN stack was used (MongoDB, Express.js, AngularJS (or Angular), and Node.js). Currently the existing code base is gradually refactored using Bootstrap/Vue.js/Laravel (PHP framework) and MySQL. As container platform Docker is used, for source code management Git and Github are used. All development is done with VSCode.

For collecting requirements, the following approach was taken: The Senior programmer conducted interviews with all AbiliCor foundation members. In a subsequent workshop the results were compiled into a single backlog. Additionally, to the interviews, other sources for requirements are Innosuisse projects, and consulting projects for inter-trade organizations (e.g. hotel, construction).

The business side documents requirements with PowerPoint slides, the IT side uses Trello (see above). Requirements are broken down into single features (formulated with user stories) up to eight hours implementation time. Checking is done with Trello (managing attributes like prioritization, writing test cases etc.). Management is also done with Trello (assigning features to releases, versioning of requirements etc.).

6.2 Kiwih

Kiwih (https://kiwih.com/) was founded by Thomas Schurter. The main goal is to offer state-of-the art IT services, mainly to big customers. Currently there are five developers (60%–100%) employed. In order to manage peaks in projects, also freelancers are available (mainly for UI design issues). The business mainly works by talking to potential customers and to discuss process optimization opportunities with them. Finding new customers is done by word-of-mouth-advertising.

Basically, there are two business models: Programming services are contracted with the customer, followed by an implementation. In addition, project management can be provided when a customer brings in own development capacities, Kiwih can the focus on the managerial part (i.e. project management). Revenues are on a project basis. Most projects are done with big customers whereas projects with smaller customers mainly deal with webpage programming.

Project management is nearly solely done with Scrum. If the customer is unsure, the advantages of Scrum are explained (i.e. appropriate when requirements are not known in advance or changing, running software from the very start, flexibility).

Kiwih is employing full-stack developers. Used technologies are.NET Core with C#, PHP, Node.js, and native Apps (Android with Java/Kotlin, iOS with Swift). Requirements engineering practices can be distinguished into the four processes collection, documentation, checking and management. Collecting requirements is mainly done with the following steps: After a telephone call to get in touch with the customer, a kick-off meeting is organized to clarify requirements. The initial backlog is created in a brainstorming session (using a whiteboard). Additionally, Kiwih offers a questionnaire to the customer which allows to answer strategic questions (like positioning the product, competitor situation etc.).

Documentation starts most often with the system specification (concept development), in order to speed up projects and not to overload the customer. For the documentation the following tools are used: A system specification template (based on

MS Office 365), Azure DevOps (formerly MS Team Foundation Server) for integrating development and operation and Jira (for Sprint planning and bug fixing). Features are documented by using the design tool InVision Studio and Use Cases. Overall, the documentation is – according to the agile nature of most products - lightweight, preferred in a wiki-style.

Checking and managing requirements is done with InVision, Jira and Azure DevOps. For projects where the frontend really matters, different versions of all screens are tracked. If there is mainly backend functionality, the versioning is done with Azure DevOps. Overall, Jira supports the management of all Scrum artefacts (like Product Backlog, Spring Backlog) and the generation of metrics (Burndown chart).

6.3 Swip.World

Swiss Innovation Pool AG aka Swip.world (https://about.swip.world/de/) was founded in 2016 by two persons. Swip.world is an open innovation platform with the mission to support innovation in a global community by connecting innovation providers and innovation seekers. Swip.world provides challenges from companies seeking for innovation. Users can contribute by participating at these challenges and may earn money from that. Swip.world uses algorithms to find experts for solving problems from a large number of expert profiles to meet the requirements of innovation challenges. Services range from pure advertising of challenges up to an all-round service related to innovation management (especially including consulting).

The procedure model of Swip.world consists of the following steps for innovation challenges:

- Step 1: Seek for companies which require solutions based on innovations (consulting for customer).
- Step 2: Formulation of a call for innovations (aka challenge) (by customer and/or Swip.world) including prize money, target group (experts or broad audience), requirements for submissions, deadlines.
- Step 3: Publication on Swip.world and invitation of persons (on platform and through other social media). Challenges can be published and seen by everyone or distributed to an invited group of experts.
- Step 4: Submission of proposals for solutions. Contributions from may be shared in a group to support collaboration or can be treated confidentially. In the first case, there is usually the possibility for other users to provide comments or assessments. Bases on this, presenters of solutions may rework their submissions.
- Step 5: Evaluations phase: determination of one or several winners, announcement and payment of winners, customer gets rights of the submission
- Step 6: Possible follow-up projects and possibly support of customer during the implementation of the innovation.

It is planned to extend the platform support to a virtual coach, but currently further coaching and consulting is provided mainly offline as a human-based process. Currently, revenues are generated from the submission of challenges and their processing through Swip.world. Further revenues are coming from consulting services which may result from the different steps of the process (see above) and further consulting services. Platform users providing innovation ideas can use the platform at no charge.

For a quick development of the first release of the platform, an established software company (Connexa Network AG) was involved (also for saving money and for acquiring further contributions by financial investors). Connexa Network AG is a spin-off of Cooltribes (cooltribes.com) located in Portugal and is known for providing a social network platform in the university and alumni area. Their social network software has been used as a starting point for the Swip.world platform, which was enriched by additional features.

As a result of limited financing of the startup, the further development of the platform was done on a very small scale, i.e. mostly by the company owner with the possibility of temporary involvement of freelancers (or an agency) in the field of software development. In addition to this, Swip.world cooperates with three inventors who are users of the platform and interested in its further development. They may provide input (i.e. requirements) for further software development, for instance with respect to a virtual innovation coach.

Software development is mainly based on agile concepts such as in the Scrum approach but not in the traditional way. The current process rather follows the Kanban logic. As the software development is currently only done to a very limited extent, specific roles of scrum such as scrum master or product owner are not defined. With further growth of the company, it is planned to employ a product owner. A suitable person for this is already known.

For managing features and bugs the visual collaboration tool Trello (trello.com) is used. Google Docs are used as a tool for formulating requirements. From those, usually stories are formulated. Jira is used as a tool for specifying these stories. The software development is mainly based on Ruby on Rails. Backbone is used as a frontend framework. Github was used during the initial software development of the platform.

Usually, requirements are first specified using Google Docs. Then mockups are used during the further specification of the software design. As a final step before software implementation, usually stories are worked out. Besides support of programming, these steps allow for a better validation of the design in cooperation with users or customers.

Requirements are formulated and collected by the company owner as a single backlog. Additionally, other sources for requirements are the cooperating inventors (see above) and possibly further users of the platform. Some prioritization of requirements is used. Usually, requirements are first specified using Google Docs. In addition, Trello and Jira are used. Jira is especially important for specifying and documenting stories based on the requirements. Checking is done mainly manually using a staging server. A new version usually comprises a packet of features resulting from two or three sprints. Automatic tests are not done currently but are planned for future development (unit and integration testing). An approval process of new releases is defined. Requirements management is mainly done with Trello.

7 Analysis of Results

The detailed data collected from the startups is shown in Table 1. The results indicate that the startups vary in terms of adopted software development practices and tools. The team sizes do vary across the startups and the selection of the tools depends on the experience of the founder and the software development team. Cost reduction and managing limited resources seems to be the biggest challenge for the startups as they also prefer involving the freelancers for UI/UX designing related works. The development practices of the startups transform from highly unstructured (for instance, startups A & B) to more structured (for instance, AbiliCor, Kiwih and SWIP) with as their maturity with the markets increases.

8 Flexibility in Startups: First Insights

Flexibility can be considered as one of the strong points of startup companies in comparison to older and larger companies which often come from more traditional industries and markets. As startup companies are often active in novel and underdeveloped (or previously nonexistent) markets, flexibility can be considered as a cornerstone for any innovative startup.

As our investigation took place during the COVID-19 period, we were, of course, also curious how the considered startup companies coped with the pandemic crisis. In most cases, the impact on the software development routines was considered as small. We observed a widespread usage of home office environments which was usually no problem as employees either already had the required infrastructure or could use mobile hardware such as notebooks. Required software was usually anyway available as server installations and due to the usage of state-of-the-art tools collaborative activities were usually well supported. Also, online conferencing systems etc. were in widespread use.

Stronger impact was observed from the market side as in some cases the offered products and services were affected by a significant drop in demand. This was related to services which were usually based on personal meetings and contacts such as consulting but could also relate to purely online services because of a reluctance of customers due to the general uncertainty regarding their economic prospects. We were informed also of some positive aspects of the crisis, e.g. regarding the temporal availability of freelancer work which may be caused by fewer opportunities for freelancers to engage in other activities. Freelancers could help the startups to foster innovations especially value proposition innovations as reported in (Gupta et al. 2020b, 2020c, 2020d, 2020e).

Apart from the pandemic crisis, we learned about other and possibly even more momentous phases in the lives of the startups. For instance, efforts to attract sufficient investor capital lead to results below expectations. Another major example is that the market targeted originally turned out as not sufficiently yielding. Typical reactions in the observed cases were, for instance, the downsizing of the business model and growth expectations or a shift and expansion of perspective concerning the targeted market and the offered types of services. It seems that these strategies have been sufficiently successful by now. In any case, the startups appear to be able to conduct further smaller or larger adaptations of their business in order to meet changed requirements related to customers, internal aspects, cooperation partners, and challenges from the further environment (including a pandemic).

Table 1. Survey of software startups.

Startup	Business model/customer situation	Maturity of SW organization	Documenting requirements	Checking requirements	Managing requirements
AbiliCor	Offering IT services based on a subscription model, consulting on top of the IT services, Joint efforts (projects) with partners	Using an ad-hoc approach of Scrum (with Trello), technology stack is currently modernized	Business: MS PowerPoint IT: Trello (features described as user stories)	Using Trello (using predefined attributes for the features)	Using Trello (assigning features to releases for the features, version management)
Kiwih	Offering programming services and project management, 5 SW developers + freelancers	Solid application of agile development, experienced SW developers	Done with Office 365, Azure DevOps and Jira	Mainly done with InVision (for screens), Azure DevOps and Jira	Mainly done with Azure DevOps and Jira
SWIP	Platform business & consulting; 1 person company	Professional skills but used to a limited amount due to limited resources	Using tools such as Google Docs. Trello and Jira	Manual checking using a staging server	Involves strongly a few platform users; Trello is used as a tool
Startup A	Freemium revenue model as the features of competitors are provided free while enhanced features are provided on subscription based	Mostly university interns. Team not well balanced. Requirement analysts not the part of team	Done informally using sticky notes and emails	Using Prototyping and MVP Verification is not much focused (due to release issues)	Tools for requirement management are not used, requirements priorities in unstructured manner and are informally documented

(*continued*)

Table 1. (*continued*)

Startup	Business model/customer situation	Maturity of SW organization	Documenting requirements	Checking requirements	Managing requirements
Startup B	The software is new and innovative (no competitor). The customers known to the founder are targeted for involvement in the software development process. The revenue model is subscription based	Team includes people with expertise in coding. People with expertise in management, requirement engineering and modelling are missing	Abstract documentation usually resides on shared documents on Google Drive; however elaborated knowledge resides as tacit knowledge in team mind	Acceptance tests are performed with users directly The software is tested using automated testing by outsourcing the task to third parties	Founder decides which requirements to release. Requirement management is very informal and requires direct communication between startup founder and the user

9 Conclusion and Future Work

The requirement management process and the infrastructure used to manage them, varies considerably across the studied startups. This is a limitation to foster knowledge sharing between startups due to heterogeneity in process models and practices but a great opportunity to foster innovations driven by learning other practices as diverse new ideas. The startup's focus is to meet product/market fit which is the biggest reason for highly informal practices but as they approach closer to the target, their practices become more structured. The previous technical experience (especially related to software engineering) of the founders and the startup team is the biggest decision factor in optimizing the practices and selection of the tools. The flexibility in the working practices, remote work nature of software engineering tasks (except those requiring interactions with the customers), availability of the freelancing working force helped the startups to cope up with the marketing fluctuations brought by the pandemic situation. Further, they also have helped the startups to diversify their product offerings to suit the changing demands of the customers in the pandemic situation.

As our study was limited regarding the number of companies and their regional distribution there is certainly much further potential regarding empirical research on software engineering practices in startup companies. Although the software business and practices are quite globalized we might still expect interesting insights from comparing startups in different parts of the world.

References

Rupp, C., Pohl, K.: Requirements Engineering Fundamentals: A Study Guide for the Certified Professional for Requirements Engineering Exam - Foundation Level - IREB Compliant, 2nd edn. Rocky Nook, Santa Barbara (2015)

Heikkilä, V., Damian, D., Lassenius, C., Paasivaara, M.: A mapping study on requirements engineering in agile software development. In: 2015 41st Euromicro Conference on Software Engineering and Advanced Applications, pp. 199–207 (2015)

Oestereich, B.: Gedanken über agiles Requirements Engineering. In: Heise Developer (2010). http://www.heise.de/developer/artikel/Gedanken-ueber-agiles-Requirements-Engineering-948348.html

Telesko, R.: Road to agile requirements engineering: lessons learned from a web app project. In: Dornberger, R. (ed.) Business Information Systems and Technology 4.0. SSDC, vol. 141, pp. 65–78. Springer, Cham (2018). https://doi.org/10.1007/978-3-319-74322-6_5

Gupta, V., Fernandez-Crehuet, J.M., Hanne, T., Telesko, R.: Requirements engineering in software startups: a systematic mapping study. Appl. Sci. **10**, 6125 (2020a)

Gupta, V., Fernandez-Crehuet, J.M., Hanne, T.: Freelancers in the software development process: a systematic mapping study. Processes **8**, 1215 (2020b)

Gupta, V., Fernandez-Crehuet, J.M., Hanne, T.: Fostering product innovations in software startups through freelancer supported requirement engineering. Results Eng. **8**, 100175 (2020c)

Gupta, V., Fernandez-Crehuet, J.M., Hanne, T.: Fostering continuous value proposition innovation through freelancer involvement in software startups: insights from multiple case studies. Sustainability **12**, 8922 (2020d)

Gupta, V., Fernandez-Crehuet, J.M., Gupta, C., Hanne, T.: Freelancing models for fostering innovation and problem solving in software startups: an empirical comparative study. Sustainability **12**, 10106 (2020e)

Runeson, P., Höst, M.: Guidelines for conducting and reporting case study research in software engineering. Empir. Softw. Eng. **14**, 131 (2009). https://doi.org/10.1007/s10664-008-9102-8

Manipulator: A Novel Collusion Attack on Trust Management Systems in Social IoT

Alper Ugur(✉)

Pamukkale University, 20020 Denizli, Turkey
augur@pau.edu.tr

Abstract. Billions of smart devices communicate with each other in the era of IoT. They share data, contribute to make decisions and sense to cooperatively perform business operations. The IoT devices implement social networking paradigms to overcome uncertainty and achieve better results. This enhances Social IoT (SIoT), a relatively new subset of IoT where smart entities form a social network of their own to provide better quality of service. Trust is what holds this social togetherness. There are various trust management protocols for managing trust between IoT entities. They were verified with known trust-related attacks such as self-promoting, bad-mouthing, ballot-stuffing etc. In this work a new trust-related attack called manipulator is introduced. It is based on historical trust consideration mechanism in trust management systems established as a safeguard for known attacks. Manipulator attack takes advantage of systems proactive interception on lack of trust and transform it to a friendly fire by deceiving that non-malicious entities were also in collusion all along. The vulnerable trust management schemes in literature against manipulator attack were discussed as well.

Keywords: Social IoT · Trust · Trust management · Vulnerability · Manipulator

1 Introduction

Machine to machine communication evolved rapidly in time became an information network and automated business systems. In this holistic system known as Internet of Things (IoT), for business objectives, large number of objects are connected to be able to collaborate without human interaction [1]. From industrial production to health technologies, variety of application areas were proposed to utilize this smart alliance of IoT objects. Smart devices manage to interact implementing a social approach. Social IoT (SIoT) extends the concept of IoT defining social relation between smart devices [2]. Through an environmental awareness, the smart devices in the SIoT, sense their surroundings, autonomously request information, interpret the collected information, and offer services to others while sustaining their individuality [3, 4].

The SIoT devices quest novel resources to improve enactment of their target services. This could be achieved by navigating a social network of "friend" objects which can scale to the billions of devices [1]. The source of this new information must be reliable,

R. Silhavy (Ed.): CSOC 2021, LNNS 230, pp. 578–592, 2021.
https://doi.org/10.1007/978-3-030-77442-4_49

dependable, and secure; otherwise, planned improvement at the end of the interaction cannot be achieved.

This endorsement requires a trust relationship where trust is defined in [5] as the presence of solid belief in the ability of an entity to act dependably, securely, and reliably within a particular context. In SIoT, trust is a state where a smart device (trustor) relies on the behaviors of another entity (trustee). The "friendly" trustor and "to be" trustee participate in a relationship only when there is adequate trust established among them in the SIoT. This constitutes relationships where devices responding only to service requests from familiar nodes, rely on them without exposure to malicious ones [6].

In a task transfer case such a road security for an automated car, or environment sensor in a nuclear plant, trustor rely on execution of the task or data obtained by trustee. It gives an opportunity to the trustee to manipulate the service. Trust management facilitates assessments for the SIoT devices to overcome this uncertainty and risk [4].

Trust management in SIoT environment conducted by several constraints on context dependent cases such as direct, indirect, local, or global trust. SIoT devices rely on a trust based on subjective or objective foundations, weights or QoS properties, composition of various reliability attributes and experience of past events [7]. Social attributes can be utilized as metrics for instance, intimacy for measuring the closeness based on the observations, honesty for anomaly behaviour during the communication [8].

Procuring trustworthiness information of IoT entities involve trust computation techniques such as trust composition, trust propagation, trust aggregation, trust update, and trust formation [9]. Trust aggregation and trust update techniques make use of self-observations (direct trust) and recommendations (indirect trust) to determine trust decision. For instance, in an event-driven update, smart devices collaborate with each other in exchanging a recommendation about behaviour of another device.

As the devices are connected and share information, a risk of vulnerability against malicious agents must be considered. Conventional security requirements, including confidentiality, integrity, availability, are essential to provide a security basis [10]. In environments like SIoT, trust attacks on trust management and recommendation systems correspondingly reveal new context-based security challenges [11–14].

In this paper a new trust attack is presented. The attack takes advantage of trust management systems historical evaluation and utilize it against system. The attack is significant that vulnerability arises in many of the current approaches proposed for trust management security.

The paper is arranged as follows. Section 2 consists of known trust attacks and countermeasures in IoT systems. In Sect. 3, detailed description of the Manipulator attack is presented. The vulnerability analysis of the current trust management methods against Manipulator attack is provided in Sect. 4. Finally, in Sect. 5, the paper is concluded with discussion and future directions.

2 Background and Motivation

This section sets the background of this work by providing an introduction towards the trust-related attacks and countermeasures presented for IoT environment. The security requirements and solutions which are available in the literature are discussed. The arguments presented are aimed to clarify the novelty of the introduced attack.

2.1 Trust

Devices in a SIoT establish trust among other devices to perform any given task. A device in SIoT can build its own social network. Once they must interact with unfamiliar devices, the requirement for trustworthiness metric will emerge. It is essential to define trust in all heterogeneous devices that are involved in achieving the right decision [15].

The decision makes trusted information available where/when it is needed and to those who need it. Trust establishment is valuable as a smart device can protect itself in hostile environments, and able to review others whether it is still trustworthy [15].

Trust in SIoT. The multifaceted term of trust is described in [16] with trust relationship where a trustor and a trustee, reliant on each other for mutual benefit. The properties of trust may fulfill the definition.

In direct, trust is based on self-observations, direct interactions, and experience. On the contrary, an indirect trust between trustor and trustee is built upon the recommendation of other devices as a function of trust transition.

- Trust may be local, differ from one relation to other, or can be global where every device in the network has same impression.
- Trust can be composition of attributes such that reliability, honesty, security, competence so on, influencing trustworthiness of a trustee [17].
- Trust should be subjective according to recommendations. This assists asymmetry property of trust, where two devices may trust each other at different levels.
- Trust is also dynamic, may change, revoked or re-established due to conditions which trust is established [5].

According to [16] the most common properties of trust in SIoT environments are subjectivity, indirect, asymmetry, composite and global.

SIoT networks differ from social networks by its characteristics such as, huge amount of device involvement, limited resources for computation and storage etc. [16].

Dynamism is another feature of the SIoT where rapid changes in network, roaming, momentarily entry and exit movements of devices are usual.

In [18] different psychological, social, or political factors such as risk, fear, delegation, cooperation, altruism etc. affecting trust assessment were defined.

Trust management protocols must consider all properties and attributes for a solution. Within numerous objectives a trust management system should achieve trust relationship evaluation, data perception trust, system security and robustness for a sustainable service [19].

2.2 Trust-Related Attacks

A malicious device would act for its own benefit targeting to break or utilize service of IoT environment. There are several attack classifications based on attacker's self-interest, targeting reputation mechanism, and intending to avoid detection of malicious behaviors [11, 20–22].

In self-promoting attack, an entity promotes its reputation dishonestly. After securing its elevated trust level (for instance promoted to be a service provider), it provides malfunction services.

Bad-mouthing attacks is one of the collusion attacks of malicious devices. They coordinate for providing bad recommendations against well behaved devices to decrease their probability to be selected for a service.

Good-mouthing or ballot-stuffing attack has similar pattern of self-promoting and bad-mouthing. The malicious group cooperate to disguise a compromised device by giving good recommendations.

Bad-mouthing or ballot stuffing attacks could be launched by a single device in SIoT. Attacker generates multiple fake identities to provide different type of recommendations for a well-behaved device. This pattern is known as sybil attack.

In on-off attack, malicious entity aiming to exhibit trusted behavior. The attacker pretends to be well behaved, hiding its harmful actions by alternating normal and malicious ones.

Opportunistic service attack is a derivative of on-off attack. Attacker decides to provide good service after experiencing certain decline in its reputation due to its former bad behavior. When good reputation is re-accomplished, device yields to its harmful deeds.

White-washing attack or re-entry problem is a self-interest attack where a malicious entity can easily remove its bad reputation in its past by registering as a newcomer in SIoT.

DoS, malicious injection, flooding, selective forwarding, and Xhole are other types of attacks can be organized against communication [23].

2.3 The Prevalence of Attacks and Countermeasures

Reviewing the prevalence and countermeasures proposed in trust management models would be an enlightening move to recognize the attacks. This section presents state-of-art of the known trust-based attacks and actions taken. Existing solutions do not consider most of the attacks due to restrictions in SIoT and thus do not provide a comprehensive one [23]. Attacks perpetrated by misbehaving devices are presented primarily for a concise background.

In [24] a trust model is proposed to protect user's security. This model uses authentication history and penalty to calculate trust values. Failed authentication attempts cause penalty points to decrease level of trust.

The trust model in [25], is the known first trust model for IoT calculating trust from behavior of the devices. Rather than using cryptographic measures the model utilizes a fuzzy trust model to identify malicious entities. The model only resilient to self-promoting among other attacks and IoT environment is confined with wireless sensors.

In [26], a dynamic trust management model is proposed considering social relationship as in SIoT environment. Hence it is not evaluated in the paper, the model supports detection of self-promoting, bad-mouth and ballot-stuff attacks [23]. A Bayesian technique is applied where both direct and indirect information collection is used in trust calculation. The model is vulnerable to behavioral changes of malicious device as in on-off and opportunistic attacks.

A trustworthiness evaluation in SIoT with both objective and subjective approaches presented in [17] by a linear computation model. Each device computes the trust based on its own experience and the recommendation of common "friends". Malicious objects modify their actions based on the social relationships as acting benevolent (to its owner or close friends) or maliciously. The objective approach tends to be affected by global decision and suffers from modified actions of adversary in the same way as in [26].

In [21], an adaptive trust management protocol for SIoT is proposed. They considered dynamical changes, used linear model and honesty, cooperativeness, and community interest parameters to calculate trust. On-off attack also poses a threat for this model [3].

Subsequent works [27–29], propose dynamic models to accommodate changes in devices. In [27], a dynamic access service recommendation scheme was proposed. The recommendation is based on the past performance, the social relationship, and the energy status of nodes. In [28], trustworthiness in SIoT systems were assessed through reputation, experience with the trustee and knowledge metrics. In [29], time-aware is considered as a new trust metric for the evaluation of the trustworthiness in SIoT.

Prior works have considered approaches to defend against self-promoting, bad-mouthing or ballot-stuffing attacks effectively but not opportunistic and on-off attacks [9].

Countermeasures taken for self-promoting attacks are weight-based or collaborated recommendations, service-feedback [25, 27, 30], for bad-mouthing and ballot stuffing attacks are recommender evaluation (honesty, trust aggregation etc.) history-based assessment (e.g., past-performance), and time-awareness mechanisms [21, 26–29].

Apart from cryptographic measures in all IoT systems which not feasible in dynamic environments [23], binding recommendations with corresponding transaction, checking social relationship between recommenders, filtering/clustering techniques to identify colluders are other techniques to identify ballot-stuffing and collusion attacks in SIoT [8].

As well as performance history of a device, storing long-term behavior, and consideration this as a trust calculation metric is often encountered type of measures for opportunistic malicious ones [8].

In [30] trust predictability method was proposed against opportunistic and on-off attacks. In each transaction, by a device's trust value was compared with the trust threshold and it is determined if a device is bad or good. The method requires backward calculations of the trust. And as a result, if a device is recognized as malicious, it will be inserted into a blacklist.

The countermeasures in the SIoT security based on trust management were intended to be performed individually. The attacks were analyzed and confronted as distinct events. As far as it is known, a unified attack is enounced for the first time in this work. The Manipulator attack presented in next section is a successive attack where collaborated devices act opportunistic to utilize exploit of defense mechanism of many up-to-date trust management systems for a ballot stuffing attack.

3 Manipulator: A Novel Collusion Attack on Trust Management Systems

Trust and recommendation-based attacks known for trust management systems are mainly self-promoting, good-mouthing, ballot-stuffing, opportunistic service, and white washing attacks. Besides attack classes such as attack on device, attack on communication, these attacks were classified as attack on service [23].

The attackers do not always strike directly to the command center on a network. As defense techniques such as intrusion detection systems become widely deployed, attackers consider how to defeat protection mechanisms when launching their attacks [31]. They infiltrate to the network with a low privileged device with trojan horses, malwares, backdoors etc. Then they exploit vulnerabilities to gain unauthorized access to the system. This requires multiple steps to a successful intrusion.

To avoid anomaly detection systems, attackers sometimes mind behaving like an ordinary device as in ballot-stuffing, on-off attacks. The defense systems were planned to react these types of dynamic changes as in [26–30]. They aimed to achieve the fact by collecting evidence. Countermeasures are designed to re-evaluate recommendations [27], experience [26, 28], explore past and present states [29] and utilize time windows for backward calculations [30].

3.1 Motivation of the Manipulator Attack

The trust management systems employ defend mechanisms for known attacks. When an attack pattern or signature is recognized, the system intends to eliminate the cause of the attack.

As mentioned before, bad-mouthing attack is a service attack, multiple malicious devices target a well-behaved device to decrease its probability to be selected for a service. Thus, one of the malicious devices will be selected for the service.

The problem is the trust management systems in literature can detect bad-mouthing attack. And most of them adapt detecting anomalies and become sensitive to behavioral variability.

And the main premise of an attacker in this case are, is there another way to seize the service for malicious devices? And any opportunity to discredit a good device without bad-mouthing?

3.2 Generation of the Manipulator Attack

In the Cambridge English Dictionary, the definition of the term manipulator is given as “a person who controls people to their own advantage, often unfairly or dishonestly”. In the case of SIoT, “a mechanical device for the remote handling of objects or materials in conditions not permitting the immediate presence of workers” definition [32] turns out to be more explanatory.

Manipulator attack can be defined in accordance with the lexical equivalent as an attack process dishonestly controlling the trust management system to attacker's own advantage for remote handling of service in conditions not permitting presence of legitimate workers.

A manipulator attack is applicable under four conditions: First, there exists a recommendation-based trust management system among devices. Many existing SIoT environments satisfy this condition. Second, the system has detection and prevention mechanism against ballot-stuffing and good-mouthing attacks. If this condition is not met, this attack is not needed for a malicious device anyway. Third, system does not know whether a device is well-behaved or malicious and must estimate it. This is also a general assumption for the environment. Four, there exists multiple malicious devices collaborating to take over a service. This condition is required for the attack methodology.

At the *Initialization*, malicious devices (M_n) opt for a benign device (b) as the target. The high probability of being selected for the service (s) will be preferred.

Mn perform attack as follows.

Step1: Friendliness. Excluding some M_{i}, m malicious devices begin ballot-stuffing b with good recommendation, where number of malicious devices $m < n$ total number of collaborating malicious entity. Recommendation must be stopped before service provider selection phase.
Step2: Self disclosure. All m acts suspicious as to be detected by anomaly detection system that behavioural variability. Modified actions should not include b.
Step3: Manipulation. Defense mechanism of the system determines malicious collaborated devices m, including b as ballot-stuffed member. The m, and b are blacklisted now and will not be interacted for a while. The system was protected itself besides it is manipulated by the collaborated devices for an undetected friendly fire.
Step3.1: White-washing. This step is not required for the attack. Devices m, re-new themselves to be in system. m would resume from the initialization to keep on harming the system service until the final step. b as a benign device stays out of the game pending reacceptance to the system.
Step4: Final. MI reputed as not part of the malicious collaboration would be selected as a s service provider hence it is the best option after elimination of the malicious (m, and all potential b s) devices.

The presented attack takes advantage of trust management system's learned actions against known basic attack types. To be clear, manipulator attack sequence diagram is given in Fig. 1.

In the following section, vulnerability analysis of the recent trust management methods against the Manipulator attack is given.

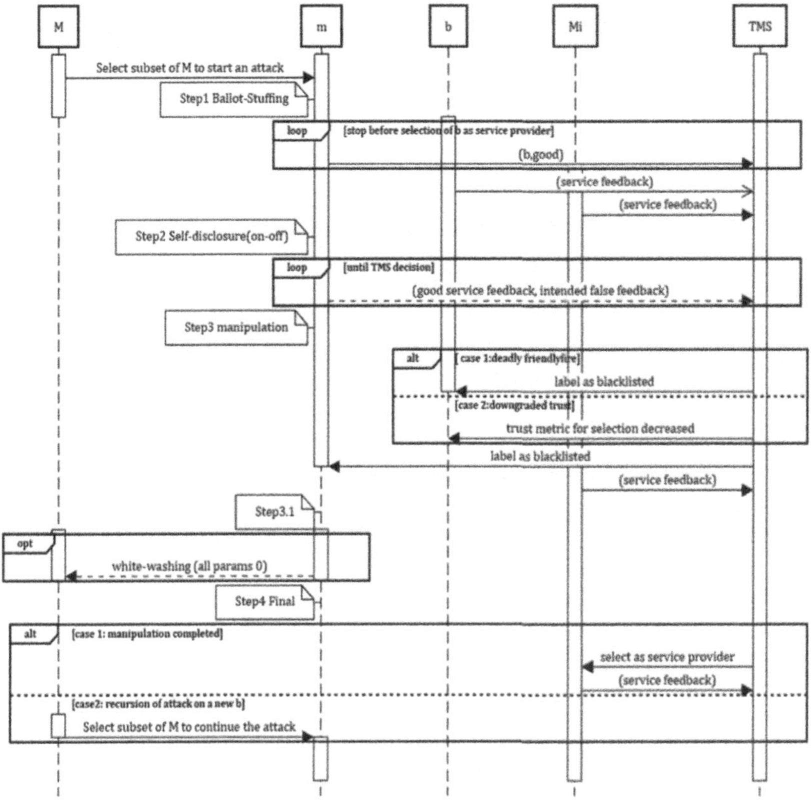

Fig. 1. Sequence diagram of manipulator attack on trust management system (TMS).

4 Vulnerabilities of Trust Management Systems Against the Manipulator Attack

Manipulator attack trade on a trust management system's security procedure against it by manipulating the mechanism. Attackers are in fact putting their own purpose into practice and while doing so, they direct the system to do what it thinks is right. Therefore, the actions taken by the trust management system is directly related with the vulnerability. The attack is successful against most of the dynamic models those supporting ballot-stuffing measures and utilizing history of a node to reveal dishonest opportunistic attacks. The manipulator attack requires a countermeasure mechanism against ballot-stuffing and indirect reputation or trust communication between devices to manipulate the trust management system. In Table 1, features of above-mentioned trust management systems are briefly presented.

Table 1. The feature analysis before attack.

Trust Management System	Measure against ballot-stuffing	Relationship
[17]	Credibility rating, long-term and short-term assessment	Direct, Indirect
[21, 26]	Honesty experience	Direct, Indirect
[27]	Positive/Negative feedback, relationship	Indirect
[28]	Experience and reputation	Direct, Indirect
[29]	Social similarity, honesty, time-aware	Direct
[30]	Evaluate towards recommender, trust predictability	Direct

In [17], devices have direct and indirect relationship. Credibility rating is utilized on devices to measure dynamic changes. Evaluations are done in long and short-term periods. Direct connection-based evaluations decrease degree of vulnerability. But it still exists hence negative feedback is given to devices those performing maliciously or providing false references, even to devices that do not act maliciously but are connected to portions of unreliable network.

Feedback is represented by f_{ij}^{l} which refers to each transaction l from device j is evaluated by device i. $f_{ij}^{l} \in \{0, 1\}$ where rate is 1 if i is satisfied by the service and 0 otherwise. Transaction factor ω_{ij}^{l} indicates the relevance of transaction l between i and j. The short-term opinion is used to evaluate the risk for oscillations of actions of a malicious device after building up its reputation. The long-term opinion is not sensitive enough to suddenly detect this situation since it requires adjustments to the accumulated score in long time. Long term opinion calculation is given as in Eq. (1).

$$\mathrm{O}_{ij}^{lon} = \frac{\sum_{i=1}^{M} \sum_{l=1}^{L} C_{ij} \omega_{ij}^{l} f_{ij}^{l}}{\sum_{l=1}^{L} C_{ijm} \omega_{ij}^{l}} \tag{1}$$

Long-term assessment will have a role in guiding the system to think that the honest device is also cooperating with the malicious ones. These characteristics would give opportunity for manipulator to decrease credibility of the honest device. Insensitivity of static weighted sums to dynamic modifications is as well a deficiency. Credibility is defined as in Eq. (2).

$$C_{ij} = \frac{(1 - \gamma - \delta)\mathrm{T_i} + \gamma\left(1 - \mathrm{F_{ij}}\right) + \delta\left(1 - \mathrm{I_j}\right)}{1 + \log\left(\mathrm{N_{ij}} + 1\right)} \tag{2}$$

Strong relations or nodes that have a high number of transactions between them, receive a lower credibility. Execution of pseudo-ballot stuffing in Step 3, affects credibility of benign device C_{ib} and therefore system's short/long opinion O_{ib} by manipulating the system to decide that device was a member of the malicious group.

The vulnerability is evaluated as medium since influence of static-weights and short-term assessment may not lower credibility of the attacked honest device as much as necessary for a friendly fire. Merely the system is yet vulnerable.

In [21] and [26], trust protocol deals with ballot-stuffing attacks by considering honesty as a trust property. The proposed protocol compares a device's recommendation toward another device. If the percentage difference between the values is above the threshold, the device is considered as suspicious and system gives it a negative honesty experience. The honesty was given as in Eq. (3).

$$D_{iq}^{Honesty}(\mathrm{t}-\Delta \mathrm{t}) \tag{3}$$

Trust recommendation T of k for i about an object j is calculated as in Eq. (4) where new recommendation is weighed according to experience.

$$T_{ij}^{X}(\mathrm{t}) = (1-\gamma)T_{ij}^{X}(\mathrm{t}-\Delta \mathrm{t}) + \gamma R_{ij}^{X}(\mathrm{t}) \tag{4}$$

Though threshold is defined as static, where a high threshold decreases false negatives (misidentifying a bad device as good) then increases false positives (misidentifying a good device as malicious) [26]. This facilitates manipulators false orientation on honest device. The honesty experience $D_{iq}^{Honesty}$ of a benevolent device can be rolled back by increasing its recommendation R_{ij}^{X} and subsequently T_{ij}^{X} above %50 by colluding malicious entities. The system is designed inevitably to give negative honesty experience for the dedicated device in case of the manipulator attack. Herewith the attack will successful after final step and vulnerability of the system is rated as high.

In [27], a reputation system is utilized to evaluate the past performance of a device based on the feedback it received. The recorded transactions may differ from one another, a weight factor is introduced to reduce the risk of malicious attacks. Feedback was calculated as in Eq. (5).

$$feedback_{uj} = 1 - \left|\mathrm{T}_{\mathrm{u,v}}(t) - R_{jv}^{dir}(t)\right| \tag{5}$$

If the recommendation from a friend is close to the actual QoS, there will be a positive feedback. Otherwise, it will be assigned with a negative one. Positive feedback is assigned not only to a device that provide quality service but also to devices that give positive recommendations. On the other hand, all nodes that provide poor service, carry out malicious attacks, or have a close relationship with the malicious nodes will be given a negative feedback.

This characteristic of the system is vulnerable to manipulator attack where attackers first advances trust making accurate feedbacks, then collaborate the attack on the honest device by giving positive feedback. Consequently, rapid drop on the feedbacks (negative) R_{jv}^{dir} alarms the system as suspicious behaviour. The irregular feedbacks given by malicious group for the target device will be considered the indication of close relationship and trigger negative feedback for the victim. The attack could be implemented indeed in a dynamic environment, so the vulnerability is rated as high.

The active attackers must white-wash to gain trust for new recommendations on a new victim. If not, new manipulator group m must be chosen for a new attack.

In [28], cooperativeness, community-interest, honesty, and similarity factors are chosen to determine trustworthiness of a device. An experience model was proposed, where many cooperative interactions in a short duration of time is required to obtain a high experience value (i.e., a strong connection between a trustor and a trustee). The longevity is related to the high value of the experience. However as in social networks when the trust is not high, uncooperative interactions can damage the experience relationship. The model utilizes reputation with experience to evaluate the trust between entities. Reputation was measured by Eq. (6) where C(i) is the total positive or negative recommendations and $Exp(i, X)$ from entity i toward X.

$$Rep(\mathrm{X}) = (1 - d)/N + \mathrm{d} \times \left(\sum\nolimits_{\forall \mathrm{i}} Rep \times \frac{Exp(i, X)}{\mathrm{C(i)}} \right) \tag{6}$$

Mathematical model of experience decrease in case of uncooperative interaction was given as in Eq. (7). It is also stated that the β parameter should be always greater than 1 because the experience relationship is hard to gain but easy to lose [28]. min_{Exp} guarantees that the experience value cannot go lower than e.g., 0.

$$Exp_{t+1} = \mathrm{Max}\langle min_{Exp}, Exp_t - \beta \times Exp_{t+1} \rangle \tag{7}$$

Malicious group may perform to achieve trust in a short period t, then practices honesty and cooperativeness against the system. The system based on cooperativeness property as the more cooperative device become more trustworthy.

Then conducting suspicious interactions after ballot-stuffing a target-honest device, will cause decrease in experience as in Eq. (7) and increase arises for negative reputation with Eq. (6) as if pseudo-cooperation is exposed by the manipulated system. The vulnerability is ranked as medium for requirement of mutual cooperativeness factor which trustworthy is measured with it.

The white-washing is optional because it causes malicious device loss of earned high reputation.

In [29], trustworthiness is calculated with metrics as knowledge: social similarity and inter-social relationship, recommendation: honesty, and reputation: time-aware. Thru time-aware approach they consider old reputation in addition to the recent reputation to deal with malicious entities who try to give reasonable service to enhance their past reputation. For the manipulator attack, honesty and reputation are keys to vulnerability once again. Cooperativeness metric could be used as the reducing factor in trust equation for the benign device. Social similarity can be employed by a manipulator to give the image of collaboration with honest device. Trustor compares trustees' vector of recommendations with average trust values of other nodes to the same device. This enables opportunistic collusion attack of manipulator devices. Reputation of a device is also produced by opinions of other devices. Knowledge is also used to include social similarity between recommenders.

In trust aggregation reputation, recommendation and knowledge are assessed as in Eq. (8).

$$T(t) = w_1 Rep + w_2 Rec + w_3 Knowledge \tag{8}$$

The manipulation will be conducted in post-evaluation phase, where system detects changing behaviour of dishonest devices within the consideration of past reputation. The 0.6 to 0.4 rate of weights for recent and past reputation values are used in simulation on cited work [29]. Closer rates will ease the manipulation where in the post assessment phase mutual direct values of victim would not turn the scale evidently.

Time-aware approach becomes an advantage for the honest device since the computation considers old reputation while calculating the new value. Further, criterion for mutual direct communication between trustor and trustee, declines the vulnerability rate from medium to low. As in [28] white-washing is optional, it requires debate on whether the old reputation is high or low by the time-aware parameter.

In [30], trust value stored in a node determines whether a node is bad or good. A window is utilized for a specific number of past transactions. In each transaction, trust value is compared with the threshold. The last value for good (0) or bad (1) is stored. Rate of good over all is defined as a prediction metric as in Eq. (9).

$$PT_i = \text{Good}/(\text{Good} + \text{Bad}) \tag{9}$$

The low prediction is evaluated as low trust or malicious opportunistic attack because of high quantity of bad records. As a result, the device is added to the blacklist.

Initially, the malicious group provide good service to gain high reputation. Then ballot-stuff the victim device as stated in the manipulator attack generation section.

The attack requires direct connection to the victim due to the system's requirement as nodes do not accept recommender unless there is no direct connection between them. However, this is not typical in SIoT, direct connection obligation makes the manipulator attack ineffective when combined with service feedback for trust evaluation. The white-washing is not applicable for the rest of the attack. The vulnerability analysis of recent trust management systems is summarized in Table 2.

Table 2. The vulnerability analysis of trust management systems against manipulator attack.

Trust Management System	Potential Cause	Malicious objects need white-washing after Manipulator attack	Vulnerability
[17]	Static-weighted sums, long-term assessment	NO	Medium
[21, 26]	Threshold	NO	High
[27]	Relationship	YES, gain trust for recommendation	High
[28]	Experience	YES/NO (optional: High reputation contributes more)	Medium
[29]	Social similarity, cooperativeness 'degree	YES/NO, (according to: time-aware recs)	Low
[30]	Blacklist, trust predictability threshold	Not applicable	None

5 Conclusion and Future Works

Identifying a dishonest collaborated attack is difficult, since instead of searching to identify cliques in a graph demonstrating their relationships, a trust management system required to identify semi-connected clusters [11]. Hence colluding attackers may not appear to have connected each other due to the varying behaviors at the phase of the attack strategy. It is possible that two attackers may have no observable interaction detected by the system and thus can be labelled as distinct devices within a window of time.

In this work, we presented a recommendation based novel attack on SIoT environments. The attack can be executed by colluding malicious entities which exploits bullet-staff countermeasure approaches. The aim of the attack is to eliminate the honest devices and take over the service as a provider.

Vulnerability of five recent approaches were analyzed and their flaws against manipulator attack were explained. One model was determined sound because of its direct service feedback characteristic. Though the attack is considered effective on many SIoT environment with many nodes that, the chance of the trustor encountering a recommender is higher than the chance of the trustor directly interacting with a trustee [26].

The presented trust management systems in SIoT have various parametric approaches. As the future works: we plan to implement and simulate the attack for the cited works. We will evaluate following features: the attack convergency, minimum number of members for the collusion, resource restrictions like energy awareness and the effect of colocation.

References

1. Atzori, L., Iera, A., Morabito, G., Nitti, M.: The Social Internet of Things (SIoT) – when social networks meet the Internet of Things: concept, architecture and network characterization. Comput. Netw. **56**(16), 3594–3608 (2012)
2. An, J., Gui, X., Zhang, W., Jiang, J.: Nodes social relations cognition for mobility-aware in the internet of things. In: 2011 International Conference on Internet of Things and 4th International Conference on Cyber Proceedings, Physical & Social Computing, pp. 687–691 (2011)
3. Chahal, R.J., Kumar, M., Batra, S.: Trust management in social Internet of Things: a taxonomy, open issues, and challenges. Comput. Commun. **150**, 13–46 (2020)
4. Lin, Z., Dong, L.: Clarifying trust in social internet of things. IEEE Trans. Knowl. Data Eng. **30**(2), 234–248 (2018)
5. Grandison, T., Sloman, M.: A survey of trust in internet applications. IEEE Commun. Surv. Tutor. **3**(4), 2–16 (2000)
6. An, J., Gui, X., Zhang, W., Jiang, J., Yang, J.: Research on social relations cognitive model of mobile nodes in Internet of Things. J. Netw. Comput. Appl. **36**(2), 799–810 (2013)
7. Senthil Kumar, J., Sivasankar, G., Selva Nidhyananthan, S.: An artificial intelligence approach for enhancing trust between social IoT devices in a network. In: Hassanien, A.E., Bhatnagar, R., Khalifa, N.E.M., Taha, M.H.N. (eds.) Toward Social Internet of Things (SIoT): Enabling Technologies, Architectures and Applications. SCI, vol. 846, pp. 183–196. Springer, Cham (2020). https://doi.org/10.1007/978-3-030-24513-9_11

8. Ahmed, A.I.A., Hamid, S.H.A., Gani, A., Khan, S., Khan, M.K.: Trust and reputation for Internet of Things: fundamentals, taxonomy, and open research challenges. J. Netw. Comput. Appl. **145**, 102409 (2019)
9. Guo, J., Chen, I.R., Tsai, J.: A survey of trust computation models for service management in internet of things systems. Comput. Commun. **97**, 1–14 (2017)
10. Frustaci, M., Pace, P., Aloi, G., Fortino, G.: Evaluating critical security issues of the IoT world: present and future challenges. IEEE Internet Things J. **5**(4), 2483–2495 (2018)
11. Hoffman, K., Zage, D., Nita-Rotaru, C.: A survey of attack and defense techniques for reputation systems. ACM Comput. Surv. **42**(1), 1–31 (2009)
12. Ning, H., Liu, H., Yang, L.T.: Cyberentity security in the Internet of Things. Computer **46**(4), 46–53 (2013)
13. Chen, I.R., Guo, J.: Dynamic hierarchical trust management of mobile groups and its application to misbehaving node detection. In: 28th IEEE International Conference on Advanced Information Networking and Applications, Victoria, Canada, 1–6 May 2014
14. Fortino, G., Fotia, L., Messina, F., Rosaci, D., Sarné, G.M.L.: Trust and reputation in the internet of things: state-of-the-art and research challenges. IEEE Access 8, 60117–60125 (2020)
15. Sfar, A.R., Natalizio, E., Challal, Y., Chtourou, Z.: A roadmap for security challenges in the Internet of Things. Digit. Commun. Netw. **4**(2), 118–137 (2018)
16. Abdelghani, W., Zayani, C.A., Amous, I., Sèdes, F.: Trust management in social internet of things: a survey. In: Dwivedi, Y.K., et al. (eds.) I3E 2016. LNCS, vol. 9844, pp. 430–441. Springer, Cham (2016). https://doi.org/10.1007/978-3-319-45234-0_39
17. Nitti, M., Girau, R., Atzori, L.: Trustworthiness management in the Social Internet of Things. IEEE Trans. Knowl. Data Eng. **26**(5), 1253–1266 (2014)
18. Cho, J.H., Chan, K., Adali, S.: A survey on trust modeling. ACM Comput. Surv. **48**(2), 1–40 (2015). Article 28
19. Yan, Z., Zhang, P., Vasilakos, A.V.: A survey on trust management for internet of things. J. Netw. Comput. Appl. **42**, 120–134 (2014)
20. De Meo, P., Messina, F., Postorino, M.N., Rosaci, D., Sarné, G.M.L.: A reputation framework to share resources into IoT-based environments. In: 2017 IEEE 14th International Conference on Networking, Sensing and Control (ICNSC), pp. 513–518 (2017)
21. Chen, I.R., Bao, F., Guo, J.: Trust-based service management for social internet of things systems. IEEE Trans. Dependable Secure Comput. **13**, 684–696 (2016)
22. Khan, W.Z., Aalsalem, M.Y., Khan, M.K., Arshad, Q.: When social objects collaborate: concepts, processing elements, attacks and challenges. Comput. Electr. Eng. **58**, 397–411 (2017)
23. Altaf, A., Haider, A., Faiza, I., Abdelouahid, D.: Trust models of internet of smart things: a survey, open issues and future directions. J. Netw. Comput. Appl. **137**, 93–111 (2019)
24. Liu, Y., Chen, Z., Xia, F., Lev, X., Bu F.: A trust model based on service classification in mobile services. In: Proceedings of the IEEE/ACM International Conference on Cyber, Physical and Social Computing, pp. 572–577 (2010)
25. Chen, D., Chang, G., Sun, D., Li, J., Jia, J., Wang, X.: TRM-IoT: a trust management model based on fuzzy reputation for Internet of Things. Comput. Sci. Inf. Syst. **8**(4), 1207–1228 (2011)
26. Bao, F., Chen, I.R.: Dynamic trust management for internet of things applications. In: Proceedings of the 2012 International Workshop on Self-aware Internet of Things, pp. 1–6. ACM, New York, USA (2012)
27. Chen, Z., Ling, R., Huang, C.M., Zhu, X.: A scheme of access service recommendation for the social Internet of Things. Int. J. Commun. Syst. **29**(4), 694–706 (2016)
28. Truong, N.B., Lee, H., Askwith, B., Lee, G.M.: Toward a trust evaluation mechanism in the social Internet of Things. Sensors **17**(6), 1346 (2017)

29. Magdich, R., Jemal, H., Ben Ayed, M.: Trust management model based on mutual evaluation method for the social internet of things. In: Themistocleous, M., Papadaki, M. (eds.) EMCIS 2019. LNBIP, vol. 381, pp. 361–375. Springer, Cham (2020). https://doi.org/10.1007/978-3-030-44322-1_27
30. Ekbatanifard, G., Yousefi, O.: A novel trust management model in the social internet of things. J. Adv. Comput. Eng. Technol. **5**(2), 57–70 (2019)
31. Fogla, P., Lee, W.: Evading network anomaly detection systems: formal reasoning and practical techniques. In: Proceedings of the 13th ACM Conference on Computer and Communications Security (CCS 2006), pp. 59–68. Association for Computing Machinery, New York (2006)
32. Dictionary.com. https://www.dictionary.com/browse/manipulator?s=t. Accessed 20 Dec 2020

Computer Science and Information Technology Students' Self-regulated Learning and Knowledge Sharing Behavior as Drivers of Individual Innovative Behavior

James K. Ngugi and Leila Goosen(✉)

University of South Africa, UNISA, Pretoria 0003, Gauteng, South Africa
GooseL@UNISA.ac.za

Abstract. Computer Science (CS), Information Technology (IT) and Information Systems (IS) students' Self-Regulated Learning (SRL) and Knowledge Sharing Behavior (KSB), together with the Course Design Characteristics (CDCs) implemented in their programs, could act as drivers to facilitate the development of their Individual Innovative Behavior (IIB). A Structural Equation Model (SEM) was therefore constructed to establish the mediating effect of KSB on the individual and contextual antecedents for the development of these students' IIB.

Keywords: Computer science students · Individual innovative behavior · Information technology · Knowledge sharing behavior · Self-regulated learning

1 Introduction

Amidst the perils and promises reported with regard to higher education in developing (especially African) countries, an international journal article a decade later by Salmi [1] still expressed concerns about quality education. This, despite the fact that, for example, things like educational technologies being used to increase students' access to an Information and Communication Technologies for Development (ICT4D) Massive Open Online Course (MOOC) [2].

More specifically related to the focus of the study reported on in this article, Chen, Li, Li, Zhang and Dong [3] lamented the inadequacy of practical methods to stimulate students' innovative tendencies. Earlier, Whetten [4] decried the lack of training of lecturers on the foundational aspect of design of courses, yet they are expected to design courses that can stimulate Individual Innovative Behavior (IIB) in Computer Science (CS) and Information Technology (IT) students. This presents a double pronged barrier to stimulating innovative tendencies.

Towards a conceptual framework for innovative education programs, Maritz, De Waal, Buse, Herstatt, Lassen and Maclachlan [5] presented a journal article in the context of European innovation management.

Literature, as cited in e.g. Ngugi and Goosen [6–8] and [9], as well as Goosen and Ngugi [10, 11] and [12], have identified multiple individual and contextual level factors or enablers, which promote individual innovative behavior among employees in

R. Silhavy (Ed.): CSOC 2021, LNNS 230, pp. 593–608, 2021.
https://doi.org/10.1007/978-3-030-77442-4_50

organizations. The problem stated in a conference paper, which argued that innovation for Computing students matters, and that the effects of Course Design Characteristics (CDC), Knowledge Sharing Behavior (KSB) and Self-Regulated Learning (SRL) on facilitating the development of IIB among undergraduate Computer Science (CS), "Information Systems (IS) and Information Technology (IT) Higher Education" Institutions (HEIs) students is not well-understood [11, p. 19]. The conference paper by Goosen and Ngugi [10, p. 376] therefore encouraged Computer Science lecturers to rethink teaching and learning in the 21st century in terms of course design characteristics towards individual innovative behavior for such students.

In a previous article, Ngugi and Goosen [7] had reported on a study, which constructed a Structural Equation Model (SEM) of course design characteristics, self-regulated learning and the mediating effect of knowledge sharing behavior as drivers of the individual innovative behavior of CS and IT students. Finally, the journal article by Ngugi and Goosen [8, p. 90] were modelling "motivation, metacognition and affective aspects of learning" towards smart innovation.

In their journal article, Janssen, Van de Vliert and West [13] **introduced** both the bright and dark sides of **individual** and group **innovation** in a special issue on organizational **behavior**.

2 A Review of the Literature

2.1 Course Design Characteristics

Job Design. In contexts of leadership, **innovation** and entrepreneurship as driving forces of the global economy, Dwivedula, Bredillet and Müller [14] looked at work motivation in temporary organizations by conducting a review of literature grounded in a **job design** perspective. A journal article related to the fields of business and psychology showed how Holman, Totterdell, Axtell, Stride, Port, Svensson and Zibarras [15] developed a model, which provided evidence of the mediating role of work-based learning strategies on both **job design** and the employee innovation process in terms of especially idea generation. The latter authors' results suggested that SRL supports the link between **job design** and favorable outcomes. While Oldham and Fried [16] investigated **job design** research and theory with regard to the past, present and future against a background of organizational behavior and human decision processes, the chapter by Parker and Zhang [17] examined future directions for **job design** research in terms of designing work that works in the contemporary world related to psycho-social factors at work in the Asia Pacific.

Barrick, Mount and Li [18] considered the roles of personality, higher-order goals and **job characteristics** in light of the theory of purposeful work behavior.

Task Characteristics. Hofmans, Gelens and Theuns [19] explored enjoyment as a mediator in the relationship between **task characteristics** and work effort through an experience sampling study in a journal article related to European work and organizational psychology. The paper by Li, Yuan and Che [20, p. 1] addressed "the importance of improving users' satisfaction and efficiency for obtaining health information through health information systems" in "order to understand how task characteristics, such as type, complexity, and difficulty, affect" users' interactions.

Autonomy. Langfred and Rockmann [21] studied the push and pull of *autonomy* as expressed in terms of the tension between individual autonomy and organizational control for knowledge work in the context of group and organization management.

Whereas De Spiegelaere, Van Gyes, De Witte, Niesen and Van Hootegem [22] researched the relationships between job insecurity, **job** *autonomy*, innovative work behavior and the mediating effect of work engagement against a background of creativity and innovation management, Giebels, De Reuver, Rispens and Ufkes [23] assessed the critical roles of **task** conflict and **job** *autonomy* in the relationship between proactive personalities and innovative employee behavior in their journal article on applied behavioral science.

Course Design. Based on the experience of the authors, most academic staff in university education join the profession without the requisite skills in curriculum and course design. Further, it takes time and effort to develop competencies in the design of effective courses, and often, thirty years later, many of them wished that they had known about learning-centered teaching much earlier. This was corroborated by Whetten [4] in terms of the principles of effective **course design**, who asserted that most CS and IT lecturers have limited training in the art of course design. Whetten, Johnson, and Sorenson [24, p. 257] observed that for CS and IT lecturers "to teach well, one must be competent in both course design and teacher-student interactions", while the journal article by Tarasewich and Nair [25] in the areas of business and economic studies indicated that **course design** should be making use of instructor and student preferences.

In the context of the university education of CS and IT students, '**job design**' can be equated to the design of a course.

2.2 Individual Innovative Behavior

Along with the rapid development of heuristics for the usability evaluation of m-commerce applications [26] and Information and Communication Technologies (ICTs), "digital technology changes how banks translate new customer demands into new products and services. To achieve this translation, banks should increase their intrapreneurship capability" and promote managers' innovative behavior at work in terms of individual factors and perceptions [27, p. 1].

With regard to the contexts of innovations and organizations as described by Zaltman, Duncan and Holbek [28], Liu and Chi [29] discussed the role of culture in e-learning against the background of **innovation** in construction education. Heukamp [30] worked on developing entrepreneurship capabilities in the Master of Business Administration (MBA) program by shaping entrepreneurial mindsets in terms of **innovation** and entrepreneurship in leadership development, while Bolton, Goosen and Kritzinger [31] showed how unified communication technologies could be used towards **innovation** at a global automotive organization.

A journal article on vocational behavior by Shanker, Bhanugopan, van der Heijden and Farrell [32] looked at the mediating effect of innovative work behavior on an organizational climate for **innovation** and organizational performance.

Chen, et al. [3] presented a study on the **innovation** capabilities of college students based on extenics and the theory of creativity, while chapters by Trošt and Škerlavaj [33], on team **innovation** through intuition and rationality, and Škerlavaj, Dysvik, Černe and Carlsen [34], appear in a book on capitalizing on creativity at work by fostering the implementation of creative ideas in organizations. Meanwhile, the management journal article by Zhang and Bartol [35] linked empowering leadership and employee creativity in terms of the influence of psychological empowerment, intrinsic motivation, and creative process engagement.

Whereas an international journal article in the field of innovation management by Sethibe and Steyn [36] studied the impact of leadership styles and the components of leadership styles on **innovative behavior**, in the context of educational management, Gkorezis [37] assessed principal empowering leadership and teacher **innovative behavior** towards a moderated mediation model. An overview of technology-supported teaching and research methods for educators [38] might also contribute.

While Hakimian, Farid, Ismail and Nair [39] surveyed the importance of commitment in encouraging employees' **innovative behavior**, the doctoral dissertation of Messmann [40] examined innovative work behavior by investigating the nature and facilitation of vocational teachers' contributions to innovation development.

2.3 Knowledge Sharing Behavior

According to Sawan [41, p. 331], **knowledge** is one of the most important resources "in an organization. Therefore, knowledge management is needed so that knowledge can be utilized effectively and efficiently. One dimension of knowledge management is knowledge sharing" behavior.

The purpose of the study by Keshavarz [42, p. 1] was aimed at investigating "the role of information literacy competencies regarding the impact of personality factors on knowledge sharing behavior in information services." In terms of the research design and methodology, a "sequential mixed-methods approach was used".

The study by Talebizadeh, Hosseingholizadeh and Bellibaş [43, p. 1] "tested the relationship between principals' learning-centered leadership and teacher professional learning in Iranian primary schools, with a focus on the mediating role of trust and knowledge sharing behavior", and with survey "data collected from 886 teachers". In such a way, a framework for partnerships promoting the continued support of schools can be built [44].

The study by Asghar and Naveed [45, p. 1] "examined the psychometric properties of the Knowledge Sharing Behavior Scale (KSBS) using" academics "at the University of Sargodha". A "quantitative research design was adopted using a survey questionnaire to collect data".

In their international journal article in the area of innovation management, Seo, Kim, Chang and Kim [46] examined the influence of interpersonal trust on innovative behavior of service workers, as well as the mediating effects of **knowledge sharing**. The latter authors found that knowledge sharing behavior fully mediated the impact of the link between trust in co-workers and innovative behavior, while the journal article by Nerstad, Searle, Černe, Dysvik, Škerlavaj and Scherer [47] offered insight into perceived mastery climate, felt trust, and **knowledge sharing** against a background of organizational behavior.

Shahsiah and Sepahvand [48] were studying the mediating role of **knowledge sharing** and market orientation to enhance the intellectual capital effect on organizational innovation in terms of the supervision of insurance companies in Lorestan province, in the context of international business management, while in their journal article related to the fields of leadership and organizational studies, Bartol and Srivastava [49] encouraged **knowledge sharing** with regard to the role of organizational reward systems.

Whereas the online information review by Chiu, Wang, Shih and Fan [50] contributed towards the understanding of **knowledge sharing behavior** in virtual communities through the integration of expectancy disconfirmation and justice theories, Liou, Chih, Yuan and Lin [51] conducted an empirical study of the antecedents of **knowledge sharing behavior** amongst the Yambol online test community in the context of internet research. A related study in the context of vocation education by Messmann [52] provided evidence of the mediating effect of **KSB** on innovative work behavior, whereas in the context of vocations and learning, Messmann and Mulder [53] looked at innovative work behavior in vocational colleges towards understanding how and why innovations are developed. Finally, Roth [54] studied **knowledge sharing behavior** intentions in wholesale distribution organizations.

Two of the existing studies, which correlated KSB with IIB, was French, McCarthy, Baraitser, Wellings, Bailey and Free [55], on young people's views and experiences of a mobile phone texting intervention to promote safer sex behavior, and Laycock, Bailie, Matthews and Bailie [56], who researched interactive dissemination for engaging stakeholders in the use of aggregated quality improvement data for system-wide change in Australian indigenous primary health care.

2.4 Self-regulated Learning

Self-Regulated Learning "is useful for understanding self-directed learning practices. However, SRL" behaviors, "despite being deemed highly context-dependent", have remained "mostly unexplored for healthcare workers in low-income countries." The study by Tuti, Paton and Winters [57, p. 1] detailed these workers' counter-intuitive self-regulated learning behaviors in such settings.

Self-regulated learning "has a predictable and instrumental effect on learning complicated knowledge." The study by Huang and Lajoie [58, p. 1] investigated "the role of SRL in acquiring" Technological Pedagogical Content Knowledge "(TPACK), an important aspect of" the effectiveness of teachers.

In an international journal article against the background of technology and design education, Barak [59] considered motivating self-regulated learning in technology education, while the journal article by Effeney, Carroll and Bahr [60] probed self-regulated learning, key strategies and their sources in a sample of adolescent Australian males with regard to educational and developmental psychology.

"During the lockdown due to SARS-CoV-2 (coronavirus lockdown), there has been a tremendous increase in the number of students taking online courses. Few studies, however, have examined the individual dispositions that influence self-regulated online learning" [61, p. 1]. One such study did examine students' perceptions of e-assessment in the context of Covid-19 in a case study of the University of South Africa (UNISA) [62].

Learning Analytics Dashboards (LADs) are intended to provide "relevant information to students and other stakeholders to inform potential next steps in the learning process." The study by Aguilar, Karabenick, Teasley and Baek [63, p. 1] examined the associations between learning analytics dashboard exposure and motivation and self-regulated learning.

"The flipped classroom model has gained prominence as advances in technology afford increasing opportunities for ubiquitous access to a variety of online resources. Despite the benefit of the flipped classroom model, flipped classrooms are not equally advantageous to" all [64, p. 1]. It was therefore necessary to design supports for promoting self-regulated learning in the flipped classroom.

According to van Alten, Phielix, Janssen and Kester [65, p. 1], Flipped Learning "(FL) makes greater use of students' self-regulated learning … skills when they regulate their online learning behavior. Previous research has shown the value of SRL support during FL to enhance students' SRL and learning outcomes."

"Academic help-seeking is a self-regulatory strategy that can have an important influence on students' learning and achievement. The primary goal of the" study by Won, Hensley and Wolters [66, p. 112] "was to investigate whether college students' sense of belonging could be used to" better understand their academic use of SRL.

"The limited instructional support in Massive Open Online Courses (MOOCs) inherently demands learners to self-regulate their learning. MOOC research shows that learners are more successful when they engage in self-regulated learning (SRL) behaviors" [67, p. 1].

Self-Regulated Learning and Knowledge Sharing Behavior. The study reported on in this paper hypothesized that self-regulated learning is a significant antecedent of KSB for CS and IT students. This was informed by the available literature, which suggested that self-regulatory skills are a significant predictor of academic achievement and, by extension, innovative behavior [15, 29, 68, 69].

3 Methods

The conference paper by Goosen and Ngugi [12, p. 333] indicated that scale "items related to Cognitive and Metacognitive Strategies (CMSs)", course design characteristics, knowledge sharing behavior and individual innovative behavior could "be gathered from literature and located within relevant theoretical conceptual frameworks."

A journal article in the area of educational psychology by Petersen, Louw and Dumont [70] examined the adjustment of questionnaires' length to accommodate university and academic performance among disadvantaged students in South Africa. During the **pilot**ing stage, some of the items in the questionnaire used in the study reported on in this article with low item-to-total correlation were eliminated to minimize the number of items within each construct, with the aid of scale reduction techniques. However, the reduction in the number of items exposed the scale to some risks. According to Creswell [71], such a reduction may have an influence on the psychometric properties of the scale and its reliability. In effect, the test length is more likely to increase the magnitude of the reliability coefficient. Thus, to obtain optimal test length, a careful balance was attained between time required and content issues. A scale reduction analysis was used to generate the inter-item total correlation, and the Cronbach [72] alpha reliability coefficients and the internal structure of tests as presented in Goosen and Ngugi [10] and [12], as well as described in Ngugi and Goosen [7]. Especially [12] further reported on the items deleted based on these reliability scores.

Another journal article against the background of psycho-educational assessment by Mowbray, Boyle and Jacobs [73] reported on an investigation using the German Test Anxiety Inventory (TAI-G) in an Australian sample, towards establishing the impact of item orientation on the structural validity of self-report measures, while, in what had been considered at the time to be a state-of-the-art assessment, Boudreau, Gefen and Straub [74] explored validation in information systems research.

With qualitative perspectives, Goosen and Mukasa-Lwanga [75] described how to take educational technologies in distance education beyond the horizon, while Goosen and Van Heerden [76] went beyond the horizon of learning programming with educational technologies.

3.1 Evaluating Model Fit and Goodness of Fit Statistics

In terms of providing some clarifications and recommendations on fit indices, Newsom [77, p. 123] indicated that the model fit indices available for use in a structural equation model can be grouped into four (4) types, namely "absolute fit indices, relative fit indices, parsimony fit indices, and those based on the" noncentrally parameter.

With regard to multivariate data analysis in SEM, Hair, Anderson, Tatham and Black [78] recommended assessing the models by comparing more than one of the available indicators. In a multidisciplinary journal article on conventional cutoff criteria versus new alternatives for fit indexes in covariance structure analysis towards structural equation modeling, Hu and Bentler [79] made specific recommendations to reduce the likelihood of Type I and Type II errors. Hence, the researcher "should use a combination of one of the … relative fit indexes and the" Standardized Root Mean Square Residual (SRMR) "(good models $< .08$) or the" Root Mean Square Error of Approximation (RMSEA) (good models $< .06$) [77, p. 127]. In contrast, Hu and Bentler [79] recommended the use of fit indices that provide different aspects of measurement properties, such as an Incremental Fit Index (IFI) and a residuals-based fit index, such as the Standardized Root Mean Square Residual.

With regard to data preparation and/or analysis, Fuller, Simmering, Atinc, Atinc and Babin [80] provided details with regard to common methods variance detection in business research. To prevent a so-called 'brain drain', Jauhar, Ghani and Islam [81] recommended that aspects with regard to **data analysis and results** should be carefully considered.

4 Results

4.1 Multicollinearity

Testing for multicollinearity in the data involved the use of regression with unweighted summated scores, as well as correlation analysis. The measure of multicollinearity in the regression analysis involved two (2) standards measures: Tolerance and the Variance Inflation Factor (VIF). The extraction of the variance inflation factors helped to examine whether collinear variables exist among the research variables, which is a necessary precursor to other forms of analysis, such as Exploratory Factor Analysis (EFA), Confirmatory Factor Analysis (CFA) and SEM, so as be sure that the assumptions of multivariate analysis were not violated. The results of the extraction of the VIF values for the explanatory variable was as indicated in Tables 1, 2 and 3, with each table having a different dependent variable from among CDC, KSB and IIB.

Table 1. Collinearity statistics for driver variables with course design characteristics as dependent variable.

Coefficients[a]	Unstandardized coefficients		Standardized coefficients	t	Sig.	Collinearity statistics	
Model	ß	Std. error	Beta			Tolerance	VIF
(Constant)	1.784	.133		13.451	.000		
Self-regulated learning	.079	.024	.190	3.274	.001	.843	1.186
Knowledge sharing behavior	.176	.040	.324	4.446	.000	.535	1.868
Innovative behavior	.078	.036	.156	2.171	.031	.553	1.810

[a]Dependent variable: Course design characteristics

Please note that collinearity statistics of the driver variables with self-regulated learning as the dependent variable were provided in Goosen and Ngugi [11].

The problem of multicollinearity can be avoided by deleting any explanatory variable with VIF > 10 from the regression model. However, the results of the study suggested that there was no problem with multicollinearity, as the largest VIF of 1.415 was lower than the cut-off value of 10 recommended by Hair, et al. [78], and was still lower that the more constrained level of 2.5 suggested by Allison [82] for multiple regression. The VIF results supported the fact that multicollinearity was not a problem in the regression models. Further, the tolerance values were all well above 0.5, and thus above the threshold value of 0.2 recommended by Field [83]. This again confirmed that there was no problem of multicollinearity in the data.

Table 2. Collinearity statistics for driver variables with knowledge sharing behavior as dependent variable.

Coefficients[a]	Unstandardized coefficients		Standardized coefficients	t	Sig.	Collinearity statistics	
Model	ß	Std. error	Beta			Tolerance	VIF
(Constant)	−.045	.270		−.168	.867		
Self-regulated learning	.089	.038	.117	2.362	.019	.826	1.210
Innovative behaviour	.483	.047	.523	10.240	.000	.773	1.293
Course design characteristics	.422	.095	.229	4.446	.000	.757	1.321

[a]Dependent variable: Knowledge sharing behavior

Table 3. Collinearity statistics for driver variables with dependent variable innovative behavior.

Coefficients[a]	Unstandardized coefficients		Standardized coefficients	t	Sig.	Collinearity statistics	
Model	ß	Std. error	Beta			Tolerance	VIF
(Constant)	.417	.305		1.368	.173		
Self-regulated learning	.065	.043	.079	1.520	.130	.816	1.226
Course design characteristics	.240	.111	.121	2.171	.031	.714	1.400
Knowledge sharing behavior	.619	.060	.572	10.240	.000	.707	1.415

[a]Dependent variable: Innovative behavior

A further test for multicollinearity involved calculating the Condition Index (CI). The results are as presented in Table 4.

All VIF values for the three (3) explanatory variables were below 2. Further, the CI was below 30 (maximum CI = 23.1). These values suggested that multicollinearity did not affect the estimations [78].

While Falk and Biesanz [84] examined two cross-platform programs for inferences and interval estimation about indirect effects in mediational models, in the context of econometric reviews, Shintani and Guo [85] provided guidelines for improving the finite sample performance of autoregression estimators in dynamic factor models when using **a bootstrap approach**.

Table 4. Collinearity diagnostics of condition index with innovative behavior as the dependent variable.

Dimension	Eigenvalue	Condition Index	Variance Proportions			
			(Constant)	Course design characteristics	Self-regulated learning	Knowledge sharing behavior
1	3.949	1.000	.00	.00	.00	.00
2	.024	12.704	.04	.00	.29	.81
3	.019	14.280	.22	.09	.70	.09
4	.007	23.100	.74	.91	.00	.10

5 Discussions

5.1 Contributions of the Present Research

Practical Implications. *Implications for Curriculum Experts.* The desire for autonomy calls for a paradigm shift in philosophy and also the procedures and processes that university Senates use to review existing programs, as well as to develop new academic programs. At the time of the study, in Kenya, the curriculum development process was guided by curriculum development policies that are developed by each university. The development of these polices is guided and aligned to the Universities Act [86], as well as standards and guidelines provided by the Commission for University Education (CUE) regulations. A paradigm shift modelled and informed by the course design characteristics could leverage these towards offering greater autonomy on the part of the student in terms of providing for increased student autonomy in scheduling of courses, decision-making, and choice of methods.

While taking a look into the innovation paradigm in the context of information diversity and a group culture of creativity, Uparna [87, p. 24] stated that "a creative idea usually requires one to do something in a new and different way". This urge to do things in a new and/or different way lays the foundation for innovation and calls for a radical shift in beliefs and values. However, a "group's goals toward deliverables may not necessarily coincide with its search for new ideas and" [87, p. 14] and creative solutions, explaining why education providers may perceive students with innovative behavior as radical, non-conformist and rabble-rousers. The antecedents and consequences of achievement goals as detailed in the PhD thesis of Kahraman [88] in the context of Turkey and the Middle East may need to be considered.

Ideas that are gauged to be viable should be allowed to progress to the third stage of being championed by students, by providing them with communities of practice, and eventually be applied and linked to industry. A potential challenge in the process would be how to measure the learning outcomes when academic programs are not prescriptive, as is the current scenario. The measurement of learning outcomes when the innovative atmosphere calls for ill-structured problems can be problematic. Ill-structured problems presented to students have no best answers, as opposed to well-structured problems. Such ill-defined problems do not "contain sufficient information to enable the designer (with) the means of meeting those requirements simply by transforming, reducing, optimizing or superimposing the given information alone" [89, p. 384]. However, with the right mix and balance of systematic methods for development in design methodology, it is possible to leverage these factors to generate learning outcomes indicative of innovative tendencies among CS and IT students.

Other Drivers of Knowledge Sharing Behavior at Individual and Contextual Level. In a journal article based in the South African context of information management, Jacobs and Roodt [90] suggested that organizational culture was the single most important driver variable, with KSB acting as a mediating effect on the turnover intentions of professional nurses. A possible enrichment of the model could incorporate suggestions from literature of constructs, such as organizational culture in terms of support and collaboration, learning and development, leadership commitment effectiveness, as well as organizational structure as defined by formalization and centralization. These variables have close correlations to KSB. Hence, the paths to benefit from such additions include:

a) Course design characteristics predicting knowledge sharing behavior;
b) Task characteristics predicting knowledge sharing behavior;
c) Knowledge characteristics predicting knowledge sharing behavior; and
d) Self-regulated learning predicting knowledge sharing behavior.

References

1. Salmi, J.: Peril and promise: a decade later. Int. J. Afr. High. Educ. **3**(1), 43–50 (2016)
2. Goosen, L.: Students' access to an ICT4D MOOC. In: Proceedings of the 47th Annual Conference of the Southern African Computer Lectures' Association (SACLA 2018), Cape Town (2018)
3. Chen, A., Li, L., Li, X., Zhang, J., Dong, L.: Study on innovation capability of college students based on extenics and theory of creativity. Procedia Comput. Sci. **17**, 1194–1201 (2013)
4. Whetten, D.A.: Principles of effective course design: what I wish I had known about learning-centred teaching 30 years ago. J. Manag. Educ. **31**(3), 339–357 (2007)
5. Maritz, A., De Waal, A., Buse, S., Herstatt, C., Lassen, A., Maclachlan, R.: Innovation education programs: toward a conceptual framework. Eur. J. Innov. Manag. **17**(2), 166–182 (2014)
6. Ngugi, J., Goosen, L.: The effects of course design characteristics, self-regulated learning and knowledge sharing in facilitating the development of innovative behaviour among technology students at universities. In: Proceedings of the Institute of Science and Technology Education (ISTE) Conference on Mathematics, Science and Technology Education, Mopani Camp, Kruger National Park (2017)
7. Ngugi, J., Goosen, L.: Modelling course-design characteristics, self-regulated learning and the mediating effect of knowledge-sharing behavior as drivers of individual innovative behavior. EURASIA J. Math. Sci. Technol. Educ. **14**(8), em1575 (2018)
8. Ngugi, J., Goosen, L.: Towards smart innovation for information systems and technology students: modelling motivation, metacognition and affective aspects of learning. In: Rocha, Á., Serrhini, M. (eds.) EMENA-ISTL 2018. SIST, vol. 111, pp. 90–99. Springer, Cham (2019). https://doi.org/10.1007/978-3-030-03577-8_11

9. Ngugi, J.K., Goosen, L.: Knowledge sharing mediating information technology student innovation. In: Khosrow-Pour, M. (ed.) Handbook of Research on Modern Educational Technologies, Applications, and Management, vol. II, pp. 645–663. IGI Global, Hershey (2021)
10. Goosen, L., Ngugi, J.K.: Rethinking teaching and learning in the 21st century: course design characteristics towards innovative behavior. In: Proceedings of the South Africa International Conference on Education, Pretoria (2018)
11. Goosen, L., Ngugi, J.K.: Innovation for computing students matter, of course! In: Proceedings of 48th Annual Conference of the Southern African Computer Lecturers' Association (SACLA 2019), Northern Drakensberg, South Africa (2019)
12. Goosen, L., Ngugi, J.K.: Towards innovative behaviour for technology students. In: Proceedings of the 10th Institute of Science and Technology Education (ISTE) International Conference on Mathematics, Science and Technology Education, Mopani Camp, Kruger National Park (2019)
13. Janssen, O., Van de Vliert, E., West, M.: The bright and dark sides of individual and group innovation: a special issue introduction. J. Organ. Behav. **25**(2), 129–145 (2004)
14. Dwivedula, R., Bredillet, C.N., Müller, R.: Work motivation in temporary organizations: a review of literature grounded in job design perspective. In: Benlamri, R., Sparer, M. (eds.) Leadership, Innovation and Entrepreneurship as Driving Forces of the Global Economy. SPBE, pp. 609–618. Springer, Cham (2017). https://doi.org/10.1007/978-3-319-43434-6_52
15. Holman, D., et al.: Job design and the employee innovation process: the mediating role of learning strategies. J. Bus. Psychol. **27**(2), 177–191 (2012)
16. Oldham, G.R., Fried, Y.: Job design research and theory: past, present and future. Organ. Behav. Hum. Decis. Process. **136**, 20–35 (2016)
17. Parker, S.K., Zhang, F.: Designing work that works in the contemporary world: future directions for job design research. In: Shimazu, A., Bin Nordin, R., Dollard, M., Oakman, J. (eds.) Psychosocial Factors at Work in the Asia Pacific, pp. 135–150. Springer, Cham (2016). https://doi.org/10.1007/978-3-319-44400-0_7
18. Barrick, M.R., Mount, M.K., Li, N.: The theory of purposeful work behaviour: the role of personality, higher-order goals, and job characteristics. Acad. Manag. Rev. **38**(1), 132–153 (2013)
19. Hofmans, J., Gelens, J., Theuns, P.: Enjoyment as a mediator in the relationship between task characteristics and work effort: an experience sampling study. Eur. J. Work Organ. Psychol. **23**(5), 693–705 (2014)
20. Li, Y., Yuan, X., Che, R.: An investigation of task characteristics and users' evaluation of interaction design in different online health information systems. Inf. Process. Manag. **58**(3), 102476 (2021)
21. Langfred, C.W., Rockmann, K.W.: The push and pull of autonomy: the tension between individual autonomy and organizational control in knowledge work. Group Org. Manag. **41**(5), 629–657 (2016)
22. De Spiegelaere, S., Van Gyes, G., De Witte, H., Niesen, W., Van Hootegem, G.: On the relation of job insecurity, job autonomy, innovative work behaviour and the mediating effect of work engagement. Creativity Innov. Manag. **23**(3), 318–330 (2014)
23. Giebels, E., De Reuver, R.S., Rispens, S., Ufkes, E.G.: The critical roles of task conflict and job autonomy in the relationship between proactive personalities and innovative employee behavior. J. Appl. Behav. Sci. **52**(3), 320–341 (2016)
24. Whetten, D.A., Johnson, T.D., Sorenson, D.L.: Learning-centered course design. In: The SAGE Handbook of Management Learning, Education and Development, pp. 255–270. SAGE, Thoasand Oaks (2009)
25. Tarasewich, P., Nair, S.K.: Course design using instructor and student preferences. J. Bus. Econ. Stud. **6**(2), 40–54 (2000)

26. Ajibola, S., Goosen, L.: Development of heuristics for usability evaluation of M-commerce applications. In: Proceedings of the South African Institute of Computer Scientists and Information Technologists (SAICSIT), Thaba Nchu, South Africa (2017)
27. Kör, B., Wakkee, I., van der Sijde, P.: How to promote managers' innovative behavior at work: individual factors and perceptions. Technovation. **99**, 102127 (2021)
28. Zaltman, G., Duncan, R., Holbek, J.: Innovations and Organizations. Wiley, New York (1973)
29. Liu, A., Hodgson, G., Lord, W.: Innovation in construction education: the role of culture in e-learning. Archit. Eng. Des. Manag. **6**(2), 91–102 (2010)
30. Heukamp, F.: Developing entrepreneurship capabilities in the MBA program. In: Canals, J. (ed.) Shaping Entrepreneurial Mindsets: Innovation and Entrepreneurship in Leadership Development, pp. 203–219. Palgrave Macmillan, London (2015)
31. Bolton, A.D., Goosen, L., Kritzinger, E.: Unified communication technologies at a global automotive organization. In: Khosrow-Pour, M. (ed.) Encyclopedia of Organizational Knowledge, Administration, and Technologies, pp. 2592–2608. IGI Global, Hershey (2021)
32. Shanker, R., Bhanugopan, R., van der Heijden, B.I.J.M., Farrell, M.: Organizational climate for innovation and organizational performance: the mediating effect of innovative work behavior. J. Vocat. Behav. **100**, 67–77 (2017)
33. Trošt, J.K., Škerlavaj, M.: Should our heart rule our head? Team innovation through intuition and rationality. In: Capitalizing on Creativity at Work: Fostering the Implementation of Creative Ideas in Organizations, pp. 64–75. Edward Elgar Publishing, Cheltenham (2016)
34. Škerlavaj, M., Dysvik, A., Černe, M., Carlsen, A.: Succeeding with capitalizing on creativity: an integrative framework. In: Capitalizing on Creativity at Work: Fostering the Implementation of Creative Ideas in Organizations, pp. 335–352. Edward Elgar Publishing Ltd., Cheltenham (2016)
35. Zhang, X., Bartol, K.M.: Linking empowering leadership and employee creativity: the influence of psychological empowerment, intrinsic motivation, and creative process engagement. Acad. Manag. J. **53**, 107–128 (2010)
36. Sethibe, T., Steyn, R.: The impact of leadership styles and the components of leadership styles on innovative behaviour. Int. J. Innov. Manag. **21**(2), 1–19 (2017)
37. Gkorezis, P.: Principal empowering leadership and teacher innovative behaviour: a moderated mediation model. Int. J. Educ. Manag. **30**(6), 1030–1044 (2016)
38. Goosen, L.: Technology-supported teaching and research methods for educators: case study of a massive open online course. In: Makewa, L., Ngussa, B., Kuboja, J. (eds.) Technology-Supported Teaching and Research Methods for Educators, pp. 128–148. IGI Global, Hershey (2019)
39. Hakimian, F., Farid, H., Ismail, M.N., Nair, P.K.: Importance of commitment in encouraging employees' innovative behaviour. Asia-Pac. J. Bus. Adm. **8**(1), 70–83 (2016)
40. Messmann, G.: Innovative work behaviour: Investigating the nature and facilitation of vocational teachers' contributions to innovation development. Unpublished Doctoral dissertation. Universität Regensburg, Regensburg (2012)
41. Sawan, F.: Impact of organizational culture on knowledge sharing behavior. In: 4th International Conference on Research of Educational Administration and Management (ICREAM 2020) (2021)
42. Keshavarz, H.: Personality factors and knowledge sharing behavior in information services: the mediating role of information literacy competencies. VINE J. Inf. Knowl. Manag. Syst. (2021)
43. Talebizadeh, S.M., Hosseingholizadeh, R., Bellibaş, M.Ş: Analyzing the relationship between principals' learning-centered leadership and teacher professional learning: The mediation role of trust and knowledge sharing behavior. Stud. Educ. Eval. **68**, 100970 (2021)

44. Vorster, J., Goosen, L.: A framework for university partnerships promoting continued support of e-Schools. In: Proceedings of the 46th Annual Conference of the Southern African Computer Lecturers' Association (SACLA 2017), Magaliesburg (2017)
45. Asghar, A., Naveed, M.A.: Psychometric evaluation of knowledge sharing behavior scale in academic environment. Libr. Philos. Pract. (e-journal) (2021)
46. Seo, H.M., Kim, M.C., Chang, K., Kim, T.: Influence of interpersonal trust on innovative behaviour of service workers: mediating effects of knowledge sharing. Int. J. Innov. Manag. **20**(2), 1–21 (2016)
47. Nerstad, C.G., Searle, R., Černe, M., Dysvik, A., Škerlavaj, M., Scherer, R.: Perceived mastery climate, felt trust, and knowledge sharing. J. Organ. Behav. **39**(4), 429–447 (2018)
48. Shahsiah, N., Sepahvand, R.: Studying the mediating role of knowledge sharing and market orientation to enhance the intellectual capital effect on the organizational innovation (Supervision of Insurance Companies in Lorestan Province). Int. Bus. Manag. **10**(20), 4776–4782 (2016)
49. Bartol, K.M., Srivastava, A.: Encouraging knowledge sharing: the role of organizational reward systems. J. Leadersh. Organ. Stud. **9**(1), 64–76 (2002)
50. Chiu, C.M., Wang, E.T., Shih, F.J., Fan, Y.W.: Understanding knowledge sharing behaviour in virtual communities: an integration of expectancy disconfirmation and justice theories. Online Inf. Rev. **35**(1), 134–153 (2011)
51. Liou, D.K., Chih, W.H., Yuan, C.Y., Lin, C.Y.: The study of the antecedents of knowledge sharing behaviour: the empirical study of Yambol online test community. Internet Res. **26**(4), 845–868 (2016)
52. Messmann, G.: Innovative work behaviour: investigating the nature and facilitation of vocational teachers' contributions to innovation development. (Unpublished Doctoral dissertation). Universität Regensburg, Regensburg (2012)
53. Messmann, G., Mulder, R.H.: Innovative work behaviour in vocational colleges: understanding how and why innovations are developed. Vocat. Learn. **4**(1), 63–84 (2011)
54. Roth, A.E.: Knowledge sharing behaviour intentions in wholesale distribution organizations, (Unpublished Doctoral dissertation). Walden University, Minneapolis (2016)
55. French, R.S., McCarthy, O., Baraitser, P., Wellings, K., Bailey, J.V., Free, C.: Young people's views and experiences of a mobile phone texting intervention to promote safer sex behavior. JMIR mHealth and uHealth **4**(2), e26 (2016)
56. Laycock, A., Bailie, J., Matthews, V., Bailie, R.: Interactive dissemination: engaging stakeholders in the use of aggregated quality improvement data for system-wide change in Australian indigenous primary health care. Front. Public Health **4**, 84 (2016)
57. Tuti, T., Paton, C., Winters, N.: The counterintuitive self-regulated learning behaviours of healthcare providers from low-income settings. Comput. Educ. **166**, 104136 (2021)
58. Huang, L., Lajoie, S.P.: Process analysis of teachers' self-regulated learning patterns in technological pedagogical content knowledge development. Comput. Educ. **166**, 104169 (2021)
59. Barak, M.: Motivating self-regulated learning in technology education. Int. J. Technol. Des. Educ. **20**(4), 381–401 (2010)
60. Effeney, G., Carroll, A., Bahr, N.: Self-Regulated Learning: Key strategies and their sources in a sample of adolescent males. Aust. J. Educ. Dev. Psychol. **13**, 58–74 (2013)
61. Hong, J.C., Lee, Y.F., Ye, J.H.: Procrastination predicts online self-regulated learning and online learning ineffectiveness during the coronavirus lockdown. Personality Individ. Differ. **174**, 110673 (2021)
62. Van Heerden, D., Goosen, L.: Students' perceptions of e-assessment in the context of Covid-19: the case of UNISA. In: Proceedings of the 29th Conference of the Southern African Association for Research in Mathematics, Science and Technology Education (SAARMSTE) (2021)

63. Aguilar, S.J., Karabenick, S.A., Teasley, S.D., Baek, C.: Associations between learning analytics dashboard exposure and motivation and self-regulated learning. Comput. Educ. **162**, 104085 (2021)
64. Yoon, M., Hill, J., Kim, D.: Designing supports for promoting self-regulated learning in the flipped classroom. J. Comput. High. Educ., 1–21 (2021)
65. van Alten, D.C., Phielix, C., Janssen, J., Kester, L.: Secondary students' online self-regulated learning during flipped learning: a latent profile analysis. Comput. Hum. Behav. **118**, 106676 (2021)
66. Won, S., Hensley, L.C., Wolters, C.A.: Brief research report: sense of belonging and academic help-seeking as self-regulated learning. J. Exp. Educ. **89**(1), 112–124 (2021)
67. Wong, J., Baars, M., de Koning, B.B., Paas, F.: Examining the use of prompts to facilitate self-regulated learning in Massive Open Online Courses. Comput. Hum. Behav. **115**, 106596 (2021)
68. Liu, M., Chi, M.: Investigating learner affective performance in web-based learning by using entrepreneurship as a metaphor. Educ. Technol. Soc. **15**(3), 202–213 (2012)
69. Za, S., Spagnoletti, P., North-Samardzic, A.: Organisational learning as an emerging process: the generative role of digital tools in informal learning practices. Br. J. Edu. Technol. **45**(6), 1023–1035 (2014)
70. Petersen, I., Louw, J., Dumont, K.: Adjustment to university and academic performance among disadvantaged students in South Africa. J. Educ. Psychol. **29**(1), 99–115 (2009)
71. Creswell, J.W.: Research Design: Qualitative, Quantitative and Mixed Methods Approaches. SAGE Publications Inc., London (2014)
72. Cronbach, L.J.: Coefficient alpha and the internal structure of tests. Psychometrika **16**(3), 297–334 (1951)
73. Mowbray, T., Boyle, C., Jacobs, K.: Impact of item orientation on the structural validity of self-report measures: an investigation using the TAI-G in an Australian sample. J. Psychoeduc. Assess. **33**(3), 278–290 (2014)
74. Boudreau, M.C., Gefen, D., Straub, D.: Validation in information systems research: a state-of-the-art assessment. MIS Q. **25**(1), 1–16 (2001)
75. Goosen, L., Mukasa-Lwanga, T.: Educational technologies in distance education: beyond the horizon with qualitative perspectives. In: Proceedings of the South Africa International Conference on Educational Technologies, Pretoria (2017)
76. Goosen, L., Van Heerden, D.: Beyond the horizon of learning programming with educational technologies. In: Proceedings of the South Africa International Conference on Educational Technologies, Pretoria (2017)
77. Newsom, J.T.: Some clarifications and recommendations on fit indices. USP **655**, 123–133 (2012)
78. Hair, J.F., Jr., Anderson, R.E., Tatham, R.L., Black, W.C.: Multivariate Data Analysis, 5th edn. Prentice-Hall, Hoboken (1998)
79. Hu, L.T., Bentler, P.M.: Cutoff criteria for fit indexes in covariance structure analysis: conventional criteria versus new alternatives. Struct. Equ. Model. Multidiscip. J. **6**(1), 1–55 (1999)
80. Fuller, C.M., Simmering, M.J., Atinc, G., Atinc, Y., Babin, B.J.: Common methods variance detection in business research. J. Bus. Res. **69**(8), 3192–3198 (2016)
81. Binti Jauhar, J., Abdul Ghani, A.B., Islam, R.: Results and data analysis. In: Binti Jauhar, J., Abdul Ghani, A.B., Islam, R. (eds.) Brain Drain, pp. 127–158. Springer, Singapore (2016). https://doi.org/10.1007/978-981-10-0977-8_5
82. Allison, P.D.: Multiple Regression: A Primer . Pine Forge Press, Thousand Oaks (1999)
83. Field, A.P.: Discovering Statistics Using SPSS, 2nd edn. Sage Publications, Thousand Oaks (2009)

84. Falk, C.F., Biesanz, J.C.: Two cross-platform programs for inferences and interval estimation about indirect effects in mediational models. SAGE Open **6**(1), 1–13 (2016)
85. Shintani, M., Guo, Z.Y.: Improving the finite sample performance of autoregression estimators in dynamic factor models: a bootstrap approach. Economet. Rev. **37**(4), 360–379 (2018)
86. Republic of Kenya, Universities Act. Kenya Gazette Supplement No. 192 (Acts No. 42). Government Printer, Nairobi (2012)
87. Uparna, J.: Information diversity and group culture of creativity: a look into the innovation paradigm. Unpublished doctoral dissertation. Rutgers University, New Jersey (2008)
88. Kahraman, N.: Antecedents and consequences of achievement goals. Ph.D. thesis. Middle East Technical University, Ankara (2011)
89. Archer, L.B.: Systematic method for designers. In: Cross, N. (ed.) Developments in Design Methodology, pp. 57–82. John Wiley & Sons, Chichester (1984)
90. Jacobs, E.J., Roodt, G.: The mediating effect of knowledge sharing between organisational culture and turnover intentions of professional nurses. S. Afr. J. Inf. Manag. **13**(1), 1–6 (2011)

The Geoimage Generation Method for Decision Support Systems Based on Natural Language Text Analysis

A. V. Vicentiy(✉)

Institute for Informatics and Mathematical Modeling, Kola Science Centre of the Russian Academy of Sciences, 14, Fersman St., Apatity 184209, Russia

Abstract. Geospatial data is a critical component of decision support systems for regional governance. The use of spatio-temporal data is most useful for rapid mapping of socio-economic phenomena and analyzing changes in the main indicators of regional development. An important advantage of using geospatial data is its versatility and applicability to a wide range of decision support tasks. Many modern decision-support information systems have their own tools for geospatial data processing. Where such tools are not available, integration with geographic information systems for visualization and analysis of geospatial data is possible. An important factor is the speed of obtaining and visualizing geospatial data. This paper describes the rapid geoimage generation method for regional governance decision support systems based on the analysis of natural language texts. The first part of this paper describes the structure of a system for recognizing named entities from texts in Russian, based on hand-crafted grammars and gazetteers. It also shows a fragment of the grammar and describes the method for populating the gazetteer. The second part of this paper briefly describes the process of geocoding and visualization of named entities extracted from natural language texts. The result of the method is an interactive geoimage (digital map). This geoimage is used by a decision-maker as an additional source of information for decision-making. A special feature of the proposed geoimage generation method for decision support systems is the high speed of obtaining geoimages displaying the named entities recognized in the text.

Keywords: Geovisualization · Decision support systems · Natural language processing · Named entity recognition · Grammar · Gazetteer · Geospatial data · Rule-based approach

1 Introduction

The management of large spatially distributed systems is a complex, challenging problem. In order to solve it effectively, it is necessary to analyze large amounts of heterogeneous data from different subject areas. The final data analysis and decision making is carried out by a decision maker (DM). Modern decision support systems make it possible to provide the decision maker with the necessary information in an easy-to-understand

R. Silhavy (Ed.): CSOC 2021, LNNS 230, pp. 609–619, 2021.
https://doi.org/10.1007/978-3-030-77442-4_51

form. One of the most important types of data used by DM for decision-making is spatiotemporal data presented in the form of thematic geoimages or specially prepared digital maps [1].

Over the last twenty years relevant geospatial data have become of great importance for decision-making support, especially for spatial planning tasks, regional socio-economic development management, marine spatial planning, and others [2, 3]. The importance of using geospatial data is due to the fact that more than half of digital data are now spatially referenced [4]. However, most of this data is unstructured or weakly structured information. This fact makes it difficult to use such data for decision-making. A significant amount of geospatial data is contained in natural language texts, such as articles, reports, regulatory documents, tweets, social media posts, comments, and so on. In order to obtain these data, it is necessary to solve the problem of information extraction [5, 6], that is, to obtain a structured representation of information from unstructured texts.

Modern Decision Support Systems (DSS) and Geographic Information Systems (GIS) have a large variety of ways and tools to retrieve, store, process, visualize and analyze data. However, despite the relevance and importance of extracting and visualizing geospatial data contained in texts, modern DSS and GIS do not have standard tools to address this problem [7, 8]. The lack of such tools limits the possibility of applying existing DSS and GIS for operational decision support in spatial planning and regional management. Therefore, we believe that the development of methods for extracting and visualizing geospatial data from natural language texts and methods for using this data for decision support systems, is an urgent scientific and applied problem. The geoimage generation method for decision support systems based on natural language text analysis described in this paper allows to provide the decision maker with an additional source of data for decision making.

2 Related Works

As mentioned above, modern decision support systems and geographic information systems generally do not have standard tools to deal with the problem of visualizing geospatial data contained in natural language texts. To retrieve geospatial data from texts, an information extraction procedure has to be performed. One of the important subtasks of information retrieval is the Named Entity Recognition (NER) problem [9]. The challenge of NER is to identify named entities in texts and determine their types (proper names, organization names, addresses, toponyms, locations, various codes, etc.) [10, 11].

Currently, there are several different approaches to solve the NER problem. The most popular approaches for recognizing named entities in natural language texts are: rule-based approach, machine learning approach, hybrid approach [12, 13]. Each of these approaches has its advantages and disadvantages. In recent years, the use of machine learning and neural networks for the recognition of named entities in natural language texts is a global trend [14]. However, the researcher has to consider the characteristics of the problem and the subject area when choosing a particular approach [15]. In some papers, the authors refer to rule-based NER algorithms as obsolete and of little promise.

However, for languages that are not as widespread as English and for languages with rich morphology [16], the rule-based and dictionary (gazetteer) approach of NER often shows good results [17, 18]. In addition, the use of rules improves the results of hybrid NER approaches [19], and also gives good results for some subject areas [20, 21].

There are now many off-the-shelf solutions, platforms, frameworks, toolkits and software libraries for natural language processing (NLP). There are now many off-the-shelf solutions, platforms, frameworks, toolkits and software libraries for natural language processing (NLP). Many of them have special facilities and tools for identifying named entities. In particular, we can mention such software tools and libraries as: GATE, Apache OpenNLP, Apache cTAKES, Apache NLPCraft, Stanford NLP, spaCy, NLTK, Google Language API, LSPL, tomita-parser, TensorFlow, and others [22]. For our work, the simplicity and speed of recognition of named entities (toponyms and organization names) in a limited and well-defined subject area, as well as the ability to take into account the features and limitations that are peculiar to the Russian language, are of great importance. We analyzed and compared the capabilities of various software tools and chose tomita-parser as the basic software tool for analyzing texts in Russian in our work.

Currently the vast majority of NER research has been done for English. The number of studies on the recognition of named entities for other languages is much smaller. This is also true for Slavic languages, to which Russian belongs. Unlike English, Russian is characterized by such phenomena as derivation, rich inflection, free word order and others. Because of these peculiarities, the recognition of named entities is a challenging task for the Russian language [23, 24]. In addition, there are far fewer corpora for Russian and other Slavic languages that can be used to develop and test NER systems. As a rule, corpora for Russian contain several times fewer tokens than corpora for English. Insufficient number of large corpora for Russian language is one of the important problems for implementation of effective NER systems based on machine learning and neural networks. Another important problem of such systems is the sharp decline of the efficiency of named entities recognition when the subject area changes. A change in the subject area may lead to the need to re-create and train neural networks that are used to identify and classify named entities. [25] Ultimately, this can lead to a significant change in the entire architecture of the NER system based on machine learning. Unfortunately, there is currently no universal solution to this problem.

3 Materials and Methods

We had to consider many factors and constraints when choosing the NER approach and the means to implement a text analysis system. In particular, it had to be taken into account that the geoimage generation method based on natural language text analysis is part of a larger decision support system for regional management. Therefore, we had to take into account the peculiarities of the DSS architecture when choosing the way to implement the text analysis subsystem. On the other hand, the subject area for our system is limited and clearly defined. In addition, to ensure satisfactory accuracy and completeness of the recognition of named entities (toponyms and organization names) it was necessary to take into account the peculiarities of the Russian language (rich morphology, free word order, etc.). Considering all the factors and limitations, we chose the NER approach based on hand-crafted rules and gazetteers. We have chosen tomita-parser as the software tool for lexico-syntactic analysis of texts. Tomita-parser is an open-source parser, distributed under the MPL 2.0 license, supports Russian language processing and allows the user to create their own context-free grammars and gazetteers.

To automate the geocoding of recognized named entities (toponyms and organization names), we have developed a software module based on the GeoPy client. This module allows us to output the geocoding results to a file in one of two formats - csv and geojson. We use the ArcGIS Online API [7] to create the geoimage. As input for geovisualization, we use the csv file with the coordinates of the objects obtained in the previous step.

For the analysis we used texts in Russian from heterogeneous sources. All the texts were divided into several main groups. The first group of texts are various normative legal acts, strategies and development programmes, laws, decrees and so on. These texts are created by official authorities. The distinctive characteristics of these texts are the presence of a clear internal structure, a large volume, references to other documents, special terminology, and a complex sentence structure.

The second group of texts are texts from various websites, including official websites of authorities, large companies and public associations, news websites, etc. These texts are created by official authorities, professional authors and ordinary Internet users. The distinguishing characteristics of these texts are the presence of some internal structure, a small volume, references to other documents, simple vocabulary, and a simple sentence structure.

The third group of texts are texts from social networks, blogs and other social media. These texts are created by official authorities or government representatives, ordinary internet users and professional authors. The vast majority of such texts are created by ordinary Internet users in the form of posts and comments. In the Russian Federation, the most popular social media are VKontakte, Facebook, TikTok, Instagram, Twitter and various channels in Telegram. Distinctive characteristics of these texts are the absence of a common internal structure, small volume, references to other documents, simple vocabulary and use of jargon, simple sentence structure, errors in the text, high "concentration of information", high speed of publication, and others.

The fourth group of texts are "all other" texts, that is, texts from various sources that do not belong to any of the other groups. For example, it can be texts of scientific articles, proceedings of various conferences, transcripts of interviews, meetings, video-conferences, and so on. These texts are created by different authors. This group of texts has no distinctive characteristics. The main criterion for inclusion of texts in this group is the relevance of the content of the texts to the context of the task of analysis.

4 Results

The workflow of the geoimage generation method for decision support systems based on natural language text analysis consists of several steps (Fig. 1).

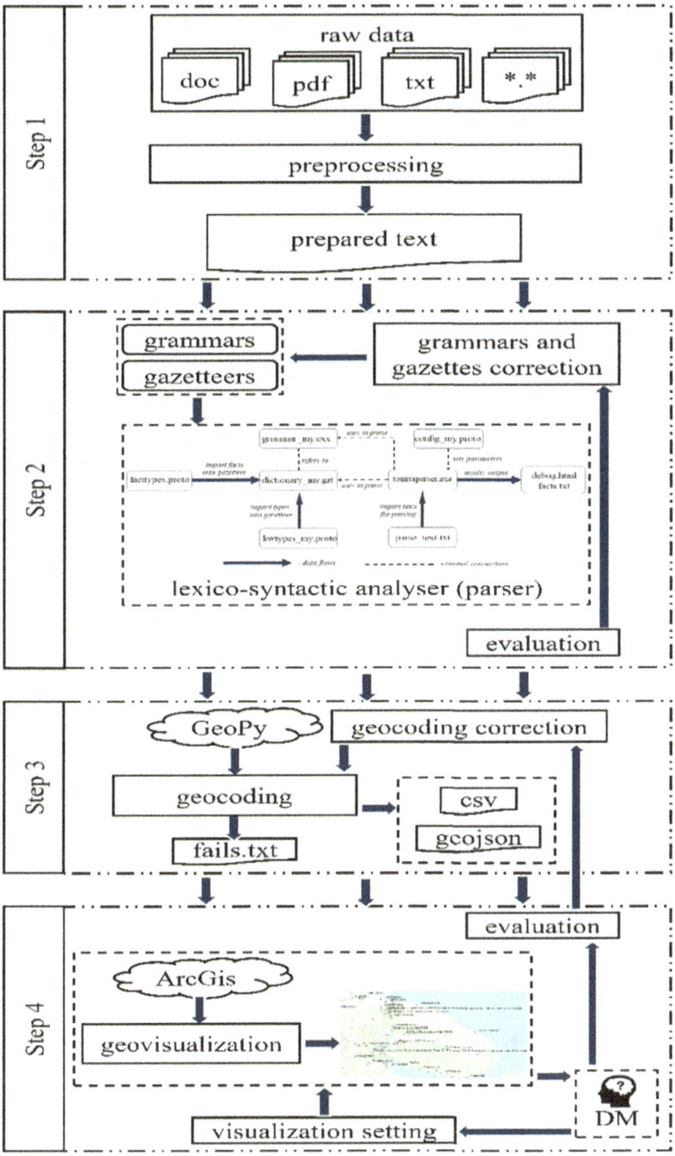

Fig. 1. The workflow of the geoimage generation method for decision support systems based on natural language text analysis

In the first step, we collect raw data and preprocess the texts. We use texts from heterogeneous sources. These texts have different formats (doc, docx, pdf with text layer, pdf without text layer, plain text, etc.), so it is necessary to do preprocessing of texts (changing encoding, text recognition, extraction of text layer and others). After preprocessing we have a txt file prepared for NER analysis. In the second step, the text is analyzed for the recognition of the named entities. Named entity recognition is implemented using a lexico-syntactic analyzer and hand-crafted context free grammars and gazetteers. Also in this step an intermediate evaluation of the recognition of the named entities can be performed. Based on the results of this evaluation, we can make changes to the rules of grammars and gazetteers. The third step is geocoding the recognized named entities. For these purposes, we have created a Python program module that uses the GeoPy client services. In the fourth step, a geoimage is generated in which the recognized named entities are marked. The geoimage allows the decision maker to change the visualization settings according to his preferences and the specifics of the problem to be solved.

In the first phase of our work, we have developed a system for the recognition of named geo-referenced entities (toponyms and organization names) in Russian texts. The system is designed for automated dynamic recognition of named entities in a limited and well-defined subject area. We use a lexico-syntactic approach to text analysis in order to recognize toponyms and organization names. Our method is based on the generalized left-to-right algorithm (GLR) for the syntactic analysis of texts, as well as hand-crafted original context free grammars and gazetteers. The use of these grammars and gazetteers allows us to take into account the peculiarities of the Russian language and the constraints of the subject area. This approach makes it possible to interpret as toponyms arbitrary named entities that have a geographical reference. To extract structured data from texts, we used the Tomita Parser. To perform geocoding, we used the GeoPy client features for python. To visualize geodata we used the Arcgis online API.

For the parser to work correctly, we have developed a specific operating environment and a certain set of additional files. The generalized scheme of the syntactical analyzer work that we have created is shown in Fig. 2. Files with the extension *.cxx are grammar files, files with the extension *.gzt are gazetteer files, *.proto are configuration files necessary for the correct operation of the syntactical analyzer, *.txt and *.html files contain source and processed data, the file tomitaparser.exe is a binary file that performs text analysis.

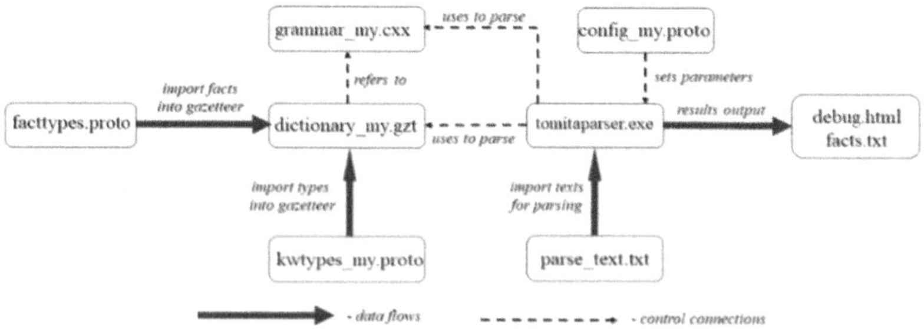

Fig. 2. The generalized scheme of the syntactical analyzer work

The greatest influence on the quality of recognition of named entities in the Russian-language text is provided by context free grammars that we have created. These grammars allow us to take into account the peculiarities of word formation in Russian and the limitations of the subject area. The Russian language is very morphologically complex and variable. In different cases, genders and numbers, nouns and adjectives, which can be a toponym or the name of an organization, have different forms. Therefore, the structure of grammars is usually very complex and the process of developing grammars is very time consuming. As an example, here is a fragment of our grammar for extracting the names of different localities from texts:

```
#encoding "utf8"
ToponymW -> 'город' | 'поселок городского типа' |
'Закрытое административно-территориальное образование' |
'сельский населённый пункт' | 'населённый пункт' |
'железнодорожная станция' | 'километр';
ToponymSokr -> 'г.' | 'пгт' | 'зато' | 'ж/д станция' |
'посёлок' | 'село' | 'км';
ToponymDescr -> ToponymW | ToponymSokr;
ToponymNameNoun -> (Adj<gnc-agr[1]>) Word<gnc-agr[1],rt>;
NumberToponym_0 -> AnyWord<wff=/[1-9]?[0-9]-
?((ый)|(ий)|(ой)|й)/>;
NumberToponym -> NumberToponym_0;
ToponymNameAdj -> NumberToponym<gnc-agr[1]> Word<gnc-
agr[0]>;
Toponym -> ToponymDescr interp (Toponym.Descr) ToponymNa-
meNoun<gram="род", h-reg1> interp (Topo-
nym.ToponymName::not_norm);
Toponym -> ToponymDescr interp (Toponym.Descr) ToponymNa-
meNoun<gram="им", h-reg1> interp (Topo-
nym.ToponymName::not_norm);
Toponym -> ToponymNameAdj<gnc-agr[1]> interp (Topo-
nym.ToponymName) ToponymW<gnc-agr[1]> interp (Toponym.De-
scr);
Toponym -> ToponymNameAdj interp (Toponym.ToponymName)
ToponymSokr interp (Toponym.Descr);
Toponym -> ToponymW<gnc-agr[1]> interp (Toponym.Descr)
ToponymNameAdj<gnc-agr[1]> interp (Toponym.ToponymName);
Toponym -> ToponymSokr interp (Toponym.Descr)
ToponymNameAdj interp (Toponym.ToponymName);
```

There is a great variety of types of settlements in the Russian language. This grammar allows us to extract from texts the names of such settlement types as: city, urban-type settlement, closed administrative-territorial formation, rural settlement, populated locality, railway station (type of settlement in Russia), kilometer (type of settlement in Russia), as well as their abbreviations.

In addition to grammars, we also use hand-created gazetteers for our subject area. These gazetteers improve the efficiency of named entity recognition. The Russian language is characterized by complex names of toponyms, which can consist of several words, including, numerals. As an example, we can cite such complex names of settlements as "The village of 25 km of the Monchegorsk-Olenya Railroad" (Поселок 25 км Железной Дороги Мончегорск-Оленья - in Russian) or "The Travel Manor 9 km of the Luostari-Nikel railroad" (Путевая Усадьба 9 км железной дороги Луостари-Никель - in Russian). Recognition of such and similar named entities without the use of special gazetteers is a very difficult problem. Given the fact that our subject area is limited and well-defined (named entities of the Murmansk region), we have created relatively compact and efficient gazetteers for automated dynamic recognition of named entities as part of our problem solving. To simplify the creation and filling of gazetteers, we have developed a python software module, which allows us to automate the filling of gazetteers. The software module allows syntactically correct addition of new records to the gazetteer, as well as correcting existing records (adding synonyms, abbreviations, former place names, etc.).

To automatically generate a geoimage (digital map) based on the recognized named entities it is necessary to geocode these entities (step three of the workflow). We have developed a python program module to geocode the recognized entities. This program module determines the coordinates of toponyms and other georeferenced named entities automatically using the capabilities of the GeoPy service. The output is a list of the named entities with their geographical coordinates. The list is output in one of the popular formats for geodata (csv and geojson). This list is used for geovisualization in the next step of the workflow.

In the last step of the geoimage generation method for decision support systems based on natural language text analysis, an automatic geovisualization of the recognized named entities is performed. To implement the automatic revisualization, we have created a software module that generates a digital map based on the data obtained in the previous step. We use the ArcGIS API to automate the geoimaging process. The output of this software module is a geoimage (interactive digital map), which is provided to the decision maker as an additional source of visual information for decision making. It can be said that this geoimage, in a sense, reflects the "geographical meaning" of the texts that have been used in the lexico-syntactic analysis using the rules and gazetteers that we have developed.

To test and demonstrate the capabilities of the proposed geoimage generation method for decision support systems based on natural language text analysis an experiment was conducted. As part of this experiment, a subject-oriented geo-image was automatically generated on which more than 70 settlements and 10 large enterprises of the Murmansk region were marked (Fig. 3).

To evaluate the efficiency of named entity recognition, we created a corpus of texts for testing, conducted an experiment and calculated precision and recall. As experimental data, we used texts obtained from the official website of the government of the Murmansk region, as well as from social networks, Twitter of the governor of the Murmansk region and other sources in Russian. According to the results of the experiment, the precision was 0.96, recall was 0.93, F-measure (2 * precision * recall/(precision + recall)) was 0.94. We believe that the results of the experiment are promising and look for ways to further improve the performance of named entity recognition.

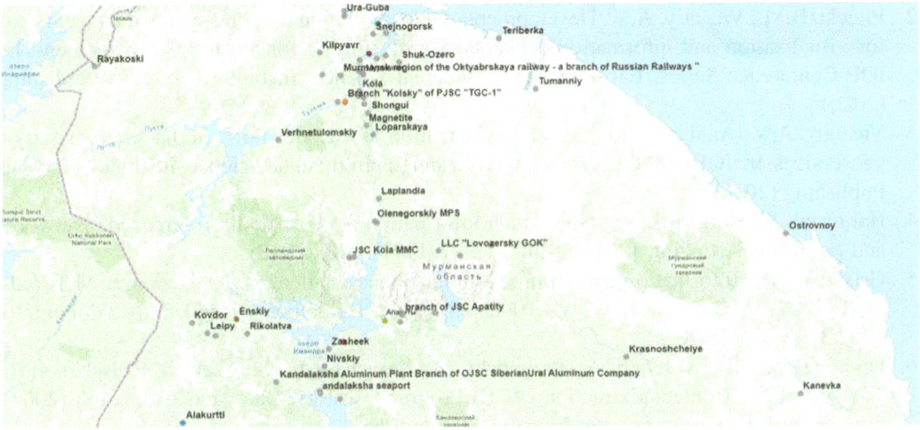

Fig. 3. The geoimage generated from analysis of natural language texts

5 Discussion and Future Work

In this paper we have proposed the geoimage generation method for decision support systems based on natural language text analysis. The method generates a digital map, which is provided to the decision maker as an additional source of information for decision making. The difference between this method and other well-known named entity recognition techniques, based, for example, on methods of distribution semantics, is that it does not require large arrays of pre-marked tests (text corpora) for operation. The NER subsystem described in the paper is implemented based on lexico-syntactic analysis. In order to improve the efficiency of named entity recognition we used rule based approach and hand-created grammars and gazetteers. This approach has shown good results for our problem with a well-defined subject area. We are currently working on improving the method. In particular, we are considering using NER systems based on machine learning. We believe that this hybrid approach will reduce the fragility of the NER system and provide acceptable recognition quality. Another direction for improving the geoimage generation method for decision support systems based on natural language text analysis is to develop an approach to take into account the individual characteristics of the end-user's perception of the information. This will allow us to generate personalized cognitive geoimages that are better suited to the needs of the user. The results will be used to create intelligent multi-subject information systems for information support for the development management of spatially-organized bio-socio-economic regional systems, including the Arctic zone of the Russian Federation.

References

1. Vicentiy, A.V.: The use of different types of geographic information systems for the construction of geoservices to support management of the Arctic territories. In: IOP Conference Series: Earth and Environmental Science. Institute of Physics Publishing (2020).

2. Pilecki, B.M., Vicentiy, A.V.: Development of a method for extracting spatial data from texts for visualization and information decision-making support for territorial management. In: IOP Conference Series: Earth and Environmental Science. Institute of Physics Publishing (2020)
3. Vicentiy, A.V.: Analysis and dynamic visualization of logistics flows in the Arctic based on geoservices. In: IOP Conference Series: Earth and Environmental Science. Institute of Physics Publishing (2020)
4. Hahmann, S., Burghardt, D.: How much information is geospatially referenced? Networks and cognition. Artic. Int. J. Geogr. Inf. Sci. **27**, 1171–1189 (2013)
5. Grishman, R.: Information extraction: techniques and challenges. In: Pazienza, M.T. (ed.) SCIE 1997. LNCS, vol. 1299, pp. 10–27. Springer, Heidelberg (1997). https://doi.org/10.1007/3-540-63438-X_2
6. Doddington, G.R., Mitchell, A., Przybocki, M., Ramshaw, L., Strassel, S., Weischedel, R.: The Automatic Content Extraction (ACE) Program – Tasks, Data, and Evaluation (2004). http://www.lrec-conf.org/proceedings/lrec2004/pdf/5.pdf
7. ArcGis official website, https://www.esri.com/en-us/arcgis/products/arcgis-pro/overview.
8. Qgis official website, https://qgis.org/
9. Nadeau, D., Sekine, S.: A survey of named entity recognition and classification Lingvisticae Investig. Int. J. Linguist. Lang. Resour. Investig. **30**, 3–26 (2007)
10. Tjong Kim Sang, E.F., De Meulder, F.: Introduction to the CoNLL-2003 Shared task: language-independent named entity recognition. In: Proceedings of the Seventh Conference on Natural Language Learning at HLT-NAACL 2003, vol. 4, pp. 142–147. Association for Computational Linguistics, USA (2003)
11. Ratinov, L., Roth, D.: Design challenges and misconceptions in named entity recognition. In: Proceedings of the Thirteenth Conference on Computational Natural Language Learning, pp. 147–155. Association for Computational Linguistics, USA (2009)
12. Yadav, V., Bethard, S.: A survey on recent advances in named entity recognition from deep learning models. In: Proceedings of the 27th International Conference on Computational Linguistics. pp. 2145–2158. Association for Computational Linguistics, Santa Fe (2018)
13. Stock, K.: Mining location from social media: A systematic review (2018). www.elsevier.com/locate/ceus
14. Collobert, R., Weston, J., Bottou, L., Karlen, M., Kavukcuoglu, K., Kuksa, P.: Natural language processing (almost) from scratch. J. Mach. Learn. Res. **12**, 2493–2537 (2011)
15. Song, H.J., Jo, B.C., Park, C.Y., Kim, J.D., Kim, Y.S.: Comparison of named entity recognition methodologies in biomedical documents. Biomed. Eng. Online. **17**, 158 (2018). https://doi.org/10.1186/s12938-018-0573-6
16. Ivanitskiy, R., Shipilo, A., Kovriguina, L.: Russian named entities recognition and classification using distributed word and phrase representations. In: Proceedings of the 3rd Annual International Symposium on Information Management and Big Data - SIMBig 2016, Cusco, Peru, 1–3 September 2016, pp. 150–156. CEUR-WS.org (2016)
17. Farmakiotou, D., Karkaletsis, V., Koutsias, J., Sigletos, G., Spyropoulos, C.D., Stamatopoulos, P.: Rule-based named entity recognition for Greek financial texts. In: In Proceedings of the Workshop on Computational Lexicography and Multimedia Dictionaries (COMLEX 2000, pp. 75–78 (2000)
18. Alfred, R., et al.: A rule-based named-entity recognition for malay articles. In: Motoda, H., Wu, Z., Cao, L., Zaiane, O., Yao, M., Wang, W. (eds.) Advanced Data Mining and Applications. Lecture Notes in Computer Science (Lecture Notes in Artificial Intelligence), vol. 8346, pp. 288–299. Springer, Heidelberg (2013). https://doi.org/10.1007/978-3-642-53914-5_25

19. Sari, Y., Hassan, M.F., Zamin, N.: Rule-based pattern extractor and named entity recognition: a hybrid approach. In: Proceedings 2010 International Symposium on Information Technology - Engineering Technology, ITSim 20'10. pp. 563–568 (2010)
20. Eftimov, T., Seljak, B.K., Korošec, P.: A rule-based named-entity recognition method for knowledge extraction of evidence-based dietary recommendations. PLoS ONE **12**, e0179488(2017)
21. I, Y.E., Çiçekli, N.: A TV content augmentation system exploiting rule based named entity recognition method. In: Abdelrahman, O.H., Gelenbe, E., Gorbil, G., Lent, R. (eds.) Information Sciences and Systems 2015. LNEE, vol. 363, pp. 349–357. Springer, Cham (2016). https://doi.org/10.1007/978-3-319-22635-4_32
22. Shelar, H., Kaur, G., Heda, N., Agrawal, P.: Named entity recognition approaches and their comparison for custom NER model. Sci. Technol. Libr. **39**, 324–337 (2020)
23. Piskorski, J., Pivovarova, L., Šnajder, J., Steinberger, J., Yangarber, R.: The first cross-lingual challenge on recognition, normalization, and matching of named entities in slavic languages. In: Proceedings of the 6th Workshop on Balto-Slavic Natural Language Processing, pp. 76–85. Association for Computational Linguistics, Stroudsburg, USA (2017)
24. Piskorski, J., et al.: The second cross-lingual challenge on recognition, normalization, classification, and linking of named entities across Slavic languages. In: Proceedings of the 7th Workshop on Balto-Slavic Natural Language Processing, pp. 63–74. Association for Computational Linguistics, Stroudsburg, USA (2019)
25. Brykina, M.M., Faynveyts, A.V., Toldova, S.: Dictionary-based ambiguity resolution in Russian named entities recognition. Case study. **1**, 163–177 (2013)

Item-Share Propagation Link Applying for Recommendation

Sahraoui Kharroubi[1](✉), Youcef Dahmani[1], and Omar Nouali[2]

[1] Faculty of Mathematics and Computer Science, Ibn Khaldoun University, Tiaret, Algeria
sahraoui.kharroubi@univ-tiaret.dz
[2] Software Laboratory, C.E.R.I.S.T, Ben Aknoun, Algiers, Algeria
onouali@cerist.dz

Abstract. The growth of social networks and various web services (e-commerce, e-learning, e-health, etc.) urgently requires recommendation techniques to satisfy users. Providing high-quality recommendations with a minimum of common feedback is a major challenge for a new recommendation algorithm. In this paper, we use more informative modelling with a weighted bipartite network around the item entity. The idea is to exploit the item-user connectivity to extract hidden information. The problem is to predict non-existent links based on existing links by double projection forward and backward. The results are promising with the implementation of real data sets.

Keywords: Ranking · Recommender systems · Prediction · Neighbourhood · Collaborative

1 Introduction

Recommendation systems have been widely used by companies to recommend relevant items to users by employing a similarity process. Large companies and websites such as Netflix, Amazon, Facebook, YouTube, Twitter, etc. integrate the techniques of recommendation in their servers. The methods of recommendation systems can be mainly classified into three approaches [1, 2]: the content-based approach which compares the content of the user profile with the content of the item; the collaborative approach which depends on the feedback of neighbour's ratings; and the hybrid approach which aims at objectively combining the two methods. In the literature, to predict a link in a bipartite network, most methods are based either on a binary history (like, dislike) or on the common neighbourhood of the active user. Currently, several in-depth studies are focused on modelling using bipartite networks [3, 4]. The structure of bipartite networks is perfectly adapted as a theoretical and practical model for several systems in the real world. The recommendation process in e-commerce or e-learning is modelled by a bipartite network that links two distinct sets of users and items. The recommendation task in the recommendation system is equivalent to a link prediction in a bipartite network. We have proposed Item Share propagation for Link Ranking in recommender systems (ISpLR). The method aims to determine the weighting of a link as a ranking value predicted by the

R. Silhavy (Ed.): CSOC 2021, LNNS 230, pp. 620–631, 2021.
https://doi.org/10.1007/978-3-030-77442-4_52

user towards an item. The idea is to accumulate the acquired information by linking the items to the users through forward-backward projection. The process can be summarized as follows:

(a) Item Share propagation for Link Ranking in a weighted bipartite network (ISpLR)
(b) Accumulation of shares linearly to pre-serve information propagated with a double projection (forward- backward)
(c) Implementation of the method without adjustment parameters and with different dense datasets.

This paper is organized as follows: the back-ground is given in Sect. 2, a detailed description of the approach is presented in Sect. 3, a description of the experimental phase and discussion of the results is given in Sect. 4 and 5 respectively. In the conclusion section, we make some suggestions for future work.

2 Background

Collaborative filtering (CF) is the most widely used and effective approach to create personalized recommendations on the Web [5]. The crucial task of CF is to form communities (neighbourhoods) of users or items through similar feedback. The community is used to predict items deemed relevant to users who have not yet judged these items. Over the past two decades, several algorithms [13–15], have been proposed and these can be classified into three main levels.

- Traditional approaches

They are based on the rating history of user's "rating matrix" and apply statistical methods to make predictions. The assumption is that users with similar ratings would have similar tastes in the future. These approaches are divided into two categories: memory-collaborative filtering (memory CF) and model-collaborative filtering (model CF) [6]. Memory CF uses the set of rating matrix scores to make predictions about the active user. In order to reduce the computational complexity, model CF uses a part of the "rating matrix" to estimate or learn a model that generates the predictions.

- Context-aware and semantic approaches

Context is additional information categorized in dimensions to better describe the user or item. These dimensions include personal information (e.g., gender, occupation, age), professional and social information (activities, interests, interactions with other users, preferences, etc.), physical conditions (weather, noise, etc.), location (geo-graphic coordinates), and time [7]. The objective is to value this information and gather similar contexts to predict an item. Adomavicius and Tuzhilin (CARS context-aware recommender systems) [8], using the mobile phone and the Global Positioning System (GPS) can recommend a restaurant or hotel to a tourist based on previous choices of its similar neighbours by context [9].

- Graph-based approaches

Each node is linked to another node forming a network structure with a particular topology. The unipartite topology represents a set of nodes and a set of edges as in the case of social networks such as Facebook, Twitter, YouTube, etc. PageRank [10] and HITS [11] are the most important algorithms for calculating the importance score of a node by a random walk or an iterative process in a network of web pages. The EdgeRank algorithm fills [12] a user's Facebook profile through the parameters of affinity, type, and freshness. The goal is to exploit the network structure to collect possible links in the future. Several studies [5, 7] are based on a bipartite structure to improve recommendation algorithms. Nesreen K. [13] implemented the SimAdapt algorithm to measure the similarity between two nodes according to the number of common neighbours in a bipartite graph. The research presented by Jung-Hun Kim et al. [14] exploits a bipartite network between textual titles and keywords in a research article in the field of physics. The paper presents a method for recommending items in a weighted bipartite network. Based on the item-rating history, the method is used to predict new links and recommend relevant items to users. For this, a cumulative share is counted on the sides of both items and users.

3 Approach

3.1 Problem and Motivation

Many complex systems are based on a net-work model that consists of two components: nodes and links. These systems focus mainly on networks with single-party topology (social net-works, links between web pages, atoms and molecules etc.). A single-party network consists of a single set of nodes that are connected by links. Network model algorithms are easy to interpret, implement, and operate. A recommendation system is a typical case of a particular network called the bipartite network.

3.2 Modelling with a Weighted Bipartite Network

A network is bipartite if the set of nodes is divided into two distinct sets I and U. Each link has one end in the first set and the other end in the second set (no links in the same set), and k determines the weight of the link. Formally, let B(N, L) be a bipartite network where N is the set of nodes and L is the set of links. B has the following properties:

$$\begin{cases} N = I \cup U \\ L = \{(i, u, R(i, j)),\ \ i \in U,\ \ R(i, j) \in \mathbb{R}^{+}\} \\ I \cap U = \emptyset \end{cases} \tag{1}$$

3.3 Item Link Prediction

Consider a recommender system which is composed of a set of items $\{I_1, I_2, ...I_N\}$ and a set of users $\{U_1, U_2, ...U_M\}$ linked by weighted links $\{(i, u, k), i \in I, u \in U, k \in N\}$.

From the previous interactions of users with the items (the rating history), let's try to deduce a non-existent link weight. In another way, can we recommend a relevant planned item to a user? This decision is equivalent to valuing the link weight between the item and the user, like (I_3, U_1) (Fig. 1).

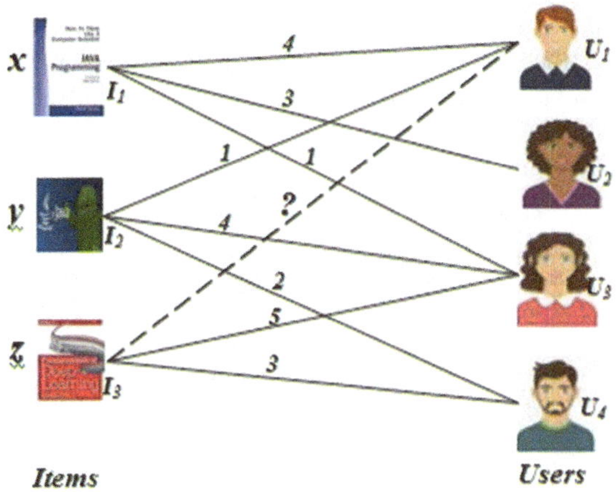

Fig. 1. Weighted link prediction

To make such a prediction we seek to determine the shares of items shared between users. In the first step (forward step), we accumulate the quotas of the users towards the items and in the second step (backward step) the shares of the items towards the users, knowing that N is the number of items, M is the number of users, and R(i, j) is the weight of the link (i, j) (rating value).

In the first step, each item has a cumulative of ratings,

$$IC_i = \sum_{j=1}^{M} R(i,j) \tag{2}$$

The share of user j,

$$q(j) = \sum_{i=1}^{M} \frac{R(i,j)}{IC_i} \tag{3}$$

In the second step, each user has a cumulative of ratings,

$$UC_j = \sum_{i=1}^{N} R(i,j) \tag{4}$$

The share of item *i*,

$$q(i) = \sum_{j=1}^{M} \frac{R(i,j)}{UC_j} \tag{5}$$

Then, the value of link prediction of the item *t* to the user *s* is given by the following formula:

$$P(t,s) = \sum_{j=1}^{M} \frac{R(t,j)}{\sum_{i=1}^{N} R(i,j)} * \left(\sum_{i=1}^{N} R(i,j)) * \frac{R(i,s)}{\sum_{j=1}^{M} R(i,j)} \right) \tag{6}$$

By replacing item cumulate (2) and user cumulate (4), we get

$$P(t,s) = \sum\nolimits_{j=1}^{M} \frac{R(t,j)}{UC_j} * \left(\sum\nolimits_{i=1}^{N} R(i,j) * \frac{R(i,s)}{IC_i}\right) \tag{7}$$

The bipartite topology offers a bidirectional reasoning item side and user side, which preserve the information shared between the nodes. The consideration of the weights of the links specifies the degree of satisfaction of the users. This method presents an item-based prediction but the same user-based reasoning can also be applied (Fig. 2).

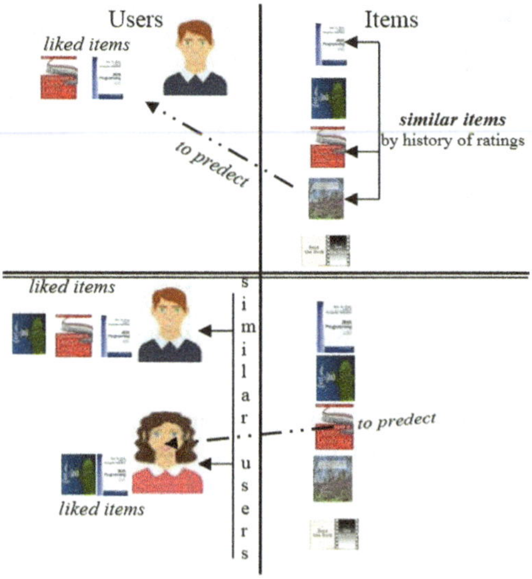

Fig. 2. Item link prediction versus user link prediction

The example below details the method proposed via item link prediction, specifying the prediction performance between them. Based on the link values presented in Fig. 1.

Either three items I_1, I_2, I_3 with initial inputs x, y, z respectively.

First step (forward): Determining user's share,

$$\begin{cases} U_1 : \frac{4}{8}x + \frac{1}{7}y \\ U_2 : \frac{3}{8}x \\ U_3 : \frac{1}{8}x + \frac{4}{7}y + \frac{5}{8}z \\ U_4 : \frac{2}{7}y + \frac{3}{8}z \end{cases}$$

Second step (backward): Determining item's share,

$$\begin{cases} I_1 : \frac{4}{5}U_1 + \frac{3}{3}U_2 + \frac{1}{10}U_3 + \frac{0}{5}U_4 \\ I_2 : \frac{1}{5}U_1 + \frac{0}{3}U_2 + \frac{4}{10}U_3 + \frac{2}{5}U_4 \\ I_3 : \frac{0}{5}U_1 + \frac{0}{3}U_2 + \frac{5}{10}U_3 + \frac{3}{5}U_4 \end{cases}$$

This gives us:

$$\begin{cases} I_1 : \frac{63}{80}x + \frac{6}{35}y + \frac{1}{16}z \\ I_2 : \frac{3}{20}x + \frac{13}{35}y + \frac{2}{5}z \\ I_3 : \frac{1}{16}x + \frac{16}{35}y + \frac{43}{80}z \end{cases}$$

So, the predictions of the links of the missing items are mentioned in Table 1.

Table 1. Item link prediction.

	U_1	U_2	U_3	U_4
I_1	**4**	3	1	**0.5304**
I_2	1	**0.45**	4	2
I_3	**0.7071**	**0.1875**	5	3

So, the prediction value of the $link(I_3, U_1) = 1/16\,(4) + 16/35\,(1) + 43/80\,(0) = 0.7071$.

Further, the prediction value of the $link(I_1, U_4) = 63/80\,(0) + 6/35\,(2) + 1/16\,(3) = 0.5304$.

To know the error of the prediction, we generalize the calculation for existing vote values (Table 2) and apply the mean absolute error measure MAE (see Sect. 4.3).

Table 2. MAE of item-based predictions.

	U_1	U_2	U_3	U_4	MAE I_i
I_1	**3.3214**	**2.3625**	**1.7857**	**0.5304**	**0.7006**
I_2	**0.9714**	**0.4500**	**3.6357**	**1.9429**	**0.1500**
I_3	**0.7071**	**0.1875**	**4.5786**	**2.5268**	**0.4473**
	MAE				0.4326

The MAE measure makes it possible to differentiate between the rating value given by the users and the prediction value generated by the formula (6).

For item I_2 and I_3:

$$MAE(I_2) = 1/3(|1 - 0.9714| + |4 - 3.6357| + |2 - 1.9429|) = 0.1500$$

$$MAE(I_3) = 1/2(|5 - 4.5786| + |3 - 2.5268|) = 0.4473$$

3.4 Recommendation Task

A recommender system processes a large number of items and selects the most relevant to recommend them to users. The decision to recommend a particular item depends on the predicted value deduced by the system. This value is compared to a δ threshold as an input parameter of the algorithm. The recommendation decision of an item t to a user s R(t, s) is given by the formula below:

$$R(t,s) = \begin{cases} t \text{ recommended to } s & \text{if } p(t,s) \geq \delta \\ t \text{ not recommended to } s & \text{otherwise} \end{cases} \tag{8}$$

p(t, s): Prediction value of item *t* to user s.

The threshold δ is bound to the rating scale as a degree of user satisfaction, for example for MovieLens and Amazon, the rating scale is limited to 5. Algorithm1 summarizes the main sequences of a recommendation task by item link prediction.

```
Algorithm item link prediction;
Input Bipartite Network B(N,L)
    N= I ∪ U  % set of nodes, I: items, U: users
    L= {(t,s), t ∈ I, s ∈ U}  % set of links
    W: L → ℝ+                 % weighted function
    (t,s) → R(i,j)
Output Weighted link
    w(t,s) = 0;
    For j=1 to M
        UC_s = 0;
    For i=1 to N
        UC_s = UC_s + R(i,j);  % s^th user cumulate
     q_s = 0 ;
    For i=1 to N
        IC_t = 0;
        For k=1 to M
          IC_t = IC_t + R(i,k);
          q_s = q_s + R(i,j)*R(i,s)/IC_t;  % t^th item cumulate

   w(t,s) = w(t,s) + R(t,j)*q_s/UC_s
   Returnw(t,s);
  If w(t,s) ≥ δ (recommend t to s)
End
```

This item-based method preserves a propagation of shared information between the two types of nodes (items and users). The recommendation task is modelled according to weighted bipartite network topology.

4 Experimentation

4.1 Datasets

To evaluate the effectiveness of the proposed approach, tests were performed on four real-world datasets:

- **Amazon-Books:** a free download dataset http://jmcauley.ucsd.edu/data/amazon/ for research purposes.
- **MovieLens:** a movie evaluation dataset https://grouplens.org/datasets/movielens/ provided by the GroupLens research group. The datasets are used for research purposes, in particular, to test the performance of collaborative filtering algorithms.
- **Yelp2018 dataset:** a set of data provided by the Yelp recommendation platform https://www.yelp.com/dataset. It includes statistics and user ratings for different items such as restaurants, hotels, shopping centers, etc.

Yahoo! Song: a dataset for user ratings of music tracks and albums. This dataset is provided by the Yahoo Research Alliance's Webscope program http://webscope.sandbox.yahoo.com/.

4.2 Comparison Methods

To see the effectiveness of the ISpLR proposal, we have implemented three methods related to this work:

- **Signed bipartite network analysis for neighbor-based collaborative ranking (SibRank)** [15]: The method exploits the relationships of a bipartite graph of positive or negative signs to infer a degree of similarity. The signs reflect trust or mistrust between users.
- **Neural Graph Collaborative Filtering (NGCF)** [16]: It uses a hierarchical layered neural graph model. This work measures an aggregation of connectivity based on user/item interactions.
- **Spectral Collaborative Filtering (SpectralCF)** [17]: Exploits the proximities of the graph and extracts connectivity information hidden in a bipartite graph. The contribution is made to reduce the cold start problem.

4.3 Metrics

In the literature on recommendation, several metrics are used such as *MAE* (Mean Absolute Error), *NMAE* (Normalized MAE), *MSE* (Mean Square Error), *RMSE*, (Root MSE). The most common being the MAE and the recall.

$$MAE = \frac{1}{n}\sum\nolimits_{i=1}^{n}\left|r_{ij} - p_{ij}\right| \quad (9)$$

r_{ij} : rating of item i by user j;
p_{ij} : prediction of item i by user j;

n: rating number
Recall: the fraction of pertinent returned items by all pertinent items.

$$R = N_{pr} / N_P \tag{10}$$

User satisfaction is strongly linked to the quality of prediction that is measured by the MAE and recall metrics.

5 Results and Discussion

We divided the datasets into two parts, 85% as a training set and 15% as a test set. The algorithm is based on the training set and generates the predictions of the test set. The difference between the generated predictions and the real votes of the learning set is measured by MAE according to the number of recommendations. The recommendation parameter (δ) is set to 3 on a scale [1–5]. Figure 3 shows the error rate according to the number of recommendations.

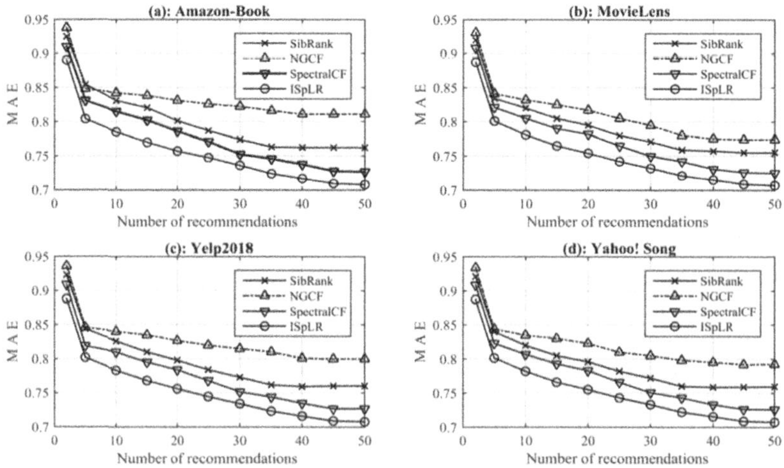

Fig. 3. Error rate depending on the length of the recommendation list

The result obtained by MAE (Fig. 3) shows a better performance in favor of the proposed method (ISpLR) compared to SpectralCF, SibRank, and NGCF respectively. We find that the MAE also varies depending on the datasets, which are better in MovieLens and Yahoo! Song and worse in Yelp2018 and Amazon-Book. The reasons for these results are as follows:

(1) The advantage of ISpLR lies in the preservation of information shared between items and users. Therefore, the information is propagated linearly without loss between each node in the network, which considerably minimizes the error rate. The SpectralCF method cumulates the hidden connectivity of the proximity nodes which slightly reduces the error;

(2) The different datasets used have an impact on the MAE according to the sparsity rate (formula 9) of each dataset.

$$sparsity\ rate = 1 - \frac{|E|}{|U| * |I|} \tag{11}$$

$|E|$: number of edges
$|U|$: number of users
$|I|$: number of items

It is clear that the sparsity rate (no rating) is proportional to the error where the quality of the recommendation system is strongly related to the number of judgments (ratings) of the users.

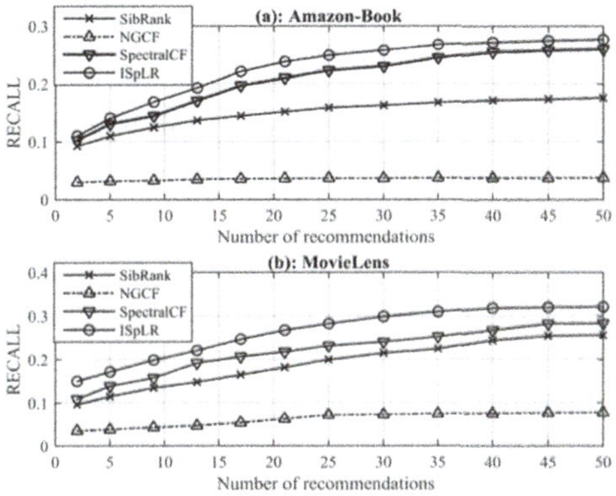

Fig. 4. Recovery rate depending on the length of the recommendation list

The recall metric as shown in Fig. 4 with the two sets Amazon-Books (higher sparsity) and MovieLens (lower sparsity). The proposed method (ISpLR) recorded a better performance; from a few recommendations (20), the recall rate reached 30%, which represents the rate of relevant items recommended (true positive rate).

NGCF has used several hierarchical layers to aggregate an order of connectivity that lowers the degree of similarity between users and closest items, which negatively affects recall and accuracy.

From these promising results, we can reason that ISpLR offers good quality predictions, even in sparse data sets and adapts well to online recommendations. On e-commerce sites, it is possible to retain only a subset of items to generate predictions for the other items.

6 Conclusion and Future Work

There is a need to model a recommendation system according to a network structure for interpretation and implementation purposes. In this paper, we addressed the problem of link prediction in a weighted bipartite network. According to the weightings of existing links, the weighting of a non-existent link can be calculated as a predictive value of an item to a user. The idea is to accumulate the shares of each user and each item in a linear and bidirectional way. Share means a part of the sharing of the item towards the user and vice versa. The advantage of this method is twofold; first, the accumulation of the shares retains the information propagated between the Items/Users nodes; second, the link prediction calculation does not require additional adjustment parameters. Classical collaborative recommendation methods based on Pearson correlation or cosine similarity [1, 2] require several steps to reach the prediction phase. Our future work will be devoted to the scalability problem (combinatorial complexity) of this method compared to existing methods and to see the prediction time for online users.

References

1. Adomavicius, G., Tuzhilin, A.: Toward the next generation of recommender systems: a survey of the state of the art and possible extensions. IEEE Trans. Knowl. Data Eng. **17**(6), 734–749 (2005)
2. Breese, J., Heckerman, D., Kadie, C.: Empirical analysis of predictive algorithms for collaborative filtering. In: 14th Conference on Uncertainty in Artificial Intelligence. Proceedings, pp.43–52. Morgan Kaufmann Publishers Inc., San Francisco (1998)
3. Musto, C., Lops P., Gemmis M. D., Semeraro G.: Semantics-aware recommender systems exploiting linked open data and graph-based features. Knowl.-Based Syst. **136**(C),1–14 (2017)
4. Yin, R., Li, K., Zhang, G., Lu, J.: A deeper graph neural network for recommender systems. Knowl.-Based Syst. **185**(1) (2019)
5. Ling, C.-Y., Zou, Y.-Z., Lin, Z.-Q., Xie, B.: Graph embedding based API graph search and recommendation. J. Comput. Sci. Technol. **34**(5), 993–1006 (2019). https://doi.org/10.1007/s11390-019-1956-2
6. Kovkov, D.V., Lemtyuzhnikova, D.V.: Decomposition in multidimensional Boolean-optimization problems with sparse matrices. J. Comput. Syst. Sci. Int. **57**(1), 97–108 (2018). https://doi.org/10.1134/S1064230718010033
7. Kharroubi, S., Dahmani, Y., Nouali, O.: A semantic layer to improve collaborative filtering systems. Int. J. Comput. Sci. Eng. **17**(4), 365–376 (2018)
8. Adomavicius, G., Mobasher, B., Ricci, F., Tuzhilin, A.: Context Aware Recommender systems. Ai Mag. **32**(3), 67–80 (2011)
9. Kashevnik, A.M., Ponomarev, A.V.: A multimodal context-aware tourism recommendation service: approach and architecture. J. Comput. Syst. Sci. Int. **56**(2), 245–258 (2017)
10. Page, L., Brin, S., Motwani, R., Winograd, T.: The PageRank citation ranking: bringing order to the Web. Stanford 12 University, CA 94305–9025, USA: Stanford InfoLab- Digital Libraries, Technical Report (1999)
11. Giannoulakis, S., Tsapatsoulis, S.: Filtering Instagram Hashtags through crowd tagging and the HITS algorithm. IEEE Trans. Comput. Soc. Syst. **6**(3), 592–603 (2019)
12. Birkbak, A., Carlsen, H.: The world of Edgerank: Rhetorical justifications of Facebook's News Feed algorithm. Computational Clture (5) (2016)
13. Nesreen, K.A., Nick, D., Liangzhen, X.: Estimating node similarity by sampling streaming bipartite graphs. In: Twenty-Seventh International Joint Conference on Artificial Intelligence, Proceeding, pp. 3286–3292. IGCAI (2018)
14. Kim, J.H., Segev, A.: Hypothesis generation using link prediction in a bipartite graph. In: IEEE International Conference on Big Data, pp. 2863–2867 (2018)
15. Shams, B., Haratizadeh, S.: SibRank: signed bipartite network analysis for neighbour-based collaborative ranking. Phys. A **458**, 364–377 (2016)

16. Wang, X., He, X., Wang, M., Feng, F., Chua, T.S.: Neural graph collaborative filtering. In: ACM SIGIR International Conference on Research and Development in Information Retrieval, pp. 165–174 (2019)
17. Zheng, L., Lu, C.T., Jiang, F., Zhang, J., Yu, P.S.: Spectral collaborative filtering. In: Twelfth ACM Conference on Recommender Systems, pp. 311–319 (2018)

Empirical Evidence in Early Stage Software Effort Estimation Using Data Flow Diagram

Vo Van Hai(✉), Ho Le Thi Kim Nhung, and Huynh Thai Hoc

Faculty of Applied Informatics, Tomas Bata University in Zlin, Nad Stranemi 4511, Zlin 76001, Czech Republic
{vo_van,lho,huynh_thai}@utb.cz

Abstract. Software effort estimation in a very early stage of the application development lifecycle is always a challenge for project managers. Many researchers have proposed different methods. They have certain advantages and limitations. This study proposed an approach called Early Effort Estimation from the data flow diagram with a complexity tag added that could estimate the effort estimation in a very early stage of the software development cycle. This method can be obtained by evaluating the customized data flow diagram, a data flow diagram with specific tag values. We also proposed a tool that implemented this method.

Keywords: Early software effort estimation · Function point (FP) · Software effort from DFD

1 Introduction

Software systems play a significant role in today's life. Building a software system is becoming more and more complex [16]. From the very beginning of the development process, software estimation is becoming increasingly important. This makes it easy for managers to plan human, financial, and other related matters. Early and accurate estimation of software project costs and efforts has long been a significant challenge for project managers. Various estimation models have been proposed in the past. COCOMO [11] is a model that has been in use for a long time, and it is still seen as a vital model today. The Function Point Analysis (FPA) [2] also is another crucial model.

With the recent development of machine learning and other techniques, estimation models based on the linear regression model, artificial neural network (ANN), fuzzy logic have been strongly proposed. Vo Van Hai et al. [22] proposed the POM model based on the linear regression model and ISBSG dataset to estimate the optimizing productivity. Xia et al. [17] suggested merging three techniques (fuzzy logic, artificial neural networks, and statistical regression) in a neuro-fuzzy function point calibration model to solve the no clear boundary between two classifications in FP counting. Ahmed et al. [18] proposed a new weights system for the FPA method based on an adapted genetic algorithm. The proposed algorithm is based on a set of initial solutions, using the biologically inspired evolution mechanisms to derive new and possibly better solutions.

R. Silhavy (Ed.): CSOC 2021, LNNS 230, pp. 632–644, 2021.
https://doi.org/10.1007/978-3-030-77442-4_53

However, classical models such as COCOMO or FPA are based on a pre-estimation of components. This takes much time before estimating results. In some cases, we would like to get results at a very early stage for user interaction or planning human resources. Machine learning-based approaches require a precise and specific amount of training data [19] for the development environment as well as the application context. Although it is the most critical problem today, it also has certain limitations [20, 21].

In addition, the prediction based on data flow diagram (DFD) has been considered before [5–7, 25]. It was mostly based on DFD combined with Entity-Relationship Diagram (ERD) to calculate the number of Data Element Type (DET), Record Element Type (RET), File Type Referenced (FTR), and then based on FPA to determine the number of FPs. Then apply other models to calculate effort, time, cost.

Our method is also based on DFD but not ERD; it adds a human factor in specifying the complexity and the boundary of the system. This may take up some extra time in the DFD design process, but not much. In return, it is very easy to estimate the function points. In the case of standard DFD (without any additional complexity tags), our method also estimates the effort based on the use of the default values. After getting the FPs, we calculate the work effort using the formula (4) with productivity based on the specific selection.

2 Research Objective

This research's main objective is to propose a new approach and a tool for effort estimation in the early stage of software development. The research questions answered by this study are as follows:

RQ1: Can DFD be used for effort estimation in the early stages of application development?

RQ2: What attributes can be added to standard DFD to help make the estimation more accurate?

To reach the goal, the following steps are performed:

- Design the DFD with and without additional complexity tags.
- Populate them to XMI standard.
- Evaluate these DFDs by using a specific algorithm.
- Compare it together.

3 Research Methods

3.1 Function Point Analysis

Function Point (FP) was first introduced by A. J. Albrecht [2] and widely accepted with many variants, from both academics and practitioners [14]. The study in this area is also known as Function Point Analysis (FPA) or Function Size Measurement (FSM). The function points measurement could be classified into FP counting and FP estimation [15].

The FPA assumes that a software program consists of a set of functions. In turn, each function can be a data transaction or transactional function. The data transaction has consisted of internal logic files (ILF) and external interface files (EIF), and the transaction function comprises External Input (EI), External Output (EO), or External Inquiry (EQ). In turn, each of these is judged as simple, average, or complex and assigned a weight accordingly (see Table 1).

Table 1. Data and transactional function complexity

		Component				
		EI	EO	EQ	EIF	ILF
Functional Complexity	Low	3	4	3	5	7
	Average	4	5	4	7	10
	High	6	7	6	10	15

The following equation can obtain the total unadjusted function points (UFP)

$$UFP = \sum_{i=1}^{n}\sum_{j-1}^{m}(W_{ij} \times S_{ij}) \tag{1}$$

where W_{ij}'s are the complexity weights and S_{ij}'s are the counts for each function component, n is the number of types, and m is the number of complexity groups.

The Value Adjustment Factor (VAF) should be aimed at before calculating the function points. The VAF is based on the rate of 14 general system characteristics (GSCs - Data Communications, Distributed Data Processing, Performance, Heavily Used Configuration, Transaction Rate, Online Data Entry, End-User Efficiency, Online Update, Complex Processing, Reusability, Installation Ease, Operational Ease, Multiple Sites, Facilitate Change). The GSCs are rated based on their degree of influence or complexity on the 0–5 scale. The following table represents the significance of the influence factors rating (Table 2).

Table 2. The GSCs Factor weights

System Influence	Rating
Not present or no influence	0
Incidental influence	1
Moderate influence	2
Average influence	3
Significant influence	4
Strong influence throughout	5

The VAF count is adjusted as

$$VAF = 0.65 + 0.01 \times \sum_{i=1}^{14} (F_i \times rating) \tag{2}$$

After all, the adjusted function points (AFP) can be reached by using the following formula

$$AFP = UFP \times VAF \tag{3}$$

3.2 The Work Effort Calculation

It is necessary to understand the metrics that are used in the estimation of the projects. Metrics are ways to measure project properties that are used, in our study, to estimate the cost and labor (work effort) of the project. To express the overall estimate, the terms person-hour or work effort is used. According to Albrecht [2], the number of person-hours is the time that is required to meet one desired result by one employee. Besides, we need to clarify the term productivity. It was defined as the amount of output per unit of input used. In the context of software, productivity measurement is often based on a simple ratio of product size to project effort [27]. In our study, we understand work effort as the ratio of the software development process output (FP size) divided by productivity.

$$Work\ Effort = \frac{Function\ Points}{Productivity} \tag{4}$$

Based on the analysis of the ISBSG dataset, Prokopova et al. [23] analyzed the influence of the selected factors in function point work effort estimation. As a result, they proposed four approaches calculating the productivity that depends on 1) FP count approach; 2) business area type; 3) industry sectors; and 4) project relative size. As this study mention, the relative size of the project is an important factor for normalized work effort. The size of the project can be categorized by Rule [24]. They provided two values, mean and median, for productivity, but they also recommended that it is preferable to use a median than the mean. In our approach, we used these values for calculating the effort.

3.3 Early Function Points

Early Function Points (EFP) and Extended Function Points (XFP) were proposed by Meli [3], with the desire to estimate the size of software in the early stages of development. This approach requires the user to understand application development's granularity such as Macro-function, Function, Micro-function, and Functional Primitive. And each level of detail, users must evaluate and assign levels of values (minimum, average, and maximum). These approaches are considered not very easy to use.

3.4 Function Points Estimation from Data Flow Diagram Method

Data Flow Diagram (DFD) was proposed as the estimator for FPA by many researchers. Rask [5, 6] introduced the algorithm for counting the Function Points using specifications from DFD and ERD data models. The automated system was also built.

O'Brien and Jones [7] proposed the counting rules to incorporate the Structured Analysis and Design Method (SSADM) into Function Point Analysis. DFD, together with the I/O structure diagram, Enquiring Access Path (EAP), and Effect Correspondence Diagram (ECD), were applied to the counting rules for the Mark II FPA.

Shoval and Feldman [8] applied Mark II Function Points with Architectural Design of Information System Based on Structural Analysis (ADISSA). This proposed method counts the attributes of all inputs and outputs from the system's data flow diagram to be built and all of the relations in the database from the database design process. It then plugs in all the numbers in the Mark II model.

Lamma et al. [9] solved the counting error; they built a system for automating the counting named FUN (FUNction points measurement). This system used the specification of a specific software system from Entity Relationship Diagram and Data Flow diagram to estimate software function points. Later, the system was automated by Grammantieri et al. [10].

Dian Pratiwi [25] calculated the FPs both on the object-oriented model (UML use case diagram, class diagram) and structure model (ERD, DFD) and found that the difference between them is not much.

4 The Proposed Method

DFD scheme is used in the gathering system requirement phase. Many applications were introduced for the design of DFD schemas, so the necessity of inter-software communication is essential. XML used as a language to represent this model is inevitable. The OMG is the general benchmarking organization for this specification language. The XML Metadata Interchange (XMI) was introduced for standardizing the format of these XML documents [13].

In this study, in order to solve the problem, we proposed that during the DFD design process, to ensure accuracy in the calculation, we must add a tag attribute in the form of <key-value> for objects as well as their relationships. We call this diagram the DFD with additional tags added (ataDFD). At a glance, ataDFD looks like the original DFD. The main difference is the attributes that were included with the original DFD. The original DFD is the DFD, which was created by DFD designing software without any additional tags. As we can see in Fig. 1, this ataDFD is the same as any original DFD. However, in Fig. 2, after exporting it to XMI format, the difference is shown (see the red-underlined lines).

One crucial issue is how to implement FPA in the early stages of system development. According to the IFPUG standard counting rule, the functional specifications must be at least 15 to 40% clear before FPs can be obtained. Otherwise, it is impossible to define EI, EO, EQ, ILF, and EIF [3, 4]. This study proposes to handle the stated problem using the DFDs schemas that appear in the requirement phase of the system development. Using

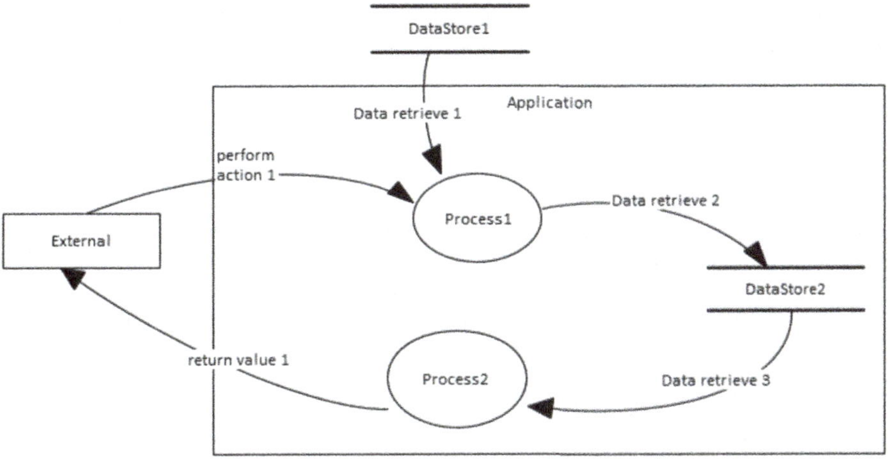

Fig. 1. A sample ataDFD

```xml
<?xml version="1.0" encoding="windows-1252"?>
<xmi:XMI xmi:version="2.1" xmlns:uml="http://schema.omg.org/spec/UML/2.1" xmlns:xmi="http://schema.omg.org/spec/XMI/2.1"
    <uml:Model xmi:type="uml:Model" name="EA_Model" visibility="public">
        <packagedElement xmi:type="uml:Package" xmi:id="EAPK_176EBFDF_A578_4b09_A161_91A386BC7400" name="Sample in Paper"
            <packagedElement xmi:type="uml:Class" xmi:id="EAID_C4EE5D6D_803B_4c40_9701_8611D12EAEEA" name="DataStore1" vi
            <packagedElement xmi:type="uml:Dependency" xmi:id="EAID_5ED66B3E_4D35_4dfb_BC52_42829F14301E" name="Data retr
            <packagedElement xmi:type="uml:Class" xmi:id="EAID_44A885CB_AABB_4781_AFB6_C05237193B7A" name="DataStore2" vi
            <packagedElement xmi:type="uml:Dependency" xmi:id="EAID_942A8B19_7DD0_4fd2_8BDA_3B607CEA9237" name="Data retr
            <packagedElement xmi:type="uml:Class" xmi:id="EAID_E370E2BC_9FB0_4165_A9A8_5CDC9DB39A22" name="External" visi
            <packagedElement xmi:type="uml:Dependency" xmi:id="EAID_0F9039E5_EBDA_42e8_8A1C_60652AA7B1F9" name="perform a
            <packagedElement xmi:type="uml:Class" xmi:id="EAID_DC733850_2390_46f2_AAE3_86E01A29A037" name="External3" vis
            <packagedElement xmi:type="uml:Activity" xmi:id="EAID_8C218347_85B0_46dd_A9DF_5FE39967EA25" name="Process1" v
            <packagedElement xmi:type="uml:Dependency" xmi:id="EAID_94F5426B_073C_490c_81B2_4AB7496C5F08" name="Data retr
            <packagedElement xmi:type="uml:Activity" xmi:id="EAID_F29F52EB_FA9E_4698_A35A_62FD1AC061CE" name="Process2" v
            <packagedElement xmi:type="uml:Dependency" xmi:id="EAID_31C5B681_A4C4_48b4_9D1C_69A87B12D069" name="return va
            <packagedElement xmi:type="uml:Dependency" xmi:id="EAID_99B0D340_DB47_4941_9C3D_041BDD74A333" visibility="pub
            <packagedElement xmi:type="uml:Class" xmi:id="EAID_8ED461F3_93A8_4aa4_BD2C_27DC17CD3152" name="Application" v
        </packagedElement>
        <DFD:DFD_DataStore base_Class="EAID_C4EE5D6D_803B_4c40_9701_8611D12EAEEA"/>
        <DFD:DFD_DataFlow base_Dependency="EAID_5ED66B3E_4D35_4dfb_BC52_42829F14301E" _Bezier="1"/>
        <DFD:DFD_DataStore base_Class="EAID_44A885CB_AABB_4781_AFB6_C05237193B7A"/>
        <thecustomprofile:Complexity base_Dependency="EAID_942A8B19_7DD0_4fd2_8BDA_3B607CEA9237" Complexity="low"/>
        <DFD:DFD_DataFlow base_Dependency="EAID_942A8B19_7DD0_4fd2_8BDA_3B607CEA9237" _Bezier="1"/>
        <DFD:DFD_External base_Class="EAID_E370E2BC_9FB0_4165_A9A8_5CDC9DB39A22"/>
        <thecustomprofile:Complexity base_Dependency="EAID_0F9039E5_EBDA_42e8_8A1C_60652AA7B1F9" Complexity="Average"/>
        <DFD:DFD_DataFlow base_Dependency="EAID_0F9039E5_EBDA_42e8_8A1C_60652AA7B1F9" _Bezier="1"/>
        <DFD:DFD_External base_Class="EAID_DC733850_2390_46f2_AAE3_86E01A29A037"/>
        <DFD:DFD_Process base_Activity="EAID_8C218347_85B0_46dd_A9DF_5FE39967EA25"/>
        <thecustomprofile:Complexity base_Dependency="EAID_94F5426B_073C_490c_81B2_4AB7496C5F08" Complexity="Low"/>
        <DFD:DFD_DataFlow base_Dependency="EAID_94F5426B_073C_490c_81B2_4AB7496C5F08" _Bezier="1"/>
        <DFD:DFD_Process base_Activity="EAID_F29F52EB_FA9E_4698_A35A_62FD1AC061CE"/>
        <DFD:DFD_DataFlow base_Dependency="EAID_31C5B681_A4C4_48b4_9D1C_69A87B12D069" _Bezier="1"/>
        <DFD:DFD_DataFlow base_Dependency="EAID_99B0D340_DB47_4941_9C3D_041BDD74A333" _Bezier="1"/>
    </uml:Model>
    <xmi:Extension extender="Enterprise Architect" extenderID="6.5">
</xmi:XMI>
```


Fig. 2. XML document from the sample DFD

DFD in FPs counting is not new, but handling it automatically helps a lot in human resources needs estimation as well as project size estimation for planning. During the design process, we must determine the complexity of the relationship. In FPA, we need to count DETs and RETs for data function counting or FTRs and DETs for transaction counting. But with the initial hypothesis of the problem that in the very early stage of the development cycle, we do not have this information. Therefore, the designer must give a subjective assessment based on their understanding of the problem they are working on. In case the designer does not provide this information, the default value will be used.

We propose four initial rules for the ataDFD designer (Table 3) and four rules (which are based on the relation between two components in the DFD diagram) for determining the function type (Table 4).

Table 3. Four initial rules for the ataDFD

• Rule 1: Dependency objects representing relationships between entities must be added with a tag named complexity, and the value must be one of the three values {Low / Average/High}. We use this triple for identifying the complex level of dependences. The Low value notates that the dependence is easy or simple. The High value means that this problem is difficult or hard. The Average value denotes that the complexity is between Low and High. The default value is Low
• Rule 2: Data-Store objects, by default, are an outside boundary, so if we want it to be in boundary, we must also add a tag named external and a value of either true or false
• Rule 3: External objects are by default located outside of the boundary; if a boundary is specified, we must add a tag named internal and a value of either true or false
• Rule 4: Because the EQ and EQ are different in terms of the computation of the rendering, so, during the DFD design process, the designer should define it as EQ or EO by adding a tag named EQ with a value of true or false. The default is EO

Table 4. Four rules for determining the function type

• Rule 1: If and the source is the DFD external and the destination is the DFD process, the function type is EI
• Rule 2: If and the source is the DFD process and the destination is the DFD external. This case can be EQ/EO. If the EO tag is true or the EQ tag is false, the function type is EO. In the opposite case, the function type is EQ
• Rule 3: If and the source is the DFD process and the destination is the DFD process, the function type is ILF
• Rule 4: If and the source is the DFD process and the destination is the DFD datastore, the function type is EIF

5 Experiments

5.1 Experimental Setup

In this section, we conduct the experiments to evaluate the effectiveness of the ataDFD method as follows.

Through case-study: We conduct the case study based on the functional requirements that are described in the warehouse management application of the software company WHS [26]. The scenario has represented that company W operates several warehouses where it stores goods for its customers. W's customers can place items into the warehouse. Items can be retrieved by their owners whenever the customer wants. The customers of W can be billed for the storage service. Company W desires the application can manage their customer data, items stored in the warehouses, and warehouse locations.

Through the empirical validation: We conduct empirical validation based on the graduate student projects in the bachelor program in Software Development at the Department of Software Engineering, Faculty of Information Technology, Industrial University of Ho Chi Minh City, Vietnam. After finishing 41 credits of study courses, the students must have a project related to developing a software application. A group of 14 students with good academic results is asked to participate in the experiment. A questionnaire is distributed to each student to ask for the following information: student id, full name, project name, start date, finished date, the number of functions, estimated FPs, the actual effort (person-hours) that is spent for the application development. Ten students have passed their project and returned the final results (Table 5).

Table 5. The dataset from the graduate student projects

No	Project name	No. functions	FPs	Effort
1	The computer controlling through mobile devices	5	146	1200
2	The online enrollment system	8	131	1100
3	Student attendance system through face recognition	5	163	1300
4	Mobile application for accessing study result and sharing events in IUH	2	86	600
5	Leftover food sharing system	3	97	730
6	Booking system for travel agencies	9	160	1150
7	Cờ-gánh, a Vietnam tradition chess	2	95	900
8	The hotel management system	10	82	600
9	A system for course rating and feedback at IUH	8	115	1050
10	A Logistic system for SME	14	112	1130

The empirical validation process of the method is shown in Fig. 3. At first, the designer designs the ataDFD diagram with an additional correspondence tag attribute. Then convert it to XML by using the standardized method. After that, our system will be parsing the XML document and counting the FPs. After getting the FPs, we use the method proposed by Prokopova [23] to calculate the effort.

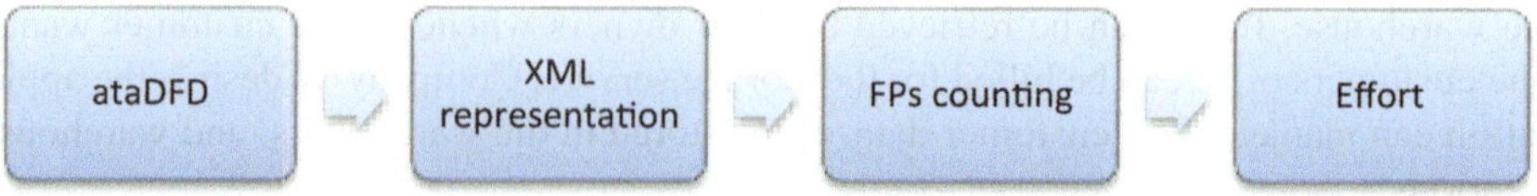

Fig. 3. The process of the method

5.2 Evaluation Criteria

The accuracy of the estimation was measured using the Magnitude of Relative Error (MRE) [28], which is defined by (5)

$$MRE_i = \frac{|Actual_i - Estimated_i|}{Actual_i} \quad (5)$$

and the Mean Magnitude of Relative Error

$$MMRE = \frac{1}{N}\sum_{i=1}^{N} MRE_i \quad (6)$$

6 Result and Discussion

6.1 Case-Study

The process of conducting a case-study is as follows: using the proposed model to estimate the number of UFP and the original DFD scheme's effort. Next, we calibrate the DFD with the additional tags, then estimate it again. Finally, we make the comparison.

With the scenario [26], we present it in the form of the DFD schema shown in Fig. 4. After counting the function point using IFPUG FPA 4.1, Fetcke [26] had got the number of FPs for company W as $UFP = 77$ and the effort for this case using [23] is 620.517. After representing the DFD schema in XML format, we test it with our proposed model, and the result shows in Table 6.

Table 6. The result of using the proposed model with the given standard DFD diagram

No	Type	Low	Average	High
1	EI	2	0	0
2	EO	3	0	0
3	EQ	0	0	0
4	ILF	8	0	0
5	EIF	0	0	0
UFP = 3 * 2 + 4 * 3 + 8 * 7 = 74				

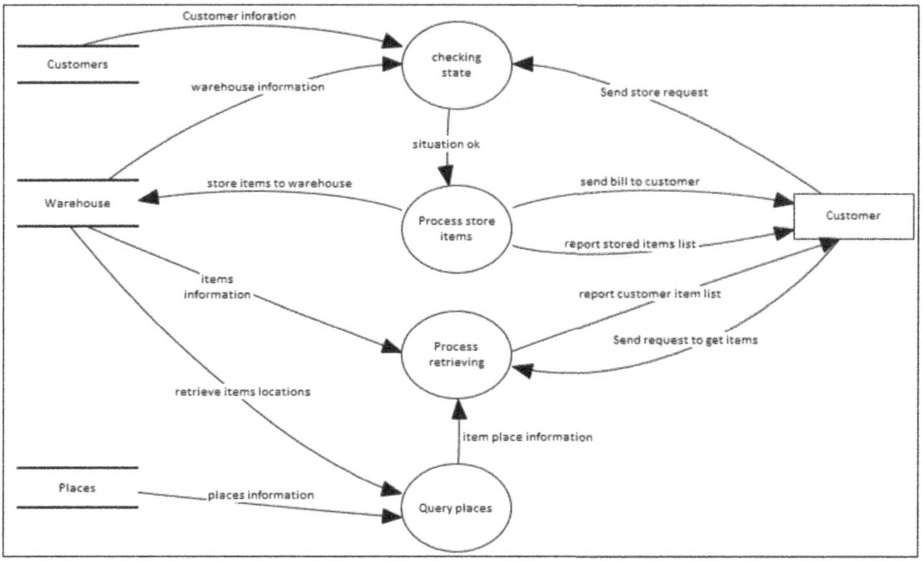

Fig. 4. The DFD of the scenario

We found that the "report customer item list" function, the "report stored items list" function, and the "item place information" related to more than one data store, the complexity of these functions should be average instead of low. Applying these changes to the DFD diagram by using the adding tags, we re-test it with our proposed model, and the result shows in Table 7.

Table 7. The result of using the proposed model with the ataDFD

No	Type	Low	Average	High
1	EI	2	0	0
2	EO	1	2	0
3	EQ	0	0	0
4	ILF	7	1	0
5	EIF	0	0	0
UFP = 3 * 2 + (4 * 1 + 5 * 2) + (7 * 7 + 1 * 10) = 79				

In this study, we assume that the VAF does not affect the calculation of AFP. It means that the VAF value is 1.0. After we have the number of AFP, we will proceed to effort estimation using the method proposed by Prokopova et al. [23]. In this experiment, we use the 'inbound logistic' as 'business type area' for effort estimation. With the mean value of inbound logistics is 0.124, we have the effort values 596.341 and 636.635 for the original DFD and ataDFD, respectively. We can found that the proposed approach effort is closer to real effort than the other one based on the original DFD.

6.2 Empirical Validation

Table 8 shows the results of the ataDFD through the MRE value.

Table 8. Results obtained by the proposed approach for tested student projects

No	Business area type	Productivity	Estimated FPs	Actual effort	Estimated Effort	MRE
1	General	0.10745	146	1200	1358.7715	0.1323
2	General	0.10745	131	1100	1219.1717	0.1083
3	Engineering	0.16815	163	1300	969.3726	0.2543
4	General	0.10745	86	600	800.3723	0.3340
5	General	0.10745	97	730	902.7455	0.2366
6	Purchasing & Supplier	0.099482	160	1150	1608.3312	0.3985
7	General	0.10745	95	900	884.1322	0.0176
8	General	0.10745	82	600	763.1456	0.2719
9	General	0.10745	115	1050	1070.2652	0.0193
10	Inbound Logistics	0.12409	112	1130	902.5707	0.2013
					MMRE	**0.1974**

The MMRE result of proposed approach on the tested dataset is *0.1974*. This result of less than *0.25* has shown that the Early Effort Estimation from the data flow diagram with a complexity tag added is acceptable for estimation models [12, 30]. Additionally, this approach can be performed at a very early stage in the application development process to bring many prospects for timing critical problems.

7 Conclusion

With the first objective, design the ataDFD with specific rules, we have a new schema with very little change compared to the standard DFD schema. Adding the tag attribute to the standard DFD schema is mostly supported by today's DFD design software. The result of ataDFD compared to the ERD-DFD scheme proposed by Rask [5, 6], the proposed model is much simpler. Even with the standard DFD model, this proposed new method can estimate the effort based on default tag attributes. This shows the ease of application of the proposed model. With the second objective, populating from DFD model to XMI standard is supported on almost all DFD design software today. This XMI standard is also easy to use for design interchange between development platforms. It is possible to see that this method does not need to calculate RETs/DETs/FTRs as the solutions suggested. This minimizes the time required to obtain the results of the estimation during software development. The proposed method brings a great advantage in the process of planning human resources for the project's future development.

Through this study's results, it is clear that we can thoroughly estimate effort estimation in the early stage of application development (RQ1). This process requires some skills in designing the ataDFD. Fortunately, most commercial applications today make it very easy to add tag attributes, as well as render documents to XMI format. Throw the experimental, and the research object can be reached. Because the standard DFD does not contain information that makes effort estimation more accurate, for the RQ2, we suggest adding these following attributes: 1) the complexity attribute that determines complexity for the relationship between objects; 2) the internal/external attribute to the data-store object in order to determine that the object is outside or inside the boundary; 3) In some cases, the external object is inside the boundary, so we add the internal/external property to determine; 4) the attributes that determine the type of EQ and EO.

In future work, we will complete our tool, as well as write the plugins for other software. Under the user's permission, we will collect the schema from the user then use these schemas for training a new system using ANN for the suggestion of a newly developed application.

Acknowledgment. This work was supported by the Faculty of Applied Informatics, Tomas Bata University in Zlín, under project IGA/CebiaTech/2021/001.

References

1. IFPUG: International Function Point Users Group. http://www.ifpug.org/. Accessed Nov 2020
2. Albrecht, A.J.: Measuring application development productivity. In: Proceedings of the IBM Applications Development Symposium, p. 83 (1979)
3. Meli, R.: Early and extended function point: a new method for function points estimation. In: IFPUG Fall Conference, 15–19, Arizona 1997
4. Meli, R., Santillo, L.: Function point estimation methods: a comparative overview. In: FESMA 1999 Conference Proceedings, Amsterdam (1999)
5. Rask, R.: Algorithm for counting unadjusted function points from dataflow diagram. Technical report, University of Joensuu (1991)
6. Rask, R.: Counting function points from SA descriptions. The Papers of the Third Annual Oregon Workshop on Software Metrics (Ed. W. Harrison), Oregon, 17–19 March 1991
7. Obrien, S.J., Jones, D.A.: Function points in SSADM. Software Qual. J. **2**(1), 1–11 (1993)
8. Shoval, P., Feldman, O.: Combining function points estimation model with ADISSA methodology for system analysis and design. In: Proceeding of ICCSSE, pp. 3–8 (1996)
9. Lamma, E., Mello, P., Riguzzi, F.: A system for measuring function points from an ER-DFD specification. Comput. J. **47**(3), 358–372 (2004)
10. Gramantieri, F., Lamma, E., Mello, P., Riguzzi, F.: A system for measuring function points from specifications. DEIS – Universita di Bologna, Bologna. and Dipartimento di Ingegneria, Ferrara, Tech. Rep DEIS-LIA-97–006 (1997)
11. Boehm, B.W.: Software Estimation with COCOMO II, Upper Saddle River, NJ, Prentice-Hall (2002)

12. Conte, S.D., Dunsmore, H.E., Shen, V.Y.: Software Engineering Metrics and Models. Benjamin/ Cummings, Menlo Park CA (1986)
13. The OMG XMI document. http://schema.omg.org/spec/XMI/2.1. accessed Nov 2020
14. Gencel, C., Demirors, O.: Functional size measurement revisited. ACM Trans. Software Eng. Methodol. **17**(3), 15.1–15.36 (2008)
15. Meli, R., Santillo, L.: Function point estimation methods: a comparative overview. In: FESMA 1999 Conference Proceedings, Amsterdam, 4–8, October 1999
16. Darcy, D.P., Kemerer, C.F., Slaughter, S.A., Tomayko, J.E.: The structural complexity of software an experimental test. IEEE Trans. Software Eng. **31**(11), 982–995 (2005). https://doi.org/10.1109/TSE.2005.130
17. Xia, W., Ho, D., Captrez, L.F.: A neuro-fuzzy model for function point calibration. WSEAS Trans. Inf. Sci. Appl. **5**(1), 22–30 (2008)
18. Ahmed, F., Bouktif, S., Serhani, A., Khalil, I.: Integrating function point project information for improving the accuracy of effort estimation. In: Proceedings of the 2nd International Conference on Advanced Engineering Computing Application Science, pp. 193–198 (2008)
19. Mange, J.: Effect of training data order for machine learning. In: 2019 International Conference on Computational Science and Computational Intelligence (CSCI), Las Vegas, NV, USA, pp. 406–407 (2019). https://doi.org/10.1109/CSCI49370.2019.00078
20. Hoffmann,A.G.: General limitations on machine learning. In: ECAI 1990: Proceedings of the 9th European Conference on Artificial Intelligence, pp. 345–347, January 1990
21. Pitt, L., Valiant, L.G.: Computational limitations on learning from examples. J. ACM, (1988). https://doi.org/10.1145/48014.63140
22. Van Hai, V., Nhung, H.L.T.K., Hoc, H.T.: Productivity Optimizing Model for Improving Software Effort Estimation. CoMeSySo2020, October 2020
23. Prokopová, Z., Šilhavý, P., Šilhavý, R.: Influence analysis of selected factors in the function point work effort estimation. In: Advances in Intelligent Systems and Computing [online]. Szczecin: Springer Verlag, 2019, s. 112–124. [cit. 2020–12–02]. ISSN 2194–5357
24. Rule, G.: 'Small project', 'medium-size project' and 'large project': what do these terms mean? PowerPoint presentation. Software Measurement Services Ltd. (SMS), copyright 2004–2005, 124 High Street, Edenbridge, Kent, United Kingdom (2005)
25. Pratiwi, D.: Implementation of function point analysis in measuring the volume estimation of software system in object-oriented and structural model of academic system. Int. J. Comput. Appl. (0975–8887) **70**(10) (2013). https://doi.org/10.5120/11995-7879
26. Fetcke, T.: A Warehouse Software Portfolio – A Case Study in Functional Size Measurement," Report No. 1999–20
27. Kitchenham, Mendes, E.: Software productivity measurement using multiple size measures. IEEE Trans. Software Eng. **30**(12), 1023–1035 (2004)
28. Briand, L.C., Emam, K.E., Surmann, D., Wieczorek, I., Maxwell, K.D.: An assessment and comparison of common software cost estimation modeling techniques. In: International Conference on Software Engineering, pp. 313–322 (1999)
29. . Albrecht, A.J, Gaffney, J.E.: Software function, source lines of code, and development effort prediction: a software science validation. IEEE Trans. Software Eng. **SE-9**(6), 639–648 (1983). https://doi.org/10.1109/TSE.1983.235271
30. Foss, T.S.: A simulation study of the model evaluation criterion MMRE. IEEE Trans. Software Eng. **29**(11), 985–995 (2003)

Analysis of the Cybersecurity Weaknesses of DLT Ecosystem

Willian Dimitrov[1](✉), Boyan Jekov[1], and Peyo Hristov[2]

[1] UNIBIT, Sofia, Bulgaria
{v.dimirov,b.jekov}@unibit.bg
[2] Cisco Systems, Inc., San Jose, CA 95134, USA
peyhrist@cisco.com
https://www.unibit.bg,
https://cisco.com

Abstract. The article examines the cybersecurity status of DLT technology in the holistic context of dynamic management of cyber protection. The advent of the DLT technologies arrived together with the continuous expansion of the surface for cyber-attacks against information systems and communication networks. The analysis is targeted to unknown dangers against the DLT ecosystem.

Keywords: DLT · Cybersecurity · Weaknesess · Vulnerabilities · Attacks

1 DLT in the Light of Cybersecurity

Consumers and organizations introducing DLT technology are exposed to unknown dangers by enemies who are constantly improving their abilities for attacks. We share the opinion that DLT technology arrived in the same manner as any other in the IT area – without security by design. From the point of view of modern concepts of cybersecurity, this raises the following questions we focus on: is this ecosystem relevant and current with (i) cybersecurity law regulations, (ii) ever-changing technologies, (iii) bad actors business practices and expertise? The current landscape in cybersecurity is characterized by the continuous expansion of the threat surface because of the raising complexity of digital space. The effectiveness of traditional cybersecurity measures reduces. The emerging new technologies do not consider the dynamics of evolving dangers. We base our work on the research literature review, information from the institutions and industrial bulletins produced by leading companies, and live security-monitoring systems from [1,2]. The research goal of the study is the real situation around DLT security in the light of a holistic view for cybersecurity in digital space

Supported by the project: CyberTwin - Reinforcing the Scientific Excellence and Innovation Capacity in Cyber Security trough Twinning.

R. Silhavy (Ed.): CSOC 2021, LNNS 230, pp. 645–655, 2021.
https://doi.org/10.1007/978-3-030-77442-4_54

and perception, based on our empirical security expertise. Blockchain security weaknesess is part of the general problem about the cyberprotection of DLT.

Europol released its 2019 Internet Organized Crime Threat Assessment (IOCTA) report, in which it painted a troublesome picture showing that cryptocurrency ransomware remains the most prominent cyber-attack that European cybercrime investigators are confronted with. According to Europol, cybercrime must be approached in a holistic sense, saying: "Countering cybercrime is as much about its present forms as it is about future projections. New threats do not only arise from new technologies but, as is often demonstrated, come from known vulnerabilities in existing technologies." Here the potential of this problem is viewed (see Fig. 1). It is generated in SAS environment (Series Plot) with the data set created from the analytics part of tr.txone-networks.com. The attacks against BlockChain protocol according to this data set for the year 2020 are November - 355695; December - 1544781; 2021 January - 1917330.

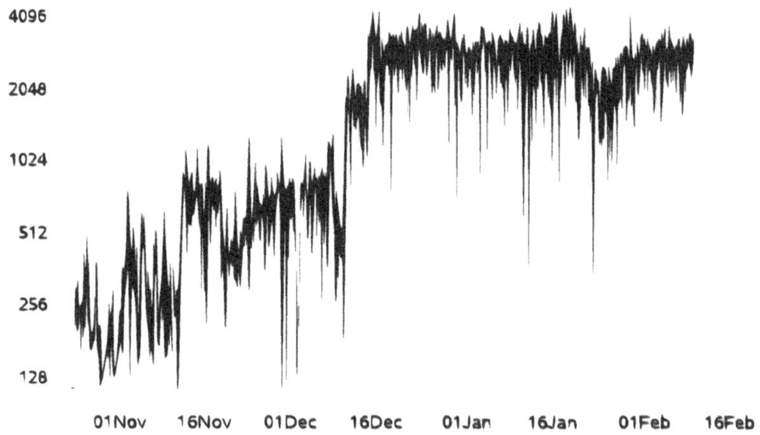

Fig. 1. Dayly global Blockchain protocol attacks. November 2020 to January 2021

2 Methodology

The analysis is designed to make a holistic summary of the many components that are related to the cybersecurity of DLT applications in different problem areas. We studied many overviews of DLT Technology. The architecture, consensus, and future trends presented in [3–5].

Our research is based on the next considerations. The scope of the DLT ecosystem includes the development process with the supply chain for software tools, distributed architecture and production lifecycle, and customer supply chain. We separate threats into architectural, organizational and social. The threats and a large part of weaknesses are enough familiar for a long time in the circles of cyber protection architects. We consider that designing a complex cybersecurity protection system for DLT is an unexplored area. The main reason is the distributed character of architecture. It leads to an unclear and complicated picture of organizational and technological aspects of the DLT system lifecycle.

The generic architectural model we consider included C2 (Command and Control) server, gateway servers [6], distributed subscribers endpoints, and customer exchanges. The DLT ecosystem uses a wide range of internet protocols like NTP, DNS and third-party platforms belonged to the supply chain. For the study, we apply (1) review of the official publications of companies that have suffered security incidents and are required to make them public following the law, (2) the adequacy of existing legal provisions, institutional opinions and standards for the required level of resistance to the capacity of malicious actors, (3) data from repositories with vulnerabilities. We consider that many security incidents with DLT is not published.

DLT implementations vary in their specific purposes. From IoT thru the digital supply chain to financial markets - the consensus mechanism used is what makes them unique - Bitcoin (PoW), Ethereum (PoW), Hyperledger Fabric (modular), Hyperledger Sawtooth (PoET), EOS (DPoS), Multichain (distributed consensus), R3 Corda (Notaries), Cardano (Ouroboros), IOTA (Tangle), Waltonchain (WPoC). So every analysed security impact should be adapted to the current context of purpose, a DLT system used and a consensus mechanism applied (if applicable). Decision generalization shouldn't be permitted.

The goal of the study is to summarize and systematize security weaknesses in DLT and explore their impact. The set of consequences are usual for cyber incidents but the main point in the area of DLT is about financial losses.

3 DLT Security Constraints

This section contains an analysis of typical incidents that show weaknesses in the architectural and organizational components and social consequences of security of DLT.

Internal Architecture. Smart contracts gave birth to distributed autonomous organizations (DAO). A DAO is a computer code through which a set of smart contracts are connected and function as a governance mechanism. On 16 June 2016, the attacker managed to retrieve approximately 3.6 million Ether from the DAO crowdfunding abusing the loophole, which is known as a "recursive call exploit" [8]. The price of ether dropped from over $20 to under $13. Several experts attempt to split The DAO to prevent more ether from being taken, but they couldn't get the votes necessary in such a short time. Because the designers didn't expect this much money, all the Ether was in a single address. Attack of a similar one could continue at any time [7] and draining Ether again [9].

Researchers from MIT found vulnerabilities in the blockchain-based voting system. It allows different kinds of adversaries to alter, stop, or expose a user's vote, including a side-channel attack in which a completely passive network adversary can potentially recover a user's secret ballot [11]. The state voting initiatives have the status of critical infrastructure.

Ecosystem Elements. Malicious actors utilize every possibility in the digital space. Large-scale crypto-mining attacks are registered in [12]. Within only two hours a malicious container was deployed on tens of Kubernetes clusters. After deployment of the crypto mining containers malicious actor enumerated the cluster resources including Kubernetes secrets. This can lead to exposure of connection strings, passwords, and other secrets which might enable lateral movement. The malicious container was deployed by the Kubernetes dashboard via a service account [12].

Another typical class of cases are related to crypto key management. One example is the Chrome extension caught stealing crypto-wallet private keys [10].

Social Issues. The following incidents show different cyber-attacks in terms of their impact. They all affect many citizens from a given area or different parts of the world.

Bitcoin giveaway scams have been around for more than two years, but a new twist in tactics has helped scammers make more than $2 million from Elon Musk's name. The new trick involves the use of Bitcoin vanity addresses to give the scam more credibility in the eyes of a suspicious user [13]. Vanity addresses are Bitcoin addresses that incorporate a custom word in the address itself – such as
"1MuskPsV7BnuvMuHGWmmXUyXKjxp3vLZX6" or
"1ELonMUsKZzpVr5Xok8abiXhhqGbdrnK5C"

Justin Lister, CEO of cyber-security firm Adaptiv, has been tracking the use of Bitcoin vanity addresses abusing Elon Musk's name. BitcoinAbuse is a website where users can report Bitcoin addresses abused in ransomware, extortions, cybercrime, and online scams. He has been sharing with ZDNet a spreadsheet that shows tracked down 66 addresses that have been reported to BitcoinAbuse by scammed users. Lister said the 66 addresses received more than 201 in Bitcoin since being created in late April 2020 [14].

In July 2020, cryptocurrency wallet company Ledger revealed a breach of 9500 customers' names and contact information [15].

Industrial cybersecurity vendor Radiflow has discovered Monero-mining malware on five servers of a water utility company located in Europe. These servers included the HMI (Human Machine Interface), which was also the control server of the physical processes of the company [16].

Cryptocurrency malware attacks involve extremely high CPU processing and network bandwidth consumption, which can threaten the stability and availability of the physical processes of a critical infrastructure operator. This new case of a cryptocurrency malware attack on an OT (Operational Technology) network poses a new threat as it runs in stealth mode and can remain undetected over time.

On February 2020, over 4,200 websites around the world started hijacking visitors' browsers to mine the Monero cryptocurrency [17].

Regulatory Measures. Smart contracts have already been used in connection with real estate transactions, bank bonds, interbank transfers, invoice financing, and homeowners, renters, pet, and insurance [18]. A lot of law regulation gaps concerning DLT are commented on in [19]. The general security of the DLT is related to the GDPR, European Cybersecurity Act, the recommendations of the ENISA and the NIST frameworks. There are also industry norms, such as PSD (Payment Service Directive) and others. The analysis of models proposed in [20] show that the scope of cybersecurity of real implementation is aligned with the next areas (see Fig. 2).

The main conclusion is that the current state of DLT regulatory landscape represents more risk than opportunity and regulators need to step in before disaster strikes [19]. The discussions of these questions take multiple volumes of articles by a vast array of experts in numerous related fields [4].

The CSA (Cloud Security Alliance) presents almost 200 weaknesses in Blockchain technology, many of which are not fully understood or documented at this time. This group investigates relevant use cases and security implications of Blockchain. Currently, they are creating a framework and glossary providing guidance and security around wallets, exchanges, cryptography and more [5].

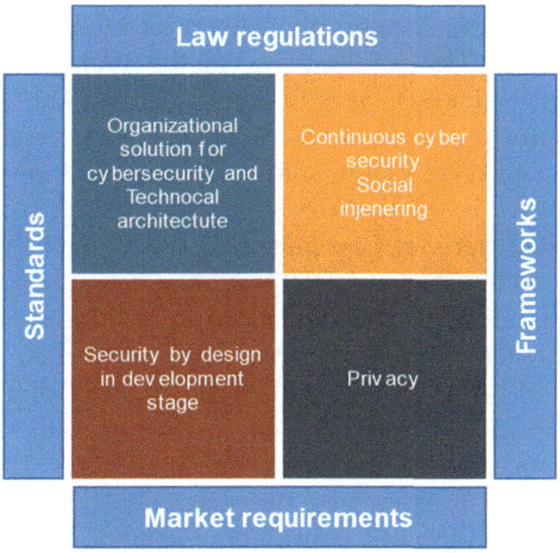

Fig. 2. A holistc view of cybersecurity aspects of DLT

4 Distribution of DLT Cybersecurity Weaknesess

The next proposed taxonomy is the research perspective. It gives a summary of risks simultaneously. Crypto crime is an emerging segment of the cybercrime ecosystem. The hacks on cryptocurrency exchanges suffered roughly $1 billion in losses during 2018. The 5 biggest bitcoin hacks of all time with the exchange name, the amount was stolen, and year occurred, according to CoinSutra: Mt. Gox, 2609 BTC, 750,000 BTC (2011); BitFloor, 24,000 BTC (2012); Poloniex, 12.3% of all BTCs – 97 BTC (2014); BitStamp, 19,000 BTC (2015); Bitfinex, 120,000 BTC (2016). The spectrum of cyber-attacks against DLT technology is wide. We systemize new ones and the main traditional attacks. The next list of attacks is partial. It consists of the most important types of attack against the DLT, that support our holistic analysis and conclusions.

4.1 Architectural Security Weaknesess

DLT also introduces different attack vectors that malicious actors may seek to exploit. For example, advanced attackers will look to influence decision-making processes around the blockchain in order to add new parts to the chain, change rules or policies or manipulate a managing entity in such a way that is not transparent or is fraudulent [21]. Our experience shows this technology, like any other, inherited a whole spectrum of environmental weaknesses: viruses, key-loggers, trojans, OS security configuration, compromised crypto keys, lack of security expertise for developing teams. In addition, DLT is functioning together with NTP, DNS, SS7 and many more protocols that have famous weaknesses existing in obsolete installations [22]. Internal security weaknesses in DLT are explored in [23,24]. Among the most vulnerable aspects of DLT technology comes from outside the chain itself. The weaknesses can be found in every architectural element. For example, this set of vulnerabilities in NAS (Network Attached Storage) [1] can compromise fragment of DLT chain, management dashboard or another important part of technical architecture. 51% of attacks by nature is an architectural security issue. In 2018, several notable cryptocurrencies such as ZenCash, Verge, and Ethereum Classic fell victim to 51% attacks [25].

The distributed architecture of DLT and the requirement for anonymity in design for cryptocurrencies create conditions for unauthorized incorporation into external systems. Attackers create crypto bots and organize the operation of Docker containers and bot networks within the Internet of Things systems. The reports with the research of specialized companies contain analyzes of incidents with millions of instances of Docker containers, which are spread on foreign platforms around the world [26]. For years, they have been extracting cryptocurrency of enormous value from endpoints, container environments and IoT systems [27].

Cryptojacking malware cause performance issues increase electricity usage, and opens the door for other hostile code. This kind of software flaws are security issues often go undiscovered for years.

4.2 Organizational Weaknesses

Organizational weaknesses in connection with DLT cybersecurity are familiar amongst cyber experts via analogy with other distributed information systems. In our opinion organizational issues for DLT cybersecurity are in the two types of supply chains - development and a production. They are consequences from the maturity stage of DLT cybersecurity. Programming tools [28] are not mature from the viewpoint of security. The development stage includes security requirements for the concept of operations, technical design, programming languages, and testing. A summary of security vulnerabilities related to two types of supply chains are presented in Table 1. There are no traces of the concept of "security by design" in the different DLT communities. The exception is R3 CORDA. The content of its site proof with parts for security procedures and documents [2].

Social engineering is an organizational problem. It has many forms, but the goal is always the same. Malicious actors try to obtain private keys, login information, or directly cryptocurrency. Hackers are targeting the employees of cryptocurrency exchanges. Then they gain access to the private networks of the exchange and steal the coins [29]. According to the case from [25], $3 million was lost due to social engineering in 2018. Crypto key management is an unsolved problem till now, although efforts of the ICT industry. The ability to generate randomness is the foundation of many security protocols. A number of applications require random values that cannot be predicted prior to being generated but are made public after generation [31]. Privacy risks and user's confidentiality could be affected by different factors. In permissive distributed ledger systems this is most clearly seen. The research has shown that a user's Bitcoin transaction can be linked to reveal user ownership and jurisdiction of information and the risk of this leakage increase when the Bitcoin addresses are reused multiple times. Legal clarity over ownership and jurisdiction of data and transactions between states is a large area of problems. They are connected with IoT too [30].

Table 1. Summary of the registered DLT vulnerabilities.

Component	Example	Amount
Swagger	CVE-2017-1000207, 8, CVSS 6.8	2
Metasploit modules	bitcoin_jacker.rb claymore_dual_miner_remote_manager_rce	2
Golang	CVE-2017-3204, 6.8; CVE-2019-11841, 4.3	2
Ethereum VM	CVE-2017-14457	1
Ethereum	cvedetails.com/vendor/17524/Ethereum.html	12
CWE – 338	cvedetails.com, Cryptographically Weak PRNG	30
Hyperledger	CVE-2018-3756	1

Exchange hacks are organizational cybersecurity issues. One of the costliest blockchain security issues isn't a problem with the blockchain technology itself. Cryptocurrency exchanges have become lucrative honeypots for hackers due to their massive crypto holdings and sometimes poor security practices. Many exchange platforms are innately centralized, so they make the decentralized benefits of blockchain obsolete. Any early incident was the 2014 Mt. Gox hack. Exchange, processing nearly 70% of all bitcoin transactions. In February 2014, the exchange revealed that a hacker had stolen approximately 850,000

BTC ($473 million) from the platform. In 2019 exchanges lost over $900 million to hackers. The cost of the 2018 Coincheck hack, the biggest cryptocurrency heist to date, was $530 million. 523 million NEM coins (known as XEM) had been stolen from a wallet connected to the Internet [32].

DLT also introduces different attack vectors that malicious actors may seek to exploit. For example, advanced attackers will look to influence decision-making processes around the blockchain in order to add new parts to the chain, change rules or policies, or manipulate a managing entity in such a way that is not transparent or is fraudulent [21].

4.3 Privacy Challenges

The core functionality of a blockchain, its immutability, seems to contradict with much of the growing global privacy legislation of today, such as the European GDPR [33]. Article 17 of the GDPR, the right to erasure, more commonly referred to as the right to be forgotten, states that a "data subject shall have the right to obtain from the controller the erasure of personal data concerning him or her without undue delay". However, the immutable nature of DLT, and arguably the core of its value proposition, prevents data from being deleted. This very contradiction between the functionality of DLT and the legal requirements of many countries poses some concerns but is an area being addressed [34] data protection by design versus tamper-proofness and transparency of blockchain (Article 25); data controller versus decentralized nodes (Article 4) [33]. Privacy issues are incorporated mainly in the CONOPS (Concept of Operations) decisions for the architecture, and development stage. Their consequences appear on the surface via the supply chain during production mode. For example, the blockchain is inappropriate for privacy requirements because every participant can see the whole data [35].

5 Conclusions and Recommendations

The problem area of the DLT is characterized by a lack of regulatory measures, standards and frameworks. The distributed architecture of DLT remains untested full scale and does not allow full coverage with security tests. The nature of this technology generates new problems with cybersecurity and confidentiality. If the blockchain utilizes a Proof-of-Work (PoW) consensus mechanism, it needs to have security measures in place to prevent a 51% attack. In this case, the vigilance of mining pools, implementing merged mining on a blockchain with a higher hash rate, or switching to a different consensus mechanism are all viable options. The safest methods of storing cryptocurrency are through either a hardware or paper wallet. These methods have minimal online touchpoints which keep coins out of reach to malicious online hackers. If the business model involves regular trading, it is recommended to use a decentralized exchange (DEX) that allows participants to trade directly from the cryptocurrency wallet [25].

Organizations must understand the potential impacts of illicit cryptocurrency mining on their operations. In its most basic form, illicit mining is a drain on the resources in anyone's enterprise, increasing the workload and the risk of physical damage on IT infrastructure, causing higher electrical bills, and decreasing the productivity of the business operations that rely on computing power. The presence of illicit cryptocurrency mining within an enterprise is indicative of flaws in their cybersecurity posture that should be addressed.

This study examines the DLT security weaknesses. The recommended measures are intended to increase the level of cyber protection of corporate IT systems. Future research directions are related to the incorporation of legislation, which is usually intended to regulate the cyber-physical world.

References

1. C. NCSC: Potential Legacy Risk from Malware Targeting QNAP NAS Devices CISA (August 2020). https://us-cert.cisa.gov/ncas/alerts/aa20-209a
2. r3: Security Issues Reporting Policy Corda (August 2020). https://www.corda.net/security
3. Zheng, Z., Xie, S., Dai, H., Chen, X., Wang, H.: An overview of blockchain technology: architecture, consensus, and future trends. In: 2017 IEEE International Congress on Big Data (BigData Congress). IEEE (June 2017)
4. Kakavand, H., Sevres, N.K.D., Chilton, B.: The blockchain revolution: an analysis of regulation and technology related to distributed ledger technologies. SSRN Electron. J. (2017). https://doi.org/10.2139/ssrn.2849251. SSRN ID 2849251
5. Blockchain/Distributed Ledger Technologies vert Cloud Security Alliance Cloud Security Alliance (2020)
6. Cimpanu, C.: Bitcoin wallet update trick has netted criminals more than $22 million (October 2020). https://www.zdnet.com/article/bitcoin-wallet-trick-has-netted-criminals-more-than-22-million
7. Siegel, D.: Understanding the DAO attack - CoinDesk (June 2016). https://www.coindesk.com/understanding-dao-hack-journalists
8. Mehar, M.I., et al.: Understanding a revolutionary and flawed grand experiment in blockchain. J. Cases Inf. Technol. **21**(1), 19–32 (2019)
9. Siegel, D.: Understanding the DAO hack for journalists (June 2018). https://medium.com/@pullnews/understanding-the-dao-hack-for-journalists-2312dd43e993.kw0ufw25q
10. Cimpanu, C.: Chrome extension caught stealing crypto-wallet private keys (January 2020). https://www.zdnet.com/article/chrome-extension-caught-stealing-crypto-wallet-private-keys
11. Specter, M.A., Koppel, J., Weitzner, D.: The ballot is busted before the blockchain: a security analysis of Voatz, the first internet voting application used in U.S. federal elections. In: 29th USENIX Security Symposium (USENIX Security 2020), pp. 1535–1553. USENIX Association (August 2020)
12. Weizman, Y.: Detect large-scale cryptocurrency mining attack against Kubernetes clusters (August 2020). https://azure.microsoft.com/en-us/blog/detect-largescale-cryptocurrency-mining-attack-against-kubernetes-clusters
13. Cimpanu, C.: Elon Musk Bitcoin vanity addresses used to scam users out of $2 million (June 2020). https://www.zdnet.com/article/elon-musk-bitcoin-vanity-addresses-used-to-scam-users-out-of-2-million

14. bitcoinglobal: Bitcoin Scam Page 2 Bitcoin Global Canada (BGC) (November 2020). https://www.bitcoinglobal.ca/category/bitcoin-scam/page/2
15. Addressing the July 2020 e-commerce and marketing data breach - A Message From Ledger's Leadership Ledger (July 2020). https://www.ledger.com/addressing-the-july-2020-e-commerce-and-marketing-data-breach
16. Radiflow: Radiflow Reveals First Documented Cryptocurrency Malware Attack on a SCADA Network Radiflow (December 2020). https://radiflow.com/-news/radiflow-reveals-first-documented-cryptocurrency-malware-attack-on-a-scada-network
17. Zorz, Z.Z.: When crypto-mining malware hits a SCADA network - Help Net Security (June 2018). https://www.helpnetsecurity.com/2018/02/08/crypto-mining-malware-hits-scada-network
18. Foreman, A.: Avoiding smart contract conflicts despite ambiguity (2020)
19. Osborne, C.: What we can expect from future cryptocurrency regulation worldwide (February 2018). https://www.zdnet.com/article/the-state-of-cryptocurrency-regulation-worldwide-and-what-the-future-will-bring
20. Hristova, T.: Blockchain application in mining management. In: E3S Web of Conferences, vol. 168, p. 00009 (2020)
21. English, E., Kim, A.D., Nonaka, M.: Advancing blockchain cybersecurity: technical and policy considerations for the financial services industry (2018)
22. Dimitrov, W.: The impact of the advanced technologies over the cyber attacks surface. In: Silhavy, R. (ed.) Artificial Intelligence and Bioinspired Computational Methods, pp. 509–518. Springer International Publishing, Cham (2020)
23. Prvulovic, M.: Hypothetical Blockchain Security Vulnerabilities and How They Can Be Addressed (June 2018). https://kapitalized.com/blog/hypothetical-blockchain-security-vulnerabilities-and-how-they-can-be-addressed
24. Praitheeshan, P., Pan, L., Yu, J., Liu, J., Doss, R.: Security analysis methods on ethereum smart contract vulnerabilities: a survey (2019)
25. LedgerOps - Elite Cybersecurity (February 2020). https://ledgerops.com/blog/2019/03/28/top-five-blockchain-security-issues-in-2019
26. Vizard, M.: Docker Hub Distributing Cryptomining Malware? - Container Journal (June 2020) https://containerjournal.com/topics/container-security/docker-hub-distributing-cryptomining-malware
27. Nadeau, M.: What is cryptojacking? How to prevent, detect, and recover from it (July 2020). https://www.csoonline.com/article/3253572/what-is-cryptojacking-how-to-prevent-detect-and-recover-from-it.html
28. Mavrevski, R., Traykov, M., Trenchev, I.: Interactive approach to learning of sorting algorithms. Int. J. Online Biomed. Eng. (iJOE) **15**(08), 120 (2019)
29. Dharshan, A.: PewDiePie incident exposes vulnerabilities in seed phrase printing - CryptoTicker (February 2020). https://cryptoticker.io/en/pewdiepie-seed-phrase-printing-hack
30. Sultan, A., Mushtaq, M.A., Abubakar, M.: IOT security issues via blockchain. In: Proceedings of the 2019 International Conference on Blockchain Technology - ICBCT 2019. ACM Press (2019)
31. Bonneau, J., Clark, J., Goldfeder, S.: On bitcoin as a public randomness source. IACR Cryptology ePrint Archive, vol. 2015, p. 1015 (2015)
32. TrendMicro: Coincheck Suffers Biggest Hack in Cryptocurrency History; Experty Users Tricked into Buying False ICO - Security News - Trend Micro USA (August 2020). https://www.trendmicro.com/vinfo/us/security/news/cybercrime-and-digital-threats/coincheck-suffers-biggest-hack-in-cryptocurrency-experty-users-buy-false-ico

33. Tatar, U., Gokce, Y., Nussbaum, B.: Law versus technology: blockchain, GDPR, and tough tradeoffs. Comput. Law Secur. Rev. **38**, 105454 (2020)
34. Patel, D.: Buzz Killer: A Realistic look at the challenges facing DLT in Trade - Trade Finance Global (April 2020). https://www.tradefinanceglobal.com/posts/challenges-dlt-in-trade
35. Binance: Summary of the Phishing and Attempted Stealing Incident on Binance (February 2020). https://www.binance.com/en/support/articles/360001547431

A Hybrid Approach with Machine Learning Towards Opinion Mining for Complex Textual Content

D. Rajeshwari[1(✉)] and D. Puttegowda[2]

[1] Department of Information Science and Engineering, ATMECE, NIE Institute of Technology, Mysore, India
[2] Computer Science and Engineering, ATMECE, Mysore, India

Abstract. Opinion Mining is increasing adopted in business enterprises where sophisticated analytical approaches are subjected to the opinion in the form of text from the consumer. Review of existing literature shows that there are still an open scope to further improve upon this technique. The proposed study considers a case study of a problem where the opinion is shared in the form of text as well as symbol/emoticon, which is quite challenging for any existing text analytics to extract the knowledge. Therefore, the proposed paper introduces a novel solution where two variants of approaches has been used for this purpose i.e. hybrid approach and machine learning approach in order to perform opinion mining from such complex textual content. The study outcome shows that proposed system offers satisfactory processing time and accuracy in large dataset of text.

Keywords: Text mining · Opinion · Sentiments · Machine learning · Unstructured data · Text analytics

1 Introduction

The present state of business harnesses the potential of opinion mining in order to understand the underlying reason for extracting an insight of specific product or service [1, 2]. Opinion mining can be traditionally carried out using emotion detection [3], multilingual analysis of sentiment [4], aspect-based analysis of sentiments [5], fine-grained analysis of sentiment [6]. However, one of the big challenges in opinion mining is the presence of unstructured data which is also not precise in many ways. This is the prime challenging factor for applying text analytics on opinion mining [7]. At present, there are various work being carried out in this perspective but there is loopholes and open-end issues in almost all the associated approaches towards opinion mining [8–10]. Another interesting fact found in existing approach is that in existing approaches are mainly focused on textual content whereas customer in modern times expresses their opinion in form of different symbols and emoticons too. Existing approaches doesn't discuss about performing text analytics on such form of complex textual contents and hence, this is the prime problem which is addressed in proposed study using machine learning approach. The organization of the paper is as follows: discussion of existing

R. Silhavy (Ed.): CSOC 2021, LNNS 230, pp. 656–666, 2021.
https://doi.org/10.1007/978-3-030-77442-4_55

opinion mining approaches is carried in Sect. 2, identified research problem in Sect. 3. Proposed methodology is discussed in Sect. 4, algorithm implementation in order to achieve the proposed aim is discussed in Sect. 5, while result analysis is carried out in Sect. 6. Finally, the summary of this paper is carried out in Sect. 7.

2 Related Work

There has been various studies carried out in order to perform analytical operation on opinion using different variants of approach. The work of Park et al. [11] suggest that structuring the respective word with their meaning is crucial to achieve accuracy like a human. The basic features like parts of speech, text colour, and font size make the corpus complex; an extensive survey is being conducted by Strobelt et al. [12].The work of Cho et al. [13] and Liu et al. [14] has discusses about different regional history analysis is possible by text data analysis such study for Roman history where various web-based visualization tool and fundamentals of visual text analytics are described. Vatrapu et al. [15] explores set theory-based visualization to complement text analysis. The sedimentation-based visualization concept of coordinated structure in text analysis has been studied by (Liu et al. [16] and Sun et al. [17], respectively. The analysis of semantic with its content plays a vital role in the content analysis as seen in study of Kurzhals et al. [18]. Ojo et al. [19] presented patient sentiment analysis using textual data. Karam et al. [20] presented a design of new hardware that supports the ecosystem of processor and memory for test analytics. The use of text-data from social media like Facebook and other feeds and surveillance camera images Neuhold et al. [21] is found. The big-data and the business complement's process management complement if the text generated is adequately analysed (Sakr et al. [22]). A tree-based visual representation of text is being practiced, but this method is not scalable (Liu et al. [23]. The bag of words, along with natural language processing and visual analytics, is being studied in the work of Benito et al. [24]. Basole et al. [25] has reviewed topic modeling on an extensive text description used in a business domain based on text analytics. Text analytics is used by Han et al. [8] in their work. Reghupathi et al. [26] has examined the use of text analytics on patent data corpora using machine learning model. Health care industries are another domain which produces a vast amount of unstructured text data Kumar et al. [27] has performed text analytics for decision support system. Text analytics has also shown its benefits in the political election campaign as seen in work of Gad et al. [28]. Text analytics facilitates benefits on the topic description of an event as in the work of Lu et al. [29]. El-Assady et al. [30] discusses a decision-making technique based on relevance feedback using text analytics. The study of sentiment analysis in crowdfunding is presented by Wang et al. [31]. Similar benefits can be achieved in further education, banking, and marketing sectors (Amalina et al. [32]). The forensic sector benefits if the complex text data from various communication sources are being analyzed by Koven et al. [33] that devices a tool that uses text analytics on the email data corpus. The use of sequence-to-sequence learning is increasing witnessed with higher accuracy and lower training time Keneshloo et al. [34].

3 Problem Description

After reviewing the existing approaches towards opinion mining, it can be seen that existing study has not considered inclusion of any non-textual object while performing mining operation. Machine learning has been applied towards text mining but without considering the complex content of the text which inevitably results in outliers or non-applicable in the present scenario of opinion. Apart from this, it is also seen that existing mechanism doesn't deals much with transformation scheme towards the complex raw structure of opinion content. Hence, the problem statement is "*it is challenging to develop a text analytics towards opinion mining which has inclusion of complex textual content.*" The next section discusses about the solution to this problem followed by discussion of series of algorithm to deal with this problem.

4 Proposed Methodology

The prime aim of the proposed study is to develop a simplified framework where complex opinion in the form of text and emoticons can be subjected to novel mining, classification using machine learning. The secondary objective of the study is to evolve up with opinion-as-a-service.

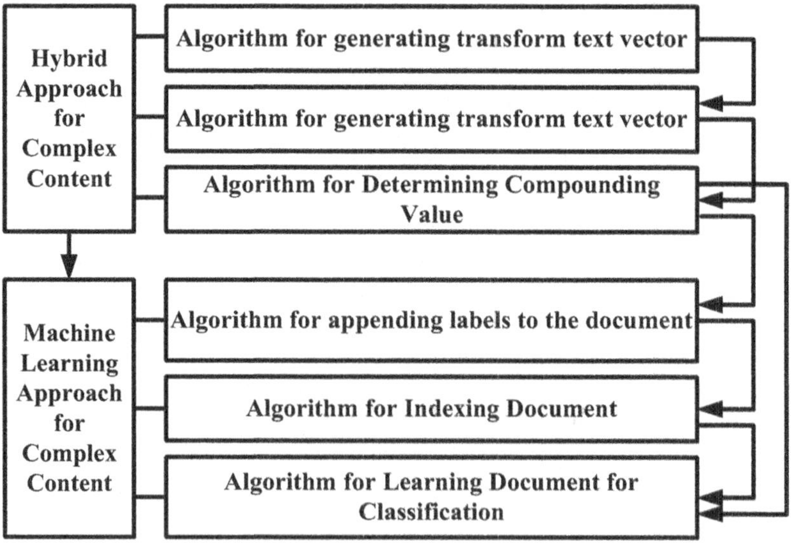

Fig. 1. Proposed implementation scheme

Figure 1 highlights that proposed system implements two essential block of operation where a hybrid approach as well as machine learning approach has been used for complex content analysis in order to support opinion mining. The proposed system also makes use of sentiment analysis concept in order to understand the degree of sentiment inclusion in the complex content shared by the consumer. The next section elaborates about the block of operation towards achieving the proposed aim.

5 Algorithm Implementation

The proposed system implements 6 sequential algorithms in order to develop opinion-as-a-service. The complete algorithmic operation is classified into two process viz. developing framework of complex context using i) hybrid approach and ii) machine learning approach. The discussions of the algorithms are as follows:

5.1 Hybrid Approach for Complex Content

This approach of algorithmic implementation focuses on constructing text vector and formulating score weightage. The first algorithm is responsible for generating transformed text vector considering the input of unstructured text *ut*. The step of the algorithm is as follow:

1. Algorithm for generating transform text vector
Input: ut (unstructured text)
Output: t_{vec} (text vector)
Start
1. init ut←Fn
2. $\beta \leftarrow f_1(ut)$
3. $t_{vec} \leftarrow f_2(t_{data})$

End

The algorithm considers unstructured text *ut* to be combination of text token *T* and symbols *S* in order to express the opinion of the consumer. The algorithm considers a variable *Fn* (Line-1) which consist of both text and symbol and is assigned to a matrix *ut* (Line-1). A tabular format is used for the transformation process using function $f_1(x)$ which when applied to *ut* gives the outcome of structured format of text β (Line-2). In the consecutive process, another function $f_2(x)$ is applied which carry out tokenization of the document process leading to generation of transformed text vector t_{vec} (Line-3) from all transformed data which is actually a subset of prior matrix β (Line-2). It should be noted that a vector of text token t_{data} consist of all the character and symbols arranged in form of an array for making it suitable for further processing. After this processing, the second algorithm is implemented whose steps are as follows:

2. Algorithm for Determining Compounding Value
Input: t_{vec} (text vector)
Output: C_{val} (compounding value)
Start
1. $\gamma \leftarrow (\tau_i)$,
2. $\lambda = \{h_j\}$
3. $\gamma \leftarrow \{\beta, \lambda\}$
4. $C_{val} = \frac{\pi}{\sqrt{\pi^2 + \rho}}$

End

This algorithm carries out the computation of the compounding value on the basis of the valence aware dictionary which is typically used for logical reasoning of sentiment analysis [35]. In this part of implementation, the algorithm considers generation of customized Vader γ that is considered as rule-based lexicon model using dependable parameter of τ_i of i_{th} type. The proposed system consider i = 3, where τ_1 represents sentiment lexicon, τ_2 represents grammar syntax, and τ_3 represents convention (Line-1). The operation is carried out in such a way that it considers a set of sentiment polarity and sentiment intensity. The matrix of Vader γ built up a list of words for all possible range of the feature vector λ (Line-2) in such a way that it represents a set with h_i components where $i = 4$ representing h_1 as words, h_2 as phrases, h_3 as symbols, and h_4 as acronym (Line-2). The proposed system offers negative score and positive score for sentiment polarity and sentiment intensity. The implementation considers γ to extract all values of β along with its associated value of λ from prior steps (Line-3). The formulation of the compounding value C_{val} using symbol π and maximum probability of the anticipated cost of value (Line-4).

5.2 Machine Learning Approach for Complex Content

This part of the implementation is carried out using machine learning approach. However, machine learning approach is not directly applied, but various processing of labelling and indexing is carried out over the document in order to make the outcome suitable to be subjected to machine learning approach. In this direction, the first implementation is carried out towards annotating labels to the data considers a dataset of University of Illinois using opinion lexicons Φ that consist of both positive and negative class of emotion. The steps of this algorithm is as follow:

3. Algorithm for appending labels to the document
Input: Φ (Opinion Lexicon)
Output: l_{doc}(labelled document)
Start
1.[pe/ne]$\leftarrow f_1(\Phi)$
2. L(mx)$\leftarrow$pe$\cup$ne
3. La(undefined: La)$\leftarrow f_2$((NaN), m, 1)
4. CLa<La(Undefined: La):W
5. $l_{doc}\leftarrow$W $\cup$Cla
End

This algorithm takes the input of Φ (Opinion Lexicon) which after processing yields l_{doc} (labelled document). The algorithm make use of a function $f_1(x)$ which assess the correctness of the data file and performs conversion of the positive and negative emotion (Line-1) followed by concatenating both of them (Line-2). This operation generates a list of words of string datatype and hence they can be represented a Not-a-number NaN (Line-3). The algorithm further obtains list of categorical variables by converting list of all NaN which is further stored in annotated labels (La) (Line-4). All the associated elements of W is now mapped with annotated labels La in the form of categorical labels (CLa) further they are concatenated in order to generate labels of data, which is further utilized for carrying out learning models (Line-5). After this operation is carried out, the next part of the operation is associated with performing filtering of the text tokens. The implemented algorithmic steps of this process are:

4. Algorithm for Indexing Document
Input:l_{doc} (labelled document)
Output:i_{doc} (indexed document)
Start
1. $i_{doc} \leftarrow f_3(l_{doc})$
2. $i_{doc} \leftarrow f_6(f_5(f_4(i_{doc=})))$

End

This algorithm takes the input of l_{doc} (labelled document) which yields an outcome of i_{doc} (indexed document). The algorithm applies a function $f_3(x)$ in order to obtain the indexed document with multiple set (Line-1). The algorithm also make use of an embedded functions where $f_4(x)$ is used for eliminating the punctuation, $f_5(x)$ is used for eliminating the stop words, and $f_6(x)$ is used for capitalize the lower case words (Line-2). The proposed system also make use of bag of words that can be also used for multiple lingual analysis. The end result of this algorithm is indexed document idoc which is finally subjected to machine learning process as discussed in algorithmic steps as follows:

5. Algorithm for Learning Document for Classification
Input: D (document)
Output: t_{Model} (classified text)
Start
1. D←fTW
2. Call Algorithm-3: ldoc(W, CLa)← Φ
3. Idx←not [ldoc(W) D]
4. Ds←num of W
5. [D-train,D-test]←f(Ds. numW)
6. D-train ←Word2vector[D-train]
7. Td←[D-train U CLa]
8. Train SVM- 1-Class-Binary Classifier
9. $t_{Model} \leftarrow F(Td)$

End

This algorithm implements Support Vector Machine in order to perform classification. The algorithm offers indexed data l_{doc} which consist of W and CLa. It perform the indexing operation of all the words that doesn't belong to the document D as constructed as Idx (Line-3). The cumulative size of the data D is basically the total number of words num W. The partitioning process of D takes place in order to carry out the cross validation of cross validate arbitrary partition on Ds in order to define the partitioned for a statistical model considering training and test data (Line-5). Applying the principle of NLP, the mapping of the word to generate vectors is carried out. Further the study implements Artificial Neural Network in order to carry out learning of the massive corpus of the text data of opinion where all the words is represented in the form of listed number in the form of vectorizing the simple mathematical function that further carry out mapping of the semantic similarity as training data obtained from function Word2vector (Line-6) while the Support Vector Machine uses training input in order to generate Td (Line-8 and Line-9).

6 Results Discussion

This section discusses about the outcome obtained after implementing the algorithm discussed in previous section. For this analysis, the proposed model considers the dataset from University of Illinois, Chicago which consists of Opinion Lexicons with 6789 list of words. All these elements in corpus belongs to two forms of sentiments i.e. positive and negative sentiments in the form of text token [36]. The words are arranged in alphabetic order. The proposed system considers a pre-trained model that offers an object named dictionary consisting of 9,99,994 word tokens in the form of string [37]

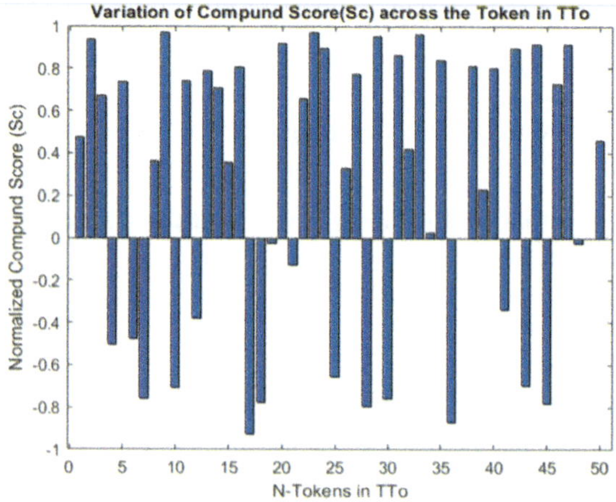

Fig. 2. Variation of normalized compound score Sc across N-token in α, N = 50

Figure 2 highlights the normalized compound score considering the number of token N as 50.

The time of processing, including all the process of tokenization, score computation, and visual presentation is shown in Fig. 3. Apart from this, extensive analysis is carried out for assessing the scalability factor by computing the processing time for increasing size of tokens (N = 50, 100, 150, 200, 250) as exhibited in Fig. 4

The algorithm is further assessed with respect to confusion matrix between the predicted class and true class associated with the test data used for the analysis. The outcome is shown in Fig. 5.

Fig. 3. Analysis of time to process

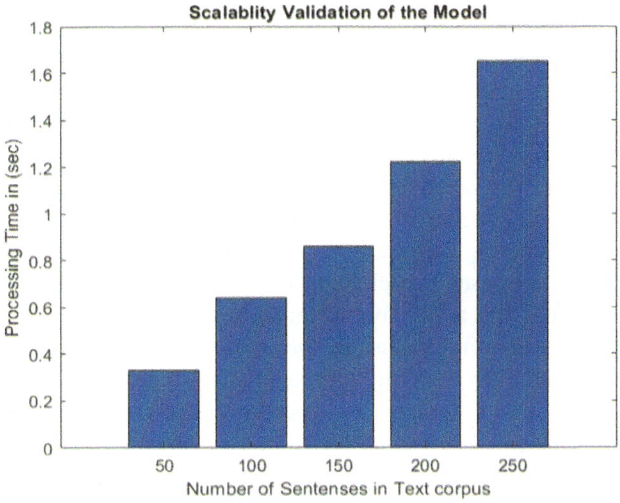

Fig. 4. Scalability assessment of proposed system

From Fig. 5, it can be seen that the proposed system offers better identification of the true positive classes for the predictive process. Hence, the algorithm is assessed to show that it offers faster data transformation operation to assist in opinion data as a service in cost effective manner.

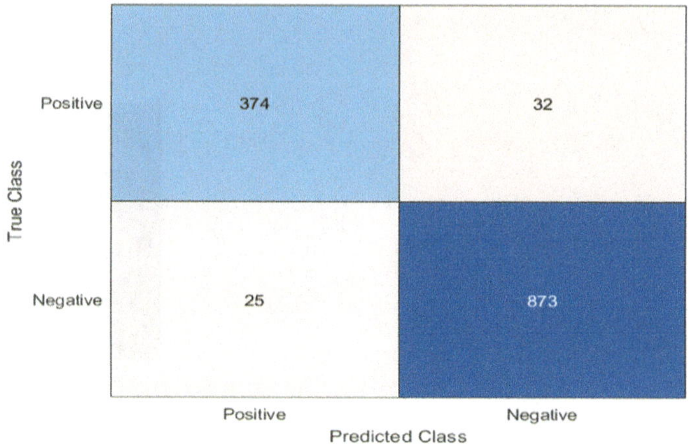

Fig. 5. Confusion matrix between the predicted Class and the true Class

7 Conclusion

This paper has discussed about a significant problem associated with opinion mining where the opinion is shared by the user in the form of textual content as well as special symbol usually in the form of emoticons. Such inclusion of opinion cannot be analyzed by conventional text analytics. The proposed system implements two novel approaches which are carried out in sequential form. The first implementation is about applying a hybrid approach while the second implementation is about applying integration machine learning approach combining both support vector machine and artificial neural network. A contribution of the proposed model are as follows: i) the proposed system is capable of performing text analytics applicable for modern form of sharing opinion by user in the form of complex opinion content, ii) the proposed system doesn't directly apply machine learning unlike any existing approaches, but it subjects the data to series of operation which makes the data suitable for performing learning operation, iii) the complete operation doesn't make use of any form of sophistical tool or technologies which makes the analytical process quite lighter, and iv) its processing time is quite satisfactory and quite scalable in contrast to size of the data. The future work will be carried out towards optimized the scheme further.

References

1. Bhatia, S., Chaudhary, P., Dey, N.: Opinion Mining in Information Retrieval. Springer, Singapore (2020). ISBN: 9789811550430, 9811550433
2. Gupta, N., Agrawal, R.: Extracting Knowledge from Opinion Mining. IGI-Global (2018). ISBN: 9781522561187, 1522561188
3. Kadiri, S.R., Alku, P.: Excitation features of speech for speaker-specific emotion detection. IEEE Access **8**, 60382–60391 (2020). https://doi.org/10.1109/ACCESS.2020.2982954
4. Ke, Z., Sheng, J., Li, Z., Silamu, W., Guo, Q.: Knowledge-guided sentiment analysis via learning from natural language explanations. IEEE Access **9**, 3570–3578 (2021). https://doi.org/10.1109/ACCESS.2020.3048088

5. Yang, J., Yang, J.: Aspect based sentiment analysis with self-attention and gated convolutional networks. In: 2020 IEEE 11th International Conference on Software Engineering and Service Science (ICSESS), Beijing, China, pp. 146–149 (2020). https://doi.org/10.1109/ICSESS49938.2020.9237640
6. Gao, Y., Liu, J., Li, P., Zhou, D.: CE-HEAT: an aspect-level sentiment classification approach with collaborative extraction hierarchical attention network. IEEE Access **7**, 168548–168556 (2019). https://doi.org/10.1109/ACCESS.2019.2954590
7. Hussein, D.M.E.-D.M.: A survey on sentiment analysis challenges. Journal of King Saud University - Engineering Sciences 30(4), 330–338 (2018)
8. Wang, R., Zhou, D., Jiang, M., Si, J., Yang, Y.: A survey on opinion mining: from stance to product aspect. IEEE Access **7**, 41101–41124 (2019). https://doi.org/10.1109/ACCESS.2019.2906754
9. Kren, M., Kos, A., Sedlar, U.: Modeling opinion of IPTV viewers based on implicit feedback and content metadata. IEEE Access **7**, 14455–14462 (2019). https://doi.org/10.1109/ACCESS.2019.2891837
10. Haq, E.U., Braud, T., Kwon, Y.D., Hui, P.: A survey on computational politics. IEEE Access **8**, 197379–197406 (2020). https://doi.org/10.1109/ACCESS.2020.3034983
11. Park, D., Kim, S., Lee, J., Choo, J., Diakopoulos, N., Elmqvist, N.: ConceptVector: text visual analytics via interactive lexicon building using word embedding. IEEE Trans. Vis. Comput. Graph. **24**(1), 361–370 (2018). https://doi.org/10.1109/TVCG.2017.2744478
12. Strobelt, H., Oelke, D., Kwon, B.C., Schreck, T., Pfister, H.: Guidelines for effective usage of text highlighting techniques. IEEE Trans. Vis. Comput. Graph. **22**(1), 489–498 (2016). https://doi.org/10.1109/TVCG.2015.2467759
13. Cho, I., Dou, W., Wang, D.X., Sauda, E., Ribarsky, W.: VAiRoma: a visual analytics system for making sense of places, times, and events in roman history. IEEE Trans. Vis. Comput. Graph. **22**(1), 210–219 (2016). https://doi.org/10.1109/TVCG.2015.2467971
14. Liu, S., et al.: Bridging text visualization and mining: a task-driven survey. IEEE Trans. Vis. Comput. Graph. **25**(7), 2482–2504 (2019). https://doi.org/10.1109/TVCG.2018.2834341
15. Vatrapu, R., Mukkamala, R.R., Hussain, A., Flesch, B.: Social set analysis: a set theoretical approach to big data analytics. IEEE Access **4**, 2542–2571 (2016). https://doi.org/10.1109/ACCESS.2016.2559584
16. Liu, S., Yin, J., Wang, X., Cui, W., Cao, K., Pei, J.: Online visual analytics of text streams. IEEE Trans. Vis. Comput. Graph. **22**(11), 2451–2466 (2016). https://doi.org/10.1109/TVCG.2015.2509990
17. Sun, M., Mi, P., North, C., Ramakrishnan, N.: BiSet: semantic edge bundling with biclusters for sensemaking. IEEE Trans. Vis. Comput. Graph. **22**(1), 310–319 (2016). https://doi.org/10.1109/TVCG.2015.2467813
18. Kurzhals, K., John, M., Heimerl, F., Kuznecov, P., Weiskopf, D.: Visual movie analytics. IEEE Trans. Multimed. **18**(11), 2149–2160 (2016). https://doi.org/10.1109/TMM.2016.2614184
19. Ojo, A., Rizun, N.: Enabling deeper linguistic-based text analytics—construct development for the criticality of negative service experience. IEEE Access **7**, 169217–169256 (2019). https://doi.org/10.1109/ACCESS.2019.2947593
20. Karam, R., Puri, R., Bhunia, S.: Energy-efficient adaptive hardware accelerator for text mining application kernels. IEEE Trans. Very Large Scale Integr. (VLSI) Syst. **24**(12), 3526–3537 (2016). https://doi.org/10.1109/TVLSI.2016.2555984
21. Neuhold, R., Gursch, H., Kern, R., Cik, M.: Driver's dashboard – using social media data as additional information for motorway operators. IET Intel. Transp. Syst. **12**(9), 1116–1122 (2018). https://doi.org/10.1049/iet-its.2018.5337

22. Sakr, S., Maamar, Z., Awad, A., Benatallah, B., Van Der Aalst, W.M.P.: Business process analytics and big data systems: a roadmap to bridge the gap. IEEE Access **6**, 77308–77320 (2018). https://doi.org/10.1109/ACCESS.2018.2881759
23. Liu, S., Chen, Y., Wei, H., Yang, J., Zhou, K., Drucker, S.M.: Exploring topical lead-lag across corpora. IEEE Trans. Knowl. Data Eng. **27**(1), 115–129 (2015). https://doi.org/10.1109/TKDE.2014.2324581
24. Benito-Santos, A., Therón Sánchez, R.: Cross-domain visual exploration of academic corpora via the latent meaning of user-authored keywords. IEEE Access **7**, 98144–98160 (2019). https://doi.org/10.1109/ACCESS.2019.2929754
25. Basole, R.C., Park, H., Chao, R.O.: Visual analysis of venture similarity in entrepreneurial ecosystems. IEEE Trans. Eng. Manag. **66**(4), 568–582 (2019). https://doi.org/10.1109/TEM.2018.2855435
26. Raghupathi, V., Zhou, Y., Raghupathi, W.: Legal decision support: exploring big data analytics approach to modeling pharma patent validity cases. IEEE Access **6**, 41518–41528 (2018). https://doi.org/10.1109/ACCESS.2018.2859052
27. Kumar, S., Singh, M.: Big data analytics for the healthcare industry: impact, applications, and tools. Big Data Min. Anal. **2**(1), 48–57 (2019). https://doi.org/10.26599/BDMA.2018.9020031
28. Gad, S., et al.: ThemeDelta: dynamic segmentations over temporal topic models. IEEE Trans. Vis. Comput. Graph. **21**(5), 672–685 (2015). https://doi.org/10.1109/TVCG.2014.2388208
29. Lu, Y., Wang, H., Landis, S., Maciejewski, R.: A visual analytics framework for identifying topic drivers in media events. IEEE Trans. Vis. Comput. Graph. **24**(9), 2501–2515 (2018). https://doi.org/10.1109/TVCG.2017.2752166
30. El-Assady, M., Sevastjanova, R., Sperrle, F., Keim, D., Collins, C.: Progressive learning of topic modeling parameters: a visual analytics framework. IEEE Trans. Vis. Comput. Graph. **24**(1), 382–391 (2018). https://doi.org/10.1109/TVCG.2017.2745080
31. Wang, W., Zhu, K., Wang, H., Wu, Y.J.: The impact of sentiment orientations on successful crowdfunding campaigns through text analytics. IET Softw. **11**(5), 229–238 (2017). https://doi.org/10.1049/iet-sen.2016.0295
32. Amalina, F., et al.: Blending big data analytics: review on challenges and a recent study. IEEE Access **8**, 3629–3645 (2020). https://doi.org/10.1109/ACCESS.2019.2923270
33. Koven, J., Felix, C., Siadati, H., Jakobsson, M., Bertini, E.: Lessons learned developing a visual analytics solution for investigative analysis of scamming activities. IEEE Trans. Vis. Comput. Graph. **25**(1), 225–234 (2019). https://doi.org/10.1109/TVCG.2018.2865023
34. Keneshloo, Y., Shi, T., Ramakrishnan, N., Reddy, C.K.: Deep reinforcement learning for sequence-to-sequence models. IEEE Trans. Neural Netw. Learn. Syst. **31**(7), 2469–2489 (2020). https://doi.org/10.1109/TNNLS.2019.2929141
35. Gilbert, C.H., Vader, H.E.: A parsimonious rule-based model for sentiment analysis of social media text. In: Eighth International Conference on Weblogs and Social Media (ICWSM-14), vol. 81, p. 82 (June 2014). http://comp.Social.gatech.edu/papers/icwsm14.vader.hutto.pdf. Accessed 20 April 2016
36. Hu, M., Liu, B.: Mining opinion features in customer reviews. In: AAAI, vol. 4, no. 4, pp. 755–760 (July 2004)
37. Risch, J., Kao, A., Poteet, S.R., Wu, Y.J.: Text visualization for visual text analytics. In: Visual Data Mining, pp. 154–171. Springer, Heidelberg (2008)

SCRUM Framework Extended with Clean Architecture Practices for Software Maintainability

Eduardo C. Arango(✉) and Omar L. Loaiza

Escuela de Posgrado, Unidad de Posgrado de Ingeniería y Arquitectura, Universidad Peruana Unión, Carretera Central Km. 19.5, Lurigancho-Chosica, Lima, Peru
{eduardo.arango,omarlj}@upeu.edu.pe

Abstract. This study consisted of the extension of the Scrum framework with Clean Architecture practices (Scrum-CA) to improve software maintainability. Methodologically, in this study: i) the requirements that Scrum-CA had to satisfy were identified; ii) identification of Scrum aspects (artifacts and/or activities) to be extended, iii) specify the extended framework; and, iv) validate the proposal through a pilot project. At the extended framework level, this proposal considers the formulation of a business context as an input to the Product Backlog. The specification of tasks in the Sprint backlog diversifies them into 4 spaces (infrastructure, controllers, application and domain) so that in the development of the Sprint the achievement of a maintainable architecture is allowed, an issue that Scrum teams must define empirically a priori. SOLID principles are added in the implementation of the product within the Sprints and coherence is maintained between the planning of the requirements and the achievement of the product architecture. Scrum-CA was validated through an implementation project in a Peruvian State entity nationwide. The results show an improvement in the maintainability of the project software, from Sprint1 to Sprint2, reducing code duplication from 4.8% to 3.5%, from 31 duplicated blocks to 9, a technical debt from 6.3% to 1.3%. Also, it is necessary to repeat Scrum-CA in other projects to generate a standard.

Keywords: Scrum · Scrum extended · Clean architecture · Software maintainability

1 Introduction

1.1 Research Context

The technological trend of Industry 4.0 is driving disruptive changes and innovations in shorter times [1]. In this sense, there is today a greater sense of urgency to adapt to changes in society and markets; such as, for example, among them to cite one, the situation of COVID 19, which forces organizations to change the way they interact with their users and/or clients, business partners [2], due to the social and sanitary conditions that are needed for its mitigation. In that order of ideas, information technologies are

R. Silhavy (Ed.): CSOC 2021, LNNS 230, pp. 667–681, 2021.
https://doi.org/10.1007/978-3-030-77442-4_56

vital for the digital transformation of organizations, having the increasing need to create and use new content, applications and/or services through the use of technologies that support their business processes, in order to maintain a competitive advantage [3]. It is also true that, if organizations grow in size and complexity over time, the software must also change to adapt to new needs [4], this means that the software, to keep being considered an adequate support, must comply with quality criteria based on standards that lengthen its operational life.

Any information system is the result of following a development process [5]; where the requirements reach the end of the life cycle converted into a product. Each activity of the life cycle must be oriented to the achievement that is expected of the product for an adequate operation in its useful life. As previously mentioned, all software changes over time, it can be during the development of the project or when it is in production [6]. So, if it is true that the software needs to be maintained over time, the life cycle activities should take into account practices that allow adequate maintainability of its architecture. An effect of not getting a scalable product over time, will definitely endanger more resources, little integration with other information systems will be evident, until having to create new software due to the fact that it is not maintainable.

It is necessary to mention that the requirements are the main element that guide all efforts in the software development processes. Based on them, the architecture of the software to be produced is formulated, which evolves as the activities of the life cycle are developed [7]. There are changes that can be introduced in any iteration under a change management protocol; They can have a different impact on the project, from changes in a class to the redefinition of functionalities, and what is clear is that all of them affect the architecture of the product, where the effects can be undesirable such as: high degree of coupling of third-party libraries and technology, not very testable components, little independence from user interfaces, external and database libraries, little ease of testing functionalities, little flexibility to add or remove functionalities, as well as to correct errors; or in any case, the effects will be seen in the future maintainability when the product is in production [8].

The foregoing places the idea of the need to think about the architecture and the necessary tasks from earlier and intermediate stages of the development of a software will be better. In this sense, what is proposed by Martin [9] with the theme of clean architectures allows us to think of architecture as a transversal theme of software development, and not as specific and discrete activities.

On the other hand, Scrum is the most widely used framework in software development today [10]. Among the benefits of Scrum, we have that it divides the work into small parts, that encourages interaction with the client and the delivery of a product in an incremental iterative way, allowing to generate value in short terms [11]. It is good to indicate a priori that Scrum does not have a focus on specific issues concerning software architecture [12]. In this sense, Martin [9] says that of all the aspects of software, maintenance is the most expensive [13] because the software changes over time in functionalities and defects appear, obsolete libraries that have to be corrected and that require of a large investment in human resources and time. In that sense, spending time to design a good architecture at the beginning of the project will tend to guarantee better scalability, maintainability and testability.

Due to the above, the objective of this article is to report the extension of the Scrum framework with Clean Architecture practices for the maintainability of the software; constituting this extension as the main contribution of this research.

This research is important because it will provide an alternative that enhances Scrum in terms of software architecture from the early stages of requirements and meets maintainability criteria, and does not postpone the topic for product analysis and design, or even worse, for the testing period. Likewise, this study will serve as a baseline for further research on requirements management, software architectures, use of artificial intelligence, maturity of software processes, etc. On the other hand, the benefits for any organization that complies with the principles of clean architecture in a software project can be evidenced indirectly, for example, in saving resources and, in the time used to understand and maintain the software. In the same way, it guarantees the possibility of postponing or rejecting technological decisions, guaranteeing a minimum impact on changes that the software may experience.

Finally, the achievements of the objective are reviewed through the execution of a pilot project in a Peruvian State entity.

1.2 Related Literature

1.2.1 Background

Taghi et al. [14], in their study "Adapting Scrum Process with 7C Knowledge Management Model" aimed to manage knowledge in SCRUM by adapting the Scrum development cycle with the 7C knowledge management model, obtaining as a result that the 7C model could be adapted to the Scrum framework.

Alotaibi [15] in its study of "Modeling Security Requirements Through Extending Scrum Agile Development Framework" aimed to develop a method to extend SCRUM, adding the best security practices in the initial requirements and during the Scrum development cycle, obtaining as a result the inclusion of security requirements and the Role of Security Owner had a positive effect on the Scrum development cycle.

According to Robert et al. [16], in their research "Agile in distress: Architecture to the rescue" which aimed to demonstrate that an early and continuous focus on software architecture allows the organization to develop the agile approach and to minimize the impact of risks in constant changes. Obtaining as a result that the architecture allows an agile development that is a key factor for the success of the project, the dedication of effort in designing an architecture that can withstand persistent changes is made in each iteration of the agile practices.

1.2.2 Maintainability

The quality of the software product can be defined as the degree to which the product satisfies the requirements of the users and also the non-functional attributes of the software. Software maintainability is defined as the ease with which functionalities can be added, correcting errors in a software product [7].

Furthermore, Martínez, et al. [13] affirm that maintenance activities constitute a large part of the effort in the software life cycle. According to Udya [17], they state that the architecture design is the first step in software development, therefore, measuring the quality of the architecture design will help to determine the quality of the software.

Alkharabsheh et al. [18], in their study of "Software Design Smell Detection a systematic mapping study" aimed to detect design defects that negatively affect the quality attributes of the software, such as comprehensibility, test capacity, extensibility, reuse and maintainability capacity, resulting in Design Smell Detection having a positive influence on the quality attributes of the software.

1.2.3 Clean Architecture

Software architecture is the way the components that make it up are organized and interrelated with each other [7].

Clean Architectures use the same principles that are the separation of responsibilities, that separates the elements of the architecture in layers with sole responsibility and it focuses on the business logic that is the center of the entire system and for communication with other parts of the system, it defines interfaces and the principle of dependency must be towards the center and the elements of the inner layer, they do not know anything about the external elements, which change frequently, it is also based on the application of the SOLID principles, the objective of CA is to design applications with low coupling and high cohesion, independent of the infrastructure, such as databases and UI, Web, etc. In this way, the application satisfies the needs of the clients providing flexibility to maintain and extend functionalities as well as to test [9].

Clean Architecture is characterized by being:

Independent of frameworks: the architecture does not depend on libraries or frameworks; these are only tools and the change of these should not impact the business logic.

Verifiable - Business rules must be tested independently of UI, database.

Independent of UI: UI can change without affecting business logic.

Database independent: You can switch databases with minimal effort because business logic does not depend on databases.

Independent of any external agency: the business rule does not know about external interfaces.

The CA is divided by layers, which are defined by external circles that form the mechanisms, the internal circles are business policies, the dependence must be towards the center and the elements of internal circles must not know anything about the external elements; likewise, the CA is based on the SOLID design principles [9].

The layers of Clean Architecture are:

Entities: it is the central circle of the CA architecture, it is the one that encapsulates the business logic of the company, where the high-level rules are defined, the same ones that change less frequently when something external changes and they can be used by different components of the architecture.

Use Cases: it is an organized step sequence where the user interacts with the system, this layer contains all the business rules of the application.

Interface Adapters: This layer prevents direct dependency on internal and external elements of the application.

Frameworks and Drivers: this layer is the outermost part of the Clean Architecture model, where are the elements that tend to change most frequently, such as databases, frameworks, that if they change, they should not affect the logic of application business.

1.2.4 Scrum Framework

Scrum is a framework for the Agile development of software products. It is based on agile principles, practices and values for the delivery of a finished product with a high value in the shortest time; it also allows fast and repeated inspection of the work carried out and, likewise, it gives autonomy to the team to determine the best way to deliver functionalities [11].

1.2.5 Clean Architecture Practices and Scrum Framework

The traditional software development model such as waterfall, prototypes, incremental iterative, unified process, etc., is characterized by following a defined plan that does not change and is rigid, where the first activity is to identify the requirements and, based on this, a plan is made setting milestones, and, for each milestone, there is a set of artifacts to deliver [7]. These development process models are complex, lack flexibility, generate many documentations and run the risk of having the documentation out of date, these processes are very bureaucratic and oriented for projects where requirements do not change frequently.

For projects that are characterized by having changing requirements, and little clarity at the beginning are the agile development processes, among them we have Scrum, which is a framework for developing complex products that has changing requirements [11]. In this sense, Staron et al. [8] tell us that the principles of agile development are increasingly used for large software development projects, but that by having self-organized teams that develop these projects, there is a risk that architectural dependencies between software components become increasingly difficult to maintain. In this sense, it is necessary to extend the Scrum framework with Clean Architecture practices to facilitate the development of an architecture with low coupling and high cohesion that adapts to these changing work environments and that does not require a high effort for its implementation as the software grows in functionalities.

In the present investigation, it was carried out with Clean Architecture because it allows the separation of responsibilities dividing into layers and promoting that each one focuses on a part of their domain and responsibility, allowing the business logic to be transferred to the product architecture [9], compared to Event Sourcing, which has a considerable storage cost, in addition to having a complexity in the learning curve, in addition to changing the paradigm of programming [19], or the traditional ways of working (3 layers, N layers, etc.) that have high coupling [19].

2 Method

This research consisted of designing 4 stages: i) identification of Clean Architecture requirements and practices; ii) identification of Scrum aspects to be extended; iii) specification of the extended framework; and, iv) validation of the framework extended in the pilot project.

Regarding the first stage, a review of reference sources was carried out in an investigatory way and that are related to Scrum and software architecture, to later consolidate the results and extract from them the requirements that are necessary for the subject extension of this Article. Likewise, a consultation with professionals was carried out for the same purpose.

Regarding stage 2, the requirements identified with the Scrum aspects (artifacts and activities) were related, in such a way that the action to be taken emerges from said analysis, such as adding a new element to an artifact or activity (A), or extending (E) the purpose by a specification.

In stage 3, it is detailed what the specification of the added and/or extended elements consists of and that they are articulated to the outline of the Scrum framework extended with Clean Architecture that this research proposes and presents. Likewise, the workflow of the extended framework is described. It should be noted that this research does not propose a new Scrum, or distort its purpose; Instead, based on agile principles and known workflow, it extends it to empower it. It is understood that for this proposal to be standard, it needs to be experienced multiple times.

Finally, in stage 4, to see the behavior of the extended framework it is necessary to see it in action in a pilot project; through a business case, which is presented in Sect. 4.4 that has a description of itself and its scope as well; then the framework runs; finally, the results obtained from its application and its effects on the maintainability of the built software are reported.

3 Results

3.1 Identification of Clean Architecture Requirements and Practices

Scrum is a framework that starts from the identification of requirements, and the business context can be understood.

Scrum is an empirical framework, not much prescriptive and highly flexible [18], so that the formation of the Product Backlog could start by wide and different paths and from the beginning not to contemplate a common business context between the contracting parties of a software development project; but what is delivered does not necessarily guarantee that antecedent information on the architecture of a software or others with which it interoperates is contemplated, producing potential future maintainability problems. So, having a minimal understanding of the context (REQ1) would be very helpful.

As it is known the result and main reason, both of a development process and a framework such as Scrum aimed at software, is the product in the hands of the client for its exploitation. It should be noted that this product is achieved based on requirements. It should be noted that all software bases its structure on an architecture guided by requirements. The above premises are reasonable to think that architecture is achieved progressively and is a transversal idea throughout a development process. It could be said in some way that the idea is not so recent, not at least in the traditional way in more ceremonious development processes such as RUP, Open Up among others; but it does in Scrum, where it seems that the variability of times and requirements tend to require recurring adjustments in the architecture, impairing its maintainability and scalability.

Then the idea that the requirements are basic and guiding remains established. If the Product backlog is the cornerstone that guides all iterative and incremental software development in Scrum, and from which sprint backlogs with identified tasks are derived, wouldn't it be better to make explicit and think about the architecture from early stages (REQ2 to REQ5), in those aspects that vary the most and those that do not? Why thinking about the variable and less variable only in the implementation within a Sprint? It is clear then, that architecture must be worked in a different way, to go from a work in layers coupled to interdependent layers; and, on the other hand, to begin with the tasks that express the least variable of the architecture and to gradually incorporate those more variable aspects.

Now, Clean Architecture, allows to think about architecture working with interdependent and better specified layers from early stages, making visible the most variant is the infrastructure layer (REQ2), since the technology that allows interaction between the user and the application or management computational information may vary due to obsolescence, technological efficiency or current use. While the least variant layer is the one that contains the corporate business rules in the domain layer (REQ5). In between are the behavioral layers of how the business rules are used in the application (REQ4) and how the results (REQ3) of these are adapted in the infrastructure layer. The previous classification of the tasks in the sprint backlog into 4 categories, reflected in the requirements REQ2 to REQ5, is called, in this research work, Architecture Backlog.

It should be noted that the implementation of an architecture must be the result of prior planning considered in the Product and Sprint backlog, what happens if this does not happen? There is a risk of achieving a product that is not very scalable and maintainable. The best option to reduce this risk is by verifying that the tasks are suitably classified and are oriented to the architecture by independent layers (REQ6). The resulting requirements are:

Table 1. Identified requirements for the Scrum extension

Code	Description
REQ01	The extended framework must make the business context explicit
REQ02	To identify tasks oriented to the infrastructure related to the product
REQ03	To identify driver-oriented tasks that are required for the product
REQ04	To identify achievement-oriented tasks of the application
REQ05	To identify tasks related to the domain related to the application

(*continued*)

Table 1. (*continued*)

Code	Description
REQ06	It must be verified that the identified tasks will allow to achieve the architecture of the product by interdependent layers
REQ07	Product programming must consider SOLID principles in the classes that make up the product architecture
REQ08	To verify adherence to SOLID principles in the development of each Sprint
REQ09	The lessons learned from each Sprint should incorporate reflection on issues related to architecture
REQ10	The technical review of the Sprint must incorporate compliance with the architectural principles in its analysis

3.2 Identification of Scrum Aspects to Extend

One of the aspects identified in Scrum is the need for the Product backlog to have as input a Business Context, due to the fact that this artifact makes more sense and mastery (control) in the team [20], it allows better planning [20] by becoming familiar and growing with it by changing the context or by explaining why changes occur in the Product Backlog [21], or helping to relate the effect of a change in the product (even if it comes late) [22] with context when communicating at daily meetings or retrospectives. This context should consist of a brief description of the business case, narrate a situation or a problematic situation related to aspects of a study reality, as well as denote the valid interlocutors, which help to understand the requirements listed in the Product backlog.

On the other hand, in the Sprint backlogs, the most pressing problems in the continuous change of a little clear Product backlog definitely redounds in the Sprint backlog and in the product. Problems such as correction of defects, improvements due to errors or inconsistencies in the design [23], conflicts in the operation of the product, etc. which clearly have to do with the maintainability of the product and its architecture [24]. This proposal enriches the Sprint backlog, by classifying the tasks in the 4 stacks (infrastructure, domain, application and business rules) of the Clean Architecture layer configuration [9]; so that the team can deal with the most and least changing tasks, in sequence or parallel, as is traditionally worked in Scrum. The detail of the tasks continues as it is traditionally worked. This division allows work focused on architecture with the usual agility, better maintainable.

Regarding the Sprint, they aim to produce increases for the client and their objectives are set as a team to meet agreed commitments. It is the stage where architecture-oriented models and their implementation are produced.

In terms of implementation within Sprint, the SOLID principles are good practices that guide the way of designing systems, which help reduce the complexity of the code as well as being more explicit and direct, reducing coupling and generating a maintainable code and easy to extend [9]. Single responsibility means that a class should have only one responsibility and not to implement other functions that are not in its domain; thus, it provides an ease of maintenance by having separate responsibilities. The open and closed principle provides a module or class that should only be open to extend functionalities but not for modification, that is why code can be added to extend functionalities, but not to modify. The Liskov substitution principle allows the use of inheritances, allowing daughter classes to be used as a father without altering their behavior. This principle helps us to have a modular system, which goes hand in hand with the other principles. The principle of interface segregation does not say that it is better to have many interfaces instead of having an interface that has many functionalities, as it is used in the traditional way, there is an interface and many functionalities are added by having many functionalities when it is required to use the interface and there will be methods that will not be used. The Dependency Inversion Principle tells us that the abstractions should not depend on the details, but the details should depend on the abstractions.

Regarding retrospectives, both in the Sprint retrospective and in the Sprint Review, an analysis is promoted based on how the product architecture that is achieved is articulated to the requirements; in the first case through lessons learned in the Scrum workflow, and in the other case, through a technical look at what was done in the Sprints. This allows to feed back the work and ensure that the product does not lose the focus of maintainability and achievement of requirements.

Table 2. Scrum aspects to be extended

Scrum aspects	Requirement	Action	Element to be added or expanded specification
Product backlog	REQ1	To insert business context in the framework	Business context
Sprint backlog	REQ2	To classify the product backlog in terms of architecture	Product backlog divided into 4 compartments: infrastructure, controllers, application and domain
	REQ3		
	REQ4		
	REQ5		
Sprint Planning	None	None	None
Sprint	REQ6	To specify procedure for modeling and implementation needs regarding SOLID principles	
	REQ7	To insert verification sub status in the to do section of Sprint	Sub status in the to do section of Scrum board
Daily meeting	None	None	None
Sprint Review	REQ8 REQ10	To add item to the Sprint review schedule	Sprint review
Sprint Retrospective	REQ9	To add item to the Sprint retrospective schedule	Sprint retrospective

3.3 Scrum Extended Framework Specification (Scrum-CA)

The extended framework begins with the formulation of the business statement by the Product Owner and which is subsequently socialized to the Scrum team. From this statement the requirements for the Product backlog are extracted, classifying them according to risk, cost and time. Next, in the Sprint Planning, the requirements of greatest value for the user were defined and according to the team's work capacity to generate the Sprint Backlog.

In the sprint backlog, the development team assumes the requirements prioritized by the Product Owner, decomposing them into the different Clean Architecture layers (domain, use cases, controllers/presenters, infrastructure and UI), as well as an estimate of time in order to complete tasks.

Then, the development team executes the implementation activities of those tasks in each Sprint, monitoring the development in a dashboard with three states: To do, Doing and Done. SOLID principles apply to definitions of software classes and methods. In parallel to the Sprints and on a daily basis, the Daily meeting is held with a focus on the clean architecture of the software, to see if there were difficulties in carrying it out; for this, the team had to answer the following questions: what have I done? What am I going to do? and what impediment have I had?

The Sprint Retrospective, is the meeting that takes place at the end of each Sprint, and the development team participates where it is evaluated what improvements must be made to the process in the next Sprint, considering the Clean Architecture practices that have been carried out. In the same way, in the Sprint Review, the Sprint is technically reviewed in terms of product with a focus on architecture (Fig. 1).

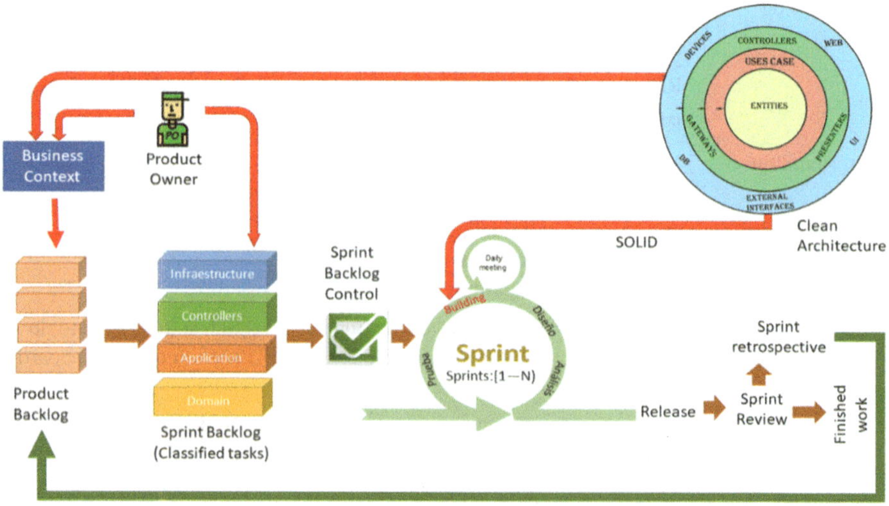

Fig. 1. Scrum + clean architecture extended framework

4 Results

4.1 Validate the Framework Extended in the Pilot Project

The implementation of the pilot test of the extended framework was carried out in a State entity in Peru. The Scrum Master carried out the inception activities, which consisted of providing an induction on the extended Scrum framework (Scrum-CA), unifying the project objectives. It should be noted that the Product Owner develops the business context, which is briefly set out below:

The State entity, where the pilot test was carried out, in the context of the health crisis caused by COVID-19, had the need to implement a virtual platform for the reception of requests from citizens via the internet, since before the reception was in physical and face-to-face form; in that sense, the State entity made the amendments to the legal regulations, whose purpose is to enable the reception of documents in digital format. The Digital Portal is a web platform, where citizens can make requests and attach the required documentation according to the type of request. Each request has a cost, the Portal must validate the payment made for said procedure according to the means of payment, whether in the state bank, own collection, in addition to interoperability with other internal systems, in this context it is required to have the system that has a modular architecture and has a separation by layers; likewise, applying the SOLID principles in the design, they helped in the generation of clean code, allowing ease of maintenance, as well as helping in interoperability with other applications for normal operation.

Activities Aimed at Discovering Clean Architecture

For the development of the Sprint Backlog, the classification of the tasks according to the Clean Architecture layers was considered; In this sense, the layers were grouped by their context in the business, which are divided by clean architecture design layers which consists of the domain layer, use case layer, controller layer, as well as the infrastructure layer [9]. Once the activities have been classified and grouped, they are defined in the following table. This Sprint backlog was verified and authorized for execution. For reasons of space, only a part of a table with more content is reproduced (Table 3).

The project for its pilot execution was divided into 2 Sprints. In the first Sprint, the portal allowed the registration of citizen requests via the internet, attaching the required documentation according to the type of request, as well as the validation of the payment made in the State entity, but in the context of the health crisis of COVID-19 it was required integration with a State payment gateway and payments made in a State bank; once the new requirement was identified, validations for payments via WebApi with the payment gateway and State bank were added to the Sprint Backlog.

For the second Sprint, the digital portal would have the functionality of review, observation, correction of observations, in addition to the validations of external payments to the State entity; for this, the development team organized the activities according to the CA layers, considering in design SOLID principles (Fig. 2).

SCRUM-CA helped in the development of the pilot project, providing the best practices for the architecture design, as well as the definition of the tasks in the sprint backlog in order to facilitate the development of the software construction.

Table 3. Definition of architecture of clean architecture in layers

Module	Layers	Use cases
Archetype of the architecture of clean architecture		
Administrative-request	Domain	
	Application	Create, Search, CustomResearch, SendProcedure, AttatRequrement, CorrectObservatio, SendEmail, CallApiRestServicesSGD, CallApiRestServicesSGD, CallApiRestServicesROP, CallApiRestServicesSGU, SearchRequirementByAdministrative ProcedureId, CreateRequirement
	Controller	GetController, PutController, PostController
	Infrastructure MySql AdministrativeRequestRepository	

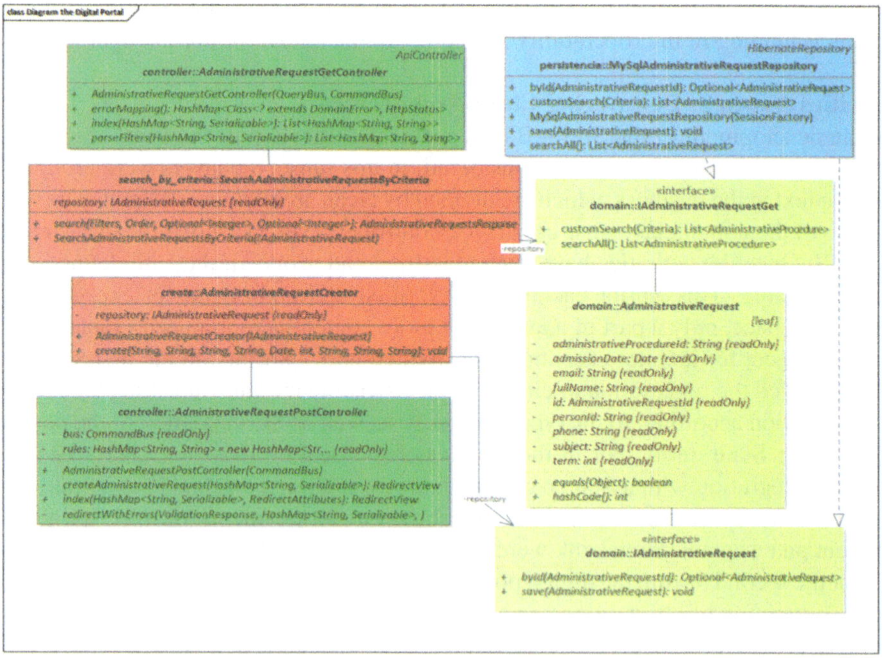

Fig. 2. Class diagram reflecting clean architecture layers

To validate the maintainability of the Digital Portal software, it was carried out with the SonarQube tool, which provides metrics for maintainability, duplication, security, etc. The evaluation considered 342 files containing 5503 lines of code, whose results issued by SonarQube were:

The first measurement on code duplication: 4.80% of duplicate lines were obtained (percentage of duplication = number of duplicate lines/number of lines * 100) as well as 31 duplicate blocks, for java projects up to 3% duplicate code is accepted and/or 10 duplicate blocks that must be reached in the next iteration. Regarding maintainability: a technical debt ratio of 6.3% was obtained, according to the SonarQube maintainability classification scale (<= 5% the rating is A, between 6 and 10% the rating is B, between 11 and 20% the rating is a C, between 21 and 50% the rating is D, any value above 50% is an E), then the lower the technical debt ratio, the better the health status of the software.

For the second measurement, the code duplication percentage increased to 3.5% and 9 duplicate blocks, and with respect to maintainability, the technical debt ratio was 1.3% of the analyzed code; therefore, SCRUM-CA influenced the maintainability of the software. These improvements in maintainability are explained by a better understanding and application of the Scrum-CA framework workflow.

At the end of the Sprint, a meeting was held to make improvements to the process carried out, as well as reviews of lessons learned (Sprint retrospective) and technical reviews (Sprint review), capturing new requirements and providing feedback on the Product Backlog, and refreshing the business context, in which a plan was created to avoid the generation of anti-patterns.

4.2 Conclusions

This research has extended the Scrum framework with Clean Architecture practices to improve the maintainability of a software developed for a Peruvian State entity. The results conclude that said improvement, from Sprint1 to Sprint2 in the project, reduces code duplication from 4.8% to 3.5%, from 31 duplicate blocks to 9 and a technical debt from 6.3% to 1.3%.

The Scrum framework extended with Clean Architecture practices, on the one hand, has followed a logical and valid methodological design in future experiences to this research or even extrapolate them to other contexts where possible; on the other hand, all its requirements have been met (Table 1), and whose extension or specification points have been achieved through the identification of those aspects that allow Scrum to reach the achievement of better maintainable software (Table 2).

This research finally concludes that it is necessary to replicate the proposed extended framework (Scrum-CA) in other projects to convert it into a standard framework.

References

1. Orellana-Daube, D.F.: El efecto global de la actual revolución tecnológica 4a revolución industrial y la industria 4.0 en acción. Rev. GEON (Gestión Organ. y Negocios **7**(2), 1–24 (2020). https://doi.org/10.22579/23463910.194

2. Isaksen, A., Trippl, M., Kyllingstad, N., Rypestøl, J.O.: Digital transformation of regional industries through asset modification. Compet. Rev. (2020). https://doi.org/10.1108/CR-12-2019-0140
3. Hsu, C.-C., Tsaih, R.-H., Yen, D.: The evolving role of IT Departments in digital transformation. Sustainability **10**(10), 3706 (2018). https://doi.org/10.3390/su10103706
4. Delgado, A.P., Velthuis, M.P.: Propuesta de marco de mejora continua de gobierno TI en entidades financieras. RISTI - Rev. Iber. Sist. e Tecnol. Inf. (15), 51–67 (2015). https://doi.org/10.17013/risti.15.51-67
5. Amaya Balaguera, Y.D.: Metodologías ágiles en el desarrollo de aplicaciones para dispositivos móviles. Estado actual. Rev. Tecnol. **12**(2) (2015). https://doi.org/10.18270/rt.v12i2.1291
6. Papamichail, M.D., Symeonidis, A.L.: A generic methodology for early identification of non-maintainable source code components through analysis of software releases. Inf. Softw. Technol. **118**, 106218 (2020). https://doi.org/10.1016/j.infsof.2019.106218
7. Sommerville, I.: Ingeniería de Software (2011)
8. Staron, M., Meding, W., Hoglund, C., Eriksson, P., Nilsson, J., Hansson, J.: Identifying implicit architectural dependencies using measures of source code change waves (2013). https://doi.org/10.1109/SEAA.2013.9
9. Martin, R.C.: Clean Architecture: A Craftsman's Guide to Software Structure and Design (2017)
10. Digital.ai, "14th State of Agile," p. 19 (2020). https://explore.digital.ai/state-of-agile/14th-annual-state-of-agile-report
11. Schwaber, K., Sutherland, J.: La Guía de Scrum TM La Guía Definitiva de Scrum: Las Reglas del Juego Español/Spanish South American. Scrum.Org, p. 22 (2017). https://www.scrumguides.org/docs/scrumguide/v2017/2017-Scrum-Guide-Spanish-SouthAmerican.pdf
12. Hossain, E., Ali Babar, M., Paik, H.Y.: Using scrum in global software development: a systematic literature review. In: Proceedings - 2009 4th IEEE International Conference on Global Software Engineering, ICGSE 2009, pp. 175–184 (2009). https://doi.org/10.1109/ICGSE.2009.25
13. Martínez, J.E., Gómez, A.F., Pino, F.J.: Generando productos software mantenibles desde el procesode desarrollo: El modelo de referencia MANTuS (2016). https://doi.org/10.4067/S0718-33052016000300007
14. Gandomani, T.J., Tavakoli, Z., Nafchi, M.Z., Najafi Sarpiri, M.: Adapting scrum process with 7C knowledge management model. In: 2019 IEEE 5th Conference on Knowledge Based Engineering and Innovation, KBEI 2019, pp. 56–59 (2019). https://doi.org/10.1109/KBEI.2019.8735008
15. Alotaibi, M.: Extending scrum framework to emphasise security: how a 'security owner' is integrated into the scrum team. In: Proceedings of the Eighth Saudi Students Conference in the UK, no. February, pp. 311–322, February 2016. https://doi.org/10.1142/9781783269150_0027
16. Nord, R.L., Ozkaya, I., Kruchten, P.: Agile in distress: architecture to the rescue. In: Dingsøyr, T., Moe, N.B., Tonelli, R., Counsell, S., Gencel, C., Petersen, K. (eds.) XP 2014. LNBIP, vol. 199, pp. 43–57. Springer, Cham (2014). https://doi.org/10.1007/978-3-319-14358-3_5
17. Dayanandan, U., Vivekanandan, K.: An empirical evaluation model for software architecture maintainability for object oriented design. In: ACM International Conference Proceeding Series, vol. 25–26-August, pp. 1–4 (2016). https://doi.org/10.1145/2980258.2980459
18. Alkharabsheh, K., Crespo, Y., Manso, E., Taboada, J.A.: Software design smell detection: a systematic mapping study. Softw. Qual. J. **27**(3), 1069–1148 (2018). https://doi.org/10.1007/s11219-018-9424-8
19. Debski, A., Szczepanik, B., Malawski, M., Spahr, S., Muthig, D.: A scalable, reactive architecture for cloud applications: CQRS and event sourcing case study. IEEE Softw. **35**(2), 62–71 (2018). https://doi.org/10.1109/MS.2017.265095722

20. Misra, S.C., Kumar, V., Kumar, U.: Identifying some important success factors in adopting agile software development practices. J. Syst. Softw. **82**(11), 1869–1890 (2009). https://doi.org/10.1016/j.jss.2009.05.052
21. Ciric, D., Lalic, B., Gracanin, D., Palcic, I., Zivlak, N.: Agile Project Management in New Product Development and Innovation Processes: Challenges and Benefits beyond Software Domain, October 2018. https://doi.org/10.1109/TEMS-ISIE.2018.8478461
22. Dan Turk, R.F., Rumpe, B.: Limitations of agile software processes. J. Environ. Prot. Ecol. **18**(3), 1259–1267 (2002)
23. Sedano, T., Ralph, P., Peraire, C.: The product backlog. In: Proceedings of International Conference on Software Engineering, vol. 2019-May, no. Section IV, pp. 200–211 (2019). https://doi.org/10.1109/ICSE.2019.00036
24. Alsalemi, A.M., Yeoh, E.T.: A survey on product backlog change management and requirement traceability in agile (scrum). In: 2015 9th Malaysian Software Engineering Conference, MySEC 2015, pp. 189–194 (2016). https://doi.org/10.1109/MySEC.2015.7475219

The Properties of Multistep Runge-Kutta Methods for Neutral Delay Differential Equations

Kewei Zhang and Haiyan Yuan(✉)

Department of Mathematics, Heilongjiang Institute of Technology, Harbin 150050, China

Abstract. This paper is devoted to the stability and convergence analysis of the multistep Runge-Kutta methods with the Lagrange interpolation of the numerical solution for a nonlinear neutral delay differential equations. Nonlinear stability and D-Convergence are introduced and proved. We discuss both the GR(l)–stability, GAR(l)-stability, and the weak GAR(l)-stability on the basis of (k, l)-algebraically stable of the multistep RK methods, we also discuss the D-convergence properties of multistep RK methods with a restricted type of interpolation procedure.

Keywords: Nonlinear neutral delay differential equations · Nonlinear stability · D-Convergence

1 Introduction

Neutral delay differential equations (NDDEs) arise in a variety of fields as biology, economy, control theory, electrodynamics and so on (see, for example, [1–5]). The stability and convergence properties of numerical methods for linear NDDEs have been widely researched by many authors (see, for example, [6–11]). For the case of nonlinear delay differential equations, this kind of methodology had been first introduced by Torelli [7, 12] and then developed by Bellen and Zennaro [13], Bellen [14] and Zennaro [15, 16]. 1997, Koto proved the asymptotic stability of natural Runge-Kutta method for a class of nonlinear delay differential equations in [17], A. Bellen [18] gave a discussion of the stability of continuous numerical methods for a special class of nonlinear neutral delay differential equations. In particular, Jackiewicz [19–21] systematically investigated the convergence of various numerical methods for more general neutral functional differential equations (NFDEs). In 2009, R. N. Yang, H. J. Gao and P. Shi, gave a novel robust stability criteria for stochastic Hopfield neural networks with time delays in [22]. R. N. Yang, Z. X. Zhang, and P. Shi [23] studied the exponential stability on stochastic neural networks with discrete interval and distributed delays in 2010. 2011, L. P. Liu [24] gave the robust stability for Neutral Time-Varying Delay Systems with Non-Linear Pe turbations. On the stability, Akio Tanikawa studied the Values of Random Zero-Sum Games in [25], and in [26] Michael Basin and Dario Calderon-Alvarez gave the delay-dependent stability studies for Vector Nonlinear Stochastic Systems with Multiple Delays.

R. Silhavy (Ed.): CSOC 2021, LNNS 230, pp. 682–700, 2021.
https://doi.org/10.1007/978-3-030-77442-4_57

However, these important convergence results are based on the classical Lipschitz conditions. The studies focusing on the stability and convergence of the numerical method for nonlinear NDDEs based on a one-sided Lipschitz condition have not yet been seen in literature until now. By means of a one-sided Lipschitz condition, in the present paper we discuss the stability and convergence of multistep Runge-Kutta(MULTISTEP) methods for nonlinear NDDEs. Thanks to the one-sided nature of the Lipschitz condition, the error bounds obtained in the present paper are sharper than those given in the references mentioned.

2 Multistep Runge-Kutta Methods for NDDEs and the Nonlinear Stability

It is the purpose of this paper to investigate the nonlinear stability and convergence properties of the following NDDEs

$$\begin{cases} y'(t) = f\big(t, y(t), y(t-\tau), y'(t-\tau)\big), & t \in [0, T] \\ y(t) = \phi(t), \quad t \in [-\tau, 0] \end{cases} \tag{2.1}$$

where $f^{[\nu]} : [t_0, T] \times C^N \times C^N \rightarrow C^N$ is a given mapping, τ is a positive delay term, and $\phi : [-\tau, 0] \rightarrow C^N$ is a continuous function. Moreover, we assume that there exist some inner product $\langle \cdot\ , \cdot \rangle$ and the induced norm $||\cdot||$ in C^N, such that

$$\mathrm{Re} < f(t, y_1, u, v) - f(t, y_2, u, v), y_1 - y_2 > \leq \alpha \|y_1 - y_2\|^2 \tag{2.2}$$

$$\|f(t, y, u_1, v) - f(t, y, u_2, v)\| \leq \beta \|u_1 - u_2\| \tag{2.3}$$

$$\|f(t, y, u, v_1) - f(t, y, u, v_2)\| \leq \gamma \|v_1 - v_2\| \tag{2.4}$$

$$\|K(t, y, u_1, v, w) - K(t, y, u_2, v, w)\| \leq q \|u_1 - u_2\| \tag{2.5}$$

where $K(t, y, u, v, w) = f(t, y, u, f(t-\tau, u, v, w)), t \in [0, T]$,
$\forall y, y_1, y_2, u, u_1, u_2, v, v_1, v_2, w \in C^N, \alpha, \beta, \gamma, q$ are constants.

In order to make the error analysis feasible, we always assume that problem (2.1) has a unique solution $y(t)$ which is sufficiently differentiable and satisfies

$$\left\| \frac{d^i y(t)}{dt^i} \right\| \leq M_i, i = 1, 2, \cdots, t \in [-\tau, T], \tag{2.6}$$

and denotes the problem class $R(\alpha, \beta, \gamma, q)$ that consists of all NDDEs with (2.2)–(2.5).

Many numerical methods have been proposed for the numerical solution of problem (2.1). In this paper, we are concerned with multistep Runge-Kutta method of the form

$$Y_i^{(n)} = y_n + h \sum\nolimits_{j=1}^{s} a_{ij} f\left(t_n + c_j h, Y_j^{(n)}\right), i = 1, 2, \cdots, s, \tag{2.7a}$$

$$Y_i^{(n-1)} = y_{n-1} + h \sum\nolimits_{j=1}^{s} a_{ij} f\left(t_{n-1} + c_j h, Y_j^{(n-1)}\right), i = 1, 2, \cdots, s, \tag{2.7b}$$

$$y_{n+1} = (1-\theta)y_n + \theta y_{n-1} + h\sum\nolimits_{i=1}^{s} b_i f\left(t_n + c_i h, Y_i^{(n)}\right) + h\sum\nolimits_{i=1}^{s} \tilde{b}_i f\left(t_{n-1} + c_i h, Y_i^{(n-1)}\right) \quad (2.7c)$$

where $\sum\limits_{i=1}^{s} b_i + \sum\limits_{i=1}^{s} \tilde{b}_i = 1+\theta, c_i = \sum_{j=1}^{s} a_{ij}, y_n$ is the numerical approximation at $t_n = nh$ to the analytic solution $y(t_n)$, $h > 0$ is a stepsize and $0 \leq \theta \leq 1$.

The above methods are studied in [11]. Now we consider the adaptation of the multistep Runge-Kutta method to (2.1)

$$Y_i^{(n)} = y_n + h\sum\nolimits_{j=1}^{s} a_{ij} f\left(t_n + c_j h, Y_j^{(n)}, \bar{Y}_j^{(n)}, \tilde{Y}_j^{(n)}\right), i = 1, 2, \cdots, s, \quad (2.8a)$$

$$Y_i^{(n-1)} = y_{n-1} + h\sum\nolimits_{j=1}^{s} a_{ij} f\left(t_{n-1} + c_j h, Y_j^{(n-1)}, \bar{Y}_j^{(n-1)}, \tilde{Y}_j^{(n-1)}\right) \quad (2.8b)$$

$$\begin{aligned} y_{n+1} = (1-\theta)y_n + \theta y_{n-1} + h\sum\nolimits_{i=1}^{s} b_i f\left(t_n + c_i h, Y_i^{(n)}, \bar{Y}_i^{(n)}, \tilde{Y}_i^{(n)}\right) \\ + h\sum\nolimits_{i=1}^{s} \tilde{b}_i f\left(t_{n-1} + c_i h, Y_i^{(n-1)}, \bar{Y}_i^{(n-1)}, \tilde{Y}_i^{(n-1)}\right) \end{aligned} \quad (2.8c)$$

where y_n is the numerical approximation to the analytic solution $y(t_n)$ with $t_n = nh$.

In particular, $y_0 = \phi(0)$. The argument $\bar{Y}_j^{(n)}$ denotes an approximation to $y(t_n + c_j h - \tau)$ and the argument $\tilde{Y}_j^{(n)}$ denotes an approximation to $y'(t_n + c_j h - \tau)$ which are obtained by a specific interpolation procedure at the point $t = t_n + c_j h - \tau$. $Y_j^{(k)} = \phi(t_k + c_j h)$ with $k < 1$, $t_k + c_j h \leq 0$.

Let $\tau = (m - \delta)h$ with integer m and $\delta \in [0, 1)$,ν, $\mu \geq 0$ be integers. Define

$$\bar{Y}_j^{(n)} = \begin{cases} \phi(t_n + c_j h - \tau) & t_n + c_j h - \tau \leq 0 \\ \sum_{i=-\mu}^{\nu} \tilde{L}_i(\delta) Y_j^{(n-m+i)} & t_n + c_j h - \tau > 0, \nu + 1 \leq m \end{cases} \quad (2.9)$$

where

$$\tilde{L}_i(\delta) = \prod\nolimits_{\substack{k=-\mu \\ k\neq i}}^{\nu} \left(\frac{\delta - k}{i - k}\right), \delta \in [0, 1), \quad (2.10)$$

and

$$\tilde{Y}_j^{(n)} = f\left(t_n + c_j h - \tau, \bar{Y}_j^{(n)}, \bar{Y}_j^{(n+1)}, \tilde{Y}_j^{(n+1)}\right) \quad (2.11)$$

We assume $m \geq \nu + 1$ is to guarantee that no (unknown) values $Y_j^{(i)}$ with $i \geq n$ are used in the interpolation procedure. It should be pointed out that the adopted interpolation procedures (2.9) is only a class of interpolation procedure for $\bar{Y}_j^{(n)}$, there also exist some other types of interpolation procedures, such as numerical schemes which use Hermite interpolation between grid points (see [27–29]). It is the aim of our future research to investigate future the adaptation of multistep Runge-Kutta methods to NDDEs by means of other interpolation procedures.

In order to consider the stability property, we also need to consider the pertubered problem of (2.1)

$$\begin{cases} z'(t) = f\left(t, z(t), z(t-\tau), z'(t-\tau)\right), & t \in [0, T] \\ z(t) = \psi(t), \quad t \in [-\tau, 0] \end{cases} \quad (2.12)$$

where $\psi : [-\tau, 0] \rightarrow C^N$ is a continuous function. The unique exact solution of the problem (2.12) is denoted as $z(t)$.

Applying the multistep Runge-Kutta method (2.7) to (2.12) leads to

$$Z_i^{(n)} = z_n + h\sum\nolimits_{j=1}^{s} a_{ij}f\left(t_n + c_jh, Z_j^{(n)}, \bar{Z}_j^{(n)}, \tilde{Z}_j^{(n)}\right), i = 1, 2, \cdots, s, \tag{2.13a}$$

$$Z_i^{(n-1)} = z_{n-1} + h\sum\nolimits_{j=1}^{s} a_{ij}f\left(t_{n-1} + c_jh, Z_j^{(n-1)}, \bar{Z}_j^{(n-1)}, \tilde{Z}_j^{(n-1)}\right), \tag{2.13b}$$

$$\begin{aligned} z_{n+1} = (1-\theta)z_n + \theta z_{n-1} &+ h\sum\nolimits_{i=1}^{s} b_i f\left(t_n + c_ih, Z_i^{(n)}, \bar{Z}_i^{(n)}, \tilde{Z}_i^{(n)}\right) \\ &+ h\sum\nolimits_{i=1}^{s} \tilde{b}_i f\left(t_{n-1} + c_ih, Z_i^{(n-1)}, \bar{Z}_i^{(n-1)}, \tilde{Z}_i^{(n-1)}\right) \end{aligned} \tag{2.13c}$$

2.1 Some Concepts

Definition 2.1. Let l be a real constant, a multistep Runge-Kutta method with an interpolation procedure is said to be $R(l)$-stable if there exists a constant C dependent only on the method and l such that $(\alpha + \beta + \gamma q)h \leq l$ and

$$\|y_n - z_n\|^2 \leq \left[2 + C(\beta + \gamma q)\right] \max_{t\leq 0} \|\phi(t) - \psi(t)\|^2, n \geq 1 \tag{2.14}$$

with stepsize h satisfying $hm = \tau$, where m is a positive integer.

$GR(l)$ - stability is defined by dropping the restriction $hm = \tau$.

Definition 2.2. Let l be a real constant, a multistep Runge-Kutta method with an interpolation procedure is said to be $AR(l)$-stable if

$$\lim_{n\to\infty} \|y_n - z_n\| = 0 \tag{2.15}$$

with stepsize h satisfying $(\alpha + \beta + \gamma q)h \leq l$ and $hm = \tau$, where m is a positive integer.

$GAR(l)$ - stability is defined by dropping the restriction $hm = \tau$.

Definition 2.3. Let l be a real constant, a multistep Runge-Kutta method with an interpolation procedure is said to be weak $AR(l)$ - stable if under the conditions of Definition 2.2, (2.15) holds when f further satisfies

$$\|f(t, u_1, v, w) - f(t, u_2, v, w)\| \leq \|Lu_1 - u_2\|^{p_0} \tag{2.16}$$

with p_0 being a positive real number and L being a nonnegative real number.

Weak $GAR(l)$ - stability is defined by dropping the restriction $hm = \tau$.

2.2 The Stability of Multistep RK Methods

Let $w_n = y_n - z_n$, $W_i^{(n)} = Y_i^{(n)} - Z_i^{(n)}$

$\bar{W}_i^{(n)} = \bar{Y}_i^{(n)} - \bar{Z}_i^{(n)}$, $\tilde{W}_i^{(n)} = \tilde{Y}_i^{(n)} - \tilde{Z}_i^{(n)}$

$Q_i^{(n)} = h\Big[f\Big(t_n + c_i h, Y_i^{(n)}, \overline{Y}_i^{(n)}, \tilde{Y}_i^{(n)}\Big) - f\Big(t_n + c_i h, Z_i^{(n)}, \overline{Z}_i^{(n)}, \tilde{Z}_i^{(n)}\Big)\Big]$,

it follows from (2.8) and (2.13) that

$$W_i^{(n)} = w_n + \sum\nolimits_{j=1}^{s} a_{ij} Q_j^{(n)}, i = 1, 2, \cdots, s \tag{2.17a}$$

$$W_i^{(n-1)} = w_{n-1} + \sum\nolimits_{j=1}^{s} a_{ij} Q_j^{(n-1)}, i = 1, 2, \cdots, s \tag{2.17b}$$

$$w_{n+1} = (1-\theta) w_n + \theta w_{n-1} + \sum\nolimits_{i=1}^{s} b_i Q_i^{(n)} + \sum\nolimits_{i=1}^{s} \tilde{b}_i Q_i^{(n-1)}. \tag{2.17c}$$

Now we will write the s-stage multistep RK methods (2.7) as a general linear method. Let $V_i^{(n)} = (W_i^{(n)}, W_i^{(n-1)})^{\mathrm{T}}$ be the internal stages and $\mu_{n+1} = (w_{n+1}, w_n)^{\mathrm{T}}$ the external vectors and $P_i^{(n)} = (Q_i^{(n)}, Q_i^{(n-1)})^{\mathrm{T}}$. Then we have a $2(s+1)$-stage partitioned general linear method

$$V_i^{(n)} = \sum\nolimits_{j=1}^{s} C_{ij}^{11} P_i^{(n)} + \sum\nolimits_{j=1}^{s} C_{ij}^{12} \mu_n, i = 1, 2, \cdots, s \tag{2.18a}$$

$$\mu_{n+1} = \sum_{j=1}^{s} C_{ij}^{21} P_i^{(n)} + \sum_{j=1}^{s} C_{ij}^{22} \mu_n,\ i = 1, 2, \cdots, s \tag{2.18b}$$

where

$$C^{11} = \begin{pmatrix} A & 0 \\ 0 & A \end{pmatrix},\quad C_{12} = \left(C_{ij}^{12}\right) = \begin{pmatrix} e & 0 \\ 0 & e \end{pmatrix},$$
$$C_{21} = \left(C_{ij}^{21}\right) = \begin{pmatrix} b & \tilde{b} \\ 0 & 0 \end{pmatrix},\quad C_{22} = \left(C_{ij}^{22}\right) = \begin{pmatrix} 1-\theta & \theta \\ 0 & 0 \end{pmatrix},$$

and $A = \left(a_{ij}\right)$, $e = (1, 1, \cdots 1)$, $b = (b_1, b_2, \cdots, b_s)$, $\tilde{b} = \left(\tilde{b}_1, \tilde{b}_2, \cdots, \tilde{b}_s\right)$.

Definition 2.4. Let k, l be real constants, a multistep RK method is said to be (k, l)-algebraically stable if there exists a diagonal nonnegative matrix G and $D = diag(d_1, d_2, \cdots, d_{2s})$ such that $M = \left(m_{ij}\right)$ is nonnegative, where

$$M(k, l) = \begin{pmatrix} kG - C_{22}^{\mathrm{T}} G C_{22} - 2l C_{12}^{\mathrm{T}} D C_{12} & C_{12}^{\mathrm{T}} D - C_{22}^{\mathrm{T}} G C_{21} - 2l C_{12}^{\mathrm{T}} D C_{11} \\ D C_{12} - C_{21}^{\mathrm{T}} G C_{22} - 2l C_{11}^{\mathrm{T}} D C_{12} & C_{11}^{\mathrm{T}} D + D C_{11} - C_{21}^{\mathrm{T}} G C_{21} - 2l C_{11}^{\mathrm{T}} D C_{11} \end{pmatrix} \tag{2.19}$$

In this paper, we use the linear interpolation procedure. Let $\tau = (m - \delta)h$ with integer $m \geq 1$ and $\delta \in [0, 1)$.

Define

$$\overline{Y}_j^{(n)} = \delta Y_j^{(n-m+1)} + (1-\delta) Y_j^{(n-m)} \tag{2.20a}$$

$$\overline{Z}_j^{(n)} = \delta Z_j^{(n-m+1)} + (1-\delta)Z_j^{(n-m)}, j = 1, 2, \cdots, s \tag{2.20b}$$

where $Y_j^{(i)} = \varphi(t_i + c_j h)$ and $Z_j^{(i)} = \psi(t_i + c_j h)$ for $i < 0$. When the stepsize h satisfies $\tau = mh$, we have

$$\overline{Y}_j^{(n)} = Y_j^{(n-m)},\ \overline{Z}_j^{(n)} = Z_j^{(n-m)},\ j = 1, 2, \cdots, s$$

Theorem 2.1. Assume that a multistep RK method is (k, l)-algebraically stable, then

$$\begin{aligned}\|\mu_{n+1}\|^2 \le k\|\mu_n\|^2 + \sum_{j=1}^{2s} d_j\Big[((2\alpha+\beta+\gamma q)h - 2l)\Big\|V_j^{(n)}\Big\|^2 \\ + h(\beta+\gamma q)\Big(\delta\Big\|V_j^{(n-m)}\Big\|^2 + (1-\delta)\Big\|V_j^{(n-m-1)}\Big\|^2\Big)\Big]\end{aligned} \tag{2.21}$$

Proof. It is well known that

$$\begin{aligned}\|\mu_{n+1}\|^2 - k\|\mu_n\|^2 - 2\sum_{j=1}^{2s} d_j\, Re < V_j^{(n)}, P_j^{(n)} - lV_j^{(n)} > \\ = -\sum_{i=1}^{2s+2}\sum_{j=1}^{2s+2} M_{ij} < r_i, r_j >\end{aligned} \tag{2.22}$$

where $r_1 = w_{n+1}$, $r_2 = w_n$,

$$r_j = Q_{j-2}^{(n)}, j = 3, 4, \cdots, s+2,\ r_j = Q_{j-s-2}^{(n-1)}, j = s+3, s+4, \cdots, 2s+2,$$

By means of (k, l)-algebraical stability of the method, we have

$$\|\mu_{n+1}\|^2 \le k\|\mu_n\|^2 + 2\sum_{j=1}^{2s} d_j\, Re < V_j^{(n)}, P_j^{(n)} - lV_j^{(n)} > \tag{2.23}$$

It follows from (2.2)–(2.5) and (2.9) that

$$2\,Re < W_j^{(n)}, Q_j^{(n)} >$$

$$\begin{aligned}
&= 2h\,Re < W_j^{(n)}, f\Big(t_n + c_j h, Y_j^{(n)}, \bar{Y}_j^{(n)}, \tilde{Y}_j^{(n)}\Big) - f\Big(t_n + c_j h, Z_j^{(n)}, \bar{Y}_j^{(n)}, \tilde{Y}_j^{(n)}\Big) > \\
&+2h\,Re < W_j^{(n)}, f\Big(t_n + c_j h, Z_j^{(n)}, \bar{Y}_j^{(n)}, \tilde{Y}_j^{(n)}\Big) - f\Big(t_n + c_j h, Z_j^{(n)}, \overline{Z}_j^{(n)}, \tilde{Y}_j^{(n)}\Big) > \\
&+2h\,Re < W_j^{(n)}, f\Big(t_n + c_j h, Z_j^{(n)}, \overline{Z}_j^{(n)}, \tilde{Y}_j^{(n)}\Big) - f\Big(t_n + c_j h, Z_j^{(n)}, \overline{Z}_j^{(n)}, \tilde{Z}_j^{(n)}\Big) > \\
&\le 2h\alpha\Big\|W_j^{(n)}\Big\|^2 + 2h\beta\Big\|W_j^{(n)}\Big\|\cdot\Big\|\overline{W}_j^{(n)}\Big\| + 2h\gamma\Big\|W_j^{(n)}\Big\|\cdot\Big\|\tilde{W}_j^{(n)}\Big\| \\
&\le 2h\alpha\Big\|W_j^{(n)}\Big\|^2 + 2h\beta\Big\|W_j^{(n)}\Big\|\cdot\Big\|\overline{W}_j^{(n)}\Big\| + 2h\gamma q\Big\|W_j^{(n)}\Big\|\cdot\Big\|\overline{W}_j^{(n)}\Big\| \\
&\le 2h\alpha\Big\|W_j^{(n)}\Big\|^2 + 2h(\beta+\gamma q)\Big\|W_j^{(n)}\Big\|\cdot\Big\|\overline{W}_j^{(n)}\Big\| \\
&\le 2h\alpha\Big\|W_j^{(n)}\Big\|^2 + 2h(\beta+\gamma q)\Big\|W_j^{(n)}\Big\|\cdot\Big(\delta\Big\|W_j^{(n-m)}\Big\| + (1-\delta)\Big\|W_j^{(n-1-m)}\Big\|\Big) \\
&\le h(2\alpha+\beta+\gamma q)\Big\|W_j^{(n)}\Big\|^2 + h(\beta+\gamma q)\Big(\delta\Big\|W_j^{(n-m)}\Big\|^2 + (1-\delta)\Big\|W_j^{(n-1-m)}\Big\|^2\Big)
\end{aligned}$$

Substitution into (2.22) gives (2.21).

Theorem 2.2. Assume that a multistep RK method is (k, l)-algebraically stable and $k \leq 1$, then the method with linear interpolation procedure is $GR(l)$-stable.

Proof. The inequality $(\alpha + \beta + \gamma q)h \leq l$ and Theorem 2.1 lead to

$$\|\mu_{n+1}\|^2 \leq \|\mu_n\|^2 + \sum_{j=1}^{2s} d_j\left[-(\beta+\gamma q)h\left\|V_j^{(n)}\right\|^2 + h(\beta+\gamma q)\left(\delta\left\|V_j^{(n-m)}\right\|^2 + (1-\delta)\left\|V_j^{(n-m-1)}\right\|^2\right)\right] \tag{2.24}$$

By induction, we have

$$\begin{aligned}
\|\mu_{n+1}\|^2 &\leq \|\mu_0\|^2 + h\sum_{j=1}^{2s} d_j(\beta+\gamma q)\left(\delta\sum_{i=-m}^{-1}\left\|V_j^{(i)}\right\|^2 + (1-\delta)\sum_{i=-m-1}^{-1}\left\|V_j^{(i)}\right\|^2\right)\\
&\leq \|\mu_0\|^2 + h\sum_{j=1}^{2s} d_j(\beta+\gamma q)[\delta(m-1)+(1-\delta)m] \max_{-m-1\leq i\leq -1}\left\|V_j^{(i)}\right\|^2\\
&= \|\mu_0\|^2 + (\beta+\gamma q)\tau\sum_{j=1}^{2s} d_j \max_{-m-1\leq i\leq -1}\left\|V_j^{(i)}\right\|^2\\
&\leq \|\mu_0\|^2 + (\beta+\gamma q)\tau\sum_{j=1}^{2s} d_j \max_{t\leq 0}\|\phi(t)-\psi(t)\|^2
\end{aligned}$$

Because $\mu_{n+1} = (w_{n+1}, w_n)^{\mathrm{T}}$, so we have

$$\begin{aligned}
\|w_n\|^2 &\leq \|w_0\|^2 + \|w_{-1}\|^2 + (\beta+\gamma q)\tau\sum_{j=1}^{2s} d_j \max_{t\leq 0}\|\phi(t)-\psi(t)\|^2\\
&\leq 2\max_{t\leq 0}\|\varphi(t)-\psi(t)\|^2 + (\beta+\gamma q)\tau\sum_{j=1}^{2s} d_j \max_{t\leq 0}\|\phi(t)-\psi(t)\|^2\\
&\leq [2 + C\,(\beta+\gamma q)\tau]\max_{t\leq 0}\|\phi(t)-\psi(t)\|^2.
\end{aligned}$$

Therefore it is $GR(l)$-stable, where $C = \sum_{j=1}^{\nu} d_j$.

Theorem 2.3. Assume that a multistep RK method is $Y^{(n)} = \left(Y_1^{(n)\mathrm{T}}, \cdots, Y_s^{(n)\mathrm{T}}, Y_1^{(n-1)\mathrm{T}}, \cdots, Y_s^{(n-1)\mathrm{T}}\right)^{\mathrm{T}} \in C^{2sN}$-algebraically stable and $\overline{Y}^{(n)} = \left(\overline{Y}_1^{(n)^T}, \cdots, \overline{Y}_s^{(n)^T}, \overline{Y}_1^{(n-1)^T}, \cdots, \overline{Y}_s^{(n-1)^T}\right)^T \in C^{2sN}$, then the method with linear interpolation procedure is $\tilde{Y}^{(n)} = \left(\tilde{Y}_1^{(n)^T}, \cdots, \tilde{Y}_s^{(n)^T}, \tilde{Y}_1^{(n-1)^T}, \cdots, \tilde{Y}_s^{(n-1)^T}\right)^T \in C^{2sN}$-stable.

Proof. Let $\zeta^{(n+1)} = (y^{(n+1)^T}, y^{(n)^T})^T \in C^{2N}$ and $F\left(t_n, Y^{(n)}, \overline{Y}^{(n)}, \tilde{Y}^{(n)}\right) = \left(f(t_n + c_1 h, Y_1^{(n)}, \overline{Y}_1^{(n)}, \tilde{Y}_1^{(n)})^T, \cdots, f(t_n + c_s h, Y_s^{(n)}, \overline{Y}_s^{(n)}, \tilde{Y}_s^{(n)})^T,\right.$

then when $f(t_{n-1} + c_1 h, Y_1^{(n-1)}, \overline{Y}_1^{(n-1)}, \tilde{Y}_1^{(n-1)})^T, \cdots, f(t_{n-1} + c_s h, Y_s^{(n-1)}, \overline{Y}_s^{(n-1)}, \tilde{Y}_s^{(n-1)})^T\Big)^T \in C^{2sN}$, we have $Y^{(n)} = hC_{11}F\Big(t_n, Y^{(n)}, \overline{Y}^{(n)}, \tilde{Y}^{(n)}\Big) + C_{12}\zeta^{(n)}$ and $\zeta^{(n+1)} = hC_{21}F\Big(t_n, Y^{(n)}, \overline{Y}^{(n)}, \tilde{Y}^{(n)}\Big) + C_{22}\zeta^{(n)}$.

The Application of Theorem 2.1 yields

$$\|\mu_{n+1}\|^2 \le \overline{k}\|\mu_n\|^2 + \sum_{j=1}^{2s} d_j\Big[\mu\Big\|V_j^{(n)}\Big\|^2 + h(\beta+\gamma q)\Big(\delta\Big\|V_j^{(n-m)}\Big\|^2 + (1-\delta)\Big\|V_j^{(n-m-1)}\Big\|^2\Big)\Big]$$

By induction, we have

$$\|\mu_{n+1}\|^2 \le \overline{k}^{n+1}\|\mu_0\|^2 + \sum_{i=0}^{n}\overline{k}^{n-i}\sum_{j=1}^{2s} d_j\Big[\mu\Big\|V_j^{(i)}\Big\|^2 + h(\beta+\gamma q)\Big(\delta\Big\|V_j^{(i-m)}\Big\|^2 + (1-\delta)\Big\|V_j^{(i-m-1)}\Big\|^2\Big)\Big]$$

$$\le \overline{k}^{n+1}\|\mu_0\|^2 + \sum_{j=1}^{2s} d_j\Big[\sum_{i=0}^{n-m}\Big(\overline{k}^{n-i}\mu + h(\beta+\gamma q)\delta\overline{k}^{n-m-i} + h(\beta+\gamma q)(1-\delta)\overline{k}^{n-m-i-1}\Big)\Big\|V_j^{(i)}\Big\|^2$$

$$+h(\beta+\gamma q)\Big(\delta\sum_{i=-m}^{-1}\overline{k}^{n-m-i}\Big\|V_j^{(i)}\Big\|^2 + (1-\delta)\sum_{i=-m-1}^{-1}\overline{k}^{n-m-i-1}\Big\|V_j^{(i)}\Big\|^2\Big)\Big]$$

$$\le \overline{k}^{n+1}\|\mu_0\|^2 + \sum_{j=1}^{2s} d_j\Big[\sum_{i=0}^{n-m}\overline{k}^{n-m-i-1}\Big(\overline{k}^{m+1}\mu + h(\beta+\gamma q)\delta\overline{k} + h(\beta+\gamma q)(1-\delta)\Big)\Big\|V_j^{(i)}\Big\|^2$$

$$Y(t) = (y(t+c_1h), \cdots, y(t+c_sh), y(t-h+c_1h), \cdots, y(t-h+c_sh)) \in C^{2sN}$$

On the other hand

$$\mu\overline{k}^{m+1} + h\overline{k}(\beta+\gamma q)\delta + h(\beta+\gamma q)(1-\delta)$$

$$\le h(\beta+\gamma q) + h\overline{k}(\beta+\gamma q)\delta + h(\beta+\gamma q)(1-\delta) \le 0$$

Considering $d_j \ge 0, j = 1, 2, \cdots, s$ and $H^h(t) = \big(H_1^h(t), H_2^h(t)\big) \in C^{2N}$, we have $Y(t) = hC_{11}Y'(t) + C_{12}H^h(t) + \Delta^h(t)$.
Because $\mu_{n+1} = (w_{n+1}, w_n)^{\mathrm{T}}$, so we have

$$Y'(t),$$

which shows that the method is $Y(t) = \big(y'(t+c_1h), \cdots, y'(t+c_sh), y'(t-h+c_1h), \cdots, y'(t-h+c_sh)\big) \in C^{2sN}$-stable.

Theorem 2.4. Assume that a multistep RK method is (k, l)-algebraically stable, $k < 1$ and $Y, \overline{Y}, \tilde{Y}, Z, \overline{Z}, \tilde{Z} \in C^{2sN}, Y^{(n)}, \overline{Y}^{(n)}, \tilde{Y}^{(n)}, Z^{(n)}, \overline{Z}^{(n)}$, then the method with linear interpolation procedure is weak $\tilde{Z}^{(n)}$-stable.

Proof. It follows from Theorem 2.1 that

$$\zeta, \varpi \in C^{2N} \tag{2.25}$$

where $\zeta^{(n+1)}$, when $\varpi^{(n+1)}$, we have $\varpi^{(n+1)} = \left(z^{(n+1)}, z^{(n)}\right)$. Analogous to Theorem 2.2, we can easily obtain

$$\|\mu_{n+1}\|^2 \leq \|\mu_0\|^2 + \sum_{j=1}^{2s} d_j\left[(\beta+\gamma q)\tau \max_{-m-1\leq i\leq -1}\left\|V_j^{(i)}\right\|^2 - \sigma\sum_{i=1}^{n}\left\|V_j^{(i)}\right\|^2\right] \tag{2.26}$$

which shows

$$\tilde{\delta} = \zeta - \varpi - hC_{21}\left(F\left(t, Y, \overline{Y}, \tilde{Y}\right) - F\left(t, Z, \overline{Z}, \tilde{Z}\right)\right), \quad p_2 \tag{2.27}$$

$$p_3, \quad \|Y - Z\| \leq p_2\left(\left\|\tilde{\Delta}\right\| + h\left\|\overline{Y} - \overline{Z}\right\| + h\left\|\tilde{Y} - \tilde{Z}\right\|\right) \tag{2.28}$$

On the other hand,

$$\|\zeta - \varpi\| \leq p_3\left(\left\|\tilde{\Delta}\right\| + \left\|\tilde{\delta}\right\| + h\left\|\overline{Y} - \overline{Z}\right\| + h\left\|\tilde{Y} - \tilde{Z}\right\|\right)$$

$$\tilde{\delta} = \zeta - \varpi - C_{21}(K_1 + K_2 + K_3)$$

$$+h\left\|f\left(t_n + c_ih, Z_i^{(n)}, \overline{Z}_i^{(n)}, \tilde{Y}_i^{(n)}\right) - f\left(t_n + c_ih, Z_i^{(n)}, \overline{Z}_i^{(n)}, \tilde{Z}_i^{(n)}\right)\right\|$$

$$\leq h\left\|f\left(t_n + c_ih, Y_i^{(n)}, \overline{Y}_i^{(n)}, \tilde{Y}_i^{(n)}\right) - f\left(t_n + c_ih, Z_i^{(n)}, \overline{Y}_i^{(n)}, \tilde{Y}_i^{(n)}\right)\right\| + h\beta\left\|\overline{W}_j^{(n)}\right\|$$
$$+h\gamma\left\|\tilde{W}_j^{(n)}\right\| \leq h\left\|f\left(t_n + c_ih, Y_i^{(n)}, \overline{Y}_i^{(n)}, \tilde{Y}_i^{(n)}\right) - f\left(t_n + c_ih, Z_i^{(n)}, \overline{Y}_i^{(n)}, \tilde{Y}_i^{(n)}\right)\right\|$$
$$+h\beta\left\|\overline{W}_j^{(n)}\right\| + h\gamma q\left\|\overline{W}_j^{(n)}\right\|$$

In view of (2.16) and (2.28), we have

$$\|\zeta - \varpi\| = \left\|\tilde{\delta} + C_{21}C_{11}^{-1}\left(W - \tilde{\Delta}\right)\right\| \tag{2.29}$$

Considering (2.2), (2.23) and (2.29) with $\leq \left\|\tilde{\delta}\right\| + p_1h\left\|C_{21}C_{11}^{-1}\right\|\left(\left\|\overline{W}\right\| + \left\|\tilde{W}\right\|\right) + (1+p_1)\left\|C_{21}C_{11}^{-1}\right\| \cdot \left\|\tilde{\Delta}\right\|$, we have

$$\leq p_3\left(\left\|\tilde{\Delta}\right\| + \left\|\tilde{\delta}\right\| + h\left\|\overline{Y} - \overline{Z}\right\| + h\left\|\tilde{Y} - \tilde{Z}\right\|\right)$$

Because $\mu_{n+1} = (w_{n+1}, w_n)^{\mathrm{T}}$, so we have
$p_3 = max\left\{1, (1+p_2)\left\|C_{21}C_{11}^{-1}\right\|\right\}$.
Which shows that the method is weak $Z^{(n)} = hC_{11}F\left(t_n, Z^{(n)}, \overline{Z}^{(n)}, \tilde{Z}^{(n)}\right) + C_{12}\varpi^{(n)}$-stable.

3 The Convergence of Multistep RK Method for NDDEs

In order to study the convergence of the method, we define

$$Y^{(n)} = \left(Y_1^{(n)^T}, \cdots, Y_s^{(n)^T}, Y_1^{(n-1)^T}, \cdots, Y_s^{(n-1)^T}\right)^{\mathrm{T}} \in C^{2sN},$$

$$\overline{Y}^{(n)} = \left(\overline{Y}_1^{(n)^T}, \cdots, \overline{Y}_s^{(n)^T}, \overline{Y}_1^{(n-1)^T}, \cdots, \overline{Y}_s^{(n-1)^T}\right)^T \in C^{2sN}$$

$$\tilde{Y}^{(n)} = \left(\tilde{Y}_1^{(n)^T}, \cdots, \tilde{Y}_s^{(n)^{\mathrm{T}}}, \tilde{Y}_1^{(n-1)^{\mathrm{T}}}, \cdots, \tilde{Y}_s^{(n-1)^{\mathrm{T}}}\right)^{\mathrm{T}} \in C^{2sN},$$

$$\zeta^{(n+1)} = (y^{(n+1)^T}, y^{(n)^T})^T \in C^{2N}$$

$$F(t_n, Y^{(n)}, \bar{Y}^{(n)}, \tilde{Y}^{(n)}) = \Big(f(t_n + c_1 h, Y_1^{(n)}, \bar{Y}_1^{(n)}, \tilde{Y}_1^{(n)})^{\mathrm{T}}, \cdots, f(t_n + c_s h, Y_s^{(n)}, \bar{Y}_s^{(n)}, \tilde{Y}_s^{(n)})^{\mathrm{T}},$$

$$f(t_{n-1} + c_1 h, Y_1^{(n-1)}, \overline{Y}_1^{(n-1)}, \tilde{Y}_1^{(n-1)})^T, \cdots, f(t_{n-1} + c_s h, Y_s^{(n-1)}, \overline{Y}_s^{(n-1)}, \tilde{Y}_s^{(n-1)})^T\Big)^T \in C^{2sN}$$

Thus, process (2.8) can be written in the more compact form

$$Y^{(n)} = hC_{11}F\left(t_n, Y^{(n)}, \overline{Y}^{(n)}, \tilde{Y}^{(n)}\right) + C_{12}\zeta^{(n)} \tag{3.1a}$$

$$\zeta^{(n+1)} = hC_{21}F\left(t_n, Y^{(n)}, \overline{Y}^{(n)}, \tilde{Y}^{(n)}\right) + C_{22}\zeta^{(n)} \tag{3.1b}$$

Definition 3.1. Multistep RK method (2.7) with an interpolation procedure is said to be D-convergent of order $\varpi^{(n+1)} = hC_{21}F\left(t_n, Z^{(n)}, \overline{Z}^{(n)}, \tilde{Z}^{(n)}\right) + C_{22}\varpi^{(n)}$ if the global error satisfies a bound of the form

$$\left\|H(t_n) - y^{(n)}\right\| \le \rho_1(t_n)\left(\left\|H(t_0) - y^{(0)}\right\| + \max_{k \le 0}\left\|Y(t_k) - Y^{(k)}\right\| + h^p\right),$$

where $Z^{(n)} = \left(Z_1^{(n)}, \cdots, Z_s^{(n)}, Z_1^{(n-1)}, \cdots, Z_s^{(n-1)}\right) \in C^{2sN}$ is defined by

$$H(t) = (y(t+h), y(t)) \in C^{2rN}$$

$Y(t) = (y(t + c_1 h), \cdots, y(t + c_s h), y(t - h + c_1 h), \cdots, y(t - h + c_s h)) \in C^{2sN}$ $\rho_1(t)$ and h_0 depend on M_i, $\overline{Z}^{(n)} = \left(\overline{Z}_1^{(n)}, \cdots, \overline{Z}_s^{(n)}, \overline{Z}_1^{(n-1)}, \cdots, \overline{Z}_s^{(n-1)}\right) \in C^{2sN}$ and τ.

Definition 3.2. Multistep RK method (2.7) is said to be algebraically stable if there exist a real symmetric, positive definite $2r \times 2r$ matrix G and a nonnegative diagonal $2s \times 2s$ matrix D such that the matrix

$$M = \begin{bmatrix} G - C_{22}^{\mathrm{T}}GC_{22} & C_{12}^{\mathrm{T}}D - C_{22}^{\mathrm{T}}GC_{21} \\ DC_{12} - C_{21}^{\mathrm{T}}GC_{22} & DC_{11} + C_{11}^{\mathrm{T}}D - C_{21}^{\mathrm{T}}GC_{21} \end{bmatrix} \tag{3.2}$$

is nonnegative definite.

Definition 3.3. Multistep RK method (2.7) is said to be diagonally stable if there exist an $2s \times 2s$ diagonal matrix $Q > 0$ such that the matrix $QC_{11} + C_{11}^{\mathrm{T}} Q$ is positive definite.

Remark 3.4. The concepts of algebraic stability and diagonal stability of multistep RK method are the generalizations of corresponding concepts of Runge-Kutta methods. Although it is difficult to examine these conditions, many results have been found, especially, there exist algebraically stable and diagonally stable multistep formulas of arbitrarily high order (cf. [30]).

Definition 3.5. Multistep RK method (2.7) is said to have generalized stage order P if P is the largest integer which possesses the following properties.

For any given problem (2.1) and $h \in [0, \overline{h}_0]$, there exist an abstract function $H^h(t)$,

$$H^h(t) = \left(H_1^h(t), H_2^h(t)\right) \in C^{2N} \tag{3.3}$$

such that

$$\left\|H(t) - H^h(t)\right\| \le p_1 h^{(p)}, \ \left\|\Delta^h(t)\right\| \le p_1 h^{p+1}, \ \left\|\delta^h(t)\right\| \le p_1 h^{p+1} \tag{3.4}$$

where the maximum stepsize $\overline{h}_0 > 0$ and the constant $p_1$depend only on the method and the bounds M_i, $\Delta^h(t)$ and $\delta^h(t)$, they are defined by the equations

$$Y(t) = hC_{11}Y'(t) + C_{12}H^h(t) + \Delta^h(t) \tag{3.5a}$$

$$H^h(t+h) = hC_{21}Y'(t) + C_{22}H^h(t) + \delta^h(t) \tag{3.5b}$$

The function $Y'(t)$ is defined by

$$Y(t) = \left(y'(t+c_1h), \cdots, y'(t+c_sh), y'(t-h+c_1h), \cdots, y'(t-h+c_sh)\right) \in C^{2sN} \tag{3.5c}$$

Especially, when $H(t) = H^h(t)$, generalized stage order is called stage order.

In this section, we focus on the error analysis of Multistep RK method for (2.1). For the sake of simplicity, we always assume that all constants h_i, γ_i, d_i and L_i are dependent only on the method, some of the bounds M_i the parameters α, β, γ and τ.

First, we give a preliminary result which will later be used several times. For any $Y, \overline{Y}, \tilde{Y}, Z, \overline{Z}, \tilde{Z} \in C^{2sN}$ for $Y^{(n)}, \overline{Y}^{(n)}, \tilde{Y}^{(n)}, Z^{(n)}, \overline{Z}^{(n)}$ $\tilde{Z}^{(n)}$, and $\zeta, \varpi \in C^{2N}$ for $\zeta^{(n+1)}, \varpi^{(n+1)}$, where $\varpi^{(n+1)} = \left(z^{(n+1)}, z^{(n)}\right)$.

Define $\tilde{\Delta}$ and $\tilde{\delta}$ by

$$\tilde{\Delta} = Y - Z - hC_{11}\left(F(t, Y, \overline{Y}, \tilde{Y}) - F(t, Z, \overline{Z}, \tilde{Z})\right) \tag{3.6a}$$

$$\tilde{\delta} = \zeta - \varpi - hC_{21}\left(F\left(t, Y, \overline{Y}, \tilde{Y}\right) - F\left(t, Z, \overline{Z}, \tilde{Z}\right)\right) \tag{3.6b}$$

Theorem 3.6. Suppose method (2.7) is diagonally stable, then there exist constants h_1, p_2 and p_3, such that

$$\|Y - Z\| \leq p_2\left(\left\|\tilde{\Delta}\right\| + h\|\overline{Y} - \overline{Z}\| + h\left\|\tilde{Y} - \tilde{Z}\right\|\right),\ h \in (0, h_1] \tag{3.7a}$$

$$\|\zeta - \varpi\| \leq p_3\left(\left\|\tilde{\Delta}\right\| + \left\|\tilde{\delta}\right\| + h\|\overline{Y} - \overline{Z}\| + h\left\|\tilde{Y} - \tilde{Z}\right\|\right),\ h \in (0, h_1] \tag{3.7b}$$

Proof. Since the method (2.7) is diagonally stable, there exists a positive definite diagonal matrix Q such that the matrix $E = QC_{11} + C_{11}^{\mathrm{T}}Q$ is positive definite. Therefore, the matrix C_{11} is obviously nonsingular and there exists an $l > 0$ depend only on the method such that the matrix

$$E_l = C_{11}^{-\mathrm{T}}EC_{11} - 2lQ \tag{3.8}$$

is also positive definite.

Define

$$W = Y - Z,\quad \overline{W} = \overline{Y} - \overline{Z},\quad \widetilde{W} = \widetilde{Y} - \widetilde{Z} \tag{3.9a}$$

$$K_1 = h\big(F(t, Y, \overline{Y}, \widetilde{Y}) - F(t, Z, \overline{Y}, \widetilde{Y})\big) \tag{3.9b}$$

$$K_2 = h\big(F(t, Z, \overline{Y}, \widetilde{Y}) - F(t, Z, \overline{Z}, \widetilde{Y})\big) \tag{3.9c}$$

$$K_3 = h\big(F(t, Z, \overline{Z}, \widetilde{Y}) - F(t, Z, \overline{Z}, \widetilde{Z})\big) \tag{3.9d}$$

then

$$\widetilde{\Delta} = W - C_{11}(K_1 + K_2 + K_3) \tag{3.10a}$$

$$\tilde{\delta} = \zeta - \varpi - C_{21}(K_1 + K_2 + K_3) \tag{3.10b}$$

Using (2.2)–(2.5), (2.9) and (3.10), we have for $h\alpha \leq l$,

$$\begin{aligned}
0 &\leq \langle W, 2lQW\rangle + 2\mathrm{Re}\langle K_1, -QW\rangle \\
&\quad - \langle W, E_1 W\rangle + 2\mathrm{Re}\left\langle W, Q(C_{11}^{-1}W - K_1)\right\rangle \\
&\leq -\lambda_l\|W\|^2 + 2\mathrm{Re}\left\langle W, Q(C_{11}^{-1}\widetilde{\Delta} + K_2 + K_3)\right\rangle \\
&\leq -\lambda_l\|W\|^2 + 2\left\|QC_{11}^{-1}\right\|\|W\|\left\|\widetilde{\Delta}\right\| + 2h\beta\|Q\|\|W\|\left\|\overline{W}\right\| + 2h\gamma\|Q\|\|W\|\left\|\widetilde{W}\right\|
\end{aligned}$$

where λ_l is the minimum eigenvalue of E_l. Therefore,

$$\|Y - Z\| \leq p_1\left(\left\|\widetilde{\Delta}\right\| + h\left\|\overline{Y} - \overline{Z}\right\| + h\left\|\widetilde{Y} - \widetilde{Z}\right\|\right),\ h \in (0, h_1]$$

where

$$p_1 = \frac{2}{\lambda_l}\max\left(\left\| QC_{11}^{-1}, \beta\|Q\|, \gamma\|Q\|\right\|\right) \tag{3.11a}$$

$$h_1 = \begin{cases} 1 & \alpha \le 0 \\ \min(1, \frac{l}{\alpha}) & \alpha > 0 \end{cases} \tag{3.11b}$$

from (3.7a) and (3.11), it follows that

$$\begin{aligned}
\|\varsigma - \varpi\| &= \left\|\tilde{\delta} + C_{21}C_{11}^{-1}\left(W - \tilde{\Delta}\right)\right\| \\
&\le \left\|\tilde{\delta}\right\| + p_1 h\left\|C_{21}C_{11}^{-1}\right\|\left(\left\|\overline{W}\right\| + \left\|\tilde{W}\right\|\right) + (1 + p_1)\left\|C_{21}C_{11}^{-1}\right\| \cdot \left\|\tilde{\Delta}\right\| \\
&\le p_3\left(\left\|\tilde{\Delta}\right\| + \left\|\tilde{\delta}\right\| + h\left\|\overline{Y} - \overline{Z}\right\| + h\left\|\tilde{Y} - \tilde{Z}\right\|\right), \\
p_3 &= max\left\{1, (1 + p_2)\left\|C_{21}C_{11}^{-1}\right\|\right\}
\end{aligned}$$

where $p_3 = max\left\{1, (1 + p_2)\left\|C_{21}C_{11}^{-1}\right\|\right\}$, which completes the proof of Theorem 3.6.

Consider the compact form of (2.13)

$$Z^{(n)} = hC_{11}F\left(t_n, Z^{(n)}, \overline{Z}^{(n)}, \tilde{Z}^{(n)}\right) + C_{12}\varpi^{(n)} \tag{3.12a}$$

$$\varpi^{(n+1)} = hC_{21}F\left(t_n, Z^{(n)}, \overline{Z}^{(n)}, \tilde{Z}^{(n)}\right) + C_{22}\varpi^{(n)} \tag{3.12b}$$

where

$$Z^{(n)} = \left(Z_1^{(n)}, \cdots, Z_s^{(n)}, Z_1^{(n-1)}, \cdots, Z_s^{(n-1)}\right) \in C^{2sN}$$

$$\overline{Z}^{(n)} = \left(\overline{Z}_1^{(n)}, \cdots, \overline{Z}_s^{(n)}, \overline{Z}_1^{(n-1)}, \cdots, \overline{Z}_s^{(n-1)}\right) \in C^{2sN}$$

$$\tilde{Z}^{(n)} = \left(\tilde{Z}_1^{(n)}, \cdots, \tilde{Z}_s^{(n)}, \tilde{Z}_1^{(n-1)}, \cdots, \tilde{Z}_s^{(n-1)}\right) \in C^{2sN},$$

$$\varpi^{(n+1)} = \left(z^{(n+1)}, z^{(n)}\right) \in C^{2N}$$

Theorem 3.7. Suppose the method (2.7) is algebraically stable for the matrices G and D, then for (3.1) and (3.12), we have

$$\left\|\varsigma^{(n+1)} - \varpi^{(n+1)}\right\|_G^2 \le \left\|\varsigma^{(n)} - \varpi^{(n)}\right\|_G^2 + p_4 h\left(\left\|Y^{(n)} - Z^{(n)}\right\|^2 + \left\|Y^{(n-m)} - Z^{(n-m)}\right\|^2 + \left\|\tilde{Y}^{(n)} - \tilde{Z}^{(n)}\right\|^2\right) \tag{3.13}$$

where $p_4 = \|D\| \cdot max\{(2\alpha + \beta + \gamma), \beta, \gamma\}$, $\|\cdot\|$ is a norm on C^{2sN} defined by

$$\|U\|_G = \langle U, GU\rangle^{\frac{1}{2}} = \left(\sum_{i,j=1}^{2s} g_{ij} < u_i, u_j >\right)^{\frac{1}{2}},$$

$$U = (u_1, u_2, \cdots, u_{2s}) \in C^{2sN},\ u_i \in C^N, u_i \in C^N.$$

Proof. Define $u^{(n)} = \zeta^{(n)} - \varpi^{(n)}$, we get from (3.9) that

$$K^{(n)} = h\Big(F(t_n, Y^{(n)}, \overline{Y}^{(n)}, \widetilde{Y}^{(n)}) - F(t_n, Z^{(n)}, \overline{Z}^{(n)}, \widetilde{Z}^{(n)})\Big)$$

and

$$W^{(n)} = C_{11}K^{(n)} + C_{12}u^{(n)},\ u^{(n+1)} = C_{21}K^{(n)} + C_{22}u^{(n)},$$

With algebraic stability, the matrix

$$M = \begin{bmatrix} G - C_{22}^{\mathrm{T}}GC_{22} & C_{12}^{\mathrm{T}}D - C_{22}^{\mathrm{T}}GC_{21} \\ DC_{12} - C_{21}^{\mathrm{T}}GC_{22} & DC_{11} + C_{11}^{\mathrm{T}}D - C_{21}^{\mathrm{T}}GC_{21} \end{bmatrix}$$

is nonnegative definite. As in [31], we have

$$\begin{aligned}\Big\langle u^{(n+1)}, Gu^{(n+1)}\Big\rangle - \Big\langle u^{(n)}, Gu^{(n)}\Big\rangle - 2\mathrm{Re}\Big\langle W^{(n)}, DK^{(n)}\Big\rangle \\ = -\Big\langle u^{(n)}, K^{(n)}\Big\rangle, M\Big\langle u^{(n)}, K^{(n)}\Big\rangle \le 0\end{aligned}$$

Using (2.2)–(2.5), we further obtain

$$\begin{aligned}\Big\langle u^{(n+1)}, Gu^{(n+1)}\Big\rangle &\le \Big\langle u^{(n)}, Gu^{(n)}\Big\rangle + 2\mathrm{Re}\Big\langle W^{(n)}, DK^{(n)}\Big\rangle \\ &\le \Big\langle u^{(n)}, Gu^{(n)}\Big\rangle + 2\mathrm{Re}\Big\langle W^{(n)}, DK_1^{(n)}\Big\rangle \\ &+ 2\mathrm{Re}\Big\langle W^{(n)}, DK_2^{(n)} + DK_3^{(n)}\Big\rangle \\ &\le \Big\langle u^{(n)}, Gu^{(n)}\Big\rangle + 2h\alpha\Big\langle W^{(n)}, DW^{(n)}\Big\rangle \\ &+ 2h\beta\Big\|D^{\frac{1}{2}}W^{(n)}\Big\| \cdot \Big\|D^{\frac{1}{2}}\overline{W}^{(n)}\Big\| + 2h\gamma\Big\|D^{\frac{1}{2}}W^{(n)}\Big\| \cdot \Big\|D^{\frac{1}{2}}\tilde{W}^{(n)}\Big\| \\ &\le \Big\langle u^{(n)}, Gu^{(n)}\Big\rangle + 2h\alpha\Big\langle W^{(n)}, DW^{(n)}\Big\rangle + h\beta\Big(\|D\| \cdot \Big\|W^{(n)}\Big\|^2 + \|D\| \cdot \Big\|\overline{W}^{(n)}\Big\|^2\Big) \\ &+ h\gamma\Big(\|D\| \cdot \Big\|W^{(n)}\Big\|^2 + \|D\| \cdot \Big\|\tilde{W}^{(n)}\Big\|^2\Big) \\ &\le \Big\langle u^{(n)}, Gu^{(n)}\Big\rangle + (2h\alpha + h\beta + h\gamma)\|D\| \cdot \Big\|W^{(n)}\Big\|^2 \\ &+ \|D\|\Big(h\beta\Big\|W^{(n-m)}\Big\|^2 + h\gamma\Big\|\tilde{W}^{(n)}\Big\|^2\Big)\end{aligned}$$

which gives (3.13). The proof is completed.

In the following, we assume that the method (2.7) has generalized stage order$P$, i.e., there exists a function $H^h(t)$ such that (3.4) holds. For any $n > 0$, we define $\hat{Y}^{(n)}$ and (k, l) by

$$(k, l) \tag{3.14a}$$

$$(k, l) \tag{3.14b}$$

where

$$\hat{\tilde{Y}}^{(n)} = \Big(f\Big(t_n + c_1 h - \tau, \overline{Y}_1^{(n)}, \overline{Y}_1^{(n+1)}, H_1^h(t_n + c_1 h - \tau)\Big), f(t_n + c_2 h - \tau, \overline{Y}_2^{(n)}, \overline{Y}_2^{(n+1)}, H_1^h(t_n + c_2 h - \tau),$$
$$\cdots, f(t_n + c_s h - \tau, \bar{Y}_1^{(n)}, \bar{Y}_1^{(n+1)}, H_1^h(t_n + c_s h - \tau), f(t_n + c_1 h - h - \tau, \bar{Y}_1^{(n-1)}, \bar{Y}_1^{(n)}, H_1^h(t_n + c_s h - h - \tau),$$
$$\cdots, f(t_n + c_s h - h - \tau, \overline{Y}_s^{(n-1)}, \overline{Y}_s^{(n)}, H_1^h(t_n + c_s h - h - \tau)\Big)$$

Theorem 3.8. Suppose the method (2.7) is diagonally stable and its generalized stage order is$P$, then there exists constants (k, l) and h_2 such that

$$\sum_{k=1}^{n}\left\|\bar{Y}^{(k)} - Y(t_k - \tau)\right\|^2 \le p_4\Big(\sum_{k=1}^{n}\left\|\zeta^{(k)} - H^h(t_k)\right\|^2 + nh^{(p+1)} + nh^{(\nu+\mu+1)}$$
$$+ \sum_{k=-m+1}^{0}\left\|Y^{(k)} - Y(t_k)\right\|^2\Big), h \in (0, h_2] \tag{3.15}$$

Proof. It follows from (2.9) that

$$\left\|\overline{Y}_j^{(k)} - y(t_k + c_j h - \tau)\right\| \le \left\|\sum_{i=-\mu}^{\nu} \tilde{L}_i(\delta)\Big(Y_j^{(k-m+i)} - y(t_{k-m+i} + c_j h)\Big)\right\|$$
$$+ \left\|\sum_{i=-\mu}^{\nu} \tilde{L}_i(\delta) y(t_{k-m+i} + c_j h) - y(t_{k-m} + c_j h + \delta h)\right\|$$

From the remainder estimation of Lagrange interpolation formula, we have

$$\left\|\sum_{i=-\mu}^{\nu} \tilde{L}_i(\delta) y(t_{k-m+i} + c_j h) - y(t_{k-m} + c_j h + \delta h)\right\| \le M_{\mu+\nu+1} h^{\mu+\nu+1}$$

Using Cauchy inequality, we further obtain

$$\left\|\overline{Y}^{(k)} - Y(t_k - \tau)\right\|^2 \le 2s(\mu + \nu + 2)\left(\tilde{L}_0^2 \sum_{i=-\mu}^{\nu}\left\|Y^{(k-m+i)} - Y(t_{k-m+i})\right\|^2 + M_{\mu+\nu+1} h^{(\nu+\mu+1)}\right)$$

where $\tilde{L}_0^2 = \underset{-\mu \le i \le \nu}{max} \ \underset{x \in [0,1]}{\text{Sup}} \left|\tilde{L}_i(x)\right|$, Hence, there exists a constant d_1 such that

$$\sum_{k=1}^{n}\left\|Y^{(k)} - Y(t_{k-\tau})\right\|^2 \le d_1\left(\sum_{k=-\mu-m+1}^{n}\left\|Y^{(k)} - Y(t_k)\right\|^2 + nh^{2(\mu+\nu+1)}\right) \tag{3.16}$$

On the other hand, a combination of (3.1a) and (3.5a) leads to

$$(k, l)$$

It follows from Theorem 3.6 that

$$\left\|Y^{(k)} - Y(t_k)\right\| = p_2\{\left\|C_{12}(y^{(k-1)}) - H^h(t_k - h) - \Delta^h(t_k)\right\|$$
$$+ h\left\|\bar{Y}^{(k)} - Y(t_k - \tau)\right\| + h\left\|\bar{Y}^{(k)} - \tilde{Y}(t_k - \tau)\right\|\}$$

which on substitution into (3.16) gives

$$(k, l)$$

Therefore, there exists p_4 and h_2 such that (3.15) holds. The proof of Theorem 3.8 is completed.

Theorem 3.9. Suppose method (2.7) is algebraically stable and diagonally stable and its generalized stage order is p. Then the method with interpolation procedure (2.9) is (k, l) convergent of order at least $\min\{p, \mu + \nu + 1\}$.

Proof. In view of (3.1) and (3.14), it follows from Theorem 3.7 that

$$(k, l) \tag{3.17}$$

Using Theorem 3.6, we have

$$(k, l)$$

$$(k, l), \ h \in (0, h_1]$$

which on substitution into (3.17) gives

$$(k, l) \tag{3.18}$$

where λ_2 is the minimum characteristic value of$G$. On the other hand,

$$\begin{aligned}\left\|y^{(n)} - H^h(t_n)\right\|_G^2 &\le \left\|y^{(n)} - \hat{y}^{(n)}\right\|_G^2 + \left\|\hat{y}^{(n)} - H^h(t_n)\right\|_G^2 + 2\left\|y^{(n)} - \hat{y}^{(n)}\right\|_G \left\|\hat{y}^{(n)} - H^h(t_n)\right\|_G \\ &\le (1+h)\left\|y^{(n)} - \hat{y}^{(n)}\right\|_G^2 + (1 + \frac{1}{h})\left\|y^{(n)} - H^h(t_n)\right\|_G^2\end{aligned}$$

In view of (3.5) and (3.14), the application of Theorem3.6 leads to

$$\left\|\hat{y}^{(n)} - H^h(t_n)\right\|_G \le p_2\left(\left\|\Delta^h(t_n)\right\| + \left\|\delta^h(t_n)\right\|\right), \ h \in (0, \overline{h}_0]$$

which gives

$$\left\|\hat{y}^{(n)} - H^h(t_n)\right\|_G^2 \le \lambda_1 p_2^2\left(\left\|\Delta^h(t_n)\right\| + \left\|\delta^h(t_n)\right\|\right)^2, \tag{3.19}$$

where λ_1 denotes the maximum eigenvalue of the matrix G. A combination of (3.4) and (3.17)–(3.19) leads to

$$(k, l)$$

$$(k, l)^{h \in (0, h_3]}$$

where

$$h_3 = \min\{\bar{h}_0, h_1\} \le 1, \ \gamma_1 = 1 + \frac{4p_3p_1^2\|C_{12}\|^2}{\lambda_2}$$

$$\gamma_2 = 2p_3(1 + 2p_1^2)(1 + 2sr^2q^2), \ \gamma_3 = 8\lambda_1 p_0^2 p_2^2$$

Therefore

$$(k, l).$$

Considering Theorem 3.8, we further obtain

$$(k, l)$$

where $h_0 = \min\{h_2, h_3\} \le 1$. Using discrete Bellman inequality, we have

$$(k, l)$$

$$(k, l) \tag{3.20}$$

Considering $\left\|H(t) - H^h(t_n)\right\| + p_0 h^p$, we obtain

$$\left\|y^{(n)} - H^h(t_n)\right\| \le \left\|y^{(n)} - H^h(t_n)\right\| + \left\|H^h(t_n) - H(t_n)\right\| \le \left\|y^{(n)} - H^h(t_n)\right\| + p_0 h^p \tag{3.21}$$

Considering (3.20) and (3.21), we can easily conclude that method (2.7) with interpolation procedure (2.9) is (k, l) convergent of order at least $\min\{p, \mu + \nu + 1\}$. The proof is completed.

4 Conclusions

In this paper we gave the stability and convergence results of multistep Runge-Kutta methods with linear interpolation procedure for solving nonlinear NDDEs (2.1). First, we gave the definitions of (k, l)-algebraically stable, algebraically stable and diagonally stable. Then we proved that if a multistep RK method is (k, l)-algebraically stable, $k < 1$ and $d_j > 0, j = 1, 2, \cdots, s$, then the method with linear interpolation procedure is weak *GAR*(l)-stable. We also proved that if a multistep RK method is algebraically stable and diagonally stable and its generalized stage order is$p$, then the method with interpolation procedure is (k, l) convergent of order at least $\min\{p, \mu + \nu + 1\}$.

We believe that the results presented in this paper can be extended to other general NDDEs. However, it is difficult to extend these results to more general neutral functional differential equations. Results extending the results presented in this paper to more general neutral functional differential equations and other delay differential equations such as delay integral differential equations will be discussed elsewhere.

Acknowledgements. The authors would like to thank the referees for giving helpful comments and suggestions. This research is supported by National Natural Science Foundation of China (Project No.11901173) and by the Heilongjiang province Natural Science Foundation (LH2019A030) and by the Innovation Talent Foundation of Heilongjiang institute of technology (2018CX17).

References

1. Weldon, T.E., Kirk, J., Finlay, H.M.: Cyclical granulopoiesis in chronic granulocytic leukemia: a simulation study. Blood **43**, 379–387 (1974)
2. Ruehli, A.E., Miekkala, U., Bellen, A., Heeb, H.: Stable time domain solutions for EMC problems using PEEC circuit models. In: Proceedings of IEEE Int. Symposium on Electromagnetic Compatibility (1994)
3. Brayton, R.K.: Small signal stability criterion for networks containing lossless transmission lines. IBM J. Res. Dev. **12**, 431–440 (1968)
4. Guglielmi, N.: Inexact Newton methods for the steady-state analysis of nonlinear circuits. Math. Models Meth. Appl. Sci. **6**(1), 43–57 (1996)
5. Kuang, Y.: Delay Differential Equations with Application in Population Dynamics. Academic Press, Boston (1993)
6. Bellen, A., Guglielmi, N., Zennaro, M.: On the contractivity and asymptotic stability of systems of delay differential equations of neutral type. BIT **39**, 1–24 (1999)
7. Bellen, A., Zennaro, M.: Numerical Methods for Delay Differential Equations. Oxford University Press, Oxford (2003)
8. Kuang, J.X., Xiang, J.X., Tian, H.J.: The asymptotic stability of one-parameter methods for neutral differential equations. BIT **34**, 400–408 (1994)
9. Qiu, L., Yang, B., Kuang, J.X.: The NGP-stability of Runge-Kutta methods for systems of neutral delay differential equations. Numer. Math. **81**(3), 451–459 (1999)
10. Tian, H.J., Kuang, J.X., Qiu, L.: The stability of linear multistep methods for linear systems of neutral differential equation. J. Comput. Math. **19**, 125–130 (2001)
11. Zhang, C.J., Zhou, S.Z.: The asymptotic stability of theoretical and numerical solutions for systems of neutral multidelay differential equations. Sci. China **41**, 504–515 (1998)
12. Torelli, L.: Stability of numerical methods for delay differential equations. J. Comput. Appl. Math. **25**, 15–26 (1989)
13. Torelli, L.: A sufficient condition for GPN-stability for delay differential equations. Numer. Math. **59**, 311–320 (1991)
14. Bellen, A.: Contractivity of continuous Runge-Kutta methods for delay differential equations. Appl. Numer. Math. **24**, 219–232 (1997)
15. Bellen, A., Zennaro, M.: Strong contractivity properties of numerical methods for ordinary and delay differential equations. Appl. Numer. Math. **9**, 321–346 (1992)
16. Zennaro, M.: Contractivity of Runge-Kutta methods with respect to forcing terms. Appl. Numer. Math. **10**, 321–345 (1993)
17. Koto, T.: The stability of natural Runge-Kutta methods for nonlinear delay differential equations. J. Ind. Appl. Math. **14**, 111–123 (1997)
18. Bellen, A., Guglielmi, N., Zennaro, M.: Numerical stability of nonlinear delay differential equations of neutral type. J. Comput. Appl. Math. **125**, 251–263 (2000)
19. Jackiewicz, Z.: One-step methods of any order for neutral functional differential equations. SIAM J. Numer. Anal **21**, 486–511 (1984)
20. Jackiewicz, Z.: Quasilinear multistep methods and variable step predictor-corrector methods for neutral functional differential equations. SIAM J. Numer. Anal **23**, 423–452 (1986)
21. Jackiewicz, Z., Kwapisz, M., Lo, E.: Waveform relaxation methods for functional differential systems of neutral type. J. Math. Appl. Anal **207**, 255–285 (1997)
22. Yang, R.N., Gao, H.J., Shi, P.: Novel robust stability criteria for stochastic Hopfield neural networks with time delays. IEEE Trans. Syst. Man Cybern. Part B Cybern. **39**(11), 467–474 (2009)
23. Yang, R.N., Zhang, Z.X., Shi, P.: Exponential stability on stochastic neural networks with discrete interval and distributed delays. IEEE Trans. Neural Netw. **21**(1), 169–175 (2010)

24. Liu, L.P.: Robust stability for neutral time-varying delay systems with non-linear perturbations. Int. J. Innov. Comput. Inf. Control **17**(10), 5749–5760 (2011)
25. Tanikawa, A.: On stability of the values of random zero-sum games. Int. J. Innov. Comput. Inf. Control **7**(1), 133–140 (2011)
26. Basin, M., Calderon-Alvarez, D.: Delay-dependent stability for vector nonlinear stochastic systems with multiple delays. Int. J. Innov. Comput. Inf. Control **7**(4), 1565–1576 (2011)
27. Stetter, H.J.: Numerische Losung von Differential gleichungen mit nacheilendem Argument. ZAMM **45**, 79–80 (1965)
28. Oberle, H.J., Pesch, H.J.: Numerical treatment of delay differential equations by Hermite interpolation. Numer. Math. **37**, 235–255 (1981)
29. Zennaro, M.: Natural continuous extensions of Runge-Kutta methods. Math. Comp. **46**, 119–133 (1986)
30. Huang, C.M.: Algebraic stability of hybrid methods. Chin. J. Num. Math. Appl. **17**, 67–74 (1995)
31. Li, S.F.: Theory of Computational Methods for Stiff Differential Equations. Hunan Science and Technology Publisher, Changsha (1997)

Comparison of Modified Object Detected Graph Cut and Hybrid Ant Colony Optimization - k-means for MRI Images Segmentation

Samer El-Khatib[1], Yuri Skobtsov[2], Sergey Rodzin[1], and Valerii Zakharov[3](✉)

[1] Southern Federal University, Rostov-on-Don, Russia
[2] St. Petersburg Polytechnic University, Saint Petersburg, Russia
[3] St. Petersburg Federal Research Center of the RAS, Saint Petersburg, Russia

Abstract. Image segmentation is the process of subdividing an image into regions that are consistent and homogeneous in some characteristics. There is neither a single segmentation model for all medical image modalities nor all methods are efficient for a specific medical image modality. In image processing and computer vision, segmentation is still a challenging problem in many real time applications and hence more research work is required. The Hybrid Ant Colony Optimization (ACO) – k-means and Modified object detected Graph Cut image segmentation algorithms for MRI images segmentation are considered in this paper. The proposed algorithms and sub-system for the medical image segmentation have been implemented. The experimental results show that the proposed algorithm has good accuracy in comparison to modified Graph cut algorithm.

Keywords: MRI images segmentation · Ant colony optimization · k-means algorithm · Swarm intelligence · Modified object detected graph cut

1 Introduction

Modern medical diagnostics are impossible without processing medical images. Medical images of magnetic resonance imaging (MRI) are complex and variable. They are used to distinguish pathological tissues from normal tissues. Nowadays, millions of medical images are regularly produced and analyzed in medical centers. One of the most difficult tasks in image processing is segmentation. All subsequent processing steps, such as classification, image retrieval and identification, directly depend on the results of segmentation.

The development of image recognition methods is one of the relevant and difficult tasks in artificial intelligence. When creating recognition systems, which contain high requirements for accuracy and performance, occurs needless to apply new methods to automate the procedure of image recognition. Despite the fact that the task of developing image recognition methods is well researched theoretically, however, there is no universal method for solving it, and the practical solution seems to be very difficult.

At computer processing and recognition of images the wide range of problems is solved. One of the main stages of recognition is the process of dividing the image into

R. Silhavy (Ed.): CSOC 2021, LNNS 230, pp. 701–708, 2021.
https://doi.org/10.1007/978-3-030-77442-4_58

non-overlapping areas (segments) that cover the entire image and are homogeneous by some criteria. Segmentation simplifies the analysis of homogeneous areas of the image, as well as brightness and geometric characteristics. This is a difficult task, the performance quality of which significantly affects the accuracy and the possibility of subsequent computer analysis of images, since there are difficulties associated with noise, blurring of images, etc.

To solve the image segmentation problem, there have been developed many methods based on the luminance, gradient and texture information of the image [1]. Numerous publications are devoted to the segmentation problem (Bajwa and Gill 2015; Christ et al. 2014; Harikumar and Kumar 2015; Maksoud et al. 2014; Mocskops et al. 2017; Nasab and Fatehi 2017; Patil and Patil 2015; Rajput et al. 2017; Rezaee and Tavakoli 2014; Saxena and Shrivastava 2017).

A suitable way to effectively solve the problem of image segmentation is to use mathematical transformations describing the collective behavior of a decentralized self-organized system that consists of a multiple agents interacting locally with each other and with the environment to achieve a predetermined goal. In nature, examples of such systems are swarm systems [2, 7]. Each agent functions autonomously, using own rules. At the same time, the behavior of the entire system is surprisingly daunting [3, 8–11].

This article is dedicated to development and evaluation for Hybrid ACO-k-means in compare to Modified Graph Cut segmentation method for MRI images.

2 Hybrid Ant Colony Optimization k-means Method for Image Segmentation

To obtain an efficient algorithm for image segmentation we have developed a method in which all the advantages of k-means and ACO algorithms are used. The first step is to set the number of clusters and initialize their centers. Then, according to the clustering algorithm k-means, we need to determine the belonging of each image pixel to a particular cluster. At this stage, the most important role is played by the ACO algorithm. It defines the relationship of each pixel with clusters of the image. This is done according to the probability, which is inversely proportional to the distance between the pixel, cluster center and the variable τ, which represents the pheromone level. The pheromone level is determined in proportion to the minimal distance between each pair of cluster centers and inversely proportional to the distance between each pixel and its center. Thus, the pheromone level value increases with increase of a distance between the centers of clusters, as well as with increase in compactness of pixels in the cluster. Under the same conditions, the probability of the pixel attachment to the cluster increases.

The evaporation of the pheromone is calculated in order to reduce the impact of the previously made choices which are of lower priority. Similarly to k-means algorithm, in the distributed state the cluster centers are updated by recalculation of the average value of pixels in each cluster. It lasts as long as the change in value of the cluster center does not vary substantially. In contrast to the k-means algorithm, the developed method does not stop at this stage. The process of clustering continues to be performed by m ants, each of which ultimately finds its solution. The criteria for finding the best solutions and the updated pheromone level are prior for the next group of m ants respectively. When the stopping criterion is reached, the clustering is completed and the best solution is found.

Hybrid ACO algorithm for image segmentation consists of the following steps:

Algorithm 1. Hybrid ACO-k-means segmentation algorithm

```
Begin
 Initialize([number clusters], [number ants]);
 Repeat
 For each ant do
   M: For each pixel do
     Calc(probability belonging pixel to cluster) (1);
   End
   Update (cluster center);
   If (NewCenter <> OldCenter) then
      goto M;
   Else
      Save(current solution);
 End
 Select Best Solution From All Ants (5);
 Update(for each pixel) (6, 7);
 Correct(common solution);
 Until criteria not reached
End
```

Software implementation of the algorithm starts with determination of the level of pheromone τ and assignment of heuristic information η for each pixel. Then, each ant determines pixel's belonging to the cluster with probability P, which can be calculated in the following way (1):

$$P_i(X_n) = \frac{[\tau_i(X_n)]^\alpha [\eta_i(X_n)]^\beta}{\sum_{j=0}^{K} [\tau_j(X_n)]^\alpha [\eta_j(X_n)]^\beta} \tag{1}$$

where $\tau_i(X_n)$ and $\eta_i(X_n)$ - information about pheromone and heuristic variable values of belonging of a pixel to the cluster i respectively, α and β - heuristic coefficients of ACO algorithm, k – number of clusters. Heuristic information is obtained as follows:

$$\eta_i(X_n) = \frac{k}{CDist(X_n, CC_i) * PDist(X_n, PC_i)} \tag{2}$$

where X_n nth pixel, CC_i - ith spectral cluster center, PC_i - ith spatial cluster center, $CDist(X_n, CC_i)$ - distance between (X_n, CC_i) according to color characteristics – (3), $PDist(X_n, PC_i)$ - the geometrical distance between (X_n, PC_i) - (4), k – constant value.

$$CDist(X_n, CC_i) = |Int(X_n) - Int(CC_i)| \tag{3}$$

where $Int(X_n)$ is the intensity of the pixel X_n.

$$PDist(X_n, PC_i) = \sqrt{(X_n.x - PC_i.x)^2 + (X_n.y - PC_i.y)^2} \tag{4}$$

where $X_n.x$ and $X_n.y$ are x and y coordinates respectively for the pixel X_n.

In the given modification we suggest to use the following set of simple rules as the target function (optimality criterion of a better solution):

- $\max_{i=1..m}\left(\sum_{k=1..K-1}\sum_{j=k+1..K} CDist\left(C_k, C_j\right)\right)$ - the maximum sum value of the color distance between the cluster centers for all ants (the distance between the clusters, in terms of color characteristics, should be at a maximum, then clusters will be different from each other), where CDist – the color distance between two pixels, C_k - center of the cluster k.
- $\min_{i=1..m}\left(\sum_{k=1..K}\sum_{p=1..S_k} PDist\left(C_k, X_p\right)\right)$ - the minimum value of the sums of geometric distances between the centers of the clusters and the pixels belonging to the cluster (the sum of the Euclidean distances between the center of the cluster and each of its pixels should be minimal,, according to the spatial characteristics, respectively, the cluster will be more homogeneous), where S_k - the number of pixels in the cluster k, PDist – Euclidian distance between two pixels, C_k - center of the cluster k.
- $\min_{i=1..m}\left(\sum_{k=1..K}\sum_{p=1..S_k} CDist\left(C_k, X_p\right)\right)$ - the minimum sum value of the color distances between the centers of the clusters and the pixels belonging to the cluster (sum of the Euclidean distances between the center of the cluster and its each pixel, according to color characteristics, should be minimal, then, the cluster will be more compact), where S_k- the number of pixels in the cluster k, CDist – the color distance between two pixels, C_k - center of the cluster k.

The fitness function of the ant colony is 3-criterial and can be obtained as follows:

$$f(m_i) = \begin{cases} f_1 = \sum_{k=1...K-1}\sum_{j=k+1...K} CDist\left(C_{m_ik}, C_{m_ij}\right) \\ f_2 = \sum_{k=1...K}\sum_{p=1...S_{m_i}k} PDist\left(C_{m_ik}, X_{m_ip}\right) \\ f_3 = \sum_{k=1...K}\sum_{p=1...S_{m_ik}} CDist\left(C_{m_ik}, X_{m_ip}\right) \end{cases} \quad (5)$$

Choice of the best solution can be presented as:

$$f(best) = \sum\nolimits_{i=1..m} \begin{cases} \max(f_1) \\ \min(f_2) \\ \min(f_3) \end{cases} \quad (6)$$

Pheromone updating is calculated as in standard ACO algorithm according to:

$$\tau_i(X_n) \leftarrow (1-\rho)\tau_i(X_n) + \sum_i \Delta\tau_i(X_n) \quad (7)$$

and pheromone evaporation is calculated as:

$$\Delta\tau_i(X_n) = \begin{cases} \frac{Q*Min(k')}{AvgCDist(k',i)*AvgPDist(k',i)}, \; if\ X_n \subset cluster\ i \\ 0, \; otherwise \end{cases} \quad (8)$$

where Q is a constant value, $Min\left(k'\right)$ - the minimum color outer cluster distance found by the most successful ant, $AvgCDist\left(k', i\right)$ and $AvgPDist\left(k', i\right)$ - the average value of the spatial Euclidean and color distances between each pixel and cluster centers for the most successful ant.

2.1 Modified Object Detected Graph Cut Image Segmentation Algorithm

The Graph Cuts algorithm has become the basis of many modern interactive segmentation algorithms [5]. Graph Cut presents a graph image. A lot of vertices consists of image pixels and two additional, artificially added vertices - source and drain.

For two adjacent vertices i and j, the edge weight is given by the expression

$$B(i,j) = \exp\left(-\frac{|C_i - C_j|^2}{2\delta^2}\right)\frac{1}{d(i,j)} \tag{9}$$

where C_i and C_j are the colors of the pixels, δ is a user-defined parameter, d (i, j) is the Euclidean distance between the pixels.

The user interactively indicates several pixels that belong to the object, and several pixels that relate to the background. The vertices belonging to the object are connected to the source by edges having infinite weight. The vertices belonging to the background are associated with the drain by ribs having infinite weight. For the obtained graph with drain and source, a minimal cut is sought, dividing the graph into 2 parts [4]. Pixels that are in the same subgraph as the source refer to the object, the rest to the background. The infinite weight of the edges connecting the pixels with the source and drain is necessary so that all the pixels of the object are in the object, and all the pixels of the background are assigned to the background. The formula for the edge weight is chosen in such a way that pixels with the most different colors are connected by an edge of minimum weight, which leads to a section of the graph along the most contrasting border. An increase in the speed of this algorithm in comparison with other algorithms has not been proven, but on the benchmarks of the reference problems of image segmentation, this algorithm works faster than its counterparts.

To improve the results of the algorithm it was proposed to use Object detection results to indicate area of interest (we obtained method that called Modified object detected graph cut). Object detection showed enhanced results and segmentation accuracy increased at average 10% in compare to original Graph Cut algorithm. Below we provided Modified object detected graph cut results in compare to ACO-k-means.

3 Testing Hybrid ACO-k-means and Modified Object Detected Graph Cut Segmentation Methods

Developed Hybrid ACO-k-means algorithm for image segmentation was investigated. Investigation parameters are: Ossirix benchmark [6], set of 150 images with various initial conditions - good quality (no noise and other artifacts) and contrast images. ACO-k-means method was compared with Modified object detected Graph Cut segmentation algorithm.

Results are presented by processing images with various initial conditions (Figs. 1 and 2). Segmentation accuracy is presented on Figs. 3 and 4. As we can observe ACO-k-means algorithm outperforms Modified object detected Graph cut's accuracy.

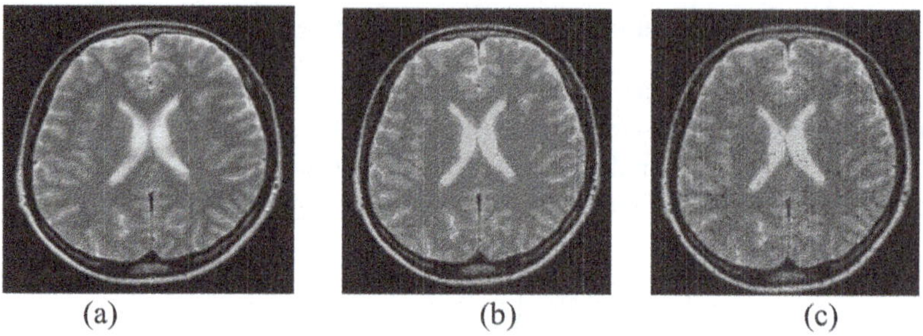

Fig. 1. (a) initial image; (b) Modified object detected Graph Cut segmentation result; (c) Hybrid ACO-k-means segmentation result

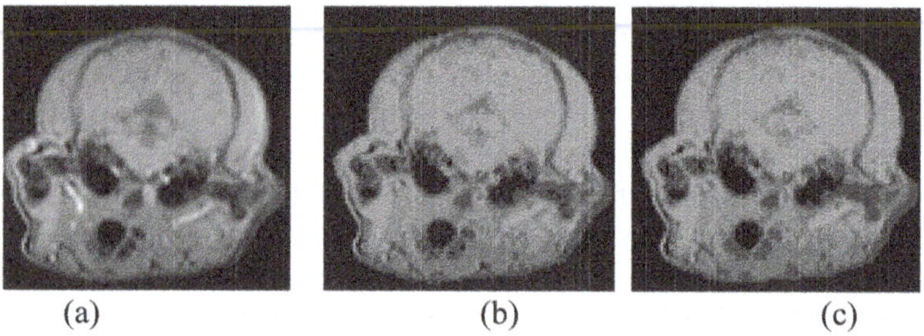

Fig. 2. (a) initial image; (b) Modified object detected Graph Cut segmentation result; (c) Hybrid ACO-k-means segmentation result

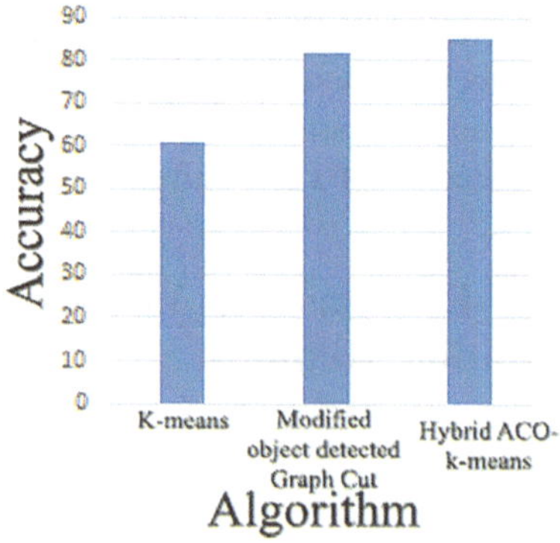

Fig. 3. Evaluation of segmentation accuracy for good quality images

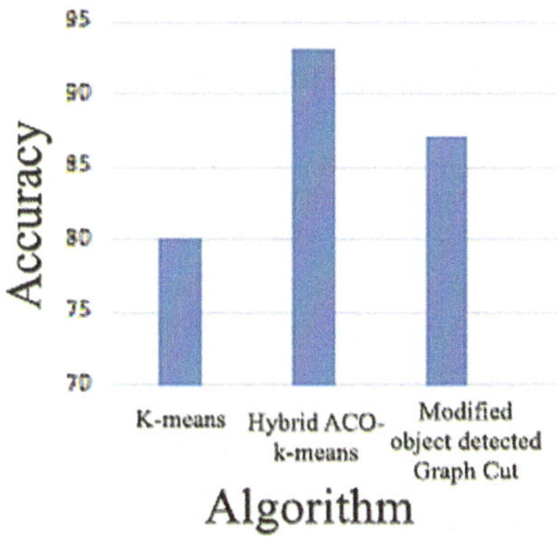

Fig. 4. Evaluation of segmentation accuracy for blurred images

4 Conclusion

In the article was presented comparison of Hybrid ACO-k-means and Modified object detected Graph Cut methods for MRI images segmentation. There were obtained comparative evaluations for the quality for the ACO-k-means segmentation method on different quality images from Ossirix benchmark. Experimentally obtained scientific data indicates the following: ACO-k-means method outperforms Modified object detected Graph cut segmentation quality at average 3%.

Acknowledgements. The reported study was funded by Russian Foundation for Basic Research according to the research project № 19-07-00570 "Bio-inspired models of problem-oriented systems and methods of their application for clustering, classification, filtering and optimization problems, including big data".

The research described in this paper is partially supported by the Russian Foundation for Basic Research (grants 20-08-01046), and state research 0073–2019–0004.

References

1. Gonzalez, R.C., Woods, R.E.: Digital Image Processing, 3rd edn. Prentice-Hall, Hoboken (2008)
2. Kennedy, J., Eberhart, R.C.: Particle swarm intelligence. In: Proceedings of the IEEE International Joint Conference on Neural Networks, pp. 1942–1948 (1995)
3. Bozhenyuk, A., El-Khatib, S., Kacprzyk, J., Knyazeva, M., Rodzin, S.: Hybrid ant fuzzy algorithm for mri images segmentation. In: Rutkowski, L., Scherer, R., Korytkowski, M., Pedrycz, W., Tadeusiewicz, R., Zurada, J.M. (eds.) ICAISC 2019. LNCS (LNAI), vol. 11509, pp. 127–137. Springer, Cham (2019). https://doi.org/10.1007/978-3-030-20915-5_12
4. Saatchi, S., Hung, C.C.: Swarm intelligence and image segmentation swarm intelligence. ARS J. (2007)
5. Boykov, Y., Veksler, O., Zabih, R.: Approximate energy minimization via graph cuts. IEEE Trans. Pattern Anal. Mach. Intell. **23**(11), 1222–1239 (2001)
6. Ossirix image dataset. http://www.osirix-viewer.com/

7. Skobtsov, Y.A., Speransky, D.V.: Evolutionary computation: hand book, 331 p. The National Open University "INTUIT", Moscow (2015). (in Russian)
8. El-Khatib, S., Skobtsov, Y., Rodzin, S.: Improved particle swarm medical image segmentation algorithm for decision making. In: Kotenko, I., Badica, C., Desnitsky, V., El Baz, D., Ivanovic, M. (eds.) IDC 2019. SCI, vol. 868, pp. 437–442. Springer, Cham (2020). https://doi.org/10.1007/978-3-030-32258-8_51
9. El-Khatib, S.A., Skobtsov, Y.A., Rodzin, S.I.: Theoretical and experimental evaluation of hybrid ACO-k-means image segmentation algorithm for MRI images using drift-analysis. Procedia Comput. Sci. **150**, 324 (2019)
10. El-Khatib, S., Skobtsov, Y., Rodzin, S., Potryasaev, S.: Theoretical and experimental evaluation of PSO-K-means algorithm for MRI images segmentation using drift theorem. In: Silhavy, R. (ed.) CSOC 2019. AISC, vol. 985, pp. 316–323. Springer, Cham (2019). https://doi.org/10.1007/978-3-030-19810-7_31
11. El-Khatib, S., Skobtsov, Y., Rodzin, S., Potryasaev, S.: Comparison of hybrid ACO-k-means algorithm and graph cut for MRI images segmentation. In: Silhavy, R. (ed.) CSOC 2020. AISC, vol. 1225, pp. 73–80. Springer, Cham (2020). https://doi.org/10.1007/978-3-030-51971-1_6

Robotic Sorting Line Model Using Coloured Petri Net

Dominik Stursa(✉), Libor Havlicek, and Libor Kupka

Faculty of Electrical Engineering and Informatics, University of Pardubice, Studentska 95, 53210 Pardubice, Czech Republic
{dominik.stursa,libor.havlicek,libor.kupka}@upce.cz

Abstract. With the increasing availability of robots and image processing systems, the automated robotic line implementations are still extensively increasing. As well, the complexity of these systems increases, which increases the demands on the determination of their function and possible analysis. For these purposes the complex robotic systems are modeled. Here, the Coloured Petri Nets were used to model the robotic sorting line. Specifically, the CPN Tools software was used to model creation, simulation, analysis and to made experiments. State space analysis was done to revelation of possible unwanted deadlocks. Furthermore, the sorting line model was used to found out the maximum number of objects that can be sorted in a defined time. As the robot manipulation time directly affects maximum number of objects in one batch, the timing procedure was created to declare robot operation dependency on types of manipulated object. The state space analysis showed that there are no unwanted deadlocks in the system. The model has been verified and declared as correct. Optimal composition and order of objects was successfully found by timed coloured Petri nets experiments.

Keywords: Colored petri net · Sorting line · Modeling

1 Introduction

With the development of robotic technologies along with computer vision approaches, automated lines are being implemented more often. Such systems are more involved in industrial applications what leads to control system improvements. With the growing number of involved components, the complexity of the control system also increases. Therefore, the state analysis and system behavior description need to defined as it simplifies the design of such system.

Modeling with event systems enjoys growing popularity because of their ability to adequately represent the world around us, especially in fields of business processes [1], production systems [2], networked control systems [3], etc. Therefore, a software for system modeling is often used to analyze system behavior. The scope of this article is the design of robotic sorting line model.

R. Silhavy (Ed.): CSOC 2021, LNNS 230, pp. 709–717, 2021.
https://doi.org/10.1007/978-3-030-77442-4_59

The event model of the robotic sorting line can help with process understanding, especially if the system is comprehensive and with its complexity the possibilities of states analysis is extensively increasing. With the complexity is also growing the possibility of the unwanted system deadlocks.

Authors' decides to use the coloured Petri net as a modeling tool, because it provides a mathematical apparatus for the possible solution and analysis of the state space of any model. Furthermore, the construction of the coloured Petri nets is capable enough to prevent unplanned deadlocks as is indicated in [4] or [5].

1.1 Motivation

The main motivation for using of the Coloured Petri Nets (CPN) is the easy creation of the state model for complex system, which can be simulated, analyzed, verified and visualized. The CPN concepts are precisely defined and clearly described in the literature [6] and [7], what makes a CPN modeling more accessible.

Furthermore, the CPN models can be mathematically described, which is allowing complex model state space analysis [8]. Moreover, there exist a lot of software for CPN model creation as is JARP [9], CPN Tools [10]. Possibilities of accurate system description by modeling is increased with use of CPN or their further improvements.

2 Problem Definition

The general automated robotic sorting systems are consisted of a several parts executing more or less important tasks. As first, the object identification must be implemented for distinguish objects that need to be sorted. Next, the control system must evaluate and calculate object positions at moment in which robot will try to pick and place object to the correct spot. As other, the selection strategy must be also defined to provide fast and accurate sorting. The robotic sorting line is captured on Fig. 1.

Sorting line could consist of a lot of different configurations. At first, the number of object types can be various. With increasing number of object types, the demands on the end effector of the robotic manipulator also increase. This is due to the reduced probability of being able to grab an object by only one effector with increasing differences in the shape of the objects. Furthermore, the number of conveyor belts can also increase when combining multiple production technologies [11].

The number of required components of robotic sorting line increases considerably with the complexity of the sorting line. Definition of the specific problem that we dealt with is stated in the next section.

2.1 Problem Statement

Problem of general sorting line is not considered here, because its solution leads to a very long duration of calculations. As such, problem was limited to the specific composition of the robotic sorting line. Specifically, the composition using two conveyor belts and three robotic manipulators were used for five object types sorting.

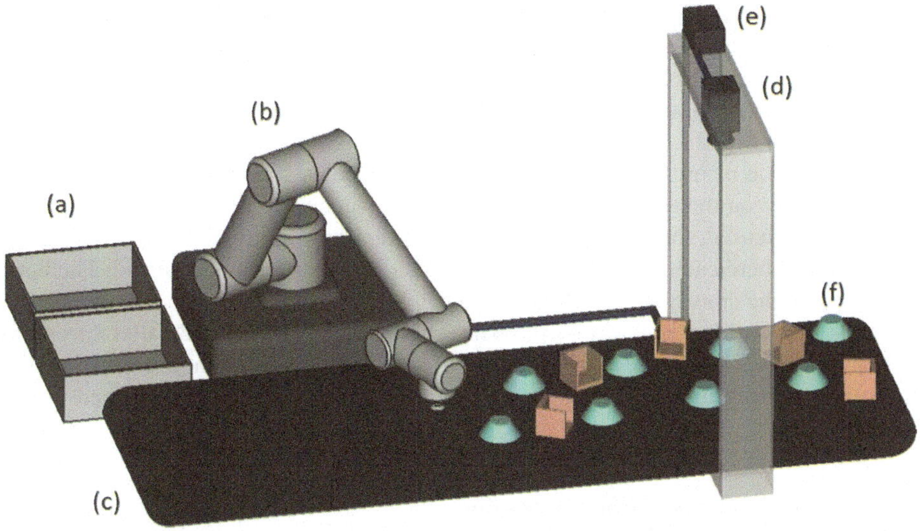

Fig. 1. Robotic sorting line illustration. Illustration contains two storage boxes (a), robotic manipulator (b), conveyor belt (c), camera sensing system (d), control system (e) and two types of object (f).

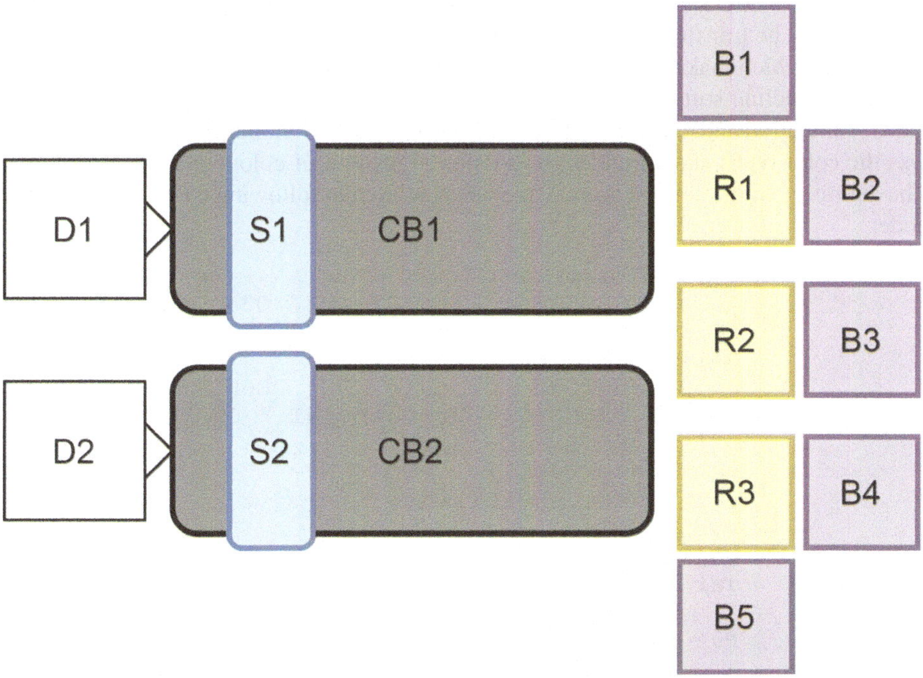

Fig. 2. Specific robotic sorting line layout.

There, these five types of objects are produced in one production. The reason for this two-belt composition is the parallel production of components of different types in two different plants. Each of robot is equipped with a different end effector and therefore can only manipulate some object types. This particular robotic sorting line composition is illustrated in Fig. 2.

The function of presented system can be described as follows. Each conveyor belt has a dispenser on the starting side, that dispenses batches of an object mixture. When the batch is released, the conveyor starts to move and transports objects mixture under the scanner, where each object in the whole batch is classified and localized. The batch is moving on the conveyor belt toward robots, which are gradually relocating objects to its boxes till no object remains on the belt. The next batch is released after successful sorting of objects into corresponding boxes from the previous batch. Presented case of the robotic sorting line was used for CPN model creation, which is described in next section.

2.2 Coloured Petri Net Model Creation

Coloured Petri net construction is a multistage process, where at least the places, transitions, edges, variables, guards and colors must be defined. Each place is defined by specific color type and can contain tokens in the same color as its type. Edges can transfer these tokens within defined conditions. When each input connection contains proper tokens, the transitions can fire (be triggered) whatever type of tokens through any output connection. The transfer of tokens is controlled by restrictive conditions on edges and transitions, which makes the correct token flow through the model.

As a modeling software was used CPN Tools [10], which allows a variety of automated simulation and analysis options. For purposes of our model was needed do define specific color types and variables for possible transition of colors through model arcs. The definition of the color sets and variables is written in following CPN Tools program code:

```
Standard declaration():
     colset OBJECT = with O1| O2| O3| O4| O5;
     colset OBJECTList = list OBJECT;
     colset INT, NO = int;
     colset BATCH = product NO * OBJECTList;
     colset PROCESS = product INT * OBJECT;
     colset BOOL = bool;
     var obj: OBJECT;
     var olist: OBJECTList;
     var n: NO;
     var e: INT;
     var b: BOOL;

     fun em(olist) =
         if lenght olist = 0
           then true
         else false;
end.
```

It can be seen from the CPN Tools standard declaration program that seven colors, five variables, and one conditional function have been defined. As these declarations are made, then the model could be presented as its on Fig. 3.

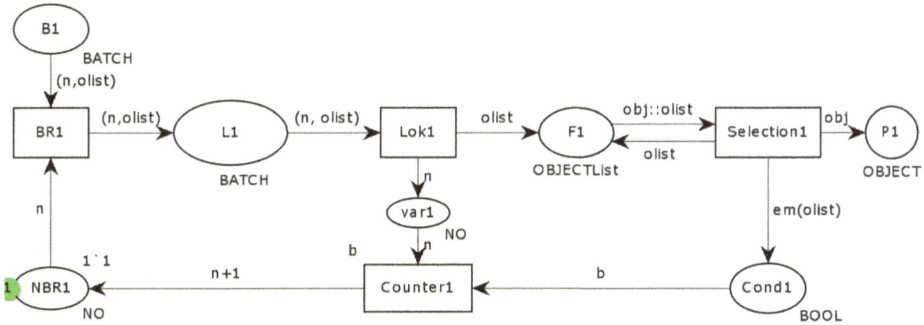

Fig. 3. CPN tools model of particular line behavior and object consumption.

The model presents the particular line behavior and object consumption process. At first, the place marked as B1 is defined as color set of BATCH type. From program declarations is obvious, that this color is product of colors NO and OBJECTList. It means that, the tokens in this place are numbered batches of objects. The color NO stands for number, which correspondents to the order of the batches. The color OBJECTList stands for batch, which is ordered set of object mixture.

The place marked as NBR1 serves as the counting token holder, which is conditioning the passing of the correct batch through the transition BR1. The number and batch tuple are then moved to the locator, which separates the number and batch. The batch is then shifted as the list of objects to the part where the gradual selection of objects is performed. This this is ensured by relation between elements F1 and Selection1. Objects are then handed over through the place P1 for manipulation by the robots. If the object list is empty after the selection, then the counter is incremented, which executes the next batch processing.

As such, the purpose of this model is to select batch, separate objects and shift them for further processing with robots. This particular model is ended by place P1, which is the first place of robot function model shown in Fig. 4.

On this particular model is showed only function of the first robot, which can manipulate O1 and O2 types of objects. The place P1 is passing objects of any type to next transitions. In this case it is the transition OR1L1. The place R1U contains token, which is signalizing that, the robot is available for manipulation task. Accordingly, to the actual robot state, the OR1L1 transition can fire. If so, the place R1 get token composed of number and object colors tuple. From place R1 leads connection to transition O1 and O2. When the type of object is O1, then the transition O1 fires object token to place OB1, which indicates that the object is placed in the box, and number token to place R1U, which shows the release of the robot resource. In case of the second branch, the workflow works the same, but with object type O2.

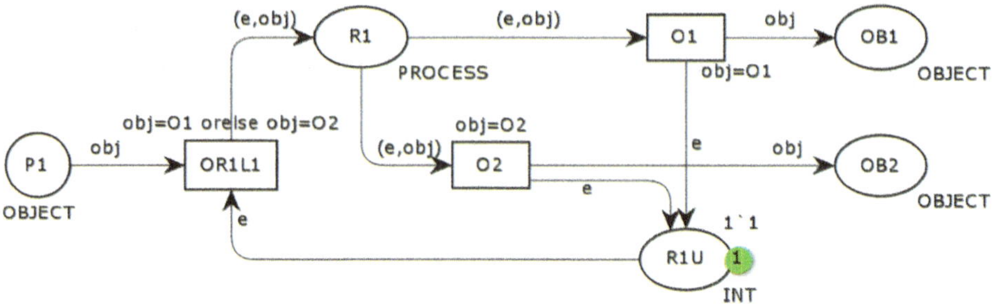

Fig. 4. CPN tools model of particular robot function for object placing into boxes.

As the robotic line was created from more similar parts defined in previous models, the entire sorting line model with two lines (model on Fig. 3.) and three robots (model on Fig. 4.) is assembled into a whole, which is captured on Fig. 5.

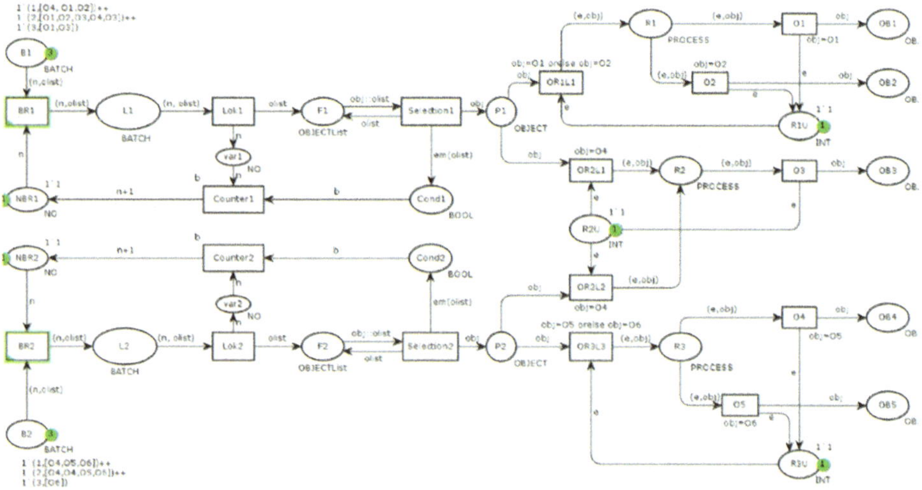

Fig. 5. CPN tools model of entire robotic sorting line

Function of complete system model should be described as follows. On each of line is processed selection of objects from batch. Each manipulated object takes its robot resource and after successful manipulation resource is released, which will allow the robot to be reused. If all objects in particular batch are placed to its boxes, then the next batch is released.

The model was extended by adding the time to color sets definitions and is necessarily described in following section.

2.3 Timed Coloured Petri Net Model

For further analysis of the dependency between batch processing time and composition of objects and their counts in the mixture, the coloured Petri net model was extended with timing. The places, where robots are manipulating objects got the time consumption stamp, which adds some duration to the operations as it works like in the real process. The duration time of each batch processing can be used for calculations for optimal queuing of objects in a batch.

In addition to coloured Petri net model, the check was placed on the placement of all objects in their boxes, which could trigger the release of another batch.

2.4 Experimental Procedure

The experimental procedure was composed from three different experiments. As first, the experiments with model proper function were done by repeated simulations with randomly generated initial batches. As second, the state space analysis was done for the same simulations as in previously mentioned experiment. Proper model function and state space analysis was tested on 100 simulation experiments.

Finally, the experiments with timing were performed. The time consumption for each object type was defined. After that, one specific batch composition, with only a given number of objects, was selected for repeated simulations in which the order of the taken objects changed. One batch composition was tested with 1000 object order changes. Total processing time was compared for each of these 1000 object orders, and best order composition was then marked as optimal. Totally, the 50 batches compositions were tested. A simplified illustration is given for a better idea on Fig. 6.

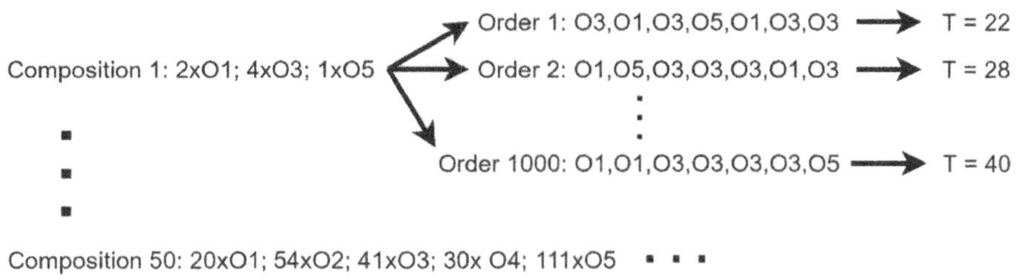

Fig. 6. Simplified illustration of experiments with batch orders and time consumption

3 Results

The model proper function was tested at first. Each of the 100 simulation experiments ended up in a state where all the objects from the batches were sorted in the appropriate boxes. The result of this experiment can be considered successful. The proposed model of the colored petri net was functional according to the requirements and therefore, can be considered as correct.

The CNP Tools provides a state space analyzing tool, which can be used to investigate the behavioral properties of coloured Petri net model. This tool was used to proper analyzation of every state space of a model for each of 100 simulations. Due to the increasing complexity of state space caused by tokens counts, the state space generation for some models lasted several tens of minutes. As an example, an average state space model contained around 15 thousand of nodes, 69 thousand of arcs between these nodes and the state space generation takes 112 s.

The analysis also showed, that the number of home markings is the same as the number of dead markings, which can be interpreted in two ways. First, this finding confirms the results of the first experiments. Second, it clearly indicates that, there were no unwanted deadlocks. Furthermore, there were no infinite occurrence sequences, which means that no infinite loops were indicated. The results of second experiment procedure also confirms the correctness of the proposed model.

The timed coloured Petri net used for time consumption measurements provided the resulting time required to sort each specific batch. From the last experiments were verified that, the objects order directly affects time consumption, so the correct sorting of objects for removal by robots should be ensured for each robotic sorting line.

4 Discussion

The coloured Petri net model and timed coloured Petri net model of specific robotic sorting line was created. The model was composed of different parts, including two lines and three robots.

With the experiments, the correctness of the model function and proper model behavior was verified. The state space analysis confirmed that, there are no unwanted deadlocks in the model.

Optimal composition and order of objects was successfully found by timed coloured Petri nets experiments. As such, this possible application of coloured Petri nets would be confirmed.

In future work, the modeling of this kind of a system should be done on more general basis, which can help with generalized mathematical analysis of the state space. Moreover, the specific parts of model should be created as a hierarchical model, which allows easier and more natural model extension.

Acknowledgments. The work has been supported by Funds of University of Pardubice (by project "SGS2021" No: SGS_2021_019), Czech Republic. This support is very gratefully acknowledged.

References

1. Liu, C., Pei, Y., Cheng, L., Zeng, Q., Duan, H.: Sampling business process event logs using graph-based ranking model. Concurr. Comput. (2020)
2. Chen, W., Liu, H., Qi, E.: Discrete event-driven model predictive control for real-time work-in-process optimization in serial production systems. J. Manuf. Syst. **55**, 132–142 (2020)
3. Zhang, X.-M., et al.: Networked control systems: a survey of trends and techniques. IEEE/CAA J. Autom. Sin. **7**(1), 1–17 (2020)
4. Huang, B., Zhou, M., Wang, C., Abusorrah, A., Al-Turki, Y.: Deadlock-free supervisor design for robotic manufacturing cells with uncontrollable and unobservable events. IEEE/CAA J. Autom. Sin. **8**(3), 597–605 (2021)

5. Pomares-Angelino, R., Lopez-Mellado, E.: Automated modelling of deadlock-free petri nets using duplicated transition labels. In: 2020 17th International Conference on Electrical Engineering, Computing Science and Automatic Control, CCE 2020, art. no. 9299169 (2020)
6. Jensen, K., Kristensen, L.M.: Coloured Petri Nets — Modeling and Validation of Concurrent Systems. Springer, Heidelberg (2009). https://doi.org/10.1007/b95112
7. van der Aalst, W.M.P., Stahl, C.: Modeling Business Processes — A Petri Net-Oriented Approach. The MIT Press, Cambridge (2011)
8. Christensen, S., Petrucci, L.: Modular state space analysis of coloured petri nets. In: De Michelis, G., Diaz, M. (eds.) ICATPN 1995. LNCS, vol. 935, pp. 201–217. Springer, Heidelberg (1995). https://doi.org/10.1007/3-540-60029-9_41
9. JARP Project. http://jarp.sourceforge.net/. Accessed 12 Feb 2021
10. CPN Tools – A tool for editing, simulating, and analyzing Colored Petri nets. https://cpntools.org/. Accessed 12 Feb 2021
11. Hino, R., Harada, M.: Development of a transportation system with multiple conveyor belts: Conditions for transportation and scheduling. Seimitsu Kogaku Kaishi/J. Jpn Soc. Precis. Eng. **78**(12), 1105–1111 (2012)

Implementation of Elliptic Curve Cryptography Between Autonomous Vehicles and Control Center

Andrej Grnac, Frederik Valocky, and Milos Orgon(✉)

Faculty of Electrical Engineering and Information Technology,
Slovak University of Technology in Bratislava, Bratislava, Slovakia
orgon@ktl.elf.stuba.sk

Abstract. This article presents a design for securing the transmission of control data between the control center and autonomous vehicle, which utilizes elliptic curve cryptography. Discussed design is using OpenSSL, an open-source software library, for all cryptographic operations. The analysis and comparison of individual types of elliptic curves suitable for ensuring the transfer of control data between the control center and the autonomous vehicle was carried out. Selected elliptic curves were used to form key pairs, i.e., to generate private and public keys for communicating nodes.

Keywords: Elliptic curve · Autonomous vehicle · Control center · Cybersecurity · Cryptography

1 Introduction

Nowadays, the smart autonomous car is a hotly discussed topic over the research as well as industrial areas. The need for heterogeneous infrastructure is obvious. The single technology, Radio Area Networks (RAN) or cable networks (such as power line, Ethernet, or fiber optic) can not be used. [1–4]. The functional parameters of these networks must come with high-reliability and availability as there is no space for mistakes due to the high-possibility of health harm if any error occurs. High secured systems are a must in these networks [5]. Several cryptography mechanisms must work together to reliably and efficiently secure communications in real-world systems [6–8]. Each encryption mechanism has its characteristics and properties that predetermine or disqualify it for use in a specific part of a comprehensive communication security system [9]. Based on the requirements resulting from the network characteristics and the properties of the communicating network nodes a set of encryption mechanisms that are best suited for the environment is then selected [10]. To obtain parameters

A. Grnac—The paper was partially supported by the grant 039STU-4/2021 and APVV-17-0190.

R. Silhavy (Ed.): CSOC 2021, LNNS 230, pp. 718–729, 2021.
https://doi.org/10.1007/978-3-030-77442-4_60

of appropriate elliptic curves designed for security applications [11], the open-source OpenSSL software library which provides support for a large number of cryptographic functions can be used. However, before proceeding with the security of data transmission between the control center and the autonomous vehicle, it was necessary to analyze and compare the different types of elliptic curves suitable for such transmission. The choice of elliptic curves was primarily focusing on the efficient formation of key pairs, i.e., to generate a private and public key for the communicating nodes. And also to ensure the authenticity and integrity of the control data, i.e., for signing and verifying an electronic signature based on the Elliptic Curve Digital Signature Algorithm (ECDSA) algorithm. Symmetric Advanced Encryption Standard (AES) encryption was selected for the encryption and decryption of the control data itself.

2 Corporate Network

Communication between two nodes (between the control center and autonomous vehicle), should be transmitted over the wireless network, i.e., standard IEEE 802.11a/b/g, even ac. Two laptops were used to verify the security of this design, simulating the two nodes in the network. One of these laptops represents a control center and the other an autonomous vehicle. As a host operating system on the first laptop, the MS Windows operating system was used, over which a virtual machine with Linux Ubuntu 18.04 operating system was created using the Oracle VM VirtualBox hypervisor. As a guest operating system on the second laptop, the MS Windows operating system was used, over which a virtual machine with Linux Ubuntu 18.04 operating system was created using the Oracle VM VirtualBox hypervisor. For all files that were created and worked with during the simulation, a directory named *controllerstorage* was created on the Control Node, which represented the control center memory. Also, a directory called *carstorage* was created on the Car Node, which contained all the files that the simulated autonomous vehicle would work with. To perform cryptographic operations, it was necessary to install the OpenSSL library on both nodes (laptops).

3 OpenSSL Library

The OpenSSL Library is an open-source software library that provides tools for securing communications over computer networks. The most widespread use is to ensure communication between clients and web servers, since OpenSSL provides a comprehensive implementation of TLS protocol, and also its now-deprecated predecessor SSL. OpenSSL is written in C and includes a cryptographic application that can implement various cryptographic functions such as key creation, certification tools, encryption or decryption itself, electronic signature signing, and verification, and all these operations are supported by a wide range of algorithms [12]. In order to work with the cryptographic library on MS Windows OS, it is necessary to utilize a UNIX-based system emulator, namely Cygwin, although in our proposed solution we were using Linux Ubuntu 18.04 operating system as VM.

OpenSSL Library was downloaded via a terminal (Linux command line). The downloaded file was in *tar* format. Extraction created a directory of the same name, which can be accessed by the commands:

```
$ wget https://www.openssl.org/source/
    openssl-1.1.0-latest.tar.gz
$ tar -zxf openssl-1.1.0-latest.tar.gz
$ cd openssl-1.1.0j
```

This directory contains many configuration and installation files. The OpenSSL manual compilation was started with the following command:

```
~/Desktop/openssl-1.1.0j$ ./config
```

After successful compilation, an executable installation file was created using the command *make*. When the executable installation file and library were obtained from the source code, the installation process was started with the command *make install*. After these steps, the OpenSSL library was installed.

```
$ make
$ make install
```

4 Choice of the Elliptic Curve

The choice of a suitable elliptic curve is very important. Dozens of different elliptic curves that are publicly known, proven, and reliable are currently available for use in cryptographic systems. Such elliptic curves have gone several years of testing to examine their properties. The OpenSSL open-source library includes support for a set of 81 predefined elliptic curves different from each other.

There are several different standards that deal with the selection of suitable and safe curves for cryptographic systems (ECC) [13]: ANSI X9.62 (1999), IEEE P1363 (2000), SEC 2 (2000), NIST FIPS 186-2 (2000), ANSI X9.63 (2001), Brainpool (2005), NSA Suite B (2005), ANSSI FRP256V1 (2011). Each of these standards tries to ensure that the solution to the Discrete Elliptic Curve Logarithm (ECDLP) problem is sufficiently complex to prevent an attacker from discovering a private key from a public key. In order to work with an elliptical curve, it is necessary to know its parameters. Specific parameters of the elliptic curve can be displayed in OpenSSL using the command, where -name is followed by the exact name of the curve whose parameters we want to print.

```
$ openssl ecparam -name prime256v1 -text
    -param_enc explicit -noout
```

Currently, elliptic curves based on the NSA Suite B standard are used in real-world systems. NSA Suite B is a set of cryptographic algorithms that were promoted in 2005 by the US National Security Agency as part of a cryptography modernization program. The elliptic curves included in NSA Suite B are called

NIST curves. These curves are among the most reliable because most of the time and money has been invested in their testing and mathematical analysis. When published in 2005, the NSA also enclosed mathematical evidence to ensure that breaking this ECDLP in a reasonable time is impossible with this type of curve. The parameters of these curves are optimized to enable the arithmetic operations required for cryptographic functions to be performed as quickly and effectively as possible.

However, there are several studies, for example [14], in which the authors hypothesized that NIST curves could include a back door in their implementation that would provide the NSA with a simple way to break the cryptographic algorithms using these curves. These theories have not yet been proven, yet there are users, who consider NIST curves to be untrustworthy and therefore implement the most common curves from the Brainpool family (2005) in their solutions. According to the authors [15], however, the Brainpool curves are not nearly as effective as the NIST curves, and the arithmetic operations on these curves are more computationally demanding, making algorithms using Brainpool curves work significantly slower. Among the components of the NSA Suite B standard are also other cryptographic algorithms that were used in our solution:

- Advanced Encryption Standard (AES) with key lengths of 128 and 256 bits,
- Elliptic Curve Digital Signature Algorithm (ECDSA) for electronic signature,
- Elliptic Curve Diffie-Hellman (ECDH) for key creation and exchange,
- Secure Hash Algorithm 2 (SHA-256 a SHA-384) hash function.

It is needed to select a curve that allows cryptographic algorithms to run quickly and effectively as possible. Because the system must be able to communicate even with a large number of vehicles. At the same time, the selected curve must not be too computationally demanding, since the computing power can be considerably limited on the side of the autonomous vehicle. Especially in electric vehicles, where battery consumption and battery life are major limiting factors.

Since we need a curve that allows fast arithmetic operations, it is obvious that the curve from the NIST family has been given priority. At the same time, the chosen curve should also be sufficiently secure, so we found a suitable curve for this solution to be the so-called "prime256v1", which provides a good compromise between performance and security. The *Weierstrass equation* of this curve is shown in Eq. (1). According to the IEEE P1363 standard, curves that satisfy Eq. (1) are most computationally efficient (both NIST and Brainpool curves have this property).

$$y^2 = x^3 + ax + b, \tag{1}$$

where a $= -3$ and b:

$$b = \begin{matrix} 5a & c6 & 35 & d8 & aa & 3a & 93 & e7 \\ b3 & eb & bd & 55 & 76 & 98 & 86 & bc \\ 65 & 1d & 06 & b0 & cc & 53 & b0 & f6 \\ 3b & ce & 3c & 3e & 27 & d2 & 60 & 4b \end{matrix}$$

5 Creating Encryption Keys

The basic element on which the entire communication security is to build is the encryption key. The OpenSSL library allows you to create different types of keys for both symmetric and asymmetric encryption methods. Since we want to use elliptic curves in the encryption process, both nodes in the network must have a key pair before communication begins, private key and public key. With OpenSSL, each node can create a key pair separately, however, the OpenSSL library needs to know the parameters of the curve that we want to use to create the key, i.e., to input curve parameters. We can use two options to input curve parameters. If we decide to use a standardized curve, which has parameters predefined in the library, we can use the following command, where after the *-name* flag we enter the exact name of the curve, and after the *-out* flag, specify the name under which to save the *.pem* file containing the key pair.

```
$ openssl ecparam -name prime256v1 -genkey -out
   controllerprivatekey.pem
```

In situations where it is desired to use a specific curve that is not predefined in the OpenSSL library, the parameters of that curve must be written to the .pem file in a defined order. After that, this file can be specified as an input variable after the *-in* command, where *ecparametersprime256v1.pem* is a file containing curve parameters.

```
$ openssl ecparam -in ecparametersprime256v1.pem
   -genkey -out controllerprivatekey.pem
```

Output file *controllerprivatekey.pem*, in both cases, it contains the name of the used curve and a key pair. Since we decided to use the NIST curve *prime256v1*, which is created over a 256-bit prime field. The private key is randomly selected 32 bytes long number. File *controllerprivatekey.pem* was created on a laptop simulating a Controller Node, since the file will be a private key of the Control Center. On the second laptop, simulating an autonomous vehicle, the key pair was obtained by the following command. The content of files in *.pem* format is encoded by Base64 encoding, see Fig. 1.

```
$ openssl ecparam -in ecparametersprime256v1.pem
   -genkey -out carprivatekey.pem
```

To actually display the private key, *.pem* files can be decoded from Base64 by the following command. Both the private key and the public key derived from it, are displayed in hexadecimal form. The public key is displayed because only one key file can be used for each cryptographic operation when working with OpenSSL. Therefore the library can automatically derive the public key if necessary. Another component of the file is the designation of the elliptic curve used to create the key.

```
$ openssl ec -in controllerprivatekey.pem -text -noout
```

The private key file contains only the name of used the elliptic curve, but not its specific parameters. This can cause a problem if there is a node in the network that does not recognize the elliptic curve by its name. This can happen when the node has an outdated OpenSSL version and this version does not have this curve defined in registers yet and does not know its parameters. This problem can be avoided by using explicit parameters instead of the curve name when creating the key. Either use the *.pem* parameter file as an input variable or use the following command:

```
$ openssl ecparam -name prime256v1 -genkey -out
   controllerprivatekeyexplicit.pem -param_enc explicit
```

This is useful when it is necessary to communicate with nodes whose specification is unknown. A difference in the length of the EC parameters can be seen. As already mentioned, it is possible to use just one key file to perform cryptographic operations that require either private our public key. To be able to encrypt communication with other nodes, these nodes must know each others public keys. In order for any node to publish a public key on the network, it is necessary to have a file that contains only the public key without the private key. The file that contains only the public key can be derived from the private key file. For the Control Node that was obtained by issuing the following command. The same procedure was used to obtain the public key for the Car Node on the given node:

```
$ openssl ec -in controllerprivatekey.pem -pubout -out
   controllerpublickey.pem
$ openssl ec -in carprivatekey.pem -pubout -out
   carpublickey.pem
```

To display the decoded contents of the public key file, following command is used. This file contains only the public key, such a file is, therefore, suitable for publication to other nodes in the network.

```
$ openssl ec -in controllerpublickey.pem -pubin -text -noout
```

The public key is 65 bytes long, representing by the coordinates (x,y) of the point on the *prime256v1* elliptic curve characterized by Eq. (1). The first byte (04) is the prefix, the next 32 bytes is the x coordinate, and the last 32 bytes is the y coordinate:

Fig. 1. Curve parameters.

$$X =$$
$$\begin{matrix} 1a & 2a & cb & 0b & 8b & 0f & 05 & fd \\ c9 & 28 & e9 & 46 & 67 & 98 & 6c & 5f \\ b7 & e9 & 27 & ed & c6 & b2 & 96 & 72 \\ 27 & 73 & 63 & 2b & 19 & 9e & 3f & c7 \end{matrix}$$
$$Y =$$
$$\begin{matrix} 8d & 96 & 09 & 12 & fc & a3 & f9 & c1 \\ 9e & 06 & 3d & 61 & f4 & 7b & b0 & 5c \\ 1e & b7 & 74 & a8 & a2 & fa & 30 & d8 \\ f0 & 1e & d4 & 64 & cc & 7e & f5 & 43 \end{matrix}$$

6 WiFi Communication Setup

The particular form of wireless data exchange between the control center and the autonomous vehicle depends on the used systems. As part of our simulation, data between nodes was sent over a WiFi network via LAN Share. This is a freely available program that allows easy and efficient exchange of files between nodes within a WLAN/LAN network, that are running on Ubuntu/Debian OS. There are also versions for other Linux distributions. There may be several problems in implementing WiFi communication between nodes. An important aspect to highlight is that the operating systems run as VMs in this simulation. Once created a Virtual Machine (VM) has, by default, only one virtual network interface that allows access only to the Internet. That means, it is not possible to communicate with other nodes on the network, not even with the guest operating system on which the VM runs. The solution to this problem was found in the Oracle manual [16]. It was necessary to assign another network interface of the *Bridged Adapter* type in the VirtualBox program (in the network settings of the VM) according to the parameters.

Another important element is the need for up-to-date packages that the Ubuntu operating system works with. When sending data, it is necessary to run the LAN Share application on both communicating nodes (laptops) in parallel. On both nodes, it is necessary to specify the directories (*controllerstorage* and *carstorage*) in which the exchanged files will be stored. This has to be done in the application settings. After that, the nodes within the LAN are able to communicate with each other.

7 Exchange of Public Keys

Once the communication channel is ready, the first step is to exchange public keys between the control center and the vehicle. The *controllerpublickey.pem* file containing the control center public key, is sent from the Control Node over the WiFi network using LAN Share, to the Car Node that automatically receives the file and stores it in the carstorage directory. Subsequently, the *carpublickey.pem* file containing the vehicles public key, is sent in the same way, from the Car Node to the Control Node. Note that public keys are sent without encryption.

8 Data Encryption and Electronic Signature Signing

When both nodes have their own key pair (in our case both Control Station Node and Car Node notebooks), it is possible to start encrypting the control data that will be transmitted between the control center and the vehicle. It is also possible to create an electronic signature of the transmitted data to ensure its authenticity. The first step was to create a file containing data to be encrypted. Because we do not have an autonomous vehicle capable of moving independently to verify the functions of this design, we used a simple text file *controldata.txt*. In this file, the *stop()* function is implemented to stop the vehicle. The control data sample on the control node is created by the command [17]:

```
$ echo 'volanie funkcie stop()' > controldata.txt
```

No encryption and decryption mechanisms are directly defined for elliptic curve-based encryption. It is possible to encrypt the data itself by asymmetric encryption using elliptic curves, but it is very computationally demanding and therefore slow. In an environment such as the VANET network, data exchange must be sufficiently secure, but at the same time, it is also essential that it is carried out as quickly as possible. For this reason, symmetric algorithms are used to encrypt and decrypt the payload. Real elliptic curves based encryption systems use the so-called Elliptic Curve Diffie-Hellman (ECDH) algorithm based on the original Diffie-Hellman (DH) algorithm [18]. The ECDH is an algorithm that allows you to use an asymmetric key pair for symmetric encryption. It is used to derive a common secret for two communicating nodes, which can be used as a password for symmetric encryption and decryption algorithms.

The ECDH principle is as follows. Suppose two nodes A (Control Node) and B (Car Node) want to exchange a secret key for symmetric encryption. Node A has a private key dA and a public key $QA{=}dAG$ (where G is the generator of the elliptic curve used). Also, the node B has a private key dB and a public key $QB{=}dBG$. Node B sends its public key QB to node A, then node A can compute $dAQb{=}dAdBG$. Similarly, node A sends its public key QA to node B, then node B can calculate $dBQA{=}dAdBG$. The common secret is the *x coordinate* of the calculated $dAdBG$ point. If an attacker eavesdropped the key exchange, he would only know the public QA and QB keys and therefore would not be able to calculate the common secret key. The common secret key on the Control Node is derived from the private key of the Control Node and the public key of the Car Node by the following command:

```
$ openssl pkeyutl -derive -inkey controllerprivatekey.pem
  -peerkey carpublickey.pem | base64 -w0 >
  sharedkeycontroller.txt
```

The derived shared key is stored in the *sharedkeycontroller.txt* file. File contents the derived key, is encoded with Base64 encoding, ensuring that the file does not contain any special characters (e.g., first byte = NULL). That could introduce an error in encryption/decryption. In the same way, a common key is derived at the Car Node, but the vehicle private key and the control center public key are used following commands. In the following figures, it can be seen that the independent *sharedkeycontroller.txt* files on the Control Node, see Fig. 2. And the *sharedkeycar.txt* files on the Car Node, see Fig. 3, contain the same common key.

```
$ openssl pkeyutl -derive -inkey carprivatekey.pem
   -peerkey controllerpublickey.pem | base64 -w0 >
   sharedkeycar.txt
```

Fig. 2. Shared secret key of Control Node.

Fig. 3. Shared secret key of Car Node.

This derived shared key can now be used as a password base for any symmetric encryption. Before the data is symmetrically encrypted, it is necessary to create an electronic signature of the data to ensure authenticity. An electronic signature guarantees that the received data has not been modified in any way along with the transmission and that the data originates from the node from which the data is expected. The electronic signature that works with EC key pairs is called Elliptic Curve Digital Signature Algorithm (ECDSA). In ECDSA, the communicating nodes must agree in advance on a hash function H to be used to create and verify the digital signature. Not the data itself is signed, but the output of the hash function resulting from that data is signed. It is important, that only the node that creates the signature, knows the private key used for signing the signature. The node that verifies the signature uses the public key of the node that created the signature. In this simulation, the SHA-256 hash function was used to create 256-bit hash output from data, which is the number of bits sufficient enough to prevent Brute Force attack. The Control Node creates a digital signature *signature.bin* from unencrypted data *controldata.txt*, using a private key *controllerprivatekey.pem* by issuing the following command:

```
$ openssl dgst -sha256 -sign controllerprivatekey.pem
   controldata.txt > signature.bin
```

Advanced Encryption Standard (AES) algorithm, the 256-bit version in CBC mode, known as *AES-256*, was used for symmetric encryption alone. The main features for which this algorithm was chosen are speed, sufficient security, and compatibility with the other algorithms and features used in this simulation, in particular ECDH and ECDSA. A detailed description of the AES symmetric encryption algorithm is not the subject of this work. Data encryption on the Control Node was performed with the following command. Where -pass flag is followed by a file containing a shared secret key that serves as the basis for the password used in AES symmetric encryption algorithm.

```
$ openssl enc -aes-256-cbc -in controldata.txt -out
   encrypteddata.txt.enc -pass file:sharedkeycontroller.txt
```

However, the *AES-256* works with 256-bit passwords and therefore it automatically calculates the hash from the shared key using the *SHA256* hash function and uses that hash output as the encryption password. The encrypted data is stored in the *encrypteddata.txt.enc* file. At this point, the Control Node has completed the process of securing the communication. All that remains is to send the encrypted data along with the digital signature to the vehicle node. The files *encrypteddata.txt.enc* and *signature.bin* are sent via LAN Share over the WiFi network to the Car Node, which automatically receives them and stores them in the carstorage directory.

9 Decrypting and Digital Signature Verification

After receiving the encrypted data at the vehicle node, the encrypted data must be decrypted by the reverse process to be read and further manipulated. Decryption at the Car Node is performed by the following command, where flag *-d* (decrypt) specifies the reverse process of encryption. The encrypted data received from the Control Node serve as input *encrypteddata.txt.enc*, the common shared key is used for decryption *sharedkeycar.txt*, and the decrypted data is saved to a file *receiveddata.txt*.

```
$ openssl enc -d -aes-256-cbc -in encrypteddata.txt.enc -out
  receiveddata.txt -pass file:sharedkeycar.txt
```

Once the data has been decrypted, all that remains is to verify the authenticity of the data. Authenticity is verified by verifying the digital signature file that was received along with the encrypted payload. The digital signature can be verified using the sender's public key. In this case, the public key of the Control Node, with the following command:

```
$ openssl dgst -sha256 -verify controllerpublickey.pem
   -signature signature.bin receiveddata.txt
```

As can be seen in Fig. 4, if the digital signature is verified by the authentic sender's public key, the signature is verified successfully. However, if an incorrect key is used for verification, or the data has been modified in some way during the transmission, signature verification will result in an error. In our performed simulation, the signature was successfully verified, therefore received data is considered to be authentic and confidential so that it can be read and further processed according to the logic of the Car Node. Based on this result it is possible to state that this design fulfilled its purpose by implementing elliptic curve cryptography to secure data during transmission over the WiFi network between remote nodes.

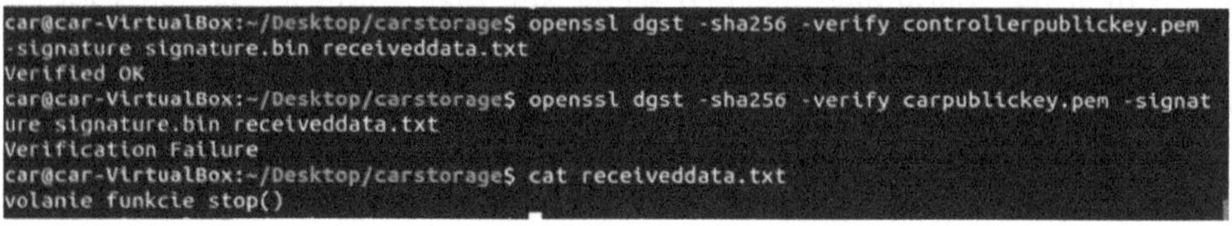

Fig. 4. Digital signature verification.

10 Conclusion

The article described the simulation of sending secure control data from the control center to an autonomous vehicle via WiFi network. For this purpose, the OpenSSL cryptographic library was used to implement elliptic curves and other cryptographic algorithms to achieve comprehensive protection and authenticity of control data when it is exchanged between the control center and an autonomous vehicle.

Prospectively, this research work will continue by implementing elliptic curves in a model that would feature an autonomous vehicle containing the Raspberry PI platform. Raspberry PI will serve as logic for the autonomous vehicle and ultrasonic radar system. The communication between the vehicle and the control center will also be processed by the Raspberry PI, whereby the control center will be able to take over vehicle control or stop the vehicle if necessary. Communication between the vehicle and the control center will be comprehensively secured. That means, transmitted data will be encrypted/decrypted on both nodes, cryptographic algorithms based on elliptic curves such ECDSA or ECDH will be implemented to provide a secure key exchange. OpenSSL library will be used to implement these cryptographic mechanisms.

References

1. Mlynek, P., Fujdiak, R., Misurec, J., Slacik, J.: Experimental measurements of noise influence on narrowband power line communication. In: 2016 8th International Congress on Ultra Modern Telecommunications and Control Systems and Workshops (ICUMT), pp. 94–100. IEEE (2016)
2. Mlynek, P., Hasirci, Z., Misurec, J., Fujdiak, R.: Analysis of channel transfer functions in power line communication system for smart metering and home area network. Adv. Electr. Comput. Eng. **16**(4), 51–56 (2016)
3. Mlynek, P., Misurec, J., Kolka, Z., Slacik, J., Fujdiak, R.: Narrowband power line communication for smart metering and street lighting control. IFAC-PapersOnLine **48**(4), 215–219 (2015)

4. Mlynek, P., et al.: Performance testing and methodology for evaluation of power line communication. Elektronika ir Elektrotechnika **24**(3), 88–95 (2018)
5. Krivtsova, I., et al.: Implementing a broadcast storm attack on a mission-critical wireless sensor network. In: International Conference on Wired/Wireless Internet Communication, pp. 297–308. Springer (2016)
6. Fujdiak, R., Mlynek, P., Blazek, P., Barabas, M., Mrnustik, P.: Seeking the relation between performance and security in modern systems: metrics and measures. In: 2018 41st International Conference on Telecommunications and Signal Processing (TSP), pp. 1–5. IEEE (2018)
7. Fujdiak, R., Misurec, J., Mlynek, P., Janer, L.: Cryptograph key distribution with elliptic curve Diffie-Hellman algorithm in low-power devices for power grids. Revue Roumaine des Sciences Techniques, pp. 84–88 (2016)
8. Fujdiak, R., Mlynek, P., Misurec, J., Slacik, J.: Simulation of intelligent public light system in smart city. In: 2017 Progress in Electromagnetics Research Symposium-Spring (PIERS), pp. 2515–2519. IEEE (2017)
9. Raso, O., Mlynek, P., Fujdiak, R., Pospichal, L., Kubicek, P.: Implementation of elliptic curve Diffie Hellman in ultra-low power microcontroller. In: 2015 38th International Conference on Telecommunications and Signal Processing (TSP), pp. 662–666. IEEE (2015)
10. Fujdiak, R., Masek, P., Hosek, J., Mlynek, P., Misurec, J.: Efficiency evaluation of different types of cryptography curves on low-power devices. In: 2015 7th International Congress on Ultra Modern Telecommunications and Control Systems and Workshops (ICUMT), pp. 269–274. IEEE (2015)
11. Fujdiak, R., Mlynek, P., Misurec, J., Masek, P.: Design of low-power random number generator using signal quantization error in smart grid. In: 2016 39th International Conference on Telecommunications and Signal Processing (TSP), pp. 7–10. IEEE (2016)
12. Henson, S., et al.: Openssl library. Open source library (1988)
13. Bernstein, D.J., Lange, T.: Safecurves: choosing safe curves for elliptic-curve cryptography (2015). https://safecurves.cr.yp.to. Citations in this document, vol. 9 (2014)
14. Bernstein, D.J., Lange, T.: Security dangers of the NIST curves. In: Invited Talk, International State of the Art Cryptography Workshop, Athens, Greece (2013)
15. Shaikh, J.R., Nenova, M., Iliev, G., Valkova-Jarvis, Z.: Analysis of standard elliptic curves for the implementation of elliptic curve cryptography in resource-constrained e-commerce applications. In: 2017 IEEE International Conference on Microwaves, Antennas, Communications and Electronic Systems (COMCAS), pp. 1–4. IEEE (2017)
16. Oracle, V.: Virtualbox user manual, Oracle Corporation.-2004.-C, vol. 357 (2011)
17. Valocky, F., Orgon, M., Fujdiak, I.: Experimental autonomous car model with safety sensor in wireless network. IFAC-PapersOnLine **52**(27), 92–97 (2019)
18. OpenSSL Wiki: Elliptic curve Diffie Hellman, OpenSSL (2014). https://wiki.openssl.org/index.php/Elliptic_Curve_Diffie_Hellman

The Hybrid Methods of Object Detection and Localization in Images

Sergey Kuleshov, Alexandra Zaytseva(✉), Ilya Shalnev, and Alexey Aksenov

St. Petersburg Federal Research Center of the Russian Academy of Sciences (SPC RAS), 14th Line, 39, 199178 St. Petersburg, Russia
cher@iias.spb.su

Abstract. The paper considers the task of watermark detection and localization in the images. Several heuristic methods for processing the characteristic images obtained from artificial neural network output for the task of identification are considered. The post-processing of characteristic images generated by the artificial neural network is used to improve the accuracy of detecting the presence of a watermark (semitransparent logo or inscription) in an image. A description of two approaches to image post-processing (the method of clustering areas and the method of selecting areas based on a template) and the results of experiments on a test sample of 500 images, artificially generated based on a set of HD resolution photos and a database of logos "UMD Logo Database for Research" are shown. The selection criteria for clusters potentially containing regions of interest can be the number of pixels in the cluster, the size of the cluster (determined based on dispersion or by using the algorithm for constructing the minimal convex hull). The use of such conditions for the selection of clusters gives a satisfactory result in the absence of noise in the form of compact artefacts.

Keywords: Artificial neural networks · Object localization · Probability map image · Classification

1 Introduction

The most convenient modern solution for tasks related to image processing is the use of artificial neural networks (ANNs) due to their wide capabilities and flexibility of use, as well as a large number of ready-made solutions [1–4]. From the literature, variants are known when post-processing of the results of neural networks is used to improve the quality and eliminate artifacts in tasks of image classification [1]. In some cases, the application of another neural network is used for the post-processing stage [2]. However, such neural network usage does not always provide a straightforward solution that does not require further post-processing. For example, in the task of object localization in image based on probability map as output of neural network the interpretation of result can be ambiguous. Solving such problems requires the implementation of hybrid methods, combining the use of neural networks and classical methods of image processing, based on additional rules for decision making.

R. Silhavy (Ed.): CSOC 2021, LNNS 230, pp. 730–737, 2021.
https://doi.org/10.1007/978-3-030-77442-4_61

The paper proposes several methods of characteristic images (ANNs output) processing on the example of automatic detection of watermark embedded to image by transparency overlay.

The given method of watermark embedding allows the copyright protection of the images submitted to photo banks, social networks and other open internet sources to prevent unauthorized reuse. Figure 1 illustrates the example of a watermarked image.

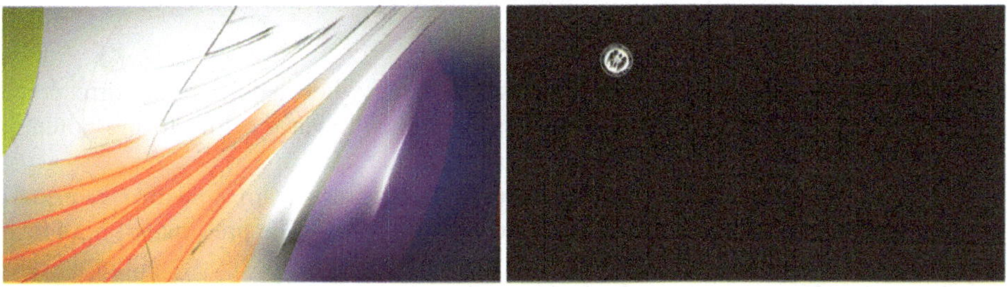

Fig. 1. The example of transparency overlay watermarking (left), the watermark (right)

The ANN architecture used in the work is based on an autoencoder (Fig. 2). The encoder consists of sequential blocks of convolution, batch normalization, and activation. Convolutional layer kernels are (3 × 3). These blocks are interleaved with layers of subsampling (pooling). In total, the encoder has 10 convolution, batch normalization and activation units and 3 subsampling layers. As the depth of the neural network increases, the number of feature maps also increases and at the same time, the size of these maps decreases. The dense representation formed by the encoder, consisting of 512 feature maps, is fed to the input of the decoder, which, in turn, based on the feature maps data, forms an output segmentation map, where the spatial localization of the watermark embedded in the digital image is displayed. The decoder consists of sequential convolution, batch normalization and activation blocks, alternating with upsampling layers, which implement the interpolation operation to increase the size of the processed feature maps. In total, the decoder has 13 convolution, batch normalization and activation units and 3 layers of increasing the dimension. At the same time, to reduce the number of processed feature maps at the stage of building a segmentation map, convolution, batch normalization and activation blocks with convolutions of size (1 × 1) are used. The output layer is a convolutional layer with (3 × 3) kernels. The output of the neural network has two channels that correspond to two classes of segmentation: pixels with watermarks and pixels without watermarks. A softmax activation function is applied to the two-channel output, which simulates the probability distribution of watermarks in the image.

We formed the training set by placing a watermark on images of various types (mainly photographs). The sample generation tool chose the specific location of the watermark, its size, as well as the degree of its visibility randomly. The ANN was trained over 100 epochs. We choose the cross-entropy function as the error function. In the course of its operation, the ANN generates a reconstructed image containing the features of connected areas with features of a watermark (characteristic image).

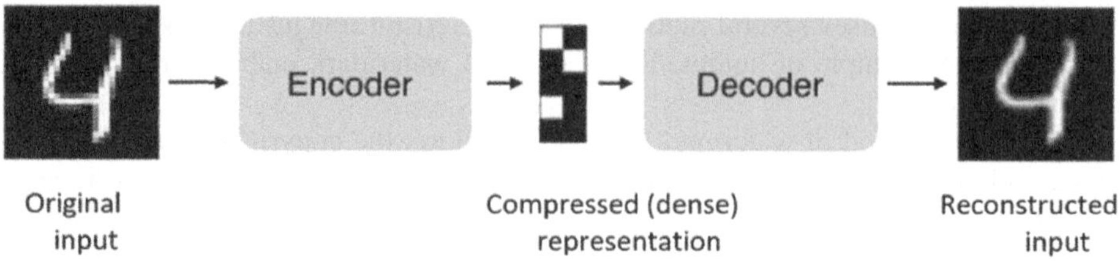

Fig. 2. ANN architecture based on a deep neural network autoencoder

The characteristic image is an 8-bit grayscale image, where a black pixel corresponds to the absence of a watermark, and the higher values towards white to the potential presence of watermark features in the corresponding area.

Figure 3 shows an example of ANN operation (characteristic image with a watermark in left and characteristic image with no watermark in right). The presence of high level noise in the characteristic image (in the form of white areas) of the ANN output does not allow using simple criteria to determine the presence of a watermark in the image (presence of white areas, exceeding the threshold by the average value of pixel brightness, searching for the center of mass of areas, etc.).

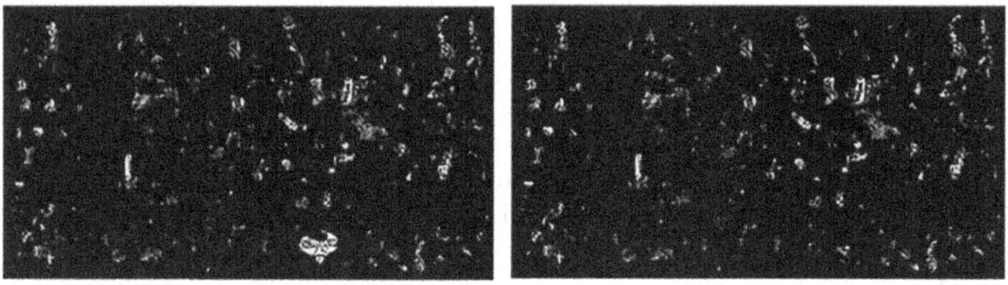

Fig. 3. An example of the result of the ANN operation (on the left – a characteristic image with an embedded logo, on the right – a characteristic image without a watermark)

2 Post-processing Method Based on Clustering Areas of Characteristic Image

The method is based on the search for clusters in the image containing compact areas of light pixels, based on the assumption that the watermark is a group of pixels bounded in the image plane (Fig. 4). The K-means method was chosen as the clustering algorithm.

The selection criteria for clusters potentially containing regions of interest can be the number of pixels in the cluster, the size of the cluster (determined based on dispersion or by using the algorithm for constructing the minimal convex hull [5]). The use of such conditions for the selection of clusters gives a satisfactory result in the absence of noise in the form of compact artifacts.

To minimize the influence of compact artifacts on the characteristic images, the criterion for detecting a watermark in a cluster was supplemented with a threshold based on the density of points in the cluster (Fig. 4). The condition for the presence of a watermark on an image expressed as follows:

$$\frac{n_k}{\pi R_k^2} \geq \gamma, \tag{1}$$

where γ – threshold value, n_k – number of points in a cluster, R_k – the radius of the cluster circle, which is calculated as $R_k = \sqrt{\max(D_{k_x}, D_{k_y})}$, where D_{k_x} – the value of the dispersion of the coordinates of points along the x axis, and D_{k_y} – the value of the dispersion of the coordinates of points along the y axis.

Those clusters that have a higher density are considered a watermark; those that have a lower density are ignored. The cluster with the maximum density is considered a watermark.

On the test sample set 74% watermark detection precision was obtained with the recall value of 56%.

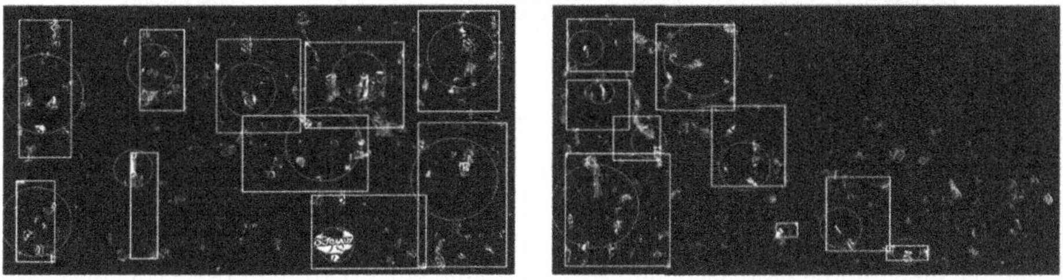

Fig. 4. An example of the result of the ANN characteristic image clustering (on the left – with an embedded logo, on the right – without a watermark)

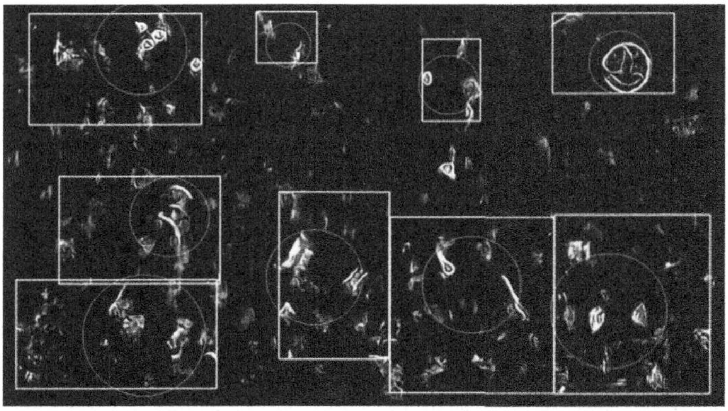

Fig. 5. An example of clustering an image with an outline watermark that gave an erroneous result

Despite the fact that the criterion based on the pixel density in the cluster gave the best results, it still has a number of weak spots. At first, the edge watermarks that have a low pixel density at significant sizes. Figure 5 shows an example of such a watermark. The algorithm correctly identified the watermark in the upper right corner, but the density of such a sign was below the threshold and because of this, such a watermark became invisible to the algorithm.

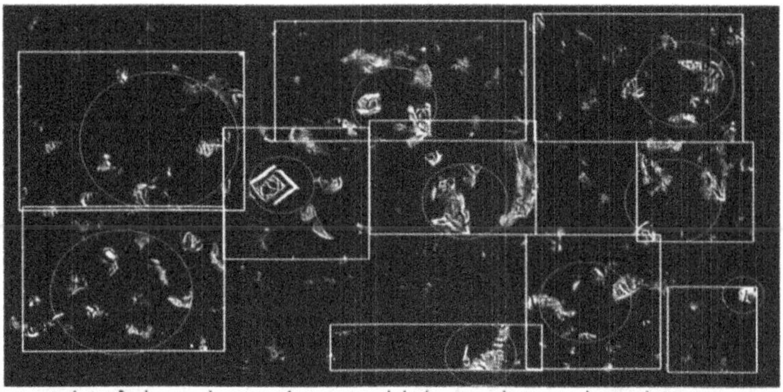

Fig. 6. An example of clustering an image with large cluster sizes that gave an incorrect result

Secondly, there is a class of images in which the size of clusters turns out to be excessive. That gives a low value of the pixel density in the cluster. In the example shown in Fig. 6, the cluster got too many pixels not related to the watermark due to which the cluster area increased, and as a result, the density of light pixels decreased. A possible reason is that the manually set maximum number of clusters was too small to adequately separate the noise from the watermark (in this case, the number of clusters was set to 10).

Using a clustering algorithm with an adaptive determination of the number of clusters could give better results.

Another problem in using characteristic image clustering is the presence of long watermarks. In Fig. 7 clustering methods detects three clusters on a single watermark. The cluster with the highest density in this method will be considered a logo. In this example, the implemented method gives a positive decision about the presence of a watermark in the image, but the actual arrangement of the clusters does not correspond to the semantics.

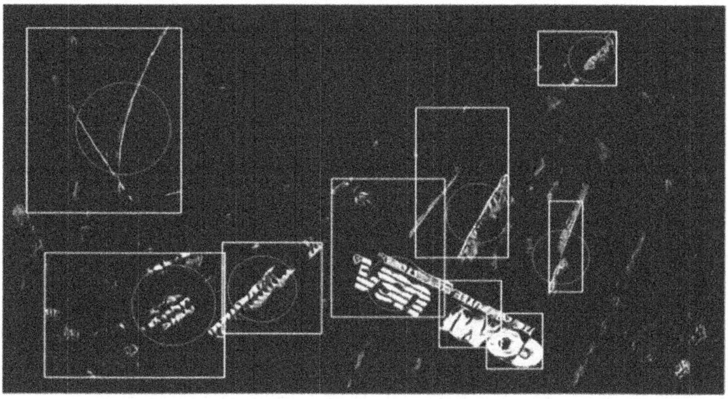

Fig. 7. Example of clustering an image with a long watermark

3 A Method for Detecting Connected Regions of a Characteristic Image Based on a Template

Another method for determining the presence of a watermark in the characteristic image generated by the ANN is based on determining the proportion of connected regions in the characteristic images. At the first stage, to select only pixels close to the maximum value we used a threshold function:

$$Y_D(x, y) = \begin{cases} 1, & \text{if } Y_S(x, y) > \gamma \\ 0, & \text{if } Y_S(x, y) \le \gamma \end{cases}, \tag{2}$$

where $Y_D(x, y)$ – pixel brightness value (x, y) after the threshold function application, $Y_S(x, y)$ – the original value of the pixel brightness of the characteristic image, γ - threshold value. Figure 8 shows an example of the result of applying the threshold function to the characteristic image.

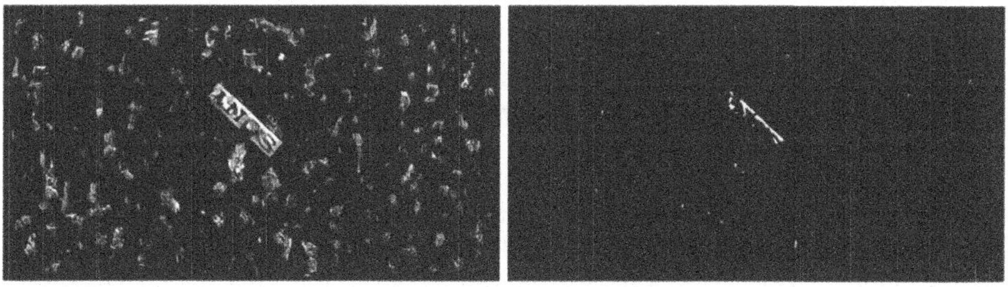

Fig. 8. The result of applying the threshold function to the characteristic image (on the left - the original characteristic image, on the right - the result)

To search for connected areas in the horizontal and vertical directions, it is proposed to use an L-shaped mask consisting of k = 5 pixels (Fig. 9). This mask in a loop moves over the image, and in the case when all pixels in the mask have a value of "1", the coincidence counter S is incremented by 1.

In the case when the condition:

$$\frac{N_1}{S} > \frac{1}{k}, \tag{3}$$

where N_1 – the number of pixels in the characteristic image having a value of "1", a decision is made on the successful watermark detection.

0	0	0	0	0	0
0	0	0	1	1	0
0	1	1	1	1	0
0	1	1	1	1	0
1	1	0	0	0	0
0	1	0	0	0	0
1	1	0	0	0	0

Fig. 9. Applying an L-shaped mask on the image

On the test sample set 77% watermark detection precision was obtained with the recall value of 85%.

4 Conclusion

This paper considers the use of post-processing of characteristic images generated by the ANN to improve the accuracy of detecting the presence of a watermark (semitransparent logo or inscription) in an image.

Authors propose an approach to solve the problem of developing decision-making algorithms based on a priori known rules after using artificial neural networks. The main issue of the considered problem is that for the processed images it is required to receive not only a binary answer about the presence of a logo (solving the classification problem), but also an answer about the spatial location of the detected logo. This imposes a limitation on the use of the neural network directly for solving the classification problem, requiring the introduction of additional processing stages.

A description of two approaches to post-processing images (the method of clustering areas and the method of selecting areas based on a template) and the results of experiments on a test sample of 500 images, artificially generated on the basis of a set of HD resolution photos and a database of logos "UMD Logo Database for Research" [6] are shown.

Table 1 shows the results of the proposed methods quality evaluation.

Table 1. Metrics of post-processing methods quality evaluation.

Method	Precision	Recall
Characteristic image clustering	0,74	0,56
Detecting connected regions of a characteristic image based on a template	0,77	0,85

The obtained results have shown that the use of post-processing can improve the precision in image classification problems on the output of the artificial neural network. Unfortunately, it is difficult to recommend the proposed methods as the generic methods for post-processing of any neural network output, since they are subject area semantics dependent. However, the development of successful solutions in particular cases is able to form a "pool of best practices".

References

1. Pan, X., Zhao, J., Xu, J.: An end-to-end and localized post-processing method for correcting high-resolution remote sensing classification result images. J. Remote Sens. **12**, 852 (2020). https://doi.org/10.3390/rs12050852
2. Chlebus, G., Schenk, A., Moltz, J.H., et al.: Automatic liver tumor segmentation in CT with fully convolutional neural networks and object-based postprocessing. J. Sci Rep. **8**, 15497 (2018). https://doi.org/10.1038/s41598-018-33860-7
3. Roth, H.R., Farag, A., Lu, L., Turkbey, E.B., Summers, R.M.: Deep convolutional networks for pancreas segmentation in CT imaging. In: Medical Imaging: Image Processing, vol. 9413, p. 94131G (2015)
4. Ghosh, A., Pal, N.R., Pal, S.K.: Image segmentation using a neural network. J. Biol. Cybern. **66**, 151–158 (1991). https://doi.org/10.1007/BF00243290
5. Cormen, T.H., Leiserson, C.E., Rivest, R.L., Stein, C.: Introduction to Algorithms. Second Edition. The MIT Press, Cambridge; London (2005)
6. DATASET: Logo Database for Research. https://lampsrv02.umiacs.umd.edu/projdb/project.php?id=47

Author Index

A
Akinpelu, Tumisang, 59, 110
Aksenov, Alexey, 730
Ali, Ammar Alhaj, 293, 307, 317
Alqassed, Ahmed, 237
Anil, G. N., 196
Antokhin, Yuri, 510
Arango, Eduardo C., 667
Avram, Anca, 221, 249
Awad, Mohamed, 237
Awwama, Emad, 307

B
Baryschnikova, E. S., 142
Berdnova, E. V., 142
Bocharov, Aleksey, 331
Bogomolov, A. S., 142
Boris, Jozef, 125
Bystrov, V. V., 418

C
Chereja, Ionela, 180, 221, 249
Chipiga, A. F., 451
Chramcov, Bronislav, 293, 307, 317

D
Dahmani, Youcef, 620
Deshko, I. P., 369
Dimitrov, Willian, 645
Dokshanin, Sergei, 88
Dranko, O. I., 142
Dubanov, A. A., 119

E
Eck, Rene van, 59
Eessaar, Erki, 375
Elhadidi, Amr, 237
El-Khatib, Samer, 701

F
Falešník, David, 351
Farana, Radim, 125
Fominykh, D. S., 142

G
Goosen, Leila, 593
Gorbunov, Alexander, 283
Grnac, Andrej, 718
Gupta, Varun, 564
Gushanskiy, Sergey, 283

H
Hahn, Sarah Myriam Lydia, 180, 221, 249
Hanne, Thomas, 564
Havlicek, Libor, 709
Hemanth Kumar, N. P., 37
Hemantha Kumar, A. R., 152
Hoc, Huynh Thai, 522, 632
Hodhod, Rania, 77
Hristov, Peyo, 645

I
Ivashchenko, V. A., 142

J
Jasek, Roman, 293, 307, 317
Jekov, Boyan, 645

R. Silhavy (Ed.): CSOC 2021, LNNS 230, pp. 739–741, 2021.
https://doi.org/10.1007/978-3-030-77442-4

K
Katta, Rasin, 293, 307, 317
Kavya, N. P., 67
Kazachenko, Aleksandr S., 275, 363
Kazachenko, Anna S., 275
Kenyeres, Jozef, 391
Kenyeres, Martin, 391
Khaliullina, D. N., 418
Kharroubi, Sahraoui, 620
Kholief, Mohamed, 77, 237
Klimeš, Cyril, 125
Kohout, Jan, 469
Komarasamy, G., 132
Koval', S. A., 451
Krayem, Said, 293, 307, 317
Kukartsev, Vladislav, 47
Kuleshov, Sergey, 730
Kumar, Karanam Sunil, 67
Kupka, Libor, 709
Kurashkin, Sergei, 47, 88, 331
Kushnikov, V. A., 142
Kushnikova, E. V., 142

L
Lapkovsky, R. Y., 142
Lazhauninkas, J. V., 142
Le Thi Kim Nhung, Ho, 632
Lei, Jianjun, 264
Leonteva, Anna, 88
Li, Li, 480
Linets, Gennadiy Ivanovich, 552
Loaiza, Omar L., 667

M
Maghawry, Ahmed, 77, 237
Mahesh Rao, K., 492
Malyar, Yuriy N., 275, 363
Malygina, S. N., 418
Mamani, Carlos R., 538
Mamani, Sol A., 538
Mareš, Jan, 469
Matei, Oliviu, 180, 221, 249
Melnik, E. V., 501
Mochalov, Valery Petrovich, 552
Murygin, Aleksandr, 47, 331

N
Nagaraja, G. S., 428
Ngugi, James K., 593
Nhung, Ho Le Thi Kim, 522
Nouali, Omar, 620
Nyandag, Bat-Erdene, 206

O
Omar, Yasser, 77
Orda-Zhigulina, D. V., 501
Orda-Zhigulina, M. V., 501
Orgon, Milos, 718

P
Palkanov, Ilya Sergeevich, 552
Panchal, Soumyashree M., 340
Pashintsev, V. P., 451
Pokorný, Pavel, 351, 406
Potapov, Viktor, 283
Prabhudeva, S., 37
Prakruthi, M. K., 132
Purnama, Aditya, 480
Puttegowda, D., 656

R
Rajeshwari, D., 656
Rezchikov, A. F., 142
Rodzin, Sergey, 701

S
Samo, Mansoor, 264
Šaur, David, 437
Seregin, Yuriy, 47, 331
Shalnev, Ilya, 730
Shashikala, J., 100
Shivaputra,, 340
Skobtsov, Yuri, 701
Skorik, Alexander D., 451
Sokolova, Irina, 510
Soria, Juan J., 538
Srinivasa, R. K., 152
Stursa, Dominik, 709
Suprun, Anton, 510
Švejda, Jaromír, 437

T
Taha, Ayman, 237
Tang, Jianlan, 480
Telesko, Rainer, 564
Thangadurai, N., 100
Tomeš, Jakub, 469
Tsvetkov, V. Ya., 369
Tsvirkun, A. D., 142
Tylová, Naďa, 469
Tynchenko, Vadim, 47, 88, 331
Tynchenko, Valeriya, 47, 88, 331

U
Ugur, Alper, 578
Ulaankhuu, Khishigdalai, 206

V
Valocky, Frederik, 718
van Eck, Rene, 110
Van Hai, Vo, 522, 632
Vanishree, K., 428
Vasilyeva, Natalya Yu., 275
Veena, S. K., 492
Vicentiy, A. V., 609
Viktor Mikhailovich, Kureichik, 170
Vladislav Ivanovich, Danilchenko, 170

W
Wang, Zhe, 1
Wijaya, Tommy Tanu, 480

Y
Yevgenia Vladimirovna, Danilchenko, 170
Yuan, Haiyan, 682

Z
Zakharov, Valerii, 701
Zapletal, Miroslav, 406
Zaytseva, Alexandra, 730
Zhang, Kewei, 682
Zimonin, Dmitry V., 363
Zuva, Tranos, 59, 110

The manufacturer's authorised representative in the EU is Springer Nature Customer Service Centre GmbH, Europaplatz 3, 69115 Heidelberg, Germany. If you have any concerns regarding our products, please contact ProductSafety@springernature.com

Printed and bound by CPI Group (UK) Ltd, Croydon, CR0 4YY
30/09/2026
02225662-0020